Lecture Notes in Computer Science 7432

Commenced Publication in 1973
Founding and Former Series Editors:
Gerhard Goos, Juris Hartmanis, and Jan van Leeuwen

George Bebis   Richard Boyle
Bahram Parvin   Darko Koracin
Charless Fowlkes   Sen Wang
Min-Hyung Choi   Stephan Mantler
Jürgen Schulze   Daniel Acevedo
Klaus Mueller   Michael Papka (Eds.)

Advances in Visual Computing

8th International Symposium, ISVC 2012
Rethymnon, Crete, Greece, July 16-18, 2012
Revised Selected Papers, Part II

Springer

Volume Editors

George Bebis, E-mail: bebis@cse.unr.edu

Richard Boyle, E-mail: richard.boyle@nasa.gov

Bahram Parvin, E-mail: parvin@hpcrd.lbl.gov

Darko Koracin, E-mail: darko@dri.edu

Charless Fowlkes, E-mail: fowlkes@ics.uci.edu

Sen Wang, E-mail: sen.wang@kodak.com

Min-Hyung Choi, E-mail: min.choi@ucdenver.edu

Stephan Mantler, E-mail: step@stephanmantler.com

Jürgen Schulze, E-mail: jschulze@ucsd.edu

Daniel Acevedo, E-mail: daniel.acevedo@kaust.edu.sa

Klaus Mueller, E-mail: mueller@cs.sunysb.edu

Michael Papka, E-mail: papka@anl.gov

ISSN 0302-9743   e-ISSN 1611-3349
ISBN 978-3-642-33190-9   e-ISBN 978-3-642-33191-6
DOI 10.1007/978-3-642-33191-6
Springer Heidelberg Dordrecht London New York

Library of Congress Control Number: 2012945624

CR Subject Classification (1998): I.3-5, H.5.2, I.2.10, J.3, F.2.2, I.3.5

LNCS Sublibrary: SL 6 – Image Processing, Computer Vision, Pattern Recognition, and Graphics

Typesetting: Camera-ready by author, data conversion by Scientific Publishing Services, Chennai, India

Printed on acid-free paper

Springer is part of Springer Science+Business Media (www.springer.com)

Preface

It is with great pleasure that we welcome you to the proceedings of the 8th International Symposium on Visual Computing (ISVC 2012) that was held in Rethymnon, Crete, Greece. ISVC provides a common umbrella for the four main areas of visual computing including vision, graphics, visualization, and virtual reality. The goal is to provide a forum for researchers, scientists, engineers, and practitioners throughout the world to present their latest research findings, ideas, developments, and applications in the broader area of visual computing.

This year, the program consisted of 11 oral sessions, one poster session, seven special tracks, and six keynote presentations. The response to the call for papers was very good; we received over 200 submissions for the main symposium from which we accepted 68 papers for oral presentation and 35 papers for poster presentation. Special track papers were solicited separately through the Organizing and Program Committees of each track. A total of 45 papers were accepted for oral presentation in the special tracks.

All papers were reviewed with an emphasis on potential to contribute to the state of the art in the field. Selection criteria included accuracy and originality of ideas, clarity and significance of results, and presentation quality. The review process was quite rigorous, involving two–three independent blind reviews followed by several days of discussion. During the discussion period we tried to correct anomalies and errors that might have existed in the initial reviews. Despite our efforts, we recognize that some papers worthy of inclusion may have not been included in the program. We offer our sincere apologies to authors whose contributions might have been overlooked.

We wish to thank everybody who submitted their work to ISVC 2012 for review. It was because of their contributions that we succeeded in having a technical program of high scientific quality. In particular, we would like to thank the ISVC 2012 Area Chairs, the organizing institutions (UNR, DRI, LBNL, and NASA Ames), the industrial sponsors (BAE Systems, Intel, Ford, Hewlett Packard, Mitsubishi Electric Research Labs, Toyota, General Electric), the international Program Committee, the special track organizers and their Program Committees, the keynote speakers, the reviewers, and especially the authors that contributed their work to the symposium. In particular, we would like to express our appreciation to BAE Systems and Riad Hammoud for their sponsorship of the "best" paper award this year.

July 2012

George Bebis
Richard Boyle
Bahram Parvin
Darko Koracin
Charless Fowlkes
Sen Wang
Min-Hyung Choi
Stephan Mantler
Jürgen Schulze
Daniel Acevedo
Klaus Mueller
Michael Papka

Organization

ISVC 2012 Steering Committee

Bebis George	University of Nevada, Reno, USA
Boyle Richard	NASA Ames Research Center, USA
Parvin Bahram	Lawrence Berkeley National Laboratory, USA
Koracin Darko	Desert Research Institute, USA

ISVC 2012 Area Chairs

Computer Vision

Fowlkes Charless	University of California at Irvine, USA
Wang Sen	Kodak Research Labs, USA

Computer Graphics

Choi Min-Hyung	University of Colorado Denver, USA
Mantler Stephan	VRVis Research Center, Austria

Virtual Reality

Schulze Jurgen	University of California at San Diego, USA
Acevedo Daniel	KAUST, Saudi Arabia

Visualization

Mueller Klaus	Stony Brook University, USA
Papka Michael	Argonne National Laboratory, USA

Publicity

Albu Branzan Alexandra	University of Victoria, Canada

Local Arrangements

Zaboulis, Xenophon	Institute of Computer Science, FORTH, Greece

Special Tracks

Porikli, Fatih	Mitsubishi Electric Research Labs, USA

ISVC 2012 Keynote Speakers

Faloutsos Petros	York University, Canada
Coquillart Sabine	INRIA, France
Schmid Cordelia	INRIA, France
Cremers Daniel	Technical University of Munich Germany
Asari Vijayan	University of Dayton, USA
Randy Goebel	University of Alberta, Canada

ISVC 2012 International Program Committee

(Area 1) Computer Vision

Abidi Besma	University of Tennessee at Knoxville, USA
Abou-Nasr Mahmoud	Ford Motor Company, USA
Agaian Sos	University of Texas at San Antonio, USA
Aggarwal J.K.	University of Texas, Austin, USA
Albu Branzan Alexandra	University of Victoria, Canada
Amayeh Gholamreza	Eyecom, USA
Agouris Peggy	George Mason University, USA
Argyros Antonis	University of Crete, Greece
Asari Vijayan	University of Dayton, USA
Athitsos Vassilis	University of Texas at Arlington, USA
Basu Anup	University of Alberta, Canada
Bekris Kostas	University of Nevada at Reno, USA
Bensrhair Abdelaziz	INSA-Rouen, France
Bhatia Sanjiv	University of Missouri-St. Louis, USA
Bimber Oliver	Johannes Kepler University Linz, Austria
Bioucas Jose	Instituto Superior Técnico, Lisbon, Portugal
Birchfield Stan	Clemson University, USA
Boufama Boubakeur	University of Windsor, Canada
Bourbakis Nikolaos	Wright State University, USA
Brimkov Valentin	State University of New York, USA
Campadelli Paola	Università degli Studi di Milano, Italy
Cavallaro Andrea	Queen Mary, University of London, UK
Charalampidis Dimitrios	University of New Orleans, USA
Chellappa Rama	University of Maryland, USA
Chen Yang	HRL Laboratories, USA
Cheng Hui	Sarnoff Corporation, USA
Cochran Steven Douglas	University of Pittsburgh, USA
Chung, Chi-Kit Ronald	The Chinese University of Hong Kong, Hong Kong
Cremers Daniel	Technical University of Munich, Germany
Cui Jinshi	Peking University, China
Dagher Issam	University of Balamand, Lebanon

Kozintsev, Igor	Intel, USA
Kuno Yoshinori	Saitama University, Japan
Kim Kyungnam	HRL Laboratories, USA
Latecki Longin Jan	Temple University, USA
Lee D.J.	Brigham Young University, USA
Li Chunming	Vanderbilt University, USA
Li Xiaowei	Google Inc., USA
Lim Ser N.	GE Research, USA
Lin Zhe	Adobe, USA
Lisin Dima	VidoeIQ, USA
Lee Hwee Kuan	Bioinformatics Institute, A*STAR, Singapore
Lee Seong-Whan	Korea University, Korea
Leung Valerie	ONERA, France
Li Shuo	GE Healthecare, Canada
Li Wenjing	STI Medical Systems, USA
Loss Leandro	Lawrence Berkeley National Lab, USA
Luo Gang	Harvard University, USA
Ma Yunqian	Honyewell Labs, USA
Maeder Anthony	University of Western Sydney, Australia
Makrogiannis Sokratis	NIH, USA
Maltoni Davide	University of Bologna, Italy
Maybank Steve	Birkbeck College, UK
Medioni Gerard	University of Southern California, USA
Melenchón Javier	Universitat Oberta de Catalunya, Spain
Metaxas Dimitris	Rutgers University, USA
Miller Ron	Wright Patterson Air Force Base, USA
Ming Wei	Konica Minolta Laboratory, USA
Mirmehdi Majid	Bristol University, UK
Monekosso Dorothy	University of Ulster, UK
Morris Brendan	University of Nevada, Las Vegas, USA
Mulligan Jeff	NASA Ames Research Center, USA
Murray Don	Point Grey Research, Canada
Nait-Charif Hammadi	Bournemouth University, UK
Nefian Ara	NASA Ames Research Center, USA
Nicolescu Mircea	University of Nevada, Reno, USA
Nixon Mark	University of Southampton, UK
Nolle Lars	The Nottingham Trent University, UK
Ntalianis Klimis	National Technical University of Athens, Greece
Or Siu Hang	The Chinese University of Hong Kong, Hong Kong
Papadourakis George	Technological Education Institute, Greece
Papanikolopoulos Nikolaos	University of Minnesota, USA
Pati Peeta Basa	CoreLogic, India
Patras Ioannis	Queen Mary University, London, UK

Pavlidis Ioannis	University of Houston, USA
Petrakis Euripides	Technical University of Crete, Greece
Peyronnet Sylvain	LRI, University Paris-Sud, France
Pinhanez Claudio	IBM Research, Brazil
Piccardi Massimo	University of Technology, Australia
Pietikäinen Matti	LRDE/University of Oulu, Filand
Pitas Ioannis	Aristotle University of Thessaloniki, Greece
Porikli Fatih	Mitsubishi Electric Research Labs, USA
Prabhakar Salil	Digital Persona Inc., USA
Prati Andrea	University IUAV of Venice, Italy
Prokhorov Danil	Toyota Research Institute, USA
Pylvanainen Timo	Nokia Research Center, USA
Qi Hairong	University of Tennessee at Knoxville, USA
Qian Gang	Arizona State University, USA
Raftopoulos Kostas	National Technical University of Athens, Greece
Regazzoni Carlo	University of Genoa, Italy
Regentova Emma	University of Nevada, Las Vegas, USA
Remagnino Paolo	Kingston University, UK
Ribeiro Eraldo	Florida Institute of Technology, USA
Robles-Kelly Antonio	National ICT Australia (NICTA), Australia
Ross Arun	West Virginia University, USA
Samal Ashok	University of Nebraska, USA
Samir Tamer	Ingersoll Rand Security Technologies, USA
Sandberg Kristian	Computational Solutions, USA
Sarti Augusto	DEI Politecnico di Milano, Italy
Savakis Andreas	Rochester Institute of Technology, USA
Schaefer Gerald	Loughborough University, UK
Scalzo Fabien	University of California at Los Angeles, USA
Scharcanski Jacob	UFRGS, Brazil
Shah Mubarak	University of Central Florida, USA
Shi Pengcheng	Rochester Institute of Technology, USA
Shimada Nobutaka	Ritsumeikan University, Japan
Singh Rahul	San Francisco State University, USA
Skurikhin Alexei	Los Alamos National Laboratory, USA
Souvenir, Richard	University of North Carolina - Charlotte, USA
Su Chung-Yen	National Taiwan Normal University, Taiwan (R.O.C.)
Sugihara Kokichi	University of Tokyo, Japan
Sun Zehang	Apple, USA
Syeda-Mahmood Tanveer	IBM Almaden, USA
Tan Kar Han	Hewlett Packard, USA
Tan Tieniu	Chinese Academy of Sciences, China
Tavakkoli Alireza	University of Houston - Victoria, USA
Tavares, Joao	Universidade do Porto, Portugal

Teoh Eam Khwang	Nanyang Technological University, Singapore
Thiran Jean-Philippe	Swiss Federal Institute of Technology Lausanne (EPFL), Switzerland
Tistarelli Massimo	University of Sassari, Italy
Tong Yan	University of South Carolina, USA
Tsechpenakis Gabriel	University of Miami, USA
Tsui T.J.	Chinese University of Hong Kong, Hong Kong
Trucco Emanuele	University of Dundee, UK
Tubaro Stefano	DEI . Politecnico di Milano, Italy
Uhl Andreas	Salzburg University, Austria
Velastin Sergio	Kingston University London, UK
Veropoulos Kostantinos	GE Healthcare, Greece
Verri Alessandro	Università di Genova, Italy
Wang C.L. Charlie	The Chinese University of Hong Kong, Hong Kong
Wang Junxian	Microsoft, USA
Wang Song	University of South Carolina, USA
Wang Yunhong	Beihang University, China
Webster Michael	University of Nevada, Reno, USA
Wolff Larry	Equinox Corporation, USA
Wong Kenneth	The University of Hong Kong, Hong Kong
Xiang Tao	Queen Mary, University of London, UK
Xue Xinwei	Fair Isaac Corporation, USA
Xu Meihe	University of California at Los Angeles, USA
Yang Ming-Hsuan	University of California at Merced, USA
Yang Ruigang	University of Kentucky, USA
Yi Lijun	SUNY at Binghampton, USA
Yu Ting	GE Global Research, USA
Yu Zeyun	University of Wisconsin-Milwaukee, USA
Yuan Chunrong	University of Tübingen, Germany
Zabulis Xenophon	Foundation for Research and Technology - Hellas (FORTH), Greece
Zhang Yan	Delphi Corporation, USA
Cheng Shinko	HRL Labs, USA
Zhou Huiyu	Queen's University Belfast, UK

(Area 2) Computer Graphics

Abd Rahni Mt Piah	Universiti Sains Malaysia, Malaysia
Abram Greg	Texas Advanced Computing Center, USA
Adamo-Villani Nicoletta	Purdue University, USA
Agu Emmanuel	Worcester Polytechnic Institute, USA
Andres Eric	Laboratory XLIM-SIC, University of Poitiers, France
Artusi Alessandro	CaSToRC Cyprus Institute, Cyprus
Baciu George	Hong Kong PolyU, Hong Kong

Balcisoy Selim Saffet	Sabanci University, Turkey
Barneva Reneta	State University of New York, USA
Belyaev Alexander	Heriot-Watt University, UK
Benes Bedrich	Purdue University, USA
Berberich Eric	Max Planck Institute, Germany
Bilalis Nicholas	Technical University of Crete, Greece
Bimber Oliver	Johannes Kepler University Linz, Austria
Bohez Erik	Asian Institute of Technology, Thailand
Bouatouch Kadi	University of Rennes I, IRISA, France
Brimkov Valentin	State University of New York, USA
Brown Ross	Queensland University of Technology, Australia
Bruckner Stefan	Vienna University of Technology, Austria
Callahan Steven	University of Utah, USA
Capin Tolga	Bilkent University, Turkey
Chaudhuri Parag	Indian Institute of Technology Bombay, India
Chen Min	University of Oxford, UK
Cheng Irene	University of Alberta, Canada
Chiang Yi-Jen	Polytechnic Institute of New York University, USA
Comba Joao	Univ. Fed. do Rio Grande do Sul, Brazil
Crawfis Roger	Ohio State University, USA
Cremer Jim	University of Iowa, USA
Crossno Patricia	Sandia National Laboratories, USA
Culbertson Bruce	HP Labs, USA
Dana Kristin	Rutgers University, USA
Debattista Kurt	University of Warwick, UK
Deng Zhigang	University of Houston, USA
Dick Christian	Technical University of Munich, Germany
DiVerdi Stephen	Adobe, USA
Dingliana John	Trinity College, Ireland
El-Sana Jihad	Ben Gurion University of The Negev, Israel
Entezari Alireza	University of Florida, USA
Fabian Nathan	Sandia National Laboratories, USA
Fiorio Christophe	Université Montpellier 2, LIRMM, France
De Floriani Leila	University of Genova, Italy
Fuhrmann Anton	VRVis Research Center, Austria
Gaither Kelly	University of Texas at Austin, USA
Gao Chunyu	Epson Research and Development, USA
Geist Robert	Clemson University, USA
Gelb Dan	Hewlett Packard Labs, USA
Gotz David	IBM, USA
Gooch Amy	University of Victoria, Canada
Gu David	Stony Brook University, USA
Guerra-Filho Gutemberg	University of Texas Arlington, USA

Habib Zulfiqar	COMSATS Institute of Information Technology, Lahore, Pakistan
Hadwiger Markus	KAUST, Saudi Arabia
Haller Michael	Upper Austria University of Applied Sciences, Austria
Hamza-Lup Felix	Armstrong Atlantic State University, USA
Han JungHyun	Korea University, Korea
Hand Randall	Lockheed Martin Corporation, USA
Hao Xuejun	Columbia University and NYSPI, USA
Hernandez Jose Tiberio	Universidad de los Andes, Colombia
Huang Jian	University of Tennessee at Knoxville, USA
Huang Mao Lin	University of Technology, Australia
Huang Zhiyong	Institute for Infocomm Research, Singapore
Hussain Muhammad	King Saud University, Saudi Arabia
Jeschke Stefan	Vienna University of Technology, Austria
Joaquim Jorge	Instituto Superior Técnico, Portugal
Jones Michael	Brigham Young University, USA
Julier Simon J.	University College London, UK
Kakadiaris Ioannis	University of Houston, USA
Kamberov George	Stevens Institute of Technology, USA
Ko Hyeong-Seok	Seoul National University, Korea
Klosowski James	AT&T Labs, USA
Kobbelt Leif	RWTH Aachen, Germany
Kolingerova Ivana	University of West Bohemia, Czech Republic
Lai Shuhua	Virginia State University, USA
Lee Chang Ha	Chung-Ang University, Korea
Levine Martin	McGill University, Canada
Lewis R. Robert	Washington State University, USA
Li Frederick	University of Durham, UK
Lindstrom Peter	Lawrence Livermore National Laboratory, USA
Linsen Lars	Jacobs University, Germany
Loviscach Joern	Fachhochschule Bielefeld, University of Applied Sciences, Germany
Magnor Marcus	TU Braunschweig, Germany
Martin Ralph	Cardiff University, UK
Meenakshisundaram Gopi	University of California-Irvine, USA
Mendoza Cesar	Natural Motion Ltd., USA
Metaxas Dimitris	Rutgers University, USA
Mudur Sudhir	Concordia University, Canada
Myles Ashish	University of Florida, USA
Nait-Charif Hammadi	University of Dundee, UK
Nasri Ahmad	American University of Beirut, Lebanon
Noh Junyong	KAIST, Korea
Noma Tsukasa	Kyushu Institute of Technology, Japan
Okada Yoshihiro	Kyushu University, Japan

Olague Gustavo	CICESE Research Center, Mexico
Oliveira Manuel M.	Univ. Fed. do Rio Grande do Sul, Brazil
Owen Charles	Michigan State University, USA
Ostromoukhov Victor M.	University of Montreal, Canada
Pascucci Valerio	University of Utah, USA
Patchett John	Los Alamos National Lab, USA
Peters Jorg	University of Florida, USA
Pronost Nicolas	Utrecht University, The Netherlands
Qin Hong	Stony Brook University, USA
Rautek Peter	Vienna University of Technology, Austria
Razdan Anshuman	Arizona State University, USA
Renner Gabor	Computer and Automation Research Institute, Hungary
Rosen Paul	University of Utah, USA
Rosenbaum Rene	University of California at Davis, USA
Rudomin, Isaac	ITESM-CEM, Mexico
Rushmeier, Holly	Yale University, USA
Sander Pedro	The Hong Kong University of Science and Technology, Hong Kong
Sapidis Nickolas	University of Western Macedonia, Greece
Sarfraz Muhammad	Kuwait University, Kuwait
Scateni Riccardo	University of Cagliari, Italy
Schaefer Scott	Texas A&M University, USA
Sequin Carlo	University of California-Berkeley, USA
Shead Tinothy	Sandia National Laboratories, USA
Sourin Alexei	Nanyang Technological University, Singapore
Stamminger Marc	REVES/INRIA, France
Su Wen-Poh	Griffith University, Australia
Szumilas Lech	Research Institute for Automation and Measurements, Poland
Tan Kar Han	Hewlett Packard, USA
Tarini Marco	Università dell'Insubria (Varese), Italy
Teschner Matthias	University of Freiburg, Germany
Umlauf Georg	HTWG Constance, Germany
Vanegas Carlos	Purdue University, USA
Wald Ingo	University of Utah, USA
Walter Marcelo	UFRGS, Brazil
Wimmer Michael	Technical University of Vienna, Austria
Woodring Jon	Los Alamos National Laboratory, USA
Wylie Brian	Sandia National Laboratory, USA
Wyman Chris	University of Calgary, Canada
Wyvill Brian	University of Iowa, USA
Yang Qing-Xiong	University of Illinois at Urbana, Champaign, USA
Yang Ruigang	University of Kentucky, USA

Ye Duan   University of Missouri-Columbia, USA
Yi Beifang   Salem State University, USA
Yin Lijun   Binghamton University, USA
Yoo Terry   National Institutes of Health, USA
Yuan Xiaoru   Peking University, China
Zhang Jian Jun   Bournemouth University, UK
Zeng Jianmin   Nanyang Technological University, Singapore
Zara Jiri   Czech Technical University in Prague,
                              Czech Republic

(Area 3) Virtual Reality

Alcañiz Mariano   Technical University of Valencia, Spain
Arns Laura   Purdue University, USA
Balcisoy Selim   Sabanci University, Turkey
Behringer Reinhold   Leeds Metropolitan University, UK
Benes Bedrich   Purdue University, USA
Bilalis Nicholas   Technical University of Crete, Greece
Blach Roland   Fraunhofer Institute for Industrial Engineering,
                              Germany
Blom Kristopher   University of Barcelona, Spain
Bogdanovych Anton   University of Western Sydney, Australia
Borst Christoph   University of Louisiana at Lafayette, USA
Brady Rachael   Duke University, USA
Brega Jose Remo Ferreira   Universidade Estadual Paulista, Brazil
Brown Ross   Queensland University of Technology, Australia
Bues Matthias   Fraunhofer IAO in Stuttgart, Germany
Capin Tolga   Bilkent University, Turkey
Chen Jian   Brown University, USA
Cooper Matthew   University of Linköping, Sweden
Coquillart Sabine   INRIA, France
Craig Alan   NCSA University of Illinois at
                              Urbana-Champaign, USA
Cremer Jim   University of Iowa, USA
Edmunds Timothy   University of British Columbia, Canada
Egges Arjan   Universiteit Utrecht, The Netherlands
Encarnao L. Miguel   ACT Inc., USA
Figueroa Pablo   Universidad de los Andes, Colombia
Fox Jesse   Stanford University, USA
Friedman Doron   IDC, Israel
Fuhrmann Anton   VRVis Research Center, Austria
Gobron Stephane   EPFL, Switzerland
Gregory Michelle   Pacific Northwest National Lab, USA
Gupta Satyandra K.   University of Maryland, USA
Haller Michael   FH Hagenberg, Austria
Hamza-Lup Felix   Armstrong Atlantic State University, USA

Herbelin Bruno	EPFL, Switzerland
Hinkenjann Andre	Bonn-Rhein-Sieg University of Applied Sciences, Germany
Hollerer Tobias	University of California at Santa Barbara, USA
Huang Jian	University of Tennessee at Knoxville, USA
Huang Zhiyong	Institute for Infocomm Research (I2R), Singapore
Julier Simon J.	University College London, UK
Kaufmann Hannes	Vienna University of Technology, Austria
Kiyokawa Kiyoshi	Osaka University, Japan
Klosowski James	AT&T Labs, USA
Kozintsev	Igor, Intel, USA
Kuhlen Torsten	RWTH Aachen University, Germany
Lee Cha	University of California, Santa Barbara, USA
Liere Robert van	CWI, The Netherlands
Livingston A. Mark	Naval Research Laboratory, USA
Malzbender Tom	Hewlett Packard Labs, USA
Molineros Jose	Teledyne Scientific and Imaging, USA
Muller Stefan	University of Koblenz, Germany
Olwal Alex	MIT, USA
Owen Charles	Michigan State University, USA
Paelke Volker	Institut de Geomàtica, Spain
Peli Eli	Harvard University, USA
Pettifer Steve	The University of Manchester, UK
Piekarski Wayne	Qualcomm Bay Area R & D, USA
Pronost Nicolas	Utrecht University, The Netherlands
Pugmire Dave	Los Alamos National Lab, USA
Qian Gang	Arizona State University, USA
Raffin Bruno	INRIA, France
Raij Andrew	University of South Florida, USA
Reitmayr Gerhard	Graz University of Technology, Austria
Richir Simon	Arts et Metiers ParisTech, France
Rodello Ildeberto	University of Sao Paulo, Brazil
Sandor Christian	University of South Australia, Australia
Santhanam Anand	University of California at Los Angeles, USA
Sapidis Nickolas	University of Western Macedonia, Greece
Sherman Bill	Indiana University, USA
Slavik Pavel	Czech Technical University in Prague, Czech Republic
Sourin Alexei	Nanyang Technological University, Singapore
Steinicke Frank	University of Münster, Germany
Suma Evan	University of Southern California, USA
Stamminger Marc	REVES/INRIA, France
Srikanth Manohar	Indian Institute of Science, India
Vercher Jean-Louis	Université de la Méditerranée, France

Wald Ingo	University of Utah, USA
Wither Jason	University of California, Santa Barbara, USA
Yu Ka Chun	Denver Museum of Nature and Science, USA
Yuan Chunrong	University of Tübingen, Germany
Zachmann Gabriel	Clausthal University, Germany
Zara Jiri	Czech Technical University in Prague, Czech Republic
Zhang Hui	Indiana University, USA
Zhao Ye	Kent State University, USA

(Area 4) Visualization

Andrienko Gennady	Fraunhofer Institute IAIS, Germany
Avila Lisa	Kitware, USA
Apperley Mark	University of Waikato, New Zealand
Balázs Csébfalvi	Budapest University of Technology and Economics, Hungary
Brady Rachael	Duke University, USA
Benes Bedrich	Purdue University, USA
Bilalis Nicholas	Technical University of Crete, Greece
Bonneau Georges-Pierre	Grenoble Université, France
Bruckner Stefan	Vienna University of Technology, Austria
Brown Ross	Queensland University of Technology, Australia
Bühler Katja	VRVis Research Center, Austria
Callahan Steven	University of Utah, USA
Chen Jian	Brown University, USA
Chen Min	University of Oxford, UK
Chiang Yi-Jen	Polytechnic Institute of New York University, USA
Cooper Matthew	University of Linköping, Sweden
Chourasia Amit	University of California - San Diego, USA
Coming Daniel	Desert Research Institute, USA
Daniels Joel	University of Utah, USA
Dick Christian	Technical University of Munich, Germany
DiVerdi Stephen	Adobe, USA
Doleisch Helmut	SimVis GmbH, Austria
Duan Ye	University of Missouri-Columbia, USA
Dwyer Tim	Monash University, Australia
Entezari Alireza	University of Florida, USA
Ertl Thomas	University of Stuttgart, Germany
De Floriani Leila	University of Maryland, USA
Fujishiro Issei	Keio University, Japan
Geist Robert	Clemson University, USA
Gotz David	IBM, USA
Grinstein Georges	University of Massachusetts Lowell, USA
Goebel Randy	University of Alberta, Canada

Görg Carsten	University of Colorado at Denver, USA
Gregory Michelle	Pacific Northwest National Lab, USA
Hadwiger Helmut Markus	KAUST, Saudi Arabia
Hagen Hans	Technical University of Kaiserslautern, Germany
Hamza-Lup Felix	Armstrong Atlantic State University, USA
Healey Christopher	North Carolina State University at Raleigh, USA
Hege Hans-Christian	Zuse Institute Berlin, Germany
Hochheiser Harry	University of Pittsburgh, USA
Hollerer Tobias	University of California at Santa Barbara, USA
Hong Lichan	University of Sydney, Australia
Hong Seokhee	Palo Alto Research Center, USA
Hotz Ingrid	Zuse Institute Berlin, Germany
Huang Zhiyong	Institute for Infocomm Research (I2R), Singapore
Jiang Ming	Lawrence Livermore National Laboratory, USA
Joshi Alark	Yale University, USA
Julier Simon J.	University College London, UK
Kohlhammer Jörn	Fraunhofer Institut, Germany
Kosara Robert	University of North Carolina at Charlotte, USA
Laramee Robert	Swansea University, UK
Lee Chang Ha	Chung-Ang University, Korea
Lewis R. Robert	Washington State University, USA
Liere Robert van	CWI, The Netherlands
Lim Ik Soo	Bangor University, UK
Linsen Lars	Jacobs University, Germany
Liu Zhanping	University of Pennsylvania, USA
Ma Kwan-Liu	University of California at Davis, USA
Maeder Anthony	University of Western Sydney, Australia
Malpica Jose	Alcala University, Spain
Masutani Yoshitaka	The University of Tokyo Hospital, Japan
Matkovic Kresimir	VRVis Research Center, Austria
McCaffrey James	Microsoft Research / Volt VTE, USA
Melançon Guy	CNRS UMR 5800 LaBRI and INRIA Bordeaux Sud-Ouest, France
Miksch Silvia	Vienna University of Technology, Austria
Monroe Laura	Los Alamos National Labs, USA
Morie Jacki	University of Southern California, USA
Mudur Sudhir	Concordia University, Canada
Museth Ken	Linköping University, Sweden
Paelke Volker	Institut de Geomàtica, Spain
Peikert Ronald	Swiss Federal Institute of Technology Zurich, Switzerland
Pettifer Steve	The University of Manchester, UK

Pugmire Dave	Los Alamos National Lab, USA
Rabin Robert	University of Wisconsin at Madison, USA
Raffin Bruno	Inria, France
Razdan Anshuman	Arizona State University, USA
Rhyne Theresa-Marie	North Carolina State University, USA
Rosenbaum Rene	University of California at Davis, USA
Santhanam Anand	University of California at Los Angeles, USA
Scheuermann Gerik	University of Leipzig, Germany
Shead Tinothy	Sandia National Laboratories, USA
Shen Han-Wei	Ohio State University, USA
Sips Mike	Stanford University, USA
Slavik Pavel	Czech Technical University in Prague, Czech Republic
Sourin Alexei	Nanyang Technological University, Singapore
Thakur Sidharth	Renaissance Computing Institute (RENCI), USA
Theisel Holger	University of Magdeburg, Germany
Thiele Olaf	University of Mannheim, Germany
Toledo de Rodrigo	Petrobras PUC-RIO, Brazil
Tricoche Xavier	Purdue University, USA
Umlauf Georg	HTWG Constance, Germany
Viegas Fernanda	IBM, USA
Wald Ingo	University of Utah, USA
Wan Ming	Boeing Phantom Works, USA
Weinkauf Tino	Max-Planck-Institut für Informatik, Germany
Weiskopf Daniel	University of Stuttgart, Germany
Wischgoll Thomas	Wright State University, USA
Wylie Brian	Sandia National Laboratory, USA
Xu Wei	Stony Brook University, USA
Yeasin Mohammed	Memphis University, USA
Yuan Xiaoru	Peking University, China
Zachmann Gabriel	Clausthal University, Germany
Zhang Hui	Indiana University, USA
Zhao Ye	Kent State University, USA
Zheng Ziyi	Stony Brook University, USA
Zhukov Leonid	Caltech, USA

ISVC 2012 Special Tracks

1. 3D Mapping, Modeling and Surface Reconstruction

Organizers

Nefian Ara	Carnegie Mellon University/NASA Ames Research Center, USA
Edwards Laurence	NASA Ames Research Center, USA
Huertas Andres	NASA Jet Propulsion Lab, USA

2. Computational Bioimaging

Organizers

Tavares João Manuel R.S.	University of Porto, Portugal
Natal Jorge Renato	University of Porto, Portugal
Cunha Alexandre	Caltech, USA

3. Optimization for Vision, Graphics and Medical Imaging

Organizers

Komodakis Nikos	University of Crete, Greece
Kohli Pushmeet	Microsoft Research Cambridge, UK
Kumar Pawan	Ecole Centrale de Paris, France
Maeder Anthony	University of Western Sydney, Australia
Carsten Rother	Microsoft Research Cambridge, UK

4. Unconstrained Biometrics: Advances and Trends

Organizers

Proença Hugo	University of Beira Interior, Covilhã, Portugal
Du Yingzi	Indiana University-Purdue University Indianapolis, Indianapolis, USA
Scharcanski Jacob	Federal University of Rio Grande do Sul Porto Alegre, Brazil
Ross Arun	West Virginia University, USA

5. Intelligent Environments: Algorithms and Applications

Organizers

Bebis George	University of Nevada, Reno, USA
Nicolescu Mircea	University of Nevada, Reno, USA
Bourbakis Nikolaos	Wright State University, USA
Tavakkoli Alireza	University of Houston, Victoria, USA

6. Object Recognition

Organizers

Scalzo Fabien	University of California at Los Angeles, USA
Salgian Andrea	The College of New Jersey, USA

7. Face Processing and Recognition

Organizers

Hussain Muhammad	King Saud Univesity, Saudi Arabia
Muhammad Ghulam	King Saud Univesity, Saudi Arabia
Bebis George	University of Nevada, Reno, USA

Organizing Institutions and Sponsors

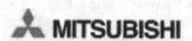

Table of Contents – Part II

ST: Intelligent Environments: Algorithms and Applications

Applications

Visualization III

Virtual Reality

ST: Face Processing and Recognition

Poster

Table of Contents – Part I

Calibration and 3D Vision

Object Recognition

Illumination, Modeling, and Segmentation

Visualization I

ST: 3D Mapping, Modeling and Surface Reconstruction

Motion and Tracking

Computer Graphics II

ST: Optimization for Vision, Graphics and Medical Imaging

HCI and Recognition

Visualization II

Iris Recognition in Image Domain: Quality-Metric Based Comparators*

Heinz Hofbauer, Christian Rathgeb, Andreas Uhl, and Peter Wild

Multimedia Signal Processing and Security Lab
Department of Computer Sciences, University of Salzburg, Austria
{hhofbaue,crathgeb,uhl,pwild}@cosy.sbg.ac.at

Abstract. Traditional iris recognition is based on computing efficiently coded representations of discriminative features of the human iris and employing Hamming Distance (HD) as fast and simple metric for biometric comparison in feature space. However, the International Organization for Standardization (ISO) specifies iris biometric data to be recorded and stored in (raw) image form (ISO/IEC FDIS 19794-6), rather than in extracted templates (e.g. iris-codes) achieving more interoperability as well as vendor neutrality. In this paper we propose the application of quality-metric based comparators operating directly on iris textures, i.e. without transformation into feature space. For this task, the Structural Similarity Index measure (SSIM), Local Edge Gradients metric (LEG), Natural Image Contour Evaluation (NICE), Edge Similarity Score (ESS) and Peak Signal to Noise ratio (PSNR) is evaluated. Obtained results on the CASIA-v3 iris database confirm the applicability of this type of iris comparison technique.

Keywords: Iris reconition, biometric comparators, image quality-metrics, image domain.

1 Introduction

Iris recognition is considered one of the most reliable biometric technologies obtaining recognition rates above 99% and equal error rates of less than 1% on several data sets. Compared to other modalities, the iris offers the advantages of being extractable at-a-distance and on-the-move [12], and numerous iris feature extraction methods have been proposed continuously over the past decade [2]. Still, the processing chain of traditional iris recognition (and other biometric) systems has been left almost unchanged, following Daugman's approach [3] consisting of (1) *segmentation and preprocessing* normalizing the iris texture by unrolling into doubly-dimensionless coordinates, (2) *feature extraction* computing a binary representation of discriminative patterns of the rectified iris texture, and (3) *biometric comparison* in feature space involving the fractional HD as dissimilarity measure, see Fig. 1.

---

* This work has been supported by the Austrian Science Fund, project no. L554-N15 and the Austrian FIT-IT Trust in IT-Systems, project no. 819382.

G. Bebis et al. (Eds.): ISVC 2012, Part II, LNCS 7432, pp. 1–10, 2012.

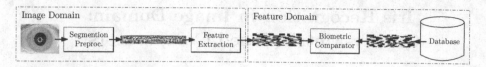

Fig. 1. Common processing chain: images are preprocessed and adaquate feature extractors generate (mostly binary) feature vectors, stored as biometric templates

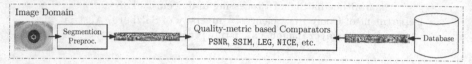

Fig. 2. Proposed processing chain: images are preprocessed and quality-metric based comparators (operating in image domain) estimate similarities between pairs of images

In accordance with the ISO/IEC FDIS 19794-6 standard image metric-based iris biometric systems are presented. ISO/IEC FDIS 19794-6 compliant databases, which store raw iris biometric data, enable the incorporation of future improvements (e.g. in segmentation stage) without re-enrollment of registered subjects. While the extraction of rather short (a few hundred bytes) binary feature vectors provides a compact storage and rapid comparison of biometric templates, information loss is inevitable. This motivates an evaluation of comparators operating in image domain, in particular image metrics, which, to our knowledge, have not been applied to iris recognition. The contribution of this work is the application of image metrics for the purpose of iris recognition, see Fig. 2. The proposed architecture involves several benefits: (1) The problem of iris recognition can be mapped to a standard image processing problem, benefiting of results in this domain. (2) Features and comparators can be easily replaced without the necessity of re-enrollment, as the entire iris image is stored for comparison and available as reference for future comparators. (3) The approach allows for easier continuous updates, e.g. by averaging iris textures each time of successful authentication. (4) Quality-based metrics in the image domain may be combined with other image domain methods, such as SIFT-based [1] or Phase-based [9] methods. Of course, the proposed technique may also be combined with traditional feature-scale methods (in which case feature extraction has to be incorporated into the comparison module), since global features used by image metrics complement the mostly localized biometric features. (5) Finally, new techniques like [6], [14] have shown, that an incremental refinement of comparison decisions saves precious computation time and can target the drawback of traditional image quality metrics being considered slow compared to trivial metrics, such as fractional HD. Regarding the security of the stored templates it is suggested to apply standard encryption algorithms (e.g. AES) in order to protect user privacy.

The following sections are organized as follows: related work is reviewed in Section 2. The proposed approach and quality metrics are introduced in Section 3. Experiments are outlined in Section 4 using an open iris database and comparing both original as well as normalized iris images. Finally, Section 5 summarizes the paper.

2 Related Work

In the context of iris biometrics, image quality metrics are largely understood as domain-specific indicators, e.g. focus assessment or measurement of pupil/iris diameter ratio, to be considered for quality checks rejecting samples if insufficiently suited for comparison [18]. Such metrics have also been applied for dynamic matcher selection in biometric fusion scenarios [20], i.e. quality is employed to predict matching performance and to select the comparator or adjust weighting of the fusion rule. Our approach is different in employing general purpose image quality metrics and their ability to measure the degree of similarity of image pairs if one of both images is subjected to a (more or less severe) degradation in quality. In our model, the degradation of a sample to be compared is not caused by compression, but by biometric noise factors (time, illumination, etc.), and the stored biometric gallery template represents the (updated) ideal representation of the biometric property of an individual.

Pursuing the idea of employing iris comparison in the image domain, the following works need to be acknowledged: Miyazawa *et al.* [13] identify the problem of feature-based iris recognition being highly dependent on the feature extraction process varying based on environmental factors, which can be avoided by computing features in the image domain. The authors suggest to apply 2D Fourier Phase components of iris images. This scheme is extended by Krichen *et al.* [9], who propose to combine global and local Gabor (i.e. wavelet instead of Fourier coefficients) phase-correlation-based iris matching directly on enhanced (using adaptive histogram equalization) iris textures for unconstrained acquisition procedures. They employ normalized cross-correlation and a Peak to Slob Ratio (PSR) as comparator, which uses mean and standard deviation of the correlation matrix. As Local correlation-based method they correlated sub-images of fixed size using correlation peak in terms of PSR and peak position of each window computing a score out of means and standard deviation. Alonso-Fernandez *et al.* [1] propose the application of Scale Invariant Feature Transformation (SIFT) for recognition, as a means of processing without transformation to polar coordinates, thus permitting less constrained image acquisition conditions. SIFT features can be extracted from original templates in scale space and matched using texture information around the feature points. Kekre *et al.* [7], [4] use the image feature set extracted from Haar Wavelets at various levels of decomposition and from walshlet pyramid for recognition. Simple Euclidean distance on the feature set is applied as the similarity measure. Furthermore, numerous advanced iris biometric comparators have been proposed [15].

3 Iris Recognition in the Image Domain

Given an image of the human eye as shown in Fig. 3 (a), the first task is the transformation into Daugman's rubbersheet model. While any accurate segmentation technique may be applied for this task, we employ the preprocessing chain in [19]. This method applies (1) reflection removal with image inpainting, (2) assessment of edge magnitude and orientation by a Weighted Adaptive Hough

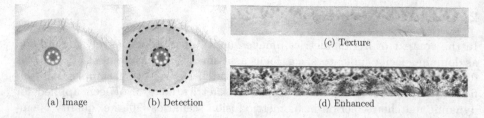

(a) Image   (b) Detection   (c) Texture
                                          (d) Enhanced

Fig. 3. Preprocessing: (a) image of eye (b) detection of pupil and iris (c) unrolled iris texture (d) preprocessed iris texture

Transform for initial center detection, followed by (3) polar and ellipsopolar transforms to detect boundary candidates, which are evaluated to (4) select the most reliable ones to be used for un-wrapping the image to a rectangular texture of 512×64 pixels. Since image-based methods are largely affected by different illumination [9], we further enhance the iris texture applying CLAHE (Contrast Limited Adaptive Histogram Equalization) [22], see Fig. 3 (b)-(d).

In image-domain iris processing, we store one full reference iris texture O per user. While template-updates can easily be handled in such a scenario, for evaluations we employ enrollment using the first eye image per user only. In order to score an authentication attempt given a claimed identity, the corresponding template image O is compared with the current sample image I. Both images of $W \times H$ pixels are compared by employing one of the following quality metrics $Q(s(I, m), O)$, where $s(I, m)$ denote a shifting of m pixels to the left or right in order to obtain a rotation invariant technique. For I and O the b bits per pixel are used with a maximum pixel value of $M = 2^b$.

All of the following image metrics[1] are full reference metrics, meaning they utilize information from the original and comparison image to calculate an assessment of the visual similarity. The following subsections describe details of applied image metrics and show, which features are used in the calculation of the quality assessment.

3.1 Peak Signal to Noise Ratio (PSNR)

The PSNR is still widely used because it is unrivaled in speed and ease of use.

The following steps are performed to calculate the PSNR.

Step 1: Calculate the mean squared error MSE $= \frac{1}{WH} \sum_{i=1}^{W} \sum_{j=i}^{H} (I(i, j) - O(i, j))^2$

Step 2: The PSNR is calculated:

$$\text{PSNR} = 10 \log_{10} \left(\frac{M^2}{\text{MSE}} \right). \tag{1}$$

---

[1] The implementation is available online at http://www.wavelab.at/sources/VQI/

3.2 Structural Similarity Index Measure (SSIM)

The structural similarity index measure (SSIM) by Wang et al. [21] uses the local luminance as well as global contrast and a structural feature to calculate a score as follows.

Step 1: Each image is transformed by convolution with a 11×11 Gaussian filter.

Step 2: The luminance, contrast and structural scores can be calculated and combined in one step as follows.

$$\text{SSIM}(I, O) = \frac{(2\mu_I\mu_O + c_1)(2\sigma_{IO} + c_2)}{(\mu_I^2 + \mu_O^2 + c_1)(\sigma_I^2 + \sigma_O^2 + c_2)}, \tag{2}$$

where μ_I is the average pixel value of image I, σ_I^2 is the variance of pixel values of image I and σ_{IO} is the covariance of I and O. The variables $c_1 = (k_1 M)^2$ and $c_2 = (k_2 M)^2$, with $k_1 = 0.01$ and $k_2 = 0.03$, are used to stabilize the division.

3.3 Local Edge Gradients Metric (LEG)

The image metric based on local edge gradients was introduced by Hofbauer and Uhl [5] and uses luminance and localized edge information from different frequency domains.

Step 1: First the global luminance difference between I and O is calculated as $\text{LUM}(I, O) = 1 - \sqrt{\frac{|\mu(O) - \mu(I)|}{M}}$, where $\mu(X) = \frac{1}{WH} \sum_{x=1}^{W} \sum_{y=1}^{H} X(x, y)$, and $X(x, y)$ is the pixel value of image X at position x, y.

Step 2: One step wavelet decomposition with Haar wavelets resulting in four sub images for each image X denoted as X_0 for the LL-subband, and X_1, X_2, X_3 for LH, HH and HL subbands, respectively.

Step 3: A local edge map is calculated for each position x, y in the image, reflecting the change in coarse structure of the image.

$$\text{LE}(I, O, x, y) = \begin{cases} 1 & \text{if } \text{EDC}(I, O, x, y) = 8, \\ 0.5 & \text{if } \text{EDC}(I, O, x, y) = 7, \\ 0 & \text{otherwise.} \end{cases}$$

$$\text{EDC}(I, O, x, y) = \sum_{p \in N(x,y)} \text{ED}(I, O, x, y, p)$$

$$\text{ED}(I, O, x, y, p) = \begin{cases} 1 & \text{if } I(x, y) < I(p) \text{ and } O(x, y) < O(p), \\ 1 & \text{if } I(x, y) > I(p) \text{ and } O(x, y) > O(p), \\ 0 & \text{otherwise.} \end{cases}$$

where $N(x, y)$ is the eight neighborhood of the pixel x, y.

Step 4: In order to assess the contrast changes a difference of gradients in a neighborhood is calculated

$$\text{LED}(I, O, x, y) = \frac{1}{8} \sum_{p \in N(x,y)} \left(1 - \sqrt{\frac{|LD(I, O, x, y, p)|}{M}} \right)^2,$$

where $\text{LD}(I, O, x, y, p) = (O(x, y) - O(p)) - (I(x, y) - I(p))$.

Step 5: The edge score is calculated by combining local edge conformity (LE) and local edge difference (LED) into

$$\text{ES}(I,O) = \frac{4}{WH} \sum_{x=1}^{\frac{W}{2}} \sum_{y=1}^{\frac{H}{2}} \left(\text{LE}(I_0, O_0, x, y) * \frac{1}{3} \sum_{i=1}^{3} \text{LED}(I_i, O_i, x, y) \right).$$

Step 6: The LEG visual quality index is calculated by combining ES and LUM.

$$\text{LEG}(I,O) = \text{LUM}(I,O)\,\text{ES}(I,O). \tag{3}$$

3.4 Natural Image Contour Evaluation (NICE)

The NICE quality index by Rouse and Hemami [17,16] uses gradient maps, adjusted for possible image shift by using a morphological dilation with a plus shaped structuring element. The actual score is computed by doing a thresholding on the image and calculating differences. The following steps are used to calculate the NICE score.

Step 1: Gradient amplitude image $\hat{I}$ is generated from I such that for $i \in [1, \ldots, W]$ and $j \in [1, \ldots, H]$ $\hat{I}$ is defined as $\hat{I}(i,j) = \sqrt{S_x(I,i,j)^2 + S_x(I,i,j)^2}$, where $\hat{I}(i,j)$ is the pixel value at location i,j and $S_x(I,i,j)$ and $S_y(I,i,j)$ are the results of a Sobel filter at position i,j in image I in direction x and y, respectively. Likewise $\hat{O}$ is generated from O.

Step 2: A binary image $B_{\hat{I}}$ is generated by thresholding with the average gradient amplitude value. That is, $B_{\hat{I}}(i,j) = 1$ if $\hat{I}(i,j) > T_{\hat{I}}$ and $B_{\hat{I}}(i,j) = 0$ otherwise, where $T_{\hat{I}} = \frac{1}{WH} \sum_{i=1}^{W} \sum_{j=1}^{H} \hat{I}(i,j)$.

Step 3: The binary image $B_{\hat{I}}$ is transformed into $B_{\hat{I}}^{+}$ by applying a morphological dilation with a plus shaped structuring element. That is, each pixel $B_{\hat{I}}^{+}(i,j)$ is set to 1 if at least one of the 4−connected neighbours of $B_{\hat{I}}(i,j)$ or $B_{\hat{I}}(i,j)$ is 1, otherwise $B_{\hat{I}}^{+}(i,j) = 0$.

Step 4: The NICE score is calculated based on the normalized Hamming distance as

$$\text{NICE}(O,I) = \frac{\sum_{i=1}^{W} \sum_{j=1}^{H} (B_{\hat{O}}^{+}(i,j) - B_{\hat{I}}^{+}(i,j))^2}{\sum_{i=1}^{W} \sum_{j=1}^{H} B_{\hat{O}}^{+}(i,j)} \tag{4}$$

3.5 Edge Similarity Score (ESS)

The ESS was introduced by Mao and Wu [11] and uses localized edge information to compare two images.

Step 1: Each image is separate into N blocks of size 8×8.

Step 2: For each image I a Sobel edge detection filter is used on each block i to find the most prominent edge direction e_I^i and quantized into one of eight directions (each corresponding to 22.5°). Edge direction 0 is used if no edge was found in the block.

Step 3: Calculate the ESS based on the prominent edges of each block:

$$\text{ESS} = \frac{\sum_{i=1}^{N} w(e_I^i, e_O^i)}{\sum_{i=1}^{N} c(e_I^i, e_O^i)}, \tag{5}$$

where $w(e_1, e_2)$ is a weighting function defined as

$$w(e_1, e_2) = \begin{cases} 0 & \text{if } e_1 = 0 \text{ or } e_2 = 0 \\ |cos(\phi(e_1) - \phi(e_2))| & \text{otherwise,} \end{cases}$$

where $\phi(e)$ is the representative edge angle for an index e, and $c(e_1, e_2)$ is an indicator function defined as $c(e_1, e_2) = 0$ if $e_1 = e_2 = 0$ and $c(e_1, e_2) = 1$ otherwise. In cases where $\sum_{i=1}^{N} c(e_I^i, e_O^i) = 0$ the ESS is set to 0.5.

4   Experiments

Experiments are carried out on the CASIA-v3-Interval iris database[2] using left-eye images only. The database consists of good quality 320×280 pixel NIR illuminated indoor images where the applied test set consists of 1307 instances, a sample is shown in Fig. 3 (a).

Recognition accuracy is evaluated in terms of false non match rate (FNMR) at a certain false match rate (FMR). The FNMR defines the proportion of verification transactions with truthful claims of identity that are incorrectly rejected, and the FMR defines the proportion of verification transactions with wrongful claims of identity that are incorrectly confirmed (ISO/IEC FDIS 19795-1), in particular, ZeroFMR defines the FNMR at a FMR of 0.1%. As score distributions overlap the Equal Error Rate (EER) of the system is defined (FNMR = FMR). At all authentication attempts 7 circular texture-shifts are performed in each direction for all comparators. A summary of obtained EERs and ZeroFMR rates at the corresponding decision thresholds for the underlying image quality metrics is given in Table 1. Receiver operating characteristics, which illustrate the tradeoff between FMR and FNMR, are plotted in Fig. 4 for experiments evaluating (a) metrics, as well as (b) impact of the used image type: original image, texture after segmentation, and enhanced texture after CLAHE normalization. Score distributions for each metric (normalized to $[0, 1]$) with respect to genuine (intra-) and impostor (inter-personal) comparisons are illustrated in Fig. 5.

4.1   Which Quality Metrics Are Useful Iris Biometric Comparators?

With respect to accuracy, the ranking of metrics is as follows: SSIM, LEG, PSNR, NICE, and ESS, with the first three metrics exhibiting EERs of less

---

[2] The Center of Biometrics and Security Research, CASIA Iris Image Database, http://www.idealtest.org

Table 1. Recognition performance of Quality Metrics

Algorithm	Type	EER	ZeroFMR	Threshold
SSIM	Enhanced	3.40%	5.34%	0.868
LEG	Enhanced	3.99%	7.72%	0.785
NICE	Enhanced	5.14%	13.32%	0.526
ESS	Enhanced	9.61%	25.97%	0.311
PSNR	Enhanced	4.21%	10.33%	0.592
PSNR	Texture	18.88%	65.37%	0.478
PSNR	Image	23.01%	80.67%	0.638
Ma *et al.*	Iris-Code	1.83%	2.02%	–
Ko *et al.*	Iris-Code	4.36%	18.45%	–

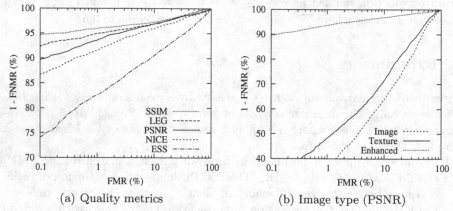

(a) Quality metrics   (b) Image type (PSNR)

Fig. 4. Receiver operating characteristics by (a) quality metric, and (b) image type

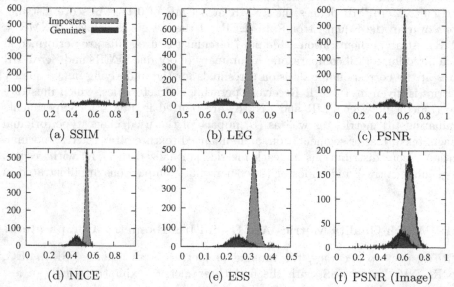

(a) SSIM   (b) LEG   (c) PSNR

(d) NICE   (e) ESS   (f) PSNR (Image)

Fig. 5. Genuine and impostor score distributions for (a) SSIM, (b) LEG, (c) PSNR, (d) NICE, (e) ESS for enhanced textures and (f) PSNR on original images

than 5%. It is interesting to see, that PSNR with 4.21% EER performs quite well on the enhanced textures although it is the most simple metric. However, for high security applications with requested low FMR, SSIM with 5.34% ZeroFMR compared to 10.33% for PSNR is clearly the better alternative. Considering recognition accuracy image metrics do not outperform feature-based techniques [2]. For instance, on the same dataset re-implementations of the approaches of Ma *et al.* [10] and Ko *et al.* [8], which extract binary iris-codes obtain EERs of 1.83% and 4.36%, respectively (see Table 1). However, image metrics are rather useful as additional features in fusion scenarios.

4.2   How Useful Is Texture Enhancement and Preprocessing?

In a second experiment, we tested the effect of texture enhancement and segmentation on iris recognition accuracy of quality metrics using PSNR as reference metric. Obtained results indicate a high degradation in case texture enhancement steps are skipped (18.88% EER instead of 4.21%). Recognition from the original eye images (without segmentation) further degraded results (23.01% EER), thus normalizaton and enhancement steps accounting for different illumination enriching the texture in the image (see Fig. 3) are extremely useful.

5   Summary

This paper applies quality metrics in image domain to the problem of iris recogniton. As opposed to the view that original iris textures exhibit too much noisy information to be used directly for comparison, we found that some metrics (SSIM, LEG, PSNR) provide quite reasonable accuracy (3.4%, 3.99% and 4.21% EER, respectively). The proposed architecture alleviates continuous template updates and enables a transparent replacement of comparators without re-enrollment. Iris texture enhancement is found to be essential to the accuracy of iris recognition in the image domain. Future work is targeted at a sophisticated analysis of fusion approaches of image-domain methods and a combination with serial comparison techniques to accelerate processing time.

References

1. Alonso-Fernandez, F., Tome-Gonzalez, P., Ruiz-Albacete, V., Ortega-Garcia, J.: Iris recognition based on sift features. In: Int'l Conf. on Biometrics, Ident. and Sec (BIdS), pp. 1–8 (2009)
2. Bowyer, K.W., Hollingsworth, K., Flynn, P.J.: Image understanding for iris biometrics: A survey. Comp. Vis. Image Underst. 110(2), 281 (2008)
3. Daugman, J.: How iris recognition works. IEEE Trans. Circ. and Syst. for Video Techn. 14(1), 21–30 (2004)
4. Kekre, H.B., Thepade, S.D., Jain, J., Agrawal, N.: Iris recognition using texture features extracted from haarlet pyramid. Int'l J. of Comp. App. 11(12), 1–5 (2010); Found. Comp. Sc.

5. Hofbauer, H., Uhl, A.: An Effective and Efficient Visual Quality Index based on Local Edge Gradients. In: IEEE 3rd Europ. Workshop on Visual Inf. Proc., p. 6 (2011)
6. Hollingsworth, K.P., Bowyer, K.W., Flynn, P.J.: The best bits in an iris code. IEEE Trans. on Pattern Anal. and Mach. Intell. 31(6), 964–973 (2009)
7. Kekre, H.B., Thepade, S.D., Jain, J., Agrawal, N.: Iris recognition using texture features extracted from walshlet pyramid. In: Prof. Int'l Conf. & Workshop on Emerging Trends in Techn (ICWET), pp. 76–81. ACM (2011)
8. Ko, J.-G., Gil, Y.-H., Yoo, J.-H., Chung, K.-I.: A novel and efficient feature extraction method for iris recognition. ETRI Journal 29(3), 399–401 (2007)
9. Krichen, E., Garcia-Salicetti, S., Dorizzi, B.: A new phase-correlation-based iris matching for degraded images. IEEE Trans. on Systems, Man, and Cyb., Part B 39(4), 924–934 (2009)
10. Ma, L., Tan, T., Wang, Y., Zhang, D.: Efficient iris recognition by characterizing key local variations. IEEE Trans. on Image Processing 13(6), 739–750 (2004)
11. Mao, Y., Wu, M.: Security evaluation for communication-friendly encryption of multimedia. In: IEEE Int'l Conf. on Image Proc, ICIP (2004)
12. Matey, J., Naroditsky, O., Hanna, K., Kolczynski, R., LoIacono, D., Mangru, S., Tinker, M., Zappia, T., Zhao, W.Y.: Iris on the move: Acquisition of images for iris recognition in less constrained environments. Proc. IEEE 94, 1936–1947 (2006)
13. Miyazawa, K., Ito, K., Aoki, T., Kobayashi, K., Nakajima, H.: An efficient iris recognition algorithm using phase-based image matching. In: IEEE Int'l Conf. on Image Proc. (ICIP), pp. 49–52 (2005)
14. Rathgeb, C., Uhl, A., Wild, P.: Incremental iris recognition: A single-algorithm serial fusion strategy to optimize time complexity. In: Proc. Int'l Conf. on Biometrics: Theory, App., and Syst (BTAS), pp. 1–6 (2010)
15. Rathgeb, C., Uhl, A., Wild, P.: Iris-biometric comparators: Minimizing trade-offs costs between computational performance and recognition accuracy. In: Proc. Int'l Conf. on Imaging for Crime Det. and Prev (ICDP), pp. 1–7 (2011)
16. Rouse, D., Hemami, S.S.: Natural image utility assessment using image contours. In: IEEE Int'l Conf. on Image Proc (ICIP), pp. 2217–2220 (2009)
17. Rouse, D., Hemami, S.S.: The role of edge information to estimate the perceived utility of natural images. In: Western New York Image Proc. Workshop (WNYIP), p. 4 (2009)
18. Tomeo-Reyes, I., Liu-Jimenez, J., Rubio-Polo, I., Fernandez-Saavedra, B.: Quality metrics influence on iris recognition systems performance. In: IEEE Int'l Carnahan Conf. on Security Technology (ICCST), pp. 1–7 (2011)
19. Uhl, A., Wild, P.: Weighted adaptive hough and ellipsopolar transforms for real-time iris segmentation. In: Proc. Int'l Conf. on Biometrics, ICB (to appear, 2012)
20. Vatsa, M., Singh, R., Noore, A., Ross, A.: On the dynamic selection of biometric fusion algorithms. IEEE Trans. on Inf. Forensics and Sec. 10(3), 470–479 (2010)
21. Wang, Z., Bovik, A.C., Sheikh, H.R., Simoncelli, E.P.: Image quality assessment: from error visibility to structural similarity. IEEE Trans. on Image Proc. 13(4), 600–612 (2004)
22. Zuiderveld, K.: Contrast limited adaptive histogram equalization. In: Heckbert, P.S. (ed.) Graphics Gems IV, pp. 474–485. Morgan Kaufmann (1994)

Gait Recognition Based
on Normalized Walk Cycles

Jan Sedmidubsky, Jakub Valcik, Michal Balazia, and Pavel Zezula

Masaryk University, Botanicka 68a, 602 00 Brno, Czech Republic

Abstract. We focus on recognizing persons according to the way they walk. Our approach considers a human movement as a set of trajectories formed by specific anatomical landmarks, such as hips, feet, shoulders, or hands. The trajectories are used for the extraction of distance-time dependency signals that express how a distance between a pair of specific landmarks on the human body changes in time as the person walks. The collection of such signals characterizes a gait pattern of person's walk. To determine the similarity of gait patterns, we propose several functions that compare various combinations of extracted signals. The gait patterns are compared on the level of individual walk cycles in order to increase the recognition effectiveness. The results evaluated on a 3D database of walking humans achieved the recognition rate up to 96 %.

1 Introduction

Human gait is defined as the manner in which a person walks. Recent studies have proven that gait can be seen as a biometric characteristic and used as a signature to recognize people. The great advantage is its possibility to be captured at a distance, even surreptitiously. However, effectiveness of gait recognition methods strongly depends on many factors such as camera view, carried accessories, person's clothes, or walking surface.

Gait recognition methods can be divided to two major categories: *model-based* and *appearance-based* approaches. Appearance-based methods [4,6,10] generally characterize the whole motion pattern of the human body by a compact representation regardless of the underlying structure. They usually combine extracted human silhouettes from each video frame into a single gait image that preserves temporal information. Nevertheless, recognition based on comparison of such gait images is restricted to one view point. To use gait features in unconstrained views, we need to adopt the model-based concept in order to estimate 3D models of walking persons.

The model-based concept fits various kinds of stick figures onto the walking human. The recovered stick structure allows accurate measurements to perform, independent of camera view. BenAbdelkader et al. [1] computed an average stride length and cadence of feet and used just both these numbers for gait recognition. Tanawongsuwan and Bobick [7] compared joint-angle trajectories of hips, knees, and feet by the dynamic time warping (DTW) similarity function, with normalization for noise reduction. Cunado et al. [5] used a pendulum model

G. Bebis et al. (Eds.): ISVC 2012, Part II, LNCS 7432, pp. 11–20, 2012.

where thigh's motion and rotation were analyzed using a Fourier transformation. The approach of Wang et al. [9] measured a mean shape of silhouettes gained by Procrustes's analysis and combined it with absolute positions of angles of specific joints. Yoo et al. [11] compared sequences of 2D stick figures by a back-propagation neural network algorithm. Recent advances in gait recognition have been surveyed in [3].

We adopt the model-based concept to recover a 3D stick figure of the human body by capturing spatial coordinates of significant anatomical landmarks, such as hands, hips, knees, or feet. The recovered stick figure is used to compute distance-time dependency signals that express how a distance between two specific joints of the human body changes in time. The collection of such signals defines a gait pattern of person's walk (Section 2). In Section 3, a novel similarity function for comparing gait patterns is introduced. To effectively compare gait patterns, we normalize them to encapsulate signals corresponding exclusively to a single walk cycle (Section 4). In Section 5, the influence of normalization and effectiveness of similarity function is deeply evaluated on a real-life 3D motion database. We distinguish from existing approaches by taking also movements of arms into account and by comparing gait patterns on the basis of normalized walk cycles.

The main contributions of this paper constitute: (1) proposal of a gait pattern that encapsulates information about a person's walk in the form of viewpoint invariant distance-time dependency signals, (2) introduction of a novel similarity function for comparing gait patterns, taking movements of legs and arms into account, and (3) experimental evaluation of recognition rate of the proposed function and its modifications, based also on diverse normalization methods.

2   Gait Representation

We introduce a structural model of a human body. This model is used for the extraction of viewpoint invariant planar signals. The collection of such planar signals forms a gait pattern of person's walk. Recognition of persons is based on comparing their gait patterns by a sophisticated similarity function.

2.1   Model Definition

We define the human model by a set of significant anatomical landmarks: clavicles C_L/C_R, elbows E_L/E_R, hands H_L/H_R, hips L_L/L_R, knees K_L/K_R, and feet F_L/F_R. The subscripts L and R express whether a given landmark is situated at the left or right side of the human body, respectively (see Figure 1). In particular, we suggest a 12-parameter model $\mathcal{M}$:

$$\mathcal{M} = (C_L, C_R, E_L, E_R, H_L, H_R, L_L, L_R, K_L, K_R, F_L, F_R),$$

where each landmark is described by a 3-dimensional *body point* $P_f = (x_f, y_f, z_f)$ captured at a given video frame $f \in F$. The domain $F = \{1, \ldots, n\}$ refers to the

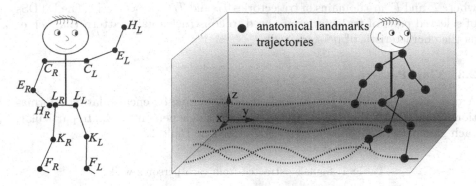

Fig. 1. Location of specific anatomical landmarks and their trajectories captured as the person walks

length of input video in terms of number of frames, i.e., to the number of times a specific body point has been captured.

A collection of consecutive points represents a motion *trajectory* (see Figure 1). Formally, each point P moving in time, as the person walks, constitutes a discrete trajectory $\mathcal{T}_P$, defined as:

$$\mathcal{T}_P = \{P_f \mid f \in F\}.$$

The discrete domain F allows us to utilize metric functions for point-by-point comparison of trajectories. Trajectories cannot be used directly for recognition because the values of their spatial coordinates depend on the calibration of system that detects and estimates particular coordinates. Moreover, persons do not walk in the same direction, which makes trajectories of different walks (even of the same person) incomparable. We rather compute distances between selected pairs of trajectories to construct distance-time dependency signals. Such signals are already independent of the walk direction and system calibration.

2.2 Distance-Time Dependency Signal

A *distance-time dependency signal* (DTDS) expresses how a distance between two trajectories changes over time as the person walks. The variation in these distances is primarily exploited as information for human recognition. The distance is always measured between two points $P_f = (x_f, y_f, z_f)$ and $P'_f = (x'_f, y'_f, z'_f)$ captured at the same video frame on the basis of the Euclidean distance L_2:

$$L_2\left(P_f, P'_f\right) = \sqrt{\left(x_f - x'_f\right)^2 + \left(y_f - y'_f\right)^2 + \left(z_f - z'_f\right)^2}.$$

A distance-time dependency signal $\mathcal{S}_{PP'}$ between two trajectories $\mathcal{T}_P$ and $\mathcal{T}_{P'}$ of points P and P' is formally defined as:

$$\mathcal{S}_{PP'} = \left\{d_f \in \mathbb{R}_0^+ \mid f \in F \cap F' \wedge d_f = L_2\left(P_f, P'_f\right)\right\},$$

where F and F' are domains of trajectories $\mathcal{T}_P$ and $\mathcal{T}_{P'}$, respectively. The DTDSs of selected pairs of trajectories are used to construct a gait pattern that serves as the characteristic of person's walk.

2.3 Gait Pattern

A *gait pattern* describes the person's style of walking by encapsulating information about DTDSs that are extracted from the single person's walk. In particular, each gait pattern consists of signals described in Table 1.

Table 1. Signals extracted from each person's walk

Notation	DTDS measured between	Notation	DTDS measured between
$\mathcal{S}_{L_L F_L}$	left hip and left foot	$\mathcal{S}_{C_R H_R}$	right clavicle and right hand
$\mathcal{S}_{L_R F_R}$	right hip and right foot	$\mathcal{S}_{F_L F_R}$	feet
$\mathcal{S}_{C_L H_L}$	left clavicle and left hand	$\mathcal{S}_{K_L F_R}$	left knee and right foot

Formally, a gait pattern $\mathcal{G}$ of person's walk is defined as the 6-parameter structure, consisting of six DTDSs:

$$\mathcal{G} = \left(\mathcal{S}_{L_L F_L}, \mathcal{S}_{L_R F_R}, \mathcal{S}_{C_L H_L}, \mathcal{S}_{C_R H_R}, \mathcal{S}_{F_L F_R}, \mathcal{S}_{K_L F_R}\right).$$

In the following, we present a methodology for measuring similarity between two gait patterns.

3 Similarity of Gait Patterns

To express similarity of gait patterns $\mathcal{G}$ and $\mathcal{G}'$, we firstly need to define the way of comparing DTDSs. We define a function Φ for measuring a distance (dissimilarity) of two signals $\mathcal{S} = \{d_f \mid f \in F\}$ and $\mathcal{S}' = \{d'_f \mid f \in F'\}$ as:

$$\Phi\left(\mathcal{S}, \mathcal{S}'\right) = \sum_{f \in F \cap F'} \left| d_f - d'_f \right|. \tag{1}$$

This function, also known as the L_1 or Manhattan distance function, sums point-by-point differences between two specific DTDSs. In case the domains F and F' are not the same, similarity is computed among their common frames only. The function returns 0 if the signals are identical and with an increasing distance their similarity decreases.

We introduce a novel similarity function D for comparing gait patterns $\mathcal{G}$ and $\mathcal{G}'$. This function is based on aggregation of four Φ functions and is formally defined as:

$$D\left(\mathcal{G}, \mathcal{G}'\right) = \Phi\left(\mathcal{S}_{L_L F_L}, \mathcal{S}'_{L_L F_L}\right) + \Phi\left(\mathcal{S}_{L_R F_R}, \mathcal{S}'_{L_R F_R}\right) +$$
$$\Phi\left(\mathcal{S}_{C_L H_L}, \mathcal{S}'_{C_L H_L}\right) + \Phi\left(\mathcal{S}_{C_R H_R}, \mathcal{S}'_{C_R H_R}\right). \tag{2}$$

Individual Φ functions compare similarity of signals between the clavicle and hand and between the hip and foot for both the arms and legs. In this way, the similarity function D expresses difference in the manner of walking of two gait patterns, taking movements of arms and legs into account. Similar to Φ, D returns 0 for identical gait patterns and with an increasing value their dissimilarity decreases.

The use of Φ functions is meaningful only in case when input signals contain the same number of footsteps and start at the same phase of a walking process, otherwise the signals are semantically incomparable. Consequently, before similarity comparison we *normalize* the signals within each gait pattern with respect to a duration and walk cycles' phase. Signals are extended or contracted to be synchronized, i.e., keeping the same phase of a walking process at every moment.

4 Normalization of Gait Patterns

We preprocess each gait pattern to contain signals aligned to 150 video frames, prior to application of the similarity function. The length of 150 frames corresponds to an average walk-cycle duration in our database used. We preprocess each gait pattern by three different normalization methods, which were proposed in [8]. These methods are briefly summarized in the rest of this section.

4.1 Simple Normalization

The most straightforward way of aligning signals is to take the first 150 video frames from each signal $S \in \mathcal{G}$, where $\mathcal{G}$ is an input gait pattern. In case the signal is shorter, it is linearly transformed to the length of 150 frames. This simple method is denoted as SN.

4.2 Footstep Normalization

In contrast to the SN approach, the footstep normalization (FN) considers speed and characteristics of walking by extracting a single walk cycle from each signal. Signals corresponding purely to a single walk cycle, which is two footsteps, can effectively be compared by the D similarity function, instead of comparing signals of unknown movements. To extract the requested walk cycle, we need to identify inceptions of individual footsteps. We select the moment when person's legs are the closest to each other as a footstep inception. Focusing on the character of one of feet signals $S_{F_L F_R} \in \mathcal{G}$ in Figure 2a, we can see a sequence of hills and valleys. Each hill represents a period of moving feet apart and their consecutive approach. The minimum of each valley expresses that both the feet are passing, i.e., the legs are the closest to each other. To determine footsteps, all the minima within the feet signal must be identified. We cannot rely on the signal to contain minima at a fixed distance and to be ideally smooth, which means without any undulation caused by measurement errors. This is the reason we utilize our specialized find-minima algorithm proposed in [8]. In particular,

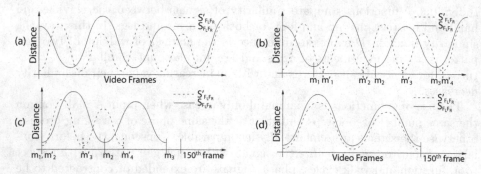

Fig. 2. Normalization of two feet DTDSs with a different number of footsteps (each hill represents a single footstep). Figure (a) represents these signals without any normalization. Figure (b) denotes identified minima of each signal. Figure (c) constitutes just the first walk cycle of the signals, which starts with the move of left foot ahead. Figure (d) shows the extracted walk cycles after linear transformation to 150 frames.

we pick the video frames m_1, m_2, m_3, m_4 where the first four minima were identified. The pairs of adjacent minima determine individual footsteps, alternately with the left or right foot in front. The requested walk cycle is formed by the first two footsteps, so each $S \in \mathcal{G}$ is cropped according to the m_1-th and m_3-th video frame. The cropped signals are linearly transformed to the standardized length of 150 video frames.

4.3   Walk Cycle Normalization

The FN approach extracts the walk cycle disregarding the fact whether the first footstep belonged to the left or right foot. However, a characteristic of some DTDSs depends on the leg which undertook a given footstep – such DTDSs are periodic on the level of walk cycles. Moreover, human walking might not be balanced, e.g., due to an injury, which even results in a different characteristic of feet signal for the left and right foot. The walk cycle normalization WN solves this problem by extracting a single walk cycle that always starts with the move of left foot ahead – the footstep of the left leg and consecutive footstep of the right leg. To identify the first footstep of the left leg, we analyze the signal $S_{K_L F_R} = \{d_f \mid f \in F\}$ that constitutes the changing distance between the left knee and right foot. If both the feet are passing, this signal achieves a higher value when the left foot is moving ahead in comparison with the opposite situation when the right foot is moving ahead. In this way, if the condition $d_{m_1} < d_{m_2}$ is met, we crop each signal $S \in \mathcal{G}$ according to the m_1-th and m_3-th video frame (m_1 and m_3 are frames where the first and third minima of the feet signal were found). Otherwise, signals are cropped according to the m_2-th and m_4-th video frame. Both the extracted footsteps form the requested walk cycle with the first footstep undertaken by the left leg. Similar to FN, the requested walk cycle is finally transformed to the length of 150 video frames. The whole normalization process is depicted in Figure 2 and described in [8] in more detail.

5   Experimental Evaluation

We evaluate effectiveness of the proposed similarity function for gait recognition and compare it with other functions and different normalization approaches. Firstly, we describe a motion-capture database used. Secondly, methodology for evaluating experimental trials is presented. Thirdly, effectiveness of examined similarity functions and influence of diverse normalization processes is reported.

5.1   Database

We utilized the Motion Capture Database (MoCap DB) [1] from the CMU Graphics Lab as a primary data source of trajectories of walking humans. This database contains motion sequences of different kinds of movements (e.g., dance, walk, box, etc.) for 144 recorded persons. We performed experiments on the subset of motion sequences that corresponded to common walking. We took all 131 walking sequences belonging to 24 recorded persons. Each person had at least two different sequences. Walking sequences are the only ones that could meaningfully be used for gait recognition.

We implemented a specialized software to extract gait patterns from 131 walking sequences. In particular, we extracted trajectories of all landmarks $P \in \mathcal{M}$ (see Section 2) for each walking sequence. The obtained trajectories were employed to compute DTDSs specified in Table 1. These DTDSs were normalized and used to construct a gait pattern for each walking sequence.

5.2   Methodology

We concentrated on verifying effectiveness of our approach by evaluating nearest-neighbors queries. To be maximally fair, we constructed one query for each person – the query object for each query was randomly chosen from gait patterns belonging to the given person. Thus 24 queries were constructed and evaluated against a database of all 131 gait patterns. The nearest found neighbor was always the same as the query gait pattern (i.e., the exact match), so it was omitted and the next closest neighbor was analyzed. If the gait pattern of the analyzed neighbor belonged to the same person as the query pattern, search was successful because of the correct person identified. Search could always be successful since at least two different gait patterns were available in the database for each person. Effectiveness – a recognition rate – was stated as a ratio between the number of correctly identified persons and the number of all persons (i.e., the number of successful queries divided by 24). Since we do not define any recognition threshold, it is not possible to calculate false positives. This is the part of our future work.

---

[1] http://mocap.cs.cmu.edu

5.3   Results

The results were deeply studied for diverse similarity functions and the three normalization approaches presented in Section 4: (1) Simple Normalization (SN), (2) Footstep Normalization (FN), and (3) Walk Cycle Normalization (WN). We expect that the SN normalization should achieve the worst recognition rate since it does not take individual footsteps into account. The WN normalization should be more effective than FN because it, furthermore, distinguishes between the left and right foot.

The normalized signals served as input parameters for computation of similarity of gait patterns. We also evaluated three different types of similarity by changing the Φ function in Equation 1. In addition to the original Manhattan distance (L_1), the Euclidean distance (L_2), and the dynamic time warping approach (DTW) [2] were used to measure similarity of two DTDSs.

We also modified Equation 2 to evaluate suitability of different DTDSs for gait recognition. Firstly, we simply modified the function D to recognize persons based purely on a single DTDS, i.e., $D = \Phi(S, S')$. This approach is denoted as *single-DTDS recognition*. Secondly, we modified the function D to combine several DTDSs with the same "weight" (e.g., the original setting with four DTDSs in Equation 2). The use of several DTDSs is referred as *multi-DTDS recognition*.

Single-DTDS Recognition. We evaluated recognition rates of all possible DTDSs that were computed for each couple of anatomical landmarks from $\mathcal{M}$. Table 2 presents the results for top-six DTDSs with the highest recognition rates. Individual rows constitute the six best signals examined and columns represent combinations of similarity functions and normalization approaches.

Table 2. Recognition rate comparison of diverse similarity functions and different normalization approaches for the single-DTDS recognition

Examined DTDS	L_1			L_2			DTW		
	WN	FN	SN	WN	FN	SN	WN	FN	SN
$S_{L_L F_R}$	**0.77**	0.54	0.27	0.75	0.52	0.21	0.67	0.58	0.44
$S_{L_L F_L}$	0.69	0.60	0.33	0.63	0.56	0.21	**0.75**	0.56	0.54
$S_{L_R F_R}$	0.71	0.58	0.25	0.67	0.50	0.25	**0.73**	0.67	0.56
$S_{C_L H_L}$	**0.73**	0.63	0.40	**0.73**	0.63	0.40	0.58	0.58	0.40
$S_{L_R F_L}$	0.56	0.52	0.23	0.56	0.54	0.25	**0.69**	0.48	0.38
$S_{C_R H_R}$	**0.67**	0.60	0.40	0.65	0.60	0.35	0.60	0.52	0.54

The best recognition rate of 0.77 (i.e., effectiveness of 77 %) was achieved by the L_1 similarity function, WN normalization, and $S_{L_L F_R}$ signal that represents the changing distance between the left hip and right foot. We can deduce that success of recognition primarily depends on a normalization approach and type of DTDS used. There is an obvious difference in distribution of recognition rates between SN and the rest of normalization approaches, which shows the usefulness of normalization. From the similarity point of view, L_1 and DTW were slightly more successful than L_2. However, the 77 % effectiveness is not still satisfactory and calls for the use of combination of more DTDSs.

Multi-DTDS Recognition. We combined several DTDSs to improve a recognition rate. The best recognition rate of 0.96 (only one query out of 24 was unsuccessful) was achieved by using the WN normalization and DTW similarity function comparing the changing distance between the shoulder and hand of the left and right arm, i.e., $D\left(\mathcal{G},\mathcal{G}'\right) = \Phi\left(\mathcal{S}_{C_L H_L}, \mathcal{S}'_{C_L H_L}\right) + \Phi\left(\mathcal{S}_{C_R H_R}, \mathcal{S}'_{C_R H_R}\right)$. The original similarity function proposed in Equation 2 achieved the 0.83 recognition rate. The lower effectiveness is caused by summing "mismatched" similarities of legs and arms. This should be solved by weighting individual functions Φ or their normalization to interval $[0, 1]$. The results of top-four similarity functions with the highest recognition rates – both the discussed similarity functions along with other two functions comparing the changing distances between (1) the hip and foot for both the legs and (2) the left hip and left foot and the left shoulder and left hand – are presented in Table 3.

Table 3. Recognition rate comparison of diverse similarity functions and different normalization approaches for the multi-DTDS recognition

Combination of examined DTDSs	L_1			L_2			DTW		
	WN	FN	SN	WN	FN	SN	WN	FN	SN
$\mathcal{S}_{C_L H_L} + \mathcal{S}_{C_R H_R}$	0.94	0.71	0.44	0.88	0.67	0.38	**0.96**	0.73	0.69
$\mathcal{S}_{L_L F_L} + \mathcal{S}_{L_R F_R}$	0.83	0.65	0.40	0.81	0.63	0.35	**0.92**	0.67	0.60
$\mathcal{S}_{L_L F_L} + \mathcal{S}_{C_L H_L}$	0.81	0.81	0.54	**0.85**	0.79	0.56	0.83	0.75	0.67
$\mathcal{S}_{L_L F_L} + \mathcal{S}_{C_L H_L} + \mathcal{S}_{L_R F_R} + \mathcal{S}_{C_R H_R}$	0.71	0.67	0.29	0.69	0.65	0.27	**0.83**	0.67	0.50

The results again confirmed the importance of normalization. The maximum recognition rate of SN was not higher than the minimal recognition rate of WN, disregarding the Φ function and combinations of DTDSs used.

6   Conclusions

We investigated the problem of gait recognition based on processing trajectories of human walking. Trajectories were used to extract distance-time dependency signals to ensure viewpoint invariant recognition. These signals were normalized in the form of walk cycles that were compared by a specialized similarity method. The results were evaluated on a real-life database and compared with diverse similarity functions and normalization approaches. The combination of signals expressing the manner of movement of left and right arm along with the walk-cycle normalization and DTW-like comparison led to the 96 % effectiveness. We demonstrated that the normalization process and the movement of arms are important characteristics to be considered for gait recognition.

We are aware of the fact that the database of 131 walking sequences is too small, so we plan to build a bigger 3D database of motion trajectories acquired by the Kinect [2] equipment. In the future, we also plan to improve a normalization approach along with similarity function, so that gait patterns could compose of more than a single walk cycle for more effective recognition.

Acknowledgements. This research was supported by the national project GACR 103/10/0886. The database used in this paper was obtained from http://mocap.cs.cmu.edu – the database was created with funding from NSF EIA-0196217.

References

1. BenAbdelkader, C., Cutler, R., Davis, L.: Stride and cadence as a biometric in automatic person identification and verification. In: 5th International Conference on Automatic Face Gesture Recognition, pp. 372–377. IEEE (2002)
2. Berndt, D.J., Clifford, J.: Finding patterns in time series: a dynamic programming approach. In: Advances in Knowledge Discovery and Data Mining, pp. 229–248. American Association for Artificial Intelligence, Menlo Park (1996)
3. Bhanu, B., Han, J.: Human Recognition at a Distance in Video. In: Advances in Computer Vision and Pattern Recognition. Springer (2010)
4. Chen, C., Liang, J., Zhao, H., Hu, H., Tian, J.: Frame difference energy image for gait recognition with incomplete silhouettes. Pattern Recognition 30(11), 977–984 (2009)
5. Cunado, D.: Automatic extraction and description of human gait models for recognition purposes. Computer Vision and Image Understanding 90(1), 1–41 (2003)
6. Han, J., Bhanu, B.: Individual recognition using gait energy image. IEEE Transactions on Pattern Analysis and Machine Intelligence 28(2), 316–322 (2006)
7. Tanawongsuwan, R., Bobick, A.F.: Gait recognition from time-normalized joint-angle trajectories in the walking plane. In: International Conference on Computer Vision and Pattern Recognition (CVPR 2001), vol. 2(C), II–726–II–731 (2001)
8. Valcik, J., Sedmidubsky, J., Balazia, M., Zezula, P.: Identifying Walk Cycles for Human Recognition. In: Chau, M., Wang, G.A., Yue, W.T., Chen, H. (eds.) PAISI 2012. LNCS, vol. 7299, pp. 127–135. Springer, Heidelberg (2012)
9. Wang, L., Ning, H., Tan, T., Hu, W.: Fusion of static and dynamic body biometrics for gait recognition. IEEE Transactions on Circuits and Systems for Video Technology 14(2), 149–158 (2004)
10. Xue, Z., Ming, D., Song, W., Wan, B., Jin, S.: Infrared gait recognition based on wavelet transform and support vector machine. Pattern Recognition 43(8), 2904–2910 (2010)
11. Yoo, J.H., Hwang, D., Moon, K.Y., Nixon, M.S.: Automated human recognition by gait using neural network. In: Workshops on Image Processing Theory, Tools and Applications, pp. 1–6. IEEE (2008)

---

[2] http://www.xbox.com/kinect

Illumination Normalization for SIFT Based Finger Vein Authentication

Hwi-Gang Kim[1], Eun Jung Lee[1], Gang-Joon Yoon[1], Sung-Dae Yang[1],
Eui Chul Lee[2], and Sang Min Yoon[3]

[1] National Institute for Mathematical Sciences, Korea
[2] Sangmyung University, Korea
[3] Yonsei University, Korea

Abstract. Recently, the biometric information such as faces, fingerprints, and irises has been used widely in a security system for biometric authentication. Among these biometric features which are unique to each individual, the blood vessel pattern in fingers is superior for identifying individuals and verifying their identities: We may obtain easily the information on blood vessels which is almost impossible to counterfeit because the pattern exists inside the body unlike the others. In this work, we propose a finger vein recognition method using an illumination normalization and a SIFT (Scale-Invariant Feature Transform) matching identification. To verify individual identification, the proposed methodology is composed of two steps: (i) we first normalize the illumination of finger vein images, and (ii) extract SIFT descriptors from the image and match them to the given data. Experimental results indicate that the proposed method is shown to be successful for authentication system.

1 Introduction

Recently, the biometric information such as faces, fingerprints, and irises has been widely used in a security system for personal identification. It is well-known fact that the blood vessel pattern is superior in identifying individuals and verifying their identities among these biometric features which are unique to each individual. As the huge network of blood vessels underneath a human's skin unlike the others, the shape of a vascular pattern is very stable over a long period of lifetime and has the distinctive structure in human body. In addition, as the blood vessels are hidden underneath the skin, it is almost impossible to copy or counterfeit the vein pattern compared to other biometric information [1]. The properties of its uniqueness, stability and safety of the vein pattern lead to in successful biometrics, and these vein features provide secure and reliable security systems for human identity verification [1].

Especially, in case of the finger vein, it is much easier to capture the information just by holding the finger on the device for a few seconds and to operate them because of a quite small amount of image data for processing, compared with other vein patterns. For this reason, the finger vein recognition has been considered as a convenient and reliable alternative to biometric signals such as faces, fingerprints, and irises. Despite these merits, the reason that the finger vein authentication technique

G. Bebis et al. (Eds.): ISVC 2012, Part II, LNCS 7432, pp. 21–30, 2012.

has not been developed well is that it is related to the poor quality of finger vein images due to illumination imbalance [2]. The vein images have been captured using the infrared lighting, and thus the illumination imbalance phenomenon occurs in vein images due to the facts that the lights from the device focus on the local areas of the finger and the rays are blocked or interrupted by the skin layer [2]. Because of this phenomenon, it is hard to extract necessary information from the obtained vein image to be used for identification. Although there is the image binarization process as a remedy of the illumination imbalance, it is difficult to detect even the approximate area of blood vessels. In applying directly to the observed image, the typical methods for illumination adjustment of the image, e.g. histogram equalization, are not proper to handle the partial illumination imbalance because they have an effect on the overall brightness of the image without considering the lighting position.

To solve this problem without manipulating or improving the device, in this paper, we propose a robust feature matching methodology by balancing the illumination of the captured finger vein images. With a captured image, we first adjust the illumination balance by subtracting the illumination components from the input images under the assumption that the low frequency of the image means the illumination components [10]. This process producing an illumination normalized image is called the illumination normalization. Then, we take a histogram equalization process on the illumination normalized image to improve the visibility of the output image, which clearly shows the vessel area. As an identification process, we use the SIFT algorithm [3] to identify the finger vein images by matching the SIFT descriptors between the illumination normalized images and the set of individual database [4].

From the experiments, this method is shown to be effective for the biometric authentication system. This results from the fact that the illumination normalized image creates a larger number of SIFT descriptors than the original image.

2   Related Work

2.1   Overview of a Vein Pattern Verification System

A flowchart of general vein pattern verification system is   provided in Fig. 1 [1].

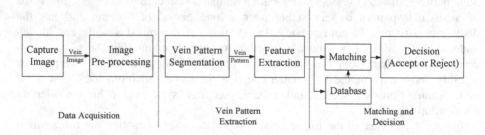

Fig. 1. Flowchart of a typical vein verification system

First, finger vein images are extracted by using the capturing device, and the captured images are normalized in the same form to facilitate the processes during the verification system. Next, to make the image segmentation easier, the images are processed through the image pre-processing step usually using the various illumination normalization or image binarization [5]. After the data acquisition step, the vein pattern images are obtained from the pre-processed images and the feature points are searched in the vein pattern images for data matching to identify individuals. Finally, the feature points are compared and matched with the collected database for the final identification decision, and the verification system determines the acceptance or the reject for individuals depending on the agreement criteria.

2.2   Finger Vein Image Acquisition

To capture figure vein images, the wavelength of the infrared lays between 700mm-1000mm to pass through most of the human tissues while the hemoglobin in the blood can absorb the infrared light fully [6]. A previously lab-made capturing device used for the finger vein images is in shown Fig. 2 [7]. In this paper, the capturing device is composed of NIR (Near Infrared light) illuminators of 850mm wavelength, a webcam, and a hot mirror. The infrared illuminators are located on the back of the finger, and infrared lights are illuminated through the finger. The light is captured by a camera underneath the person's finger. The NIR cutting filter inside the webcam is removed, and a NIR passing filter is attached in front of the camera lens. And the hot mirror is tilted at a 45° angle in front of camera. The hot mirror reflects NIR light evenly over the finger while allowing visible light to pass through the finger. It can also minimize the size of the capturing device as shown in Fig. 2 (a) [2][7]. A user can simply touch an index-finger tip at a fixed position on the device as shown in Fig. 1 (b), which improves user convenience. Examples of acquired images from this device are shown in Fig. 3 [7].

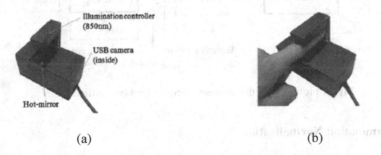

(a)   (b)

Fig. 2. Prototype camera for capturing finger vein image [7]. (a) Structure of the finger vein capture system, (b) example of using the system

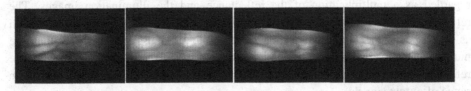

Fig. 3. Examples of captured finger vein images by the vein capturing device

2.3 Localizing and Resizing the Finger Vein Image

The captured images include shaded regions at the top and the bottom of the images. And the shapes of the finger images are different for each person as shown in Fig. 3. Therefore, we need to localize the captured images by cutting out the shaded regions, and then, and then resize into the same size (150×60 pixel) by stretching the cropped images. Localized and resized images from this process are shown in Fig. 4.

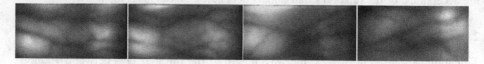

Fig. 4. Localized and resized images from Fig. 3

3 Proposed Method

In this section, we propose a finger vein authentication method by matching SIFT descriptors obtained from illumination normalized images unlike general systems as shown in Fig. 1. The overall flowchart of the proposed method for a vein verification system is summarized in Fig. 5.

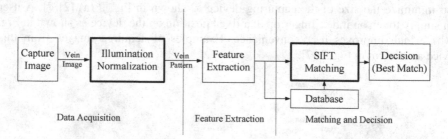

Fig. 5. Flowchart of the proposed method for vein verification

3.1 Illumination Normalization

In order to recover and enhance visual information from low-contrast images, Retinex [8], which was originally proposed by Edwin Land [9] in 1986, has been used popularly. The core assumption on Retinex process is that an observed image is regarded as a multiplication of the reflectance and the illumination images. The reflectance image contains the significant information of the objects in the image such as edges, and the illumination image acts like a Gaussian filter, so that the multiplication operation has a smoothing effect on the image. In Retinex process, the reflectance components are obtained by acting on logarithm to compress the dynamic range of output signals.

Because the shape of the blood vessels is projected poorly due to the effect of the interruption caused by the skin surface and other organs inside the finger, it is hard to directly detect significant information from the projected image. In this work, we assume the interruption effect to be a filtering one. Hence, we will regard the finger vein image as those obtained by taking logarithm already. With the assumption, we preprocess the images by directly subtracting the illumination component from the captured images [10]. Then, applying the histogram equalization, we obtain the illumination normalized image on the basis of our previous research [10]. Empirically, we observe that the illumination components distort the image seriously. Since it is, however, well known that the SIFT descriptors are invariant under the scale and illumination, we suggest using the SIFT descriptors at the identification process step.

Let $I(x, y)$ be a captured finger vein gray image of which the pixel values lie in [0, 255]. We may assume that an image is a function of $\Omega = \{(x, y) | 1 \le x \le n, 1 \le y \le m\}$ into [0, 255]. To subtract the illumination component $R(x, y)$ from the image $I(x, y)$, we first generate $R(x, y)$ by the convolution of the original image $I(x, y)$ with an 11×11 moving average filter $F_{ave}(x, y)$ in order to smooth the input image,

$$R(x, y) = F_{ave}(x, y) * I(x, y). \tag{1}$$

To avoid the negative values appearing in subtracting the illumination image $R(x, y)$, we take the following operation to obtain the desired image $I_N(x, y)$,

$$I_N(x, y) = \frac{1}{2}(I(x, y) - R(x, y) + 255). \tag{2}$$

In Eq. (2), we multiply by 1/2 to make the pixel values lie in the range of 0~255 because some values are exceeded over 255 in subtracting the illumination image directly. The image $I_N(x, y)$ appears blurredly in gray because the image $I(x, y)$ was smoothed already as mentioned earlier and the specific areas of $I(x, y)$ close to the light bulbs are seen in white but almost all of the rest areas appear vague, which we call the illumination imbalance. For these reasons, the smoothing effect remains very slight and thus the differences $I(x, y) - R(x, y)$ are very small. The image $R'(x, y) := 255 - R(x, y)$ is called the inverse image of $R(x, y)$.

After overcoming the illumination imbalance effect, we apply the histogram equalization to $I_N(x, y)$, which improves the image visibility because the normalized output image is not distinguished clearly between the veins and background areas:

$$H(x, y) = \frac{255}{m \times n} \Phi(I_N(x, y) + 1), \quad (x, y) \in \Omega. \tag{3}$$

Here Φ is the cumulative histogram function and $H(x, y)$ is the final illumination normalized image. The block diagram of the illumination normalization method is shown in Fig. 6.

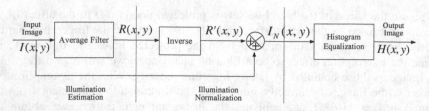

Fig. 6. Block diagram of the proposed method

As examples, the images at each step are shown in Fig. 7 [10]. We can see in Fig 7 that the blood vessel of the output image is more distinguishable than the input image.

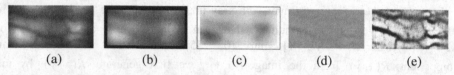

(a)   (b)   (c)   (d)   (e)

Fig. 7. Examples of finger vein images in the proposed method: (a) Input image $I(x, y)$, (b) illumination image $R(x, y)$, (c) inverse image $R'(x, y)$, (d) illumination normalized image $I_N(x, y)$, and (e) output image $H(x, y)$

3.2   Data Matching Using SIFT Algorithm

Now, we are in position to identify the individual from whom the finger vein image $I(x, y)$ was captured. In this paper, to match the finger vein images with the collected database, we use the SIFT algorithm by extracting the local feature points which are invariant to image scale, rotation, illumination, and so on. The major stages for extracting image features are scale-space extrema detection, keypoint localization, orientation assignment, and keypoint descriptor [3]. We apply the SIFT algorithm to $H(x, y)$ to generate the set S of SIFT descriptors. By matching the SIFT descriptors in S to the given data of SIFT descriptors, we take the best matching image in the database as an identification individual [4].

4   Experimental Results

We experiment to confirm the matching performance of the proposed method using the SIFT matching algorithm. For performance comparison with original and illumination normalized ones, we find the SIFT descriptors of the original input images and the illumination normalized output images obtained by the proposed method. In this experiment, the original images are the images before the illumination normalization process, and the modified images are those obtained through the illumination normalization process. In Fig. 8, we figure out the apparent difference in numbers of the SIFT descriptors for a captured finger vein image.

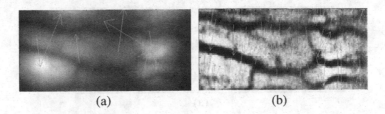

(a)   (b)

Fig. 8. Comparison of SIFT descriptors: (a) The original input, (b) the modified output image

We can see in Fig. 8 (b) that the output image by the proposed method has much more descriptors than the original image in Fig. 8 (a). As we may see, the more SIFT descriptors we have, the higher the possibility of identification increases in matching descriptors for the personal identification.

Next, we perform the SIFT matching test for the captured original image and the prior collected images in database which is shown in Fig. 9. The top image in Fig. 9 shows that the original image has 10 matching points to the same person. This result can be considered as a successful matching, but it is insufficient to meet the needs of the identification system. In addition, the image shares a matching point with a different image as shown in Fig. 9 (b). When we perform the SIFT matching test between the illumination normalized images, the number of matching points increases dramatically and the images obtained from different persons share no matching points as shown in Fig. 10. (The right image in Fig. 10 (b) is the illumination normalized image of the right image in Fig .9 (b).)

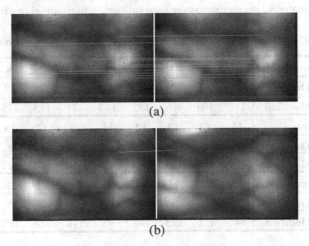

(a)

(b)

Fig. 9. Matching performance of original images obtained from (a) the same person, and (b) different persons

In Fig. 10, there are much more matching points compared with Fig. 9(a). Furthermore, in case of the different persons, the related images have no matching point as shown in Fig. 10 (b).

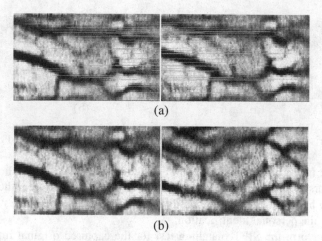

(a)

(b)

Fig. 10. Matching performance of modified images obtained from (a) the same person, and (b) different persons

Lastly, we experiment the SIFT matching test with more samples in order to evaluate the performance for security systems. For this test, we capture 10 images (of index-finger) from 30 participants using the capturing device. And we save 9 images per each person in the database. For a recognition test about the normalized images, we also normalize the images in the database by the proposed method. Then we compare the 30 remaining original (modified, resp.) images to 270 original (modified, resp.) images in the database through the SIFT matching. First, we check the numbers of descriptors for the 30 remaining finger vein images, which are given in Table 1.

Table 1. The number of the SIFT descriptors for the 30 participants

Participants		1	2	3	4	5	6	7	8	9	10
Keypoints	Original	11	12	2	8	11	16	7	12	5	7
	Modified	110	108	108	135	112	131	113	115	130	122
Participants		11	12	13	14	15	16	17	18	19	20
Keypoints	Original	7	8	6	2	7	3	9	8	5	3
	Modified	117	137	124	109	125	110	110	122	123	153
Participants		21	22	23	24	25	26	27	28	29	30
Keypoints	Original	5	9	4	10	14	6	5	5	17	11
	Modified	141	126	133	108	95	118	127	100	117	98

We can see from the table that the modified images by the proposed normalization have much more keypoints compared with the original images.

And, to confirm the identification performance for each person, we have done the matching experiment using the SIFT matching.

To test a recognition performance, we estimate the recognition probability of identifying the participant by finding the best matching person from the database.

Table 2. The recognition probability (OI: Original Image, MI: Modified Image)

Person	9		10		13		14		16	
Rank	OI	MI	OI	MI	OI	MI	OI	MI	OI	MI
1	1.00	1.00	1.00	1.00	1.00	1.00	0.00	1.00	1.00	1.00
2	1.00	1.00	1.00	1.00	1.00	1.00	0.24	1.00	0.43	1.00
3	1.00	1.00	1.00	1.00	0.50	1.00	0.03	1.00	0.03	1.00
4	0.35	1.00	0.90	1.00	0.43	1.00	0.03	1.00	0.03	1.00
5	0.19	1.00	0.82	1.00	0.33	1.00	0.03	1.00	0.03	1.00
6	0.03	1.00	0.69	1.00	0.03	1.00	0.03	1.00	0.03	1.00
7	0.03	1.00	0.15	1.00	0.03	1.00	0.03	1.00	0.03	1.00

After choosing 5 participants randomly, then the recognition probability is calculated as

$$p = \frac{\#(A \cap \sum_R)}{\#(\sum_R)} \tag{4}$$

where $\#(\cdot)$ denotes the number of elements in a set, A is the set of all (relevant) images in the database captured from the corresponding participant, and $\sum_R$ is the set of the retrieved images in the database which are the best possible matches up to rank R. We rank the participants according to the number of matching descriptors. From Table 2, the recognition probability stays perfect when we use the modified images, while the probability decreases seriously for the original images. We also note that in the experiment, we have the perfect matches for all the participants in using modified images but there occur 3 false identifications when we identify by the best match in using illumination free images.

5   Conclusion

As the importance of personal privacies increases, the biometrics has been significant for securing personal information. In these circumstances, the finger vein image has been spotlighted as a means of a personal identification system using biometrics such as faces, fingerprints, irises, and finger veins. In spite of its many advantages as a biometric authentication due to the convenience of acquisition and the reliability of finger vein features, the finger vein recognition techniques are still to be more developed regarding the issues of the poor quality of vein images. This is deeply related to the illumination imbalance caused by the several limitations like as light scattering, optical blur, and its existence inside a finger. In this work, we propose an illumination normalization method to overcome the illumination imbalance in the finger vein images. With a finger vein image, we fulfill the illumination normalization process as follows. After we estimate the illumination components by the convolution of the captured image and a moving average filter, we balance the image by subtracting the estimated illumination components. Then, we equalize the histogram of the illumination-free image to improve the image visibility. Finding SIFT descriptors from the illumination normalized image, we apply the SIFT matching algorithm to confirm the matching accuracy that is an indispensable condition for a security system to be successful. The experimental results show that the normalized images by the proposed method have much better performance than the original images in aspects of the number of descriptors and of the identification accuracy. Consequently, the proposed method enhances the identification for the finger vein authentication. With the proposed method, the identification step using the SIFT algorithm can be applied for many other verification systems and can be used widely.

Acknowledgement. The author S. M. Yoon was supported by the MKE(The Ministry of Knowledge Economy), Korea, under the "IT Consilience Creative Program" support program Supervised by the NIPA(National IT Industry Promotion Agency)(NIPA-2012-H0201-12-1001). The author E. C. Lee was supported by 2012 Research Grant from Sangmyung University.

References

1. Wang, L., Leedham, G., Cho, D.S.-Y.: Minutiae feature analysis for infrared hand vein pattern biometrics. In: Pattern Recognition. Part Special Issues in The Journal of the Pattern Recognition Society, vol. 41, pp. 920–929 (2008)
2. Lee, E.C., Park, K.R.: Image restoration of skin scattering and optical blurring for finger vein recognition. Optics and Lasers in Engineering 49, 816–828 (2011)
3. Lowe, D.G.: Distinctive Image features from Scale-Invariant Keypoints. International Journal of Computer Vision 60, 97–110 (2004)
4. Lowe, D.G.: Demo Software, SIFT Keypoint Detector, http://www.cs.ubc.ca/~lowe/keypoints/
5. Ladoux, P.-O., Rosenberger, C., Dorizzi, B.: Palm Vein Verification System Based on SIFT Matching. In: Tistarelli, M., Nixon, M.S. (eds.) ICB 2009. LNCS, vol. 5558, pp. 1290–1298. Springer, Heidelberg (2009)
6. Mulyono, D., Woodell, G.A., Jobson, D.J.: A Study of finger vein biometric for personal identification. In: International Symposium on Biometric and Security Technologies, Taipei, pp. 1–8 (2008)
7. Lee, E.C., Lee, H.C., Park, K.R.: Finger Vein Recognition Using Minutia-Based Alignment and Local Binary Pattern-Based Feature Extraction. Imaging Systems and Technology 19, 179–186 (2009)
8. Rahman, Z., Woodell, G.A., Jobson, D.J.: Retinex Image Enhancement: Application to Medical Images. In: The NASA Workshop on New Partnerships in Medical Diagnostic Imaging, Greebelt, Maryland (2001)
9. Land, E.H.: An alternative technique for the computation of the designator in the retinex theory of color vision. In: PNAS. Proceedings of the National Academy of Sciences of the United States of America, vol. 83, pp. 3078–3080 (1986)
10. Lee, E.J., Lee, E.C.: Illumination normalization method of infrared vein image to enhance localization performance of vein region. Journal of Korea Multimedia Society (submitted, 2012)

Higher Rank Support Tensor Machines

Irene Kotsia[1], Weiwei Guo[2], and Ioannis Patras[1]

[1] School of Electronic Engineering and Computer Science
Queen Mary, University of London
{irene.kotsia,i.patras}@eecs.qmul.ac.uk
[2] College of Electronic Science and Engineering
National University of Defense Technology, China
weiweiguo@nudt.edu.cn

Abstract. This work addresses the two class classification problem within the tensor-based large margin classification paradigm. To this end, we formulate the higher rank Support Tensor Machines (STMs), in which the parameters defining the separating hyperplane form a tensor (tensorplane) that is constrained to be the sum of rank one tensors. The corresponding optimization problem is solved in an iterative manner utilizing the CANDECOMP/PARAFAC (CP) decomposition, where at each iteration the parameters corresponding to the projections along a single tensor mode are estimated by solving a typical Support Vector Machine (SVM)-type optimization problem. The efficiency of the proposed method is illustrated on the problems of gait and action recognition where we report results that improve, in some cases considerably, the state of the art.

1 Introduction

Tensors are multidimensional objects, whose elements are indexed by more than two indices. Images, grayscale and color videos can be regarded as 2nd, 3rd and 4th order tensors, respectively. The advantages of using tensorial representations have recently attracted significant interest from the research community. Within the last decade, several works proposed tensor-based extensions of fundamental methods. Principal Component Analysis was extended to Multilinear Principal Component Analysis (MPCA) [1], Linear Discriminant Analysis (LDA) to Multilinear Discrimininant Analysis (MDA) [2], two class Support Vector Machines (SVMs) to Support Tensor Machines (STMs) [3], Non-negative Matrix Factorization (NMF) to Non-negative Tensor Factorization (NTF) [4]-[6] and Vector Correlation to Canonical Analysis Correlation of tensors (CAC) [7].

The extension of the above mentioned algorithms within the tensorial framework has led to considerable performance improvements. This is due to the fact that the proposed algorithms can better retain and utilize information about the structure of the high dimensional space the data lie in. By contrast, vector schemes discard this information since the input vectors that they use are created by stacking the rows or columns of the original data tensors in a rather

G. Bebis et al. (Eds.): ISVC 2012, Part II, LNCS 7432, pp. 31–40, 2012.

arbitrary way. This may lead to overfitting, especially for small sample size problems. Tensor-based algorithms on the other hand decompose the whole problem in a number of smaller and simpler ones, each of which is typically defined over a certain tensor mode and is of lower dimensionality.

In this paper, we exploit the advantages of tensor-based framework in order to address the classification problem within the large margin paradigm. We formulate the higher rank STMs, in which the parameter defining the separating hyperplane from a tensor (tensorplane) is constrained to be the sum of rank one tensors. The benefit of the above mentioned proposed scheme is a twofold one. First, the use of tensor representations for processing tensorial input data, instead of using their vectorized formes is intuitively closer to the nature of the data as in that way we retain their topology. The second theoretical benefit lies in the use of the CP decomposition. This is a general decomposition that decomposes a tensor into a sum of component rank-one tensors. Our framework exploits the benefits offered by the CP decomposition, allowing in that way multiple projections of the input tensor along each mode leading to considerable improvement in the discriminative ability of the resulting classifier.

The rest of the paper is organized as follows. In Section 2, we briefly organize and review related works in tensor-based formulations, as well as related works in action and gait recognition. In Section 4, we introduce the novel algorithm that is able to handle tensorial representations of the data. The power of the proposed classifier is demonstrated on the gait and actions recognition problems in Section 5. Finally, conclusions are drawn in Section 6.

2   Related Work

In [3], the SVMs learning framework was extended to handle tensorial input. More precisely, a two class STM formulation was proposed, where the weights parameters were defined as rank one tensors, one for every mode of the input tensor. This is in contrast to the approach followed in this work, where the weights parameters are defined as a higher rank tensor, that can be written as a sum of rank one tensors following the rationale behind the CP decomposition, creating in that way a novel higher rank two class STMs classifier.

So far as action recognition is concerned, we focus on methods that rely on viewpoint dependent, template-based representations. Early approaches utilize holistic representations such as temporal images, motion history images [8], optical flow templates [9] or action shapes [10] used with Mahalanobis distances, Low-Rank SVMs, or NN, respectively. Lately, features that are robust to illumination and small viewpoint variations, such as Histograms of Oriented Gradients (HOGs) and optical flow are used in template representations. In [11], the activity patterns correlation was used to build a Markov Chain Monte Carlo (MCMC) Bayesian network for classification. In [7] the canonical correlation of tensorial input data is defined by considering the canonical correlations along each tensor mode. The sum of the several selected canonical correlations is used as a similarity measure in a NN classifier.

Regarding gait recognition, most of the methods developed during the past few years use sequences of silhouettes [12][13], acquired by background subtraction followed by Linear Time Normalization (LTN), HMMs and similarity measures. In [14], Gait Energy Images (GEI) are calculated by the extracted silhouettes and classified using a minimum Euclidean distance classifier on the distances calculated between a test sequence and the gallery sequences. More recently, tensors have been introduced (3rd order tensors created by sequences of silhouettes) so as to better capture the information along the time dimension. In [15], the average gait images are used to extract tensors by applying three different Gabor representations. General Tensor Discriminant Analysis is then applied for gait recognition. In [1], the dimensionality of the input gait tensors is reduced using MPCA and the eigentensors acquired are rearranged to form a feature vector that is subsequently used to estimate the matching score between a test sequence and a gallery set of sequences. Similarly to [7], in [1] the tensor-based analysis is confined to the unsupervised dimensionality reduction stage, that is separated from the classification stage.

3   Useful Notations in Multilinear Algebra

The notations of tensorial algebra that will be used in this paper are consistent with the ones presented in [16]. Tensors of order 3 or higher will be denoted by boldface Euler script calligraphic letters (e.g. $\mathcal{X}$). Let us define the inner product of two tensors $\mathcal{X}, \mathcal{Y} \in \mathbb{R}^{I_1 \times \cdots \times I_M}$ as:

$$\langle \mathcal{X}, \mathcal{Y} \rangle \triangleq \sum_{i_1}^{I_1} \cdots \sum_{i_M = 1}^{I_M} x_{i_1 \ldots i_M} y_{i_1 \ldots i_M} \tag{1}$$

and the Frobenius norm of a tensor as

$$\|\mathcal{X}, \mathcal{X}\| = \sqrt{\langle \mathcal{X}, \mathcal{X} \rangle}. \tag{2}$$

If Kruskal tensor representations are used, as in the case of the CANDE-COMP/PARAFAC (CP) decomposition, then a tensor $\mathcal{X}$ can be written as a sum of component rank-one tensors. For example, an M order tensor $\mathcal{X} \in \mathbb{R}^{I_1 \times I_2 \cdots \times I_M}$, can be written as:

$$\mathcal{X} = \sum_{r=1}^{R} \mathbf{u}_r^{(1)} \circ \mathbf{u}_r^{(2)} \circ \cdots \circ \mathbf{u}_r^{(M)} \tag{3}$$

where R is a positive integer and $\mathbf{u}_r^{(1)} \in \mathbb{R}^{I_1}, \mathbf{u}_r^{(2)} \in \mathbb{R}^{I_2} \ldots \mathbf{u}_r^{(M)} \in \mathbb{R}^{I_M}$, for $r = 1, \ldots, R$.

In the classifier that will be presented below, the weights are regarded to form a tensor $\mathcal{W}$, that can be decomposed in a sum of R rank-one tensors as (according to the CP decomposition):

$$\mathcal{W} = \sum_{r=1}^{R} \mathbf{u}_r^{(1)} \circ \mathbf{u}_r^{(2)} \circ \cdots \circ \mathbf{u}_r^{(M)} \tag{4}$$

where $\mathbf{u}_r^{(j)} \in \mathbb{R}^{I_j}, j = 1, 2, \ldots M$. The weights tensor defined above is regarded to be the generalization of the weights vector $\mathbf{w}$ defined by the SVMs formulation (the normal vector that is perpendicular to the separating hyperplane).

If the $\mathbf{u}_r^{(j)}$, $r = 1, \ldots, R$ considered for a mode j are stacked in a matrix $\mathbf{U}^{(j)}$ as

$$\mathbf{U}^{(j)} = \left[\mathbf{u}_1^{(j)}, \mathbf{u}_2^{(j)}, \ldots, \mathbf{u}_R^{(j)} \right], j = 1 \ldots, M. \tag{5}$$

then the j-th matricization of $\mathbf{W}$ can be written as:

$$\mathbf{W}_{(j)} = \mathbf{U}^{(j)} \left[\mathbf{U}^{(M)} \odot \ldots \odot \mathbf{U}^{(j+1)} \odot \mathbf{U}^{(j-1)} \odot \ldots \odot \mathbf{U}^{(1)} \right]^T = \mathbf{U}^{(j)} [\mathbf{U}^{(-j)}]^T. \tag{6}$$

Moreover, we can further note that:

$$\langle \mathcal{W}, \mathcal{W} \rangle = \mathrm{Tr}[\mathbf{W}_{(j)} \mathbf{W}_{(j)}^T] = \mathrm{vec}(\mathbf{W}_{(j)})^T \mathrm{vec}(\mathbf{W}_{(j)}) \tag{7}$$

where $\mathrm{Tr}[\mathbf{A}]$ and $\mathrm{vec}(\mathbf{A})$ are the trace and the vectorization of the matrix $\mathbf{A}$, respectively.

4   Higher Rank Support Tensor Machines

In this paper, $\mathcal{X} \in \mathbb{R}^{I_1 \times I_2 \times \cdots \times I_{M+1}}$ represents a dataset consisting of L samples. Every dataset sample is a tensor of order M denoted by $\mathcal{X}_{i_n:} \in \mathbb{R}^{I_1 \times I_2 \times \cdots \times I_M}$, $i_n = 1, 2, \ldots, L$, that is indexed by M indices $(i_1, i_2, \ldots, i_M)$ and belongs to one of the two classes $\mathcal{Q}_1$ and $\mathcal{Q}_2$.

4.1   Higher Rank Support Tensor Machines

We aim at learning a multilinear decision function $g : \mathbb{R}^{I_1 \times I_2 \times \cdots \times I_M} \to [-1, 1]$, that during testing, classifies a test tensor $\mathcal{Y}\mathbb{R}^{I_1 \times I_2 \times \cdots \times I_M}$ and has the following form:

$$g(\mathcal{Y}) = \mathrm{sign}\left(\langle \mathcal{Y}, \mathcal{W} \rangle + b \right). \tag{8}$$

The weights tensor $\mathcal{W}$ is estimated by solving the following soft STMs problem:

$$\min_{\mathcal{W}, b, \boldsymbol{\xi} \geq 0} \frac{1}{2} \langle \mathcal{W}, \mathcal{W} \rangle + C \sum_{i=1}^{I_{M+1}} \xi_i \tag{9}$$

$$\text{s.t.} \quad y_i \left[\langle \mathcal{W}, \mathcal{X}_{i:} \rangle + b \right] \geq 1 - \xi_i, 1 \leq i \leq I_{M+1}, \xi_i \geq 0 \tag{10}$$

where b is the bias term, $\boldsymbol{\xi} = [\xi_i, \ldots, \xi_w]$ is the slack variable vector and C is the term that controls the relative importance of penalizing the training errors. The training is performed in such a way that the margin of the Support Tensors is maximized while the upper bound on the misclassification errors in the training set is minimized. The above problem however is not convex with respect to all parameters in $\mathcal{W}$. For this reason we adopt an iterative scheme in which at each iteration we solve only for the parameters that are associated with the j-th mode of the parameters tensor $\mathcal{W}$, while keeping all the other parameters fixed. At the iterations for the j-th mode we solve the following optimization problem:

$$\min_{\mathbf{W}_{(j)},b,\boldsymbol{\xi}\geq 0} \frac{1}{2}\mathrm{Tr}\left[\mathbf{W}_{(j)}\mathbf{W}_{(j)}^T\right] + C\sum_{i=1}^{I_{M+1}} \xi_i \tag{11}$$

$$\text{s.t.}\quad y_i\left[\mathrm{Tr}(\mathbf{W}_{(j)},\mathbf{X}_{(j)i:}^T) + b\right] \geq 1 - \xi_i, \quad 1 \leq i \leq I_{M+1},\ \xi_i \geq 0. \tag{12}$$

Under our assumption/constraint that the tensor $\mathcal{W}$ is written as a sum of rank one tensors, we replace $\mathbf{W}_{(j)}$ in the above equation. We should mention here that the initial values for the rank-one tensors $\mathbf{u}$, and subsequently for their matricized forms $\mathbf{U}$, are randomly chosen. Then (11) is rewritten as:

$$\min_{\mathbf{U}^{(j)},b,\boldsymbol{\xi}\geq 0} \frac{1}{2}\mathrm{Tr}\left[\mathbf{U}^{(j)}(\mathbf{U}^{(-j)})^T(\mathbf{U}^{(-j)})(\mathbf{U}^{(j)})^T\right] + C\sum_{i=1}^{I_{M+1}} \xi_i \tag{13}$$

$$\text{s.t.}\quad y_i\left[\mathrm{Tr}\left(\mathbf{U}^{(j)}(\mathbf{U}^{(-j)})^T\mathbf{X}_{(j)i:}^T\right) + b\right] \geq 1 - \xi_i, \quad 1 \leq i \leq I_{M+1},\ \xi_i \geq 0, \tag{14}$$

that is at each iteration we solve for the parameters $\mathbf{U}^{(j)}$ for the mode j while keeping the parameters $\mathbf{U}^{(k)}$ for all other modes $k, k \neq j$, fixed.

In what follows we will bring the optimization problem defined in (13) in a form that can be solved using a classic vector-based SVM implementation. Let us define $\mathbf{A}_{\mathrm{STMs}} = \mathbf{U}^{(-j)^T}\mathbf{U}^{(-j)}$. By definition, $\mathbf{A}_{\mathrm{STMs}}$ is a positive definite matrix. Also, let $\widetilde{\mathbf{U}}^{(j)} = \mathbf{U}^{(j)}\mathbf{A}_{\mathrm{STMs}}^{\frac{1}{2}}$. Then,

$$\mathrm{Tr}\left[\mathbf{U}^{(j)}(\mathbf{U}^{(-j)})^T(\mathbf{U}^{(-j)})(\mathbf{U}^{(j)})^T\right] = \mathrm{Tr}\left[\widetilde{\mathbf{U}}^{(j)}(\widetilde{\mathbf{U}}^{(j)})^T\right] = \mathrm{vec}(\widetilde{\mathbf{U}}^{(j)})^T\mathrm{vec}(\widetilde{\mathbf{U}}^{(j)}). \tag{15}$$

By letting $\widetilde{\mathbf{X}}_{(j)i:} = \mathbf{X}_{(j)i:}\mathbf{U}^{(-j)}\mathbf{A}_{\mathrm{STMs}}^{-\frac{1}{2}}$ we have

$$\mathrm{Tr}\left[\mathbf{U}^{(j)}(\mathbf{U}^{(-j)})^T\mathbf{X}_{(j)i:}^T\right] = \mathrm{Tr}\left[\widetilde{\mathbf{U}}^{(j)}\widetilde{\mathbf{X}}_{(j)i:}^T\right] = \mathrm{vec}(\widetilde{\mathbf{U}}^{(j)})^T\mathrm{vec}(\widetilde{\mathbf{X}}_{(j)i:}). \tag{16}$$

Then, (13) is written as

$$\min_{\mathbf{U}^{(j)},b,\boldsymbol{\xi}\geq 0} \frac{1}{2}\mathrm{vec}(\widetilde{\mathbf{U}}^{(j)})^T\mathrm{vec}(\widetilde{\mathbf{U}}^{(j)}) + C\sum_{i=1}^{I_{M+1}} \xi_i \tag{17}$$

$$\text{s.t.}\quad y_i\left[\mathrm{vec}(\widetilde{\mathbf{U}}^{(j)})^T\mathrm{vec}(\widetilde{\mathbf{X}}_{(j)i:}) + b\right] \geq 1 - \xi_i, \quad 1 \leq i \leq I_{M+1},\ \xi_i \geq 0. \tag{18}$$

Thus, we can see that the optimization problem for the j-th mode in (13) can be formulated as a vector SVM problem (17) with respect to $\widetilde{\mathbf{U}}^{(j)}$. This can lead to a straightforward implementation in which we solve (17) with respect to $\widetilde{\mathbf{U}}^{(j)}$ using a standard SVM implementation and then solve for $\mathbf{U}^{(j)}$ as $\mathbf{U}^{(j)} = \widetilde{\mathbf{U}}^{(j)}\mathbf{A}_{\mathrm{STMs}}^{-\frac{1}{2}}$.

Table 1. Alternating projection for higher rank STMs

Input:	The set of training tensors $\mathcal{X}_i|_{i=1}^N$ and their corresponding labels $y_i = \{+1, -1\}$.
Output:	The parameters of the classification tensorplane $\mathcal{W}$ and $b \in \mathbb{R}$.
Step 1:	Initialize $\mathcal{W}$ written as a sum of random rank-one tensors.
Step 2:	Repeat **steps 3-5** until the convergence condition is satisfied.
Step 3:	For j=1 to M (number of modes):
Step 4:	Calculate $\mathcal{W}$ by optimizing: $$\max_{0 \le \alpha_i^j \le C} -\frac{1}{2} \sum_{i=1}^{I_M+1} \sum_{k=1}^{I_M+1} \alpha_i^j \alpha_k^j y_i y_k \mathrm{Tr}\left(\mathbf{U}^{(-j)} \left[\mathbf{U}^{(-j)^T} \mathbf{U}^{(-j)} \right]^{-1} \mathbf{U}^{(-j)^T} \mathbf{X}_{(j)i:}^T \mathbf{X}_{(j)i:} \right)$$ $$+ \sum_{i=1}^{I_M+1} \alpha_i^j$$ $$\text{s.t.} \quad \boldsymbol{\alpha}^{j^T} \mathbf{y} = 0.$$
Step 5:	Finish calculation of $\mathbf{U}^j$.
Step 6:	Check if the convergence rule $\|\mathcal{W}^{(t)} - \mathcal{W}^{(t-1)}\| / \|\mathcal{W}^{(t-1)}\| \le \varepsilon$ holds. If it is valid, the calculated $\mathcal{W}$ has converged.
Step 7:	Finish.

5   Experimental Results

In this section, we address the gait recognition and the human action recognition problems using publicly available databases. For both problems we study the influence of the rank R.

Note that both the gait and the action recognition problems are multiclass classification problems. We address them by combining multiple two class classifiers that are trained in an one-against-all manner. The final decision is taken either using a voting scheme or by assigning each test sample to a class k, such that the distance of the sample in question to the separating tensorplane of the class k is minimum over the distances to the tensorplanes of all other classes.

5.1   Gait Recognition

The database used for the gait recognition experiments is the USF HumanID Gait Challenge data sets version 1.7. This is a database that has been widely used in the literature (e.g. [1]), a fact that allows easy comparison to previous works. The database has 452 sequences of 74 subjects walking in elliptical paths in front of the camera. Three variations are provided for each subject: viewpoint (left/right), shoe type (two different types) and surface type (grass/concrete). Seven experiments (referred to as Probe Sets) are available, each one of which contains 71 sequences of each subject. The Probe Sets are of increasing complexity/difficulty with probe set A being the easiest and probe set G being the most difficult one. There are no common sequences between the gallery sets and any of the probe sets. Each probe set is unique.

The input features for our algorithms consisted of tensors constructed from silhouettes extracted from the gait sequences. The silhouettes are provided by [1] and are also used as features in [1]. The classifiers were trained using the Gal dataset and tested using each one of the A, B, C, D, E, F and G datasets, following the protocol described in [13]. In order to calculate the accuracy of gait recognition for a probe sequence *probe* consisting of N_{probe} samples (gait periods) against a gallery sequence *gallery* consisting of $N_{gallery}$ samples, we calculate the distance of that particular probe sample from each of the separating tensorplanes of each class of the gallery samples. In order to calculate a symmetric dissimilarity measure between a probe sequence and a gallery sequence (class) we also calculate of each gallery sample to the separating tensorplane of each class of each probe sample, as well. The dissimilarity measure between a probe sequence *probe* and a gallery sequence *gallery* is chosen to be the minimum of the calculated distances. Finally, the probe sequence *probe* is assigned to the gallery sequence class to which is has the smallest dissimilarity.

We first discuss the effect of rank R, i.e. the number of simultaneous projections along each mode, on the recognition accuracy. In Fig. 1, we plot the recognition accuracy against the rank R for the most difficult probe set, namely Probe Set G. It is clear that while there are flunctuations in the performance depending on the value of R, higher ranks (i.e. $R > 1$) give better results, especially for the more difficult probe sets. The best recognition accuracy rate is achieved when the rank R is equal to 18, that is when 18 simultaneous projections along each mode of the tensor are used. The same value for the Rank R is also the optimal for the rest of the Probe Sets.

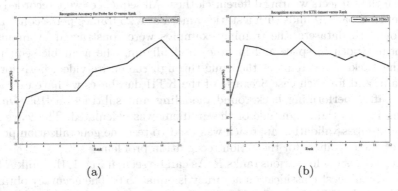

(a)   (b)

Fig. 1. Rank R versus recognition accuracy for (a) Probe Set G and (b) KTH database

In Table 2 we present the recognition accuracy for the various probe sets obtained by our method and the recognition accuracy reported elsewhere in the literature on the same datasets. We also present results when the tensorial input is vectorized to be fed into SVMs and when rank one formulations are used for STMs, for comparison reasons. We report the Cumulative Scores 1 and 5 (CS-1 / CS-5, respectively) similarly to [1][12]. As we can see, tensors largely outperform vectors. This was expected given the fact that the dimensionality is large and only few training examples are available. The best accuracies achieved for each Probe Set are highlighted in bold.

Table 2. Comparison of proposed method with state of the art (CS-1/CS-5)

Probe Set	Baseline [13]	HMM [17]	LTN [12]	GEI [14]	ETG [1]	ETGLDA [1]	SVMs	STMs $R = 1$	$R = 18$
A	79/96	99/100	94/99	**100/100**	92/96	99/100	80/97	92/100	99/100
B	66/81	**89/90**	83/85	85/85	85/90	88/93	79/93	81/90	85/93
C	56/76	78/90	78/83	80/88	76/81	**83/88**	68/85	73/88	79/93
D	29/61	35/65	33/65	30/55	39/55	36/71	30/54	47/67	**53/72**
E	24/55	29/65	24/67	33/55	29/52	29/60	23/46	48/79	**62/88**
F	30/46	18/60	17/58	21/41	21/58	21/59	24/49	29/49	**41/71**
G	10/33	24/50	21/48	29/48	21/50	21/60	12/37	31/71	**50/88**
Mean	42/64	53/74	50/72	54/67	52/69	54/76	45/62	57/68	**67/86**

5.2 Human Action Recognition

For action recognition, we used the KTH [18] dataset. The KTH Action Dataset depicts 25 subjects performing 6 different activities namely: "boxing", "hand-clapping", "handwaving", "jogging", "running" and "walking". The data were collected in four different settings corresponding to indoors recordings, outdoor recordings with scale variations (camera zoom in and out) and outdoor recordings with the subjects wearing different clothes. All sequences were recorded over homogeneous backgrounds with a static camera at a 25 frames-per-second frame rate. For both datasets, the training examples were constructed by manually segmenting (both in space and in time) and aligning the available sequences. Since the background remains the same through the entire video, both for the Weizmann and for each case Scenario of the KTH databases we have extracted features after performing background modeling and subtraction [19]. On the resulting images the magnitude of the gradient was calculated. The leave-one-person-out cross-validation approach was used to test the generalization performance of the classifiers for the action recognition problem.

We report results for various ranks R. As can be seen in Fig. 1, the rank R that provides the highest classification accuracy is equal to 6. The accuracy obtained for higher rank STMs is equal to 93.3% . The equivalent accuracy obtained for STMs when the rank was set equal to 1 are significantly lower, namely 89.3% Thus, we notice a considerable improvement in comparison to plain STMs when higher ranks R are used. In Table 3 we present previously reported results - note however that the various methods may use different feature sets.

Table 3. Accuracies achieved by various methods for the KTH action database

Method	[20]	[21]	[22]	[23]	[24]	[25]	[26]	[27]	[7]
Accuracy	88.0%	88.3%	81.2%	81.5%	90.5%	91.7%	88.3%	87.7%	95.3%
Method				SVMs	STMs (Rank 1)	STMs (Rank 6)			
Accuracy				88.5 %	89.3%	93.3%			

6   Conclusions

In this work, we addressed the two class classification problem within the tensor-based large margin classification framework. More precisely, we proposed the higher rank Support Tensor Machines (STMs), in which the parameters defining the separating hyperplane form a tensor (tensorplane) that can be written as a sum of rank one tensors, according to the CANDECOMP/PARAFAC (CP) decomposition. The corresponding optimization problem of the proposed classifier was solved in an iterative way, with the estimation of the parameters corresponding to the projections along a single tensor mode achieved using a typical SVM formulation. The efficiency of the proposed method was illustrated on the problems of gait and action recognition where the results reported improved, in some cases considerably, the state of the art.

Acknowledgment. This work was supported by the EPSRC grant 'Recognition and Localization of Human Actions in Image Sequences' (EP/G033935/1).

References

1. Lu, H., Plataniotis, K.N., Venetsanopoulos, A.N.: Mpca: Multilinear principal component analysis of tensor objects. IEEE Trans. on Neural Networks 19 (2008)
2. Yan, S., Xu, D., Yang, Q., Zhang, L., Tang, X., Zhang, H.J.: Multilinear discriminant analysis for face recognition. IEEE Trans. on Image Processing 16 (2007)
3. Tao, D., Li, X., Wu, X., Hu, W., Maybank, S.J.: Supervised tensor learning. Knowledge and Information Systems 13 (2007)
4. Zafeiriou, S.: Discriminant nonnegative tensor factorization algorithms. IEEE Trans. on Neural Networks 20 (2009)
5. Zafeiriou, S., Petrou, M.: Nonnegative tensor factorization as an alternative csiszar tusnady procedure: algorithms, convergence, probabilistic interpretations and novel probabilistic tensor latent variable analysis algorithms. Data Mining and Knowledge Discovery 22 (2011)
6. Zafeiriou, S.: Algorithms for nonnegative tensor factorization. Tensors in Image Processing and Computer Vision (2009)
7. Kim, T.-K., Cipolla, R.: Canonical correlation analysis of video volume tensors for action categorization and detection. IEEE Trans. on Pattern Analysis and Machine Intelligence 31 (2009)
8. Bobick, A., Davis, J.: The recognition of human movements using temporal templates. IEEE Trans. on Pattern Analysis and Machine Intelligence 23 (2001)
9. Wolf, L., Jhuang, H., Hazan, T.: Modeling appearances with low-rank svm. In: Proceedings of IEEE Conference on Computer Vision and Pattern Recognition (2007)
10. Gorelick, L., Blank, M., Shechtman, E., Irani, M., Basri, R.: Actions as space-time shapes. IEEE Trans. on Pattern Analysis and Machine Intelligence 29 (2007)
11. Loy, C.C., Xiang, T., Gong, S.: Multi-camera activity correlation analysis. In: IEEE Conference on Computer Vision and Pattern Recognition (2009)
12. Boulgouris, N.V., Plataniotis, K.N., Hatzinakos, D.: Gait recognition using linear time normalization. Pattern Recognition 39 (2006)

13. Sarkar, S., Phillips, P.J., Liu, Z., Vega, I.R., Grother, P., Bowyer, K.W.: The humanid gait challenge problem: Data sets, performance, and analysis. IEEE Trans. on Pattern Analysis and Machine Intelligence 27 (2005)
14. Han, J., Bhanu, B.: Individual recognition using gait energy image. IEEE Trans. on Pattern Analysis and Machine Intelligence 28 (2006)
15. Tao, D., Li, X., Wu, X., Maybank, S.J.: General tensor discriminant analysis and gabor features for gait recognition. IEEE Trans. on Pattern Analysis and Machine Intelligence 29 (2007)
16. Kolda, T.G., Bader, B.W.: Tensor decompositions and applications. SIAM Review 51 (2009)
17. Kale, A., Rajagopalan, A.N., Sunderesan, A., N. Cuntoor, A.R.C., Krueger, V., Chellappa, R.: Identification of humans using gait. IEEE Trans. on Image Processing (2004)
18. Schuldt, C., Laptev, I., Caputo, B.: Recognizing human actions: A local svm approach (2004)
19. Tzimiropoulos, G., Zafeiriou, S., Pantic, M.: Subspace learning from image gradient orientations. IEEE Trans. on Pattern Analysis and Machine Intelligence (2012)
20. Oikonomopoulos, A., Patras, I., Pantic, M.: Spatiotemporal localization and categorization of human actions in unsegmented image sequences. IEEE Trans. on Image Processing 20 (2011)
21. Ahmad, M., Lee, S.: Human action recognition using shape and clg-motion flow from multi-view image sequences. Pattern Recognition 41 (2008)
22. Dollar, P., Rabaud, V., Cottrell, G., Belongie, S.: Behavior recognition via sparse spatio-temporal features. In: 2nd Joint IEEE International Workshop on Visual Surveillance and Performance Evaluation of Tracking and Surveillance (2005)
23. Niebles, J., Wang, H., Fei-Fei, L.: Unsupervised learning of human action categories using spatial-temporal words. International Journal of Computer Vision 79 (2008)
24. Fathi, A., Mori, G.: Action recognition by learning mid-level motion features. Proceedings of IEEE Conference on Computer Vision and Pattern Recognition 41 (2008)
25. Jhuang, H., Serre, T., Wolf, L., Poggio, T.: A biologically inspired system for action recognition. In: Proceedings of IEEE Conference on Computer Vision (2007)
26. Rapantzikos, K., Avrithis, Y., Kollias, S.: Dense saliency-based spatiotemporal feature points for action recognition. In: IEEE Conference on Computer Vision and Pattern Recognition (2009)
27. Ali, S., Shah, M.: Human action recognition in videos using kinematic features and multiple instance learning. IEEE Trans. on Pattern Analysis and Machine Intelligence 1 (2010)

Multi-scale Integral Modified Census Transform for Eye Detection

Inho Choi and Daijin Kim

Department of Computer Science and Engineering
Pohang University of Science and Technology (POSTECH)
{ihchoi,dkim}@postech.ac.kr

Abstract. In this paper, we propose a multi-scale integral modified census transform (MsiMCT) for eye detection. Modified census transform shows a good classification performance. However, it does not represent block properties. We therefore consider a method for overcoming this limitation. First, we propose the integral modified census transform (iMCT) using integral images, which can compute the mean intensity of the rectangular region rapidly. Using iMCT, we propose the multi-scale integral modified census transform (MsiMCT), which is structured as a concatenation of various iMCTs. The proposed MsiMCT can describe from pixel features to block features, and can therefore be implemented in many applications, such as face detection and human body detection, without modification. Our experimental results using various images show that the proposed method provides good detection accuracy, in terms of the detection rate of the number of selected weak classifiers.

1 Introduction

The detection of facial features is among the most important and interesting research topics in the field of automated face analysis. It is used in the pre-processing of face analysis by face normalization fiducial points. In particular, human eyes are stable and stand out among the other facial feature points. In this paper, we discuss the eye detection algorithm and a new texture descriptor: the multi-scale integral modified census transform (MsiMCT).

We consider a natural and intuitive method that learns and detects human facial components. Freund et al. introduced the AdaBoost algorithm and showed that it had a good generalization capability [1]. Viola et al. proposed a robust face detection method using AdaBoost with simple features, which showed a good performance in locating the face region [2]. Following Viola's method, the AdaBoost algorithm has been widely used for the detection problem. Froba and Ernest used AdaBoost with a modified version of the census transform (MCT) [3] for face detection. This combination was very robust to changes in illumination, and found the face very quickly. Wang et al. tried to overcome the limitation of the Haar feature using RNDA features [4]. Niu et al. used AdaBoost classifiers, while bootstrapping on both positives and negatives, in a method that consists of two cascade classifiers [5]. Choi and Kim used AdaBoost

G. Bebis et al. (Eds.): ISVC 2012, Part II, LNCS 7432, pp. 41–50, 2012.

with an MCT, incorporating eye verification and a correction, to improve the eye detection rate [6]. Zhang et al. described a multi-block local binary pattern (MB-LBP), based on a local binary pattern operator for rectangular regions [7]for face detection. Choi and Kim included eye detection and eye blink detection in a smart phone application using AdaBoost with an MCT [8].

As shown in the above previous works, many object detection algorithms use the boosting algorithm with texture descriptions, such as the Haar-like feature, LBP, and MCT. In addition, using the boosted learning algorithm, some features that have been shown to result in a good classification performance can be selected automatically.

We consider a new descriptor that is based on an MCT and integral images. It can represent block properties that are similar to the Haar-like feature and the MB-LBP. Using these properties, we propose a new descriptor: the multi-scale integral MCT (MsiMCT). It is a non-parametric local transform of 3 3 neighbor pixels for any rectangular region that is computed by integral images. It can represent any ordered pattern, the pixel intensity of which is greater than the mean of 3 3 pixels. It is built by concatenating several blocks that have different normalized sizes for the same image. The MsiMCT can describe from pixel properties to block properties. Because of this characteristic, the MsiMCT can be implemented in many applications, such as face detection and human body detection, without modification.

This paper is organized into six sections. In the second section, we will describe various works related to object description. In the third section, our proposed new descriptor, which is called MsiMCT, will be explained. In the fourth section, we will describe the eye detection process using the proposed detector with the MsiMCT feature. In the fifth section, we will show the experimental results of the proposed eye detection over various eye databases. Finally, we will present our conclusions.

2 Previous Work

2.1 Modified Census Transform

MCT is a non-parametric local transform that modifies the census transform developed by [3]. It is an ordered set of comparisons of pixel intensities in a local neighborhood that represents which pixels have greater intensity than the mean pixel intensity. Let $N(x)$ be a spatial neighborhood of the pixel at $\mathbf{x}$ and $N'(x)$ include the pixel at $\mathbf{x}$, i.e., $N'(\mathbf{x}) = \{\mathbf{x} \cup N(\mathbf{x})\}$. Let $I(\mathbf{x})$ and $\bar{I}(\mathbf{x})$ be the pixel intensity at $\mathbf{x}$ and the mean of pixel intensity at $\mathbf{x}$ over $N'(x)$, respectively. Then, the MCT of the pixel at $\mathbf{x}$ is defined as

$$\Gamma(\mathbf{x}) = \bigotimes_{\mathbf{y} \in N'(\mathbf{x})} C(\bar{I}(\mathbf{x}), I(\mathbf{y})), \qquad (1)$$

where C is a comparison function defined as

$$C(\bar{I}(\mathbf{x}), I(\mathbf{y})) = \begin{cases} 1 \text{ if } \bar{I}(\mathbf{x}) < I(\mathbf{y}), \\ 0 \text{ otherwise}, \end{cases} \qquad (2)$$

and $\otimes$ denotes a concatenation operator. In a 3×3 neighborhood, the MCT provides 511 different local structure patterns.

2.2 Multi-Block Local Binary Pattern

MB-LBP encodes rectangles using a local binary pattern operator [7] (Fig. 1). The MB-LBP can be calculated rapidly using integral images, and it can compute larger scale structures than the original LBP. The MB-LBP is defined by comparing the center rectangle's average intensity with those of its neighborhood rectangles. An output value of the MB-LBP is obtained as

$$MB - LBP = \sum_{i=1}^{8} s(g_i - g_c) 2^i \tag{3}$$

where g_c is the average intensity of the center rectangle, g_i is the average intensity of the center pixel's neighborhood rectangles,

$$s(x) = \begin{cases} 1 \text{ if } x > 0, \\ 0 \text{ otherwise.} \end{cases} \tag{4}$$

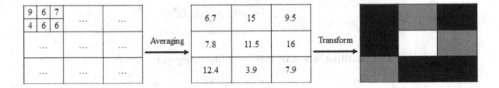

Fig. 1. Example of MB-LBP

3   Integral Modified Census Transform

3.1 Integral Modified Census Transform

In this section, we describe the integral modified census transform (iMCT). Fig. 2 shows a graphical representation of iMCT.

The iMCT encodes rectangular regions by an intensity comparator using integral images. To compute a specific rectangular region, four access points to the integral image are required (Fig. 3).

Each rectangular region is calculated as

$$R(1,1) = ii(2,2) - ii(2,1) - ii(1,2) + ii(1,1) \tag{5}$$
$$R(2,1) = ii(3,2) - ii(3,1) - ii(2,2) + ii(2,1) \tag{6}$$
$$R(x,y) = ii(x+1,y+1) - ii(x+1,y) - ii(x,y+1) + ii(x,y) \tag{7}$$

where $ii(x,y)$ is the integral image.

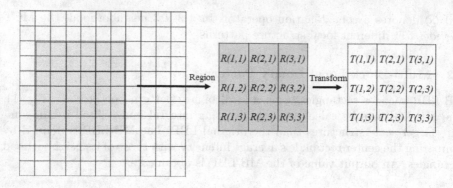

Fig. 2. A graphical representation of iMCT

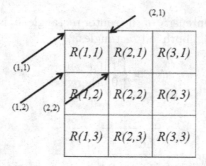

Fig. 3. Illustration of iMCT using integral images

The iMCT is defined as

$$T(x', y') = \begin{cases} 1 \text{ if } R(x', y') > \bar{R}, \\ 0 \text{ otherwise,} \end{cases} \tag{8}$$

where

$$\bar{R} = \frac{1}{9} \sum_{x'=x-1}^{x'=x+1} \sum_{y'=y-1}^{y'=y+1} R(x', y'). \tag{9}$$

3.2   Multi-scale Integral Modified Census Transform

The MsiMCT is a concatenation of various iMCTs (Fig. 4). It can represent various sizes of block for the input data. It is built by concatenating several blocks that have different normalized sizes for the same image. The MsiMCT can describe from pixel properties to block properties.

In the AdaBoost training phase, Fig. 4-(b) and (c) are selected. The block features facilitate the training phase.

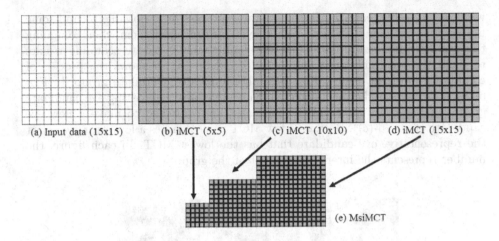

(a) Input data (15x15)   (b) iMCT (5x5)   (c) iMCT (10x10)   (d) iMCT (15x15)

(e) MsiMCT

Fig. 4. Example of MsiMCT

3.3   AdaBoost Training with MsiMCT

We present the AdaBoost training with the MsiMCT features. In AdaBoost training, we select weak classifiers that separate the positive and negative eyes and then construct a strong classifier that is a linear combination of the weak classifiers. The weak classifier tries to find the lookup table that represents the error for each MCT at a specific location (scale and pixel position).

4   Eye Detection Process

4.1   Grouping in Scanning Process

In face detection, the sizes of the detected faces are very different. In addition, the human eye has a variety of shapes. We therefore use a scanning method that uses various sizes of the sliding window (Fig. 5).

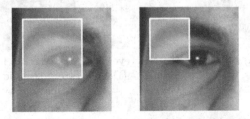

Fig. 5. Scanning using various sizes of sliding window

To scan eye candidates, the detected face is divided into two sub-regions that contain the left and right eye candidate regions. In each scanning step, the sliding window has a different size. In the scanning process, it is possible to find some non-eyes. If we know that they are non-eyes, we can reject them when compiling the eye detection results [8]. Thus, we combine and separate scanning results to build the eye candidate group (Fig. 6 and 7). Fig. 6 shows the process of combining two scan results. Fig. 6-(a) shows one eye candidate group. In Fig. 6-(b), a new eye candidate is detected, which overlaps the existing eye candidate group. In this case, the two eye candidates are combined to make a new eye candidate group (6-(c)). Because a low MCT value denotes a low error, we select the representative eye candidate that has the lowest MCT. In each figure, the number represents the intersection count of the group.

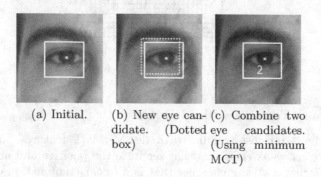

(a) Initial.   (b) New eye can- (c) Combine two
                 didate. (Dotted eye   candidates.
                 box)   (Using minimum
                                  MCT)

Fig. 6. Grouping of scanning results: case 1

Fig. 7 shows the process of separating two scanning results. Fig. 7-(a) shows one eye candidate group. In 7-(b), a new eye candidate is detected but does not overlap the first eye candidate. In this case, the two eye candidates are separated and form two distinct eye candidate groups (7-(c)).

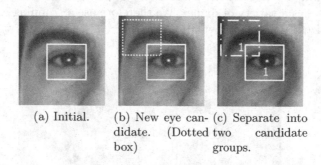

(a) Initial.   (b) New eye can- (c) Separate into
                 didate. (Dotted two   candidate
                 box)   groups.

Fig. 7. Combining and separating scanning results: case 2

After scanning an input image and executing the combining and separating steps, we select an eye candidate group according to the maximum intersection count of each eye candidate group (Fig. 8). Fig. 8 shows three eye candidate

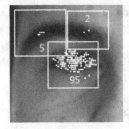

Fig. 8. Choosing an eye candidate group

groups that have intersection counts of five, two, and ninety-five, respectively. In this figure, we choose the white box that has the highest intersection count. This step facilitates the elimination of outliers of scanning the input image.

4.2   Refinement of Scanning Results

Using the average MCT value of a selected eye candidate group, we calculate the final center position of two eyes. First, we calculate the average MCT value $\bar{\Gamma}$ of a selected eye candidate group. Second, we compute the average position of the candidates with an MCT value lower than the average MCT value $\bar{\Gamma}$. Because the MCT value refers to the difference between the detected region and the real eye, a low MCT value indicates proximity to the real eye. By using the Gaussian estimation, the detection results will be more stable and reliable. This refinement process is summarized in Table. 1.

Table 1. Refinement of eye candidate scanning results

1. Calculate average the MCT value Γ of the selected eye candidate group.
2. For each eye candidate, if the MCT value is lower than $\bar{\Gamma}$, then it is cumulated to the final result.
3. Calculate the average position of the final detection result group and this locates the center of the eye.

5   Experimental Results

5.1   Experimental Setup

To compare the training result, we used many training databases: the Asian Face Image Database PF01 ([9]), POSTECH Face Database PF07 ([10]) , and some images from the Internet and prepared 25,000 eye images and 50,000 non-eye images. To evaluate the proposed eye detection method, we used two face databases: the BioID and the AR564 face database. As a measure of eye detection, we define an eye detection error as

$$err = \frac{max(d_l, d_r)}{d_{lr}}, \tag{10}$$

where d_l and d_r are the Euclidean distance between the ground truth and the detected position of the left eye and right eye, respectively, and d_{lr} is the Euclidean distance between the ground truth of the left eye and right eye. To estimate the detection accuracy over a number of selected weak classifiers, we set the normalized error to 0.1 (10%).

5.2   Eye Detection Results

The BioID face database consists of 1,521 gray level images, where the pixel resolution of each image is 384x286 pixels. Fig. 9 shows the detailed eye detection rate over the number of selected weak classifiers in the BioID database images with 10% normalized error.

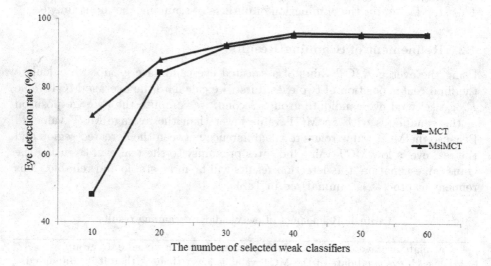

Fig. 9. Eye detection rate over the number of weak classifiers in the Bio ID database (10% normalized error)

We used a subset of the AR face database (AR-564) [11]. The AR-564 face database consists of 564 images (94 peoples × 3 facial expressions × 3 illuminations).

Our proposed MsiMCT has good detection accuracy at the same number of weak classifiers in the Bio and AR564 databases. From these results, it can be seen that block features are useful for representing texture description. Our proposed descriptor therefore facilitates obtaining reliable eye detection results.

6   Conclusion

In this paper, we propose a new descriptor for object representation: the multi-scale integral modified census transform (MsiMCT). The proposed iMCT can

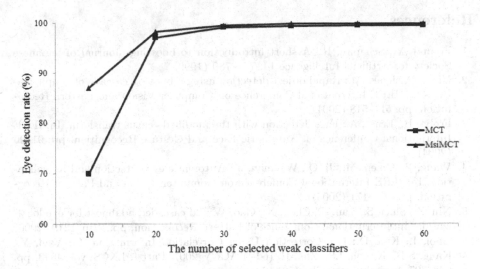

Fig. 10. Eye detection rate over the number of weak classifiers in the AR564 database (10% normalized error)

Table 2. Comparison of eye detection rate as a function of the number of selected weak classifiers in the BioID and AR564 database

Descriptors	Number of weak classifiers					
	10	20	30	40	50	60
MCT(Bio)	48.19	84.22	91.98	95.27	95.27	95.4
MsiMCT(Bio)	71.6	88.03	92.77	96.06	95.86	95.99
MCT(AR564)	69.80	96.98	99.11	99.29	99.47	99.47
MsiMCT(AR564)	87.03	98.22	99.47	99.82	99.82	99.82

represent rectangular properties using integral images. The MsiMCT can describe from pixel features to block features by various iMCTs. Because block features are useful for representing texture description, the proposed eye detection with MsiMCT is more accurate than MCT. From these features, the proposed descriptor can be implemented in many applications without modification, for instance, face detection, facial feature point detection, entire human body detection, eye blinking detection, and so on.

Acknowledgements. This work was partially supported by the MKE (The Ministry of Knowledge Economy), Korea, under the Core Technology Development for Breakthrough of Robot Vision Research support program supervised by the NIPA (National IT Industry Promotion Agency)(NIPA-2012-H1502-12-1002). Also, this work was supported by the IT R&D program of MKE/KEIT. [10040246, Development of Robot Vision SoC/Module for acquiring 3D depth information and recognizing objects/faces]

References

1. Freund, Y., Schapire, R.: A short introduction to boosting. Journal of Japanese Society for Artificial Intelligence 14, 771–780 (1999)
2. Viola, P., Jones, M.: Rapid object detection using a boosted cascade of simple features. In: IEEE International Conference on Computer Vision and Pattern Recognition, pp. 511–518 (2001)
3. Froba, B., Ernst, A.: Face detection with the modified census transform. In: IEEE International Conference on Automatic Face and Gesture Recognition, pp. 91–96 (2004)
4. Wang, P., Green, M., Ji, Q., Wayman, J.: Automatic eye detection and its validation. In: IEEE International Conference on Computer Vision and Pattern Recognition, pp. 164–171 (2005)
5. Niu, Z., Shan, S., Yan, S., Chen, X., Gao, W.: 2d cascaded adaboost for eye localization. In: International Conference of Pattern Recognition, pp. 1216–1219 (2006)
6. Choi, I., Kim, D.: Eye Correction Using Correlation Information. In: Yagi, Y., Kang, S.B., Kweon, I.S., Zha, H. (eds.) ACCV 2007, Part I. LNCS, vol. 4843, pp. 698–707. Springer, Heidelberg (2007)
7. Zhang, L., Chu, R., Xiang, S., Liao, S., Li, S.Z.: Face Detection Based on Multi-Block LBP Representation. In: Lee, S.-W., Li, S.Z. (eds.) ICB 2007. LNCS, vol. 4642, pp. 11–18. Springer, Heidelberg (2007)
8. Choi, I., Han, S., Kim, D.: Eye detection and eye blink detection using adaboost learning and grouping. In: IEEE International Conference on Computer Communications and Networks, pp. 1–4 (2011)
9. Je, H., Kim, S., Jun, B., Kim, D., Kim, H., Sung, J., Bang, S.: Asian Face Image Database PF01, Technical Report. Intelligent Multimedia Lab, Dept. of CSE, POSTECH (2001)
10. Lee, H., Park, S., Kang, B., Shin, J., Lee, J., Je, H., Jun, B., Kim, D.: The postech face database (pf07) and performance evaluation. In: IEEE International Conference on Automatic Face and Gesture Recognition, pp. 1–6 (2009)
11. Martinez, A., Benavente, R.: The AR Face Database, CVC Technical Report #24 (1998)

A Comparative Analysis of Thermal and Visual Modalities for Automated Facial Expression Recognition

Avinash Wesley, Pradeep Buddharaju, Robert Pienta, and Ioannis Pavlidis

University of Houston and Georgia Institute of Technology

Abstract. Facial expressions are formed through complicated muscular actions and can be taxonomized using the Facial Action Coding System (FACS). FACS breaks down human facial expressions into discreet action units (AUs) and often combines them together to form more elaborate expressions. In this paper, we present a comparative analysis of performance of automated facial expression recognition from thermal facial videos, visual facial videos, and their fusion. The feature extraction process consists of first placing regions of interest (ROIs) at 13 fiducial regions on the face that are critical for evaluating all action units, then extracting mean value in each of the ROIs, and finally applying principal component analysis (PCA) to extract the deviation from neutral expression at each of the corresponding ROIs. To classify facial expressions, we train a feed-forward multilayer perceptron with the standard deviation expression profiles obtained from the feature extraction stage. Our experimental results depicts that the thermal imaging modality outperforms visual modality, and hence overcomes some of the shortcomings usually noticed in the visual domain due to illumination and skin complexion variations. We have also shown that the decision level fusion of thermal and visual expression classification algorithms gives better results than either of the individual modalities.

1 Introduction

The detection and recognition of human facial expressions is a challenging task. Among different individuals the geometry, size, and color of the face vary greatly. Furthermore, a single expression can be formed at many different intensities and speeds, sometimes so subtle that it goes unnoticed to a human observer. This intense variance compounded with the subtlety of expressions necessitates more detailed and automated approaches to facial expression detection.

Visual cameras are most commonly used to capture facial data due to their low cost and ubiquitous availability. Several automated facial expression recognition algorithms were proposed in the recent years from visual imagery [1], [2], [3]. Bartlett et. al reported 93% accuracy of automated facial recognition on the Cohn-Kanade expression dataset [4], and recently Kotsia and Pitas reported classification accuracy of 99.7% and 95.1% on the same dataset [5]. Visual approaches, while shown to be quite effective on particular databases, have a few

G. Bebis et al. (Eds.): ISVC 2012, Part II, LNCS 7432, pp. 51–60, 2012.

unaddressed obstacles. A major drawback is their tendency to lose accuracy when classifying subjects of darker skin tones. The OpenCV face detection system, which has become a basis for comparison shows a significant disparity in the accuracy of classifying dark- versus light-skinned subjects [6]. Furthermore, many databases used to test visual-based expression recognition systems have a narrow variety of positions, textures, and intensities of light. This usually simplifies the task of classification and result in higher accuracy measurements. Hence visual approaches tend to perform well under sterile lab conditions, but under varied light conditions they may operate at lower accuracies [6].

Thermal imaging is a well known alternative to visual imagery because of its illumination invariance [7]. A thermal camera measures the radiations emitted from the surface of the skin, which is a result of heat dissipation from core body due to blood flow, metabolic activities, subcutaneous tissue structure and the sympathetic nervous activities. Though study has been done in the area of thermal face recognition [8], few have attempted to explore facial expression recognition using this modality. Khan et al. explored and proved through statistical analysis the feasibility of automated facial expression classification through thermal imaging [9]. Yoshitomi et al. reported success rates of 90% [10]. An unsupervised local and global feature localization algorithm for facial expression classification was proposed by Trujilo [11].

Despite solving the illumination problem encountered in visual imagery, thermal imaging poses a major challenge as facial thermograms may change depending on ambient temperature and the physical condition of the subject. This renders difficult the task of acquiring similar features for the same expressions. Past studies in face recognition noticed that the thermal face recognition performance deteriorates over time [12], posing a necessity to perform similar studies for automated facial expression recognition. In this paper, we collected simultaneous visual and thermal data during both ideal and challenging conditions, and further present comparative results from both modalities as well as their fusion. To the best of our knowledge, this is the first time such comparative study is being reported.

FACS, developed by psychologists Ekman and Friesen [13], is most commonly used to classify human facial expressions through analysis of possible contortions of facial geometry. FACS breaks down the development of expressions into particular action units, each of which is derived from a muscle or muscle group in the head. In this paper, we present a feature extraction and classification algorithm for a total of 8 action units (AU 1+2, 4, 6+12, 9, 10, 12, 15, 17). We selected these specific action units because they are the exemplary when forming any of the 6 universal emotions: surprise (AU1+2), fear (AU4), sadness (AU15), disgust (AU9, AU10), anger (AU4) and happiness (AU6+12, AU12) [14].

2 Methodology

Our automated facial expression recognition algorithm mainly contains three steps - face acquisition, facial feature extraction, and expression classification.

In this section, we will explain in detail our experimental setup to collect simultaneous visual and thermal facial data, local facial feature extraction algorithm, and expression classification methods.

2.1   Experimental Setup

A snapshot of our experimental setup can be seen in the figure 1. A total of 8 subjects participated in our experiments with age range from 20 to 30 years, both genders, and varying ethnicities. To facilitate comparison, we collected simultaneous data from both midwave thermal infrared and a monochrome CCD visual cameras as shown in figure 1. The room is equipped with low, medium, and high intensity fluorescent lighting to simulate the effect of illumination variation on visual imagery. We used a portable heater fan to simulate the effect of variable atmospheric air conditions on thermal imagery. The subjects were instructed to rinse their face and apply a small amount of 70% isopropyl alcohol. In order to ensure that the evaporation of the volatile alcohol mixture did not adversely affect the data, each subject waited a mandatory period of 15 minutes before beginning the data collection. A FACS encoder trained each subject regarding the facial expressions they were supposed to make during the data collection by showing them the videos of each expression. Each subject was allowed as much time as they needed to practice each expression, and they also have an option to skip any expression if they so desired. For each subject, we first record their relaxed and neutral expression for 25 seconds, followed by a visual instruction on a screen in front of them regarding the next expression they are supposed to make. The subjects were asked to repeat each expression 14 times at any intensity of their choice in order to simulate the variety of natural expression in everyday formulation.

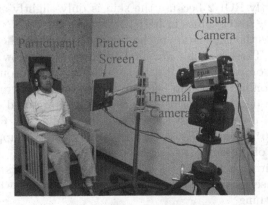

Fig. 1. The experimental setup used to simultaneously collect thermal and visual facial data from subjects

2.2   Local Feature Extraction

The typical feature extraction algorithms in automated facial expression recognition can be categorized as holistic (where the face is processed as a whole), and local (where only facial features or areas that are prone to change with facial expressions are processed) [1]. Our feature extraction algorithm falls in the latter category with regions of interest (ROIs) placed at 13 fudicial points on the face (as shown in figure 2). The ROIs are carefully chosen according to various facial muscles involved in different FACS action units as explained below:

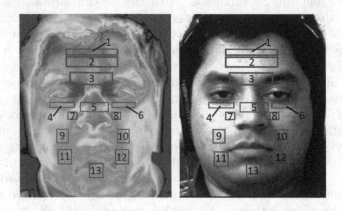

Fig. 2. The 13 regions of interest used to capture facial movement and deformation

ROIs 1 and 2: measures contraction of the frontalis muscles which raise the eyebrows. The raising of the eyebrows, present in FACS action units 1 and 2, are most common associated with expressions of surprise. The vertical placement of ROIs 1 and 2 distinguish between action unit combination 1+2 and 4. Action unit 4 affects mostly ROI 2 because the skin is only slightly stretched on the forehead, producing lower values in ROI 1.

ROI 3: captures the translation of the tissue actuated by the corrugator and procerus muscles. These muscles are used to furrow the brow, action unit 4, when one is angry or sad. This ROI detects both the translation of the eyebrow and the deformation in the skin in between the eyebrows generate signal.

ROIs 4 and 5: detects the orbicularis oculi. These are used to detect action unit 6, the critical difference between a Duchenne smile (AU 6 and 12) and a simple smile (AU 6). These ROIs detect the subtle raising of upper-cheek tissue and the wrinkling of the outer eye-edge.

ROI 6: measures the quadratus labii superioris which is responsible for scrunching the nose tissue. This is most commonly formed when a person is disgusted at something.

ROIs 7 and 8: additional measures to detect the lower set of elevator muscles, used to raise the tissue surrounding the nose. These attempt to measure action unit 10, a secondary expression of disgust. These measure the translation of new tissue from around the nose, just above the periorbital region.

ROIs 9 and 10: detects the contraction of the zygomaticus muscles, used most strongly in smiles. These measure action unit 12, the widening of the lips. These detect specifically the translation of cheek tissue as well as the crease formed at the edges of the mouth during a smile.

ROIs 11 and 12: measures the contraction of the triangularis, which lowers the other edges of the mouth into a frown. Action unit 15 is necessary for expressing sadness. These two ROIs measure both tissue translation and crease formation around the bottom edges of the mouth.

ROI 13: measures the change in the tissue attached to the mentalis. This allows for the measurement of any chin flexion, especially used to raise the lower lip.

The thermal and visual facial videos were recorded at 25 and 30 fps respectively. We computed the neutral ROIs by computing the mean values in each ROI from first 25 seconds of the video, when the subjects made neutral expression. Then the principal components were computed for each ROI by treating each pixel within the ROI as a variable. The frames corresponding to greatest change from the neutral ROI will have the largest principal component values during the expression as depicted in figure 3.

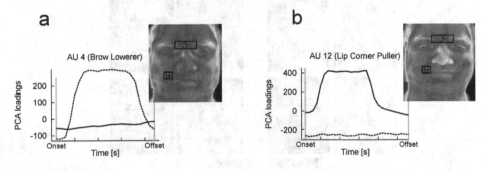

Fig. 3. PCA values from ROIs 3 and 9 while the subject is making **(a)** angry expression, AU 4 (Brow lowerer) and **(b)** happy expression, AU12 (Lip corner puller). It can be clearly seen that PCA values for ROI 3 has large values during angry expression, while it has large values for ROI 9 during happy expression.

After the principal components have been found for all ROIs, a profile for each expression is determined by computing the standard deviation of each ROI-principal component. To do this, we first annotate the onset (marking the start) and offset (marking the end) frames for each expression as shown in figure 3. The standard deviation expression profiles are generated by computing the standard deviation of each of the 13 ROIs during the window between the onset and offset. These expression profiles denote the amount of deviation found over the course of the expression, and hence are used to train the classifier.

2.3   Classification

The standard deviation expression profiles computed in the feature extraction step are used to train feed-forward multilayer perceptrons [15] for both visual and thermal modalities. Each multilayer perceptron utilizes 14 input nodes, 10 sigmoid nodes in the hidden layer and 8 output nodes to classify expressions. Thermal and visual perceptron classifiers were generated separately by training them with expressions that were coded by a certified FACs encoder to determine a ground truth. In order to study the effect of fusion of thermal and visual modalities, we use a simple decision level fusion scheme, where for each test expression, the result from the perceptron with maximum confidence is considered.

3   Experimental Results and Discussion

In order to test the performance of each of the thermal and visual modalities during both ideal and challenging conditions, each subject was asked to participate in two sessions - Phase I and Phase II. In this section, we will present results from each of these sessions.

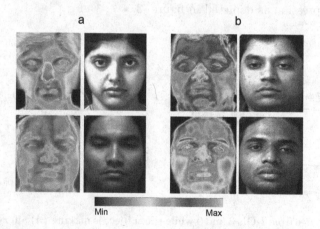

Fig. 4. (a) A sample from Phase I (illumination variance) dataset. The top row shows the thermal and visual images acquired from a subject under bright lighting, middle row shows corresponding images from another subject acquired under low lighting. **(b)** A sample from Phase II (temperature variance) dataset. The top row shows the thermal and visual images acquired while warm air is blown under high setting, middle row shows corresponding images from another subject while air is blown under low setting. The bottom row shows the color map used for thermal images.

3.1   Phase I Experiments - Illumination Variance

In the first session (Phase I), we introduced variability in visual imagery by using different lighting (low, medium, and high intensity) in the room for different subjects during the data collection. This resulted in a considerable variability in visual imagery (as shown in figure 4a) and hence posed a challenging condition for the visual perceptron classifier. However, the room temperature was maintained constant throughout the session, maintaining an ideal condition for thermal imagery. The Phase I dataset consisted of a total of 448 expressions from each of the thermal and visual modalities.

We used 10-fold cross validation and percentage split in order to test the classification accuracy. Table 1(left) shows the confusion matrix and Table 2 shows the accuracy for all the test action units from thermal and visual modalities, as well as their decision-level fusion. As we expected, thermal modality performed better than visual modality because visual imagery is affected by the illumination variance introduced in the dataset. However, the decision-level fusion of thermal and visual perceptron classifiers performed better than either of them.

3.2   Phase II Experiments - Temperature Variance

In the second session (Phase II), we introduced variability in thermal imagery by blowing a heater fan (at different speeds of low, medium, and high), affecting the subject's thermal signature. This introduced a considerable variability in thermal imagery (as shown in figure 4b), posing challenging conditions for the thermal perceptron classifier. However, the lighting in the room was maintained constant throughout the session — an ideal condition for visual imagery. The Phase II dataset consisted of a total of 448 expressions from each of the thermal and visual modalities.

Table 1 (right) shows the confusion matrix and Table 2 shows the accuracy for all the test action units from thermal and visual modalities, as well as their decision-level fusion. As we expected, the visual modality has better results in Phase II than in Phase I, since constant lighting is maintained during the data collection. However, an interesting observation is that despite the temperature variance introduced in the dataset, thermal modality remains unaffected in Phase II and has almost similar performance to that in Phase I. Also, the decision-level fusion of thermal and visual perceptron classifiers again performed better than either of them.

As explained in section 2.2, the features fed to the classifiers are the principal components computed in each of the ROIs, which actually measures the change from neutral ROI during the expression. In the visual imagery much of this change is a result of the formation of shadows on portions of face depending on the particular expression being made. It is possible that no new shadows are formed in the case of planar deformations or poor lighting. This is the reason why the classifier performance was poor on Phase I dataset where different lighting conditions were used during data collection. The thermal data however captures not only the translation, but also the deformation of the tissue due to the unique heat patterns generated on face during the expression. These deformations introduce variability that can always be measured by principal components. Hence the classifier performance was same on both Phase I and Phase II datasets, even though considerable variability was introduced on the thermal data in Phase II

Table 1. Confusion matrices of Phase I (illumination variance) and Phase II (temperature variance) experiments; for thermal and visual modalities, and their fusion

Test AUs		Phase I Classified AUs								Phase II Classified AUs							
		1+2	4	6+12	9	10	12	15	17	1+2	4	6+12	9	10	12	15	17
1+2	T	96	2	0	2	0	0	0	0	100	0	0	0	0	0	0	0
	V	92	4	0	2	0	0	0	2	98	0	0	0	0	0	2	0
	F	96	2	0	2	0	0	0	0	100	0	0	0	0	0	0	0
4	T	2	98	0	0	0	0	0	0	0	97	0	0	0	0	3	0
	V	8	86	0	6	0	0	0	0	0	100	0	0	0	0	0	0
	F	2	98	0	0	0	0	0	0	0	100	0	0	0	0	0	0
6+12	T	0	0	98	0	2	0	0	0	0	0	97	0	0	3	0	0
	V	0	0	72	0	10	12	2	4	0	0	97	0	0	3	0	0
	F	0	0	98	0	0	0	0	2	0	0	97	0	0	3	0	0
9	T	0	2	0	92	2	2	0	2	0	0	0	100	0	0	0	0
	V	2	2	0	96	0	0	0	0	0	0	2	98	0	0	0	0
	F	0	0	0	98	0	0	0	2	0	0	0	100	0	0	0	0
10	T	0	0	2	0	96	0	2	0	0	0	0	0	100	0	0	0
	V	0	0	0	4	96	0	0	0	0	0	0	0	100	0	0	0
	F	0	0	0	0	100	0	0	0	0	0	0	0	100	0	0	0
12	T	0	0	6	0	2	90	0	2	2	0	6	0	0	82	8	4
	V	0	0	8	0	0	92	0	0	4	0	2	0	0	88	6	0
	F	0	0	4	0	0	94	0	2	2	0	0	0	0	94	4	0
15	T	0	5	0	0	3	3	90	0	0	0	3	0	0	5	90	3
	V	0	0	0	0	0	0	90	10	0	0	0	0	0	15	82	3
	F	0	0	0	0	0	0	95	5	0	0	0	0	0	8	92	0
17	T	0	0	2	0	0	0	2	96	0	0	2	0	0	0	0	98
	V	4	0	0	0	2	5	5	84	0	0	0	0	0	0	0	100
	F	2	0	0	0	2	0	0	96	0	0	0	0	0	0	0	100

Table 2. Accuracy of Phase I (illumination variance) and Phase II (temperature variance) experiments for thermal and visual modalities, and their fusion

	Thermal	Visual	Fusion
Phase I Accuracy	94.81%	88.64%	97.03%
Phase II Accuracy	94.6%	94.6%	98.01%

by using a heater fan. As one would expect, the fusion of the two modalities always performed better than either of the individual modalities.

There are a few challenges in classification of certain action units that were noticed in both modalities. The largest type of misclassification in the thermal domain is between action units 1+2 and 4. This error is caused largely by low intensity action unit 1+2, which develops a weak signal in the topmost ROI 1, which mostly resembles the low signal generated by action unit 4, and hence confuses the perceptron classifier. Hence in these cases the perceptron misclassified the lower signal action unit 1+2 as action unit 4. Similarly there is considerable misclassification between action units 1+2 and 4 in the visual approach, although the reason is slightly different. Medium to strong contraction of the frontalis (au1+2) creates wrinkles on the forehead, which casts shadows and in turn affects the PCA output. In a few instances the intensity was so low that very few shadows were generated, and therefore it was classified as action unit 4.

The second largest source of misclassification in both modalities is between action unit 12 and 15. Au 12 pulls the corner of the lips back and upwards (obliquely) creating a wide U shape to the mouth while au 15, the lip corner depressor pulls the lip corners down. Both of these action units produce strong signals in the ROIs placed in the buccal region (ROIs 9, 10, 11 and 12), which in turn confuses the perceptron classifier in some cases, and hence leads to misclassification.

The third largest source of misclassification in thermal is between action unit combination 6+12 and 12. This error is caused when the two ROIs measuring the orbicularis oculi do not detect the subtle deformation of the skin around the eye socket.

4 Conclusion

The visual approach has long been considered the most powerful approach to facial expression recognition. We have shown through pilot experiments that the thermal modality can be an alternative or a strong addition to visual modality that can overcome some of its shortcomings, such as illumination dependency. We have collected two sessions of simultaneous thermal and visual facial expression datasets, with each session comprising a challenging variability in each modality. We noticed that the visual modality has best performance when the lighting conditions are kept constant, but the performance degraded considerably when illumination variance was introduced in the dataset. However, the thermal modality performed equally well even in the presence of heat variability in the dataset. The decision-level fusion of thermal and visual modalities performed better than either of the individual modalities. To the best of our knowledge this is the first comparative study between the two modalities for automated facial expression recognition. As a future work, we plan to extend the dataset considerably, and also investigate more sophisticated fusion techniques for thermal and visual modalities.

References

1. Fasel, B., Luettin, J.: Automatic facial expression analysis: A survey. Pattern Recognition 36, 259–275 (1999)
2. Pantic, M., Member, S., Rothkrantz, L.J.M.: Automatic analysis of facial expressions: The state of the art. IEEE Transactions on Pattern Analysis and Machine Intelligence 22, 1424–1445 (2000)

3. Bartlett, M.S., Littlewort, G., Lainscsek, C., Fasel, I., Movellan, J.: Machine learning methods for fully automatic recognition of facial expressions and facial actions. In: Proc. IEEE Int'l Conf. Systems, Man and Cybernetics, pp. 592–597 (2004)
4. Bartlett, M.S., Littlewort, G., Frank, M., Lainscsek, C., Fasel, I., Movellan, J.: Recognizing facial expression: Machine learning and application to spontaneous behavior. In: IEEE Computer Society Conference on Computer Vision and Pattern Recognition, CVPR 2005, San Diego, California, pp. 568–573. IEEE Computer Society (2005)
5. Kotsia, I., Pitas, I.: Facial expression recognition in image sequences using geometric deformation features and support vector machines. IEEE Transactions on Image Processing 16, 172–187 (2007)
6. Whitehill, J., Littlewort, G., Pasel, I., Bartlett, M., Movellan, J.: Toward practical smile detection. IEEE Transactions on Pattern Analysis and Machine Intelligence 31, 2106–2111 (2009)
7. Socolinsky, D.A., Wolff, L.B., Neuheisel, J.D., Eveland, C.K.: Illumination invariant face recognition using thermal infrared imagery. In: IEEE Computer Society Conference on Computer Vision and Pattern Recognition, vol. 1, p. 527 (2001)
8. Kong, S.G., Heo, J., Abidi, B.R., Paik, J., Abidi, M.A.: Recent advances in visual and infrared face recognition - a review. Computer Vision and Image Understanding 97, 103–135 (2005)
9. Khan, M.M., Ingleby, M., Ward, R.D.: Automated facial expression classification and affect interpretation using infrared measurement of facial skin temperature variations. ACM Trans. Auton. Adapt. Syst. 1, 91–113 (2006)
10. Yoshitomi, Y., Miyawaki, N., Tomita, S., Kimura, S.: Facial expression recognition using thermal image processing and neural network. Robot and Human Communication, 380–385 (1997)
11. Trujillo, L., Olague, G., Hammoud, R., Hernandez, B.: Automatic feature localization in thermal images for facial expression recognition. In: Proceedings of the 2005 IEEE Computer Society Conference on Computer Vision and Pattern Recognition (CVPR 2005) - Workshops, vol. 03, pp. 14–21. IEEE Computer Society, Washington, DC (2005)
12. Socolinsky, D.A., Selinger, A.: Thermal face recognition over time. In: Proceedings of 17th International Conference on of the Pattern Recognition, ICPR 2004, vol. 4, pp. 187–190. IEEE Computer Society, Washington, DC (2004)
13. Ekman, P., Friesen, W.: Facial Action Coding System: A Technique for the Measurement of Facial Movement. Consulting Psychologists Press (1978)
14. Ekman, P., Friesen, W., Hager, J.C.: Facial Action Coding System Investigator's Guide (2002)
15. Hall, M., Frank, E., Holmes, G., Pfahringer, B., Reutemann, P., Witten, I.H.: The weka data mining software: An update. SIGKDD Explorations 11 (2009)

Vertebrae Tracking in Lumbar Spinal Video-Fluoroscopy Using Particle Filters with Semi-automatic Initialisation

Hammadi Nait-Charif[1], Allen Breen[2], and Paul Thompson[3]

[1] National Centre for Computer Animation, Bournemouth University, UK
[2] Anglo-European College of Chiropractic, Bournemouth, UK
[3] School of Health and Social Care, Bournemouth University, UK
hncharif@bournemouth.ac.uk

Abstract. Vertebrae tracking in lumbar spinal video-fluoroscopy is the first step in the analysis of vertebrae kinematic in patients with lower back pain. This paper presents a technique to track the vertebrae using particle filters with image gradient based likelihood measurement. In the first X-ray frame, the vertebrae are semi-automatically segmented and a bi-spline curve is fitted to the landmark points to construct the vertebrae outlines; then a particle filter is used to track the vertebrae through the sequence. The proposed technique is able to track the vertebrae in both lateral and frontal video-fluoroscopy sequences. The tracking results compare well with the ground truth data obtained by manually segmenting the vertebrae.

1 Introduction

Abnormal kinematic behaviour of the lumbar spine has been associated with low back pain [1, 2]. Intervertebral kinematic can provide useful diagnostic and follow up of back pain [3]. Hence the measurement of inter-vertebral motion has been investigated and many techniques have been developed to measure inter-vertebral motion [4] [3, 5] as well as many techniques to automatically segments the vertebrae [6–10]. Recording continuous spinal motion was first introduced by Breen et al. in [11]. This technique captures dynamic frames of spinal motion with low X-ray dosage than the normal single X-ray images [12]. In these X-ray videos the amount of radiation is such that the quality of a single image is much lower that of standard single X-rays. Although it is possible to see the similarities between vertebrae, there still a large variation in a single patient and between different patients.

Many researchers have focused on the segmentation of vertebrae and many techniques have been proposed to achieve better segmentation. Benjelloun et al. proposed a framework for vertebra segmentation using active shape models [13]. Statistical models are created after a training stage and the vertebrae are segmented using vertebrae detected contours. Lecron et al. also used active shape models and edge polygonal approximation to segment the vertebra in high resolution X-ray images. To speed up the segmentation, parts of their scheme were

G. Bebis et al. (Eds.): ISVC 2012, Part II, LNCS 7432, pp. 61–69, 2012.
© Springer-Verlag Berlin Heidelberg 2012

processed on multi-CPU/multi-GPU architecture. Zhen et al. in [12] presented a Hough transform (HT) based technique to segment the vertebrae within an image sequence; where they used Fourier descriptors to describe the vertebral body shape. Klinder at al. presented a two-scale framework for the modelling and segmentation of the spine [14]. The global spine shape is expressed as a consecution of local vertebra coordinate systems while individual vertebrae are modelled as triangulated surface meshes.

The majority of These techniques were mainly applied to the segmentation/-tracking of vertebrae in lateral view sequences. In this paper we present more general tracking technique which is applied to both lateral and frontal video-fluoroscopy sequences. This technique uses few particle filters with an image gradient based likelihood measurement to track the vertebrae in parallel (i.e. each vertebra is tracked by a corresponding particle filter). The tracking is semi-automatically initialised as the user selects few land mark points in the first frame on which the detected edges are superposed.

The remainder of this paper is organised as follows. Section 2 briefly presents the particle filter used in this paper. Section 3 describes the vertebra model and the likelihood measurement based on image gradient and edge cues. The semi-automatic initialisation is described in Section 4. Section 5 presents the tracking results and compares them with ground truth data obtained but hand annotation. Finally, section 6 concludes the paper.

2   Particle Filters in Visual Tracking

Visual tracking is often formulated from a Bayesian perspective as a problem of estimating some degree of belief in the state $\boldsymbol{x}_t$ of an object at time t given a previous observations $\boldsymbol{z}_{1:t}$ [15]. Bayesian filtering recursively computes a posterior density that can be written using Markov assumption:

$$p(\boldsymbol{x}_{t+1} \mid \boldsymbol{z}_{1:t+1}) \propto p(\boldsymbol{z}_{t+1} \mid \boldsymbol{x}_{t+1})p(\boldsymbol{x}_{t+1} \mid \boldsymbol{z}_{1:t}) \tag{1}$$

Applying a Markov assumption, the prior density is the posterior density propagated from the previous time step using a dynamic model given by

$$p(\boldsymbol{x}_{t+1} \mid \boldsymbol{z}_{1:t}) = \int p(\boldsymbol{x}_{t+1} \mid \boldsymbol{x}_t)p(\boldsymbol{x}_t \mid \boldsymbol{z}_{1:t})dx_t \tag{2}$$

The posterior in (1) cannot be computed analytically unless linear-Gaussian models are adopted. Isard and Blake suggested particle filtering for visual tracking in the form of Condensation [16] which is adopted in this paper. In Condensation, the posterior density $p(\boldsymbol{x}_t \mid \boldsymbol{z}_{1:t})$ is estimated at each time step t by a set of N particles $\{\boldsymbol{x}_t^n, w_t^n\}_{n=1}^N$ where each particle is a weighted random sample and $\sum_{n=1}^N w_t^n = 1$. The posterior is then

$$p(\boldsymbol{x}_{t+1} \mid \boldsymbol{z}_{1:t+1}) \propto p(\boldsymbol{z}_{t+1} \mid \boldsymbol{x}_{t+1}) \sum_{n=1}^N w_p^n p(\boldsymbol{x}_{t+1} \mid \boldsymbol{x}_t^n) \tag{3}$$

where the prior is now a mixture with N components. The Condensation involves (a) selecting the nth mixture component with probability w_t^n, (b) drawing a sample from it, and (c) assigning to the sample a weight proportional to its likelihood. Resampling is used to obtain samples with equal weights. The algorithm is given in Table 1. The dynamic (motion) model is encapsulated by the transition density $p(\boldsymbol{x}_{t+1} \mid \boldsymbol{x}_t^n)$. Typically, a sample can be drawn from it by adding random process noise and then applying deterministic dynamics (drift).

Table 1. Condensation Particle Filter

Draw samples $\boldsymbol{x}_{t+1}^n$ from $p(\boldsymbol{x}_{t+1} \mid \boldsymbol{x}_t^n)$
Assign weights $w_{t+1}^n = p(\boldsymbol{z}_{t+1} \mid \boldsymbol{x}_{t+1}^n)$
Normalise weights so that $\sum_{n-1}^N w_t^n = 1$
Resample with replacement to obtain samples $\boldsymbol{x}_t^n$

---

3 Vertebrae Models

In order to apply the particle filters, both the state vector and the likelihood models have to be defined. As this research is about tracking different vertebrae viewed from two different angles, we adopt two distinct models for each vertebra. Benjelloun et al. used a local model and global model to segment cervical vertebrae from a single X-ray scan using active shape models [13]. Using global models is not possible in our case as the variation in the vertebrae shape in the lower spine is very large and single model would not be able to capture this variation. However since the vertebrae are moving in the image plane, the outline of each vertebra is expected to remain the same in each video-fluoroscopy sequence. Therefore, in this paper we adopt a rigid contour model for each vertebra in any given sequence. For the frontal view, a vertebra shape is represented by a closed contour; while for the lateral view the vertebra shape is represented by an open contour as show in the Fig. 1

Although the movement of the patient (i.e. that of the vertebrae) is controlled, the motion model of each vertebrae can only be estimated. This is due partially to the fact that the calibration is practically impossible as each patient is unique and also the inter-vertebrate motion is specific to each patient especially when there are abnormalities in the lumbar spine.

While the shape of a vertebrae is assumed to be invariant in each sequence, the position and the orientation do change. Hence the sate of a vertebra model at time t is given by $e_t = (x_t; y_t; \theta_t)$; where x_t and y_t are the image coordinates of the contour centre and θ_t is the orientation relative to the centre of mass of the contour.

The likelihood measurement is based on the aggregation of intensity gradient information along each vertebra boundary. The gradient-based measurement $\psi(p)$ involves searching for maximum gradient magnitude points along short normal search line segments to the vertebrae model. In this paper, there are 100 such lines, each one is 7 pixels long. As the maximum gradient should be ideally on the contour model, the distance between the maximum gradient point and the contour is used to penalise the contours which are further away from the maximum gradient. The gradient-based likelihood measurement $\psi(p)$ is given by

$$\psi(x_t^n) = \sum_{n=1}^{N} \frac{\text{Max}_{i=1}^{M}\{\lambda_{i,n}\}}{1 + \eta D_n} \tag{4}$$

where N is the number of normal search line segments, $\lambda_{i,n}$ is the gradient at the ith point/pixel along the line n, D_n is the Euclidian distance between the point with the maximum gradient and the vertebrae contour model, and η is a weighing factor. In this paper the weighting factor is kept constant at $\eta = 0.1$. Then likelihood is computed as follows

$$p(z_t \mid x_t^n) = \frac{\psi(x_t^n)}{\sum_{n=1}^{N} \psi(x_t^n)} \tag{5}$$

This likelihood should give a clear maximum in the correct location which corresponds to the model being aligned with maximum gradient.

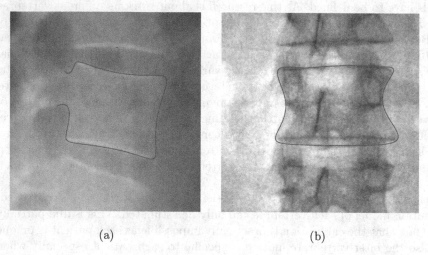

(a)   (b)

Fig. 1. Side view and frontal view of a Vertebrae: (a) Vertebra contour model in side view (b) Vertebra contour model in frontal view

4   Semi-automatic Initialisation

Tracking initialisation is an important step in any object tracking scheme. In this paper we adopt a semi-automatic approach where the user is guided by the detected edges in the first frame to specify few land mark points along the vertebra outline. Canny edge detector is used to detect the edges in the first

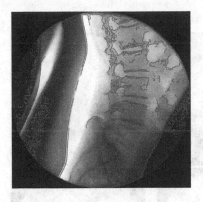 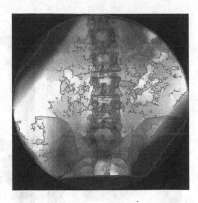

Fig. 2. Edge detected using Canny Edge detector

frame and these edges are superposed on the frame image. Since the X-rays are of law quality, the detected edges are very noisy as can be seen in figure 2. It is also clear from 2 that some edge curves appear to be aligned with vertebrae outlines; but some parts of the vertebrae outlines have no detected edge on them. In this paper, the user manually selects few land marks along the vertebra outline where no edge is detected and only selects a start and an end points on each edge curve which are considered to be aligned with vertebrae outlines. This edge segment then is automatically sampled, and sampled point are added to the manually selected landmarks to form the initial land mark sequence $S = [x_1, x_2, .., x_K; y_1, y_2, .., y_K]$ along the outline of the vertebra. Then a parametric spline is fitted to these points to form the vertebra contour model. Parametric splines are fitted independently to both $X = [x_1, x_2, .., x_K]$ and $Y = [y_1, y_2, .., y_K]$ using the parametric splines $x = f(t)$ and $y = g(t)$. The parametric splines would also help to filter out the out layers. Figure 1 shows the obtained vertebrae outline using the fitted spline.

5   Evaluation

5.1   Tracking Results

The proposed tracking method was implemented using a Gaussian transition density with a diagonal covariance matrix. Specifically, the variance parameters were $\sigma^2 = 25$ pixels for the vertebra model centre of gravity and $\sigma^2 = 4^o$ degree for the model rotation relative to the vertebra centre of gravity. In the sequences, the image size was resampled down to 870×870 pixels.

The proposed tracking technique is evaluated on two lumbar spinal video-fluoroscopy sequences of sagital and lateral flexions. 3 shows the tracking results for the side view with sagital flexion, while 4 shows the tracking results

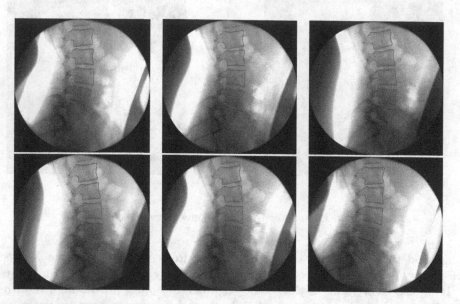

Fig. 3. Tracking results for the side view with sagital flexion: Frames 1, 40, 80, 120, 160, and 200

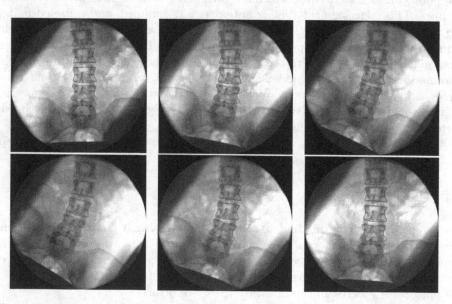

Fig. 4. Tracking results for the frontal view with lateral flexion to the right: Frames 1, 40, 80, 120, 160, and 200

for the frontal view with lateral flexion. As we can see, the tracking was successful and all the vertebrae were correctly tracked throughout the sequences.

5.2 Comparison with Ground Truth Data

To get some quantitative measurement of the quality of the tracking, we have manually annotated the vertebrae in the frames 150 from each sequence. Figure 4 shows the manually segmented vertebrae in blue and the tracked vertebrae in red. Although we did not get a perfect match, the overall tracking is very close to the segmented contours. The error here can be attributed to the segmentation in both the ground truth data and the initialisation stage of the tracker rather than the tracking technique.

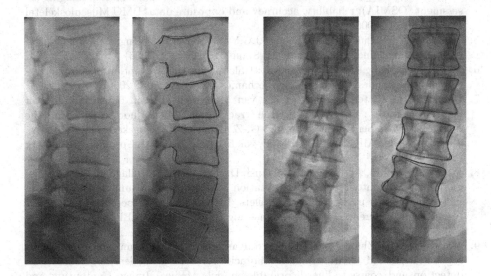

Fig. 5. The manually segmented vertebrae in blue and the tracked vertebrae in red in Frame 150 in both frontal and lateral views. Frames with no contours on are displayed for reference.

6 Conclusion

A particle filter based technique for tracking vertebrae in lumbar spinal video-fluoroscopy has been proposed. In the first X-ray frame, the vertebrae are semi-automatically segmented and a bi-spline curve is fitted to the landmark points to construct the vertebrae outlines; then a particle filter is used to track the vertebrae through the sequence. The proposed technique was able to track the vertebrae in both lateral and frontal video-fluoroscopy sequences. Compared with the ground truth data obtained by manually segmenting the vertebrae in a given frame showed that the proposed technique would be a good starting point for vertebrae kinematic analysis in patients with lower back pain.

References

1. Abbott, J.H., Fritz, J.M., McCane, B., Shultz, B., Herbison, P., Lyons, B., Stefanko, G., Walsh, R.M.: Lumbar segmental mobility disorders: comparison of two methods of defining abnormal displacement kinematics in a cohort of patients with non-specific mechanical low back pain. BMC Musculoskeletal Disorders 7, 45 (2006)
2. Barrett, C.J., Singer, K.P., Day, R.: Assessment of combined movements of the lumbar spine in asymptomatic and low back pain subjects using a three-dimensional electromagnetic tracking system. Manual Therapy 4, 94–99 (1999)
3. Holt, C.A., Evans, S.L., Dillon, D., Ahuja, A.S.: Three-dimensional measurement of intervertebral kinematics in vitro using optical motion analysis. Proceedings of the Institution of Mechanical Engineers Part H Journal of Engineering in Medicine 219, 393–399 (2005)
4. Breen, A.C., Muggleton, J.M., Mellor, F.E.: An objective spinal motion imaging assessment (OSMIA): reliability, accuracy and exposure data. BMC Musculoskeletal Disorders 7, 1 (2006)
5. Wong, K.W.N., Luk, K.D.K., Leong, J.C.Y., Wong, S.F., Wong, K.K.Y.: Continuous dynamic spinal motion analysis. Spine 31, 414–419 (2006)
6. McCane, B., King, T.I., Abbott, J.H.: Calculating the 2D motion of lumbar vertebrae using splines. Journal of Biomechanics 39, 2703–2708 (2006)
7. Reinartz, R., Platel, B., Boselie, T., van Mameren, H., van Santbrink, H., ter Haar Romeny, B.: Cervical Vertebrae Tracking in Video-Fluoroscopy Using the Normalized Gradient Field. In: Yang, G.-Z., Hawkes, D., Rueckert, D., Noble, A., Taylor, C. (eds.) MICCAI 2009. LNCS, vol. 5761, pp. 524–531. Springer, Heidelberg (2009)
8. Wong, A., Mishra, A., Fieguth, P., Clausi, D., Dunk, N.M., Callaghan, J.P.: Shape-guided active contour based segmentation and tracking of lumbar vertebrae in video fluoroscopy using complex wavelets. In: Conference Proceedings of the International Conference of IEEE Engineering in Medicine and Biology Society, pp. 863–866 (2008)
9. Ma, J., Lu, L., Zhan, Y., Zhou, X., Salganicoff, M., Krishnan, A.: Hierarchical segmentation and identification of thoracic vertebra using learning-based edge detection and coarse-to-fine deformable model. Medical Image Computing and Computer-Assisted Intervention 13, 19–27 (2010)
10. Zamora, G., Sari-Sarraf, H., Long, L.R.: Hierarchical segmentation of vertebrae from x-ray images. In: Medical Imaging 2003 Image Processing, vol. 5032, pp. 631–642 (2003)
11. Breen, A.: A computer/x-ray method for measuring spinal segmental movement: a feasibility study. In: 2nd Annual Conference on Research and Education of the Pacific Consortium for Chiropractic Research, pp. 13–14 (1987)
12. Zheng, Y., Nixon, M.S., Allen, R.: Automatic Segmentation of Lumbar Vertebrae in Digital Videofluoroscopic Imaging. IEEE Transactions on Medical Imaging 23, 45–52 (2004)
13. Benjelloun, M., Mahmoudi, S., Lecron, F.: A framework of vertebra segmentation using the active shape model-based approach. International Journal of Biomedical Imaging (2011) 621905

14. Klinder, T., Wolz, R., Lorenz, C., Franz, A., Ostermann, J.: Spine segmentation using articulated shape models. Medical Image Computing and Computer-Assisted Intervention 11, 227–234 (2008)
15. McKenna, S.J., Nait-Charif, H.: Tracking human motion using auxiliary particle filters and iterated likelihood weighting. Image and Vision Computing 25, 852–862 (2007)
16. Isard, M., Blake, A.: ICONDENSATION: Unifying Low-Level and High-Level Tracking in a Stochastic Framework. In: Burkhardt, H.-J., Neumann, B. (eds.) ECCV 1998. LNCS, vol. 1406, pp. 893–908. Springer, Heidelberg (1998)

Mutual Information for Multi-modal, Discontinuity-Preserving Image Registration

Giorgio Panin

German Aerospace Center (DLR)
Institute for Robotics and Mechatronics
Münchner Straße 20, 82234 Weßling

Abstract. Multi-sensory data fusion and medical image analysis often pose the challenging task of aligning dense, non-rigid and multi-modal images. However, optical sequences may also present illumination variations and noise. The above problems can be addressed by an invariant similarity measure, such as mutual information. However, in a variational setting convex formulations are generally recommended for efficiency reasons, especially when discontinuities at the motion boundaries have to be preserved. In this paper we propose the TV-MI approach, addressing for the first time all of the above issues, through a primal-dual estimation framework, and a novel approximation of the pixel-wise Hessian matrix, decoupling pixel dependencies while being asymptotically correct. At the same time, we keep a high computational efficiency by means of pre-quantized kernel density estimation and differentiation. Our approach is demonstrated on ground-truth data from the Middlebury database, as well as medical and visible-infrared image pairs.

1 Introduction

An important problem in computer vision is to find visual correspondences between two views of a scene, possibly acquired by multi-modal sensors, or under different illumination conditions. The former is a preliminary step for multi-sensory data fusion, as well as medical image analysis and visualization. However, robustness to illumination and image noise is also a vital requirement for motion estimation in optical sequences.

In the optical flow literature, we can first distinguish between global and local methods, dating back to [1] and [2] respectively, or combinations of both [3]. The former minimize a global energy, that combines a pixel-wise *data term*, assessing the quality of matching, with a *regularization* prior, coping with the ill-posedness of the problem. The others extend data terms to local windows of a given aperture, increasing robustness to noise and avoiding further regularization, but usually limited to a sparse set of features in textured areas, roughly undergoing planar homographies.

Global energies are efficiently minimized through locally convex approximations of the nonlinear cost function, typically obtained by linearizing residuals, under an L^p-norm or a convex M-estimator. For differentiable cost functions, discretized Euler-Lagrange equations are employed: for example, in [1] a linearized

G. Bebis et al. (Eds.): ISVC 2012, Part II, LNCS 7432, pp. 70–81, 2012.

L^2-norm data term is regularized by the L^2-norm of the motion field $|\nabla f|$, and the resulting quadratic problem is solved by Jacobi iterations. These algorithms are also suitable for graphics hardware implementation, because of their highly parallel structure.

For preserving motion discontinuities at the surface boundaries, the total variation (TV) regularizer employs instead the L^1-norm, that allows non differentiable solutions, however adding non-trivial issues to the optimization procedure. Earlier works in this direction [4] use the approximate L^1 regularizer $\sqrt{|\nabla f|^2 + \epsilon^2}$, where ϵ is a small positive constant, thus keeping the Euler-Lagrange framework. However, this procedure introduces ill-conditioning, especially for small ϵ.

More recently, careful studies have shown how to directly and efficiently address convex TV-L^1 problems [5], including optical flow [6], by means of *primal-dual* formulations, that introduce a dual variable and solve a *saddle-point* problem in two alternate steps (min-max), coupled by a quadratic penalty.

Considering the data term, the simplest and most common assumption is the *brightness constancy*, that may be violated in presence of photometric changes. This happens in case of a variable camera exposure, as well as environment light variations, and especially for multi-modal data (such as medical, or multi-spectral images), that bear nonlinear and many-to-one relationships. Since the L^p-norm is not robust to such variations, several alternatives have been proposed.

To cope with smooth, additive illumination fields, in [6] both images are pre-processed by a structure-texture decomposition [7], which amounts to a L^1 denoising (the *ROF* model [8]), producing a *structure* image, that is afterwards removed so that only texture components are used for matching.

Other works introduce additional terms such as image gradients [4], which are robust to additive changes, but also more noisy and requiring a proper relative weighting; while others estimate smooth, additive illumination fields [5], or complex parametrized models [9].

A different class of approaches looks instead for more robust and invariant matching indices. For example, normalized cross-correlation (NCC) is invariant to brightness mean and variance, thus allowing linear photometric relationships; it has been recently included into the convex variational framework [10], through local correlation windows, and a second-order Taylor expansion with numerical differentiation. Another index is the correlation ratio (CR) [11], which is invariant to a class of nonlinear, one-to-one relationships.

So far, the most general index is mutual information (MI), defined in information theory to express the statistical *dependency* between two random variables, in this case the corresponding grey pairs: in this way, any photometric relationship is enforced, also nonlinear and many-to-one.

Due to this property, as well as a higher robustness to outliers and noise, MI has been initially proposed for medical image registration [12,13]. Later on, it has been applied to stereo, in [14] and in the semi-global matching (SGM) algorithm [15], for object tracking [16] and visual navigation [17].

Notably, [11] considered a unified variational formulation of global NCC, CR and MI, as well as their local counterparts, for multi-modal and non-rigid registration. This approach only relies on gradient descent, through the nonlinear Euler-Lagrange equations.

Although MI has been used also for variational registration, in this case we are not aware of any locally convex formulation, which, as we have seen, is the key for an efficient optimization using discontinuity-preserving priors.

Our main contribution is, therefore, the integration of global MI into the primal-dual TV framework through locally convex, second-order Taylor expansion. Furthermore, we adopt a particular approximation of the Hessian matrix, motivated by the following insights.

In fact, it is well-known that MI is a cascade of two mappings: one at the level of grey-value statistics (Sec. 3.1) and one at pixel-level (Sec. 3.2), where both Hessian contributions contain first- and second-order terms. We choose to retain at the upper level only second-order terms, while keeping only first-order ones at the lower level. This leads to a block-diagonal, negative-semidefinite approximation, resulting in directional searches along image gradients, while being asymptotically correct.

By contrast, the traditional approximation first proposed in [18], intuitively following the Gauss-Newton approach, neglects second-order terms everywhere. However, this has been put recently under discussion [17],while already confirmed by a seldom usage even in a few dimensions (e.g. Levenberg-Marquardt strategies [19] show less efficiency than the LSE counterpart).

At pixel level, instead, (2×2) rank-1 structure tensors are consistent with the *aperture problem* of global approaches. By comparison, the second-order approximation of local NCC [10] neglects off-diagonal terms, further decoupling the horizontal and vertical flow components, by assuming in most places to have diagonally-dominant, full-rank blocks, due to the extended sampling windows. In our case, this assumption would be clearly incorrect.

The remainder of the paper is organized as follows: in Sec. 2 we review the primal-dual variational approach. Sec. 3 describes our formulation for the MI data term and optimization strategy, finally resuming the TV-MI algorithm. Sec. 4 shows experimental results on the Middlebury training dataset and multi-modal images, and Sec. 5 proposes future developments.

2   TV-Regularized Motion Estimation

Given two images I_0, I_1, a motion field $f = (u(x,y), v(x,y))$ is sought in order to match corresponding points $I_0(x,y), I_1(x+u, y+v)$ with possibly sub-pixel accuracy, such that some "similarity" index is maximized, at the same time keeping a smooth field, while preserving discontinuities at the motion boundaries.

The first requirement can be expressed, omitting the x, y coordinates for brevity, by a global data term $E_{data}(I_0, I_1(u,v))$. The other constraints are usually incorporated into a smoothness (or soft penalty) term $E_{smooth}(u,v)$, which is a function of the local behaviour of the field, typically through the spatial gradients

$$\arg\min_{(u,v)} E_{smooth}\left(\nabla u, \nabla v\right) + \lambda E_{data}\left(I_0, I_1(u,v)\right) \tag{1}$$

with a proper weighting factor λ.

Following [5], let $F = E_{smooth}, G = \lambda E_{data}$, we have the general problem

$$\arg\min_{f\in X} F\left(Df\right) + G\left(f\right) \tag{2}$$

where $f : \Omega \to \mathbb{R}^2$ belongs to an Euclidean space X of functions with open domain, $D : X \to Y$ is a linear operator such as the component-wise gradient, mapping onto another space Y, $F : Y \to \mathbb{R}^+$ and $G : X \to \mathbb{R}^+$ are the prior and data terms, for example given by an integral over Ω of the respective L^p-norm. Both spaces are endowed with the scalar product $\langle \cdot, \cdot \rangle$ and induced norm

$$\langle f, g \rangle = \sum_i \int_\Omega f^i g^i \, dx \, dy; \quad \|f\| = \sqrt{\langle f, f \rangle} \tag{3}$$

summed over the vector field components $i = \{1, 2\}$.

If both F, G are convex and lower semi-continuous [5], then (2) can be cast into a saddle-point problem

$$\min_{f\in X} \max_{p\in Y} \langle Df, p \rangle + G\left(f\right) - F^*\left(p\right) \tag{4}$$

where $p \in Y$ is the dual variable, and F^* is the Legendre-Fenchel *conjugate*

$$F^*(p') \equiv \sup_{p\in Y} \langle p', p \rangle - F\left(p\right) \tag{5}$$

In order to solve (4), first-order algorithms alternate descent and ascent steps in the respective variables f, p, by defining the resolvent, or *proximal* operators

$$f = (I + \tau\partial G)^{-1}\left(\tilde{f}\right); \quad p = (I + \sigma\partial F^*)^{-1}\left(\tilde{p}\right) \tag{6}$$

where τ, σ are two parameters, I is the identity mapping, and ∂F is the *subgradient* of F, which extends the (variational) gradient to non-differentiable but convex functions, being well-defined over the whole domain Y.

This operator is given by

$$(I + \tau\partial G)^{-1}\left(\tilde{f}\right) = \arg\min_f \frac{1}{2\tau} \left\|f - \tilde{f}\right\|^2 + G\left(f\right) \tag{7}$$

and similarly for F^*. Then, an efficient algorithm (Alg. 1 in [5], with $\theta = 1$) iterates the following steps

  – Initialization: choose $\tau, \sigma > 0$ s.t. $\tau\sigma \|D\|^2 \leq 1$, set initial values f^0, p^0, and the auxiliary variable $\bar{f}^0 = f^0$

− Iterate: for $n = 1, 2, \ldots$

$$\begin{cases} p^n = (I + \sigma \partial F^*)^{-1} \left(p^{n-1} + \sigma D \bar{f}^{n-1} \right) \\ f^n = (I + \tau \partial G)^{-1} \left(f^{n-1} - \tau D^* p^n \right) \\ \bar{f}^n = 2 f^n - f^{n-1} \end{cases} \tag{8}$$

where D^* is the dual operator: $\langle Df, p \rangle_Y = \langle f, D^* p \rangle_X$.

In particular, the total variation regularizer

$$F_{TV} = \int_\Omega |Df| \, \mathrm{d}x \mathrm{d}y \tag{9}$$

is the isotropic L^1-norm of the *distributional* derivative, that is defined also for discontinuous fields, and reduces to the gradient $D = \nabla$ when f is sufficiently smooth, so that $|Df| = \sqrt{f_x^2 + f_y^2}$. The corresponding dual operator is the divergence, $\nabla^* p = -\mathrm{div} p$.

Thus, proximal operators in (8) are applied to

$$\tilde{p}^n \equiv p^{n-1} + \sigma \nabla \bar{f}^{n-1}; \ \tilde{f}^n \equiv f^{n-1} + \tau \mathrm{div} p^n \tag{10}$$

In the following, we will consider the problem in a discrete setting, where f, p are defined on pixel grids, and the discretized operators are given in [5]. Then, it can be shown that $\|D\|^2 \le 8$, and a common choice is $\tau = \sigma = 1/\sqrt{8}$. Furthermore, $(I + \tau \partial F_{TV}^*)^{-1}$ is the point-wise Euclidean projection

$$p = (I + \tau \partial F_{TV}^*)^{-1} (\tilde{p}) \iff p_{x,y} = \frac{\tilde{p}_{x,y}}{\max (1, |\tilde{p}_{x,y}|)} \tag{11}$$

3   Mutual Information Data Term

Formally, MI is the Kullback-Leibler divergence between $P(i_0, i_1)$ and the product of marginals $P(i_0) P(i_1)$

$$MI(I_0, I_1 | f) = H(I_0) + H(I_1 | f) - H(I_0, I_1 | f) \tag{12}$$

$$= \int_0^1 \int_0^1 P(i_0, i_1 | f) \log \frac{P(i_0, i_1 | f)}{P(i_0) P(i_1 | f)} di_0 di_1$$

where H are the marginal and joint entropies, and we emphasize the dependency of the I_1 sample on f.

This quantity must be *maximized* with respect to f, so we can write $E_{data} = -MI(I_0, I_1 | f)$. In order to introduce our Hessian approximation, we will first consider the statistical dependency of MI on grey values, and then the lower-level dependency upon flow vectors.

3.1 Approximating the Hessian: Grey-Value Statistics

For a given a density estimate $P(i_0, i_1)$, obtained from a sample of grey pairs $I_{0,h}, I_{1,h}; h = 1, \ldots, N$, let us consider the dependency of MI on the I_1 sample[1] (suppressing the 1 index)

$$\frac{\partial MI}{\partial I_h} = \iint_{i_0, i_1} \frac{\partial P(i_0, i_1)}{\partial I_h} \log \frac{P(i_0, i_1)}{P(i_1)} \tag{13}$$

$$\frac{\partial^2 MI}{\partial I_h \partial I_k} = \iint_{i_0, i_1} \frac{\partial^2 P(i_0, i_1)}{\partial I_h \partial I_k} \log \frac{P(i_0, i_1)}{P(i_1)} +$$
$$\frac{\partial P(i_0, i_1)}{\partial I_h} \frac{\partial P(i_0, i_1)}{\partial I_k} \left(\frac{1}{P(i_0, i_1)} - \frac{1}{P(i_1)} \right)$$

This Hessian is generally not diagonal since, although sampling schemes for $P(i_0, i_1)$ ensure that mixed partials are zero, the last term is generally non zero for $h \neq k$, leading to a problem of untractable complexity.

In order to reduce MI to a sum of independent terms, [14] and [15] linearize $P \log P$ around the previous density estimate $\tilde{P} = P(I_0, I_1 | \tilde{f})$, leading to $P \log P \approx P \log \tilde{P}$. Although these methods are derivative-free, this corresponds to neglecting *first-order terms* in the Hessian, that cause the undesired coupling.

We can see that the resulting accuracy is mainly related to the finite sample size N, and to the kernel bandwidth: in fact, because of the products, first order terms decay as $1/N^2$, while second order terms as $1/N$. Moreover, we observed that the approximation is always best at the optimum, i.e. when the joint density is maximally clustered. Finally, the eigenvalues of our approximation have always a larger magnitude than those of the true Hessian, that can be seen from the fact that first-order terms on the diagonal are always non-negative.

Among the many existing non-parametric procedures for entropy estimation, we decided to follow the efficient strategy used in [15,14], extended to our derivative-based framework. Briefly resumed, it consists of a Parzen-based estimation, with pre-quantized kernels assigned to the cells of a (256×256) joint histogram P. The density is estimated, after warping $I_1(f)$, by collecting the histogram of $I_{0,h}, I_{1,h}$, and subsequently convolving it with an isotropic Gaussian K_w of bandwidth w. Afterwards, a further convolution of $\log P$ with the same kernel, evaluated at the same sample points[2], produces the desired data terms, whose sum is the entropy.

$$H(I_0, I_1) = -\frac{1}{N} \sum_h \left[K_w * \log \left(K_w * P \right) \right] (I_{0,h}, I_{1,h}) \tag{14}$$

and similarly for the marginal entropy $H(I_1)$, this time with mono-dimensional convolutions, and a possibly different bandwidth w_1.

---

[1] Notice that we write log instead of $(1 + \log)$ as it is often found, because derivatives of a twice-differentiable density integrate to 0.

[2] In order to keep a sub-pixel/sub-grey precision, we perform bilinear interpolation at non-integer histogram positions.

From (14) we obtain derivatives in a straightforward way, again by convolution of the $\log P$ table

$$\frac{\partial H}{\partial I_{1,h}} = -\frac{1}{N} \left[K'_w * \log\left(K_w * P\right)\right] \left(I_{0,h}, I_{1,h}\right) \tag{15}$$

$$\frac{\partial^2 H}{\partial I_{1,h}^2} \approx -\frac{1}{N} \left[K''_w * \log\left(K_w * P\right)\right] \left(I_{0,h}, I_{1,h}\right)$$

where K', K'' are first- and second-derivatives along I_1, and the last equation comes from the previously explained approximation to (13). All of these operations are efficiently carried out by the FFT. The bandwitdths w, w_1 for (14) are estimated in a maximum-likelihood way, according to *cross-validation* rules [11], that can be shown to require a convolution by $\partial K / \partial w$.

In practice, performing the above convolutions is still an expensive operation; therefore, we update those tables only once per pyramid level, while interpolating them at new values of $I_1(f)$ for computing the Hessian and gradient. The latter operations are performed in an intermediate "warp loop" (Fig. 1), while the innermost loop alternates primal-dual steps (8) until convergence.

3.2 Approximating the Hessian: Directional Derivatives

At the pixel level, the *aperture* problem results in rank-deficient (2×2) diagonal blocks of the overall Hessian. In fact, after decoupling pixel-wise dependencies, we can compute derivatives of MI w.r.t. the flow

$$\frac{\partial MI}{\partial f_h} = \frac{\partial MI}{\partial I_{1,h}} \nabla I_{1,h} \tag{16}$$

$$\frac{\partial^2 MI}{\partial f_h^2} = \frac{\partial^2 MI}{\partial I_{1,h}^2} \nabla I_{1,h} \, \nabla I_{1,h}^T + \frac{\partial MI}{\partial I_{1,h}} \frac{\partial^2 I_{1,h}}{\partial f_h^2}$$

where, hereafter dropping the h index

$$\nabla I \, \nabla I^T = \begin{bmatrix} I_x^2 & I_x I_y \\ I_x I_y & I_y^2 \end{bmatrix}; \; \frac{\partial^2 I}{\partial f^2} = \begin{bmatrix} I_{xx} & I_{xy} \\ I_{xy} & I_{yy} \end{bmatrix} \tag{17}$$

are the (rank-1) structure tensor, and the Hessian of I_1, respectively.

At the optimum, in absence of noise, $\partial (MI)/\partial I_1$ vanishes from (16), so we approximately keep the rank-1 term, scaled by the second derivatives of MI. The image Hessian is seldom used in the literature, because it may be indefinite, and consists of possibly noisy values.

However, we have to further check the factor $\partial^2 MI / \partial I_1^2$ in order to ensure a negative-semidefinite matrix. In fact, during the initial stages the density is spreaded out, and some places may have a positive (or almost zero) curvature. Therefore, we threshold each factor to a maximum value $D_{max}^2 < 0$.

In order to cope with the rank deficiency, the primal step thus relies on the regularizing prior, whose strict convexity ensures a unique minimum. Since the

Initialization: Let I_0, I_1 be two images, set $f \equiv 0$ and an initial guess for w, w_1. Compute the two pyramids at L levels, including sub-octaves and related subsampling.
Outer loop: let f_{l-1} be the result at the previous level

1. Upsample $f_{l-1} \to f_l$ (and the dual field $p_{l-1} \to p_l$)
2. Warp I_1 and ∇I_1 at f_l, and collect the joint histogram
3. Adapt w, w_1 with maximum-likelihood ascent
4. Compute the entropy tables for MI (14)(15)
5. **Warp loop:** initialize $f^0 = f_l$, and repeat
   (a) Warp I_1 at f^0 and compute MI gradient and Hessian, by interpolating the tables at (I_0, I_1)
   (b) **Inner loop:** iterate $n = 1, 2, \ldots$
      i. Perform the dual step (10)(11) to obtain $\tilde{f}^n$
      ii. Solve (20) and update the primal variable f^n
   (c) Apply median filtering to f^n, and update the expansion point $f^0 = f^n$

Fig. 1. The TV-MI algorithm

prior $\left\| f - \tilde{f} \right\|^2$ is isotropic in (x, y), the problem reduces to a mono-dimensional search, along $\mathbf{n} = \nabla I_1 / |\nabla I_1|$.

For this purpose, first- and second-order *directional* derivatives are given by

$$\frac{\partial MI}{\partial \mathbf{n}} = \frac{\partial MI}{\partial I_1} |\nabla I_1| \tag{18}$$

$$\frac{\partial^2 MI}{\partial \mathbf{n}^2} = \frac{\partial^2 MI}{\partial I_1^2} |\nabla I_1|^2 + \frac{\partial MI}{\partial I_1} \frac{\partial^2 I_1}{\partial \mathbf{n}^2}$$

where, once again, the last term of the second derivative is neglected.

Thus, we look for $\rho \equiv \mathbf{n}^T \left(f - \tilde{f} \right)$, the projected motion field along $\mathbf{n}$, and conversely $f = \tilde{f} + \rho \mathbf{n}$. Several primal-dual steps (Fig. 1) are needed for the TV-regularized optimization, so that prior values $\tilde{f}^n$ will be different from the initial expansion point f^0.

Therefore, by defining $\tilde{\rho}^n = \mathbf{n}^T \left(f^0 - \tilde{f}^n \right)$, dropping the n index, the primal step becomes

$$\arg\min_{\rho} \left\{ \frac{\rho^2}{2\tau} - \lambda \left[\frac{\partial MI}{\partial I_1} |\nabla I_1| (\rho - \tilde{\rho}) + \frac{1}{2} \frac{\partial^2 MI}{\partial I_1^2} |\nabla I_1|^2 (\rho - \tilde{\rho})^2 \right] \right\} \tag{19}$$

where derivatives are computed at f^0, that is solved by

$$\rho = \frac{\dfrac{\partial MI}{\partial I_1} |\nabla I_1| - \dfrac{\partial^2 MI}{\partial I_1^2} |\nabla I_1|^2 \tilde{\rho}}{\dfrac{1}{\lambda \tau} - \dfrac{\partial^2 MI}{\partial I_1^2} |\nabla I_1|^2} \tag{20}$$

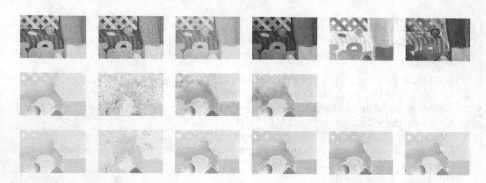

Fig. 2. Photometric variations of different types (see text), added to the second image of the *RubberWhale* sequence. First row: image; Second row: result of TV-L^1 with illumination field estimation [5] ($\beta = 0.05$); Third row: result of TV-MI.

Table 1. Ground-truth comparison on the Middlebury dataset. On each entry, results for TV-L^1 (left) and TV-MI (right) are shown. Optimization failures are marked.

Dataset	Dimetrodon	Grove2	Grove3	Hydrangea	RubberWhale	Urban2	Urban3	Venus
Average angular error (AE)								
Original	3.39, 3.16	2.95, 2.66	7.86, 6.64	2.83, 2.57	4.95, 4.43	3.17, 2.65	5.83, 5.14	4.74, 4.59
Noise	26.09, 26.52	19.47, 18.33	27.72, 16.19	44.39, 10.41	33.34, 31.05	24.84, 11.59	27.47, 15.72	30.69, 16.86
Linear	6.08, 3.13	4.49, 2.57	8.96, 6.53	3.48, 2.52	7.32, 4.29	5.24, 2.62	17.77, 5.47	7.03, 4.67
Square	5.80, 3.15	5.14, 2.68	9.38, 6.71	4.35, 2.58	8.68, 4.42	10.34, 2.73	15.61, 5.53	7.47, 4.38
Neg. square	80.79, 3.15	53.44, 2.71	115.67, 6.73	125.83, 2.59	72.65, 4.43	105.00, 2.74	133.18, 5.66	92.03, 4.45
Two-to-one	88.56, 3.31	59.12, 3.01	115.10, 7.54	96.92, 4.24	66.74, 4.61	88.64, 2.92	118.66, 5.54	96.18, 11.32
Average end-point error (EE)								
Original	0.18, 0.17	0.21, 0.19	0.77, 0.66	0.23, 0.22	0.16, 0.14	0.40, 0.35	0.82, 0.67	0.32, 0.30
Noise	1.14, 1.18	1.32, 1.19	2.19, 1.47	2.73, 1.07	0.98, 0.87	2.36, 1.09	3.42, 1.86	2.17, 1.16
Add. field	0.22, 0.53	1.71, 0.44	0.79, 1.00	0.20, 0.77	0.14, 0.56	1.52, 1.81	1.61, 1.63	0.38, 0.63
Linear	0.30, 0.17	0.31, 0.18	0.85, 0.65	0.39, 0.21	0.24, 0.13	0.55, 0.36	1.44, 0.69	0.43, 0.30
Square	0.28, 0.17	0.36, 0.19	0.89, 0.67	0.44, 0.23	0.27, 0.13	1.03, 0.37	1.67, 0.75	0.48, 0.31
Neg. square	31.04, 0.17	32.40, 0.19	77.91, 0.67	66.03, 0.23	43.24, 0.13	24.18, 0.37	35.27, 0.77	43.85, 0.31
Two-to-one	22.65, 0.18	32.21, 0.21	44.61, 0.73	27.52, 0.42	32.51, 0.14	46.93, 0.67	75.56, 0.69	40.79, 0.71

4   Experimental Results

In order to assess the quality of the TV-MI algorithm, we tested it first on optical sequences with ground-truth, using the Middlebury datasets[3], and compared with the illumination-robust TV-L^1 algorithm [5], that estimates additive fields $q(x, y)$

$$I_t \approx \nabla I_1^n \cdot (u - u^n, v - v^n) + I_t^n + \beta q \qquad (21)$$

with an additional coefficient β, so that f is augmented to $f = (u, v, q)$. This over-parametrization leads to a compromise between robustness and precision: a high β tends to estimate strong brightness variations and suppress motion, while a low β cannot deal with the actual illumination changes, increasing the risk of divergence.

For this comparison, we run the TV-L^1 Matlab implementation available at the TU-Graz computer vision website[4]. Our algorithm is currently in Matlab

---

[3] http://vision.middlebury.edu/flow/
[4] http://www.gpu4vision.org

code, showing roughly the same timing: for example, the *RubberWhale* sequence takes about 45 sec. for TV-MI and 51 sec. for TV-L^1.

Throughout all sequences, parameters were set as follows: data term weight $\lambda = 1$ for TV-MI ($\lambda = 50$ for TV-L^1), initial guess for kernel size $w = 5$, pyramid levels ≈ 30 (with reduction factor 0.9), primal-dual coefficients $\tau = \sigma = 1/\sqrt{8}$, 1 warp iteration and 50 inner-loop iterations.

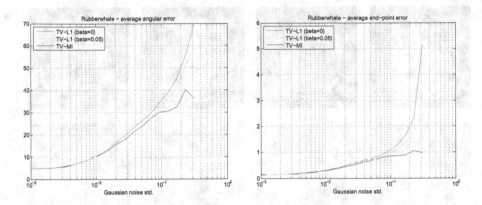

Fig. 3. Average estimation errors at different levels of additive noise

In the first set of experiments, we also set $\beta = 0$, obtaining the result marked *Original* in Table 1. As we can see, for a constant illumination, our algorithm shows similar performances or slight improvements.

Subsequently, we create more challenging conditions, by making photometric changes to the second image I_1 of each sequence (Fig. 2 shows the *RubberWhale* example) in the following order: additive Gaussian noise ($\sigma = 0.1$), linear map $0.7I_1 + 0.3$, nonlinear one-to-one map I_1^2, with color inversion $1 - I_1^2$, and two-to-one map $2|I_1 - 0.5|$.

In order to cope with these changes, we set $\beta = 0.05$ for TV-L^1. We can see how MI can cope with linear and nonlinear maps, outperforming L^1 most of the times, and showing an improved robustness to random noise (see also Fig. 3).

Examples of MRI/CT and near infrared (NIR)/optical pairs, bearing more complex photometric relationships, are shown in Fig. 4.

5   Conclusions

In this paper, we presented the TV-MI approach for multi-modal and discontinuity-preserving variational image registration.

Future developments may follow several directions. For example, the TV regularizer can be replaced by a more robust, anisotropic Huber term [10]. Moreover, as for any global data term, MI performances degrade in presence of a slowly varying illumination field, that creates a *one-to-many* relationship by spreading

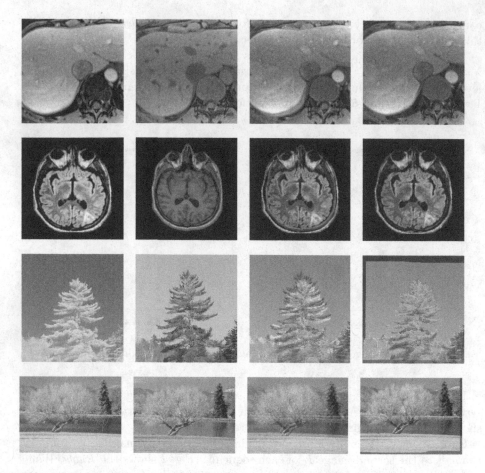

Fig. 4. Multi-modal registration of medical and infrared-optical images. From left to right: original images; superimposed images, before and after warping. Optical/NIR pictures re-printed with permission (©Jeremy McCreary, www.dpFWIW.com).

out the joint histogram. For this purpose, here one may resort either to a local formulation of statistics [11], or to an additional parametric field.

Finally, a GPU-based implementation can largely improve the speed of histogram sampling, FFT convolution, gradient and Hessian interpolation, and solution to the primal problem.

References

1. Horn, B.K.P., Schunk, B.G.: Determining optical flow. Artificial Intelligence 17, 185–203 (1981)
2. Lucas, B.D., Kanade, T.: An iterative image registration technique with an application to stereo vision (darpa). In: Proceedings of the 1981 DARPA Image Understanding Workshop, pp. 121–130 (1981)

3. Bruhn, A., Weickert, J., Schnörr, C.: Lucas-kanade meets horn-schunck: combining local and global optic flow methods. International Journal of Computer Vision 61, 211–231 (2005)
4. Brox, T., Bruhn, A., Papenberg, N., Weickert, J.: High Accuracy Optical Flow Estimation Based on a Theory for Warping. In: Pajdla, T., Matas, J. (eds.) ECCV 2004, Part IV, LNCS, vol. 3024, pp. 25–36. Springer, Heidelberg (2004)
5. Chambolle, A., Pock, T.: A first-order primal-dual algorithm for convex problems with applications to imaging. Journal of Mathematical Imaging and Vision 40, 120–145 (2011)
6. Wedel, A., Pock, T., Zach, C., Bischof, H., Cremers, D.: In: An Improved Algorithm for TV-L1 Optical Flow, pp. 23–45. Springer, Heidelberg (2009)
7. Aujol, J.F., Gilboa, G., Chan, T.F., Osher, S.: Structure-texture image decomposition - modeling, algorithms, and parameter selection. International Journal of Computer Vision 67, 111–136 (2006)
8. Rudin, L.I., Osher, S., Fatemi, E.: Nonlinear total variation based noise removal algorithms. Phys. D 60, 259–268 (1992)
9. Haussecker, H.W., Fleet, D.J.: Computing optical flow with physical models of brightness variation. IEEE Trans. Pattern Anal. Mach. Intell. 23, 661–673 (2001)
10. Werlberger, M., Pock, T., Bischof, H.: Motion estimation with non-local total variation regularization. In: IEEE Computer Society Conference on Computer Vision and Pattern Recognition (CVPR), San Francisco, CA, USA (2010)
11. Hermosillo, G., Chefd'Hotel, C., Faugeras, O.D.: Variational methods for multimodal image matching. International Journal of Computer Vision 50, 329–343 (2002)
12. Gaens, T., Maes, F., Vandermeulen, D., Suetens, P.: Non-rigid Multimodal Image Registration Using Mutual Information. In: Wells, W.M., Colchester, A.C.F., Delp, S.L. (eds.) MICCAI 1998. LNCS, vol. 1496, pp. 1099–1106. Springer, Heidelberg (1998)
13. Wells, W., Viola, P., Atsumi, H., Nakajima, S., Kikinis, R.: Multi-modal volume registration by maximization of mutual information. Medical Image Analysis 1, 35–51 (1996)
14. Kim, J., Kolmogorov, V., Zabih, R.: Visual correspondence using energy minimization and mutual information. In: 9th IEEE International Conference on Computer Vision (ICCV 2003), Nice, France, October 14-17, pp. 1033–1040 (2003)
15. Hirschmüller, H.: Stereo processing by semiglobal matching and mutual information. IEEE Trans. Pattern Anal. Mach. Intell. 30, 328–341 (2008)
16. Panin, G., Knoll, A.: Mutual information-based 3d object tracking. International Journal of Computer Vision 78, 107–118 (2008)
17. Dame, A., Marchand, E.: Accurate real-time tracking using mutual information. In: IEEE Int. Symp. on Mixed and Augmented Reality, ISMAR 2010, Seoul, Korea, pp. 47–56 (2010)
18. Thevenaz, P., Unser, M.: Optimization of mutual information for multiresolution image registration. IEEE Transactions on Image Processing 9, 2083–2099 (2000)
19. Pluim, J.P.W., Maintz, J.B.A., Viergever, M.A.: Mutual-information-based registration of medical images: a survey. IEEE Transactions on Medical Imaging 22, 986–1004 (2003)

Mass Detection in Digital Mammograms
Using Optimized Gabor Filter Bank

Muhammad Hussain[1], Salabat Khan[3], Ghulam Muhammad[2], and George Bebis[4]

[1] Department of Computer Science
[2] Department of Computer Engineering,
College of Computer and Information Sciences,
King Saud University, Riyadh 11543, Saudi Arabia
[3] National University of Computer and Emerging Sciences,
Islamabad, Pakistan
[4] Department of Computer Science and Engineering,
University of Nevada at Reno
mhussain@ksu.edu.sa

Abstract. Breast cancer is the second major type of cancer that causes mortality among women. This can be reduced if the cancer is detected at its early stage but the existing methods result in a large number of false positives/negatives. Detection of masses is more challenging. A new method for mass detection is proposed that uses textural properties of masses. A Gabor filter bank is used for this purpose. The decision of how many Gabor filters must be there in the bank and the selection of the appropriate parameters of each individual Gabor filter is critical. Particle swarm optimization (PSO) and a clustering technique are used to design and select the optimal Gabor filter bank. Support vector machine (SVM) is used as an application oriented fitness criteria. The empirical evaluation of the method over 512 ROIs from DDSM database depicts that it yields better performance (99.41%) than the traditional Gabor filter bank and other state-of-the-art methods that exploit texture properties of masses.

1 Introduction

Breast cancer is the most common form of cancer that affects women all over the world and is considered the second major type of cancer that causes mortality among women. According to the statistics of National Cancer Institute, Surveillance, Epidemiology, and End Results (SEER) program, lifetime risk of developing breast cancer among American women is 12.2%, exceeded only by the lung cancer [1, 2]. In the European Community, breast cancer represents 19% of cancer deaths and the 24% of all cancer cases [3, 4]. Women diagnosed between the ages of 40-49 years are the major victims having about 25% of all breast cancer deaths. In the United States, for instance, breast cancer remains the leading cause of death for women in their forties [3]. The World Health Organization's International Agency for Research on Cancer (IARC) has estimated more than one million cases of breast cancer to occur annually and reported that more than 400,000 women die each year [5].

G. Bebis et al. (Eds.): ISVC 2012, Part II, LNCS 7432, pp. 82–91, 2012.

Masses are one of the signs of breast cancer and mammography is effective and cheap means of mass detection, but the number of false positives/negatives is high. In this paper, we propose a new method for mass detection using digital mammograms exploiting texture properties of masses. The texture properties are useful to correctly detect the masses [3] at their early stage. For the texture properties, we use Gabor filter bank that characterizes micro-patterns (e.g. edges, lines, spots and flat areas) at different scales. Gabor filters were used for the breast cancer detection earlier (see e.g. [2] and references therein), our emphasis is on improving the detection performance using especially optimized Gabor filters. There are two main optimization concerns; first, how many filters are appropriate to be used in the bank and second, what should be the parameter values of each Gabor filter in the bank [6]. Clearly, both of these issues are application oriented and a generic Gabor filter bank [7, 8] doesn't perform well in different application scenarios.

We use a systematic approach that unifies the filter selection and design process employing Particle Swarm Optimization (PSO) and an incremental clustering algorithm. PSO is used to search for optimal parameters of Gabor filters. On the other hand, incremental clustering algorithm removes the redundant Gabor filters from the bank by putting together similar filters in the same cluster. The filters in the same cluster are represented with a single filter which is equal to the centroid of the cluster. This approach helps in two ways: first, mass detection accuracy is improved (99.41%) and second, the computational cost is reduced.

The rest of the paper is organized as follows. In Section 2, the proposed method is presented. Section 3 discusses the experimental setup. Results are presented and discussed in Section 3.2. Finally, Section 4 concludes the paper.

2   Proposed Method

In this section, we discuss in detail the proposed method for mass detection. First, a brief overview of Gabor filter bank is given, then Gabor filters selection and design using PSO and incremental clustering is discussed followed by feature extraction mechanism.

2.1   Gabor Filter Bank

Texture is an important part of the visual world of humans and animals [9]. Texture properties can be used to collect information about micro-patterns like edges, lines, spots and flat areas, which help in detection. Masses in a mammogram do contain edges and local spatial patterns; Gabor filters initialized with different scales and orientations are helpful to detect these patterns and thus provide powerful statistics which could be very useful for breast cancer detection. Gabor filters are biologically motivated convolution kernels [10] that have enjoyed wide usage in a myriad of applications [11, 12]. Two dimensional Gabor filter family is defined as follows [11]:

$$g(x,y) = \frac{1}{2\pi\sigma_x\sigma_y}e^{\left[-\frac{1}{2}\left(\frac{\tilde{x}^2}{\sigma_x^2}+\frac{\tilde{y}^2}{\sigma_y^2}\right)\right]}e^{(2\pi jW\tilde{x})} . \qquad (1)$$

$$\tilde{x} = x\cos\theta + y\sin\theta \quad and \quad \tilde{y} = -x\sin\theta + y\cos\theta . \qquad (2)$$

where σ_x and σ_y are the scaling parameters and describe the neighborhood of a pixel, W is the central frequency of the complex sinusoidal and $\theta \in [0,\pi)$ is the orientation of the normal to the parallel stripes of the Gabor function. A single Gabor filter is therefore fully described with four parameters $\psi = \{\theta, W, \sigma_x, \sigma_y\}$. A bank contains several Gabor filters initialized with different parameters setting of ψ in order to extract the micro-patterns that might be present at different scales and orientation in the image.

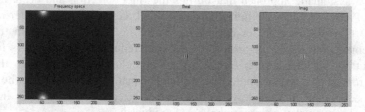

Fig. 1. Gabor Filter with frequency = 0.2, orientation = 0 degree

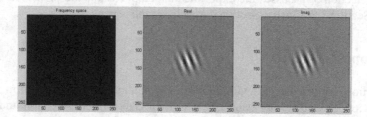

Fig. 2. Gabor Filter with frequency = 0.05, orientation = 157.5 degree

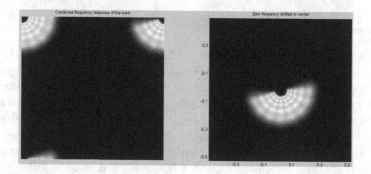

Fig. 3. Combined frequency response of the 40 filters in the Gabor bank

The orientations and frequencies for a bank are calculated using the following equations:

$$orientation\,(i) = \frac{(i-1)\,\pi}{m} \text{ where } i = 1, 2,..., m \;,$$

(3)

$$frequency\,(i) = \frac{f_{max=0.2}}{(\sqrt{2})^{i-1}} \text{ where } i = 1, 2,......, n,$$

(4)

where m is the total number of orientations and n is the total number of frequencies. For instance, the bank with 5 scales and 8 orientations has the frequencies (0.2, 0.14, 0.1, 0.07 and 0.05) for a given number of scales and orientation angles (0, 22.5, 45, 67, 90, 112.5, 135 and 157 in degrees). In Figures 1 and 2, two filters from this bank with ($f = 0.2$, o=0°) and ($f = 0.05$, o=157.5°) are shown for an image with resolution 256×256. Figure 3 plots the combine frequency responses of the entire 40 filters in this bank.

2.2   Gabor Filters Optimization

Different parameter settings of a Gabor filter bank lead to quite different filter responses. We use PSO [13] to select the Gabor filter bank with optimal response. PSO starts with an initial population of particles with random positions and velocities. The position of a particle is an n-dimensional vector and represents a candidate solution to the n-dimensional problem. In the standard PSO, velocities of the particles are stochastically updated based on two factors. First, the personal best position p_b found so far by the particle itself and second, the global best position g_b found so far by the entire swarm. The quality of a particle position is determined using an application specific fitness function. The velocity and position of ith particle are updated using following equations:

$$x^i_{t+1} = x^i_t + v^i_{t+1}.$$

(5)

$$v^i_{t+1} = w_i v^i_t + c_1 r_1 (p^i_b - x^i_t) + c_2 r_2 (p_{gb} - x^i_t).$$

(6)

where, x^i_t and v^i_t are n-dimensional vectors and represent the position and velocity of the ith particle in iteration t, respectively, w_i is known as inertia weight, $w_i v^i_t$ is the weighted current velocity, $c_1 r_1 (p^i_b - x^i_t)$ is the weighted deviation from self-best position and $c_2 r_2 (p_{gb} - x^i_t)$ is the weighted deviation from global best position. If w_i is large, the particle will move faster and the search becomes less refined and if it is too low, the search takes longer. The values of c_1 and c_2 (known as acceleration coefficients) are set to some predefined values at initialization and r_1, r_2 are calculated randomly in the range from 0 to 1.

Particle Encoding

The four parameters in ψ represent a single Gabor filter to be optimized. In order to optimize N filters, the particle will contain $4N$ dimensions where each dimension is a continuous value. It is to be noted that the encoding scheme is quite flexible and any number of pre-specified filters could be easily encoded by simply increasing/ decreasing the dimensions of the particle.

Each of the parameter in ψ has its own constraints and ranges which could be imposed using some prior knowledge. We have adopted the ranges as used in [6]; the orientation $\theta \in [0, \pi)$, radial frequency is allowed to be in the range $[0 \geq W \leq 0.5]$ and σ_x, σ_y should be in the range of $[0.796 > \sigma < w/5]$ where w is the width of the mask in the filter. In our experiments, we used a swarm size of 10 particles that are initialized randomly within the given ranges of each parameter. The velocities of the particles quickly attain very large values and the particles might take longer jumps in the search space. In this case the search becomes less refined. In order to ensure a finer search in the close proximities of the particles' current positions, velocity clamping is used. The velocity of a particle in any of its dimensions if exceeds $\pm V_{max}$, the value at that dimension is clamped to $\pm V_{max}$ where $\pm V_{max}$ is the maximum velocity in both positive and negative directions. In our approach, V_{max} is equal to the maximum range values of the parameters in the respective dimensions. In order to ensure that the particle must not overfly from the given ranges of parameters, velocity resetting is used. After applying Equation 5, if the position value of the particle at a particular dimension doesn't remain in the given range, older position value at that dimension is used (new position value which is out of bound is ignored) and velocity at that dimension is reset to zero (known as velocity resetting).

The particles keep searching for the optima until the stopping criterion is met. Two stopping criteria are used; first, the total iterations are executed or the optimal is found. Second, the convergence is detected. The search process is assumed to converge given that the global best fitness value of the entire swarm doesn't improve over last few iterations.

Incremental Clustering

During the execution of PSO, some filters encoded in the particle may become similar. Such similar filters generate similar responses and thus introduce redundancy in the feature space. To overcome this problem, incremental clustering is used that group together redundant filters of a particle in a cluster. Later, the centroid of the cluster is used as a single filter. The clustering is performed in the parameter space of ψ using the algorithm proposed in [6]. The first filter in the particle encoding is assigned to the first cluster. The *nth* filter $\psi^n = \{\theta^n, W^n, \sigma_x^n, \sigma_y^n\}$ in the particle encoding is assigned to the *ith* cluster, only if the following conditions are satisfied.

$$\theta^i - Thr_\theta \leq \theta^n \leq \theta^i + Thr_\theta$$
$$W^i - Thr_W \leq W^n \leq W^i + Thr_W$$
$$\sigma_x^i - Thr_\sigma \leq \sigma_x^n \leq \sigma_x^i + Thr_\sigma$$
$$\sigma_y^i - Thr_\sigma \leq \sigma_y^n \leq \sigma_y^i + Thr_\sigma$$

where $i \in [1 \ N]$ and N represents the total existing clusters. The $\{\theta^i, W^i, \sigma_x^i, \sigma_y^i\}$ is the centroid of the *ith* cluster and *Thr* represents the threshold values used to quantify the similarity between a filter and the centroid of the cluster and calculated as follows based on the minimum and maximum range values of the parameters in ψ:

$$Thr_\theta = \frac{\pi}{K} \times 0.5, \quad Thr_W = \frac{(W_{max} - W_{min})}{K} \times 0.5$$

$$\& \ Thr_\sigma = \frac{(\sigma_{max} - \sigma_{min})}{K} \times 0.5$$

where K is a user defined parameter and used to handle the extent of threshold values over the tradeoff between model compactness and accuracy [6] e.g. large values of K result in more compact models.

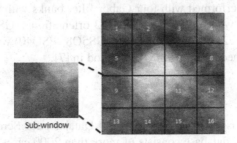

Fig. 4. ROI image partitioning into sub-windows

2.3 Feature Extraction

For feature extraction, an ROI of a mammogram is first partitioned into sub-images, called as sub-windows. In particular, each ROI (e.g. 512x512) is first divided into square patches of equal sizes (e.g. 128x128), see Figure 4. In this way, sixteen patches are created, labeled 1-16. The patches are then combined to create overlapping sub-windows e.g. patches 1, 2, 5 and 6 comprise the first sub-window, 2, 3, 6 and 7 the second, and so on. With this formation, 9 overlapping sub-windows are created. It may be noted that by increasing/ decreasing the size of a patch, an ROI can be partitioned in different sizes and numbers of sub-windows. Feature extraction is performed by convolving sub-windows with a Gabor filters bank. A slightly modified design strategy is already used for texture based feature extraction [8, 11]. The magnitude responses of each Gabor filter in the bank are collected from all sub-windows and represented by three moments: the mean $\mu_{i,j}$, the standard deviation $\sigma_{i,j}$ and the skewness $k_{i,j}$, where i means ith filter in the bank and j represents jth sub-window. These moments correspond to the statistical properties of a group of pixels in a sub-window and positioning of pixels is essentially discarded which compensates for any errors that might occur during partitioning of ROIs into sub-windows. With a Gabor bank of 40 filters (5 scales × 8 orientations) applied on 9 sub-windows (Figure 4) of a single ROI, a feature vector of size 1080+1(class label) is obtained:

$$[\mu_{1,1}, \sigma_{1,1}, k_{1,1}, ..., \mu_{1,2}, \sigma_{1,2}, k_{1,2}, ..., \mu_{40,9}, \sigma_{40,9}, k_{40,9}, class] \ . \tag{7}$$

2.4 Fitness Function (SVM)

For classification, we used SVM [14] with RBF kernel. The error rate of SVM classifier is used as fitness value to evaluate the effectiveness of the filters encoded as a particle. SVM formulation is based on statistical learning theory and has attractive generalization capabilities in linear as well as non-linear decision problems [14]. SVM classifiers [14] are the most advanced ones, generally, designed to solve binary classification problems; thus perfectly suite our requirements.

3   Experimental Setup

The experiments are performed with four Gabor filter banks without (with) optimization: GS2O3 (PSO06) with 6 filters (2 scales × 3 orientations), GS3O5 (PSO15) with 15 filters, GS4O6 (PSO24) with 24 filters and GS5O8 (PSO40) with 40 filters. These four banks were designed by following the method in [7].

3.1   Dataset

The proposed method is evaluated using Digital Database for Screening Mammography (DDSM) [15]; this database consists of more than 2000 cases where each case is labeled by expert radiologists and locations of masses are encoded as code-chains. We randomly selected 512 (256 normal and 256 mass) cases from the database. Using code chains, we extracted 256 ROIs which contain true masses. In addition, we extracted 256 ROIs containing normal but suspicious tissues. Later each ROI is resized to a power of 2. Some sample ROIs are shown in Figure 5.

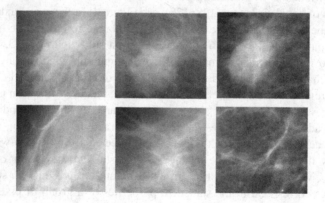

Fig. 5. Sample mass ROIs (top row) and normal but suspicious ROIs (bottom row)

3.2   Results and Discussions

We used 150 ROIs (75 normal and 75 mass) as validation set during PSO based Gabor filter bank optimization. Once the optimized filter bank is found, all the 512 ROIs are used for comparison and 10-fold cross validation is used for evaluation. In particular, the data set is randomly partitioned into 10 non-overlapping and mutually exclusive subsets. One subset is used as testing set and the remaining nine subsets are used to train the classifier. The performance metric is the average percentage accuracy of SVM. Optimal parameters of SVM with RBF kernel are selected using the method of Chih-Wei Hsu et al. [16].

For fair comparison, the PSO is initialized with the same size of bank as in the non-optimized Gabor filter bank. For instance, in Table 1, PSO15 is obtained by initializing the PSO with the same 15 filters as are used for GS3O5; the value (13) on the right side of PSO15 is the number of filters finally selected using incremental clustering algorithm.

The following PSO parameters were used during all the experiments: *swarm size* = 10, *total generations* = 50, *iterations_for_converge* = 5, c_1 & c_2 = 1.4 and *inertia weight* = 0.7. These parameter values have been chosen because they seem to provide reasonable performance as reported earlier in the literature [13]. The optimization of the parameter values is avoided since the goal is to evaluate the parameter settings' generalization ability across a wide range of experimental configurations used for the classification task. The results in Table 1 are generated for the ROI images scaled to 64×64, *patch size* = 16×16 and clustering threshold K = 3. It is obvious from Table 1 that the optimized Gabor filter bank performs better in all the cases. The best results are found for PSO06 (**99.21±1.37**). After analyzing the results in Table 1, two points are noteworthy: first, the optimized filter bank is more compact, which results in the smaller feature space (good for computational efficiency and better generalization capability of the SVM). Second, the accuracy is remarkable even when some of the information might go due to rescaling of ROI.

Table 1. Recognition results with PSO Gabor filter banks and non-optimized Gabor filters banks (ROI size: 64×64)

Method	Sensitivity	Specificity	Accuracy (%)
GS2O3	94.85±4.38	98.11±2.03	96.27±2.35
PSO06 (6)	**98.89±2.54**	**99.66±1.10**	**99.21±1.37**
GS3O5	96.97±4.73	97.33±.91	97.06±2.65
PSO15 (13)	98.54±1.89	98.81±2.57	98.62±2.08
GS4O6	95.55±3.75	98.84±1.89	97.06±2.12
PSO24 (22)	96.40±4.06	98.35±2.90	97.45±1.86
GS5O8	97.24±3.79	99.55±1.44	98.43±1.80
PSO40 (34)	98.32±2.20	99.67±1.05	99.02±1.39

In order to observe the effect of image size and patch size, we repeated the experiments with the best filters bank i.e. PSO06 and ROIs scaled to 128×128 with patch size of 32×32 pixels, keeping all other parameters the same as used for the results in Table 1. From Table 2, it is clear that when ROIs are rescaled to bigger size, accuracy is improved. ROI sizes 256x256 and 512x512 can further improve the accuracy.

Table 2. Performance with different image and patch sizes

Method	Image Size	Patch Size	Accuracy (%)
PSO06 (6)	64×64	16×16	99.21±1.37
PSO06 (6)	128×128	32×32	**99.41±0.95**

Comparison with Other Methods

We compared the proposed method with two state-of-the-art methods which also exploit the texture properties of masses. One of these uses LBP descriptor (LBP method) [18] and the other uses a variant of WLD descriptor [17] – MSWLD (Multiscale Spatial WLD). We implemented these methods using the same software and hardware environment, and tested on the same database, which we used for the evaluation of our method. The comparison results are shown in Table 3. It is obvious that the proposed method performs better than LBP and MSWLD based methods. Further to test whether the classification results of the proposed method are significantly higher than those of LBP and MSWLD based methods, we used 2-tail t-test. The null hypothesis is that the difference between the mean measures of performance (accuracy and Az) is zero and the alternative hypothesis is that the difference is positive. The test results with 5% significance level are given in Table 4. The null hypothesis is rejected at 0.05 significance level for the two methods. It indicates that the classification performance obtained using the proposed method is higher than that obtained by other two methods and the difference is statistically significant. All p-values are much less than 0.05, which indicates that the differences are actually statistically highly significant.

Table 3. Comparison between Gabor, MSWLD, and LBP

Method	Accuracy
Gabor (G)	**99.41±0.95**
MSWLD (M) [17]	98.00±0.55
LBP (L) [18]	92.00±0.99

Table 4. Statistical significance 2-tail t-test results

Method	Accuracy	
	t-val	p-val
G vs M	4.061866	**0.000731758**
G vs L	17.07809	**1.44352E-12**

4   Conclusion

A new method for mass detection in mammograms is proposed. In the proposed method, features are extracted using the Gabor filters bank optimized using Particle Swarm Optimization (PSO) and an incremental clustering algorithm. PSO is used for the design of Gabor filters whereas clustering algorithm handles the filters selection problem. The resultant optimized filters are evaluated over ROIs rescaled to different sizes using an application oriented fitness function based on SVM. The Gabor filters bank become more compact and the recognition rate of cancerous tissues in the digital mammograms has improved. For future work, we plan to investigate the performance over large sizes of ROIs and the optimization of PSO parameter values with stagnation being handled. It will also be interesting to see the effect of different sizes of patches and sub-windows.

Acknowledgement. This work is supported by the National Plan for Science and Technology, King Saud University, Riyadh, Saudi Arabia under project number 08-INF325-02.

References

1. Altekruse, S.F., Kosary, C.L., Krapcho, M., et al.: SEER Cancer Statistics Review, pp. 1975–2007. National Cancer Institute, Bethesda (2010)
2. Yufeng, Z.: Breast Cancer Detection with Gabor Features from Digital Mammograms. Algorithms 3(1), 44–62 (2010)
3. Lladó, X., Oliver, A., Freixenet, J., Martí, R., Martí, J.: A textural approach for mass false positive reduction in mammography. Comp. Medical Imag. and Graph. 33(6), 415–422 (2009)
4. Esteve, J., Kricker, A., Ferlay, J., Parkin, D.: Facts and figures of cancer in the European Community. Tech. rep., International Agency for Research on Cancer (1993)
5. Mohamed, M.E., Ibrahima, F., Brahim, B.S.: Breast cancer diagnosis in digital mammogram using multiscale curvelet transform, Comp. Medical Imag. and Graph. 34(4), 269–276 (2010)
6. Zehan, S., George, B., Ronald, M.: On-road Vehicle Detection Using Evolutionary Gabor Filter Optimization. IEEE Trans. on Intell. Transp. System 6(2), 125–137 (2005)
7. Ville, K., Joni-Kristian, K.: Simple Gabor feature space for invariant object recognition. Pattern Recognition Letter 25(3), 311–318 (2004)
8. Manjunath, B., Ma, W.: Texture features for browsing and retrieval of image data. IEEE Tran. on Pattern Analysis and Machine Intelligence 18(8), 837–842 (1996)
9. Peter, K., Nikolay, P.: Nonlinear Operator for Oriented Texture. IEEE Tran. on I. Proc. 8(10), 1395–1407 (1999)
10. Daugman, J.G.: Two-dimensional spectral analysis of cortical receptive field profiles. Vis. Res. 20, 847–856 (1980)
11. Zehan, S., George, B., Ronald, M.: Monocular Precrash Vehicle Detection: Features and Classifiers. IEEE Trans. on Image Proc. 15(7), 2019–2034 (2006)
12. Yu, S., Shiguan, S., Xilin, C., Wen, G.: Hierarchical Ensemble of Global and Local Classifiers for Face Recognition. IEEE Trans. on Image Proc. 18(8), 1885–1896 (2009)
13. Eberhartand, R.C., Kennedy, J.: Particle Swarm Optimization. In: Proc. of IEEE International Conference on Neural Networks, pp. 1942–1948 (1995)
14. Vapnik, V.: Statistical Learning Theory. Springer, New York (1995)
15. Heath, M., Bowyer, K., Kopans, D., Moore, R., Kegelmeyer, P.J.: The digital database for screening mammography. Int. Work. Dig. Mamm., 212–218 (2000)
16. Hsu, C.W., Chang, C.C., Lin, C.J.: A Practical Guide to Support Vector Classification, Technical report, Department of Computer Science and Information Engineering, National Taiwan University (2010)
17. Hussain, M., Khan, N.: Automatic mass detection in mammograms using multiscale spatial weber local descriptor. In: Proc. IWSSIP 2012, Austria, April 11-13 (to appear, 2012)
18. Lladó, X., Oliver, A., Freixenet, J., Martí, R., Martí, J.: A textural approach for mass false positive reduction in mammography. Computerized Medical Imaging and Graphics 33(6), 415–422 (2009)

Comparing 3D Descriptors for Local Search of Craniofacial Landmarks

Federico M. Sukno[1,2], John L. Waddington[2], and Paul F. Whelan[1]

[1] Centre for Image Processing & Analysis,
Dublin City University, Dublin 9, Ireland
[2] Molecular & Cellular Therapeutics,
Royal College of Surgeons in Ireland, Dublin 2, Ireland

Abstract. This paper presents a comparison of local descriptors for a set of 26 craniofacial landmarks annotated on 144 scans acquired in the context of clinical research. We focus on the accuracy of the different descriptors on a per-landmark basis when constrained to a local search. For most descriptors, we find that the curves of expected error against the search radius have a plateau that can be used to characterize their performance, both in terms of accuracy and maximum usable range for the local search. Six histograms-based descriptors were evaluated: three describing distances and three describing orientations. No descriptor dominated over the rest and the best accuracy per landmark was strongly distributed among 3 of the 6 algorithms evaluated. Ordering the descriptors by average error (over all landmarks) did not coincide with the ordering by most frequently selected, indicating that a comparison of descriptors based on their global behavior might be misleading when targeting facial landmarks.

1 Introduction

We address the comparison of local geometry descriptors for highly accurate localization of 3D facial landmarks, in the context of craniofacial research [1,2]. In contrast to applications on facial biometrics, which emphasize robustness to challenging conditions [3], medical applications tend to have a greater focus on the highly accurate localization of landmarks, as they constitute the basis for the analysis, often aimed at detecting quite small shape differences. Depending on the author, localization and repeatability errors are considered clinically relevant when they exceed 1 mm [4] or 2 mm [5]. Acquisition conditions are therefore carefully controlled to minimize holes and other artifacts.

In this context, we aim to evaluate the performance of a variety of local descriptors for the purpose of facial landmark localization. There are two key elements that motivate an interest in such a study: 1) the popularity of local descriptors for the detection of 3D facial landmarks, as opposed to global cues, and 2) the fact that previous comparisons of 3D descriptors have been focused on the detection and reproducibility of generic keypoints, rather than on the accuracy of specific ones (e.g. landmarks).

G. Bebis et al. (Eds.): ISVC 2012, Part II, LNCS 7432, pp. 92–103, 2012.
© Springer-Verlag Berlin Heidelberg 2012

The first element derives from the difficulty in detecting individual points globally on the human face. In general, facial landmarks can be distinguished from their neighboring points based on some geometric properties. For example, the nose tip can be detected as a curvature *peak* or *cap*, while eye corners are *pits* or *cups*. However, the chin tip is also a *peak* and the mouth corners are also *pits* [6,7]. Unfortunately, as we will show, these coincidences are not exclusive of simple descriptors, such as curvature, but they persist for more elaborate ones. Thus, landmark descriptors do not work satisfactorily on a global basis and one usually combines local search with higher level constraints based on the spatial relationships between sets of landmarks [8,9,10]. Hence, it is of interest to assess how different descriptors perform when constrained to a local search.

On the other hand, prior work on the evaluation of 3D descriptors has not focused on the accuracy of specific points. It is common practice to report the detection and reproducibility of *keypoints* defined generically as those that are most distinct from the rest for the descriptor of choice. The number of detected keypoints and their reproducibility are used as measures of quality [11]. Accuracy is secondary, indirectly addressed by means of the acceptance radius (the maximum distance at which two different points are considered to match).

Bronstein et al. [12] put more emphasis on accuracy by evaluating the identification of dense correspondences with different keypoint detection algorithms. They report the average geodesic distances to the true correspondences; this is possible because they constrain the matching pairs to different instances of the same object, after some synthetic transformations.

Closer to the present study are the evaluations reported by Romero & Pears [14] and Creusot et al. [13]. However, in both cases the evaluation is performed based on the joint search of all points under the global constraints provided by a graph-matching scheme. Furthermore, descriptors are not compared individually but, rather, combined together. While Creusot et al. provide the resulting weights for the descriptor combinations using Linear Discriminant Analysis (targeting 14 landmarks on a set of 200 scans), these do not constitute an optimal criteria to assess the performance of each descriptor individually.

In this work we present a comparison of local descriptors for a set of 26 facial landmarks relevant in the context of craniofacial dysmorphology [1]. The evaluation is performed on a per-landmark basis with a focus on the accuracy that can be achieved when the descriptors are constrained to search in a local neighborhood of the targeted landmark. We empirically show that, for a good descriptor, the curves of overall accuracy against the search radius have a plateau that is indicative of the descriptor's accuracy and usable range (Section 2). The set of descriptors that are evaluated are detailed in Section 3, results are presented in Section 4 and conclusive remarks are provided in Section 5.

2   Analysis of Local Accuracy

We start from a set of annotated facial surfaces in 3D, organized in meshes M described by sets of vertices and triangles. We will indicate that a vertex $\mathbf{v}$

belongs to the vertices of mesh $\mathcal{M}$ simply by $\mathbf{v} \in \mathcal{M}$. Evaluation will be based on the (Euclidean) distance from a given vertex $\mathbf{v}$ to the ground truth (manual location of the considered landmark) and will be denoted by $d(\mathbf{v})$.

The model-instance of a descriptor for a given landmark will be referred to as the *template*; for example, the average of the spin images [16] of the nose tip (for a set of *training* scans) is a descriptor template for the nose tip using the spin image descriptor. The value resulting from the evaluation of a descriptor *template* at a given vertex $\mathbf{v}$ will be denoted as descriptor *score* $s(\mathbf{v})$.

We define the expected local accuracy $\bar{e}_L(r_S)$ over the distances from $\mathbf{v}_i^{max}$ (the vertices that obtain the maximum score in each mesh) to the ground truth position of the targeted landmark, evaluated only on a neighborhood composed of vertices whose distances do not exceed the search radius parameter r_S,

$$\bar{e}_L(r_S) = E[d(\mathbf{v}_{i,r_S}^{max})] \tag{1}$$

$$\mathbf{v}_{i,r_S}^{max} = \{\mathbf{v} \in \mathcal{M}_i \,|\, d(\mathbf{v}) \leq r_S \,\wedge$$
$$\forall \mathbf{w} \neq \mathbf{v}, d(\mathbf{w}) \leq r_S, \mathbf{w} \in \mathcal{M}_i \,:\, s(\mathbf{v}) \geq s(\mathbf{w})\} \tag{2}$$

where $E[x]$ is the expected value of x. That is, given a target landmark, for each mesh $\mathcal{M}_i$ we consider a neighborhood of radius r_S around the ground truth position of the landmark and select $\mathbf{v}_i^{max}$ as the vertex with the maximum score in this neighborhood. We are interested in the expected distance of these maximum-score vertices to the targeted landmark.

It is evident that $\bar{e}_L(r_S) \leq r_S$. However, a useful descriptor should also beat chance (i.e. random selection), which would be equivalent to a uniform distribution of the scores over the neighborhood (i.e. a fully uninformative descriptor):

$$\bar{e}_L^{rand}(r_S) = \int_{A(r_S)} r \, P(r) \, dA = \int_{A(r_S)} \frac{r}{A(r_S)} \, dA$$

$$A(r) = \pi r^2, \quad dA = 2\pi r \, dr \quad \Rightarrow \quad \bar{e}_L^{rand}(r_S) = \int_0^{r_S} \frac{2\pi r^2}{\pi r_S^2} dr = \frac{2}{3} r_S \tag{3}$$

Hence, we require that $\bar{e}_L(r_S) < \frac{2}{3} r_S$. Fig. 1 shows three examples of $\bar{e}_L(r_S)$, selected to illustrate the different behaviors observed in the landmarks used for this study:

- In the first example, $\bar{e}_L(r_S)$ initially increases with r_S until reaching a flat region or *plateau*. This means that, except for relatively small r_S, the descriptor produces, on average, a maximum score consistently at the same distance from the target.
- In the second example, $\bar{e}_L(r_S)$ behaves similarly to the first example up to a certain r_S, after which there is a sudden increase, which typically reaches a second plateau.
- In the third example, $\bar{e}_L(r_S)$ does not show any plateau (at least for the range of interest). Although it is below the theoretical limit of $\frac{2}{3} r_S$, its value constantly increases.

The descriptor from the first example is the most useful one, because it could be used for global search; however the second one is the most frequent. The reason for this is the presence of highly similar points from the viewpoint of the descriptor that is used. For example, both inner eye corners are evidently similar to each other (*twin* points), hence when targeting one of them we will find an increase in $\bar{e}_L(r_S)$ at the average distance between inner eye corners. Any descriptor will show such an increase of the error due to twin points but there can also be strong similarities between different landmarks (e.g. this is typical between the mouth and nose corners).

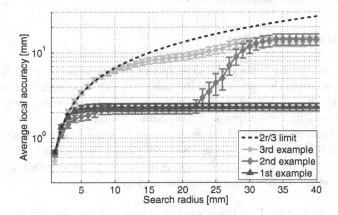

Fig. 1. Expected local accuracy, $\bar{e}_L(r_S)$, for some facial landmarks. The theoretical value for random choice ($\frac{2r_S}{3}$) is also provided for reference.

The third example is the least useful one, because it indicates somewhat erratic behavior: the descriptor finds confusing points roughly at any distance, indicating that the targeted landmark is not distinctive enough.

Thus, the curves of expected local accuracy can be very informative about the performance of a descriptor for a given landmark, allowing us to compare between different alternatives. Nonetheless, proceeding this way would require one plot per landmark, each containing several curves (one per descriptor). A more practical solution is to focus on the analysis of the first plateau, which is the most important part of the curve. This can be done with just three numbers: the value of $\bar{e}_L(r_S)$ and the plateau limits, in terms of r_S.

Two important considerations when identifying the plateau are the stability of $\bar{e}_L(r_S)$ and the range of interest. Due to the presence of outliers, we used the median to estimate $\bar{e}_L(r_S)$. Regarding the range of interest, it is evident from the second example that we should focus our analysis before the transition from the first to the second plateau (around 25 mm in this case), as it indicates the presence of a strong source of false positives, possibly due to a twin point. However, sometimes there is no plateau before this sudden increase, and it is useful to analyze the curve generated by the difference to the theoretical limit (random choice), which we might regard as the accuracy gain, $\overline{G}_L(r_S)$:

$$\overline{G}_L(r_S) = \frac{2}{3}r_S - \overline{e}_L(r_S) \tag{4}$$

In general, the accuracy gain is a monotonically increasing curve. However, in the presence of a sudden increase of $\overline{e}_L(r_S)$, the accuracy gain will drop and we will find a local maximum. Therefore, the search for the plateau should be constrained up to the first local maximum of $\overline{G}_L(r_S)$. Note that this criterion applies only to curves like the one in example 2, as the curves in examples 1 and 3 will not produce local maxima on $\overline{G}_L(r_S)$.

3   Evaluated Descriptors

In this section we briefly review the descriptors that were evaluated, including the default parameters that were used in each case. We choose three descriptors based on histograms of distances and three based on signatures of orientations (histograms of relative orientations of the normal vectors).

The descriptors are computed for each vertex $\mathbf{v}$, whose normal vector is $\mathbf{n}_v$, considering a neighborhood $\mathcal{N}_v = \{\mathbf{w} \mid \|\mathbf{w} - \mathbf{v}\| \leq r_N\}$, namely all points within a radius r_N. Except for the spin image approach, the implementations used in this paper are based on the Point Cloud Library [15].

Spin Images (SI) [16]: This descriptor is computed as a bi-dimensional histogram of distances. One axis encodes the unsigned distance to the normal vector and the other encodes the signed distance projected in the direction of the normal vector. That is, the histogram is generated from the following pairs:

$$(\alpha, \beta) = \left(\sqrt{\|\mathbf{w} - \mathbf{v}\|^2 - (\mathbf{n}_v \cdot (\mathbf{w} - \mathbf{v}))^2}, \ \mathbf{n}_v \cdot (\mathbf{w} - \mathbf{v}) \right) \tag{5}$$

By default, the histograms contain 15×15 bins and the contribution of each point is calculated with bilinear interpolation. While this descriptor has been proposed more than a decade ago, it is still very widespread and is often used as an indicator of baseline performance.

3D Shape Contexts (3DSC) [17]: This descriptor is based on a 3D-histogram computed on a spherical support region centered at the interest point, $\mathbf{v}$, and with its North pole oriented with the normal, $\mathbf{n}_v$. The default structure has 11 elevation bins and 12 azimuth bins, both uniformly spaced, and 15 radial bins logarithmically spaced so that more importance is assigned to shape changes that are closer to the interest point. Similar to spin images, these descriptors are based on distances, as the value for each bin is based on the number of points that fall within its boundaries. The contribution of each point is weighted by 1) the inverse of its local density (to account for uneven sampling) and 2) the inverse of the cube root of the bin volume, due to the large difference between bin sizes, especially along radius and elevation.

As the spherical support region is defined based only on $\mathbf{v}$ and $\mathbf{n}_v$, there is an ambiguity on the azimuth origin. This is dealt with by calculating as many descriptors per point as the number of azimuth bins, covering all possible shifts. The computation of multiple descriptors is done for the model (i.e. during training), so that during matching only one descriptor is computed and matched to multiple descriptors by choosing the one that yields the highest score.

Unique Shape Context (USC) [18]: This descriptor is analogous to the 3DSC but without the ambiguity in the azimuth direction, thanks to the definition of a local reference frame consisting of 3 unit vectors that replace the orientation of the North pole with the normal and are computed by a distance-weighted eigen-decomposition, followed by a sign disambiguation step [19].

The use of a local reference frame reduces the computational complexity during point matching and might also improve accuracy by reducing false positives due to spurious similarities that arise from inconvenient azimuth rotations (i.e. those that make the descriptors of different points become too similar). Nonetheless, errors or instabilities in the computation of the reference frame (e.g. due to noise) might impair the performance of this descriptor.

Signature of Histograms of OrienTations (SHOT) [19]: This descriptor is based on a histogram of orientations, rather than distances as for the three previously presented. A spherical domain is defined based on the local reference frame as described above for USC, with a coarse and isotropic grid of 8 azimuth, 2 elevation and 2 radial divisions. For each of these grid divisions, an 11-bin histogram is defined encoding the cosine of the angle between the normals of points within the grid division and the reference point, $\mathbf{v}$.

The use of the cosine, divided in equally spaced bins, results in a non-uniform division of the angular space, favoring directions nearly perpendicular to $\mathbf{n}_v$. Tombari et al. argue that such directions are more informative than those close to $\mathbf{n}_v$, which are more likely to appear in nearly-planar regions (e.g. due to noise). Quadrilinear interpolation is used to construct the local histograms to avoid boundary effects and the whole descriptor is normalized to sum the unit for robustness with respect to sampling density.

Point Feature Histograms (PFH) [20]: This descriptor is based on histograms of 3 angles and, optionally, one distance (not used in this paper). Given the points in the considered neighborhood $\mathbf{w} \in \mathcal{N}_v$, all possible pairs of points are analyzed. For each pair $(\mathbf{w}_i, \mathbf{w}_j), i \neq j$ with normals $(\mathbf{n}_i, \mathbf{n}_j), i \neq j$, the following angles are computed:

$$\alpha = \left((\mathbf{w}_j - \mathbf{w}_i) \times \mathbf{n}_i\right) \cdot \mathbf{n}_j, \qquad \phi = \frac{\mathbf{n}_i \cdot (\mathbf{w}_j - \mathbf{w}_i)}{\|\mathbf{w}_j - \mathbf{w}_i\|}$$

$$\theta = \arctan \frac{\left(\mathbf{n}_i \times \left((\mathbf{w}_j - \mathbf{w}_i) \times \mathbf{n}_i\right)\right) \cdot \mathbf{n}_j}{\mathbf{n}_i \cdot \mathbf{n}_j} \tag{6}$$

The resulting angles are used to construct a three-dimensional histogram (5 bins for each angle were used, yielding a descriptor of length 125). Note that, due to the evaluation of all possible pairs within $\mathcal{N}_v$, the computational complexity for this descriptor is much higher than all other ones evaluated here.

Fast Point Feature Histograms (FPFH) [21]: This is the fast variant of PFH, and is computed in two steps. Firstly, a Simplified Point Feature Histogram (SPFH) is constructed for each point, as described for PFH but considering only the relations between the reference point and each of the neighbors. Later on, this estimation is refined to obtain the final descriptor by weighting the simplified histograms by the inverse of their Euclidean distance to the reference point. The authors also point out the sparseness of the resulting three-dimensional histogram and propose, instead, to compute separate histograms for each of the three angles and simply concatenate them (11 bins per angle are used, resulting in a descriptor of length 33).

4   Performance Comparison

4.1   Data

Our test dataset consisted of 144 facial scans acquired by means of a hand-held laser scanner[1]. This type of scanner allows acquisition of a three dimensional surface by smoothly sweeping a scanning wand over an object, in a manner similar to spray painting. The whole facial surface was acquired, up to (and including) the ears. Special care was taken to avoid occlusions due to facial hair. There is some heterogeneity regarding the extent to which neck and shoulders were included.

A unique surface is reconstructed by combining the different sweeps, which allows coverage of multiple viewpoints. Thus, the complete facial surface can be obtained irrespective of the head pose and possible self-occlusions. This is an important advantage compared to single-view scanners used in other databases.

The dataset contains exclusively healthy volunteers acting as controls in the context of craniofacial dysmorphology research. The mesh resolution varies between 1.2 and 2.4 mm and, on average, there are 24.2 thousand vertices per mesh. Each scan was annotated with 26 anatomical landmarks, in accordance with definitions in [22] (based on [23]), as indicated in Fig. 2.

4.2   Results

We computed the expected local accuracy curves, as defined in Section 2, for each descriptor-landmark pair, varying the search radius r_S from 1 to 200 mm. Landmarks with bilateral symmetry (left and right) were merged together by considering each as a separate instance of the same test.

---

[1] Cobra Wand 192 (FastSCAN^{TM}, Colchester, VT, USA).

Table 1. Expected local accuracy for neighborhood radius $r_N = 30$ mm. If a plateau is found, its value and limits are indicated, otherwise (n.p - no plateau) only the limit based on the first peak of $\overline{G}_L$ is indicated. For each landmark (rows), the best descriptor is highlighted in boldface and those that do not differ significant from it are indicated with an asterisk. The neighborhood radius for which we obtained the best performance is also indicated with a symbol: 20 mm ($\downarrow$), 30 mm ($-$) or 40 mm ($\uparrow$).

Landmark	SI	3DSC	USC	SHOT	PFH	FPFH
en	**1.7** $\uparrow$	2.2 $\uparrow$	2.8 $\downarrow$	6.2 $\downarrow$	4.6 $-$	2.4 $\uparrow$
(2)	**(5 - 23)**	(5 - 24)	(6 - 23)	(8 - 21)	(8 - 21)	(5 - 23)
ex	4.0 $\uparrow$ *	3.8 $-$ *	n.p $-$	**3.8** $-$	6.1 $-$	6.6 $\downarrow$
(2)	(11 - 37)	(11 - 86)	(< 23)	**(6 - 23)**	(11 - 68)	(14 - 52)
n	3.4 $\uparrow$	**1.9** $\uparrow$	3.4 $\uparrow$	3.1 $-$	5.1 $\downarrow$	2.4 $-$
	(6 - 200)	**(5 - 200)**	(6 - 18)	(5 - 17)	(7 - 200)	(5 - 200)
a	**1.5** $-$	1.6 $\downarrow$ *	2.5 $\downarrow$	4.8 $\downarrow$	6.2 $\downarrow$	5.7 $\downarrow$
(2)	**(4 - 25)**	(3 - 27)	(4 - 16)	(6 - 26)	(12 - 17)	(9 - 14)
ac	**2.5** $\uparrow$	3.9 $\downarrow$	n.p $-$	4.3 $\downarrow$	5.6 $-$	6.4 $\downarrow$
(2)	**(14 - 23)**	(10 - 25)	(< 105)	(6 - 21)	(12 - 22)	(13 - 22)
nt	n.p $\uparrow$	n.p $\uparrow$	13.2 $\downarrow$	8.5 $\uparrow$	7.6 $\downarrow$	**7.0** $-$
(2)	(< 8)	(< 9)	(14 - 200)	(15 - 200)	(11 - 200)	**(12 - 200)**
prn	2.8 $\uparrow$	**1.3** $\uparrow$	1.9 $\uparrow$	3.0 $-$	4.3 $\downarrow$	1.8 $-$
	(4 - 200)	**(2 - 200)**	(3 - 200)	(4 - 200)	(5 - 200)	(3 - 200)
sn	2.0 $-$	**1.7** $-$	n.p $\downarrow$	2.6 $-$	6.7 $\downarrow$	2.6 $-$
	(5 - 60)	**(3 - 200)**	(< 111)	(4 - 200)	(10 - 200)	(5 - 200)
ch	2.7 $\downarrow$	3.1 $\uparrow$	3.4 $-$	**2.2** $\downarrow$	5.7 $\uparrow$	3.7 $\uparrow$
(2)	(8 - 23)	(7 - 14)	(12 - 30)	**(4 - 43)**	(10 - 24)	(9 - 40)
cph	n.p $\downarrow$	n.p $\uparrow$	13.9 $\downarrow$	8.5 $\uparrow$ *	**7.9** $-$	8.3 $-$ *
(2)	(< 9)	(< 10)	(21 - 30)	(16 - 39)	**(12 - 200)**	(15 - 200)
li	n.p $\downarrow$	2.7 $\uparrow$	2.5 $-$	**1.9** $\uparrow$	8.1 $-$	5.1 $\uparrow$
	(< 11)	(12 - 42)	(8 - 30)	**(4 - 49)**	(9 - 200)	(7 - 200)
ls	10.8 $\uparrow$	**2.6** $\uparrow$	10.5 $\uparrow$	3.6 $\uparrow$	5.0 $-$	4.1 $-$
	(21 - 38)	**(13 - 200)**	(20 - 34)	(13 - 123)	(6 - 200)	(8 - 200)
sto	6.3 $\uparrow$	2.7 $\uparrow$ *	**2.4** $-$	2.8 $\uparrow$	6.1 $-$	5.3 $\uparrow$
	(14 - 84)	(7 - 58)	**(6 - 27)**	(7 - 15)	(8 - 200)	(9 - 200)
sl	9.4 $\uparrow$	2.9 $\downarrow$ *	n.p $\uparrow$	**2.6** $\uparrow$	5.9 $-$ *	6.3 $-$ *
	(13 - 21)	(9 - 200)	(< 16)	**(4 - 200)**	(10 - 97)	(11 - 18)
pg	18.7 $\uparrow$	**4.7** $\uparrow$	13.4 $\uparrow$	5.3 $\uparrow$ *	7.1 $-$	4.9 $\uparrow$ *
	(23 - 200)	**(9 - 200)**	(19 - 200)	(9 - 200)	(9 - 200)	(12 - 200)
t	n.p $-$	n.p $-$	n.p $-$	**7.4** $-$	13.4 $\downarrow$	8.1 $\downarrow$ *
(2)	(< 58)	(< 125)	(< 142)	**(20 - 96)**	(23 - 89)	(25 - 100)
oi	**7.9** $\uparrow$	n.p $-$	n.p $-$	12.3 $\uparrow$	15.1 $\downarrow$	9.1 $\uparrow$
(2)	**(17 - 22)**	(< 17)	(< 129)	(23 - 30)	(25 - 37)	(17 - 27)

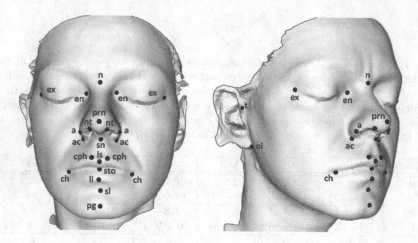

Fig. 2. The 26 landmarks used in this study: en = endocanthion; ex = exocanthion; n = nasion; a = alare; ac = alar crest; nt = nostril top; prn = pronasale; sn = subnasale; ch = cheilion; cph = crista philtrum; li = labiale inferius; ls = labiale superius; sto = stomion; sl = sublabiale; pg = pogonion; t = tragion; oi = otobasion inferius [22]

The descriptor template for each landmark was computed as the median descriptor from a training set, created by means of 6-fold cross-validation. To compute the scores $s(\mathbf{v})$, the descriptor template was compared with the one of each vertex using (minus) the Euclidean distance (i.e. considering each descriptor as a point in N-dimensional space, N being the descriptor length). The only exception was the case of spin images, where we used the (2D) cross-correlation, as suggested in the original paper [16]. Nonetheless, we should mention that results using Euclidean distance (with the descriptor normalized to sum the unit) were similar to those using cross-correlation.

Table 1 summarizes the results. Each cell describes the first plateau of the expected local accuracy curve: the number on the top indicates its median value and the ones below (in parentheses) indicate its limits. Recall that the plateau is only searched for r_S values below the first peak of $\overline{G}_L$. The plateau range was determined as the region for which $\bar{e}_L$ did not vary by more than 10%.

The best descriptor for each landmark is highlighted in boldface and those not significantly different from it are indicated with an asterisk[2]. For example, the best descriptor for the inner-eye corners (en) is SI; if we constrain the search to a radius below 22 mm we can expect to locate each of the inner-eye corners at 1.7 mm from their correct (ground truth) position. Clearly, the great majority of landmarks must be constrained to a local search range for all six descriptors and only a few of them could be used globally (e.g. n, prn, pg, sn).

It is interesting to analyze the consistency of the plateau ranges. We observed strong agreement for the different descriptors on symmetric landmarks for which the twin point is relatively nearby (e.g. en, a, ac), especially for the upper limit,

---

[2] $p > 0.05$ on a paired Wilcoxon signed rank test.

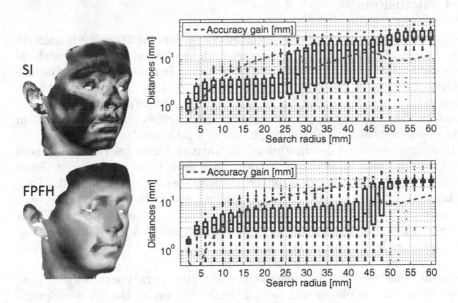

Fig. 3. Left: a facial scan with descriptor scores color-coded (red = high, blue = low) for SI and FPFH, targeting the mouth corners (ch). Right: boxplots of the distances of the highest-score vertices (for the whole set of 144 meshes) for different search radii, r_S. The expected local accuracy, $\bar{e}_L$ is estimated as the median; the discontinuous lines indicate the accuracy gain, $\overline{G}_L$.

which is the most important one. On the other hand, symmetric landmarks that are further apart showed significant variations in their plateau limits across descriptors. This is due to the presence of strong sources of false positives different from the twin points. To illustrate this, Fig. 3 shows the descriptor scores obtained by SI and FPFH for the mouth corners (ch) color-coded on a facial scan. While SI tends to produce high scores on the nose corners (ac), which are approximately 20 mm apart from the mouth corners, FPFH does not show high scores on the nose and the upper plateau limit is therefore extended up to the twin point (the other mouth corner), at about 40 mm, as indicated on the right of the figure. Note also that both descriptors show a peak of $\overline{G}_L$ at about 45 mm, but for SI there is an earlier one at 27 mm, in coincidence with the narrower usable local range of the descriptor for this particular landmark.

We also explored the influence of the neighborhood size used for the computation of the descriptors, testing for $r_N = 20$, 30 and 40 mm. Table 1 shows the results for $r_N = 30$ mm and, for each cell, it is also indicated whether that neighborhood size or one of the two other options was optimal (see table caption). The full tables are available on-line[3].

---

[3] http://fsukno.atspace.eu/Research.htm

5   Conclusions

We present a comparison of local geometry descriptors for 3D facial landamrks on a dataset of 144 scans at moderately high resolution, annotated with 26 anatomical points relevant for craniofacial research. To facilitate the analysis we explored the patterns generated when computing the local accuracy at different search radii. It was found that the most useful descriptors present a flat region or *plateau* that can be used to characterize the descriptor's behavior, both in terms of accuracy and maximum range for the local search.

Six histograms-based descriptors were evaluated: three describing distances and three describing orientations. No descriptor dominated over the rest. From the point of view of the overall error (i.e. the average over all landmarks), 3DSC was the best, followed by SHOT, FPFH, SI, USC and PFH. However, 3DSC was best only for 5 out of 26 landmarks (+6 that did not differ significantly from the best), while SHOT did so for 8(+3) landmarks and SI for 8(+2) landmarks. This illustrates how a comparison of descriptors based on their global behavior might be misleading if targeting facial landmarks.

Finally, while for some landmarks the expected accuracy was below 2 mm, some others did not obtain satisfactory results for any of the six descriptors. This is the case for the ear points (t and oi), typically difficult due to their very complex geometry, but also for symmetric points very close to each otehr (nt, ls), where the accuracy of the best descriptor was similar to the separation between the twin points and therefore not good enough to distinguish between them.

Acknowledgment. The authors would like to thank their colleagues in the Face3D Consortium (www.face3d.ac.uk), and the financial support provided for it from the Wellcome Trust (grant 086901/Z/08/Z).

References

1. Hennessy, R., Baldwin, P., Browne, D., Kinsella, A., Waddington, J.: Frontonasal dysmorphology in bipolar disorder by 3D laser surface imnaging and geometric morphometrics: Comparison with schizophrenia. Schizophr Res. 122, 63–71 (2010)
2. Sharifi, A., Jones, R., Ayoub, A., et al.: How accurate is model planning for or-thognathic surgery. Int. J. Oral. Max. Surg. 37, 1089–1093 (2008)
3. Bowyer, K., Chang, K., Flynn, P.: A survey of approaches and challenges in 3D and multi-modal 3D + 2D face recogn. Comput. Vis. Image Und. 101, 1–15 (2006)
4. Plooij, J., Swennen, G., Rangel, F., et al.: Evaluation of reproducibility and relia-bility of 3D soft tissue analysis using 3D stereophotogrammetry. Int. J. Oral. Max. Surg. 38, 267–273 (2009)
5. Aynechi, N., Larson, B., Leon-Salazar, V., et al.: Accuracy and precision of a 3D anthropometric facial analysis with and without landmark labeling before image acquisition. Angle Orthod. 81, 245–252 (2011)
6. Dibeklioglu, H., Salah, A., Akarun, L.: 3D faceial landmarking under expression, pose, and occlusion variations. In: Proc. BTAS, pp. 1–6 (2008)

7. Segundo, M., Silva, L., Bellon, et al.: Automatic face segmentation and facial landmark detection in range images. IEEE T. Syst. Man Cy B: Cybernetics 40, 1310–1330 (2010)
8. Gupta, S., Markey, M., Bovik, A.: Antopometric 3D face recognition. Int. J. Comput. Vision 90, 331–349 (2010)
9. Szeptycki, P., Ardabilian, M., Chen, L.: A coarse-to-fine curvature analysis-based rotation invariant 3D face landmarking. In: Proc. BTAS, pp. 1–6 (2009)
10. Passalis, G., Perakis, N., Theoharis, T., et al.: Using facial symm to handle pose variations in real-world 3D face recogn. IEEE T. Pattern Anal. 33, 1938–1951 (2011)
11. Salti, S., Tombari, F., Stefano, L.: A performance evaluation of 3D keypoint detectors. In: Proc. 3DimPVT, pp. 236–243 (2011)
12. Bronstein, A., Bronstein, M., Castellani, U., et al.: SHREC 2010: robust correspondence benchmark. In: Proc. 3DOR (2010)
13. Creusot, C., Pears, N., Austin, J.: Automatic keypoint detection on 3D faces using a dictionary of local shapes. In: 3DimPVT, pp. 204–211 (2011)
14. Romero-Huertas, M., Pears, N.: Landmark localisation in 3D face data. In: Proc. AVSS, pp. 73–78 (2009)
15. Rusu, R., Cousins, S.: 3D is here: Point cloud library (PCL). In: Proc. ICRA, pp. 1–4 (2011)
16. Johnson, A., Hebert, M.: Using spin images for efficient object recognition in cluttered 3D scenes. IEEE T. Pattern Anal. 21, 433–449 (1999)
17. Frome, A., Huber, D., Kolluri, R., Bülow, T., Malik, J.: Recognizing Objects in Range Data Using Regional Point Descriptors. In: Pajdla, T., Matas, J(G.) (eds.) ECCV 2004. LNCS, vol. 3023, pp. 224–237. Springer, Heidelberg (2004)
18. Tombari, F., Salti, S., Stefano, L.D.: Unique shape context for 3D data description. In: Proc. 3DOR, pp. 57–62 (2010)
19. Tombari, F., Salti, S., Stefano, L.D.: Unique signature of histograms for local surface description. In: Proc. ECCV, pp. 356–369 (2010)
20. Rusu, R., Blodow, N., Marton, Z., et al.: Aligning point cloud views using persistent feature histograms. In: Proc. IROS, pp. 3384–3391 (2008)
21. Rusu, R., Blodow, N., Beetz, M.: Fast point feature histograms (FPFH) for 3D registration. In: Proc. ICRA, pp. 3212–3217 (2009)
22. Hennessy, R., Kinsella, A., Waddington, J.: 3D laser surface scanning and geometric morphometric analysis of craniofacial shape as an index of cerebro-craniofacial morphogenesis: initial applic to sexual dimorph. Biol. Psychiat. 51, 507–514 (2002)
23. Farkas, L.: Anthropometry of the head and face, 2nd edn. Raven Press, New York (1994)

Vision-Based Tracking of Complex Macroparasites for High-Content Phenotypic Drug Screening

Utsab Saha[1,2] and Rahul Singh[1,*]

[1] Department of Computer Science
[2] Open University Program, San Francisco State University,
San Francisco, CA 94132, USA
rahul@sfsu.edu

Abstract. This paper proposes a method for vision-based automated tracking of schistosomula, the etiological agent of schistosomiasis, a disease which affects over 200 million people worldwide. The proposed tracking system is intended to facilitate high-throughput and high-content drug screening against the schistosomula by taking into account their complex phenotypic response to different candidate drug molecules. Our method addresses the unique challenges in tracking schistosomula, which include temporal changes in morphology, appearance, and motion characteristics due to the effect of drugs, as well as behavioral specificities of the parasites such as their tendency to remain stagnant in dense clusters followed by sudden rapid fluctuations in size and shape of individuals. Such issues are difficult to address using current bio-image tracking systems that have predominantly been developed for tracking simpler cell movements. We also propose a novel method for utilizing the results of tracking to improve the accuracy of segmentation across all images of the video sequence. Experiments demonstrate the efficacy of the proposed tracking method.

1   Introduction

The World Health Organization (WHO) has declared schistosomiasis to be amongst the diseases for which new treatments are urgently needed [1]. An important technical challenge in this context is to develop technologies for high-throughput drug screens (HTS) against the parasite. A crucial sub-goal lies in targeting the juvenile schistosomula, as they exhibit greater resistance to the only available current drug, praziquantel (PZQ). Current efforts in drug discovery against schistosomiasis are based on human mediated whole organism screens, where the effect of a drug on the parasite is manually evaluated. Such efforts are costly, resource-intensive, and fundamentally incapable of exploring the chemical space effectively. Thus, the development of an automated computer vision-based system capable of analyzing the phenotypic responses of schistosomula to drug compounds represents a highly promising direction of research. An automated vision system also has significant advantages over human assays, such as detecting subtle phenotypic responses to a drug (e.g. slight changes in color, texture, motion) and offering quantitative and rigorous phenotype descriptors.

---

* Corresponding author.

G. Bebis et al. (Eds.): ISVC 2012, Part II, LNCS 7432, pp. 104–114, 2012.

In this paper, we build upon our prior research [2] [11] [7] and propose a method for the automated tracking of juvenile schistosomula in video microscopy. This work represents a significant foundational step towards our ultimate goal, namely the creation of a complete vision based system for high-throughput drug screening against parasitic diseases in general and schistosomiasis in particular and its application in the discovery of new and effective therapeutics against these diseases.

1.1   Problem Formulation

The tracking of complex and highly evolved parasites like schistosomula presents several unique challenges. First, the movement of the parasites is based on contracting and expanding motions which can cause rapid fluctuations in their size and shape. For instance, the size of a parasite can increase by a factor of two to three within a few frames in a video sequence. Second, the parasites have a tendency to remain stagnant in dense clusters of touching or partially overlapping individuals. A cluster of parasites can often remain touching in this manner through a significant duration of a video sequence. This can result in frequent segmentation and tracking error. Further, parasites often exhibit significant shape deformation, but little translational movement (i.e. the total distance traveled by the parasite throughout a recording can be small). Moreover, parasites that have been stationary for a long time can suddenly exhibit motion. Third, the translucent nature of the parasite's membranes can reveal the outlines of its anatomy, which can sometimes be mistaken as boundaries of the parasite. This causes most segmentation methods to incorrectly segment a parasite into multiple regions which misleads tracking. Figure 1 presents instances of these challenges.

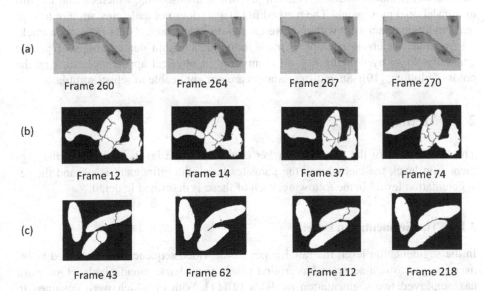

Fig. 1. Examples image sequences which illustrate the challenges of tracking schistosomula. (a) Closely spaced cluster of three touching parasites. (b) Example of erroneous splitting in segmentation.   (c) Example of erroneous splitting in frames 43 and 112 and erroneous merging in frame 62.

1.2   Related Work

Many tracking methods exist for video microscopy, most notably in cell tracking, where active contours are widely used. However, they typically break down in videos which contain dense clusters of cells [6]. Additionally, contour evolution is not well suited for schistosomula, which can exhibit high variability in size and shape between frames. Methods such as [8] do focus on separating a cluster of cells but often make extensive use of contextual specificities, such as the elliptical shape common to many types of cells. For schistosomula a priori assumptions about expected shape changes induced by different compounds are hard to make. Another class of methods involves the use of stochastic filters, which can be very effective if the motion characteristics of the object can be modeled [9]. The motion patterns of schistosomula, however, are difficult to model as they can suddenly contract or expand after remaining stagnant for long periods. *C. elegans*, a nematode and a well-established modal organism, has also been the subject of considerable research effort that is related to our research [3] [4]. Notably, [4] models the motion characteristics of *C. elegans*, utilizing a computer-controlled tracker to keep a target worm in the center of the field of view. However, this approach restricts the tracked sample to a single worm and does not address the challenges of tracking large groups of worms in high throughput settings. The initial motivation of our research is provided by work in pedestrian tracking. The graph-theoretic approach of [5] in particular offers useful insights into dealing with segmentation errors and complex scenes and forms the foundation for the proposed method. However, the use of stochastic filters in [5] to model pedestrians was not suitable to the motion characteristics of juvenile schistosomula, which exhibit minimal translational movement. The method in [5] also does not make any attempt to separate individual members within dense groups of pedestrians. Other pedestrian tracking methods which do attempt to track individuals within dense groups of people generally do so by utilizing key differentiators in physical appearance, such as the color of clothing [10]. Such differentiators are not applicable to schistosomula.

2   Method

The architecture of the proposed tracker consists of five levels. These are: the segmentation level, the blobs level, the parasites level, the refinement level, and the re-segmentation level. In the following, each of these is described in detail.

2.1   The Segmentation Level

In the segmentation level, the raw images of the video sequence are translated to binary images, which separate foreground pixels from background pixels. Our work has employed two segmentation methods [2][11], both of which were designed to segment schistosomula. However, we note that the segmentation method can be varied so long as the output is a single binary image for each frame of the video sequence. The latter stages of the tracker will generalize to any such segmentation method.

2.2   The Blobs Level: Determining Temporal Correspondences between Blobs

Our method in the blobs level closely mirrors the formulation from [5]. In the following, we briefly review this formulation. Each foreground region in the segmented image represents a *blob*. The goal of the blobs level is to find an association between each blob in frame i-1 and the corresponding blob(s) in frame i. An undirected bipartite graph, G_i (V_i, E_i) is used to model the blob associations, where $V_i = B_i \cup B_{i-1}$. B_i and B_{i-1} represent the set of blobs in frame i and frame i-1, respectively. A blob in frame i-1 that is associated with a blob in frame i is connected by an edge in G_i. Two blobs connected by an edge in this manner are said to be *neighbors*. Figure 2c shows an example blob graph generated for two frames in a video sequence.

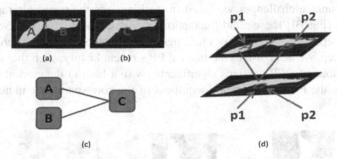

Fig. 2. (a) Blobs in frame i-1 (b) Blobs in frame i (c) Blob graph showing blob associations (d) blob-parasite graph

Based on [5], we make use of two simplifying assumptions when constructing G_i: the *locality constraint* (L-constraint) which requires that two blobs must overlap sufficiently with each other in order to be considered neighbors and the *blob evolution constraint* (E-constraint) which disallows a blob from simultaneously splitting and merging in G_i. The resulting set of candidate graphs are then weighted according to a cost function. The cost function $C(G_i)$ defines two sets of blobs, *parents*, P_i, and *children* C_i, where $P_i \cup C_i = V_i$. The parent set, P_i, contains all blobs which have degree greater than 1 in G_i, all blobs which have degree 0 in G_i, and all blobs which have degree one that are only in B_i. The descendant set includes all neighbors of the blobs in the parent set: $C_i = \cup N_i(u)$, $u \in P_i$, where $N_i(u)$, is the set of all neighbors of u. We then formulate the cost function as:

$$C(G_i) = \sum_{b=P_i} \frac{|A(b) - S_i(b)|}{\max(A(b), S_i(b))}$$

(1)

In Eq.(1), $A(b)$ is the area of the blob b and S_i (b) $= \sum_{n \in N_i(b)} A(b)$

We can see from the above equation that the cost function penalizes graphs where the area of corresponding blobs changes significantly between consecutive frames.

2.3   The Parasite Level: Determining Blob-Parasite Correspondence

In the parasites level and subsequent levels of the tracker, our method offers an alternative approach to [5] and attempts to solve the unique challenges associated with crowded scenes and the peculiar motion characteristics of schistosomula.

A *parasite* is modeled as a rectangular region with a unique identifier (ID). The parasites level begins by initially assigning one putative parasite (PP) to each blob in the first frame of the video. On subsequent frames of the video, if a PP is associated with a blob in frame i-1, then it is also associated with all neighbors of that blob in frame i (recall that neighbors are given by the blob graphs). The relationship between PP's and blobs is represented as a directed bipartite graph, which we refer to as a *blob-parasite graph* (BPG). Figure 2d shows an example of a BPG for two frames. One of the unique challenges we faced in tracking was the parasite's rapid fluctuations in size (Fig. 3). These rapid fluctuations in size coupled with putative segmentation errors represent a significant challenge to the method in [5]. Therefore, as an additional step, we make use of the tracked PP's recent history, such that if one of the recent positions of a PP overlaps significantly with a blob in the current frame, then we associate the PP with the blob, regardless of the correspondence indicated by the BPG.

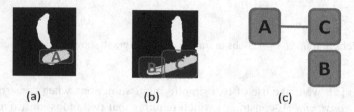

(a)   (b)   (c)

Fig. 3. (a) Blobs in frame i-1 (b) Blobs in frame i. Blob A in frame i-1 simultaneously expands to more than twice its previous size and is erroneously split into two regions. (c) Blob graph showing blob associations, failing to capture the true relationship between blobs A and B due to insufficient overlap.

2.4   Refinement of Blob-Parasite Correspondences

The completion of the parasite level leads to the generation of a BPG for each frame of the video. Next, in the refinement level, we improve the precision of tracking and resolve cases where erroneous tracking may have occurred. To do so, we revisit the assumption made in the first processing step of the parasites level, that there is a one-to-one relationship between parasites and blobs in the first frame. There are two cases which can confound this assumption: (1) Two or more parasites get erroneously merged into a single region in the first frame of the video (Fig. 4). (2) A single parasite is erroneously split into two or more regions in the first frame of the video (Fig. 5). In the following, we describe our method for handling the two cases.

2.4.1 Dealing with Erroneously Merged Parasites

Two observations guide the design of our method for detecting situations where more than one parasite is erroneously merged into a single region in the first frame. First, the erroneously merged putative parasite (EMPP) will likely be larger in size than the average sized PP in the well. Second, the EMPP will likely exhibit frequent splitting behavior throughout the video sequence (i.e. the EMPP may split into a variety of sub-regions in various frames of the video). In Figure 4a we present an EMPP which splits into sub-regions in Figure 4b, 4c, and 4d. We developed the notion of a *split volatility factor* to formalize these observations. A PP is considered *volatile* if it splits into more than one blob in a large number of frames, given by

$$N_{split} = S + K\sigma(\frac{1}{2})^{\frac{A}{B}} \qquad (2)$$

where N_{split} is the minimum number of frames in which the PP must split in order to be deemed volatile, S is the average mode number splits for all parasites, K is a constant, σ is the splits standard deviation, A is the area of the parasite, and B is the average mode of all blob areas. After we conclude that a given PP is an EMPP, we use a *cumulative overlay* to impute the most likely locations of parasites which are contained within the original EMPP. The following steps represent one complete cycle of the cumulative overlay technique for a given EMPP shown in Figure 4a:

1. All of the blobs associated with the EMPP throughout all frames of the video sequence are overlaid on top of one another (Figure 4e). We restricted the size of acceptable blobs to a range of $.75 B < A_{Blob} < 2 B$, where A_{Blob} is the area of the blob, and B is the average mode of all blob areas.
2. Bounding boxes of blobs which overlap by at least 50% are grouped and assigned a frequency count, defined as the number of grouped bounding boxes.
3. The final remaining regions are called *intensity boxes* (Figure 4f). Each intensity box is considered a candidate location for a parasite which may be in the original EMPP. The final number of intensity boxes we select is given by $R = A_{EMPP}/B$, where A_{EMPP} is the area of the original EMPP and B is the average mode of all blob areas. The intensity boxes are ranked by frequency count, and we select the top R intensity boxes as the locations of new imputed parasites.
4. The original EMPP from the first frame is removed, and the newly imputed parasites are inserted in its place. (Figure 4f).

2.4.2 Dealing with Erroneously Split Parasites

To detect cases of erroneous splitting, the intuition is converse of that used for detecting erroneous merges. First, an ESPP is likely to be smaller in size than the average PP's. Second, an ESPP will likely exhibit frequent merging behavior throughout the video. Figure 5 shows an example of a parasite which was erroneously split into two regions on frame 1 and then merges on frames 2 and 4. We can see from the BPG's that a merge occurs when more than one parasite share an edge with the same blob.

To measure the probability that a given parasite is an ESPP, we again define the notion of a *merging volatility factor*, given by:

$$N_{merge} = M + K\sigma \tag{3}$$

In Eq.(3) N_{merge} is the minimum number of frames in which the putative parasite must merge, M is the average mode number of merges of all parasites, K is a constant, and σ is the merges standard deviation.

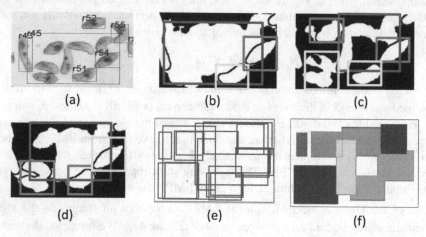

Fig. 4. (a) Raw image of a cluster of parasites (b), (c), and (d) are segmented binary images of three frames of the video sequence. (e) cumulative overlay of all bounding boxes associated with the putative parasites in the cluster (f) Final intensity boxes.

One key observation about ESPP's is that they cannot occur in isolation. The existence of one ESPP necessarily entails the existence of at least one neighboring ESPP with which it frequently merges. The set of all neighboring ESPP's is given by the BPG's, in which the neighboring ESPP's will frequently share an edge with the same blob. After we determine that a given PP is an ESPP, we backtrack to the first frame of the video and merge the ESPP's with the neighboring ESPP's.

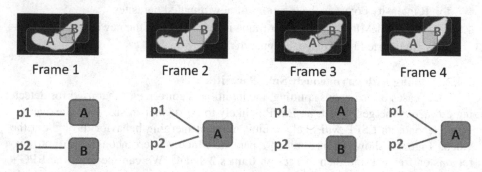

Fig. 5. Example erroneously split putative parasite (ESPP) and corresponding parasite graphs for four frames of a video sequence

After the above strategies remove the erroneous PP's and insert the newly imputed parasites into the first frame, we repeat another iteration of the parasites level, which repositions the newly inserted parasites appropriately in each frame of the video.

2.5   Re-segmentation

Although the refinement of blob-parasite correspondences does improve the precision of tracking when there is segmentation error, the original segmentation inside the rectangular regions is still erroneous. The close-up views of the imputed parasites in Figure 6c show that the shape of each parasite ends abruptly at the corner of its bounding box and includes portions of surrounding parasites. Such distortions signif-icantly undermine our attempts to accurately measure changes in the parasite's shape and other features that characterize a parasite's response to a drug. In the following, we describe a method which utilizes the results of the tracking to improve the under-lying segmentation.

We can detect segmentation errors by examining the BPG's. When a parasite be-comes associated with more than one blob on a certain video frame, we can conclude that a segmentation error occurred and led to erroneous splitting of the parasite into multiple regions on that frame (see for instance, Figure 2d). Figure 7a illustrates the process of extrapolating information from two correctly segmented images (i.e. the PP is connected to exactly one blob in the BPG) in the PP's previous and forward history to improve the erroneously segmented image. An analogous operation is performed to correct the segmentation for erroneously merged parasites (Figure 7b).

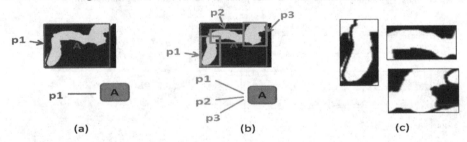

(a)   (b)   (c)

Fig. 6. (a) First frame of video prior to refinement level and corresponding BPG below, show-ing one EMPP associated with blob A. (b) First frame of video after refinement level and cor-responding BPG below, with EMPP removed and three new imputed parasites inserted, all associated with blob A. (c) Close-up views of the three new imputed parasites.

3   Experimental Results

It is critical that the method can effectively track schistosomula exhibiting a wide range of phenotypes and motion characteristics which may occur as a result of drug insult. In the following, we present results of the tracking system on five videos (Ta-ble 1). In each video, the parasites in the well were exposed to a different drug com-pound, as shown in the first column of Table 1. Our definition of a correctly tracked parasite requires that it be tracked accurately in 99% of the total frames in the video sequence. The method was able to consistently track over 90% of the parasites throughout the entirety (99% of total frames) of the video sequence, even though only 63%-89% of the parasites were correctly segmented throughout the entirety (99% of total frames) of the video sequence.

The results of the re-segmentation level are shown for a sample video containing 57 parasites, out of which 54 were correctly tracked throughout the video. Figure 8b shows the first frame of the video sequence. The number of corrected segmentation errors for each of the 54 tracked parasites is shown in Fig. 8a. The method was able to detect and correctly resolve 1198 segmentation errors in the 440 frames of the video sequence.

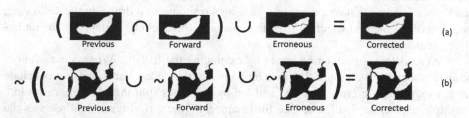

Fig. 7. (a) Process for correcting erroneously split regions. If a pixel is turned on in either of the correctly segmented images, then we also turn on that pixel in the erroneous segmentation. (b) Process for correcting erroneously merged regions. If a pixel is turned off in either of the correct segmentations, then we also turn off that pixel in the erroneous segmentation.

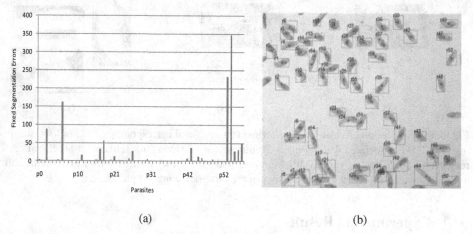

(a)   (b)

Fig. 8. (a) Number of corrected segmentation errors for each of 54 tracked parasites in a video sequence of 440 frames. (b) First frame of the video sequence showing all parasites.

Table 1. Segmentation and tracking results for five videos

Compound	Total Parasites	Segmented Parasites (99% of frames)	Tracked Parasites (99% of frames)
Doxepin	57	39 (68%)	54 (95%)
Control	45	36 (80%)	44 (98%)
Chlorprothixene	44	39 (89%)	44 (100%)
Amitriptyline	27	17 (63%)	25 (93%)
Cyclobenzaprine	39	25 (64%)	36 92%)

4   Conclusions

We have presented a tracking system which can robustly track schistosomula exhibiting a wide range of phenotypes. This works represents a solution to a unique tracking problem since none of the many trackers designed to track fast movement or well defined biological entities are applicable to the problem faced in HTS of complex parasites. Results from the proposed research bridge this gap and can lead to development of promising new technologies for high throughput and high content drug screening.

Acknowledgements. The authors thank C. Caffrey and B. Suzuki for the Schistosomiasis data. This research was funded in part by the NIH-NIAID through grant 1R01A1089896-01 and the NSF through grant IIS-0644418.

References

1. Nwaka, S., Hudson, A.: Innovative lead discovery strategies for tropical diseases. Nat. Rev. Drug Discov. 5, 941–955 (2006)
2. Lee, H., et al.: Quantification and clustering of phenotypic screening data using time-series analysis for chemotherapy of schistosomiasis. BMC Genomics 12(suppl. 1), S4 (2012)
3. Geng, W., Cosman, P., Berry, C.C., Feng, Z., Schafer, W.R.: Automatic track-ing, feature extraction and classification of C. elegans phenotypes. IEEE Transactions on Biomedical Engineering 51(10), 1811–1820 (2004)
4. Baek, J.-H., Cosman, P., Feng, Z., Silver, J., Schafer, W.R.: Using machine vision to analyze and classify Caenorhabditis elegans behavior phenotypes quantitatively. Journal of Neuroscience Methods 118(1), 9–21 (2002)
5. Masoud, O., Papanikolopoulos, N.: A novel method for tracking and counting pedestrians in real-time using a single camera. IEEE Trans Vehicular Technology 50(5), 1267–1278 (2001)
6. Li, K., Miller, E.D., Chen, M., Kanade, T., Weiss, L.E., Campbell, P.G.: Cell population tracking and lineage construction with spatiotemporal context. Med. Image Anal. 12(5), 546–566 (2008)
7. Singh, R., Pittas, M., Heskia, I., Xu, F., McKerrow, J.H., Caffrey, C.: Automated Image-Based Phenotypic Screening for High-Throughput Drug Discovery. In: IEEE Symposium on Computer-Based Medical Systems, pp. 1–8 (2009)

8. Al-Kofahi, O., Radke, R.J., Goderie, S.K., Shen, Q., Temple, S., Roysam, B.N.: Automated Cell Lineage Construction: A Rapid Method to Analyze Clonal Development Established with Murine Neural Progenitor Cells. Cell Cycle 5(3), 327–335 (2006)
9. Blake, A., Isard, M.: Condensation – conditional density propagation for visual tracking. International Journal of Computer Vision 28(1), 5–28 (1998)
10. Heisele, B., Kressel, U., Ritter, W.: Tracking nonrigid, moving objects based on color cluster flow. In: Proc. IEEE Conf. Computer Vision and Pattern Recognition, pp. 257–260 (June 1997)
11. Moody-Davis, A., Mennillo, L., Singh, R.: Region-Based Segmentation of Parasites for High-throughput Screening. In: Bebis, G. (ed.) ISVC 2011, Part I. LNCS, vol. 6938, pp. 43–53. Springer, Heidelberg (2011)

Cell Nuclei Detection Using Globally Optimal Active Contours with Shape Prior

Jonas De Vylder[1], Jan Aelterman[1], Mado Vandewoestyne[2], Trees Lepez[2],
Dieter Deforce[2], and Wilfried Philips[1]

[1] Department of Telecommunications and Information Processing,
IBBT - Image Processing and Interpretation, Ghent University, Ghent, Belgium
[2] Laboratory of Pharmaceutical Biotechnology, Faculty of Pharmaceutical Sciences,
Ghent University, Ghent, Belgium
jonas.devylder@telin.ugent.be
http://telin.ugent.be/~jdvylder

Abstract. Cell nuclei detection in fluorescent microscopic images is an important and time consuming task for a wide range of biological applications. Blur, clutter, bleed through and partial occlusion of nuclei make this a challenging task for automated detection of individual nuclei using image analysis. This paper proposes a novel and robust detection method based on the active contour framework. The method exploits prior knowledge of the nucleus shape in order to better detect individual nuclei. The method is formulated as the optimization of a convex energy function. The proposed method shows accurate detection results even for clusters of nuclei where state of the art methods fail.

1 Introduction

Cell nuclei are one of the most studied objects in microscopic biology. This is because they are easily visualized independent of the type of cells, typically using a fluorescent staining, while containing relevant biological information for a wide range of applications, e.g. cell division in tumours, root growth in plants, full embryonic development, etc. [1, 2].

Due to the biological importance of cell nuclei, several automated detection methods have been proposed in the past. These methods can generally be categorized in two groups: edge based and intensity based. The first group starts by detecting edges, both binary and continuous edge maps have been used, and tries to fit a specific shape model to them [3–6]. The robustness of these methods strongly depends on the output of the edge detector, which is not always sufficient to detect individual nuclei in case of clustered nuclei. The second group first segments the nuclei from the background, this is typically done based on intensity, e.g. by some sort of automatic thresholding [7, 8]. For isolated nuclei this is rather straightforward, but for touching nuclei this requires an extra step to detect each individual nucleus. This is mainly done by assuming that the detected segments should have a convex shape [2, 9–11]. These methods are non optimal since both steps are independent: the second step can only use the result of the first step, instead of all information available, e.g. the complete image. Incorrectly segmented pixels can have a big influence on the individual detection.

G. Bebis et al. (Eds.): ISVC 2012, Part II, LNCS 7432, pp. 115–124, 2012.

In contrast too state of the art methods, this paper proposes an intensity based segmentation method which tackles the individual nuclei detection in a single step. The decision of segmenting a pixel as foreground is based both on intensity and on the likelihood that a nucleus is located at that location. This combined approach will result in more accurate nuclei detection as will be shown in the results section. This is achieved using the active contour framework. The proposed method has a convex energy function, thus is invariant of the initialization [12, 13]. This paper is arranged as follows. The next section provides a brief description of convex energy active contours. In section 4 our proposed algorithm is presented, while section 5 elaborates on the optimization of our proposed method. Section 6 shows the results of our technique and the convergence of our method is studied. Section 7 recapitulates and concludes.

2   Notations and Definitions

In the remaining of this paper we will use specific notations. To make sure all notations are clear, we briefly summarize the notations and symbols used in this work.

We will refer to an image, F in its vector notation, i.e. $\mathbf{f}(i * m + j) = F(i, j)$, where $m \times n$ is the dimension of the image. In a similar way we will represent the contour in vector format, $\mathbf{u}$. If a pixel (i, j) is part of the segment, it will have a value above a certain threshold, all background pixels will have a value lower than the given threshold. Note that this is similar to level-sets. The way these contours are optimized however is different than with classical level-set active contours, as is explained in the next section. We will use the gradient image operators in combination with this vector notation, however the semantics of this operator remains the same as if it was used with the classical matrix notation:

$$\nabla(\mathbf{f}(i * m + j)) = \big(F(i + 1, j) - F(i, j), F(i, j + 1) - F(i, j)\big) \tag{1}$$

Further we will use the following inner product and norm notations:

$$\langle \mathbf{f}, \mathbf{g} \rangle = \sum_{i=1}^{mn} f(i)g(i)$$

$$\|\mathbf{f}\|_0 = \sum_{i=1}^{mn} 1 - \delta_{0,f(i)}$$

$$|\mathbf{f}| = \sum_{i=1}^{mn} | f(i) |$$

$$\|\mathbf{f}\|_2 = \sqrt{\sum_{i=1}^{mn} f(i)^2}$$

Where the weights $g(i) \geq 0$ and $\delta_{i,j}$ represents the Kronecker delta, which is equal to one if i and j are equal and is zero in all other cases. The $l0$ norm, i.e. $\|\mathbf{f}\|_0$, counts the non-zero elements of the vector. Note that the $l0$ norm is not a real norm in the mathematical sense, since the triangle inequality does not hold.

3 Convex Energy Active Contours

In [12] an active contour model was proposed which has global minimisers. This active contour is calculated by minimizing the following convex energy:

$$E[\mathbf{u}] = |\nabla \mathbf{u}| + \beta \langle \mathbf{u}, \mathbf{r} \rangle \tag{2}$$

with

$$\mathbf{r} = (\mu_f - \mathbf{f})^2 - (\mu_b - \mathbf{f})^2 \tag{3}$$

where μ_f, μ_b are respectively the expected foreground and background intensity and β is a weighting parameter tuning the influence of the data-fit in relation to the total variation regularization. Note that this energy is convex, only if μ_f and μ_b are constant. The expected intensities μ_f, μ_b can be calculated from a training dataset. Chan et al. [12] proved that $\mathbf{u}$ is well-defined as the solution of a convex energy function over a convex domain if $\mathbf{u} \in [0,1]^{mn}$, i.e.

$$\hat{\mathbf{u}} = \operatorname*{arg\,min}_{\mathbf{u} \in [0,1]^{mn}} |\nabla \mathbf{u}| + \beta \langle \mathbf{u}, \mathbf{r} \rangle \tag{4}$$

Furthermore, the steady state of the gradient flow corresponding to this energy function coincides with the steady state of the gradient flow of the original active contours without edges (ACWE) [12, 14], i.e. an optimum of this convex energy is also an optimum of the original ACWE energy function. Note that $\hat{\mathbf{u}}[i]$ can have any value between 0 and 1, thus the found active contour does not have to represent a crisp segmentation. A binary segmentation result can be given by thresholding $\hat{\mathbf{u}}$, i.e.

$$\Phi_\alpha(\hat{\mathbf{u}}[i]) = \begin{cases} 1 & \text{if } \hat{\mathbf{u}}[i] > \alpha \\ 0 & \text{otherwise} \end{cases} \tag{5}$$

for some $\alpha \in [0,1]$. In [13] it is shown that $\Phi_\alpha(\hat{\mathbf{u}})$ itself is a global minimizer for the energy in eq. (2) and by extension for the energy function of the ACWE model.

4 Active Contours with Sparse Shape Prior

The energy function in eq. (2) tries to remove noisy segmentation pixels by regularizing the energy function using total variation. This regularization is useful if pixels are incorrectly classified, i.e. background pixels detected as foreground or vice versa, due to noise in the image. In microscopic images however, incorrectly detected nuclei are often caused by clutter in the image, e.g. dead cells or bleed-through from other fluorescent channels. This is not solved using total variation since these incorrectly detected nuclei correspond to natural objects. Therefore, we propose a regularization term that exploits the regular shape of cell nuclei, penalizing segments with strongly deviate from the expected shape. In this work, we model a nucleus as a disk. Thus, given a predefined radius, r, and location, (x,y), we can calculate the ideal u, i.e. a binary image where the pixels within a distance r of (x,y) are equal to one and all other pixels are equal to zero. For a discrete range of nuclei diameters it is possible to enumerate all possible u's, this is of course under the assumption that there is only one nucleus in the image. We will refer to each of these possible u's as *words*. In most applications however the image does contain multiple nuclei, where it is unknown how many nuclei are present, this requires a superposition of multiple words to represent u. The optimal active contour can be represented using a minimal number of words, i.e. one word per nucleus. This representation of the active contour in terms of a superposition of words can be used as a new regularization term:

$$\hat{u}, \hat{t} = \underset{t, u \in [0,1]^{mn}}{\arg\min} \|t\|_0 + \beta \langle u, r \rangle \qquad \text{such that} \qquad u = Dt \qquad (6)$$

where D a matrix of dimension $mnr \times mn$, representing a dictionary of words, with mn corresponding to the number of pixels, and r corresponding to the number of possible diameters of a nucleus. Thus we can represent this dictionary as $D = [d_{a,1} \; d_{a,2} \; d_{a,3} \; ... \; d_{a+1,1} \; d_{a+1,2} \; d_{a+1,3} \; ... \; d_{b,n}]$ with $d_{i,j}$ the vector form of a word, i.e. a binary representation of a disk with radius $i \in [a, b]$ and centroid at pixel j. Thus, t represents a weighting vector for the words. The energy term minimized by u is based on a l_0 norm which is non-convex. Fortunately the l_1 norm can be used as a good approximation instead [15]. This new prior results in the following active contour:

$$\hat{u}, \hat{t} = \underset{t, u \in [0,1]^{mn}}{\arg\min} |t| + \beta \langle u, r \rangle \qquad \text{such that} \qquad u = Dt \qquad (7)$$

By using the l_1 norm as an approximation of the l_0 norm, we penalize representations which use more words than necessary. This sparsity constraint results in isolated peaks in t which can easily be detected using a peak detector such as is used for peak detection after the Hough transform. These detected peaks represent the centroids of the detected nuclei.

5   Optimization

In order to optimize the constrained problem in eq. (7) the problem is approximated by the following unconstrained optimization problem:

$$\hat{u}, \hat{t} = \underset{t, u \in [0,1]^{mn}}{\arg\min} |t| + \beta \langle u, r \rangle + \gamma \|u - Dt\|_2^2 \qquad (8)$$

Where γ is a weighting parameter. Note that this only approximates the constraint $u = Dt$. Although there exists efficient techniques to enforce this constraint exactly, e.g. augmented Lagrangian or Bregman methods, we propose to use the approximation in eq. (8) instead. This allows the active contours to detect nuclei whose shape slightly deviates from the circular model or to detect partially overlapping nuclei. Given the convexity of eq. (8), this problem can be solved by iteratively optimizing for t and u independently, i.e.

$$\hat{t} = \underset{t}{\arg\min} |t| + \gamma \|u - Dt\|_2^2 \qquad (9)$$

$$\hat{u} = \underset{u \in [0,1]^{mn}}{\arg\min} \beta \langle u, r \rangle + \gamma \|u - Dt\|_2^2 \qquad (10)$$

The problem in eq. (9) is a typical $l_1 - l_2$ optimization problem for which a variety of optimization methods have been proposed, e.g. FISTA, primal-dual methods, etc. [16–18]. In the next section we evaluate several of these methods for the optimization of eq. (9). The problem in eq. (10) can be optimized by solving a set of Euler-Lagrange equations. For an optimal $\mathbf{u}$, the following optimality condition should be satisfied:

$$\mathbf{u} = -2\frac{\beta}{\gamma}\mathbf{r} + D\mathbf{t} \tag{11}$$

The solution of eq. (11) is unconstrained, i.e. $\mathbf{u}$ does not have to lie in the interval $[0, 1]^{mn}$. However minimizing eq. (10) is equivalent to minimizing a set of quadratic functions. So if $\mathbf{u}[i] \notin [0, 1]$ then the constrained optimum is either 0 or 1, since a quadratic function is monotonic in an interval which does not contain its extremum. So the constrained optimum is given by:

$$\hat{\mathbf{u}}[i] = \max\left(\min\left(\mathbf{u}[i], 1\right), 0\right) \tag{12}$$

6 Results

6.1 Detection

To validate the proposed method, a synthetic dataset is analyzed [19]. These synthetic images were proposed as a common benchmark for nuclei segmentation. The synthetic images show the same intrinsic properties of real microscopic images of cell nuclei: blurred nuclei, non uniform intensity in a nucleus, touching nuclei, non uniform background, etc. In Fig. 1(a) an example of such a synthetic image is shown. Fig. 1(b) shows the detection result using an edge based detection method [6]. The result of cellProfiler, i.e. an intensity based method [9], is shown Fig. 1(c), whereas Fig. 1(d) depicts nuclei detection using the proposed method. The proposed method shows more accurate results for the detection of isolated nuclei even if nuclei are clustered together.

To quantitatively validate the result, a dataset of 20 images, each containing 300 nuclei, were analyzed. The results are shown in Table 1. The first row represents the results of an edge based method [6], whereas the next two rows correspond with two intensity based images [9, 10]. Note that the proposed method, the last row, get the best results both for cell detection metrics, in the first 4 columns, as for the Dice coefficient, which is a metric for segmentation quality. Mark that lacking ground truth for individual nuclei, the Dice coefficient measures the similarity of all segmented nuclei compared with the ground truth for all nuclei instead of for individual nucleus.

In Fig. 2 an example of nuclei detection in a real microscopic image is shown. These microscopic image shows nuclei of peripheral blood mononuclear cells, captured with a PALM MicroBeam system. In Fig. 2.a shows the ground truth detection, which corresponds to manually annotations. Fig. 2.b and Fig. 2.c correspond to state of the art detection methods. Note that they both falsly detect nuclei at places where there is some smeared staining, i.e. when a nucleus ruptures it releases its staining, which results in bright smears. CellProfiler does not only suffer from false detections, but also merges touching nuclei. The proposed method does not suffer from false detection due to dye smears, while still being able to detect touching nuclei, as can be seen in Fig. 2.d.

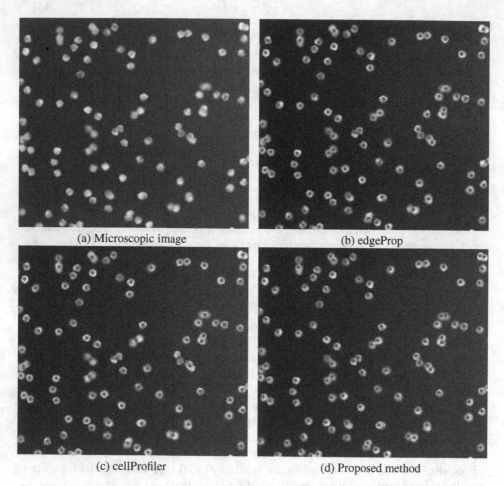

(a) Microscopic image

(b) edgeProp

(c) cellProfiler

(d) Proposed method

Fig. 1. Example of cell detection using different methods on an image of the broad Bioimage Benchmark Collection

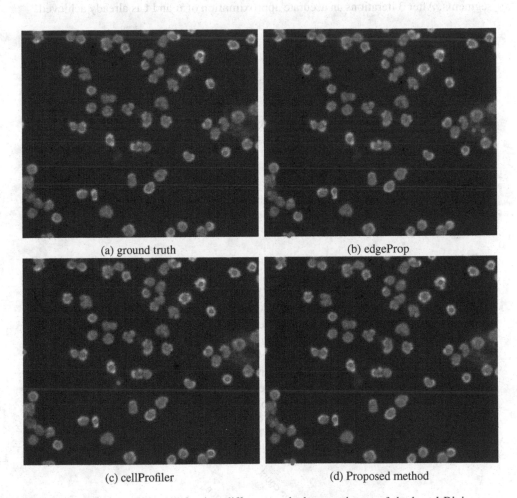

(a) ground truth

(b) edgeProp

(c) cellProfiler

(d) Proposed method

Fig. 2. Example of cell detection using different methods on an image of the broad Bioimage Benchmark Collection

6.2   Convergence

The optimal **u** and **t** are calculated by alternating optimizing eq. (8) for **u** and **t**. This results in fast convergence as can be seen in Fig. 3, where the mean squared error of **u** and **t** are plotted after each iteration. Note that this error is in comparison with the optimal active contour, i.e. **u** and **t** after 10 iteration, and not to the ground truth segments. After 3 iterations an accurate approximation of **u** and **t** is already achieved.

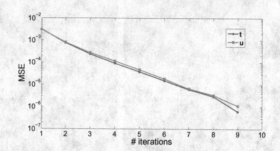

Fig. 3. Convergence of **u** and **t** in function of the number of iterations, the MSE is calculated between **u** and **t** and there optimal values, i.e. there values after convergence

Solving the problem in eq. (9) requires to iteratively optimize an $l_1 - l_2$ problem. For this type of problems several optimization methods have been proposed in literature [16–18]. In Table 2 we compare the performance of several of these methods: DALM[17], FISTA [18], L1LS [16], PALM [16] and SpaRSA[17]. The last column shows the computational time of a single iteration for the analysis of a 256×256 image. These measurements where done using an Intel i7 Q720 1.6 GHz CPU with 4GB RAM. The L1LS optimizer needs significantly less iterations than the other tested methods, however a single iteration is much more time consuming. Therefore FISTA turns out to be the fastest method, with an average convergence after 6.10 s. This results in a total processing time of 18.05s for the complete segmentation of an image.

Table 1. A comparison of different cell nuclei detection and segmentation methods

	average count	δ	precision	recall	Dice
edgeProp	299.85	1.785	0.9973	0.9968	0.941
cellProfiler	298.85	2.231	0.9923	0.9885	0.939
cellc	315.85	3.631	0.9445	0.9945	0.929
proposed	300.05	0.224	0.9998	1	0.981

Table 2. A comparison of convergence of different optimization methods used for the inner optimization step, i.e. eq. (9)

	mean # iterations	min # iterations	max # iterations	time (s)
DALM	1819	529	4000	0.013
FISTA	281	270	293	0.026
L1LS	12	11	14	19.778
PALM	84	73	100	0.292
SpaRSA	1101	867	1311	0.012

7 Conclusion

This paper proposed a novel segmentation technique to detect and segment cell nuclei in fluorescent microscopic images. The method fits within the active contour framework has a convex energy function. The method uses prior knowledge about the shape of cell nuclei, which is done by representing the segmentation result using a dictionary. The proposed method was tested on a benchmark dataset, specifically proposed for this type of application. The method results in accurate nuclei detection and outperforms state of the art methods, e.g. precision and recalls of respectively 0.9998 and 1.

Acknowledgment. Jonas De Vylder is funded by the Institute for the Promotion of Innovation by Science and Technology in Flanders (IWT). The authors would like to thank the following institutes and researcher for sharing their codes and data: Tampere University of Technology for the use of CellC Cell Counting, Broad institute for providing both the benchmark dataset and the cellProfiler software and Allen Y. Yang (University of California, Berkeley) for the l_1 optimization benchmarking toolkit.

References

1. Keller, P.J., Schmidt, A.D., Wittbrodt, J., Stelzer, E.H.K.: Reconstruction of zebrafish early embryonic development by scanned light sheet microscopy. Science 322, 1065–1069 (2008)
2. Chen, X.W., Zhou, X.B., Wong, S.T.C.: Automated segmentation, classification, and tracking of cancer cell nuclei in time-lapse microscopy. IEEE Transactions on Biomedical Engineering 53, 762–766 (2006)
3. Gladilin, E., Goetze, S., Mateos-Langerak, J., Van Driel, R., Eils, R., Rohr, K.: Shape normalization of 3d cell nuclei using elastic spherical mapping. Journal of Microscopy-Oxford 231, 105–114 (2008)
4. Hukkanen, J., Hategan, A., Sabo, E., Tabus, I.: Segmentation of cell nuclei from histological images by ellipse fitting. In: The 2010 European Signal Processing Conference (2010)
5. Li, G., Liu, T., Nie, J., Guo, L., Chen, J., Zhu, J., Xia, W., Mara, A., Holley, S., Wong, S.T.C.: Segmentation of touching cell nuclei using gradient flow tracking. Journal of Microscopy-Oxford 231, 47–58 (2008)

6. De Vylder, J., Philips, W.: Computational efficient segmentation of cell nuclei in 2d and 3d fluorescent micrographs. In: Proceedings of SPIE Photonics West: Conference on Imaging, Manipulation and Analysis of Biomolecules, Cells, and Tissues (2011)
7. Gudla, P.R., Nandy, K., Collins, J., Meaburn, K.J., Misteli, T., Lockett, S.J.: A high-throughput system for segmenting nuclei using multiscale techniques. Cytometry Part A 73A, 451–466 (2008)
8. Sezgin, M., Sankur, B.: Survey over image thresholding techniques and quantitative performance evaluation. Journal of Electronic Imaging 13, 146–168 (2004)
9. Kamentsky, L., Jones, T.R., Fraser, A., Bray, M.A., Logan, D.J., Madden, K.L., Ljosa, V., Rueden, C., Eliceiri, K.W., Carpenter, A.E.: Improved structure, function and compatibility for cellprofiler: modular high-throughput image analysis software. Bioinformatics 27, 1179–1180 (2011)
10. Selinummi, J., Seppala, J., Yli-Harja, O., Puhakka, J.A.: Software for quantification of labeled bacteria from digital microscope images by automated image analysis. Biotechniques 39, 859–863 (2005)
11. Cloppet, F., Boucher, A.: Segmentation of overlapping/aggregating nuclei cells in biological images. In: 19th International Conference on Pattern Recognition, vols. 1-6, pp. 789–792 (2008)
12. Chan, T.F., Esedoglu, S., Nikolova, M.: Algorithms for finding global minimizers of image segmentation and denoising models. Siam Journal on Applied Mathematics 66, 1632–1648 (2006)
13. Bresson, X., Chan, T.F.: Active contours based on chambolle's mean curvature motion. In: IEEE International Conference on Image Processing, vols. 1-7 , pp. 33–36 (2007)
14. Chan, T., Vese, L.: An active contour model without edges. In: Scale-Space Theories in Computer Vision, vol. 1682, pp. 141–151 (1999)
15. Baraniuk, R.G.: Compressive sensing. IEEE Signal Processing Magazine 24, 118–121 (2007)
16. Kim, S.J., Koh, K., Lustig, M., Boyd, S., Gorinevsky, D.: An interior-point method for large-scale l1-regularized least squares. IEEE Journal of Selected Topics in Signal Processing 1, 606–617 (2007)
17. Yang, A.Y., Sastry, S.S., Ganesh, A., Yi, M.: Fast l1-minimization algorithms and an application in robust face recognition: A review. In: 2010 17th IEEE International Conference on Image Processing (ICIP), pp. 1849–1852 (2010)
18. Zibulevsky, M., Elad, M.: L1-l2 optimization in signal and image processing. IEEE Signal Processing Magazine 27, 76–88 (2010)
19. Ruusuvuori, P., Lehmussola, A., Selinummi, J., Rajala, T., Huttunen, H., Yli-Harja, O.: Set of synthetic images for validating cell image analysis. In: Proc. of the 16th European Signal Processing Conference, EUSIPCO 2008 (2008)

A Novel Gait Recognition System
Based on Hidden Markov Models

Akintola Kolawole and Alireza Tavakkoli

University of Houston–Victoria

Abstract. The advances in computing power, availability of large-capacity storage devices and research in computer vision have contributed to recent developments in gait recognition. The ease of acquiring human videos by low cost equipments makes gait recognition much easier and less intrusive than other biometric systems. In this paper, a gait recognition system using a space-model-based approach is proposed. The proposed mechanism is able to detect moving object of interest, while track them and analyzing their gait for recognition. The system captures videos of subjects in front a stationary camera. The identification module makes use of the shape and dynamics of the system using HMM. Then it models the gait properties by accepting the feature vectors as input and model the dynamics through state transitions and observation probabilities. The experimental results show that the proposed gait recognition system successfully recognizes humans using their gait.

1   Introduction

Gait is a spatio-temporal phenomenon that typifies the motion characteristics of an individual [1]. Gait recognition is aimed to recognize human beings from distance, using the way they walk [2]. In kinesiology (human kinetics) the goal has been to understand human motion with applications in sports, medicine, elderly care and early detecting of movement disorders [3].

Analyzing human gait has found considerable interest in recent computer vision research [4]. The application areas of gait recognition are enormous. The use of gait analysis as a means of deducing the physical well-being of people is examined in [4]. This method is based on transforming the joint motion trajectories using wavelets to extract spatio-temporal features which are then fed as input to a vector quantizer; a self-organizing map for classification of walking patterns of individuals with and without pathology. It was shown that the algorithm is successful in extracting features that successfully discriminate between individuals with and without locomotion impairment. Rohila, [5], recognized that abnormal gait can be used to identify certain behaviors that might generate an alert.

Gait also has been analyzed for use as a biometric system to identify people from a distance [6]. Human gait recognition works from the observation that an individual style is unique and can be used for identification [6]. Pushpa and Arumugam observed that natural biometric systems fails in two ways; Failure to match in low resolution images, and User co-operation to obtain accurate results, [6]. These shortcomings might be a plus to using gait for biometric purposes.

G. Bebis et al. (Eds.): ISVC 2012, Part II, LNCS 7432, pp. 125–134, 2012.

Although an intrusive approach to identification, a successful gait recognition biometric system must overcome a number of challenges. In [5], the following challenges were identified and categorized into two classes:

– External factors: Such factors mostly impose challenges to the recognition approach. e.g.

  • Viewing angles like frontal view, side view etc.
  • Lighting conditions like day/night, shadow etc.
  • Outdoor/indoor environments e.g. sunny, rainy days
  • Walking surface conditions e.g. hard/soft, dry/wet, grass/concrete, etc.

– Internal factors: Such factors cause changes of the natural gait due to internal changes in body.

  • Foot injury
  • Lower limb disorder
  • Physiological changes in body due to aging, drunkenness, pregnancy, etc.

Various Gait recognition algorithms have been proposed in the literature and compete to achieve better recognition results. These approaches can broadly be classified as model-based and modeless algorithms. The model-based techniques are motivated by bio-mechanical and input/output modeling approaches, while the modeless techniques use shape and motion information for gait analysis [7].

In this research we propose to use a space-model algorithm for gait modeling and recognition based on [1] and [8]. Kernel-based density estimation is used in the proposed system for moving object detection [9]. The object detection algorithm can be used in indoor and outdoor environments. In the object detection method, a reference background model is built initially by using about 200 frames of the video. At each new frame, foreground pixels are detected by comparing the Intensity value histogram with a threshold. The detected foreground noise is filtered by the use of a sequence of morphological operations closing and opening operations. Then, a contour of the connected components is extracted. Small contours are removed by finding contour whose size is less than a predetermined threshold. Finally, each detected foreground object is represented with its spatio-spectral model to be used for tracking and feature extraction. A tracking algorithm tracks the detected objects in successive frames by using a color histogram matching technique [10].

1.1 Motivation

There are many challenges facing the recognition of human gait and we seek daily to improve the gait recognition by carrying out research on ways to improve the existing algorithms. With several gait algorithms, no algorithm has been able to work in different environment. Some algorithms are view invariants [11] and [12], while others are variant multiple viewpoint and demand constraints on the direction of walk cycles. Gait recognition may also be impacted by quality of the silhouettes extracted during foreground extraction.

One of the motivations for this research is to investigate more robust gait properties (i.e. distance feature vectors). Another motivation for our approach is to study how a better silhouette extraction from the scene background can effectively impacts the recognition results. Several Gait images have been used in the literature for gait recognition, e.g gait energy images(GEI) [13] and gait exemplars [1], but none have been able to effectively use gait representatives. Also to our knowledge none of the approaches in the literature have been able to efficiently use Bhattacharyya clustering for this purpose. In this paper we investigate the use of gait representative and Bhattacharyya clustering in modeling gait dynamics for HMM gait recognition.

1.2   Research Scope

The research in Gait usually includes two stages, "low-level" and "high-level". The low-level stage involves extracting features from images such as the boundary of object silhouettes. The task of high-level is then to recognize these objects with the extracted features. This paper is concerned with both "low-level" and "high-level" stages. The low level is concerned with object localization, while the high level performs the object tracking and gait recognition. To this end, we focus on gait recognition in a modeled background environment. Walking direction is also constrained to be perpendicular to the direction of projection on the image plane. The subjects walk pass the camera in 0 degree of view. All direction of walk is not considered. Also, other forms of people behaviors like running, jumping, etc. are not considered in this work.

1.3   Contributions

The main contributions of this thesis can be summarized as follows:

1. The use of Gait representative
2. Detection of gait cycle using new MiniMax algorithm on *dfeet* .
3. Development of a gait recognition system for human recognition.

2   Methodology

In this paper an approach based on Hidden Markov Models (HMM) is adopted for gait recognition. The block diagram of the proposed method is shown in figure 1. The distance from the centroid of the bounding box to the boundary of the silhouette is used as the feature vector. The temporal shape and dynamics of the silhouette walking characteristics are modeled using the HMM. It is shown that the proposed method brings the high rate of identification in [8].

After the feature extraction stage, the gait cycle is detected using *dfeet* and MiniMax algorithm. In order to clarify the starting point and the ending point of the object for identification, one Gait cycle (1GC) is used for the unit of identification focusing on the periodicity of a walking person [8].

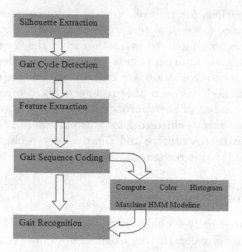

Fig. 1. The Gait Recognition Block Diagram

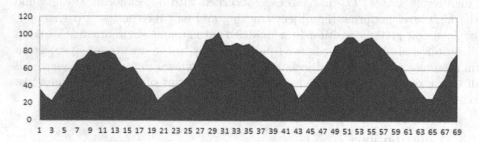

Fig. 2. The *dfeet* measure, plotted as shown above in order to determine the number of sequences in the video frame that falls within a cycle

Hidden Markov Models are robust to temporal changes in a pattern. Therefore, the temporal transitions of feature parameter are captured using HMMs. Using HMMs requires a training and an identification phase. Every 1 gait cycle (1GC) in both phases need to be employed, according to [1] and [8]. In the proposed method, however we adopted half gait cycle as our gate cycle to achieve a more efficient recognition benchmark. Half gate cycle is the distance covered by the movement of one leg minimum position to the maximum position and back to the minimum position as shown in figure 2.

2.1   Feature Extraction

Sequences of silhouette distance signals are used as inputs to the training and recognition phases. We present each shape instance (silhouette) as a 1-D signal of the distance from the centroid of the bounding box of the image to the outer

Fig. 3. Key frames of sequence of gaits in one gait cycle showing the feature vectors extracted

Fig. 4. Key frames of sequence of gaits showing *dfeet* calculation. The *dfeet* is the distance between the two feet at the position shown in figure 3 (the position is take as 1/5 of human height from ground up).

contour of the silhouette as depicted in figure 3. This feature is then normalized. It is found that this feature represents the shape and dynamic features of the system and can discriminate each silhouette, according to [8].

2.2 Gait Cycle Detection

Detection of gait cycle is essential for training and recognition. Typically, for side or frontal views, cycles can be detected using features such as width or height of bounding box [14]. However, in this paper we made use of the *dfeet* measure, shown in figure 2. We calculate the width of the lower leg regions in sequence of silhouettes as depicted in the colored region in figure 6.

Kale *et al.* in [1] used half gait cycle for as a cycle for gait recognition. They argue that there is periodicity in a person's walking pattern, and it can be considered that walking operation is a periodic signal. According to [8], the advantage obtained by making 1-Gait Cycle the unit of identification is that:

Shape changes of the walking silhouette over time is an important cue in determining its internal motion. Much like [8], we convert a two-dimensional silhouette shape into a one-dimensional distance signal from its associated centroid. This significantly reduces the subsequent computational cost. A distance

```
1. Set n=0;
2. For feature item [i] i=1 T
2.1.    Get feature item[i]
2.2.    Set finmin to false
2.3.    if( not finmin)
            if  feature [i] < feature[i-1]  // item is less than next
                                            //  feature item
                if fetitem < fetvect[i+1]) && (fetitem < fetvect[i+2]) &&
                (fetitem < fetvect[i+3]) && (fetitem < fetvect[i+4])  &&
                (fetitem < fetvect[i+5]))
                    set minvalue[n] to fetvect[i];
                    set mini index to i;
                    minindex[n] = i;
                    set n to n+1;
        if (findmin)
            if ((fetitem > fetvect[i+1]) && (fetitem > fetvect[i+2]) &&
            (fetitem > fetvect[i+3]) && (fetitem >  fetvect[i+4])  &&
            (fetitem > fetvect[i+5]))
                set maxvalue[n] to fetvect[i];
                set max index to i;
                maxindex[n] = i;
                set n to n+1;
            Set findmin = false;
```

Fig. 5. Gait Cycle detection using the proposed MiniMax Algorithm

signal sequence is used to approximately represent a temporal gait pattern. After the silhouette of a walking figure has been extracted, its boundary can be easily obtained using a border following algorithm based on connectivity [2]. Automatic detection of number of gaits within a cycle is obtained using the MiniMax Algorithm. The gait cycle's MiniMax algorithm is shown in figure 5.

2.3   Feature Vector Quantization

A discrete HMM is taken into consideration for the recognition process of human gait. The feature vector of distance signals needs to be converted into a finite set of symbols. The VQ technique plays a reference role in HMM based approach in order to convert continuous sequence of gait signals into a discrete sequence of symbols for discrete HMM. The VQ concept is entirely determined by calculating the Bhattacharyya distance between a gait representative and each gait within the gait cycle. M A single cycle is used for gait feature extraction. We just pick one of the features within a cycle as a representative gait feature. And the distance between the chosen representative feature and each of the other features is calculated using Bhattacharyya distance to obtain a code for each feature vectors to be used for training the HMM. Bhattacharyya distance has been used for feature clustering in [15] and has proven to be quite effective.

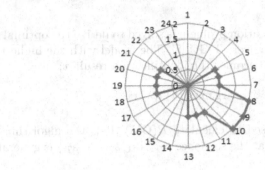

Fig. 6. Normalized gait distance from representative gait for one cycle (Gait signature). The representative gait is in the middle with value 0.

2.4 Gait Classification Using HMM

The Hidden Markov Model is a finite set of states. These states are associated with probability distribution functions. the transitions among the states are governed by a set of probabilities called transition probabilities. The states are not visible to an external observer and therefore are "hidden". Only the outcome of the state probability distributions are observable by the outside world.

RG-HMM. This section presents the gait identification process using the observation feature vector, containing shape information. The temporal transition of a person's silhouette encodes the individual's characteristics. We use Hidden Markov Model (HMM) as the primary mode of identification. HMM is robust to temporal changes in the patterns and is mainly used for recognition of voice or signature. However, it is also used for sequential image recognition such as gesture recognition [1].

Since the walking motion can be assumed as a temporal series, it is modeled by an HMM using the observed feature vector. We call this model RG-HMM, since a single gait is used to represent the gait cycle from which all other gaits distances are computed and coded. The RG-HMM is modeled for every person.

2.5 Training

In this stage, the state transitional and the output symbol probabilities of RG-HMM are initialized randomly for every person to generate the RG-HMM. Given the model λ and the feature vector $\mathbf{O}$, the model parameters A and B are calculated to maximize the probability $P(\mathbf{O}|\lambda)$.

Each model is trained using a single cycle sequence of gaits. Each gait is the observed symbol. To find the optimal values for the model λ, the Baum-Welch algorithm is used. In the Baum-Welch algorithm, recursively calculates the transitional and model probabilities after proper values are initialized.

2.6   Identification

In this stage, the HMM evaluation problem is solved to derive the optimal model from the feature vector $\mathbf{O}$ in every 1-GC [8]. The model with the highest probability of the feature vector becomes the identification result $\hat{x}$:

$$\hat{x} = \arg\max_x P(\mathbf{O}|\lambda_x) \tag{1}$$

The forward algorithm is used to calculate $P(\mathbf{O}|\lambda)$. It is the algorithm which calculates the probability that the sequence $S = \{s_1, s_2, \cdots, s_T\}$ is generated by the given HMM model.

To test a particular gait, we extract the feature vectors, calculate a gait cycle, and then quantize the sequence of gaits within the cycle using Bhattacharyya distance. We then forward this gait sequence to the HMM as input sequence. A forward algorithm is then applied to the gait sequence to obtain the probability of the model given the sequence. This sequence is passed to all the models and the model that maximizes the probability is recognized as the gait that the sequence belongs.

3   Experimental Results

In this section we present the experiments carried out using the design philosophy highlighted above. We also present the results of our gait development system using videos of walking people. The application is implemented using Microsoft Visual C++ and the OpenCV Library.

The proposed detection algorithm was tested on a number of videos with the frame size of 320x480 pixels. We examined the training and detection time of our technique. To calculate the training speed of the algorithm, 200 frames with the background and no foreground objects were used. The total training time in the original algorithm on one processor with 1.5GHz clock speed was about 40 ms per frame. When the algorithm was run on a dual and quad core processors the training time dropped 20 and 10 ms, respectively. The foreground detection time was even faster than the training stage.

On average the proposed technique reached the foreground detection time of about 2ms per frame. The shadow detection and removal runs at about 200 μs. We implemented a hidden right to left Hidden Markov Model to test our model. Videos of five subjects walking in front of the camera at angle 0 degree to the camera were used to train each of the HMM.

We extracted the 1-D signal distance of the silhouettes from these videos. To quantize these signals we use a representative gait within a cycle. We calculate the distance between a feature vector of a particular gait within the cycle to the representative gait feature vector using Bhattacharyya distance.

We calculate the cycle of these subjects using the *dfeet* , [8], and the MiniMax algorithm. After obtaining the cycles of these gait sequences, we used the gaits that are within each cycle to train the HMM.

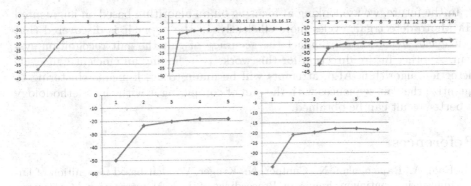

Fig. 7. model2-model6 Baulm-welch training graphs. These training graphs show the accuracy of the gait model trained for each individual.

Table 1. Recognition Results: The log-likelihood of the identification process in the form of a confusion matrix

	Data 1	Data 2	Data 3	Data 4	Data 5	Data 6
Model 1	-20.74	-11.69	-24.80	-60.70	-37.00	-18.66
Model 2	-27.51	-12.81	-10.55	-40.37	-44.00	-21.35
Model 3	-48.04	-23.86	-9.19	-29.45	-46.01	-46.38
Model 4	-36.70	-23.02	-19.79	-19.95	-24.69	-25.74
Model 5	-24.12	-16.99	-22.46	-42.10	-21.05	-23.05
Model 6	-28.95	-18.99	-17.74	-28.16	-44.68	-17.13

In this experiment, we train each gait sequences using a four state Left-to-Right Hidden Markov Model. The transition probabilities of each of the HMM were set to $\{0.25, 0.25, 0.25, 0.25\}$ while the state transition probabilities and the observation probabilities were randomly selected. Graphs in figure 7 show the training of the models to optimal values using Baum-Welch algorithm.

4   Discussions and Conclusions

In this project we articulated an automatic gait recognition system with an online object detection/tracking mechanism. The detection algorithm is working quite efficiently. The gait recognition is working and has been able to recognize the majority of the objects efficiently.

As discussed earlier, the gait videos from model 2 to model 6 were downloaded from a benchmark gait database while gait video for model 1 is our own lab tested video. The proposed system was able to recognize all of the downloaded videos. Only for model 1, which matches its own data (-20.74 for data 1) and also identifies as data 2(-11.69 for data 2), did we have the erroneous identification. However, even for this model the correct match had the second best matching score (-12.81 log-likelihood compared with -11.69 log-likelihood with model 1). All other models are uniquely identified by the system.

In the future, we are looking at focusing on an algorithm based on augmented HMMs to detect gait cycles more efficiently. In the paper it is discovered that gait cycles detection can impact the accuracy of gait the gait recognition system. Another future direction for this works is in finding an efficient gait cycle detection algorithm. Also, attempts will be made to use k-means clustering to quantize the gait sequences with the aim of comparing it with this methodology if better result can be obtained.

References

1. Kale, A., Rajagopalan, N., Cuntoor, N., Kruger, V.: Gait-based recognition of humans using continuous hmms. In: Proceedings of the Fifth International Conference on Automatic Face and Gesture Recognition (2002)
2. Bouchrika, I., Nixon, M.: Gait recognition by dynamic cues. In: Proceedings of the International Conference on Pattern Recognition, Tampa, FL (2008)
3. Harris, G., Smith, P.: Human Motion Analysis: Current Applications and Future Directions. IEEE (1996)
4. Katiyar, R., Pathak, V.K.: Clinical gait data analysis based on spatio-temporal features. International International Journal of Computer Science and Information Security 7 (2010)
5. Rohila, N.: Abnormal gait recognition. International Journal on Computer Science and Engineering (2), 1544–1551
6. Rani, R., Arumugam, G.: An efficient gait recognition system for human]identification using modified ICA. International Journal of Computer Science and Information Technology 2 (2010)
7. Ding, T.: A robust identification to gait recognition. In: Proceedings of the IEEE Conference on Computer Vision and Pattern Recognition, Anchorage, AK (2008)
8. Iwamoto, K., Sonobe, K., Komatsu, N.: A gait recognition method using hmm. In: Proceedings of the SICE Annual Conference, pp. 1936–1941 (2003)
9. Tavakkoli, A., Kelley, R., King, C., Nicolescu, M., Nicolescu, M., Bebis, G.: A visual tracking framework for intent recognition in videos. In: Proceedings of the International Conference on Pattern Recognition, Tampa, FL (2008)
10. Tavakkoli, A., Nicolescu, M., Bebis, G.: A Spatio-Spectral Algorithm for Robust and Scalable Object Tracking in Videos. In: Bebis, G., Boyle, R., Parvin, B., Koracin, D., Chung, R., Hammound, R., Hussain, M., Kar-Han, T., Crawfis, R., Thalmann, D., Kao, D., Avila, L. (eds.) ISVC 2010. LNCS, vol. 6455, pp. 161–170. Springer, Heidelberg (2010)
11. Lee, C.-S., Elgammal, A.: Towards Scalable View-Invariant Gait Recognition: Multilinear Analysis for Gait. In: Kanade, T., Jain, A., Ratha, N.K. (eds.) AVBPA 2005. LNCS, vol. 3546, pp. 395–405. Springer, Heidelberg (2005)
12. Spencer, N., Carter, J.N.: Viewpoint invariance in automatic gait recognition. In: Proc. AutoID 2002, pp. 1–6 (2002)
13. Bhanu, B., Han, J.: Human Recognition at a Distance in Video. Springer (2010)
14. Wang, L., Hu, W., Tan, T.: A new attempt to gait-based human identification. In: Proceedings of the 16th International Conference on Pattern Recognition, pp. 115–118 (2002)
15. Mak, B., Barnard, E.: Attention-based target localization using multiple instance learning. In: Proceedings of the 4th International Conference on Spoken Language, Philadelphia, PA, pp. 2005–2008 (1996)

Motion History of Skeletal Volumes
for Human Action Recognition

Abubakrelsedik Karali and Mohamed ElHelw

Ubiquitous Computing Group, Center for Informatics Sciences,
Nile University, Egypt
a.karali@ieee.org, melhelw@nileuniversity.edu.eg

Abstract. Human action recognition is an important area of research in computer vision. Its applications include surveillance systems, patient monitoring, human-computer interaction, just to name a few. Numerous techniques have been developed to solve this problem in 2D and 3D spaces. However most of the existing techniques are view-dependent. In this paper we propose a novel view-independent action recognition algorithm based on the motion history of skeletons in 3D. First, we compute a skeleton for each volume and a motion history for each action. Then, alignment is performed using cylindrical coordinates-based Fourier transform to form a feature vector. A dimension reduction step is subsequently applied using Principle Component Analysis and action classification is carried out by using Euclidian distance, Mahalonobis distance, and Linear Discernment analysis. The proposed algorithm is evaluated on the benchmark IXMAS and i3DPost datasets where the proposed motion history of skeletons is compared against the traditional motion history of volumes. Obtained results demonstrate that skeleton representations improve the recognition accuracy and can be used to recognize human actions independent of view point and scale.

1   Introduction

Human motion analysis is a key area in computer vision research. Human motion analysis is divided into action recognition and activity recognition. The former typically deals with identifying actions each represented by short, and occasionally periodic, motion patterns such as walking, jumping, running, jogging, .etc. In the latter, long and complex motion patterns are used to identify human activity and group interactions. This paper focuses on human action recognition systems where the goal is to automatically classify ongoing activities in unlabeled videos. The applications of human action recognition systems include surveillance systems, patient monitoring systems, and a variety of systems that involve human-computer interfaces. In the simple case where a video is segmented to contain only one human action, the objective of the system is to correctly classify the video into its action category.

Action and activity recognition approaches are classified into two categories [1] single-layered approaches and hierarchical approaches. Single-layered approaches are approaches that represent and recognize human actions directly based on sequences of

G. Bebis et al. (Eds.): ISVC 2012, Part II, LNCS 7432, pp. 135–144, 2012.

images. On the other hand, hierarchical approaches recognize high-level human activities by describing them in terms of actions, which they generally call sub-events. Single-layered approaches are classified into two types depending on how they model human activities: space-time techniques and sequential techniques. Space-time techniques represent action videos in three-dimensions, 2D video images over time domain, while sequential techniques interpret an action video as a sequence of observations. Space-time techniques are further divided into three categories: space-time spaces themselves, trajectories, or local interest point descriptors.

The proposed approach belongs to space-time spaces techniques. Space-time approaches are first introduced by [2] it was scale invariant, however the algorithms was limited to cyclic actions. Then [3, 4] introduced a view based template approach using Motion Energy Images (MEI) and motion history images (MHI) to indicate presence of motion and the order respectively but this approach was view variant and limited to 2D cases. In [5, 6] they extends the MHI into the 3D space and introduced the Motion History Volumes (MHV). MHV characterizes the body shape and topology as it takes into account the entire volume occupied by the human, and hence better captures the pose than other representations.

In this research, we propose the use of motion history skeletal volumes (**MHSVs**) where skeletonized volumes are used to compute 3D motion history features. MHSVs are view- and scale-invariant and experimental results demonstrate they can be used to build enhanced view-invariant action recognition systems. The motivation beyond using skeletons is that, as will be shown, skeletons provide robust way to encode geometric topology and surface information [7]. The key contribution of the paper is the introduction of a novel motion representation leading to a metric that is used for improved human action recognition in videos. The paper is organized as follows: Section 3 provides necessary definitions. Section 4 describes the proposed work while Section 5 presents obtained results and discussion whereas future work is provided in Sections 5 and 6, respectively.

2   Definitions

2.1   Skeletons and Skeletonization

Skeletons are geometric shape-descriptors that are centered within the shape to capture shape's topology and geometry in a compact manner [8]. They have gained increasing interest in computer science since 1967 when they were used in [8]. They are currently being used for surface smoothing, volumetric animation, and feature-based information in numerous application domains such as computer-aided design, computer graphics, scientific visualization, and medical analysis, just to name a few. The process of computing skeletons is called skeletonization. There are several techniques for skeletonization including thinning [9], Voronoi-diagrams [10], distance fields [11], general fields [12], in addition to a number of hybrid approaches (survey can be found in [13]).

2.2   Motion History and Energy Volumes

A motion history volume (MHV) and motion energy volume (MEV) are 3D representations that inherit from the motion history image (MHI) and motion energy im-
ages(MEI)[4] 2D space representations respectively. MEV and MHV [5] capture
motion information in 3D space. To this end, voxel values are binary values encoding
motion occurrence at a voxel in case of MEV, or multiple-values encoding how recently motion occurred at a voxel in case of MHV. Consider the function $D(x,y,z,t)$,
where $D = 1$, indicating motion at time t and location (x,y,z), then the MHV function
is given by:

$$MHV_T = \begin{cases} if\ (D(x,y,z,t) = 1) & T \\ else & max\ (0, MHV_T(x,y,z,t-1) - 1) \end{cases} \quad (1)$$

where T is the 'motion duration. The associated MEV encodes the occurrence of the
voxel during motion duration in binary volume and can be easily computed by thresholding $MHV > 0$.

3   Methodology

The proposed MHSV-based motion classification algorithm is composed of two stages: (1) MHSV feature extraction, and (2) processing and classification, as illustrated
in Figure 1. In the first stage, actions are input in the form of 3D binary blobs called
visual hulls [14]. Visual hulls are then skeletonised and a motion history volume is
subsequently computed for each action from the skeletonised samples to obtain
MHSV features. In the second stage, MHSVs are processed by applying cylindrical
coordinates Fourier transform to produce rotation-invariant feature vectors. Before
classification, Principle component analysis (PCA) is applied for dimension reduction
by projecting the data along the principle axes that maximize the discrimination between samples. A single Gaussian model is used to represent each action where the
model is described by the distribution mean and variance. Action classification is

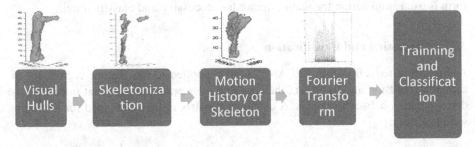

Fig. 1. Flow Diagram of the proposed action recognition algorithm. Starting from a 3D visual
hull, a skeleton is computed, and then MHSV is generated. Fourier transform is applied to
obtain feature vector to be used in the action classification procedure.

then carried out by using 3 different techniques: Euclidian distance, Mahalanobis distance, and Linear Discriminant Analysis (LDA). Details of the proposed algorithm are next described.

3.1 Feature Extraction

Skeletons are computed from three dimensional visual hull following the algorithm presented in [11] where a skeleton is defined as the singularities in the Euclidean distance-to-boundary field, *i.e.* the distance transform. Singularities are the points where the field is non-differentiable. When the distance transform is seen as a height map, the singularities can be seen as the local peaks of the distance transform. In order to extract skeletons, we calculate the divergence of the distance transform and a threshold is used to extract the skeleton and preserve the connectivity of the skeleton parts, Samples of the output of the skeletonization are shown in Figure 2.

Afterwards, for each action we compute the motion history of the skeletal volume (MHSV) based on Equation 1. Generating 64*64*64 element volumes, each motion history is then scaled in time and spatial domains and centered to the occupying volume for the next step. In order to get a rotation invariant feature, MHSVs are then transformed into cylindrical coordinates [6] in which a rotation is represented as translation using:

$$v\left(\sqrt{x^2 + y^2}, \tan\left(\frac{x}{y}\right), z\right) \to v(r, \theta, z), \tag{2}$$

where (x,y,z) and (r, θ, z) are the basis of the original coordinates and the cylindrical coordinates, respectively. The resulting MHSV is centered and scaled to obtain a scale invariant feature. A 3D Fourier transformation is carried out using Equation 3. Details of the use of cylindrical coordinates Fourier transformation are found at [5].

$$V(k_r, k_\theta, k_z) = \int\limits_{-\infty}^{\infty} \int\limits_{-\pi}^{\pi} \int\limits_{-\infty}^{\infty} V(r, \theta, z) \, e^{-j(K_r r + k_\theta \theta \, k_z z)} dr d\theta dz, \tag{3}$$

where k_r, k_θ, k_z are the bases of the Fourier coordinates. The output of Forier transform is used as an action the feature vector for processing and classification.

3.2 Processing and Classification

A dimension reduction using PCA[15] is then applied to the feature vectors for dimension reduction and to project the data along the principle axes that maximize the discrimination between samples. A covariance matrix is first computed for N training samples using:

$$\Sigma(i, j) = \frac{1}{N}(f_i - \mu)\left(f_j - \mu\right)^{\mathrm{T}} \tag{4}$$

where μ represents the average of all feature vectors f, and the superscript T donates the transpose of the matrix. The covariance Σ is then decomposed into the Eigen

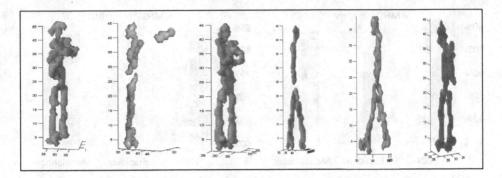

Fig. 2. Samples skeletonization output for from IXMAS dataset, from left to right: "cross arms", "punch" "check watch", "turn around", "walk" and "scratch head "

values and the corresponding Eigen vectors. The Eigen values are then sorted decreasingly and the Eigen vectors matrix is reformed according to the sorted Eigen values resulting into a projection matrix that is used to define a new projective space.

Each action is defined by a single Gaussian model represented by the mean μ and variance σ over all training set for the same action. All testing samples are then projected into the new projective space using the projection matrix and assigned to the class of minimum distance based on:

Euclidian distance, given by:

$$d_e\left(\mu_j, f\right) = \sqrt{\sum_i (\mu_{j_i} - f_i)^2}, \tag{5}$$

where μ_j is the mean of the model represent the class j, f is the test feature vector after projection and the subscript i index for the elements inside of the feature vector.

Mahalanobis distance, given by:

$$d_m\left(\mu_j, f\right) = \sqrt{\sum_i \frac{(\mu_{j_i} - f_i)^2}{\sigma_i}} \tag{6}$$

where μ_j is the mean of the model represent the class j, f is the test feature vector after projection, μ_i is the ith element in the mean μ and f, and σ_i is the variance for this element over all training samples.

Linear Discriminant Analysis (LDA) technique, such as the Fisher Linear Discriminant [16], can be used for classification. According to Fisher discriminant analysis, the separation between classes is the ratio of the variance between classes to the variance within the classes. In our case, if y is the feature vector after PCA, μ_j is the mean of each action model and μ is the average of all samples, the variance matrix within class S_w and between classes S_b is given by:

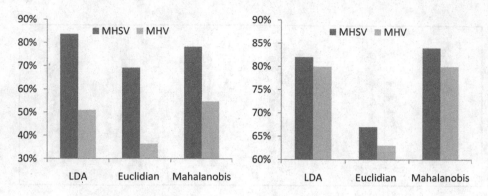

Fig. 3. Classification accuracy using MHSV and MHV on i3DPost (left) and IXMAS (right) datasets using three classification techniques

$$S_w = \sum_{i=1}^{c} \sum_{j}^{ni} (y_i - \mu_i)(y_i - \mu_i)^T \tag{7}$$

$$S_b = \sum_{i=1}^{c} (\mu_j - \mu)(\mu_j - \mu)^T \tag{8}$$

where c is the number of classes and ni is the number of samples per class i. A projection matrix W that maximize Sb and minimize Sw would be equal to the largest Eigen values of $S_w^{-1} * S_b$ [5] and sample z is assigned to the class j that makes Equation 9 minimum.

$$d(\alpha_j, z) = \|\alpha_j - z\|^2 \tag{9}$$

where α is the class mean and z is the feature vector after the LDA.

4   Results and Discussion

We evaluate our approach using two benchmark datasets the IXMAS and the i3DPost. Details of each dataset are as following:

INRIA's IXMAS Motion Acquisition Sequences dataset [17] contains 13 actions each performed 3 times by 10 actors (5 males / 5 females). The acquisition was achieved using 5 cameras. To demonstrate view-invariance, the actors freely changed their orientation for each acquisition. Silhouettes were extracted from the videos using standard background subtraction techniques. Afterwards, visual hulls are carved from a discrete space of voxels of resolution of 64*64*64.

The I3DPost dataset [18] is an action database that has been created by using eight camera setup to produce multi-view videos. Each video depicts one of eight persons (2 females and 6 males) performing one of twelve different human motions, six actions and six activities. The subjects have different body sizes, clothing and are of different sex and nationality. The database contains 104 multi-view videos or 832 (8 × 104) single-view videos. The multi-view videos have been further processed to produce a 3D mesh at each frame describing the respective 3D human body surface. We voxelized the 3D human meshes using mesh rasterization followed by hole filling and then resized it into 64*64*64 volume.

The performance of the proposed MHSV approach is compared against the well-established MHV approach using the above two datasets. MHVs generate view- and scale-invariant features and are thus used in many action recognition systems. For each dataset, action classification is carried out by using LDA, Euclidian distance, and Mahalanobis distance over all actors. The average accuracy of classification is computed and a pooled confusion matrix is subsequently created for each classification technique.

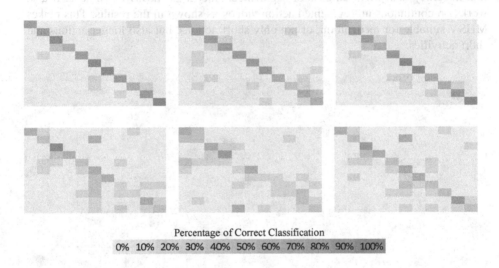

Percentage of Correct Classification

0% 10% 20% 30% 40% 50% 60% 70% 80% 90% 100%

Fig. 4. The confusion matrices between the 11 actions in i3DPost dataset for MHSV (top) and MHV (bottom) employing LDA (right), Euclidian distance (center) and Mahalonobis distance (left)

A leave-one-out routine is used to split action data into learning and testing where one actor is iteratively selected out of the training data and used for testing. As shown in Figure 3, the action recognition accuracy achieved by applying the proposed MHSV approach on the I3DPost dataset is 83.6%, 69%, 78.1% for the LDA, Euclidian distance, and Mahalanobis distance classifications, respectively. The corresponding recognition accuracy for the MHV approach for the same dataset is 50.9%, 36.4%, 54.5%, for the same classification techniques. For the IXMAS dataset, the action recognition accuracy of MHSV is 81.66%, 66.94% and 83.61%, in case of LDA,

Euclidian and Mahalonobis classifications, respectively. The corresponding values for the MHV approach are 79.722%, 62.778% and 79.722%. The confusion matrices in Figure 4 and Figure 5 illustrate the performance of the three action classification techniques when MHSV and MHV are used with the I3DPost and the IXMAS datasets, respectively.

Table 1. Performance of MHSV vs. MVH in case of manually segmented action videos

Classification	LDA	Euclidian	Mahalanobis
MHSV	87.576	74.242	88.788
MHV	82.424	66.061	83.939

As demonstrated in the above results, the proposed MHSV action recognition approach offer improved recognition accuracy than the MHV approach when used on two of the benchmark datasets. This can be attributed to the fact that skeletons provide robust action representation since they encode topology geometry and surface information. The proposed MHSV approach produces better human action recognition accuracy than MHV in case of segmented videos, as shown in Table 2. It also works on continuous un-segmented action videos as shown in the results. This makes MHSV suitable for recognition, of not only short actions, but also longer actions and short activities.

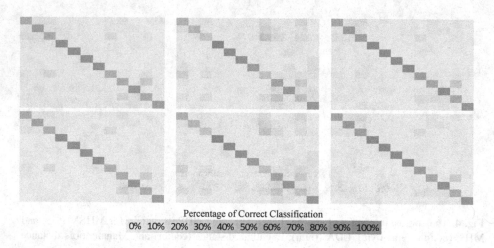

Percentage of Correct Classification

0% 10% 20% 30% 40% 50% 60% 70% 80% 90% 100%

Fig. 5. The confusion matrices between the 12 actions in IXMAS dataset for MHSV (top) and MHV (bottom) employing LDA (right), Euclidian distance (center) and Mahalonobis distance (left)

In IXMAS we reached a better accuracy using 12 actions. And we tested our algorithm against 11 actions in i3DPost and reached a higher accuracy. Table 2 shows the accuracy of the proposed approach against the corresponding accuracies of the state of art approaches on IXMAS and I3DPostdatasets. In [5, 19, 20], it should be noted that only subsequences that maximally represent the action were used in their experiments, and these subsequences were selected manually. These subsequences are generated by splitting each action into sub actions using the change in motion energy. However higher accuracies they achieved, it cannot be generalized because of the lack of automation. From the tables we can see that the accuracy drops by increasing the number of actions.

Table 2. Comparison between our approach against the state of arts aproachs on IXMAS and i3DPost Datasets

IXMAS		Number of actions			I3DPost		Number of actions			
Year	Author	11	12		Year	Author	5	8	10	11
2006	Wienland[5]	93.91%			2009	Gkalelis[25]	90%			
2007	Wienland[20]	81.27%			2010	Iosifidis[26]		91%		
2007	Lv[21]		80%							
2008	Turaga[22]	98.33%			2011	Holte[27]	84%		80%	
2008	Yan[23]	78%			2012	Karali				84%
2011	Liu[24]		82.8%							
2012	Karali		83.61%							

5 Conclusions

In this paper we discussed motion history of 3D skeletons as a novel feature for human action classification. First, we compute the skeleton for each volume, then a motion history for each action. Then alignment is performed using cylindrical coordinates based Fourier Transform forming feature vector. A dimension reduction step is then applied using Principle Component Analysis and finally classification is performed by using Euclidian distance, Mahalonobis distance and Linear Discernment analysis. The proposed algorithm is evaluated on the benchmark IXMAS and i3DPost datasets where the proposed motion history of skeletons is compared against the traditional motion history of volumes. Obtained results demonstrate that skeleton representations improve the recognition accuracy and can be used to recognize human actions independent of view point and scale.

References

1. Aggarwal, J.K., Ryoo, M.S.: Human activity analysis: A review. ACM Comput. Surv. 43(3), 1–43 (2011)
2. Polana, R., Nelson, R.: Recognizing activities. In: Proceedings of the 12th IAPR International Conference on Pattern Recognition, Conference A: Computer Vision & Image Processing, vol. 1 (1994)
3. Bobick, A.F., Davis, J.W.: The recognition of human movement using temporal templates. IEEE Transactions on Pattern Analysis and Machine Intelligence 23(3), 257–267 (2001)
4. Davis, J.W.: Hierarchical motion history images for recognizing human motion. In: Proceedings of IEEE Workshop on Detection and Recognition of Events in Video (2001)
5. Weinland, D., Ronfard, R., Boyer, E.: Free viewpoint action recognition using motion history volumes. Comput. Vis. Image Underst. 104(2), 249–257 (2006)
6. Weinland, D., Ronfard, R., Boyer, E.: Motion history volumes for free viewpoint action recognition. In: IEEE International Workshop on Modeling People and Human Interaction, vol. 104(2) (2005)
7. Tangelder, J.W.H., Veltkamp, R.C.: A survey of content based 3D shape retrieval methods. In: Proceedings of Shape Modeling Applications (2004)

8. Blum, H.: A transformation for extracting new descriptors of shape. In: Models for the Perception of Speech and Visual Form, pp. 362–380. W. Wathen-Dunn

9. Svensson, S., di Baja, G.S.: Simplifying curve skeletons in volume images. Comput. Vis. Image Underst. 90(3), 242–257 (2003)

10. Ogniewicz, R.L., Kübler, O.: Hierarchic Voronoi skeletons. Pattern Recognition 28(3), 343–359 (1995)

11. Bouix, S., Siddiqi, K.: Divergence-Based Medial Surfaces. In: Vernon, D. (ed.) ECCV 2000, Part I. LNCS, vol. 1842, pp. 603–618. Springer, Heidelberg (2000)

12. Ahuja, N., Jen-Hui, C.: Shape representation using a generalized potential field model. IEEE Transactions on Pattern Analysis and Machine Intelligence 19(2), 169–176 (1997)

13. Reniers, D.: Skeletonization and Segmentation of Binary Voxel Shapes, Technische Universiteit Eindhoven (2008)

14. Laurentini, A.: The Visual Hull Concept for Silhouette-Based Image Understanding. IEEE Trans. Pattern Anal. Mach. Intell. 16(2), 150–162 (1994)

15. Abdi, H., Williams, L.J.: Principal component analysis. Wiley Interdisciplinary Reviews: Computational Statistics 2(4), 433–459 (2010)

16. Swets, D.L., Weng, J.J.: Using discriminant eigenfeatures for image retrieval. IEEE Transactions on Pattern Analysis and Machine Intelligence 18(8), 831–836 (1996)

17. 4D Repository, INRIA Xmas Motion Acquisition Sequences (IXMAS). INRIA Xmas Motion Acquisition Sequences (IXMAS),
http://4drepository.inrialpes.fr/public/viewgroup/6 (cited 2012)

18. Gkalelis, N., et al.: The i3DPost Multi-View and 3D Human Action/Interaction Database. In: Proceedings of the 2009 Conference for Visual Media Production, pp. 159–168. IEEE Computer Society (2009)

19. Turaga, P., Veeraraghavan, A., Chellappa, R.: Statistical analysis on Stiefel and Grassmann manifolds with applications in computer vision. In: IEEE Conference on Computer Vision and Pattern Recognition, CVPR 2008 (2008)

20. Weinland, D., Boyer, E., Ronfard, R.: Action Recognition from Arbitrary Views using 3D Exemplars. In: IEEE 11th International Conference on Computer Vision, ICCV 2007 (2007)

21. Fengjun, L., Nevatia, R.: Single View Human Action Recognition using Key Pose Matching and Viterbi Path Searching. In: IEEE Conference on Computer Vision and Pattern Recognition, CVPR 2007 (2007)

22. Turaga, P., et al.: Machine Recognition of Human Activities: A Survey. IEEE Transactions on Circuits and Systems for Video Technology 18(11), 1473–1488 (2008)

23. Pingkun, Y., Khan, S.M., Shah, M.: Learning 4D action feature models for arbitrary view action recognition. In: IEEE Conference on Computer Vision and Pattern Recognition, CVPR 2008 (2008)

24. Liu, J., et al.: Cross-view action recognition via view knowledge transfer. In: Proceedings of the 2011 IEEE Conference on Computer Vision and Pattern Recognition, pp. 3209–3216. IEEE Computer Society (2011)

25. Gkalelis, N., Nikolaidis, N., Pitas, I.: View indepedent human movement recognition from multi-view video exploiting a circular invariant posture representation. In: Proceedings of the 2009 IEEE International Conference on Multimedia and Expo, pp. 394–397. IEEE Press, New York (2009)

26. Iosifidis, A., Nikolaidis, N., Pitas, I.: Movement recognition exploiting multi-view information. In: 2010 IEEE International Workshop on Multimedia Signal Processing, MMSP (2010)

27. Holte, M.B., et al.: 3D Human Action Recognition for Multi-view Camera Systems. In: Proceedings of the 2011 International Conference on 3D Imaging, Modeling, Processing, Visualization and Transmission, pp. 342–349. IEEE Computer Society (2011)

Compressive Matting

Sang Min Yoon[1] and Gang-Joon Yoon[2]

[1] Yonsei Institute of Convergence Technology, Yonsei University, Korea
[2] National Institute for Mathematical Science, Korea

Abstract. Image matting may be defined as the extraction and composition of foreground pixels from a given image using color or opacity estimation. The alpha matting still have their own problems due to statistical optimization from small number of features. In this paper, a novel alpha matting methodology using compressive sensing, which is less dependent on the given trimap and scribbles, is proposed. The experimental results obtained for many complex natural images show that our proposed matting method can provide good mattes from either scribbles or large parts of unknown regions in a trimap.

1 Introduction

Image matting, which involves the extraction and composition of foreground elements from a given image via the estimation of pixels with color or opacity, has been extensively studied over the last 20 years. State of the art alpha matting techniques try to efficiently extract a high degree of mattes from a still image or video sequences that are to be used in film production and broadcasting systems. Here, image matting defines a given input image as the composite of a foreground layer and a background layer, and combines those by using a linear blending of opacity value in each pixel. It is mathematically represented as an observed image I as a convex combination of a foreground image F and a background image B by using the alpha matte α:

$$I = \alpha F + (1 - \alpha)B \qquad (1)$$

where α is a value within [0,1]. From this Eq. (1), the matting process must solve an inverse problem with several unknowns when only given three constraints are given. Thus, the task of alpha matting is to recover the value of α, B, F at every pixel. To properly extract meaningful foreground objects from natural input image, several matting approaches start to segment the input image into three regions which is referred to as a *trimap* which is shown in second image of Figure 1: roughly segmentation of three regions like "definitely foreground", "definitely background", and "unknown" regions. If $\alpha = 1$ or $\alpha = 0$, we classify a pixel within an image as 'definitely foreground' or 'definitely background' respectively. The trimap reduces the dimension of the solution space for the matting problem, and leads the matting algorithm to generate the user-desired result. Given this trimap, these methods typically solve for F, B, and α simultaneously. This is typically done by iterative nonlinear optimization, alternating the estimation of its true (F, B, α) triplets. However, the alpha matting is very dependent on color measurement and opacity optimization within a small sliding window. It also relies on

G. Bebis et al. (Eds.): ISVC 2012, Part II, LNCS 7432, pp. 145–154, 2012.

Fig. 1. Alpha matting from a complex image. From left to right; a complex natural image, user-drawn trimap which is used for alpha matting, high quality matte extracted by using compressive sensing, composite image of the extracted foreground in a new background.

its configuration of *"trimap"* and *"scribbles"*. Although the reported performance of previous alpha matting has improved significantly over the years, their improvements have partly contributed to sophisticated feature extraction procedure, complex statistical model suggestion, and costly optimization techniques. However, most previous alpha matting techniques typically experience difficulty handling image with a significant portion of mixed pixels or when the foreground object has many holes like the example trimap as shown in Figure 1. Such a complex natural image which has rough trimap requires a great deal of user interaction and experience to extract a good matte. They could not extract the meaningful features and optimize its model with well defined statistical analysis.

In this paper, we propose alpha matting using compressive sensing, which can be approximated by compressing sparse data from a given image and trimap. Such a scheme allows for the robust estimation of the unknown alpha value from only a small amount of entities while reducing the ambiguity in whom or what they are.

2   Related Work

Previous matting approaches can be separated into three categories like sampling-based approach, propagation-based approach, and other approaches by automatically reducing the unknowns in matte estimation.

An early sampling-based matting algorithm was proposed by [1], whose approach was based on manifold, connecting the 'frontiers' of each object's color distribution. Based on their approach, Bayesian matting [2] used a continuously sliding window for neighborhood definition, which much inward from the foreground and background area. In order to build color distributions, the algorithm used foreground and background samples additionally to compute F, B, and α. Thus, every pixel within a neighborhood region will contribute to model the foreground and background Gaussian. However, the sampling-based matting is weak when the background color is non-Gaussian. The "knockout matting" [3] method has been developed to avoid the disadvantages of parametric sampling matting by a weighted average of known foreground and background pixels. Generally, these sampling-based matting approaches are very dependent on the nearby known pixels in the sliding window from a tripmap.

For independence of foreground and background color, propagation-based approaches are proposed. The Poisson matting [4] solves a Poisson equation for the matte by assuming that the foreground and background are slowly varying comparing to the matte. This algorithm interacts closely with the user by beginning from a hand-painted trimap offering painting tools to correct errors within the matte. Random walk based alpha matting [5] proposed a technique by minimizing a quadric cost function. GrabCut approach [6] relies on graph cut optimization to extract foregrounds from images. Given a small amount of user-defined bounding boxes, this approach erodes and dilates the extracted foreground region. The closed-from matting techniques [7] solved the problem also by minimizing a cost function from careful analysis of the problem.

3   Our Approach

Alpha matting is identical to an optimization problem: given the mixed color of each pixel, its aim is to reconstruct the optimal value of F, B, and α. To solve this ill-posed problem, we propose a new matting methodology using *compressive sensing technique* to reconstruct its alpha value using small amount of measurements. The compressive sensing that has attracted a great deal of research interest by numerous mathematicians and computer scientists is very efficient to solve the problem of recovery of an unknown vector from the information provided by linear measurement [8] [9]. As inspired by compressive sensing and this based classification theory, we shall explain the details how we extract the alpha values from the natural input image. The sub-scheme to extract the α value can be separated into two steps in this Section; we firstly predict the label of unknown pixel where it is near to background or foreground, and then we measure the alpha value by using sum of mean-squared-errors over all labels.

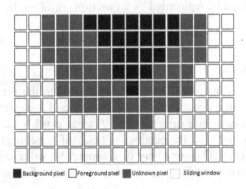

Background pixel   Foreground pixel   Unknown pixel   Sliding window

Fig. 2. Procedure for alpha matting in a sliding window using compressive sensing

Given the mixed color of each pixel, the aim of alpha matting is to assign the optimal value of α to separate the foreground from the image. As inspired by compressive sensing and basic classification theory, we shall explain the details on how we extract the alpha values from the natural image. To know the alpha value of the unknown pixel (painted in red), we slide the window (painted in yellow) as shown in Figure 2.

We estimate its alpha value by using definitely foreground (painted in white) and definitely background (painted in black) color information around the unknown (painted in red).

Given the input image and a specified trimap where the two areas of foreground (F) and background (B) in the image are roughly estimated, to assign an α value to each pixel p ($\notin F \cup G$), our algorithm first assembles a small number of foreground and background samples as candidates for estimating the resemblance of the color information at this position to the nearby foreground and background colors. Let w-neighborhood W_p be the set of all definitely foreground (F_p) (painted in white within a yellow box in Figure 2) and definitely background pixels (B_p)(painted in black within a yellow box in Figure 2) in a rectangle window (Ω_P) centered at pixel p (painted with red). Using the information of foreground and background pixels in W_p, we estimate where the color at the pixel p is closer between F_p and B_p. To the end, we use RGB information as well as the luminance (YUV) data to enhance the accuracy, since we are more sensitive to the changes of brightness and intensity.

To be more precise, let $\mathbb{I}$ be a given color image consisting of $N \times M$ pixels. We can regard $\mathbb{I}$ as a function f defined on $\Omega = \{(n,m) \in \mathbb{Z}_+ : 1 \leq n \leq N, 1 \leq m \leq M\}$ into $\mathbb{Z}_+^3$. That is, for each $x \in \Omega$, there exist three non-negative integers $0 \leq f_R(x), f_G(x), f_B(x) \leq 255$ such that

$$f(x) = (f_R(x), f_G(x), f_B(x)) \tag{2}$$

Here $\mathbb{Z}_+ = \{0, 1, 2, \cdots\}$. Now we assume that F and G are preassigned subsets of Ω corresponding to the foreground and background images in $\mathbb{I}$ (trimap), respectively.

Before going to alpha matting of the unknown pixel, the definitely foreground and definitely background can be segmented by a function $\alpha : \Omega \to [0, 1]$ such that

$$\alpha(x) = \begin{cases} 1, & \text{if } x \in F \\ 0, & \text{if } x \in B. \end{cases} \tag{3}$$

The main issue on alpha matting is how to assign $\alpha(p)$ to the pixel $p \in \Gamma := \Omega \setminus (F \cup B)$. We may presume that the data in foreground and background have their own similarity. In mathematical language, the data $f(p)$ at pixel p can be represented in terms with finite data in $f(F_p)$ (respectively $f(B_p)$) if p belongs to the foreground (resp. background) image. Thus one way to estimate the value $\alpha(p)$ is to check the projections of $f(p)$ into the spaces $\mathcal{F}_p$ spanned by the data in F_p and $\mathcal{B}_p$ spanned by the data in B_p.

Based on this approach, we propose a method to define such a function α via compressive sensing technique. For each $p = (n, m) \in \Gamma$, we construct a rectangular neighborhood Ω_p denoted by

$$\Omega_p = \{(n+i, m+j) \in \Omega : -k_1 \leq i \leq k_2, -\ell_1 \leq j \leq \ell_2\} \tag{4}$$

The positive integers k_1, k_2, ℓ_1 and ℓ_2 are determined to be minimum integers satisfying

$$a := |\Omega_p \cap F| \geq Th_{min} \qquad b := |\Omega_p \cap B| \geq Th_{min} \tag{5}$$

where $|G|$ means the number of the elements in a set G (i.e., the cardinality of G). Th_{min} is the minimum pixel number of the definite foreground and the background images in Ω_p, which is dependent of the pixel p. That is, the size of the sliding window centered at pixel p varies as the pixel p is changed. Let $F_p = \Omega_p \cap F$ and $B_p = \Omega_p \cap B$, and we number the elements in F_p and G_p, which are denoted by $F_p = \{x_1, x_2, \ldots, x_a\}$ and $B_p = \{y_1, y_2, \ldots, y_b\}$. Now, we define two sets of column vectors $\{v_1, \ldots, v_a\}$ and $\{w_1, \ldots, w_b\}$ in $\mathbb{R}^6$ with $v_k = f(x_k)$ for $k = 1, \ldots, a$ and $w_k = f(y_k)$ for $k = 1, \ldots, b$. Let $\mathcal{F}_p$ and $\mathcal{B}_p$ be the subspaces of $\mathbb{R}^6$ spanned by $\{v_1, \ldots, v_a\}$ and $\{w_1, \ldots, w_b\}$, respectively. In this work, we may assume from the similarity that each vector $f(x)$ for $x \in \Omega$ can be well represented as a linear combination of few vectors in $\{f(v) : v \in F\}$ or in $\{f(w) : w \in B\}$. That is, $f(p)$ can be represented as

$$f(p) = \alpha_1 v_1 + \cdots + \alpha_a v_a + \beta_1 w_1 + \cdots + \beta_b w_b \qquad (6)$$

and one of $\{\alpha_k\}_{k=1}^a$ and $\{\beta_k\}_{k=1}^b$ are all zeros or far close to zero. In this aspect, we try to find sparse coefficients $\{\alpha_k\}_{k=1}^a$ and $\{\beta_k\}_{k=1}^b$ satisfying the representation Eq. 6. The vector relation Eq. 6 can be written in the matrix form as

$$f(p) = \begin{pmatrix} f_1(p) \\ f_2(p) \\ \vdots \\ f_6(p) \end{pmatrix} = \begin{pmatrix} v_{11} & \cdots & v_{a1} & w_{11} & \cdots & w_{b1} \\ v_{12} & \cdots & v_{a2} & w_{12} & \cdots & w_{b2} \\ \vdots & \ddots & \vdots & \vdots & \ddots & \vdots \\ v_{16} & \cdots & v_{a6} & w_{16} & \cdots & w_{b6} \end{pmatrix} \begin{pmatrix} \alpha_1 \\ \vdots \\ \beta_1 \\ \vdots \\ \beta_b \end{pmatrix} \qquad (7)$$

By A, we denote the $6 \times (a + b)$ matrix appearing in the right hand side of the Eq. 7 which has the vectors v_k and w_k as its column vectors. In this situation, we want to find the sparest vector $z_0 = (\alpha_1, \ldots, \alpha_a, \beta_1, \ldots, \beta_b)^T \in \mathbb{R}^{a+b}$ satisfying Eq.7:

$$z_0 = \operatorname*{argmin}_{z \in \mathbb{R}^{(a+b)}} \|z\|_0 \qquad \text{subject to} \quad Az = f(p) \qquad (8)$$

where $\|z\|_0$ is the pseudo-norm of a vector $z = (z_k)^T \in \mathbb{R}^{a+b}$ given as the number of nonzero components in z,

$$\|z\|_0 = |\{k : z_k \neq 0\}|. \qquad (9)$$

In this case, it is well-known [8], [10], [11] that compressive sensing is much powerful to find a sparse vector z_0 satisfying Eq.9 by solving the optimization problem

$$z_0 = \operatorname*{argmin}_{z \in \mathbb{R}^{(a+b)}} \|z\|_1 \qquad \text{subject to} \quad Az = f(p) \qquad (10)$$

where the quantity $\|z\|_1$ is given as $\|z\|_1 = \sum_{i=1}^{a+b} |z_i|$.

With the vector $z_0 = (\alpha_1, \ldots, \alpha_a, \beta_1, \ldots, \beta_b)^T \in \mathbb{R}^{a+b}$ obtained by solving the optimization problem Eq. 10, we are now ready to define $\alpha(p)$. We assign to $\alpha(p)$ the value given by

$$\alpha(p) := \frac{(\sum_{k=1}^a |\alpha_k|^2)^{\frac{1}{2}}}{(\sum_{k=1}^a |\alpha_k|^2)^{\frac{1}{2}} + (\sum_{k=a+1}^{a+b} |\beta_k|^2)^{\frac{1}{2}}}. \qquad (11)$$

The alpha value is differently estimated according its label as shown in Eq.11, which is the ratio of the similarity between two classes. Note that $0 \leq \alpha(p) \leq 1$ and if p comes from the foreground image, then $\beta_k = 0$, $k = 1, \ldots, b$, and hence $\alpha(p) = 1$. And if p belongs to the background image, then $\alpha_k = 0$, $k = 1, \ldots, a$, and then $\alpha(p) = 0$. This shows the definition of the quantity $\alpha(p)$ in this situation seems reasonable.

4   Experiments

In this Section, we conduct several experiments to evaluate the effectiveness and robustness of our proposed alpha matting method. This Section is separated into three parts: experimental setup (Section 4.1), alpha matting results from various images (Section 4.2), and comparison of our method against remarkable previous approaches (Section 4.3).

4.1   Experimental Setup

We conducted several experiments to evaluate the alpha matting performance to the various complex natural images. Most previous approaches provided a few successful mattes to verify their methods using their own database. For fair comparison and evaluation test, we used several complex images from alpha matting evaluation website [1]. The foreground objects have a variety of hard and soft boundaries and different boundary lengths. We will show the extracted alpha matte using our proposed method and compare with previous approaches.

For our experiments, we initiated the size of sliding window as 25×25, but we increased the window size when we could not extract known color information within a sliding window because of big holes in a trimap. The processing time to estimate the alpha values was also dependent on resolution of natural input image, number of unknown pixels in trimap, and size of sliding window.

4.2   Evaluation of Our Proposed Alpha Matting

This experiment addresses the problem of alpha matting using our proposed compressive sensing. Figure 3 is the examples of natural input image from the image set (Figure3(a)), user-drawn trimap (Figure3(b)), alpha matte by using our proposed approach (Figure3(c)), and ground-truth alpha (Figure3(d)). This Figure represents that our proposed method is very robust against complex background and rough trimap. Even though the background color is very near to foreground object and their trimaps have numerous big holes, the foreground objects are sharply separated from the background and extract its opacity using compressive sensing technique. For example of third natural image from Figure 3, the the unknown region of the trimap is 31.71% from a given image, so it is very hard to extract its exact alpha value from the natural image like thin stalk of the plant because its sliding window has only few color information near to unknown pixel.

---

[1] http://www.alphamatting.com/

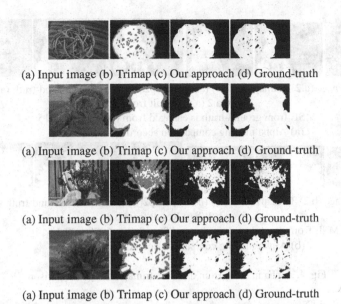

(a) Input image (b) Trimap (c) Our approach (d) Ground-truth

(a) Input image (b) Trimap (c) Our approach (d) Ground-truth

(a) Input image (b) Trimap (c) Our approach (d) Ground-truth

(a) Input image (b) Trimap (c) Our approach (d) Ground-truth

Fig. 3. Alpha matting using our proposed approach and comparison with ground truth

Figure 4 shows the change of its extracted alpha matte according to its trimap. The unknown regions from trimap have increased from 9.7% to 17.0% (Figure 4(a)), but the MSE (Mean Squared Error) from ground-truth is a little bit changed from 0.0012 to 0.0015. In the case of second example, the MSE has only changed from 5.0×10^{-4} to 8.1×10^{-4} when the unknown regions are increased from 10.98% to 18.08%. This Figure proves that our proposed approach is not depedent on the change of user-drawn trimap. We also conduct the alpha matting by using scribbles using only few lines instead of user-drawn trimap. Figure 5 is examples of alpha matting from scribbles. The scribbles using "red" and "yellow" lines in Figure 5 provide less color information than trimap, so some previous approaches failed to extract correct alpha values from few input information. We first extract foreground and background scribbles and then train the given foreground and background pixels using compressive sensing based classification. The total color information of the scribbles are used for input vector to reconstruct matrix and measure the similarity instead of known pixel information from trimap. The opacity of other regions within an input image is determined by compressive sensing based classification and its total errors from all classes.

4.3 Matting Comparison

To evaluate and prove the robustness of our proposed method, we compare our proposed method to previous 15 remarkable matting methods like Poisson matting (Figure 6(c)) [4], bayesian matting (Figure 6(d)) [2], easy matting (Figure 6(e)) [12], iterative BP matting (Figure 6(f)) [13], geodesic matting (Figure 6(g)) [14], random walk matting (Figure 6(h)) [5], high-resolution matting (Figure 6(i)) [15], robust matting (Figure

(a.1) Input image (a.2) Trimap1 (9.7%) (a.3) Trimap2(17.0%) (a.4) Ground-truth (a.5) Result
from a.2 (a.6) Result from a.3

MSE from ground-truth is changed from 0.0012 to 0.0015

(a) Alpha matting comparison according to its trimap

(b.1) Input image (b.2) Trimap1(10.98%) (b.3) Trimap2(18.08%) (b.4) Ground-truth (b.5) Result
from b.2 (b.6) Result from b.3

MSE from ground-truth is changed from 5.0×10^{-4} to 8.1×10^{-4}

(b) Alpha matting comparison according to its trimap

Fig. 4. Alpha matting from different trimap and its comparison

Fig. 5. Alpha matting using scribbles. From input image, we extract foreground and background by using simple lines. The "red" line is foreground, and "yellow" line is the background. Right images show the extracted alpha matte from scribbles.

6(j)) [16], large kernel matting (Figure 6(k)) [17], closed-form matting (Figure 6(l)) [7], learning based matting (Figure 6(m)) [18], real-time shared matting (Figure 6(n)) [19], improved color matting (Figure 6(o)) [20], segmentation based matting (Figure 6(p)) [21], shared matting (Figure 6(q)) [22], and our approach (Figure6(r)), respectively. The red box area from natural input image is zoomed to efficiently visualize the difference between our approach and previous approaches.

As shown in Figure 6, most sampling based matting approaches like Bayesian matting (Figure 6(d)) failed to extract robust foreground pixel because the color of unknown region is ambiguous because its color is very close to foreground object and background. In particular, the "elephant" image, Figure 6(a), has similar color information between "definitely foreground" and "definitely background" areas. Other approaches like propagation based approach have better results than sampling based approaches, but the main problem of their approaches is the blurring effect of the complex area. However, our compressive sensing based alpha matting provide the sharp separation from the ambiguous background like the zoomed area of the input image.

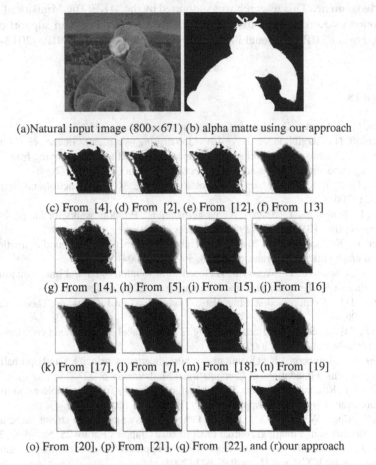

(a)Natural input image (800×671) (b) alpha matte using our approach

(c) From [4], (d) From [2], (e) From [12], (f) From [13]

(g) From [14], (h) From [5], (i) From [15], (j) From [16]

(k) From [17], (l) From [7], (m) From [18], (n) From [19]

(o) From [20], (p) From [21], (q) From [22], and (r)our approach

Fig. 6. Comparison of alpha matting between our approaches and previous remarkable approaches. The zoom in area from an input image is compared to each other. The box which is painted in "red" is our approach.

5 Conclusion

The success of the proposed compressive sensing based alpha matting is attributed to superb capability in feature extraction and similarity measure. The compressive sensing based approach was also very efficient to solve the optimization problem by suppressing the noisy features while preserving its structure in a complex natural image. The alpha value which is estimated by approximating its errors from definitely background and definitely foreground classes was robust in a complex input image with rough trimap or scribbles. In the experimental section, we have demonstrated that our proposed alpha matting is less dependent on its trimap or scribbles. Moreover, we also compared our proposed approach with previous approaches to prove that compressive sensing based alpha matting is more robust where it has complex background and rough trimap.

Acknowledgement. This research was supported by the MKE(The Ministry of Knowledge Economy), Korea, under the IT Consilience Creative Program support program supervised by the NIPA(National IT Industry Promotion Agency) (NIPA-2012-H0201-12-1001).

References

1. Ruzon, M.A., Tomasi, C.: Alpha estimation in natural images (2000)
2. Apostoloff, N., Fitzgibbon, A.: Bayesian video matting using learnt image priors (2004)
3. Berman, A., Vlahos, P., Dadourian, A.: Comprehensive method for removing from an image the background surrounding a selected object. U.S. Patent 6, 134–135 (2000)
4. Sun, J., Jia, J., Tang, C.K., Shum, H.Y.: Poisson matting. ACM Transactions on Graphics 23, 315–321 (2004)
5. Grady, L., Schiwietz, T., Aharon, S., Westermann, R.: Random walks for interactive alpha-matting. In: IN VIIP, pp. 423–429. ACTA Press (2005)
6. Rother, C., Kolmogorov, V., Blake, A.: "grabcut": interactive foreground extraction using iterated graph cuts. ACM Trans. Graph 23, 309–314 (2004)
7. Levin, A., Lischinski, D., Weiss, Y.: A closed-form solution to natural image matting. IEEE Trans. Pattern Analysis and Machine Intelligence 30, 228–242 (2008)
8. Donoho, D.L.: Compressed sensing. IEEE Transactions on Information Theory 52, 1289–1306 (2006)
9. Hsu, D., Kakade, S.M., Langford, J., Zhang, T.: Multi-label prediction via compressed sensing. CoRR abs/0902.1284 (2009)
10. Candès, E.J., Romberg, J.K.: Quantitative robust uncertainty principles and optimally sparse decompositions. Foundations of Computational Mathematics 6, 227–254 (2006)
11. Candès, E.J., Romberg, J.K., Tao, T.: Stable signal recovery from incomplete and inaccurate measurements. Comm. Pure Appl. Math. 59, 1207–1223 (2006)
12. Guan, Y., Chen, W., Liang, X., Ding, Z., Peng, Q.: Easy matting - a stroke based approach for continuous image matting. Journal of Computer Graphics Forum 25, 567–576 (2006)
13. Wang, J., Cohen, M.F.: An iterative optimization approach for unified image segmentation and matting. In: ICCV, vol. II, pp. 936–943 (2005)
14. Bai, X., Sapiro, G.: Geodesic matting: A framework for fast interactive image and video segmentation and matting. International Journal of Computer Vision 82 (2009)
15. Rhemann, C., Rother, C., Acha, A.R., Sharp, T.: High resolution matting via interactive trimap segmentation. In: CVPR, pp. 1–8 (2008)
16. Wang, J., Cohen, M.F.: Optimized color sampling for robust matting. In: CVPR, pp. 1–8 (2007)
17. He, K., Sun, J., Tang, X.: Fast matting using large kernel matting laplacian matrices. In: CVPR, pp. 2165–2172 (2010)
18. Zheng, Y., Kambhamettu, C.: Learning based digital matting. In: ICCV (2009)
19. Gastal, E.S.L., Oliveira, M.M.: Shared sampling for real-time alpha matting. Computer Graphics Forum 29, 575–584 (2010); Proceedings of Eurographics
20. Rhemann, C., Rother, C., Gelautz, M.: Improving color modeling for alpha matting (2008)
21. Rhemann, C., Rother, C., Kohli, P., Gelautz, M.: A spatially varying PSF-based prior for alpha matting. In: CVPR, pp. 2149–2156. IEEE (2010)
22. Moh, H.P.N.C.: Segment-based image matting using inpainting to resolve ambiguities. Master's thesis, Massachusetts Institute of Technology. Dept. of Electrical Engineering and Computer Science (2008)

A Template-Based Completion Framework
for Videos with Dynamic Backgrounds

Tatsuya Yatagawa and Yasushi Yamaguchi

The University of Tokyo / JST CREST
Komaba 3-8-1, Meguro, Tokyo, Japan, 153-8902

Abstract. This paper presents a video completion framework with a novel foreground extraction method based on templates. Video completion is a process of filling in missing regions of a video with appropriate fragments from the input video. In existing video completion techniques, appropriate fragments are effectively detected by separating the video into foreground and background in advance. Any existing foreground/background separation used in the completion methods assumes that the moving objects are foreground. Accordingly, undesirable moving objects are detected as foreground. These moving objects are called "dynamic background". This paper proposes a novel foreground extraction method exploiting templates which are originally used to accelerate video completion itself. Both the efficiency and accuracy of the process are demonstrated over existing methods.

1 Introduction

Completion is a process to fill in a portion of an image or video, which is used for photo editing and movie post production, in particular, for the removal of unnecessary objects in images and videos. Old damaged movies are usually recovered by completion methods as well.

The early work on video completion employs PDEs (Partially Differential Equations) for filling in missing regions of an input video [1]. Colors on the boundary of a missing region are propagated by solving PDEs and the missing region is filled in with the colors gradually. This method is efficient when the missing regions are small and do not occlude moving objects. Another work based on Bayesian motion interpolation restores stationary background occluded partly by moving objects [2]. This method copies a part of other frames into missing region with the interpolated motion fields. However, it cannot restore moving objects occluded by other objects or constantly occluded background.

The first example-based video completion method [3] solves completion problems as global optimization. Missing regions are filled in with the appropriate spatio-temporal fragments detected from the input video. Although this method is able to fill in large missing regions, it is inherently time-consuming.

A recent video completion method involves derivative based local motion estimates [4]. In this method, each appropriate pixel color is interpolated by tracing back the estimated motion field. Although this method is efficient to restore

G. Bebis et al. (Eds.): ISVC 2012, Part II, LNCS 7432, pp. 155–165, 2012.

non-periodic motions, it is strongly affected by noises because the derivatives are used to estimate the motion field. Another completion method has used a tracking based approach [5]. This method is tolerant of noises because it utilizes mean shift tracking. However, this tracking based search is not efficient enough to complete larger size of videos.

Video completion has been accelerated by a precomputation, that is, foreground/background subtraction. One method which consists of the separation is based on summarized images of foreground and background, which is called "mosaic images" [6]. For foreground completion, a mosaic image is used to reduce the search space of fragments. Furthermore, this method restores the static background by simply copy pasting fragments from the background mosaic.

Another method makes use of "object templates," which are small images containing foreground objects [7]. A sequence of object templates represents a motion of foreground objects and is used to restore foreground objects by dynamic programming. In the latest work [8], the foreground/background separation also plays an important role to address video completion problems.

These methods involving foreground/background separation have succeeded in reducing computational cost remarkably and in providing natural completion results. However, the video samples of these methods include only static background such as a grass field, a street or a room scene. Unfortunately, it is difficult for these methods to handle videos with moving objects in a background scene because the foregrounds in their context are the objects with periodic motions. Therefore, irregularly moving objects such as waves, blown tree branches or swaying curtains cannot be dealt with by these methods.

In this paper, we present a novel video completion framework which is able to handle videos with dynamic backgrounds. This framework takes advantage of the object templates to extract foreground objects. In addition, it introduces further speed up into the existing completion method. Finally, some experiments are made to compare our results to those of conventional methods. The contributions of our work are as follows:

- A novel approach for efficiently extracting foreground objects with periodic or locally similar motions by modeling and tracking these objects (in Section 3). This extraction is performed only within the templates for the purpose of improving its computational efficiency and accuracy.
- A technique to speed up the existing completion method by using multi-scale computation of template dissimilarities (in Section 4.1).
- A simple solution based on the work [6] for dynamic background completion (in Section 4.2).

2   Algorithm Overview

Our framework is associated with the work [7]. Their completion framework makes use of object templates to restore moving foreground objects. The completion problem is treated as an energy minimization problem which can be solved by dynamic programming. Each object template consists of a window

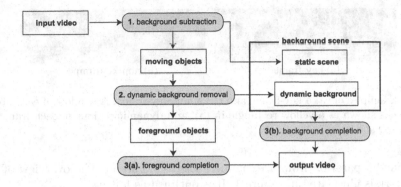

Fig. 1. An overview of our framework. Rounded rectangles represent the processes.

surrounding foreground objects and a binary mask that represents a shape of the objects. The proposed method utilizes the object templates also for foreground extraction. An object template has rich information of foreground objects such as their shapes, positions, and colors. This information is helpful for determining whether a pixel is a part of foreground objects.

In the existing work, foreground objects are extracted with the statistics-based background subtraction algorithm [9]. It is not applicable to videos which include moving objects in background, namely dynamic backgrounds, because the subtraction method assumes that foreground objects are moving and a background scene is static. Consequently, a certain criterion is necessary to distinguish foreground objects from dynamic backgrounds. For this purpose, a guide of a user is often helpful. For example, the video snapcut [10] is one of the most beneficial methods to extract an object which is intended by a user. However, the amount of a user's task increases as the number of frames contained in a video. In addition, it is not obvious for a user whether the selected foreground objects are desirable for current completion methods. In other words, foreground objects extracted by user interaction do not always lead a good result.

Our framework aims at extracting moving objects with periodic motions as foreground objects because objects with such motions are efficiently restored by existing completion methods. In the original work [7], a user has to assign IDs to moving objects for computing object templates. By contrast, a user only has to set two kinds of parameters to do it in our framework. One is for the size of foreground objects and the other is for the extent of foreground objects over a video frame. These two values are employed for evaluating the accuracy of foreground extraction. When the extracted region is a cluster of pixels with a certain area size, it is considered to consist of only foreground objects. A frame which yields this successful extraction is regarded as a "key-frame." Key-frames are used as seeds to track moving objects, and it accomplish automatic

| (a) key-frame | (b) non-key-frame |

Fig. 2. Examples of (a) a key-frame and (b) a non-key-frame. A window of each object template is shown as a yellow rectangle in (a), and dynamic regions in each frame are colored red and green.

foreground separation from a dynamic background scene. The overview of our framework is illustrated in Figure 1. It is outlined as follows.

1. Moving objects are extracted by a simple background subtraction. The input video is separated into moving objects and a static scene.
2. Actual foreground is extracted by the template-based dynamic background removal. In this step, key-frames that contains only foreground objects are detected from input video frames. At the beginning, object templates are computed for the key-frames. Then, dynamic backgrounds contained in the other frames are removed by using the object templates.
3. After foreground extraction, object templates computed for all input frames are directly adopted to foreground completion (a) with the method [7]. Mosaic images used in [6] are applied to background completion (b).

3   Template-Based Dynamic Background Removal

Moving objects are separated with a conventional background subtraction. A background image is estimated by averaging a series of pixels on the same spatial position in all input frames. For a video recorded with a moving camera, a global motion is computed to align frames. A set of pixels is separated as moving objects when the pixels have either different colors from the estimated background or optical flows of a certain magnitude. In order to reduce noise, both the color difference and the flow magnitude are convoluted with Gaussian filters in advance. Tiny motions of objects in the scene are ignored by this convolution.

After separating moving objects, some frames are selected as key-frames which contains only actual foreground objects. Here, we assume there exist several frames which do not include dynamic backgrounds. Key-frames are selected for each moving objects. The examples in this paper contain more than one moving object, and they have to be identified. In our implementation, moving objects are identified with the cues of optical flows by a simple clustering algorithm k-means++ [11].

To select key-frames, the following constraints proposed in the work [12] are employed:

$$\text{Size}(I_i^{dyn}) > minArea, \tag{1}$$

$$\text{Var}(\{(x,y)|(x,y) \in I_i^{dyn}\}) < maxSpan, \tag{2}$$

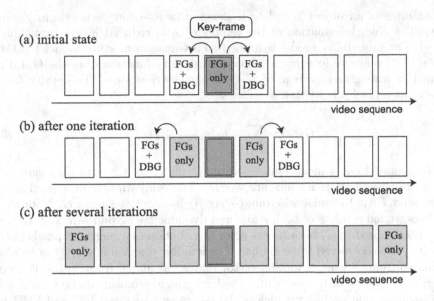

Fig. 3. The scheme of template based dynamic background removal. As the computation proceeds, dynamic backgrounds (DBGs) are removed from moving objects. As a result, only foreground objects (FBs) remain in all frames.

where I_i^{dyn} is the dynamic region in the i-th frame. Size($\cdot$) represents the number of pixels in the region ($\cdot$), Var($\cdot$) represents the determinant of the covariance matrix calculated from the input set of coordinates. The parameters, $minArea$ and $maxSpan$, are specified by a user. The former represents a lower bound of the area size while the latter represents an upper bound of the variance value of a foreground object. In the original work, key-frames represent the frames that have moving parts of target objects, contrarily, those in our framework represent the frames that have only foreground. Examples of a key-frame and a non-key-frame are shown in Figure 2. Moving objects in a key-frame distribute only around a foreground object while those in a non-key-frame spread over the frame.

An input video has several key-frames. An object template is computed for each key-frame. Each object template consists of a window that surrounds foreground objects and a binary mask that represents the shape of foreground objects. A window of each object template is determined to have enough size to include foreground objects. In our experiment, we use 1.5–2.0 times as large size as the average size of moving objects in all key-frames.

Dynamic backgrounds in non-key-frames are removed by a segmentation-based method using a probability model. This process is proceeded frame by frame and is starting from the key-frames. The scheme of this process is illustrated in Figure 3. Let O_i be an object template in the i-th frame which is one of the key-frames, and M_i be the corresponding binary mask of O_i.

An example of an object template is denoted as a yellow rectangle in Figure 2 (a). The color information of the template O_i is reduced into a probability model. The probability model is formed as a Gaussian mixture model (GMM) of the RGB colors of foreground objects. A density function G of the GMM is defined as a weighted average of K Gaussian distributions. The density for a color c_p at a pixel p is written as follows:

$$G(c_p) = \sum_{k=1}^{K} \omega_k g_k(c_p | \mu_k, \Sigma_k), \tag{3}$$

where μ_k and Σ_k is a mean and a covariance matrix of the k-th Gaussian distribution g_k, and ω_k is its mixture weight. The compatibility of a pixel p for foreground $Pr(p|\mathcal{F})$ and background $Pr(p|\mathcal{B})$ is calculated from G. Note that the background consists of both static and dynamic backgrounds in key-frames. G_{i+1} is initialized to G_i because the color distribution of foreground pixels in O_i and O_{i+1} are considered to be similar. The window position of O_{i+1} can be also estimated from O_i using a motion vector that is calculated from optical flows of foreground objects with mean shift. The background removal can be formulated as a segmentation problem, which is the assignment of a label $\mathcal{L}_p \in \{\mathcal{F}, \mathcal{B}\}$ to each pixel p. This problem can be solved by minimizing the Gibbs energy for segmentation [13]. The energy is defined as follows:

$$E_{segm}(O) = U(O) + \lambda V(O), \tag{4}$$

where U and V are called "data term" and "smoothness term" respectively, and λ is balancing the two terms. They are written as follows:

$$U(O) = \sum_{p \in O} Pr(p|\mathcal{L}_p), \tag{5}$$

$$V(O) = \sum_{p \in O} \sum_{q \in N_p} \exp(-\beta \|c_p - c_q\|^2), \tag{6}$$

where N_p denotes the neighboring pixels of p , and β is the constant parameter. By this segmentation, the final binary mask M_{i+1} is computed. After that, the template segmentation proceeds to $O_{i+2}, O_{i+3}, ...$, finally, dynamic background is removed from entire video frames. Overall algorithm of the template-based dynamic background removal is outlined in Algorithm 1.

4   Video Completion

4.1   Foreground Completion

Object templates have been computed for all video frames after the dynamic background removal. The template-based completion framework in [7] is seamlessly adapted. In their framework, a dissimilarity function $d(O_i, O_j)$ between two templates O_i and O_j is defined as follows:

$$d(O_i, O_j) := \min_{\mathbf{m}}[E_O(O_i, O_j; \mathbf{m}) + E_N(O_i, O_j; \mathbf{m})], \tag{7}$$

Algorithm 1. Template-based dynamic background removal

Input: the list of templates generated from key-frames.
Output: foreground extraction result.

1: **while** the template list is not empty **do**
2:      Pop a template O_i from the list.
3:      Initialize the foreground model G_{i+1} to G_i.
4:      The position of O_i is translated into that of O_{i+1} by the motion vector.
5:      Determine the binary mask M_{i+1} by minimizing the Gibbs energy defined in
        Equation 4.
6:      Update G_{i+1} and the positions of O_{i+1}.
7:      Add templates O_{i+1} to the template list.
8:      Repeat 3–7 for O_{i-1}.
9: **end while**

---

where $\mathbf{m}$ is an alignment vector between O_i and O_j. Two energy functions E_O and E_N represent the energy for the overlapping area and the non-overlapping area, respectively. The energy functions E_O and E_N are given as below:

$$E_O(O_i, O_j; \mathbf{m}) := \sum_{\mathbf{p} \in O_i} [O_i(\mathbf{p}) - O_j(\mathbf{p} + \mathbf{m})]^2 M_i(\mathbf{p}) M_j(\mathbf{p} + \mathbf{m}), \tag{8}$$

$$E_N(O_i, O_j; \mathbf{m}) := 255^2 \sum_{\mathbf{p} \in O_i} M_i(\mathbf{p})[1 - M_j(\mathbf{p} + \mathbf{m})]. \tag{9}$$

In these definitions, the object template O_i represents a mapping of an intensity value of a pixel at a position $\mathbf{p}$. The binary mask M_i takes the value 1 when a pixel at a position $\mathbf{p}$ is foreground, otherwise takes 0.

In order to compute the dissimilarity $d(O_i, O_j)$, the alignment vector $\mathbf{m}$ has to be determined. This dissimilarity measure was a bottleneck of the foreground completion. The problem of determining the alignment vector can be regarded as a pattern matching problem. Here, we introduce a multi-scale solution into this pattern matching problem for acceleration. The dissimilarity function is firstly computed in a coarse level of a Gaussian pyramid of the templates. The alignment vector in a coarse scale is used as a cue in finer scales. As a result, the computation of the dissimilarity is accelerated. By the multi-scale solution, computation of the foreground completion is about 2^l times accelerated with a Gaussian pyramid of l levels.

4.2   Background Completion

At the end of our framework, the background scene is restored, and this process is based on the work [6]. Their foreground completion technique which allows moving objects to be restored is adopted to the background completion in our framework.

A background mosaic image is calculated by averaging color values of pixels in both static and dynamic backgrounds. In addition, an optical flow mosaic

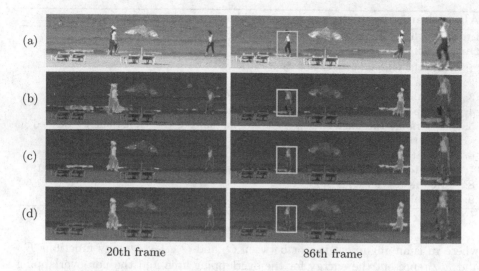

(a)

(b)

(c)

(d)

20th frame   86th frame

Fig. 4. Comparisons of the foreground extractions. Input frames are shown in the row (a), results of the background subtraction method [14] are in (b), results of the simple background subtraction before the dynamic background removal are in (c), and our extraction results are in (d).

is also calculated by averaging flows of the pixels. An appropriate fragment for restoring a missing region is detected with these mosaic images. Pixels in mosaic images are more or less influenced by the color or optical flow of original pixels in the input video frames. The possibility of a spatial position of an appropriate video fragment can be limited to the positions where the dissimilarities for pairs of image fragments in the mosaics are small. The appropriate video fragment is chosen from those on the limited positions of the input video.

For the static regions, we can copy and paste a fragment in the estimated background, however, the estimated background is often over-smoothed. In order to avoid over-smoothed completion results, static regions are also completed by the same manner as that for dynamic backgrounds.

5   Experimental Results

The dynamic background removal results are shown in Figure 4. The comparison is made toward the background subtraction method [14] which is improved solution of the work [9]. The top row shows input frames, the second row shows the results of the background subtraction [14], the third row shows the results of a simple subtraction before the dynamic background removal, and the bottom row shows the results of our template-based background subtraction.

According to Figure 4, dynamic backgrounds, namely waves, are accurately removed in only the result of our approach. Therefore, we may claim that our template-based approach is powerful to remove dynamic backgrounds.

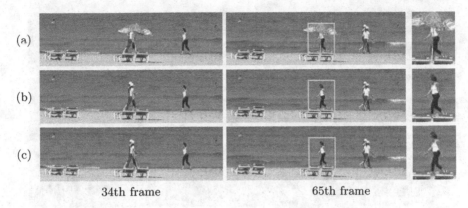

(a)

(b)

(c)

34th frame   65th frame

Fig. 5. Results of video completion are compared. Input video frames are shown in the row (a), and an umbrella in the center is removed with the method [3] in (b), and with our method in (c).

The results of video completion are shown in Figure 5 and 6. In both figures, the top row shows input frames, the second row shows the results of the completion method [3] and the bottom row shows the results of our framework. The sample shown in Figure 5 has a missing region which covers an umbrella in the center, thus the missing region is static. In contrast, the other sample shown in Figure 6 has a missing region which covers a man jogging from left to right, thus the missing region is moving. For both static and dynamic missing regions, our framework has achieved as accurate completion results as the method [3]. In our implementation, the computational time for the example in Figure 5 is less than 6 minutes for the entire process. On the other hands, the method [3] takes at least 4 hours. For both examples, each experiment has been executed by a single core process of the 2.67 GHz Core i7 CPU.

6   Discussion

Our framework can provides as accurate results as those of the method [3], in addition, the overall computational cost of our framework is much less than that. For the best of our knowledge, the method [3] was only the method that can handle videos including dynamic backgrounds, whereas it costs much time for the computation. Our framework have enabled the video completion method [7] to process videos with dynamic backgrounds. Hence, completion for such videos is remarkably accelerated.

Although our framework is helpful to restore videos with dynamic backgrounds, unnatural completion result might be provided in a particular situation. Background completion in our framework is based on mosaic images proposed in the work [6]. Information in each input frame is sometimes over attenuated in these mosaic images, which results in the difficulty of estimating accurate positions of

48th frame	58th frame	68th frame

Fig. 6. In this example, the man jogging from left to right is removed from (a) input frames. The frames in (b) show the results of the method [3], and the frames in (c) show the results of our method.

appropriate fragments. Eventually it would be hard to restore dynamic backgrounds even though the dynamic backgrounds are separated properly.

Moreover, our framework spends about 90% of computational time for background completion. Although the mosaic images can reduce the search space of fragments, this reduction is not enough to restore videos of a larger size.

Limitations. There are several limitations in our framework. One of the limitations is about background scenes. Our framework does not accept videos over which dynamic backgrounds are constantly moving, because we assume that at least a few frames do not include dynamic backgrounds. Although the seashore scene in our examples can be regarded to contain a constantly dynamic background, most of the motions are tiny enough to be ignored with the Gaussian convolution. Another limitation is about foreground objects. The foreground objects have to repeat a particular motion, in other words, an appropriate motion for missing region has to be copied from somewhere in the input video. This limitation is common in other completion algorithms.

7   Conclusion

In this paper, we present a template-based foreground extraction method and a completion framework for videos with dynamic backgrounds. Accordingly, the proposed framework succeeds in accelerating completion for videos with dynamic backgrounds. As mentioned earlier, our framework still have challenging problems. One is about computational cost and the other is about robustness of the result. We will address these problems for future work.

Acknowledgement. This work was partially supported by JSPS Grant-in-Aid for Scientific Research (B) (22300030).

References

1. Bertalmio, M., Bertozzi, A.L., Sapiro, G.: Navier-stokes, fluid dynamics, and image and video inpainting. In: Proc. of IEEE Conference on Computer Vision and Pattern Recognition, vol. 1, pp. I-355-I-362 (2001)
2. Kokaram, A., Collis, B., Robinson, S.: Automated rig removal with bayesian motion interpolation. IEE Proc. on Vision, Images and Signal Processing 152, 407-414 (2005)
3. Wexler, Y., Shechtman, E., Irani, M.: Space-time completion of video. IEEE Trans. on Pattern Analysis and Machine Intelligence 29, 463-476 (2007)
4. Shiratori, T., Matsushita, Y., Tang, X., Kang, S.B.: Video completion by motion field transfer. In: Proc. of IEEE Conference on Computer Vision and Pattern Recognition, vol. 1, pp. 411-418 (2006)
5. Jia, Y.T., Hu, S.M., Martin, R.R.: Video completion using tracking and fragment merging. The Visual Computer 21, 601-610 (2005)
6. Patwardhan, K., Sapiro, G., Bertalmio, M.: Video inpainting under constrained camera motion. IEEE Trans. on Image Processing 16, 545-553 (2007)
7. Venkatesh, M., Cheung, S., Zhao, J.: Efficient object-based video inpainting. Pattern Recognition Letters 30, 168-179 (2009)
8. Ling, C.H., Lin, C.W., Su, C.W., Chen, Y.S., Liao, H.Y.: Virtual contour guided video object inpainting using posture mapping and retrieval. IEEE Trans. on Multimedia 13, 292-302 (2011)
9. Horprasert, T., Harwood, D., Davis, L.S.: A statistical approach for real-time robust background subtraction and shadow detection. In: Proc. of IEEE Frame-Rate Applications Workshop, vol. 99, pp. 1-19 (1999)
10. Bai, X., Wang, J., Simons, D., Sapiro, G.: Video snapcut: robust video object cutout using localized classifiers. ACM Trans. on Graphics 28, 70:1-70:11 (2009)
11. Arthur, D., Vassilvitskii, S.: k-means++: the advantages of careful seeding. In: Proc. of Annual ACM-SIAM Symposium on Discrete Algorithms, pp. 1027-1035 (2007)
12. Liu, F., Gleicher, M.: Learning color and locality cues for moving object detection and segmentation. In: Proc. of IEEE Conference on Computer Vision and Pattern Recognition, pp. 320-327 (2009)
13. Boykov, Y.Y., Jolly, M.-P.: Interactive graph cuts for optimal boundary & region segmentation of objects in n-d images. In: Proc. of IEEE International Conference on Computer Vision, vol. 1, pp. 105-112 (2001)
14. Zivkovic, Z.: Improved adaptive gaussian mixture model for background subtraction. In: Proc. of International Conference on Pattern Recognition, vol. 2, pp. 28-31 (2004)

3D Action Classification Using Sparse Spatio-temporal Feature Representations

Sherif Azary and Andreas Savakis

Computing and Information Sciences and Computer Engineering
Rochester Institute of Technology, Rochester, NY 14623

Abstract. Automatic action classification is a challenging task for a wide variety of reasons including unconstrained human motion, background clutter, and view dependencies. The introduction of affordable depth sensors allows opportunities to investigate new approaches for action classification that take advantage of depth information. In this paper, we perform action classification using sparse representations on 3D video sequences of spatio-temporal kinematic joint descriptors and compare the classification accuracy against spatio-temporal raw depth data descriptors. These descriptors are used to create overcomplete dictionaries which are used to classify test actions using least squares loss L_1-norm minimization with a regularization parameter. We find that the representations of raw depth features are naturally more sparse than kinematic joint features and that our approach is highly effective and efficient at classifying a wide variety of actions from the Microsoft Research 3D Dataset (MSR3D).

1 Introduction

Action classification in video is the process of labeling a sequence of movements generated by humans performing a task or function. Automatic classification of human actions is a fundamental but challenging task in computer vision research for applications such as surveillance and human computer interaction. To complicate the difficult task of automatic action classification, human actions are unconstrained and can drastically vary among individuals in terms of their shapes, sizes, timing of action executions, viewing angles, occlusions, lighting conditions, and other noise.

One of the challenges with action classification is establishing an accurate and efficient descriptor given the vast amount of data in a video sequence of the action being performed. Spatial action representations, such as body models [1], body pose estimations [2], kinematic joint models [3] and stick figures [4] tend to be intuitive and descriptive but have difficulty in accounting for the human body's high degree of variability in shape and size. Spatial parametric image features such as contour/silhouette representations [5], optical flow [6], and motion history images/motion energy images [7] don't require body part labeling or tracking, but are computationally intensive and have difficulties with occlusions. Spatial statistical models such as Bag of Features [8], Support Vector Machines (SVM's) [9], Principal Component

G. Bebis et al. (Eds.): ISVC 2012, Part II, LNCS 7432, pp. 166–175, 2012.

Analysis (PCA) [10], and Manifold Learning such as supervised locality preserving projections (sLPP) [11] account for occlusions, noise, and outliers but require carefully selecting features that describe actions in the temporal domain.

Some approaches towards temporal description of human actions revolve around the idea of feature concatenation in the time domain. The work of Junejo et al. [12] tackles a view independent approach to action recognition on 2D video sequences using Self Similarity Matrices (SSM). The approach is to capture temporal histograms of gradient orientations in the spatial domain and concatenate the features descriptors into one large local SSM feature vector descriptor. Other approaches are based on spatio-temporal interest points such as cuboids using temporal Gabor filters [13], Harris 3D detectors as a 3D extension of the Harris corner detector for detecting significant local variations on both space and time [14], and Hessian detectors that are scale and affine invariant across the space and time domains [15]

One common theme across the many different approaches to action classification is the issue of processing large amounts of data to describe a human action. In many real-world systems, data is often sparse which means that a small portion of the data can describe the entire system, which would be beneficial in reducing high-dimensional data. The theory stems from the *Pareto Principle*, a phenomenon that in any population contributing to some common effect only a few members of the population actually contributes to the majority of the effect [16]. This phenomenon can be observed in a wide variety of applications including economics [17], biology [18], and social networks [19]. In the computer vision field, sparse representations have been used for face recognition [20], super-resolution [21], denoising [22], and image classification [23].

Zhang et al. [24] use sparse representations and Bag of Words of spatio-temporal feature descriptors which are projected into a lower dimensional space using PCA and apply L_1-minimization to classify actions. Their approach has an accuracy of 86.5% on the 2D KTH action dataset. Liu et al. [25] use motion context descriptors to represent frame description and motion context and find sparse representations for training and testing. When applied on the 2D Weizmann dataset they achieve an action classification accuracy of 96.77%.

In this paper, we utilize sparse representations to efficiently represent high dimensional signals of human actions. We apply sparse representations on 3D video sequences using L_1-minimization because of its effectiveness in approximating linear models and use this approach for classifying actions. The remainder of this paper is organized as follows: Section 2 discusses feature descriptors for 3D action sequences and sparse representations and error minimization methods. In Section 3, we describe the MSR3D dataset, the experimental setup, and the recognition results. Finally, Section 4 concludes the paper.

2   Sparse Representation of Actions in Depth Video Sequences

2.1   Feature Descriptors of Actions

Our goal is to define feature descriptors which represent an over-complete dictionary of human actions from depth data, meaning that the dimension of the feature vector is larger than the dimension of the input. We selected two distinct feature descriptors for comparison and evaluation: kinematic joint coordinates and raw depth data.

The 3D video sequences allow for the calculation of kinematic joint positions of the individual conducting an action using an approach inspired by Shotton et al [26]. They present the core component of the Kinect gaming platform, where 3D positions of body joints are predicted from a single depth camera using randomized decision forest classifiers for body part labeling. Specifically, mean-shift is used to classify each pixel in an image using spatial mode distribution along with the randomized decision forests to propose 3D joint positions. The approach is invariant to pose, body shape, and clothing. The MSR3D dataset used for our experiment provides 20 kinematic joint coordinates per frame of the forehead, neck, shoulders, chest, torso, elbows, hips, hands, wrists, knees, and feet. These kinematic coordinates are captured into a feature vector after the joint coordinates are subtracted from the center torso of the human to define relative data to account for localization invariance. The difference between the coordinates and the torso coordinates are then normalized to define features which are scale invariant.

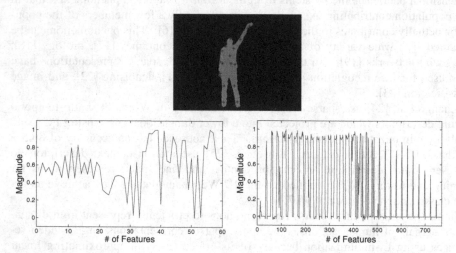

Fig. 1. Example frame of a test subject performing a waving action in 3D space with kinematic coordinates. The bottom left plot shows the kinematic feature descriptor for the frame instance while the bottom right plot shows the depth surface descriptor for the frame instance.

We utilize features extracted from raw depth data by determining the largest connected object in the scene and defining a bounding box around that region of interest. The raw data is read from the scene, scaled to a constant feature size, and normalized to obtain a feature descriptor that is invariant to scale and localization.

Given an image representing a single frame of a video sequence, we extract a feature descriptor a of m-dimensions for a single video frame $a \epsilon R^m$ as shown in Fig. 1. A spatio-temporal surface feature descriptor is generated taking into account the temporal aspect of the action $a_{spatio-temporal} \epsilon R^{m \times t}$ where t is the time duration of the action. To account for variance in action execution time, the surfaces features are resized to 40 frames using bicubic interpolation. An example of an individual executing a hand wave is shown in the surface plots of Fig. 2 for kinematic joint coordinates and raw 3D surface data respectively. The surface descriptors are then vectorized to represent a single human action feature descriptor with a fixed executing time while maintaining invariance to scale and localization in both cases.

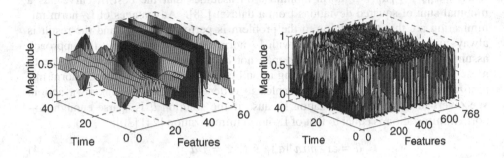

Fig. 2. The left feature action surface plot shows the kinematic surface plot for which there are 60 features per frame over 40 frames. The right feature action surface plot shows the raw 3D depth information of 768 features across 40 frames.

2.2   Sparse Representation Classification

Sparse representations are very useful because of their dimensionality reduction properties and resulting efficient representation of high dimensional signals. Sparse representations uncover semantic information following the principle that high dimensional data associated with the same class exhibit a degenerate structure and therefore share common low-dimensional properties [27].

We are given a sparse matrix $D = [D_1, D_2, ..., D_p] \epsilon R^{m \times n}$ representing an overcomplete dictionary for training of n-action samples, each of m-dimensions, with p-separate action classes. A test sample $x \epsilon R^m$ of an action class in the dictionary lies in the linear range of the training samples of that same action class p and the problem becomes a linear equation. Given a test sample x and the sparse matrix D, a linear representation is defined as:

$$x = Da \tag{1}$$

where $a \epsilon R^n$ is a coefficient vector for which non-zero coefficients represent the pth action class in the linear range. Corruption and occlusions can complicate the action classification process affecting the coefficient vector representation [27] by either providing no unique solution or allowing many solutions. Least squares minimization approaches can be used to address the issue. If there is a large number of action classes p, the coefficient representation is naturally sparse [24] and ideally we can find the sparsest solution using L_0-norm minimization:

$$\hat{a} = \arg min \, \|a\|_0 \,\, s.t. \,\, x = Da \tag{2}$$

where $\|a\|_0$ counts the number of non-zeros in vector a. However, the system is underdetermined and finding the sparsest solution is NP-hard. L_2-norm minimization or Euclidean norm is a least squares minimization approach based on:

$$\hat{a} = \arg min \|a\|_2^2 \ s.t. \ x = Da \tag{3}$$

where $\|a\|_2^2 = \sum_i a_i^2$. L_2-norm minimization assumes that the best-fit curve has a minimal sum of squared deviations from a dataset [28]. Advantages of L_2-norm minimization are that the solution to the problem is performed easily and the result is always unique. However, an issue with L_2-norm minimization is that the approach assumes a normal distribution which may not be the case for collected data due to noise and errors in the dataset resulting in outliers [29]. In other words, L_2-norm minimization utilizes all available examples in order to identify the unknown action. We revert to L_1-norm minimization because if the solution $\hat{a}$ is sparse enough, L_0-norm minimization is equal to that of L_1-norm minimization [24] [30]:

$$\hat{a} = \arg min \|a\|_1 \ s.t. \ x = Da \tag{4}$$

where $\|a\|_1 = \sum_i |a|$. L_1-norm minimization promotes sparse solutions and can be reformed as a convex linear programming optimization method. Furthermore, L_1-norm minimization is an effective technique for solving underdetermined systems of linear equations [31] and concentrates on few non-zero coefficients making the approach robust with built-in outlier detection. The reconstruction coefficients from L_2-norm minimization and L_1-norm minimization are shown below to demonstrate the promotion of sparse solutions using L_1-norm minimization.

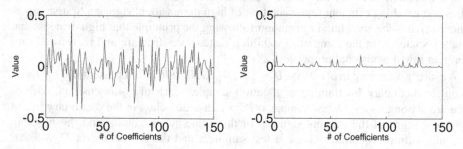

Fig. 3. These plots show the reconstruction coefficients of an action on a dictionary of training examples. The left plot shows the L_2-norm minimization coefficients and the right plots shows the L_1-norm minimization coefficients which are sparse.

There are many methods for L_1-norm minimization and we use the following *Least Squares Loss* method with an added regularization term:

$$\hat{a} = \arg min \|Da - x\|_2^2 + \lambda \|a\|_1 \ s.t. \ x = Da \tag{5}$$

where λ is L_1-norm regularization parameter or a penalty parameter. The role of the regularization term is to select the solution that achieves sparser solutions by providing low variance feature selection, improved approximations, and more interpretable solutions [32].

3   Experiments

3.1   Experimental Setup for MSR3D Dataset and Cross Validation Methods

The dataset used for our experiments is the MSR Action 3D (MSR3D) Dataset proposed by Li et al. [33]. It consists of depth map sequences with a resolution of 320x240 pixels recorded with a depth sensor at 15 FPS. There are ten subjects performing twenty actions two to three times with a total of 567 depth map sequences. The dataset actions are: high arm wave, horizontal arm wave, hammer, catch, tennis swing, forward punch, high throw, draw X, draw tick, tennis serve, draw circle, hand clap, two hand wave, side boxing, golf swing, side boxing bend, forward kick, side kick, jogging, and pick up and throw. No corresponding RGB information is available, however 3D joint positions are provided as shown in Fig. 1. The formats of the data provided are JPG images of each frame. All silhouettes of the individuals conducting the actions have already been segmented out from the background as demonstrated in the sample action frames of Fig. 4.

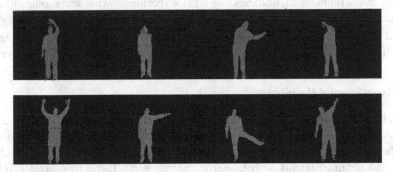

Fig. 4. Sample frames from the MSR3D dataset with plotted kinematic joints of a high arm wave, horizontal arm wave, golf swing, draw X, two-hand wave, side boxing, side kick, and a tennis serve

Table 1. List of actions included in the four subsets used in the experiments. Subset 1 and Subset 2 group actions with similar characteristics. Subset 3 groups actions that are dissimilar. The Full Set includes all actions.

Subset 1	Subset 2	Subset 3	Full Set
Hor. Arm Wave	High Arm Wave	High Throw	
Hammer	Hand Catch	Forward Kick	
Forward Punch	Draw X	Side Kick	
High Throw	Draw Tick	Jogging	All Actions
Hand Clap	Draw Circle	Tennis Swing	
Bend	Two Hand Wave	Tennis Serve	
Tennis Serve	Forward Kick	Golf Swing	
Pickup & Throw	Side Boxing	Pickup & Throw	

The work of Li et al. [33] use action graphs to model the dynamics of the actions and a Bag of Features (BoF) to encode the action and classify test samples against a training set. To compare against their results, we provide the same experimental setup, where 20 actions were divided into three subsets consisting of 8 actions each as presented in Table 1. Additionally, in this paper we test against the entire set of activities which we labeled as Full Set. The Subsets 1 and 2 were designed to group activities with similar movements while Subset 3 was designed to group actions that are more dissimilar, and therefore more suitable for sparser solutions.

In the work of Li et al. [33] three types of tests were conducted as follows: training with 1/3 of the training samples and testing with 2/3 of the samples, training with 2/3 of the samples and testing against 1/3, and training with half of the samples and testing against the other half. Since cross validation was not used, and without knowing which samples were used for testing and training for each test, we felt the best method of comparison was to use 2-Fold cross validation (2FCV) and compare against the third test. In 2FCV, we randomly select half the subjects for testing and half the subjects for training, and additionally we train and test on both sets allowing for each action sample to be used for either training or validation on each fold. To ensure large and over-complete dictionaries, we also experiment with leave one out cross validation (LOOCV) where each test subject is validated against the remaining subjects and repeated for all subjects until all subjects have been used for training and testing. The results from each subject are averaged to give a final performance result.

3.2   Experimental Results

The results of our approach are presented in Tables 2 and 3 for the kinematic joint feature descriptor and the raw depth data extrapolated from the 3D video sequences. We performed leave one out cross validation (LOOCV) and 2-Fold cross validation (2FCV) and obtained results with L_1-norm minimization, L_2-norm nearest neighbor, and the Bag of Features method presented by Li et al [33].

Table 2. Results using kinematic joint feature descriptors and cross validation methods, L_1-norm minimization, nearest neighbor, and the Bag of Features method by Li et al.

Subset	Cross Validation Approach	L_1-norm Minimization	L_2-norm Nearest Neighbor	Bag of Features [33]
Subset 1	LOOCV	80.73%	80.21%	
	2FCV	77.66%	76.60%	72.90%
Subset 2	LOOCV	77.11%	78.78%	
	2FCV	73.17%	75.61%	71.90%
Subset 3	LOOCV	93.89%	89.29%	
	2FCV	91.58%	89.47%	79.20%
Full Dataset	LOOCV	72.11%	72.32%	
	2FCV	63.23%	73.54%	

Table 3. Results using raw depth feature descriptors and cross validation methods against L_1-norm minimization, nearest neighbor, and the Bag of Features method by Li et al.

Subset	Cross Validation Approach	L_1-norm Minimization	L_2-norm Nearest Neighbor	Bag of Features [33]
	LOOCV	67.79%	61.76%	
Subset 1	2FCV	74.47%	69.15%	72.90%
	LOOCV	84.50%	71.50%	
Subset 2	2FCV	84.15%	67.07%	71.90%
	LOOCV	82.37%	74.99%	
Subset 3	2FCV	88.42%	86.32%	79.20%
	LOOCV	71.05%	58.82%	
Full Dataset	2FCV	76.23%	71.75%	

As suspected, the best action classification accuracies came from Subset 3 because the dissimilarity between the grouped actions naturally encourages sparser solutions. We also find that with kinematic joint descriptors, L_1-norm minimization does not drastically outperform L_2-norm minimization. This indicates that the normal distribution and utilization of all available training examples is sufficient and that the kinematic descriptor is not sparse enough to accurately classify actions. However, the kinematic joint descriptor is very powerful in action classification accuracy and is comparable to the Bag of Features approach.

When examining the raw depth data feature descriptor, we begin to see that the natural sparse representation of each action sequence results in an improvement over nearest neighbor classification. We obtain an accuracy of 76.23% on all 20 3D video actions using 2FCV which performs 5.18% better than LOOCV, indicating that the training dictionaries are over-complete without training a majority of the action samples. This is even more apparent when noticing that in almost all cases 2FCV's outperform LOOCV for both L_2-norm minimization and L_1-norm minimization.

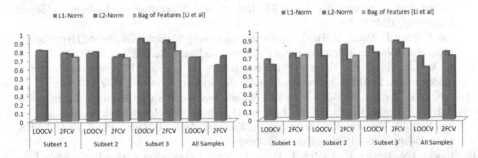

Fig. 5. Plots showing bar charts from Table 2 and Table 3 for action classification. The left chart represents the results from using kinematic joint descriptors and the right chart represents the results from using raw depth feature descriptors.

4   Conclusion

In this paper, we present a novel approach to action classification of 3D video sequences using sparse representations of spatio-temporal kinematic joint features and raw depth features which are invariant to scale and localization. We create overcomplete dictionaries and take advantage of the sparse nature of the feature descriptors to classify actions using least squares loss L_1-norm minimization with parameter regularization. We find that the representations of raw depth features is naturally more sparse than kinematic joint features as a result of comparing L_1-norm minimization with L_2-norm nearest neighbor classification.

References

1. Ghasemzadeh, H., Loseu, V., Jafari, R.: Collaborative Signal Processing for Action Recognition in Body Sensor Networks: A Distributed Classification Algorithm Using Mo-tion Transcripts. In: Proc. 9th ACM/IEEE Int. Conf. Inf. Process. (2010)
2. Raja, K., Laptev, I., Perez, P., Oisel, L.: Joint pose estimation and action recognition in image graphs. In: 18th IEEE International Conference on International Conference on Image Processing, ICIP (2011)
3. Weinland, D., Boyer, E., Ronfard, R.: Action Recognition from Arbitrary Views using 3D Exemplars. In: IEEE ICCV (2007)
4. Maji, S., Bourdev, L., Malik, J.: Action recognition from a distributed representation of pose and appearance. In: IEEE Conference on Computer Vision and Pattern Recognition, CVPR (2011)
5. Wang, Y., Zhang, Z.: View-invariant action recognition in surveillance videos. In: First Asian Conference on Pattern Recognition, ACPR (2011)
6. Imtiaz, H., Mahbub, U., Ahad, M.A.R.: Action recognition algorithm based on optical flow and RANSAC in frequency domain. In: Proceedings of SICE Annual Conference, SICE (2011)
7. Ahad, M.A.R., Tan, J., Kim, H., Ishikawa, S.: Action recognition by employing combined directional motion history and energy images. In: IEEE Computer Society Conference on Computer Vision and Pattern Recognition Workshops, CVPRW (2010)
8. Lopes, A. P. B., Oliveira, R. S., de Almeida, J. M., de A Araujo, A.: Comparing alternatives for capturing dynamic information in Bag-of-Visual-Features approaches applied to human actions recognition. In: IEEE International Workshop on Multimedia Signal Processing, MMSP (2009)
9. Liu, J., Yang, J., Zhang, Y., He, X.: Action Recognition by Multiple Features and Hyper-Sphere Multi-class SVM. In: 20th International Conference on Pattern Recognition, ICPR (2010)
10. Ji, X., Liu, H., Li, Y.: Human actions recognition using Fuzzy PCA and discriminative hidden model. In: IEEE International Conference on Fuzzy Systems, FUZZ (2010)
11. Azary, S., Savakis, A.: View Invariant Activity Recognition with Manifold Learning. In: Bebis, G., Boyle, R., Parvin, B., Koracin, D., Chung, R., Hammound, R., Hussain, M., Kar-Han, T., Crawfis, R., Thalmann, D., Kao, D., Avila, L. (eds.) ISVC 2010. LNCS, vol. 6454, pp. 606–615. Springer, Heidelberg (2010)
12. Junejo, I.N., Dexter, E., Laptev, I., Perez, P.: View-Independent Action Recognition from Temporal Self-Similarities. IEEE Transactions on Pattern Analysis and Machine Intelligence (2011)

13. Gall, J., Yao, A., Razavi, N., Gool, L.V., Lempitsky, V.: Hough Forests for Object Detection, Tracking, and Action Recognition. IEEE Transactions on Pattern Analysis and Machine Intelligence (2011)
14. Laptev, I.: On Space-Time Interest Points. International Journal of Computer Vision (2005)
15. Willems, G., Tuytelaars, T., Van Gool, L.: An efficient dense and scale-invariant spatio-temporal interest point detector. In: Forsyth, D., Torr, P., Zisserman, A. (eds.) ECCV 2008, Part II. LNCS, vol. 5303, pp. 650–663. Springer, Heidelberg (2008)
16. Juran, J.M.: The non-Pareto Principle: Mea culpa. Quality Progress, 8–9 (May 1975)
17. Farmer, J.D., Geanakoplos, J.: Power laws in economics and elsewhere. Santa Fe Institute, Santa Fe (2008)
18. West, G.B.: The Origin of Universal Scaling Laws in Biology. Oxford University Press, New York (1999)
19. Mislove, A., Marcon, M., Gummadi, K.P., Druschel, P., Bhattacharjee, B.: Measurement and analysis of online social networks. In: IMC 2007 Proceedings of the 7th ACM SIGCOMM Conference on Internet Measurement, New York, NY (2007)
20. Wright, J., Yang, A., Ganesh, A., Sastry, S., Ma, Y.: Robust Face Recognition via Sparse Representation. In: IEEE Transactions on Pattern Analysis and Machine Intelligence, PAMI (2009)
21. Qiu, F., Xu, Y., Wang, C., Yang, Y.: Noisy image super-resolution with sparse mixing estimators. In: 4th International Congress on Image and Signal Processing, CISP (2011)
22. Bao, L., Liu, W., Zhu, Y., Pu, Z, Magnin: Sparse representation based MRI denoising with total variation. In: 9th International Conference on Signal Processing, ICSP (2008)
23. Zuo, Y., Zhang, B.: General image classification based on sparse representation. In: 9th IEEE International Conference on Cognitive Informatics, ICCI (2010)
24. Zhang, J., Wang, Y., Chen, J., Li, Q.: Sparse Representation for Action Recognition. In: 3rd International Congress on Image and Signal Processing, CISP 2010 (2010)
25. Liu, C., Yang, Y., Chen, Y.: Human Action Recognition using Sparse Representation. In: IEEE International Conference on Intelligent Computing and Intelligent Systems, ICIS (2009)
26. Shotton, J., Fitzgibbon, A., Cook, M., Sharp, T., Finocchio, M., Moore, R., Kipman, A., Blake, A.: Real-Time Human Pose Recognition in Parts from Single Depth Images. In: CVPR (2011)
27. Wright, J., Ma, Y., Maira, J., Sapiro, G., Huang, T.S., Yan, S.: Sparse Representation for Computer Vision and Pattern Recognition. Proceedings of the IEEE 98(6), 1031–1044 (2010)
28. Miller, S.J.: The Method of Least Squares. Mathematics Department Brown University, Providence, RI (2006)
29. Bektaş, S., Şişman, Y.: The comparison of L1 and L2-norm minimization methods. International Journal of the Physical Sciences, IJPS (2010)
30. Donoho, D.L., Elad, M., Temlyakov, V.: Stable recovery of sparse overcomplete representations. IEEE Transactions on Information Theory (2005)
31. Donoho, D.L., Tsaig, Y.: Fast Solution of L1-norm Minimization Problems When the Solution May Be Sparse, Stanford CA, 94305, Department of Statistics, Stanford University (2006)
32. Schmidt, M.: Least Squares Optimization with L1-Norm Regularization. University of British Columbia (2005)
33. Li, W., Zhang, Z., Liu, Z.: Action recognition based on a bag of 3D points. In: CVPR Workshop, San Fransisco, CA (June 2010)

SCAR: Dynamic Adaptation for Person Detection and Persistence Analysis in Unconstrained Videos*

George Kamberov, Matt Burlick, Lazaros Karydas, and Olga Koteoglou

Department of Computer Science
Stevens Institute of Technology
Hoboken, NJ 07030, USA
{gkambero,mburlick,lkarydas,okoteogl}@stevens.edu

Abstract. In many forensic and data analytics applications there is a need to detect whether and for how long a specific person is present in a video. Frames in which the person cannot be recognized by state of the art engines are of particular importance. We describe a new framework for detection and persistence analysis in noisy and cluttered videos. It combines a new approach to tagging individuals with dynamic person-specific tags, occlusion resolution, and contact reacquisition. To assure that the tagging is robust to occlusions and partial visibility the tags are built from small pieces of the face surface. To account for the wide and unpredictable ranges of pose and appearance variations and environmental and illumination clutter the tags are continuously and automatically updated by local incremental learning of the object's background and foreground.

1 Introduction

In this work address a basic persistence analysis task in understanding unconstrained videos: detect if a specific person is present and estimate the length of her presence. The objective is to detect the faces of specific individuals and to retrieve as many as possible of the frames containing at least a part of the individuals' faces. Unconstrained videos often involve illumination and environment clutter, frequent and hard to analyze shot transitions (camera motions and zooms, stitching together multiple camera views and scenes), noisy low quality videos, and lossy compression. In such videos there are many frames where only small and often featureless parts of the face are visible. We do not need to recover the full face, in fact it is often impossible to do that. However, we do need to be able to keep track of the face in the presence of significant occlusions, image deterioration and environmental and scene clutter. The key of the proposed approach is to develop dynamic tags, see Figure 1, that can be used to verify the subjects' presence. Persistence analysis is an important task combining recognition and tracking but, to our knowledge, there is no published work addressing it specifically. The state of the art in face detection, recognition, and tracking provides useful insights and tools but appears inadequate to our persistence analysis task. In a set of experiments reported in Section 3 we evaluated all the state of the art (SOA) tracking systems for which the authors have provided publicly available code: [1,2,4,7,9]. The evaluation shows that the

* This research was partially supported by: the National Science Foundation, Award # 0916610; two Gifts from the Gerondelis Foundation; the Robert Crooks Stanley Fellowship Fund.

G. Bebis et al. (Eds.): ISVC 2012, Part II, LNCS 7432, pp. 176–187, 2012.
© Springer-Verlag Berlin Heidelberg 2012

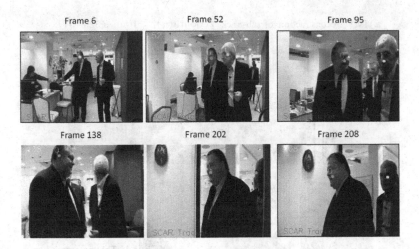

Fig. 1. Automatic detection, pose variations, camera motions, exiting and re-entering and occlusions of the tagged persons. Yellow tags indicate initialization/on-line relearning of a new person-specific tag mask and (local) appearance model. Green tags indicate that the persons tag was updated through mean shift. Frame 6: Initial recognition of person 0. Frame 52: Initialization of person 1, continued tracking of person 0. Frame 95: Continued persistence of person 0, a new foreground model and a tag for person 1 is re-acquired upon re-entry in the field of view (FOV) – SCAR detected that person 1 exited the FOV in video frame 71. Frames 138 and 202: pose variations and partial occlusion of person 1. Frame 208: re-initialization (new appearance model) for person 1 while person 0 is still present.

state-of-the-art trackers are struggling when tracking faces in unconstrained videos with difficult backgrounds, fast moving characters, occlusions, exits and re-entries, and camera motion. We also ran the OpenCV CAMSHIFT: not surprisingly on unconstrained videos it is by far the worst performer. The adaptive background enhancement reported in [9] at least gives results in the same ball-park as the rest of the SOA trackers. See Table 1 and Table 2. In addition, we evaluated a baseline method *Picasa BL* for "detection, recognition, and tracking by per-frame detection and recognition". Picasa BL is a straight forward approach based on submitting to Google's Picasa the frames in each video and then considering each set of recognized people's faces as a track. One would have expected that the baseline method should have been severely disadvantaged since it does not use at all the time coherency or the spatial-temporal relationships inherent in video data. The experiments show that this is not the case in videos with substantial clutter (See Table 2.). Still, as Table 1 indicates tracking by per-frame recognition also fails at very high rates on videos involving occlusions and pose and illumination variations. The performance of SOA trackers can be improved sometimes by careful per-video manual tuning. This is not an option in persistence analysis of real world forensics collections of video snippets and collages whose volumes often exceed tera bytes of data. The evaluation results and the need to develop self-tuning methods show that a new approach is needed to tackle persistence analysis in un-constrained videos.

a.) IMT Deconstruction of sub-track $\tau(\text{Boutaris})_1 = \text{frames}[52 - 64]$

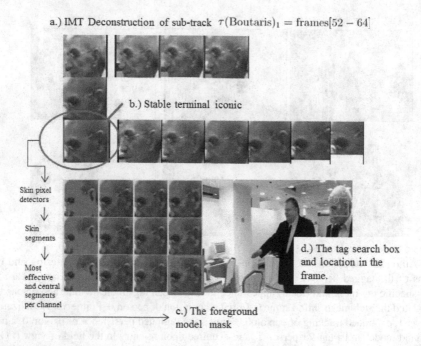

b.) Stable terminal iconic

Skin pixel
detectors

Skin
segments

Most
effective
and central
segments
per channel

c.) The foreground
model mask

d.) The tag search box
and location in the
frame.

Fig. 2. The IMT re-initialization pipeline: a.) History decomposition into initial, mature, and terminal tracklets; b.) Stable terminal iconic used for the re-initialization model; c.) The mask used for building the re-initialization foreground model; d.) The tag location and its search box in the iconic's frame.

2   The SCAR Algorithm

We propose a new approach based on the following novel ideas and observations: (**1.**) Use small, informative, central descriptors: that is, build descriptors from small informative central patches instead of elaborate representations encapsulating the whole object and its surrounding supporters/distractors. See Figure 2 c.) and d.). Using small central patches makes the tracking more robust to occlusions and also to the effects of camera motions. (**2.**) Use just enough context: couple the foreground object descriptor patch with some background context but use as little as possible context to flesh out the object while encompassing only a few possible distracters. See Figure 5. This approach is also aimed at minimizing the probability of occlusion. (**3.**) Detect when the current foreground model is not adequate or when a contact is occluded or exits from the FOV. (**4.**) Contact re-acquisition is a hard cognitive task and it is handled by either: (i) Detection and recognition using recognition memory and a high level recognition engine (we use a plug-in face detection and recognition engine); (ii) If the recognition fails because the engine is not trained to handle the hard poses, partial views, or illumination clutter, then engage a secondary mechanism called IMT-based re-initialization which looks at the track up to date and extracts a probable foreground model. See Figure 2. (5.) Use an information theoretic approach to handle occlusion resolution even in the absence of scene depth and elaborate scene context and semantics models.

The proposed method applies to persistence analysis of arbitrary objects. The only place where we specialize to faces is where we select the actual masks (for intialization/reinitialization) for building the foreground models. Different color models can be used but for faces we will use 1D Hue.

2.1 Dynamic Tags and Persistence Analysis

In our framework, in each frame f the tag of an object o is built from an appearance model for the object and an axis aligned square bounding an object segment. The tag is marked on the video frame as the center of the axis aligned bounding square. The appearance model is represented as a class conditional distribution. Following [3] we call the tag square the search box and denote by $P(C|o; f)$ the class conditional distribution representing the probability for any pixel color C given that the pixel belongs to the object segment enclosed by the search box in this frame. To study the persistence of an object (person) we build an initial tag from a crop and then update the tag, the search box, and the supporting foreground appearance model dynamically.

Without loss of generality we may assume that for each object o the class conditional distribution $P(C|o; f)$ is locally stable with respect to time. Therefore, the sequence of video frames $f_1, \ldots, f_T$ is broken into consecutive intervals of adjacent frames $\tau(o)_1, \ldots, \tau(o)_{m(o)}, m(o) \leq T$, so that

$$P(C|o; f) \approx \text{const}_j, \forall f \in \tau(o)_j, j = 1, \ldots, m(o). \tag{1}$$

In fact this is always true, even for the most violent and unstable appearances in which case each subinterval will contain a single frame. The fundamental observation that underlines our method is that within a CAMSHIFT framework we can estimate this temporal partition of the video for each object. (See Figure 3.) Indeed for each frame f and object tag o let $P(o; f) = 1/r(o; f)$ be the prior probability that a pixel belongs to the object and so let $P(b; f) = (r(o; f) - 1)/r(o; f)$ be the prior probability that a pixel belongs to the background. In order to deal successfully with both background changes and object appearance changes we will have to estimate $r(o; f)$ for each frame and for each object. In the original CAMSHIFT both priors were postulated to be constant and equal to 0.5. As the authors point out this is the least informed choice and it is not surprising that it can not account for most object appearance changes and environment clutter. As observed in [9] the selection of $r(o; f)$ should be connected to the relative sizes of the area

$$s(o; f) = \text{Area of the search box}(o; f) \tag{2}$$

and the area $\beta(o; f)$ of the *calculation region* for the object in the frame, [3]. The calculation region is an axes-aligned bounding box where the object is supposed to reside. It is centered at the center of the selection box but is slightly larger so that it encompasses some background pixels. See Figure 5. Thus following [9] we define

$$r(o; f) = \frac{\beta(o; f)}{s(o; f)} > 1. \tag{3}$$

The video partition can be chosen so that $r(o; f)$ is locally constant (in time). That is, both (1) and

$$r(o; f) = \text{const}_j^*, \forall f \in \tau(o)_j, j = 1, \ldots, m(o) \tag{4}$$

hold.

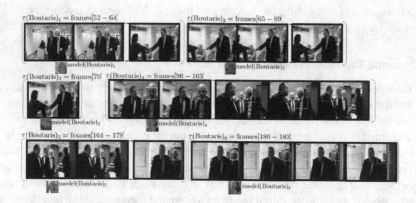

Fig. 3. Detecting and tagging of a contact (Yiannis Boutaris) in a 184 frames video. The contacts' search boxes are color coded to distinguish model initialization/reinitialization (yellow and using recognition, pink when using the IMT mechanism) vs CAMSHIFT with background adaptation (green). $\tau(\text{Boutaris})_1$: SCAR automatically builds an initial tag based on a face crop produced by Picasa in frame 52; it then dynamically updates the tag until frame 64 by mean shifting with a continuous background adaptation. $\tau(\text{Boutaris})_2$ and $\tau(\text{Boutaris})_3$: New IMT based foreground models are built in frame 65 and then again in frame 70 to cope with the progressive occlusion as the contact exits camera FOV. SCAR detects the contact's ultimate exit from the FOV in frame 71. $\tau(\text{Boutaris})_4$: The contact is re-acquired in frame 96 after it re-enters the FOV; $\tau(\text{Boutaris})_5$ and $\tau(\text{Boutaris})_6$: new foreground models are reacquired as needed to account for the pose variations and the severe visibility occlusions. The appearance model for Yiannis Boutaris for the time interval $\tau(\text{Boutaris})_i$ is built using the crop and the mask shown in model($\text{Boutaris})_i, i = 1, \ldots, 6$.

The basic steps in the SCAR framework are illustrated in Figure 3. They are: (i) generation of temporally stable foreground models and initial tags from image crops (See Section 2.2.); (ii) Tag and foreground models dynamic updates including: (ii.a) tag adaptation using the temporally stable foreground models and locally constant object priors and mean shifting with background adaptation, (ii.b) re-initialization and contact reacquisition (See Section 2.3.); (iii) Handling contact overlaps. (See Section 2.4.)

2.2   Initial Tag Selection

To obtain a tag from a facial crop we combine appearance models (skin models in the case of faces, using the skin detectors introduced in [5]), fragment-based shape cues; and centrality cues expressed in the weights of the fragments. To obtain the fragments encoding the shape of the person's facial area detected in a given frame we need to segment the image crop enclosing the facial area. The shape fragments are just the segments of the segmented facial area. So we will refer to fragments as segments. In a unconstrained video the facial crop can cover only small and/or severely occluded parts of the face. Further complications can be caused by environmental clutter, compression, and abnormal viewing angles. To deal with segmentation bias we use an automatically selected approximation of the central segmentation of the crop. The central segmentation (CS) was introduced in [10]. By design this is the segmentation whose entropy equals the mean of the entropies of all possible segmentations. It is obtained through gradient descend. In practice we can only compute approximations of the CS by following a fixed number of steps down the gradient descent path. Instead of manually selecting thresholds to stop the gradient descend we chose as approximation the segmentation whose most effective segments explain most of the content of the crop. To do this we measure the effectiveness $p(\sigma, S)$ of a fragment σ in a segmentation S of a crop C as the relative area of the fragment. Thus

$$p(\sigma, S) = \mathrm{Area}(\sigma)/\mathrm{Area}(C).$$

Under the single hypothesis that the crop *does enclose just the object of interest* the quality of the segmentation is measured by its *effective number of different fragments*:

$$R(S) = \mathrm{int}(1/\sum_{\sigma \in S} p(\sigma, S)^2).$$

To encode the shape of the object (facial area) enclosed by a crop we chose an approximate central segmentation S whose top $R(S)$ effective segments cover most of the crop area.

Once an effective approximate segmentation is chosen we can proceed to build a facial tag by ranking segments according to their effectiveness and centrality. In particular, the weight of a fragment σ, is $\omega(\sigma) \propto e^{-\mathrm{dist}(\sigma, P)}\,\mathrm{Area}(\sigma)$, where P is the centroid of the object or object tag crop. See Figure 4.

The crops are selected by one of two mechanisms – recognition when possible, or the memory-based IMT mechanism (see Figure 2).

2.3   Foreground Models and Tag Dynamics

The first question we should address is how do we recognize a good tag initialization – i.e., a mask identifying a selected set of pixels that belong to the object o. Without loss of generality we may assume that this has happened at frame 0. Can we now mean-shift to the next frame? If the initialization is good, then we should be able to select a tight calculation box around the mask which also contains some background pixels, otherwise it is known that a mean shift algorithm will fall in an unstable mode. (The extreme case is when the object is flat color and takes the whole image.) Thus if the initialization is good, one must be able to choose a calculation region such that the object and the objects background distributions are different $P(C|o; 0)$ and $P(C|b; 0)$, that is

$$\mathrm{dist}(P(C|o; 0), P(C|b; 0)) > 0. \tag{5}$$

Central segmentation approximation via gradient
descent. Stop as the effective segmentation
components cover the majority of the object crop

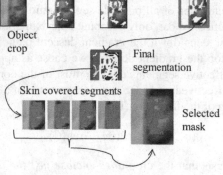

Object
crop

Final
segmentation

Skin covered segments

Selected
mask

Fig. 4. Automatic mask generation combining appearance models (skin models in the case of faces), and fragment-based shape cues and centrality cues expressed in the weights of the segments. The final tag is a minimal axis aligned square bounding box of the mask.

Theoretically it does not matter what metric is used in Equation (5). Recall that for a chosen $r = r(o; 0) > 1$, that is, a chosen calculation region, we have

$$P(C; 0) = P(C|o; 0)P(o; 0) + P(C|b; 0)P(b; 0)$$
$$= \frac{P(C|o; 0)}{r} + \frac{(r-1)P(C|b; 0)}{r}.$$

Thus we have,

$$P(C|b; 0) = \frac{P(C; 0)r - P(C|o; 0)}{r - 1}. \tag{6}$$

Once $r = r(o; 0)$ is chosen, we compute $P(C; 0)$ and $P(C|o; 0)$ by computing the color histograms in the calculation window and the color histogram of the object. We reached to the criterion for a good initialization: we must be able to chose $r > 1$ such that

$$0 \le P(C; 0)r - P(C|o; 0) \le r - 1 \tag{7}$$

and (5) hold. If such an $r > 1$ cannot be chosen, we can not use this initialization to start a mean-shifter. It is informative to think how can we go about to find r and what could go wrong. The obvious procedure is, start with $r = 1$, in which case (5) fails, and keep increasing r until the inequalities (5) and (7) hold. If the initialization is badly chosen, then this may not happen – eventually r will become so large that the calculation box will overflow the video frame.

On the other hand, suppose that the initialization is chosen well and so we can find the appropriate $r = r(o; 0) > 1$. At this point we can start mean shifting, which gives us the location of the object center in the next frame. First we need to check whether the model of the object appearance is still valid. That is, we check if

$$0 \leq P(C; 1)\, r(o; 0) - P(C|o; 0) \leq r(o; 0) - 1 \tag{8}$$

holds. If it does, then frames 0 and 1 are in the time interval where both $r(o; 1) = r(o; 0)$ and $P(C|o; 0) \approx P(C|o; 1)$ and we proceed with adapting the object background as in [9]. We will discuss how to do this in Section 2.3. If this condition fails, this means that the object appearance model is no longer valid, that is the object has changed too much. Essentially either the object is not visible anymore or we must find a way to compute the new model. Next we will discuss the methods to deal with this situation. The important moral is that we have gleaned a method to diagnose substantial changes in the appearance models and that this is done without using any hand tuned thresholds, or making the assumption that we can somehow estimate and list all possible appearances of an object or person.

Updating the Foreground and Background Models: At the beginning of the processing of each new frame f and for each tracked object we must: (i) check if the object tag appearance model is still valid in the new frame; (ii) update the object tag background model.

The discussion in the previous section dealt with the case $f = 1$. In the general case we have to check if

$$0 \leq P(C; f) r - P(C|o; f - 1) \leq r - 1, \tag{9}$$

where $r = r(o; f - 1)$. As long as (9) holds the foreground model is valid, and following [9], the background model is automatically updated since

$$P(C|b; f) = \frac{P(C; f) r - P(C|o; f)}{r - 1}.$$

The only subtlety is that to account for computational precision when verifying and adjusting the axes aligned boxes so that the calculation box contains enough background pixels we check that

$$\mathrm{dist}(P(C|o; f), P(C|b; f)) > \tau(o),$$

where

$$\tau(o) = \mathrm{dist}(P(C|\text{object mask}; 0), P(C|\text{selection box}'; 0)),$$

where

$$\text{selection box}' = \text{selection box}(0) \setminus \text{object mask}.$$

The object tag mask is the initial skin pixels mask. In contrast, in [9] the threshold $\tau(o)$ was hand-tuned. If (9) fails, then the object might have become invisible or we may have to re-build its appearance model, i.e., re-initialize the object.

Re-acquisition: The optimal method to re-initialize an object/person model is to have a good detection and recognition engine which will find a correct mask for the object. When this is impossible we attempt to re-initialize using the incremental knowledge acquired up to date. We perform recursive initial-mature-terminal appearance analysis (IMT) [6] using the Hellinger distance on the tracklet terminating at the frame where the object appearance model became invalid. The tracklet is split into three clusters (the initial, the mature, and the terminal clusters, respectively). And we use the stable final appearance in the terminal tracklet of the object/person to generate the re-initialization mask. The stable final appearance is obtained by selecting the centroid of stable sub-cluster of the terminal portion of the tracklet. See Figure 2.

Sometimes a tracked object/person becomes invisible. It could have left the FOV, or it could be occluded or the tracker may have lost it. Such tracks could still be re-activated as the object/person is recognized in a future frame. To reactivate tracks we keep a log of the suspended tracks and the individuals associated with them. Each future frame is checked by a plug in recognition engine and if the individual object is found, then the track is reactivated using a re-initialization mask.

2.4 Handling Overlaps

To handle target overlaps when tagging multiple objects we propose a mechanism for early detection of possible overlaps and a related mechanism for adjusting the target masks to separate the objects if possible. A possible target overlap is detected when the search area of a one tracked object o_i intersects with the calculation area of another tracked object o_j. Without actual depth ordering it is not clear who is the occluding and who is the occluded. To reduce the chance for an person-person occlusion we try to locate a sub-area of o_i that is still representative of the object appearance. The selection area of o_i is shrunk until the it does not overlap o_j's calculation region. To avoid the loss of appearance information we need to compare the total information loss as we shrink the search area and the object mask information loss. The resizing is invalid if full info loss < mask info loss. To compute these losses we use object specific masks $\sigma(S(o))$. For a face tracker $\sigma(S(o))$ is the collection of skin pixels inside the search box of the object $S(o) = \text{scale}(\frac{1}{r}, \frac{1}{r})C(o)$ where $C(o)$ is the calculation box of our object in this frame. Thus for each object o and for each frame we have the chain $\sigma(S(o)) \subset S(o) \subset C(o)$. And the detection of a possible occlusion event involving o_i and o_j in frame f is equivalent to $S(o_i) \cap C(o_j) \neq \emptyset$. If this event is detected we begin to shrink the search (and hence the calculation box) of o_i to $S'(o_i)$ such that $S'(o_i) \cap C(o_j) = \emptyset$. We use the Kullback-Leibler divergence to model information loss

$$\text{full info loss} = \text{KL}(P(C|S(o_i))||P(C|S'(o_i)))$$

$$\text{mask info loss} = \text{KL}(P(C|\sigma(S(o_i)))||P(C|\sigma(S'(o_i)))).$$

If the resizing is invalid, i.e., full info loss < mask info loss, the object is considered occluded.

3   Evaluation

As person trackers have improved there has been additional interest in *unconstrained, ad-hoc* data sets. Several of these do exist [11], however they tend to focus on eas-ier examples: talking heads style interviews and fixed camera surveillance videos. The SCAR framework is evaluated on a set of videos involving large camera transitions, low

Fig. 5. The search boxes - marked by the inner yellow or green squares, and the corresponding calculation regions – marked by the outer, red rectangles

resolution imagery (or distant subjects), and complicated scenarios involving object occlusions and fast appearance changes. The data set consists of both single-subject videos as well as multiple-person videos. Additionally it includes videos where a subject may leave and then re-enter the scene. The videos are:

The full Trellis and David Indoor: http://www.cs.toronto.edu/~dross/ivt;
Clips 0:0-1:19 from http://www.youtube.com/watch?v=Bl51HtRvJcI;
Clip 0:0:05.75-1:17 from http://www.youtube.com/watch?v=tC3Q5mOGlDA.

The results for the tagging task are shown in Table 1 and Table 2. For reference we also show the performance of several state-of-the-art trackers [9,4,1,7,2] and the baseline Picasa tagging-by-per-frame-recognition. To avoid bias and implementation-based variations the evaluation tables include only comparisons with systems whose code was made available by the original authors. For all trackers the original code was used. The default parameters were used for all runs. For each video and for each person in the video the trackers and Picasa BL are seeded with the first recognition of the person in the video. Typically tracker and face detection performance measures involve the distance between the centers of the tracker/detection box $D_k(t)$ and the hand-labeled ground truth box $G_k(t)$. For testing tagging performance this measure is not relevant since the SCAR tags are already small by design and can cover very small patches of

Table 1. Tracker Evaluation on Single-Person Videos Trellis and David Indoor

Tracker	Precision	Recall	Accuracy
SCAR	97.9%	100.0%	97.9%
IVT[7]	92.9%	100.0%	92.9%
USC_CT[4]	87.0%	100.0%	87.0%
MIL[2]	82.4%	100.0%	82.4%
FaceTrack[8]	77.9%	99.2%	77.7%
Picasa BL	**100%**	**69.8%**	**69.8%**
FRAGTrack[1]	86.0%	53.4%	49.5%
abcSHIFT[9]	85.1%	28.4%	25.8%

Table 2. Evaluation while tracking the persistence of single individuals in unconstrained videos involving multi person scenarios, occlusions, pose and illumination variations, large camera transitions

Tracker	Precision	Recall	Accuracy
SCAR	65.6%	81.8%	66.4%
Picasa BL	**100%**	**49.5%**	**58.6%**
IVT[7]	55.2%	68.0%	44.6%
MIL[2]	43.8%	100.0%	43.8%
FRAGTrack[1]	41.0%	71.1%	34.4%
USC_CT[4]	34.2%	100.0%	34.2%
abcSHIFT[9]	55.8%	31.4%	24.8%

the face. We consider a valid detection to occur in frame t if the two windows intersect. Indeed, this choice gives a slight advantage to the competing trackers since SCAR tends to have very small boxes. In fact, in the evaluation tests the median size of the SCAR detection boxes is 1.4% of a video frame. As a face recognition engine we used a plug-in to Picasa.

4   Summary, Conclusions, and Future Work

We described a new fully automatic approach to address the persistence analysis task. The development of a new approach is motivated by: (i) The need to obtain high accuracy rates; (ii) The need to develop a completely self-tuning framework that does not require manual tweaking and tuning. The contributions in this paper are: (i) formulation the persistence analysis task as a tagging problem; (ii) a novel method to select a representative tag from an object (facial) crop box in an image; (iii) a method to evaluate whether an initial tag can be used to initiate a dynamic sequence of continuously varying tags that can be used to mark the individual in subsequent frames; (iv) a CAMSHIFT based method to update a tag; (v) a method to detect abrupt changes in the scene (e.g., the person is occluded or exits the field of view) or in the person's appearance which is equivalent to detecting that the current foreground model is no more valid; (vi) a dual-mechanism method to re-acquire contacts and in effect re-initialize persons' tags; (vii) a method to resolve person-person occlusions. An evaluation on publicly available data and code shows that SCAR outperforms the current SOA methods. The proposed method applies to persistence analysis of arbitrary objects. The only place where we specialize to faces is when we select the actual mask (for intialization/reinitialization), in which case we use skin color models (and a face recognition engine if possible). In our future work we will extend the tagging framework to tackle a wider category of objects.

Acknowledgements. We are very grateful to the authors of [1,2,4,7,9,10] for providing their code.

References

1. Adam, A., Rivlin, E., Shimshoni, I.: Robust Fragments-based Tracking using the Integral Histogram. In: IEEE Conference on Computer Vision and Pattern Recognition (CVPR), vol. 1, pp. 798–805 (2006), 1, 10, 11, 12
2. Babenko, B., Yang, M.-H., Belongie, S.: Visual Tracking with Online Multiple Instance Learning. In: CVPR (2009), 1, 10, 11, 12
3. Bradski, G.R.: Computer vision face tracking for use in a perceptual user interface. Intel. Technology Journal (Q2) (1998), 4
4. Dinh, T.B., Vo, N., Medioni, G.: Context tracker: Exploring supporters and distracters in unconstrained environments. In: IEEE Conference on Computer Vision and Pattern Recognition (CVPR), pp. 1177–1184 (2011), 1, 10, 11, 12
5. Gomez, G., Morales, E.F.: Automatic feature construction and a simple rule induction algorithm for skin detection. In: In Proc. of the ICML Workshop on Machine Learning in Computer Vision, pp. 31–38 (2002), 6
6. Kamberov, G., Burlick, M., Luczinski, B., Karydas, L., Kamberova, G.: Collaborative track analysis, data cleansing, and labeling. In: Internatioal Symoposium on Visual Computing. LNCS. Springer (2011), 9
7. Lim, J., Ross, D., Lin, R.-S., Yang, M.-H.: Incremental learning for visual tracking. In: Advances in Neural Information Processing Systems, pp. 793–800 (2005), 1, 10, 11, 12
8. Saragih, J., Lucey, S., Cohn, J.: Deformable model fitting by regularized landmark Mean-Shift. International Journal of Computer Vision 91(2), 200–215 (2011), 11
9. Stolkin, R., Florescu, I., Kamberov, G.: An adaptive background model for camshift tracking with a moving camera. In: Proceedings of the 6th International Conference on Advances in Pattern Recognition, pp. 147–151 (2007), 1, 2, 4, 5, 8, 10, 11, 12
10. Wang, H., Oliensis, J.: Rigid shape matching by segmentation averaging. IEEE Trans. Pattern Anal. Mach. Intell. 32(4), 619–635 (2010), 6, 12
11. Wolf, L., Hassner, T., Maoz, I.: Face recognition in unconstrained videos with matched background similarity. In: IEEE Conference on Computer Vision and Pattern Recognition (CVPR), pp. 529–534 (June 2011), 10

Exploiting 3D Digital Representations of Ancient Inscriptions to Identify Their Writer

Georgios Galanopoulos, Constantin Papaodysseus,
Dimitiris Arabadjis, and Michael Exarhos

National Technical University of Athens,
School of Electrical and Computer Engineering,
{ggalan,cpapaod}@cs.ntua.gr, alphad.d@gmail.com,
mexarhos@mail.ntua.gr

Abstract. The paper introduces a methodology for the automatic classification of ancient Greek inscriptions to cutters by exploiting the three dimensional digital representation of each inscription. In particular, the authors employed surface information features extracted from 3D datasets of the letters depicted on each inscription. Therefore, implementations of various alphabet symbols are used to extract a three dimensional "ideal" prototype of the symbol for each inscription separately. Next, statistical criteria are introduced so as to reject the hypothesis that two inscriptions have been carved by the same writer, determining thus the distinct number of cutters who carved a given set of inscriptions. The remaining inscriptions are then classified to the (be) determined by the previous step cutters by maximizing resemblance likelihood of their underlined alphabet symbols "ideal" prototypes. The methodology has been applied to twenty eight Ancient Athenian inscriptions and classified them to eight different cutters. The classification results have been fully confirmed by expert epigraphists.

1 Introduction

Ancient inscriptions form an important information carrier for great civilizations of the past, since they provide to archaeologists direct access to them. Identifying the writer who carved an inscription can be very helpful to its dating. Suppose that one can identify, via digital processing and pattern analysis, the writer who carved a set of inscriptions and at least any of these inscriptions can be dated. Then, all other inscriptions of the same writer can be immediately dated. Thus, the development of a system that automatically identifies writers and classifies inscriptions to carvers is of major significance for their dating and consequently it is of immense importance for archaeology, history and other Classical disciplines.

Writer identification methods employing handwritten text have been proposed by various researchers. In [9] the authors use feature vectors formed by morphologically processing of horizontal projection profiles of words. In [10], [11] the authors combine edge based features of uppercase letters along and/or their connected component contours to create stochastic patterns characterizing each writer. In [12] authors use identification methods for Chinese letters based on generalized Gaussian density on

G. Bebis et al. (Eds.): ISVC 2012, Part II, LNCS 7432, pp. 188–198, 2012.

wavelet domain that characterize wavelet coefficients of preprocessed handwriting image. Writer identification systems for ancient inscriptions have been proposed in [1] and [2]. In [1], mean values and variance for specific geometric characteristics of each letter are employed and the statistical distribution of their differences is used for the classification. In [2], "ideal" prototypes estimations are computed for the letters of each inscription separately and statistical criteria and likelihood considerations are used to compare these prototypes.

There are various reasons for which the authors tackled the problem of writer identification by using three dimensional representations of inscriptions. In fact: a) 3D representation of an inscription carries practically all available information about the writer, since writing is achieved by carving letters on marble/stone b) consequently, while in images writer identification is based on letter shapes extracted from shading diversifications, in 3D representations one may employ information associated with the bed of carved letters which is a far more rich of information than shading and it is actually the cause of shading diversifications c) it is well founded [8] that if one develops different statistical methods for solving a specific problem, then if the results of these methods coincide, then the degree of confidence about the correctness of the obtained results is asymptotically increased reaching certainty d) a side gain is the preservation of cultural heritage by 3D digitization of inscribed documents.

2 Extracting Representative Geometrical Models for Each Alphabet Symbol and Each Inscription

The range sensor used for acquiring the three dimensional mesh datasets of each inscription was a state of the art structured light 3D scanner (Imetric, IScan M). The average distance between neighboring points of each inscription letter was less than 0.15mm (Fig. 1). The reference system of the points' coordinates was defined by the 3D scanner.

If the carved letters and the inscription material roughness could be excluded, the remaining surfaces of the inscriptions would be quite planar and the curvature at each surface point too low and almost negligible.

2.1 Inscription Local "Ideal" Plane and Surface Representation

We first create a coordinate system whose xy-plane approximates the "ideal" plane that locally best fits the inscription upper surface. Initial estimation of parameters A, B, C and D of this plane are calculated via: a) a linear method including calculation of the principal component axes of the initial letter point coordinates and b) non-linear method including the minimization [5] of the error function E- 1.

$$err = \frac{1}{n}\sum_{i=1}^{n}(z_i - z_{est})^2 = \frac{1}{n}\sum_{i=1}^{n}\left(z_i + \frac{A}{C}x_i + \frac{B}{C}y_i + \frac{D}{C}\right)^2 \qquad \text{E- 1}$$

The initial estimation results for the plane from both methods are quite similar as expected. For simplicity and better inspection, a reference system J is selected on the plane that has the following features: a) the system center coincides with the average of the x, y coordinates of the projections of the $[x_i, y_i, z_i]$ on the plane and b) the direction of axis x is parallel to the writing direction of the letter's inscription row. Feature (a) is automatically calculated while feature (b) is selected by user. Afterwards, the necessary parameters of the transformation function F including rotation and translation operators are calculated and the new coordinates $[x_{Ji}, y_{Ji}, z_{Ji}] = F([x_i, y_i, z_i])$ are also calculated (Fig. 2)

In order to obtain a better estimation of the "ideal" plane the following syllogism is made: Consider the hypothetical surface S_m which remains if one could ideally remove all letters of the inscription together with their beds. The remaining surface S_m is as a rule quite rough, due to many factors such as: the wear the inscription suffered, the initial roughness of the inscription material (marble/stone), chemical interaction of this material with the environment, etc. Exactly because this roughness is due to many parameters, it is logical to assume that central limit theorem applies; hence the z-axis coordinates of the points of surface S_m, are expected to follow a normal distribution around the "ideal" plane zero level. The performed Kolmogorov-Smirnoff tests did not reject this hypothesis (a=0.001). Since we cannot "a priori" remove the carved letters from the inscriptions' relief, the normal distribution parameters cannot be calculated explicitly.

Therefore we follow an alternative approach: If z_{max} and z_{min} are the upper and lower limits of the values of z-axis coordinates of $[x_{Ji}, y_{Ji}, z_{Ji}]$, $E_z = \{z_{min}, z_{max}, 100\}$ is a partition of the $[z_{min}, z_{max}]$ interval into 100 equally sized sections, E_{zi} is the median value of each section and $H(E_{zi})$ is the histogram distribution of the z-axis coordinates among E_z, the $H(E_z)$ is very well approximated by the normal distribution function $f(a,\mu,\sigma) = a \cdot e^{-(\mu-z_i)^2/(2\sigma^2)}$ (Fig. 3)

The parameters (a,μ,σ) can be calculated by minimizing [5] the following error function

$$E = \frac{1}{n}\sum_{i=1}^{n}\left(H(E_{z_i}) - a \cdot e^{-\frac{(\mu-E_{z_i})^2}{2\sigma^2}}\right)^2 \qquad n=1..100 \qquad \text{E-2}$$

Since 95,4% of the normal distribution values lie in the interval $[\mu-2\sigma, \mu+2\sigma]$ the parameters of the plane are recalculated according to E-1 by excluding any point whose z-axis coordinate does not lie in the above interval. As a result the final estimation of the "ideal" plane is calculated and the points $[x_i, y_i, z_i]$ are finally transformed to $[x_{fi}, y_{fi}, z_{fi}]$, where I is the reference system adjusted to the local "ideal" plane of this inscription segment The implementation of Delaunay triangulation [4] on the $[x_{fi}, y_{fi}]$, data points offers the primary faces of the actual surface of the inscription (Fig. 4).

2.2   Letter Extraction

Letters are considered as trenches on the actual surface and we proceed as follows:

A) Reject points that obviously do not belong to the "trenches"

The histogram distribution of the z-axis coordinates of point set $[x_{fi}, y_{fi}, z_{fi}]$, is firstly recreated and the params (a,μ,σ) of the normal distribution function are evaluated as before. Every point with $z>z_{th}$ where $z_{th}=\mu-2\sigma$ is considered as point that belongs to the roughness of the inscription and is rejected. The value of threshold z_{th} is based on the fact that 97,7% of the population of the normal distribution lies in the interval ($\mu-2\sigma$, ∞).

B) Select trenches' beds and steep walls

The elevation angle θ of the surface normal on each point is calculated as the average of normals of all facets whose points belong to a sphere centered at the point in hand of a properly chosen diameter, eliminating thus possible small undulations on the surface of the trenches. The choice of the magnitude of the radius of the sphere is evaluated by trial and error asymptotic convergence in the sense that local increase of the radius of the sphere slightly affects the obtained average of normals. For the inscriptions in hand 1mm was a very good value for this radius.

Points are sorted in ascending order according to their z-axis value and to every point, say P_i the attribute "not checked" is given. Then, the following algorithm is executed starting from the first available point in the list.

i   Select all neighbouring points of P_i with distance less than 1mm and z-axis coordinate value satisfying $z_{Pj+}>z_{Pi}$. These points form set P_{j+}.
ii Select all neighbouring points of P_i with distance less than 1mm and z-axis coordinate value satisfying $z_{Pj-}<z_{Pi}$. These points form set P_{j-}.
iii If P_{j-} set is empty or every point is attributed as "rejected" or "terminal", attribute "rejected" is given to P_i and we get to step (v)
iv If P_{j+} set is empty or at least one point exists for which $\theta_{Pj+}>\theta_{Pi}$, then attribute "valid" is given to P_i else the attribute "terminal" is given to P_i.
v   Next point P_{i+1} is examined from step (i).

Every point attributed as "valid" or "terminal" is finally selected as letter point. These points along with facets extracted previously and containing them make up the geometrical model of the letter.

3   Three Dimensional Ideal Letter Prototype Evaluation

It is assumed that the writer has in his mind an "ideal" prototype for each alphabet symbol he carves on the inscription material. Let $r=r(u,v)$ the surface of the ideal prototype. Due to various reasons, such as the instability of the writer's hand, the mechano-elastic properties of inscription base material, the variations on the placement of the employed tools and the writer's mood, each implementation is a distorted realization of the ideal prototype. Therefore, if $n(u,v)$ integrates all kinds of possible distortions, the surface of each actual letter is $r=r(u,v)+n(u,v)$. Ideal letter prototype extraction includes the suppression of the $n(u,v)$ term from actual letter realizations from each inscription and is performed in two steps.

3.1   Potential Ideal Prototypes Creation

Let R_i, $i=1...n$, the n geometrical models of the realizations of an alphabet symbol (e.g. A) appearing in inscription E as they have been extracted with the methodology stated in section 2.

Sub step 1: A realization R_i is selected randomly as a reference one and weight $w_i=1$ is attributed to all its points.

Sub step 2: Another available realization R_j is randomly selected and ICP algorithm [6], [7] is used to fit R_j with R_i. In order to achieve high quality match, ICP algorithm is performed against various resized versions of R_j that are placed in different initial positions relative to R_i. The closest point operator used for each point A_i in the implementation of ICP is affected by the surface normals (shooting) of point A_i and its neighboring points B_j. If A_{iv} and B_{jv} the unit surface normals of points A_i and B_j, B_j is the closest point to A_i only if $|B_{jv} \cdot A_{iv}| > \gamma$, $\gamma = \sqrt{3}/2$ and geometrical distance of A_i and B_j is minimum. The maximum angle deviation between surface normals A_{iv} and B_{jv} corresponding to the selected threshold γ is $\pi/6$. High quality match between m points A_i of letter R_i and n points B_j of letter R_j is achieved when the following quantity is minimized

$$E = \frac{1}{n}\sum_{i=1}^{n} d(xB_j, A_i) + \frac{1}{m \cdot W}\sum_{i=1}^{m}\left(w_i \cdot d(A_i, xB_j)\right)$$ E- 3

where xB_j are the transformed points of P_j at each step of ICP algorithm, d(A_i, B) is the square distance between A_i and the previously defined closest point B, w_i the points weight and W is the sum of w_i. If either of the two terms in E- 3 was missing, optimization output would be erroneous since the algorithm would force R_j either to shrink or expand endlessly.

Sub step 3: R_i and R_j points are combined so as to keep a simpler reference model. The combination is performed in the following way. For every point A_i of R_i we find all points B_j from R_j which satisfy the following criteria:

$$|(A_i - B_j) \cdot A_{iv}| > \alpha/2$$ E- 4

$$d(A_i, B_j) < \beta$$ E- 5

$$|B_{jv} \cdot A_{iv}| > \gamma$$ E- 6

where $\alpha=0.86$ (angle 30°), $\beta=3$mm and $\gamma=\sqrt{2}/2$ (angle 45°). If m the number of points B_j and B_{jp} their traces on the line defined by A_{iv}, the new position A_{inew} of A_i is

$$A_{inew} = w_i \cdot A_i + \frac{1}{m}\sum_{j=1}^{m} B_{jp}$$ and then value of w_i is increased by one.

After exhausting all points A_i, we remove every point B_j that has satisfied all three afore mentioned criteria E- 4, E- 5 and E- 6. The remaining points from R_j are given weight $w_i=1$ and together with the points of R_i form a new reference set of points. Then, we continue with next letter realization from sub step 2.

After exhausting all realizations R_j of the considered alphabet symbol, we obtain a set of points, which is considered to be a potential "ideal" prototype of the specific alphabet symbol and for the given inscription E.

3.2   Final Selection of the Ideal Prototype

The methodology described in section 3.1 is applied with a different letter realization as reference among the n available and a random permutation of the remaining $n-1$ letters, leading to n potential "ideal" prototypes I_i. However, only one of them should be selected. The selection is based on the potential "ideal" prototype that best matches every letter realization of the alphabet symbol in hand. Therefore the methodology of the sub step 2 of section 3.1 is used to match all realizations R_j to a specific I_i each time. If $E(I_i,R_j)$ the fitting error as defined in E- 3, then we assume that mean value $E_m(I_i)$ of quantites $E(I_i,R_j)$ $(j=1...n)$ is an adequate estimation of the quality of the specific potential "ideal" prototype. The potential prototype with the minimum $E_m(I_i)$ is considered to be the finally selected "ideal" prototype for the specific alphabet symbol.

We would like to emphasize that "ideal" prototype I_{iL} is, in fact, an estimation of the actual prototype for the alphabet symbol L, the writer had in mind at least in the period he carved the given inscription. The smaller the overall error $E_m(I_i)$ is, the better we assume that this estimation is.

4   Statistical Analysis and Classification to Cutters

Comparison of "ideal" prototypes of same symbols among different inscriptions will be used in order to classify the inscriptions to cutters.

Quantifying the assumptions previously stated, it is quite sound to consider that if two inscriptions were carved by the same writer, the extracted "ideal" prototypes of same alphabet symbols will match and their fitting error will be quite small. On the other hand, if two inscriptions were carved by different writers, it is quite expectable that there will be at least one symbol whose extracted "ideal" prototypes will not match and their fitting error will be significant.

Let I_{iL}, I_{jL} be the extracted "ideal" prototypes for alphabet symbol L for inscriptions E_i and E_j and R_{1R}, R_{2R}, $..R_{mR}$ be the letter L of the inscription i placed at the position where best fit I_{iL}. Let ε^k_{iL} $(k=1..n_i)$ be the minimum signed distance between point P_k of I_{iL} and its closest points from R_{1R}, R_{2R}, $..R_{mR}$. The closest point is defined in the region of P_k fulfilling E- 4, E- 5 and E- 6. The sign of ε^k_{iL} is equal to the sign of the internal product defined by the formula $(P_k - P_c) \cdot P_{kV}$, where P_{kV} is surface normal on point P_k and P_c its closest point.

According to our assumptions and due to the central limit theorem, ε^k_{iL} will follow normal distribution with population mean value 0 and variance σ^2_{iL}. If

$$\varepsilon_{iL} = \sum_{k=1}^{n_i} \left(\varepsilon^k_{iL} \right)^2$$ is the total square error, quantity $V_{iL} = \varepsilon_{iL}/\sigma^2_{iL}$ will follow chi-square

(X^2). Similarly to definition of ε^k_{iL}, ε^k_{jiL} $(k=1...n_i)$ is the minimum signed distance between point P_k of I_{iL} and its closest point from I^R_{jL} (I^R_{jL} is I_{iL} at the position that best fits I_{iL}). According to our assumptions, quantity ε^k_{jiL} will follow normal distribution with mean value μ_{ji} and variance σ^2_{jiL}, while quantity $V_{jiL} = \frac{1}{\sigma^2_{jiL}} \sum_{k=1}^{n_j} \left(\varepsilon^k_{jiL} - \mu_{ji} \right)^2$

will follow an analogous chi-square (X^2).

It is proved in [3] that quantity

$$F_{ji} = \left(\frac{\varepsilon_{jiL}}{\sigma^2_{jiL}} \right) \cdot \left(\frac{\varepsilon_{iL}}{\sigma^2_{iL}} \right)^{-1} = \left(\frac{1}{\sigma^2_{jiL}} \sum_{k=1}^{n_j} \left(\varepsilon^k_{jiL} \right)^2 \right) \cdot \left(\frac{1}{\sigma^2_{iL}} \sum_{k=1}^{n_i} \left(\varepsilon^k_{iL} \right)^2 \right)^{-1} \qquad \text{E- 7}$$

follows the Snedecor distribution with $(n_j\text{-}1, n_i\text{-}1)$ degrees of freedom.

We state the hypothesis that E_i and E_j have been carved by the same writer. If the hypothesis is valid, then a) population mean value μ_{ji} should be zero and b) population variances σ^2_{jiL}, σ^2_{iL} should be equal. Therefore, E- 7 becomes

$$F_{ji} = \left(\sum_{k=1}^{n_j} \left(\varepsilon^k_{jiL} \right)^2 \right) \cdot \left(\sum_{k=1}^{n_i} \left(\varepsilon^k_{iL} \right)^2 \right)^{-1} \qquad \text{E- 8}$$

If $f(x, n_1, n_2)$ is the Snedecor probability density function for (n_1, n_2) degrees of freedom, the following value is calculated:

$$G_{ji} = \int_{F_{ji}}^{\infty} f\left(x, n_j - 1, n_i - 1 \right) \cdot dx \qquad \text{E- 9}$$

Quantity G_{ji} tests the likelihood of the stated hypothesis. If BE a properly selected threshold for confidence interval and $G_{ji} < BE$, the stated hypothesis that E_i and E_j are carved by the same writer is rejected. In any other case the hypothesis cannot be rejected. Quantity G^L_{ji} is calculated for all possible inscription pairs E_i, E_j among the N available and for the selected symbols L. For every pair of inscriptions E_i, E_j and tested alphabet symbols L_1, L_2, L_n the value $G_{ji} = \min(G^{L1}_{ji}, G^{L2}_{ji,...} G^{Ln}_{ji})$ is kept.

The initial confidence interval threshold is $BE = 10^{-8}$. The inscription pair that rejects with the greatest confidence interval the hypothesis that has been carved from the same writer is found. These two inscriptions are automatically classified into two different cutters X_1 and X_2. The remaining inscriptions are tested and the one that rejects the hypothesis that has been carved either from X_1 or X_2 is found. This inscription is automatically classified to writer X_3. The process is repeated for all remaining inscriptions until the stated hypothesis of common origin cannot be rejected. Then, the confidence interval is decreased to 10^{-7}, 10^{-6} and 10^{-5}. After the completion of the current process, X inscriptions have been classified to X different cutters. The remaining $N\text{-}X$ inscriptions are classified to the above cutters with the aid of the following quantity:

$$U_{ji} = \sqrt[n]{\prod_{k=1}^{n} f\left(F^{L_k}_{ji}, n^{L_k}_j - 1, n^{L_k}_i - 1 \right)} \qquad \text{E- 10}$$

where $f(x, n_1, n_2)$ the Snedecor density function with (n_1, n_2) degrees of freedom, L_k the tested alphabet symbol and $F^{L_k}_{ji}$ the quantity defined in E- 7 for the "ideal" prototypes of the symbol L_k of inscriptions E_i and E_j In this case, E_i belongs to the set of already classified to different cutters inscriptions X, while E_j belongs to the remaining $N\text{-}X$ inscriptions. Therefore, for each unclassified inscription E_j we calculate X number values, each one depicting the likelihood that E_j is carved by the same writer as E_i. The maximum likelihood value determines hand that most probably carved E_j.

5   Application of the Introduced Methodology

Expert epigraphist Prof. Stephen Tracy indicated to the authors 28 inscriptions located at the premises of either the Epigraphical Museum of Athens or the Museum of the Ancient Agora of Athens. Apart from an arbitrary identification code, Prof. Tracy revealed no other information about the inscriptions. A 3D mesh dataset for each inscription was acquired and 3D mesh datasets for depicted alphabet symbols A, M, N and Σ were isolated. These four letters have been chosen by the authors due to their high frequency of appearance and the convenience in the recognition of their appearance on the inscriptions.

Subsequently, the methodology described in previous sections was implemented. Firstly, geometrical models of letters have been generated and "ideal" prototypes were extracted for each available alphabet symbol and each inscription separately. The average number of letter realizations per inscription used for "ideal" prototype extraction was 15. Deviation to lower values of employed realizations was due to the overall number of letter realizations on the inscriptions.

Then, "ideal" prototypes and statistical analysis of section 0 has been applied. In the first stage, the employed methodology extracted the eight inscriptions of column 2 of Table 2, rejecting with maximum likelihood the hypothesis that any two of them were carved by the same writer. As a result, those inscriptions were classified to eight different cutters. The ordering of inscriptions listed in column 2 of Table 2 corresponds to the order with which the statistical analysis rejected the above hypothesis. with confidence interval greater than 10^{-5}. For each one of the remaining twenty inscriptions, pair wise calculations of quantity U_{ji} defined in E- 10 have been performed with each one of the inscriptions listed in column 2 of Table 2, leading to the classification listed in third column of Table 3, based on consideration of maximum value of U_{ji} (Fig. 5). The classification results were in full accordance with the established opinion of Prof. Tracy and several prominent epigraphists.

Table 1. List of inscriptions employed

E_i	Identification Code	E_i	Identification Code	E_i	Identification Code
1	I0247	11	I6295	21	I7457
2	I1640	12	I7188	22	I7481
3	I4033	13	I7190	23	I7482
4	I4266	14	I7233_7335	24	I7519
5	I4917	15	I7237	25	I7566
6	I5039	16	I7237a	26	I7567_7568_7569
7	I5297	17	I7245	27	I7587
8	I6006	18	I7400	28	I7723
9	I6053	19	I7405		
10	I6124	20	I7446		

Table 2. Final classification to cutters of employed inscriptions

Carver	E_i	Inscriptions
X_1	13	-
X_2	22	E_{23}, E_2, E_3
X_3	20	E_{21}, E_{28}, E_{17}
X_4	10	E_{18}, E_{19}
X_5	7	E_{11}, E_9, E_8, E_{25}, E_{24}, E_{26}
X_6	1	E_5, E_{12}
X_7	16	E_{14}, E_{15}
X_8	6	E_4, E_{27}

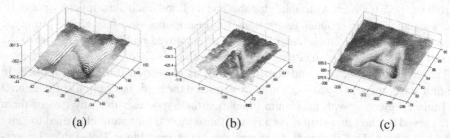

(a)   (b)   (c)

Fig. 1. (a), (b), (c) 3d meshes visualization of various letter realizations from three different inscriptions. The reference system is defined by the 3D scanner. The color of each point is proportional to its z-component in the corresponding coordinate system.

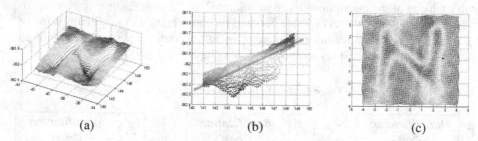

(a)   (b)   (c)

Fig. 2. (a) Letter at initial reference system, (b) Calculated plane along with point mesh and (c) Point mesh in the new reference system. The color of each point is proportional to its z-component in the corresponding coordinate system.

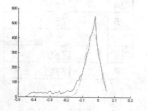

Fig. 3. Histogram distribution $H(E_z)$ (continuous line) and normal distribution function $f(\alpha,\mu,\sigma)$ (dotted line) approximating $H(E_z)$

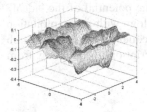

Fig. 4. Letter points after transformation and primary faces extracted by Delaunay triangulation

Fig. 5. Points within conical volumes defined by criteria E- 4, E- 5 and E- 6

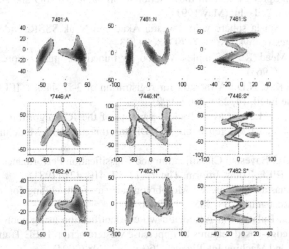

Fig. 6. Visualization of esitimated "ideal" prototypes of letters A. N and Σ from inscription E_{22} (I7481) matched with their corresponding prototypes from inscriptions E_{20} (I7446) and E_{23} (I7482). The hypothesis that E_{20} and E_{22} have been carved by the same cutter is rejected *and therefore E_{20} and E_{22} are attributed to different carvers.* After determining the count of cutters, equation E- 10 attributes E_{23} to the same cutter with E_{21}.

6   Conclusion

A methodology for identifying the writer who carved an ancient inscription has been presented. The methodology employs the acquired three dimensional datasets of the realizations of alphabet symbols to create an estimation of an ideal prototype surface of the specific symbol. Then, statistical criteria are introduced for the classification process. Application of the method revealed that 8 distinct cutters carved the 28 inscriptions in hand. The obtained classification is in full accordance with the established opinion of prominent epigraphists and signify the potential of the "ideal" prototype as introduced in [2] and used also in this methodology with full exploitation of the 3D representation of the archaeological find. Consequently, further research will include application of the methodology to more inscriptions originating from several ancient civilizations, to ancient coins, to vessels, etc.

References

[1] Papaodysseus, C., Roussopoulos, P., Panagopoulos, M., Fragoulis, D., Dafi, D., Panagopoulos, T.: Identifying hands on ancient Athenian inscriptions: First step towards a digital approach. Archaeometry 49(4), 749–764 (2007)

[2] Panagopoulos, M., Papaodysseus, C., Roussopoulos, P., Dafi, D., Tracy, S.: Automatic writer identification of ancient Greek inscriptions. IEEE Pattern Analysis and Machine Intelligence 31(8), 1404–1414 (2009)

[3] Kokolakis, G., Spiliotis, J.: «An Introduction to probability theory and statistics,» Symeon Publications, 2nd edn. (May 1991)

[4] Delaunay, B.: Sur la sphère vide, Izvestia Akademii Nauk SSSR. Otdelenie Matematicheskikh i Estestvennykh Nauk 7, 793–800 (1934)

[5] Nelder, J.A., Mead, R.: A Simplex Method for Function Minimization. Computer Journal 7, 308–313 (1965)

[6] Besl, P.J., McKay, N.D.: A method for registration of 3-D shapes. IEEE Trans. Pattern Anal. Machine Intell. 14, 239–256 (1992)

[7] Rusinkiewicz, S., Levoy, M.: Efficient variants of the ICP algorithm. In: Proceedings Third International Conference on 3-D Digital Imaging and Modeling, pp. 145–152 (2001)

[8] Kokolakis, G.E.: Bayesian Classification and Classification Performance for Independent Distributions. IEEE, Trans. Inform. Theory, IT-27, 419–421 (1981)

[9] Zois, E.N., Anastassopoulos, V.: Morphological waveform coding for writer identification. Pattern Recognition 33, 385–398 (2000)

[10] Schomaker, L., Bulacu, M.: Automatic writer identification using connected-component contours and edge-based features of uppercase western script. IEEE Transactions on Pattern Analysis and Machine Intelligence 26(6), 787–798 (2004)

[11] Bulacu, M., Schomaker, L.: Text-Independent Writer Identification and Verification Using Textural and Allographic Features. IEEE Transactions on Pattern Analysis and Machine Intelligence 29(4) (April 2007)

[12] Heb, Z., Youb, X., Tanga, Y.Y.: Writer identification using global wavelet-based features. Neurocomputing 71(10-12), 1832–1841 (2008)

What the Eye Did Not See – A Fusion Approach to Image Coding

Ali Alsam, Hans Jakob Rivertz, and Puneet Sharma

Department of Informatics & e-Learning(AITeL)
Sør-Trøndelag University College(HiST)
Trondheim, Norway
er.puneetsharma@gmail.com

Abstract. The concentration of the cones and ganglion cells is much higher in the fovea than the rest of the retina. This non-uniform sampling results in a retinal image that is sharp at the fixation point, where a person is looking, and blurred away from it. This difference between the sampling rates at the different spatial locations presents us with the question of whether we can employ this biological characteristic to achieve better image compression. This can be achieved by compressing an image less at the fixation point and more away from it. It is, however, known that the vision system employs more that one fixation to look at a single scene which presents us with the problem of combining images pertaining to the same scene but exhibiting different spatial contrasts. This article presents an algorithm to combine such a series of images by using image fusion in the gradient domain. The advantage of the algorithm is that unlike other algorithms that compress the image in the spatial domain our algorithm results in no artifacts. The algorithm is based on two steps, in the first we modify the gradients of an image based on a limited number of fixations and in the second we integrate the modified gradient. Results based on measured and predicted fixations verify our approach.

1 Introduction

From the very beginning of photography, cameras were designed and iteratively improved with the aim of mimicking the human visual system. From this perspective, a camera is thought of as a machined eye–a device that is sensitive to illumination. Equally, we normally think of algorithms such as white-balancing [1], adaptation [2] and tone mapping [3] as being similar to the biological processes of the vision system.

A camera is of course not a human visual system. The two are different in a number ways some of which are relevant to the work presented in this article. Primarily, while digital camera manufacturers are striving to produce devices with progressively higher resolution, the human brain has evolved to be efficient, i.e. use less information to reach greater conclusions. Thus while the camera sensor has a uniform number of pixels per unit area, the human eye has a much

G. Bebis et al. (Eds.): ISVC 2012, Part II, LNCS 7432, pp. 199–208, 2012.
© Springer-Verlag Berlin Heidelberg 2012

higher resolution in the fovea which is the center part of the retina [4]. It is well known that the fovea is responsible for our central, sharpest vision while the cone distribution in the rest of the retina results in blurred vision [4].

In the process of exploring a scene, the brain directs the eyes to different spatial locations. At those locations, known as fixations the eyes pause and gather the visual information [5]. Due to the concentration of photo-receptors at the fovea, we can think of each pause as the time taken to capture an image that is sharp at the fixation point and blurred away from it. Given that the average distribution per unit area and spatial location of the cones in the retina is known, it is possible to model the spatial contrast of the retinal image at each fixation.

For a given scene, the number of fixations and their locations vary. The question of whether fixations are guided by image features has been addressed extensively in vision research; and some conclusions are widely accepted. Specifically, experiments have shown that for a given image, people tend to look at the same regions [6,7], they tend to look at the central part [8,7] and that certain image attributes such as luminance and colour contrasts tend to attract fixations [9,10]. Furthermore, fixations can be measured using eye trackers and the experimental data shows conclusively that for a general image the human visual system employs more than one fixation [6].

Based on a given digital image and a number of measured or predicted fixations, we can model the foveation effect, i.e a sharp region at the fixation point and blurring away from it. The result of such a model would be a number of images with different spatial contrast. As an example, see figure 1 where we have modeled the foveation effect based on 3 different fixations. Given such an image series we might wonder how the vision system integrates the different foveation results into a seamless visual experience; and subsequently how we can design signal processing algorithms that offer such functionality.

In this article, we present an algorithm which integrates a number of differently foveated images in the gradient domain. The algorithm starts by calculating the gradients of the input image. Having done that a number of fixation locations are used to calculate the corresponding foveated gradients. Here we use the foveation function described by Geisler and Perry [11]. As a second step, the gradients are combined using the fast colour to gray algorithm by Alsam and Drew [12]. The Alsam and Drew algorithm [12] combines the gradients from n channels into a single gradient by arguing that the maximum horizontal and vertical differences over all the channels result in the maximum contrast. Thus the gradient fusion step is guaranteed to result in a gradient where the maximum differences pertaining to the fixations locations are maintained. As a final step the resultant gradient is integrated using the modified Frankot-Chellappa-algorithm [13] proposed by Alsam and Rivertz [14].

The need for a fast algorithm to combine foveated images is best motivated in the image compression domain where improvements in statistically based image compression, i.e. methods that are based on data analysis have long slowed down. The use of human vision steered compression is seen by researchers as the most promising path toward further improvements. In this regard, the algorithm presented in this article can be used as part of an image compression pipeline with very promising results. From our initial tests, we have noticed that the algorithm results in reduced storage requirements without the added artifacts associated with frequency based compressions in the wavelets domain.

Like other foveation driven algorithms, our method is dependent on accurate estimation of the fixation points. Thus in our experimental section, we present results based on measured fixation data as well as predictions based on the visual saliency algorithm by Itti et al. [15].

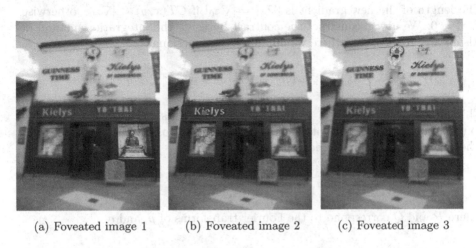

(a) Foveated image 1   (b) Foveated image 2   (c) Foveated image 3

Fig. 1. Figures show the foveated images for three fixations, here the fixation points are represented as red dots

2   The Filter and the Integration

Experiments for measuring the contrast sensitivity of the human eye have been carried out [16,17]. Based on these experiments, the contrast threshold has been modeled through the function

$$CT(f, \theta) = CT_0 \exp \left(\alpha f \frac{\theta + \theta_2}{\theta_2} \right).$$

Here, f is the spatial frequency measured in degrees, θ is the retinal eccentricity. CT_0 is the minimal contrast threshold, θ_2 is the half-resolution eccentricity constant, and α is the spatial frequency decay constant. The values used in [18] are $\alpha = 0.106$, $\theta_2 = 2.3$, and $CT_0 = 1/64$.

Given a normalized gray scale image $z_0 : \Omega \to [0,1]$. Denote its width by w, measured in pixels. An observer views the image from a distance d, measured in pixels. The maximal spatial frequency of the image is given by $f_d = \frac{w}{4 \arctan \frac{w}{2d}}$. If r is the distance measured in pixels from a fixation point, then $\theta(r) = \arctan \frac{r}{d}$.

The gradient ∇z_0 is modified by setting the its magnitude to zero if its is less than $CT(f_d, \theta)$ for some of the fixation points.

We make a new contrast threshold function based on $f = f_d$ and the fixation points, $(x_1, y_1), (x_2, y_2), \ldots, (x_n, y_n)$.

$$CT(x, y) = \min(CT_1(x, y), CT_2(x, y), \ldots, CT_n(x, y)),$$

where $CT_k(x, y) = CT\left(f_d, \theta\left(\sqrt{(x - x_k)^2 + (y - y_k)^2}\right)\right)$, $k = 1, 2, \ldots, n$. This step is equivalent to the Alsam and Drew method [12].

The direction of the original and modified gradients are $\hat{u} = \nabla z_0 / |\nabla z_0|$. The length of the new gradient is $|\nabla z| = |\nabla z_0|$ if $CT(x, y) < |\nabla z_0|$, otherwise $|\nabla z| = 0$. We now reconstruct the contrast by using the integration method of Alsam and Rivertz [14] where we minimize the functional:

$$W(z) = \lambda \int_\Omega |z - z_0|^2 \, dx \, dy + \int_\Omega \left(|z_x - p|^2 + |z_y - q|^2\right) dx \, dy.$$

This minimization results in an image whose gradients are as close as possible to (p, q), under the constraint that the luminance is close to the original image. The image z in the Fourier domain can be taken as

$$Z(u, v) = \frac{\lambda Z_0 - i(uP + vQ)}{\lambda + u^2 + v^2},$$

where P and Q correspond to the Fourier transforms of p, and q.

3　Results

To test the proposed method, we used images and corresponding fixations data from the study by Judd et al. [6]. The results for two images and the associated fixations are shown in figures 2 to 3. In the left column the foveated images for three fixations are shown. Here, the fixation points are represented as red dots. In agreement with the predicted results for the application of the contrast function by Wang and Bovik [18], we notice that the regions around the fixation points are sharper than the rest. The images in the right column show the original image, the result obtained by combining the foveated images using the proposed method, and the difference between the result and the original image. We notice that the result image is sharp in the regions corresponding to the three fixation points, we further notice that the image represents a good approximation of the original with greater differences in the parts that the observer deemed to be less salient. Here we remark that the difference between the original and the result can be optimized by controlling the λ parameter defined in the previous section.

In figure 4, the left column contains the foveated images obtained by using the first three salient points from the visual saliency algorithm by Itti et al. [15] and the right column contains the original image, the result obtained by using

(a) Foveated image 1   (d) Original image

(b) Foveated image 2   (e) Result

(c) Foveated image 3   (f) Difference

Fig. 2. In the left column the foveated images for three fixations are shown. Here, the fixation points are represented as red dots. The images in the right column show the original image, the result obtained by combining the foveated images using the proposed method, and the difference between the result and the original image. We notice that the result image is sharp in the regions corresponding to the three fixation points, we further notice that the image represents a good approximation of the original with greater differences in the parts that the observer deemed to be less salient. In the difference image, the dark regions indicate the locations where the differences are higher.

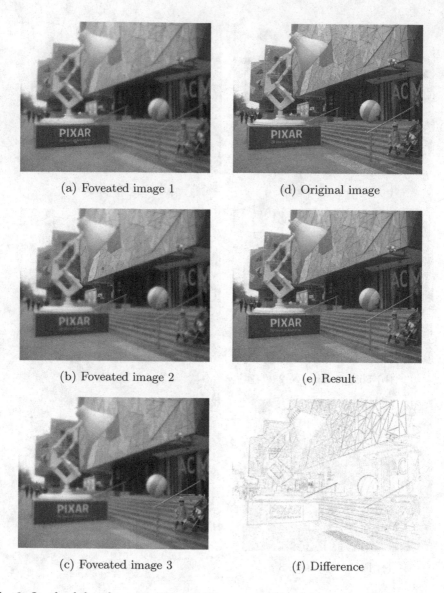

(a) Foveated image 1   (d) Original image

(b) Foveated image 2   (e) Result

(c) Foveated image 3   (f) Difference

Fig. 3. In the left column the foveated images for three fixations are shown. Here, the fixation points are represented as red dots. The images in the right column show the original image, the result obtained by combining the foveated images using the proposed method, and the difference between the result and the original image. We notice that the result image is sharp in the regions corresponding to the three fixation points, we further notice that the image represents a good approximation of the original with greater differences in the parts that the observer deemed to be less salient. In the difference image, the dark regions indicate the locations where the differences are higher.

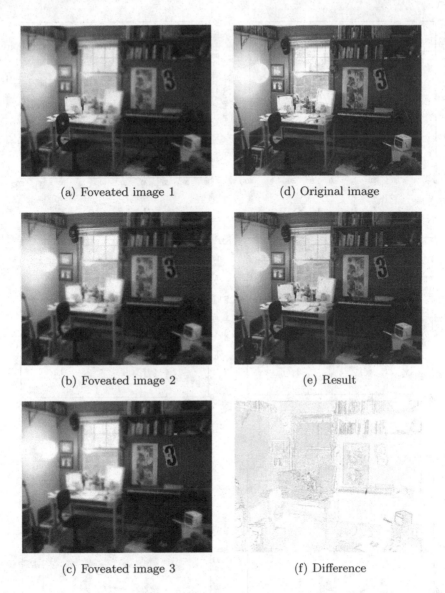

(a) Foveated image 1   (d) Original image

(b) Foveated image 2   (e) Result

(c) Foveated image 3   (f) Difference

Fig. 4. In the left column the foveated images obtained by using first three salient points from the visual saliency algorithm by Itti et al. [15] are shown. Here, the fixation points are represented as red dots. The images in the right column show the original image, the result obtained by combining the foveated images using the proposed method, and the difference between the result and the original image. We notice that the result image is sharp in the regions corresponding to the three fixation points, we further notice that the image represents a good approximation of the original with greater differences in the parts that the observer deemed to be less salient. In the difference image, the dark regions indicate the locations where the differences are higher.

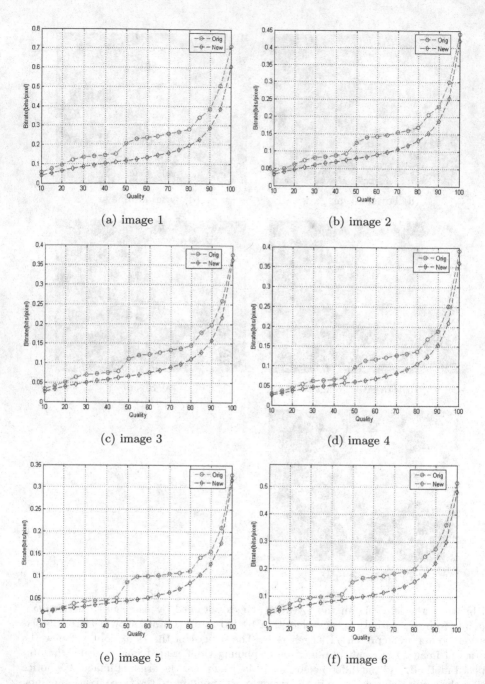

(a) image 1

(b) image 2

(c) image 3

(d) image 4

(e) image 5

(f) image 6

Fig. 5. Figures show the bitrates for saving the original image and corresponding result image in JPEG format with different quality values, ranging from 10 to 100 based on six different images. Here we notice that for the same compression quality the new images require lower storage space.

the proposed method, and the difference between the result and the original image. For this experiment, we notice that the results are very similar to those obtained for the first test image. We underline, however, that the choice of fixation locations and the number of salient regions is clearly related to the results that we obtain, where the higher the number of fixations and the more spread they are in the image plane the closer the result is going to resemble the original.

Finally, in figures 5(a) to 5(f), we show the bitrates obtained by saving the original image and corresponding result image in JPEG format with different quality values, ranging from 10 to 100 based on six different images. Here we notice that for the same compression quality the new images require lower storage space. Given that the foveation function reduces the high frequency elements of the original image, we can argue that this result is not surprising. The advantages of this approach are, however, more subtle than a simple removal of high frequency elements- we have removed high frequencies locally- in regions where the foveation function predicts that the observer couldn't see with the sharp part of their vision.

4   Conclusion

This article presents an algorithm to combine a series of differently foveated images pertaining to an identical scene. This is achieved by using image fusion in the gradient domain. The advantage of the algorithm is that unlike other algorithms that compress the image in the spatial domain our algorithm results in no artifacts. The algorithm is based on two steps, in the first we modify the gradients of an image based on a limited number of fixations and in the second we integrate the modified gradient. Results based on measured and predicted fixations verify our approach. The need for a fast algorithm to combine foveated images is best motivated in the image compression domain where improvements in statistically based image compression, i.e. methods that are based on data analysis have long slowed down. The use of human vision steered compression is seen by researchers as the most promising path toward further improvements. In this regard, the algorithm presented in this article can be used as part of an image compression pipeline with very promising results. From our initial tests, we have noticed that the algorithm results in reduced storage requirements without the added artifacts associated with frequency based compressions in the wavelets domain.

References

1. Chikane, V., Fuh, C.S.: Automatic white balance for digital still cameras. Journal of Information Science and Engineering 22, 497–509 (2006)
2. Hurley, J.B.: Shedding light on adaptation. Journal of General Physiology 119, 125–128 (2002)

3. Qiu, G., Guan, J., Duan, J., Chen, M.: Tone mapping for hdr image using optimization a new closed form solution. In: 18th International Conference on Pattern Recognition, ICPR 2006, vol. 1, pp. 996–999 (2006)

4. Cormack, L.K.: Computational models of early human vision. In: Handbook of Image and Video Processing, pp. 325–345. Elsevier Academic Press (2005)

5. Rajashekar, U., van der Linde, I., Bovik, A.C., Cormack, L.K.: Gaffe: A gaze-attentive fixation finding engine. IEEE Transactions on Image Processing 17, 564–573 (2008)

6. Judd, T., Ehinger, K., Durand, F., Torralba, A.: Learning to predict where humans look. In: International Conference on Computer Vision, ICCV (2009)

7. Alsam, A., Sharma, P.: Analysis of eye fixations data. In:Proceedings of the IASTED International Conference, Signal and Image Processing (SIP 2011), pp. 342–349 (2011)

8. Tatler, B.W.: The central fixation bias in scene viewing: Selecting an optimal viewing position independently of motor biases and image feature distributions. Journal of Vision 7, 1–17 (2007)

9. Itti, L., Koch, C.: Computational modelling of visual attention. Nature Reviews Neuroscience 2, 194–203 (2001)

10. Meur, O.L., Callet, P.L., Barba, D., Thoreau, D.: A coherent computational approach to model bottom-up visual attention. IEEE Transactions on Pattern Analysis and Machine Intelligence 28, 802–817 (2006)

11. Geisler, W.S., Perry, J.S.: A real-time foveated multiresolution system for low-bandwidth video communication. In: SPIE Proceedings, vol. 3299, pp. 1–13 (1998)

12. Alsam, A., Drew, M.S.: Fast colour2grey. In: 16th Color Imaging Conference: Color, Science, Systems and Applications, Society for Imaging Science & Technology (IS&T)/Society for Information Display (SID) Joint Conference, pp. 342–346 (2008)

13. Frankot, R.T., Chellappa, R.: A method for enforcing integrability in shape from shading algorithms. IEEE Transactions on Pattern Aanalysis and Machine Intelligence 10, 439–451 (1988)

14. Alsam, A., Rivertz, H.J.: Constrained gradient integration for improved image contrast. In: Proceedings of the IASTED International Conference, Signal and Image Processing (SIP 2011), pp. 13–18 (2011)

15. Itti, L., Koch, C., Niebur, E.: A model of saliency-based visual attention for rapid scene analysis. IEEE Transactions on Pattern Analysis and Machine Intelligence 20, 1254–1259 (1998)

16. Banks, M., Sekuler, A., Anderson, S.: Peripheral spatial vision: limits imposed by optics, photoreceptors, and receptor pooling. J. Opt. Soc. Am. A 8, 1775–1787 (1991)

17. Arnow, T.L., Geisler, W.S.: Visual detection following retinal damage: Predictions of an inhomogeneous retino-cortical model. In: Human Vision and Electronic Imaging. Proceedings of SPIE, vol. 2674 (1996)

18. Wang, Z., Bovik, A.C.: Embedded foveation image coding. IEEE Transactions on Image Processing 10, 1397–1410 (2001)

Knot Detection in X-Ray CT Images of Wood

A. Krähenbühl[1], B. Kerautret[1], I. Debled-Rennesson[1],
F. Longuetaud[2], and F. Mothe[2]

[1] Université de Lorraine, LORIA, Adagio Team, UMR 7503, 54506, Nancy, France
[2] INRA, UMR1092 LERFoB, 54280 Champenoux, France

Abstract. This paper presents an original problem of knot detection in 3D X-ray Computer Tomography images of wood stems. This image type is very different from classical medical images and presents specific geometric structures. These ones are characteristic of wood stems nature. The contribution of this work is to exploit the original geometric structures in a simple and fast algorithm to automatically detect and analyze the wood knots. The proposed approach is robust to different wood qualities, like moisture or noise, and more simple to implement than classical deformable models approaches.

1 Introduction

Detect and identify automatically digital objects in 3D volumetric images is a challenge, especially in medical imaging applications. Various methods are proposed to extract and measure internal characteristics of the human, like brain or blood vessel. Currently, the most effective techniques use deformable models [1].

(a)   (b)

Fig. 1. Example of wood image and 3D volumic rendering

In this work, we focus on a similar problem with 3D digital images of wood. The images are obtained by medical X-Ray Computer Tomography (CT) scanners and are illustrated on Fig. 1. On Fig. 1(a), a raw image is projected in an intensity interval which highlight four knots starting from the pith[1]. The related volume obtained by thresholding is rendered on Fig. 1(b).

A knot as in Fig. 1 and Fig. 2 is the first part of a branch, included within the tree stem due to the radial growth of the stem. The frequency and size of the knots are the first depreciation factors considered by wood suppliers for estimating the price of timber. Knottiness is also one of the main criteria considered in the visual grading of lumber. Knowledge of knot geometry and location would be valuable at sawmills for optimising cutting decisions or improving the grading of logs. X-ray CT has been recognised as being the most promising method to non-destructively analyse the internal structure of wood [2]. CT scanners designed expressly for the wood industry are now available and some of

---

[1] Pith : central part of stem.

G. Bebis et al. (Eds.): ISVC 2012, Part II, LNCS 7432, pp. 209–218, 2012.

the largest sawmills are now equipped with them. Moreover, such detailed data about knottiness are needed in the field of forest research to study for example the relationship between tree growth conditions and wood quality.

Several methods were proposed for measuring knottiness on the basis of CT images (see [3] for a review). But no method is entirely automatic and well adapted to all species of wood and to wood containing wet areas such as sapwood[2]

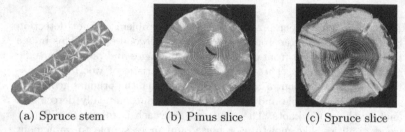

(a) Spruce stem   (b) Pinus slice   (c) Spruce slice

Fig. 2. X-Ray CT images of stem : (a) 3D volumetric rendering , (b) dry and (c) with sapwood

We proposed in [3] an entirely automated method to identify knots from X-ray CT images of a piece of wood (log or beam) and obtain data on knot geometry without any human intervention. However, the first step of the algorithm, image segmentation, was not studied in details. A simple thresholding operation was sufficient enough for processing the air-dried wood that were used in this study. For an application in industrial conditions, there is a need to improve the segmentation algorithm, especially for wet wood containing areas of sapwood (see Fig. 2(c)).

The main contribution of this paper is to automatically and quickly detect all knots of a given log based on the analysis of X-ray CT images. Various softwood species are studied, various moisture contents were considered leading to have large sapwood areas visible in the CT images (see Fig. 2(c)), and our method identifies all knots in all cases.

The proposed knot detection is based on a very precise detection of the localisation of each knot. To do this, we use intensity variation between two consecutive slices and study a specific intensity histogram. We proceed in two steps. The first one is to select slice intervals where intensity variations are important around the pith. In the second step, we keep a set of angular sectors also centered on the pith in each interval from the same criterion of intensity variation.

In the next section we list all tools needed by the proposed method. Section 3 introduces the definition of the cumulative z-motion histogram which allows to identify the meaning slice intervals and to deduce the angular sector of each knot. The last section shows results and presents perspectives of this work.

---

[2] Sapwood : wood area located at the periphery of the stem.

2   Suitable Tools for X-Ray CT Images of Logs

2.1   Pith Detection

Pith detection is an essential first step to wood knot detection. In particular, in the proposed method it will allow to focus on a restricted area around the pith. In this area, we are sure that intensity variations observed in images indicate the presence of knots. Indeed, the knots are connected to the pith where they have their origin. They are oriented radially in a log, from the pith to the bark, and generally upward. Some knots are totally included into the log and covered by the radial growth rings[3] but in general, thay are still visible at the outside of the log in the form of branches.

To detect pith, we use the algorithm proposed by Fleur Longuetaud [4] and improved by Boukadida et al. [5]. The idea is to detect the pith successively in each slice. Its location in slice n allows to steer and speed up pith computing on slice $n + 1$.

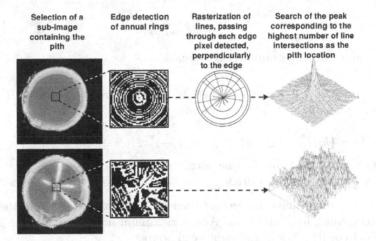

| Selection of a sub-image containing the pith | Edge detection of annual rings | Rasterization of lines, passing through each edge pixel detected, perpendicularly to the edge | Search of the peak corresponding to the highest number of line intersections as the pith location |

Fig. 3. Pith detection in two slices, with (below) and without (above) knot

The first step of the algorithm uses the roundness of growth rings. They are detected in each slice I_k with a Sobel filter. This one computes G_x and G_y images, respectively the gradients of I_k along x and y.

A new image I'_k is computed with a thresholding of G norm (see Fig. 3). I'_k allows to define a set of lines. For that, we are computing Θ, the direction of gradient in each pixel of I'_k not null. After, we are building the map of the intersection locations for all lines taken two by two. The intersection locations are computed with an Hough transform [6]. The pith position is then defined like the location of the largest number of intersecting lines.

In a second step, the pith position is corrected in slices where pith neighbourhood is noisy. It is usually the slices with knots. These ones disturb the roudness

---

[3] Growth rings: concentric circles in stem. They are centered on pith.

of growth rings and this noise hugely affects the gradient direction and therefore Hough transform. In these slices, the pith is recomputed by linear interpolation of the previous and the next slice.

2.2 Intervals Detection in an Histogram

The method that we propose detects intervals centered on maximums. It is a problem that can be very difficult to solve according to noise level on histogram. The detection that we are now presenting is suitable for the histograms we have to deal with.

Let H be an N-class histogram with $H = (H_k)_{k \in [1,N]}$.

Smoothing. The first step consists to smooth H with an averaging filter of radius r_f. The aim is to improve the computing of the discrete derivative during the following step. We get H^l histogram where the k class is :

$$H_k^l = \frac{1}{2r_f + 1} \sum_{i=k-r_f}^{k+r_f} H_i$$

with r_f : radius of histogram averaging mask.

Maximums Computing. We defined M as the set of the index of H^l maximums from a neighbourhood of r_M radius.

$$M = \{k | H_k^l > t_M \text{ and } \forall i \in [-r_M, -1] \cup [1, r_M], H_k^l > H_{k+i}^l\}$$

with t_M: minimum value for one maximum.

r_M: width of neighbourhood on which the maximums are defined.

Value of a maximum must be greater than t_M. This allows to avoid the small maximums coming from the noise. And a maximum must have a value greater than the ones of its $\frac{r_M}{2}$ left and right neighbours.

Intervals Computing. Let V be the set of searched intervals. We determine V from the M set of maximum indexes. An interval of classes is computed from each maximum, starting from each side of this maximum.

$$V = \{ \ [k-i, k+j] \mid k \in M$$

$$\text{and } \exists \ t_i \in [k-i,k] \quad \text{such that } \begin{cases} \forall t \in [t_i, k], & H_t^l > t_V \\ \forall t \in]k-i, t_i[, \ H_t^l \leq t_V \text{ and } (H_t^l)' > 0 \\ (H_{k-i}^l)' \leq 0 \end{cases}$$

$$\text{and } \exists \ t_j \in [k+1, k+j] \quad \text{such that } \begin{cases} \forall t \in]k, t_j], & H_t^l > t_V \\ \forall t \in]t_j, k+j[, \ H_t^l \leq t_V \text{ and } (H_t^l)' < 0 \\ (H_{k+j}^l)' \geq 0 \end{cases}$$

$$\text{and } j - i > w_V \ \}$$

with $(H_t^l)' = H_t^l - H_{t-1}^l$
and t_V: derivative threshold.
     w_V: minimum width of an interval.

Derivatives are tested on each side of maximums only, from the nearest neighbour with a value lower than t_V. An interval is kept if its magnitude is larger than w_V.

3   Z-Motion Histogram and Knot Detection

With detection methods of pith and intervals, presented in previous section, we can determine knot areas. They take the form of angular sectors of cylinder and our goal is to get one area by knot. Let us specify that we detect these areas in image sequences from medical X-Ray scanner. Each image corresponds to a circular slice of log (see Fig. 2). Pixel values corresponds to density on Hounsfield scale. This density is linearly correlated to the wood density. We referred either to wood density or to images intensity.

On X-Ray images of log, pixel intensity is approximately between −3000 and 3000. These bounds changed few from an image to an other. On Hounsfield scale, wood density is usually greater than −900 and always lesser than 530. All image values out of this interval will be considered as outlier and ignored.

Our method consists in applying twice the process of interval detection. The two processed histograms will be constructed from the same notion : the z-motion.

The Z-Motion. We decided to use the z-motion by observing successively an image sequence. By scrolling an image sequence of one log, egg-shaped shapes are observed in motion. These movements occur from pith to bark, orthogonally to the growth rings, when the images are scrolled from the bottom to the top of the tree. These are the knot movements that have their origin at the pith location and have grown radially and upwards.

The z-motion Z of a slice I_k is defined like the intensity variation between the two successive slices I_{k-1} and I_k :

$$Z_k = |I_k - I_{k-1}|$$

With help of this notion, we now present the two steps of our method.

Before this two steps, it is necessary to detect pith (see Sec. 2.1) on all slices. It allows to define a C circle centered on pith in each slice of log. Then for the first step, the histogram is constructed by cumulating the z-motion of each slice on the entire circle C. We then obtain the H_{slices} histogram on which we apply the process of interval detection (see Sec. 2.2). During the second step, we subdivided the C circle in a set of angular sectors with same size. For each interval of H_{slices}, we combine the z-motion separately on each angular sector.

Hereafter are more details about the construction of these two histograms.

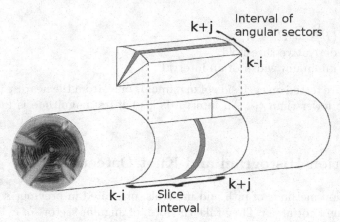

Fig. 4. Angular sesctors intervals on a given slice interval. The green sections represents the slice and the angular sector where the z-motion cumulation is maximum. We can see in initial log section the considered circle C and the localisation of pith.

3.1   The Histogram of Cumulative Z-Motion

The Slice Histogram. Let I be a set of images of a log slice. Let C be a circle of radius r_C. We defined the $H_{slices}(k)$ value of H_{slices} histogram for the I_k slice as follows :

$$H_{slices}(k) = \sum_{(i,j)\in C} Z_k(i,j)$$

where

- C is centered on the pith position in I_k slice.
- $\forall (i,j), I_k(i,j) \in [i_{min}, i_{max}]$ et $I_{k-1}(i,j) \in [i_{min}, i_{max}]$
- $\forall (i,j), Z_k(i,j) \in [z_{min}, z_{max}]$

with r_C : radius of the cumulative z-motion circle.
        $[i_{min}, i_{max}]$: intensity interval.
        $[z_{min}, z_{max}]$: interval of intensity variations.

The z-motion is only computed inside the C circle centered on pith. According to experts, the minimum wood density can be fixed to -900. It is our value for i_{min}. For i_{max}, we fixed its value to 530, a biological constant for the maximum wood density. Variations lesser than z_{min} due to growth rings, sapwood, etc. are considered as noise. Variations in $[z_{min}, z_{max}]$ interval are potential movements of knots. Variations greater than z_{max} are considered as outliers.

In this first step, each histogram value corresponds to the z-motion sum in a slice. Intervals are therefore slice intervals where the cumulative z-motion is important (see Fig. 5).

One of the most important parameters is the r_C radius of C circle. It is determine on each slice. We start from pith in ten directions as long as pixel values are greater than -900. In fact, values of X-ray scanned images fall drastically

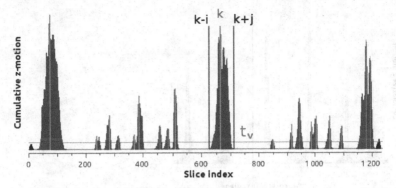

Fig. 5. Histogram after detection of slice intervals in a spruce log

below -900 when we leave the log. Then we compute the average radius r and r_C is fixed to $0.8 \cdot r$.

The z-motion interval $[z_{min}, z_{max}]$ is fixed to $[200, 500]$ to just cumulate important z-motion. After histogram construction, it is smoothed with an averaging mask where $r_f = 2$. Each histogram value becomes candidate to the maximum set if its value is greater than $t_M = 5\%$ of the maximum value of H_{slices}.

To compute intervals, we fix the t_V threshold to the same value as t_M. All the first neighbors of a maximum greater than t_V belong to the interval of this maximum. From the first neighbor lower than t_V, following neighbors are added to the initial interval until the discrete derivative goes through 0. Finally we just keep slice intervals with a width greater than $w_V = 10$. We obtain a set of slice intervals $V_{H_{slices}}$ with all knots.

The Histogram of Angular Sectors. This second step is applied on each slice interval obtained at previous step. For a given slice interval, we cumulate z-motion separately in each angular sector. We obtain the histogram of cumulative intensity by angular sector (see Fig. 6). Angular sectors are a partition of C circle in N sectors with a same angle. We choose $N = 360$.

Let $[a, b] \in V_{slices}$ be an interval of slice indexes. Let C be a circle with a r_C radius split in 360 sectors with same angle. We defined the value $H_{sectors}(\alpha)$ of $H_{sectors}$ histogram restricted to angle of α index as follows :

$$H_{sectors}(\alpha) = \sum_{k \in [a,b]} \sum_{(i,j) \in C_\alpha} Z_k(i,j)$$

with

- C centered on pith coordinates in I_k
- C_α the C angular sector of α index $\in [0, 359]$
- $\forall k, \forall (i,j), I_k(i,j) > i_{min}$ et $I_{k-1}(i,j) > i_{min}$
- $\forall k, \forall (i,j), Z_k(i,j) \in [z_{min}, z_{max}]$

For an interval V_q of V_{slices}, we obtain an histogram H_{sector} of z-motion distribution according to angular sector. Like in the first step, we use algorithm of interval detection to obtain a set of angular intervals which contains one knot.

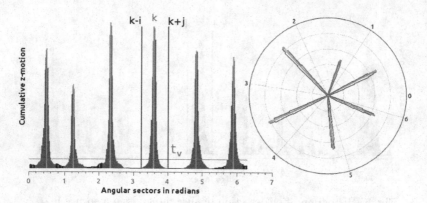

Fig. 6. Histogram and circular projection after detection of sector interval

We keep the same parameters as at the first step, t_M and t_V excepted. In fact, the majority of parameters was fixed in studying scanned picture properties and in this second step, images are the same.

However, t_M and t_V correspond to histogram properties. On histograms of angular sectors, peaks are usually the same size, unlike the slice histograms. In fact, the knots of a slice interval have about about same volume. We fix rationally t_M and t_V at 50% of the $H_{sectors}$ maximum value without risking to forget a maximum and therefore a knot sector.

4   Results

The presented method is able to quickly isolate each log knot in angular sectors. The idea is to select an efficient segmentation method without introducing a long and complex processing step. It is important for sawmills that hope to scan logs at 2 meters per second. For biologists, the most important criterion is the precision. They want to obtain accurate measurements such as length, inclination, maximal diameter and volume of knots. The biologists and sawmills are both very interested by the exact number of knots in a log.

To fulfil this dual purpose, we have chosen to use the extraction of connected components in each angular sector based on a simple threshold. The connected components of one slice are first computed and are then merged with the ones of the previous slice.

The Fig. 7 presents several results of the connected component extraction. The main connected component of one slice is represented in light yellow to the left images and the corresponding 3D reconstruction is given in green on the right image. On logs without sapwood (see Fig. 7(a)), we obtain all knots with an high precision. In contrast, when logs contain sapwood (see Fig. 7(b)), the connected component extraction produces a component where the knot is connected to sapwood and growth rings. This happens when the density of sapwood and growth rings is close to the knot. Thresholding before extraction does not allow to remove them whithout distort the knot.

(a) Without sapwood   (b) With sapwood

Fig. 7. Detected angular sectors and connex components with 3D visualization

The first parameter that our algorithm allow to compute, even with sapwood, is the number of knots. It is the number of detected angular sectors. We can also determine the inclination of all knots by extracting angle corresponding to the maximum of histogram for each angular sector (see Fig. 6). The length is determined in logs without sapwood. When there are sapwood and when a knot is connected to it, experts estimate that node leaves the stem and length can be estimate by the radius of log. The diameter can be exactly computed at any position in the 3D reconstruction of knots without sapwood. In the others cases, we can compute the maximum diameter because it is computed from the middle of knot and sapwood is not present at this place. Without sapwood, the volume can also be determined from the 3D reconstruction of knots. The sapwood induce an overestimation due to the set of parts connected to the knot.

To compare our method with other usual approaches of deformable model, we experiment one of them able to deal with automated topology changes [7]. For this purpose, the initial model was created by using a cylindrical mesh defined from the outer shape contour (see Fig. 8(a)). Note that we use the euclidean metric model implementation. The model is efficient to extract knots in logs without sapwood (See Fig. 8(a-c)). However it is not able to isolate the knot parts from sapwood areas (See Fig. 8(d)).

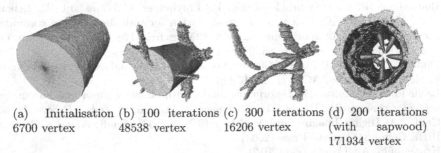

(a)   Initialisation (b) 100 iterations (c) 300 iterations (d) 200 iterations
6700 vertex   48538 vertex   16206 vertex   (with   sapwood)
                                                             171934 vertex

Fig. 8. Experiments of deformable models [7] on wood with and without sapwood

If the segmentation results are comparable for the logs without sapwood, the real advantage of our method is the computation time. We use histogram computation and extraction of connected components. The knot detection time is about the second with our approach whereas deformable model takes really more time, more than one hour. The proposed method is also able to detect and analyze the knots for the case of sapwood while the deformable fails to isolate such a part in this case. The source code of the proposed method is available online [8].

5   Conclusion

The method proposed in this paper automatically detects and counts all the knots of a given piece of wood based on the analysis of X-Ray CT images. It allows to easily deduce several measures such as inclination, length, maximal diameter and volume of knots. All its steps are based on elementary operations such as histogram computation and therefore the method is very fast. In future works, we will improve our method with objective to free ourselves of sapwood.

Acknowledgements. The authors would like to thank Jacques-Olivier Lachaud and Benjamin Taton for kindly providing the source code of the deformable model [7].

References

1. Mcinerney, T., Terzopoulos, D.: Deformable models in medical image analysis: A survey. Medical Image Analysis 1, 91–108 (1996)
2. Schad, K.C., Schmoldt, D.L., Ross, R.J.: Nondestructive methods for detecting defects in softwood logs. USDA Forest Service Research paper FPL-RP-546 (1996)
3. Longuetaud, F., Mothe, F., Kerautret, B., Krhenbhl, A., Hory, L., Leban, J., Debled-Rennesson, I.: Automatic knot detection and measurements from x-ray ct images of wood: A review and validation of an improved algorithm on softwood samples. Computers and Electronics in Agriculture 85, 77–89 (to appears, 2012)
4. Longuetaud, F., Leban, J.M., Mothe, F., Kerrien, E., Berger, M.O.: Automatic detection of pith on ct images of spruce logs. Computers and Electronics in Agriculture 44, 107–119 (2004)
5. Boukadida, H., Longuetaud, F., Colin, F., Freyburger, C., Constant, T., Leban, J.M., Mothe, F.: Pithextract: a robust algorithm for pith detection in computer tomography images of wood - application to 125 logs from 17 tree species. Computers and Electronics in Agriculture 85, 90–98 (2012)
6. Duda, R.O., Hart, P.E.: Use of the hough transformation to detect lines and curves in pictures. Commun. ACM 15, 11–15 (1972)
7. Lachaud, J.-O., Taton, B.: Deformable model with adaptive mesh and automated topology changes. In: Rioux, M., Boulanger, P., Godin, G. (eds.) Proc. 4th Int. Conf. 3-D Digital Imaging and Modeling (3DIM 2003), Banff, Alberta, Canada. IEEE Computer Society Press (2003)
8. Krähenbühl, A.: Tkdetection (2012),
   https://github.com/adrien057/TKDetection

Diffusion-Based Image Compression
in Steganography

Markus Mainberger[1], Christian Schmaltz[1], Matthias Berg[2],
Joachim Weickert[1], and Michael Backes[2]

[1] Mathematical Image Analysis Group,
Faculty of Mathematics and Computer Science, Campus E1.7
Saarland University, 66041 Saarbrücken, Germany
{mainberger,schmaltz,weickert}@mia.uni-saarland.de
[2] Information Security and Cryptography Group,
Faculty of Mathematics and Computer Science, Campus E1.1
Saarland University, 66041 Saarbrücken, Germany
{berg,backes}@cs.uni-saarland.de

Abstract. We demonstrate that one can adapt recent diffusion-based image compression techniques such that they become ideally suited for steganographic applications. Thus, the goal is to embed secret images within arbitrary cover images. We hide only a small number of characteristic points of the secret in the cover image, while the remainder is reconstructed with edge-enhancing anisotropic diffusion inpainting. Even when using significantly less than 1% of all pixels as characteristic points, sophisticated shapes of the secret can be clearly identified. Selecting more characteristic points results in improved image quality. In contrast to most existing approaches, this even allows to embed large colour images into small grayscale images. Moreover, our approach is well-suited for uncensoring applications. Our evaluation and a web demonstrator confirm these claims and show advantages over JPEG and JPEG 2000.

1   Introduction

In a time that is characterized by devices that allow anyone at anytime to take and share high resolution digital images, the need for image compression codecs with high compression rates is more important than ever. A promising recent class of image compression codecs relies on inpainting techniques based on partial differential equations (PDEs); see e.g. [1]. In contrast to conventional methods such as JPEG, PDE-based image compression does not require transformations to the frequency domain: It simply stores a small fraction of characteristic pixels. In the decoding step, the missing pixels are reconstructed by the inpainting effect of the PDE. PDE-based approaches can outperform JPEG and even its sophisticated successor JPEG 2000 [2,3]. This oberservation holds all the more for perceptual quality metrics as shown in [4]. However, they have not been adapted to specific application scenarios outside classical image compression so far.

   In this paper, we consider one of those scenarios, namely *steganography*. Steganography is the art of hiding the *presence* of an embedded message (called

G. Bebis et al. (Eds.): ISVC 2012, Part II, LNCS 7432, pp. 219–228, 2012.

secret) within a seemingly harmless message (called *cover*). It can be seen as the complement of *cryptography*, whose goal is to hide the *content* of a message. Media data such as images currently constitute the most widely used cover types for practical steganographic systems, mainly because they contain much redundant data. However, most existing approaches only embed single words or a short text within a cover image. We show that, by exploiting the potential of PDE-based image compression codecs, our approach can hide complete images.

Apart from steganography, embedding messages within other messages is also useful in applications such as digital content protection and copyright management: A content provider can distribute redacted media data already containing the missing information. Afterwards the missing information can be selectively disclosed to paying customers by simply supplying the password.

Example algorithms to embed data in images are *Jsteg* [5], *F5* [6], and *YASS* [7]. *Jsteg* replaces least significant bits with those to embed, which is vulnerable to histogram attacks [8]. *F5* implements matrix encoding to improve the embedding efficiency and a permutation to spread the changes over the whole steganogram. *YASS* employs error correcting codes to be robust against *JPEG* compression. Often, steganographic methods also focus on hiding information inside specific file formats such as JPEG [7,9] or JPEG 2000 [10,11].

Our Contribution. In this article, we show how PDE-based image compression must be adapted such that it meets the requirements occurring in steganography. We call the resulting novel method *Steganograhy with Diffusion Inpainting (SDI)*. First we select a very small number of characteristic points of the given secret image. Thereby, we carefully combine insights from several recent papers to obtain a fast compression codec which yields high quality results and small file sizes at the same time. Then, we encode the obtained bitstream by employing a state of the art entropy coder. Before embedding this data, we encrypt it to avoid detectable patterns that might be a result of the compression.

Note that, for steganographic purposes, it is desirable to keep the number of characteristic points – and thus the amount of data that must be hidden – as small as possible. Since experiments have demonstrated that diffusion inpainting is well-suited to extremely high compression rates [2], it fits perfectly to steganography. Even with less than 1% of all pixels as characteristic points, the shapes of the secret image can be clearly identified.

Interestingly, our SDI approach even allows to black out a region of an image and embed its former content in the remainder of the image. Using the correct password, the censored part can be uncovered at any time. Diffusion-based inpainting is particularly well-suited for this application, as the boundary of the censored region provides additional information which facilitates the reconstruction process. A web demonstrator for hiding and censoring is available at http:// stego.mia.uni-saarland.de.

Related Work. To the best of our knowledge, inpainting methods have not been used for steganographic purposes before. In [12] techniques are developed to prevent an unexpected user from eliminating objects in videos by inpainting.

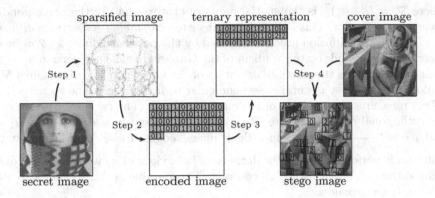

Fig. 1. Schematic overview of the main steps of our steganography method: Steps 1 and 2 illustrate the encoding of the PDE-based image compression codec. Steps 3 and 4 depict the steganographic preparation and embedding phases.

This is achieved by segmenting the objects, followed by a steganographic embedding. However, this has nothing to do with our approach, which tailors diffusion inpainting for compact data representation in steganography.

Paper Structure. Section 2 illustrates our diffusion-based compression algorithm. Based on this algorithm, we explain how to embed secret images within arbitrary cover images in Section 3. In Section 4, we describe our performance measurements before concluding the paper in Section 5.

2 PDE-Based Image Compression

2.1 Basic Concept

While anisotropic diffusion processes have been popular for denoising images for many years, their use in image compression is relatively new and a topic of ongoing research. Compared to conventional compression methods such as *JPEG* or *JPEG 2000*, diffusion-based image compression offers a number of advantages: It has a very intuitive physical interpretation, it is well-suited for extremely high compression rates, where it can outperform *JPEG* and even *JPEG 2000* [2,3], and it can be used in a generic way for various types of data – including 2-D images, 3-D data sets, surface data and videos.

For simplicity, we consider a greyscale image, i.e. a function $f : \Omega \to \mathbb{R}$, where $\Omega \subset \mathbb{R}^2$ denotes a rectangular image domain. The extension to colour images does not create specific problems. The idea behind diffusion-based image compression is to store only the grey-values of a small subset $K \subset \Omega$ [1]. Using the pixels in K as Dirichlet boundary data, we reconstruct f in the unspecified domain $\Omega \setminus K$ as steady state ($t \to \infty$) of the partial differential equation (PDE)

$$\partial_t u = \operatorname{div}(\boldsymbol{D}(\boldsymbol{\nabla} u_\sigma)\,\boldsymbol{\nabla} u) \,, \tag{1}$$

where $\nabla = (\partial_x, \partial_y)^\top$ is the spatial nabla operator, and div the corresponding divergence operator. This PDE is known as *edge-enhancing anisotropic diffusion (EED)* [13]. The diffusion process is steered by the positive definite 2×2 *diffusion matrix D*. It depends on the gradient of the Gaussian-smoothed version u_σ of the image u, where σ is the standard deviation of the Gaussian. The direction of ∇u_σ is along the steepest ascent, i.e. perpendicular to image edges, and its magnitude $|\nabla u_\sigma|$ measures the strength of the edge (contrast). The two eigenvectors of D are orthogonal resp. parallel to ∇u_σ, and their eigenvalues are given by $\mu_1 = 1$ and $\mu_2 = \frac{1}{\sqrt{1+|\nabla u_\sigma|^2/\lambda^2}}$. Thus, along image edges, $\mu_1 = 1$ ensures that full diffusion is performed, while the decreasing behaviour of μ_2 w.r.t. $|\nabla u_\sigma|$ reduces diffusion across edges with high contrast. The parameter $\lambda > 0$ allows to steer this contrast dependence.

2.2 Step 1: Finding Characteristic Pixels

Since steganographic methods are often applied in situations with limited resources, the compression algorithm has to be fast. Compared to methods such as [2], which are tuned to yield optimal *quality*, we thus give greater priority to the *running time*. However, real-time capable methods are often designed for more restricted application areas, e.g. to cartoon-like images [3]. Since our codec aims at combining interactive running times with suitability for a large class of images, we have to modify ideas from [2] and [3] and combine them with results from [14]. Moreover, our compression codec should work particularly well for high compression rates such that we are also able to hide a large colour image within a much smaller greyscale image. This also allows not to exploit the maximal available space such that the secret information is only stored in a fraction of all pixels, in order to prevent an observer from detecting the secret information.

To fulfil these requirements, we first extract a set of characteristic pixels which will be used for inpainting in the recovering step (Figure 1, Step 1), followed by saving these points in a compact representation, see next section (Figure 1, Step 2). We rely on an adaptive subdivision method similar to the one chosen in [2], but refrain from time-consuming optimisation steps. Instead, we exploit the theoretical results obtained in [14] to speed up the method.

As in [2], our subdivision method starts with the rectangle defined by the whole image domain, or the censored image part, respectively. The rectangle is recursively split into smaller rectangles until a predefined stopping criterion is fulfilled. The final inpainting mask K, i.e., the pixel locations used in the reconstruction, is given by the four corner points and the centre of each rectangle. This allows to efficiently encode pixel positions; see Section 2.3. Moreover, the algorithm can adaptively store more pixels in significant structures by using a finer subdivision instead of investing many points in unimportant image areas.

In [2], splitting a rectangle is stopped as soon as the interpolation error within the rectangle is smaller than a threshold. Thus, a rectangle is split whenever decoding would lead to an inadequate reconstruction in this area. The drawback of this natural stopping criterion is that its evaluation requires to inpaint image parts while compressing. Thus, this method is too slow for our purpose.

Therefore, we exploit the analytical result obtained in [14] for homogeneous diffusion inpainting to speed up the subdivision. In this paper, it is proven that the pixel positions should be chosen depending on $|\Delta f|$, the magnitude of the Laplacian of the original image f. Even though we prefer EED over homogeneous diffusion due to its superior inpainting quality, experiments indicate that this choice still yields good results. Our new subdivision scheme exploits this idea and works as follows: During the subdivision process, we assign each rectangle the sum over the Laplace magnitudes $|\Delta f_\sigma|_i$ of all pixels i within the rectangle. We denote these sums, which can be computed efficiently using the idea of integral images [15], by $D(R)$. Starting with the rectangle defined by the whole image domain, the rectangle that corresponds to the largest sum is subdivided in the middle of the x or y direction, whichever is larger. This rectangle can be found efficiently by utilising a priority queue, where the rectangle R has the priority $-D(R)$. After each subdivision step, the two new rectangles are included into the queue.

Obviously, the number of subdivisions s is related to the size of the encoded image. Therefore, we can adapt s to the desired compression ratio. We perform a binary search for s to approximate this compression rate.

2.3   Step 2: Compact Representation

To store the inpainting mask K obtained in the last section, it is sufficient to represent the splitting decisions with a binary tree structure. We encode this tree in the image header: First the depth range of the tree is stored, followed by single bits indicating the splitting decisions. The file header also contains a few bytes for properties such as image size and number of channels.

The grey or colour values of the selected points are quantised regularly into q possible values. In contrast to [2], where a time-consuming optimisation of q is advocated to achieve best quality, we use the constant value $q = 32$. To further speed up the algorithm, we employ the $LPAQ$ [16] coder to compress the brightness data instead of the slower PAQ [16] coder proposed in [2]. Finally, the binary strings containing the compressed pixel values and the file header are concatenated to obtain a bit stream representation of the secret.

To restore the image, the saved grey values are placed at the locations indicated by the stored splitting decisions, and the remainder of the image is inpainted using EED. For uncensoring, we can exploit additional information in the reconstruction step, namely the fact that (a part of) the boundary of the censored image is known. This has two consequences: First of all, characteristic points lying on known boundaries do not have to be saved, which reduces the size of the compressed file. Secondly, the known boundary pixels can be used as known points in the inpainting step, which improves the reconstruction quality.

3   Steganographic Preparation and Embedding

In this section we present the steganographic part our SDI algorithm, depicted as Step 3 and 4 in Figure 1. It allows to hide a bitstream of a restricted size, as obtained at the end of the previous section, in a set of cover image pixels. An observer cannot detect whether the resulting image contains a hidden secret or not. Moreover, the secret bitstream can only be extracted with a password p.

3.1 Step 3: Steganographic Preparation

As a first steganographic step we need to add the size of the bitstream (in bytes) to its beginning to be able to separate secret data from cover data when extracting the bitstream again. Then we encrypt this concatenation using the *Advanced Encryption Standard* (*AES*) in *Cipher-block Chaining* (*CBC*) mode. This step further ensures the confidentiality of the secret data. The key used for this encryption is derived from password p by using the hash function SHA-256 as a random oracle [17]. Following the approach proposed in [18], we split the encrypted binary stream into blocks of 11 bits. Each block is transformed to 7 ternary bits (note that $2^{11} < 3^7$). This conversion is necessary to use *mod-3 matching* later on, which embeds a ternary bit in a cover pixel.

3.2 Step 4: Embedding

Next, we need to select the cover image pixels in which we want to hide the secret data. The simplest way would be to start at the beginning of the cover's data and to sequentially embed the ternary bits into the pixels of the cover. However, this approach simplifies the detection of the secret image since all modifications of the cover occur at its beginning, except the ternary stream is large enough to exhaust the cover's capacity. Therefore, we use a pseudo-random permutation ρ, inferable by anybody who knows the password p, to spread the ternary bits homogeneously over the cover's data. This ensures that each pixel of the cover image carries secret information with the same probability. Our embedding of secret data into an image relies on the assumption that the distribution of the least significant bits for each byte representing a cover pixel essentially corresponds to random noise. Hence, changing the pixel values by ±1 does not change an image in a manner that is noticeable by the user. However, *LSB replacement*, i.e. overwriting the *least significant bits* with those that should be embedded, is vulnerable to statistical detection methods, as even values are never decreased and odd values are never increased [8]. *LSB matching* [19] instead randomly decides whether to increase or decrease a value, which bypasses these statistical tests. We use mod-3 matching which embeds a ternary bit t by modifying a pixel value v, such that $v \bmod 3 = t$. More precisely, if $v - t \bmod 3 = 1$, we set $v := v - 1$; if $v - t \bmod 3 = 2$, we set $v := v + 1$; otherwise we leave v unchanged. In the special cases where v already is 0 or 255, we skip over v as modifying completely saturated values might be noticeable in some scenarios. Furthermore, an embedding of t in a value $v \in \{1, 254\}$ might result in shrinkage with a completely saturated byte $\in \{0, 255\}$; see [6]. The extraction cannot distinguish such saturated stegobytes from skipped saturated image bytes. Therefore, we re-embed the respective ternary bit in the subsequent cover byte, until the result is not a saturated byte.

Neglecting those two restrictions, we can approximately embed one ternary bit per cover image pixel. That means if the cover image's data consist of n pixels, we can embed up to $t = 11 \cdot \lfloor \frac{n}{7} \rfloor$ bits, which amounts to up to $b - \lfloor \frac{t}{128} \rfloor$ AES blocks. Assuming 4 bytes of length information at the beginning of the stream, we can embed secret data of size up to $\frac{128 \cdot b}{8} - 4$ bytes in the cover image.

To extract the secret image from the cover, we simply invert the operations explained above: Given the password p, we can infer the pseudo-random permutation ρ. This allows us to extract the bits of the ternary representation in the correct order by calculating mod-3 for each value and skipping over completely saturated values as explained above. To determine how many pixels need to be processed, we decrypt the first 128 extracted bits with the key derived from password p. Then, by using the obtained length information we extract the remaining bits of the secret bitstream and decrypt it. Finally we apply the decompression algorithm from Section 2.3 to reconstruct the secret image.

3.3 Uncensoring Images

SDI as described above can immediately be applied to hide an image into another one. However, there are various modifications and extensions that allow the usage of SDI in other interesting fields of application.

For instance, SDI can be used for uncensoring applications where censored parts of an image are still contained in the same image and accessible by a password. However, it should not be detectable that there is any other information available than the visible one. In this scenario, only some minor adjustments are necessary. First of all, we store the location of the part of the image that is to be censored using a few additional bits in the header of the compressed image. Moreover, as explained in Section 2.3, we can exploit the fact that the boundary of the censored region is known. Additionally, the file header of our approach is much smaller than those of JPEG and JPEG 2000, which is an additional advantage over these approaches. This makes our diffusion-based image compression algorithm particularly well-suited for such uncensoring scenarios.

Even in cases where a significant part of the image is censored and only a small surrounding frame of the censored region is available for embedding, we still achieve good reconstruction results. We encourage the reader to verify this claim with our web demonstrator (see `http://stego.mia.uni-saarland.de`).

4 Experiments

For a quantitative evaluation of SDI, we consider the *mean squared error (MSE)*, which is defined as

$$\text{MSE} := \frac{1}{M \cdot N} \sum_{m=1}^{M} \sum_{i=1}^{N} (f_{m,i} - u_{m,i})^2 , \tag{2}$$

where N is the number of image pixels, M is the number of channels, and $(f_{m,i})_{i=1..N}$ and $(u_{m,i})_{i=1..N}$ are the pixel values of the original secret image in channel m, pixel i and its reconstructed version, respectively.

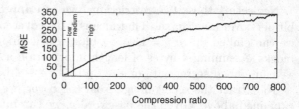

Fig. 2. Colour test images: **(a) Left:** Cover image *cake*, 256 × 256 pixels. **(b) Right:** Secret image *knife*, 512 × 512 pixels.

Fig. 3. Compression ratio versus mean squared error when our compression codec is applied to the test image *knife* (see Figure 2). The green rectangles mark the automatically computed ratios corresponding to the quality settings "high", "medium", and "low" of our web-based prototype.

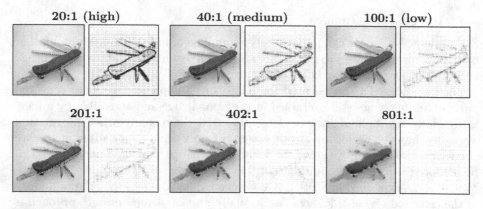

Fig. 4. Reconstructed images and corresponding inpainting masks for increasing compression ratios of the test image *knife* (see Figure 2). The first row depicts the results obtained with different quality settings offered by our web-based prototype.

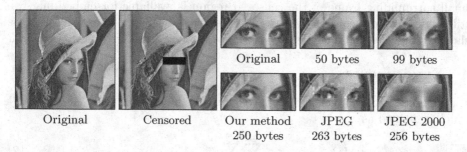

Fig. 5. The information removed from the original image was hidden in the censored image. The magnifications on the top of the right side illustrate the reconstruction quality with different amounts of data. The images in the bottom row compare the performance of our codec with JPEG and JPEG 2000.

In Figure 2, we apply our method to two test images. Note that the secret image is four times larger than the cover image. Figures 3 and 4 reflect the respective results of the recovered secret depending on the number of stored secret pixels. As expected, the higher the compression ratio, the lower the quality. However, it is typically only necessary to recognise what was originally depicted. With SDI this is possible: Even for compression ratios as high as 800 : 1, we can clearly identify the secret to be a knife. This means we could hide this secret in the given cover roughly 40 times.

For the colour images from Figure 2, the embedding and extracting time on a standard PC lies in the order of seconds. If the speed of the hiding process is negligible, it is also possible to replace the suggested compression codec by the more sophisticated one from [2], which yields results of even higher quality at the same compression ratio.

The benefit of our codec over JPEG and JPEG 2000 can be seen in Figure 5: Storing the censored image part with JPEG or JPEG 2000 yields a much worse quality, even though our compression algorithm additionally stored the position of the censored image part, while the saved JPEG and JPEG 2000 image parts were manually pasted onto the right position.

5   Conclusions and Future Work

We have established a novel application field for recent PDE-based image compression techniques by adapting them to the needs of steganographic problems. In order to compress the image data in short time, high quality, and with small file sizes, it was necessary to combine a number of concepts and insights. This was integrated with carefully engineered steganographic preparation and embedding processes to end up with SDI, a novel technique for hiding secret image data in a cover image in a secure manner. Moreover, we have shown the usefulness of the SDI approach for uncensoring applications.

The success of these applications benefits heavily from the fact that diffusion-based image compression offers several properties that makes it preferable over classical codecs such as JPEG or JPEG 2000. This includes for instance its excellent performance for very high compression ratios, and its ability to exploit known boundary data in uncensoring applications. It appears that PDE-based compression and steganography constitute an ideal, hitherto unexplored match.

We are convinced that our work is not the end of the road: Due to the generic concept of diffusion-based inpainting, the SDI approach can be extended to a variety of different secret and cover formats, including 2-D images, 3-D data sets, surface data and videos. Even hiding small videos within a single image seems within range given our experiments. This is part of our ongoing research.

References

1. Galić, I., Weickert, J., Welk, M., Bruhn, A., Belyaev, A., Seidel, H.P.: Image compression with anisotropic diffusion. Journal of Mathematical Imaging and Vision 31, 255–269 (2008)

2. Schmaltz, C., Weickert, J., Bruhn, A.: Beating the Quality of JPEG 2000 with Anisotropic Diffusion. In: Denzler, J., Notni, G., Süße, H. (eds.) Pattern Recognition. LNCS, vol. 5748, pp. 452–461. Springer, Heidelberg (2009)
3. Mainberger, M., Bruhn, A., Weickert, J., Forchhammer, S.: Edge-based image compression of cartoon-like images with homogeneous diffusion. Pattern Recognition 44, 1859–1873 (2011)
4. Galić, I., Zovko-Cihlar, B., Snježana, R.D.: Computer image quality selection between JPEG, JPEG 2000 and PDE compression. In: Proc. 19th International Conference on Systems, Signals and Image Processing, pp. 437–441. IEEE Computer Society (2012)
5. Upham, D.: JSteg (1997), http://zooid.org/~paul/crypto/jsteg/
6. Westfeld, A.: F5-A Steganographic Algorithm. In: Moskowitz, I.S. (ed.) IH 2001. LNCS, vol. 2137, pp. 289–302. Springer, Heidelberg (2001)
7. Solanki, K., Sarkar, A., Manjunath, B.S.: YASS: Yet Another Steganographic Scheme That Resists Blind Steganalysis. In: Furon, T., Cayre, F., Doërr, G., Bas, P. (eds.) IH 2007. LNCS, vol. 4567, pp. 16–31. Springer, Heidelberg (2008)
8. Westfeld, A., Pfitzmann, A.: Attacks on Steganographic Systems. In: Pfitzmann, A. (ed.) IH 1999. LNCS, vol. 1768, pp. 61–76. Springer, Heidelberg (2000)
9. Jókay, M., Moravcík, T.: Image-based JPEG steganography. Tatra Mountains Mathematical Publications, 65–74 (2010)
10. Su, P.C., Ku, C.C.J.: Steganography in JPEG 2000 compressed images. IEEE Transactions on Consumer Electronics 49, 824–832 (2003)
11. Zhang, I., Wand, H., Wu, R.: A high-capacity steganography scheme for JPEG 2000 baseline system. IEEE Transactions on Image Processing 18, 1797–1803 (2009)
12. Chou, Y.-C., Chang, C.-C.: Restoring Objects for Digital Inpainting. In: Pan, J.-S., Huang, H.-C., Jain, L.C. (eds.) Information Hiding and Applications. SCI, vol. 227, pp. 47–61. Springer, Heidelberg (2009)
13. Weickert, J.: Theoretical foundations of anisotropic diffusion in image processing. Computing Supplement 11, 221–236 (1996)
14. Belhachmi, Z., Bucur, D., Burgeth, B., Weickert, J.: How to choose interpolation data in images. SIAM Journal on Applied Mathematics 70, 333–352 (2009)
15. Crow, F.C.: Summed area tables for texture mapping. Computer Graphics 18, 207–212 (1984)
16. Mahoney, M.: Data compression programs (2011), http://mattmahoney.net/dc/ (last visited July 22, 2011)
17. Bellare, M., Rogaway, P.: Random oracles are practical: A paradigm for designing efficient protocols. In: Denning, D.E., Pyle, R., Ganesan, R., Sandhu, R.S., Ashby, V. (eds.) Proc. First ACM Conference on Computer and Communications Security, pp. 62–73 (1993)
18. Zhang, X., Wang, S.: Efficient steganographic embedding by exploiting modification direction. IEEE Communications Letters 10, 781–783 (2006)
19. Sharp, T.: An Implementation of Key-Based Digital Signal Steganography. In: Moskowitz, I.S. (ed.) IH 2001. LNCS, vol. 2137, pp. 13–26. Springer, Heidelberg (2001)

Video Analysis Algorithms for Automated Categorization of Fly Behaviors

Md. Alimoor Reza[1], Jeffrey Marker, Siddhita Mhatre[2], Aleister Saunders[2],
Daniel Marenda[2], and David Breen[1,*]

[1] Department of Computer Science
[2] Department of Biology
Drexel University, Philadelphia, PA 19104, USA
david@cs.drexel.edu

Abstract. The fruit fly, *Drosophila melanogaster*, is a well established
model organism used to study the mechanisms of both learning and memory *in vivo*. This paper presents video analysis algorithms that generate
data that may be used to categorize fly behaviors. The algorithms aim
to replace and improve a labor-intensive, subjective evaluation process
with one that is automated, consistent and reproducible; thus allowing
for robust, high-throughput analysis of large quantities of video data.
The method includes tracking the flies, computing geometric measures,
constructing feature vectors, and grouping the specimens using clustering techniques. We also generated a Computed Courtship Index (CCI),
a computational equivalent of the existing Courtship Index (CI). The
results demonstrate that our automated analysis provides a numerical
scoring of fly behavior that is similar to the scoring produced by human
observers. They also show that we are able to automatically differentiate between normal and defective flies via analysis of their videotaped
movements.

1 Introduction

The fruit fly, *Drosophila melanogaster*, is a well-established model organism used
to study the mechanisms of both learning and memory. The ability to study
learning and memory behavior in *Drosophila* has significantly increased our understanding of the underlying molecular mechanisms of this behavior, and allows
for the rational design of therapeutics in diseases that affect cognition. The existing human-observer-based techniques used to assess this behavior in flies are
powerful. However, they are time-consuming, tedious, and can be subjective. An
automated method based on video analysis of fly movements can replace and
improve a labor-intensive, possibly inconsistent evaluation process. Automating
the process will provide a reliable and reproducible analysis of fruit fly courtship
behavior. Additionally, it also enables quantification of many aspects of the
behaviors that are not easily perceived by a human observer. Moreover, the

---

* Corresponding author.

G. Bebis et al. (Eds.): ISVC 2012, Part II, LNCS 7432, pp. 229–241, 2012.

automation promises the possibility of robust, high-throughput analysis of substantial quantities of video data, which would be useful for large-scale genetic and drug screens, and disease modeling.

In addition to the extensive genetic tools available for the analysis of fly neural circuitry associated with learning and memory [1,2], both adult and larval flies exhibit a number of behaviors that can be altered with training [3]. One of these behaviors that is well established in the field is courtship conditioning. Courting behavior by male flies in *Drosophila* follows a linear, stereotyped, and well documented set of behaviors [4], and this behavior is modified by previous sexual experience [5]. Courtship conditioning is a form of associative learning in *Drosophila*, where male courtship behavior is modified by exposure to a previously mated female that is unreceptive to courting [5,6]. Thus, after 1 hour of courting a mated female, males suppress their courtship behavior even towards subsequent receptive virgin females for 1-3 hours [5,7]. This courtship suppression is measured by the Courtship Index (CI), which is calculated by dividing the total amount of time each male fly spends courting by the total duration of a testing period [5,6]. CI is the standard metric used to assess learning and memory in courtship suppression analysis.

A computational approach to fly behavior quantification and characterization has been developed based on the analysis of videos of courting fruit flies [8]. The approach includes identifying individual flies in the video, quantifying their size (which is correlated with their gender), and tracking their motion, which also involves computing the flies' head directions. Geometric measures are then computed, for example distance between flies, relative orientation, velocities, contact time between the flies, and the time when one fly looks at another. This data is computed for numerous experimental videos and produces high-dimensional feature vectors that represent the behavior of the flies. We formulated a computational equivalent (Computational Courtship Index) of the existing CI, based on the feature vector values, and compared it with CI. Clustering techniques, e.g., k-means clustering, are then applied to the feature vectors in order to computationally group the specimens based on their courtship behavior. Our results show that we are able to reproduce CI values and automatically differentiate between normal and memory/learning defective flies using only the data derived from our video analysis.

2   Previous Work

Significant research has been conducted on tracking objects (cars, humans, animals, and insects) in video frames [9]. Isard et al. developed a real-time tracking system, BraMBle (A Bayesian framework for multiple blob tracker) that combines blob tracking and particle filtering approaches to develop a Bayesian multiple-blob tracker [10]. Khan et al. identify objects entering and leaving a tracking arena and keep track of the trail of the targets using particle filter based approaches [11]. Branson et al. address tracking multiple targets (i.e., three mice) even with occlusion [12]. Tao et al. present a CONDENSATION-like sampling algorithm [13] for tracking multiple objects which supports addition, deletion, and occlusion of objects [14]. Comaniciu et al. focus on tracking based on target representation and localization [15].

Several studies have been done on fly behavior modeling and analysis. Dankert et al. describe an automated social behavior (aggression and courtship) measurement of *Drosophila* [16] using a vision-based tracking system. Though the system has a courtship analysis component, it has not been used to assess courtship suppression in learning and memory. Branson et al. present a method for high-throughput computer-vision-based tracking that quantifies individual and social behaviors of fruit flies [17] and showed that an ethogram-based behavior quantifications can differentiate between gender and genotype. Valente et al. developed a system based on spatial and temporal probabilistic measures for analyzing the trajectory of *Drosophila melanogaster* in a circular open field arena [18].

Though many aspects of fly behaviors have been previously studied [16,17,18], to the best of our knowledge, no work has been done to computationally quantify and categorize the learning and memory capabilities of individual fly specimens by analyzing videos of their courtship behavior. Additionally, the technical contributions of our work include: 1) automated fly tracking, 2) determination of fly head direction, 3) formulation of a computational equivalent of the existing Courtship Index, and 4) automated grouping of fly specimens.

3   Computational Pipeline

We developed a computational pipeline to analyze videos of courting fruit flies in order to characterize the memory and learning behaviors of the flies. A sample video of 10-minute duration is processed in the pipeline. Images are extracted from the video and processed by the pipeline components: segmentation, filtering, fly tracking, and geometric measures computation. A feature vector is generated for each video segment using the computed geometric measures. The feature vectors are used to generate a Computed Courtship Index (CCI). A clustering method (k-means clustering) is used to then group the samples according to their behaviors as defined by the feature vectors.

3.1   Extract Chamber Images from the Video

A fly pair is video-taped inside a circular chamber. An input video may contain 3 to 6 of these chambers. Individual frames are extracted from the video at a rate of 30 frames-per-second. The extracted images are cropped to isolate each chamber. The rectangular coordinates surrounding a chamber are manually specified in the first frame of a segment. Each chamber is isolated using the same coordinates from the subsequent frames. Two example frames are presented in Figure 1 A.

3.2   Segmentation

A binary image is produced from each cropped image, where a group of white pixels represents a fly object and black pixels represent the background. This stage utilizes background detection, image subtraction and thresholding.

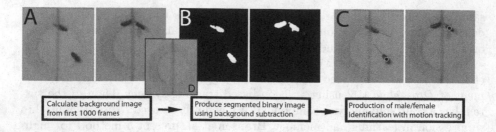

Fig. 1. Steps in the fly identification and tracking process

Background Detection. This component processes a set of N images to find the most frequent color that appears in each pixel, producing a single background image. In general, using the first 1000 cropped images was found to be sufficient for accurately detecting the background. One such background image is presented in Figure 1 D.

Image Subtraction and Thresholding. Each of the approximately 18,000 frames in a video segment are subtracted from the background image. The resulting images are thresholded to compute a corresponding segmented binary image. If the intensity difference between the background and foreground pixel exceeds a threshold, the binary pixel is set to white. Empirically, a threshold value of 0.15 (in a 0 to 1 range) was found to be sufficient for identifying fly pixels in the sample videos.

3.3 Filtering

The filtering stage removes noise and fills voids from the binary segmented images, and identifies the white-region representation(s) of the fly pair, as seen in Figure 1 B.

Filling Voids. The first step in the filtering stage fills voids inside any collections of white pixels. The voids inside the white regions are removed using a flood-filling process, starting from the top left corner pixel. The 4-connected black region is found. The white regions are defined as the complement of the flood-filled black region; thus removing interior voids in the white regions.

Fly Object Identification. We assume that the two flies will produce two sets of white pixel regions that have approximately the same area. If the areas of the two largest white regions differ greatly we assume that the flies are touching each other and form one large white pixel region. Collections of white pixels are identified in the void-filled binary images. The area of each white region is calculated. One or two white regions are selected in each image based on the following conditions:

1. If the ratio between the second largest white region and largest white region is less than or equal to a given value, e.g., $\frac{1}{15}$, the largest white region

Image Y (labels unknown)   Image X (labels known)   Image Y (labels transferred)   Image X (labels known)

Fig. 2. Propagating fly object labels from Image X to Image Y

is identified as a single object. This single object represents the two flies in contact with each other. In our experience using this ratio consistently allowed us to identify when flies are in contact.

2. If the current image only has one white region, it is selected as the fly object, and indicates that once again the flies are together.

3. Otherwise, the largest two white pixel regions are selected to represent the two flies.

3.4 Fly Object Tracking

The two fly objects are tracked for all of the frames of a video segment. The first step of the process is to divide the 18,000 filtered, binary images into smaller subsequences. A *subsequence* is a contiguous series of images where each image contains two fly objects (white regions). A *subsequence* with fewer than 15 (i.e., half of the extraction frame-rate) images is not processed. Let us define a subsequence starting from image number M to image number N (in increasing order of extraction). The images at the start and end of this subsequence are denoted by I_M and I_N respectively. The Euclidean distance between the centroids of the two white regions in each image is computed. The frame with the maximum inter-centroid distance is identified as a starting image for this subsequence. Let us assume that the starting image is frame S and the image is denoted by I_S. Each white region in image S is labeled and these labels are propagated from image S both forwards and backwards in the subsequence. A minimum distance criterion is used for establishing the label correspondences between images.

A four-way distance calculation is performed between the two fly objects in the source frame and the two fly objects in an adjacent frame. The minimum of the four distances determines one of the label correspondences, i.e. the label of the object in a source image that has the smallest Euclidean distance is transferred to the closest fly object in the adjacent image. The second fly object in the adjacent image is given the label of the other fly object in the source image, as seen in Figure 2. The labels for the first and second white regions of image I_S are initialized to be A and B respectively. The fly objects in the immediately adjacent images, I_{S-1} and I_{S+1}, are assigned labels from the fly objects in I_S. This labeling information is then propagated out to the remaining images in the subsequence.

Once the white regions are tracked and labeled (A or B) in all of the images in a subsequence, we attempted to also identify the gender of the fly objects by

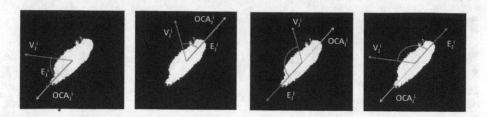

Fig. 3. Definition of a fly object's Oriented Central Axis OCA^i_J from its velocity vector V^i_J and eigen vector E^i_J

assuming that the larger object is female. Our experiments showed that this assumption was only valid approximately 80% of the time. Since this identification was not deemed sufficiently reliable, we did not use this size/gender information during subsequent calculations.

3.5 Fly Object Head Direction Determination

Several geometric quantities are needed for the head direction calculation. The **Velocity Vector** of fly^i in frame J is denoted by V^i_J. The centroid of fly^i in frame J is denoted by C^i_J, a 2-dimensional point. The velocity vector V^i_J of frame J for fly^i is calculated using the following equation,

$$V^i_J = \frac{C^i_{J+1} - C^i_J}{J+1-J} \ pixels/frame.$$

The denominator in the previous equation equals 1, since we compute the velocity vector from two consecutive frames. The velocity vectors $\{V^i_M, V^i_{M+1}, , V^i_N\}$ are calculated for all frames of a subsequence for fly^i.

The **Eigen Vector** of a frame J for fly^i is denoted by E^i_J. E^i_J is the eigen vector that corresponds to the largest eigen value of the covariance matrix generated from the locations of the white pixels representing fly^i.

The **Oriented Central Axis** OCA^i_J is the Eigen Vector E^i_J of fly^i pointing in the direction of the fly's velocity vector V^i_J. It is determined from the sign of the dot product between V^i_J and E^i_J with the following equation,

$$OCA^i_J = \begin{cases} E^i_J & if(V^i_J \cdot E^i_J) > 0 \\ -E^i_J & if(V^i_J \cdot E^i_J) < 0 \\ 0 & if(V^i_J \cdot E^i_J) = 0 \end{cases} \tag{1}$$

OCA^i_J is calculated for all images in the subsequence. The three quantities, V^i_J, E^i_J and OCA^i_J, are presented in Figure 3 for a single fly object.

The **Head Direction Vector** of fly^i in frame J is denoted by HD^i_J. HD^i_J can be defined either as E^i_J or $-E^i_J$. The head direction vector of the starting frame, HD^i_S, in the subsequence is first calculated. Head direction vectors for the remaining frames in the subsequence are derived from HD^i_S utilizing the following calculations.

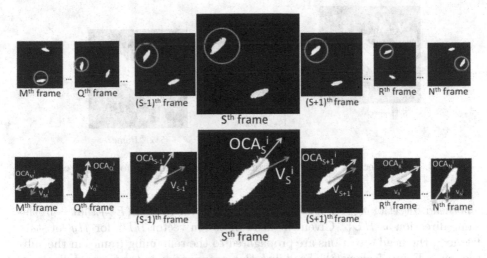

Fig. 4. Longest positive dot product sequence identification

Longest Positive Dot Product Series. In a courtship assay, a fly moves in the forward (i.e. head direction) for most of the time in the assay chamber. Given this observation, we search for the longest time period when a fly moves in a consistent direction. The computation of the longest series of consecutive positive dot products between the adjacent OCA_j^is will give us the greatest number of frames where the angle between each of the adjacent pairs of OCA_j^is is less than 90 degrees. In other words, the adjacent pairs of OCA_j^is in the series are oriented in the same direction. The blue circles in Figure 4 identify the series of fly objects for which we are computing head directions. The bottom row of Figure 4 shows the vectors OCA_j^i (yellow arrows) and V_j^i (blue arrows) for the subsequence from frame M to frame N. The *longest positive dot product series* extends from frame Q to frame R. We compute the pairwise dot product between each of the adjacent OCA_j^is and select the longest series of dot products in which all the products are positive. In this series (from frame Q to frame R) the adjacent pairs of OCA_j^is are oriented in the same direction.

Start Frame Head Direction. Given the series of longest positive dot products of the OCA_j^is in the subsequence, we choose the frame with the maximum speed in the series as a start frame. Let us denote this start frame to be frame number S. The head direction vector HD_S^i of fly^i in frame number S is defined as OCA_S^i, since it is the central axis vector pointing in the same direction as V_S^i.

Propagate the Head Direction Vectors. The head direction vector HD_X^i for fly^i in frame number X immediately adjacent to frame number Y, whose head direction vector HD_Y^i for fly^i is known, is computed by

$$HD_X^i = \begin{cases} E_X^i & if\ (HD_Y^i \cdot E_X^i) \geq 0 \\ -E_X^i & Otherwise. \end{cases}$$

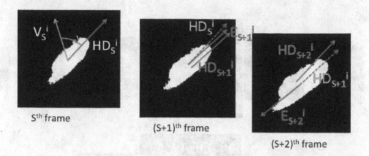

Fig. 5. Head direction propagation from start frame S in the forward direction

The head direction vector HD_X^i is the Eigen vector of fly^i (E_X^i) that is in the same direction as HD_Y^i. Given the head direction vector HD_S^i for fly^i of start frame S, the head directions are propagated to the remaining frames in the subsequence. Figure 5 shows the head direction propagation in the forward direction from start frame S. Figure 1 C presents the centroids and head directions for two flies overlaid on the original processed image.

3.6   Compute Geometric Measures

Five geometric measures are computed for each video segment. They are 1) *percentage of frames when one fly is looking at the other*, 2) *percentage of frames when flies are together*, 3) *distribution of the distances between the fly centroids*, 4) *distribution of the head direction angles between the flies*, and 5) *distribution of the flies' speeds*. For the first measure the head direction vector of fly object A is intersected with the pixels of fly object B, and the head direction of fly object B is intersected with fly object A. The number of frames where these intersections are non-null are counted and divided by the number of frames of the video segment. For the second measure the number of frames that contain one fly object are divided by the number of frames of the video segment. For the third measure the Euclidean distance between fly object centroids is calculated in each frame and accumulated in a histogram. For the fourth measure the angle between fly object head directions is calculated in each frame and accumulated in a histogram. For the fifth measure V_S^i is calculated in each frame and its length is accumulated in a histogram.

3.7   Create Feature Vector

Four types of video clips of 10-minute duration are processed. They are: 1) *first 10 minutes*: first 10 minutes in the normal training period of a male fly, 2) *last 10 minutes*: last 10 minutes in the normal training period of a male fly,

3) *immediate recall 10 minutes*: immediate recall 10 minutes[1] of a normal-trained male fly, and 4) *sham 10 minutes*: immediate recall 10 minutes of a sham-trained[2] male fly. From the geometric measures an 8-dimensional feature vector that represents the behavior of the flies is computed for each video segment. The elements of the feature vector are: 1) *percentage of frames when one fly is looking at the other*, 2) *percentage of frames when flies are together*, 3,4) *mean and standard deviation of inter-centroid distances*, 5,6) *mean and standard deviation of head direction angles*, and 7,8) *mean and standard deviation of speeds multiplied by 100*. The last two numbers in the feature vector are multiplied by 100 so that all eight numbers have the same order of magnitude.

3.8   Generate Computed Courtship Index

A computational equivalent of the Courtship Index (CI), the Computed Courtship Index (CCI) is computed based on the values of the 8-dimensional feature vectors extracted from the N sample videos. The matrix $\mathbf{A}$ is an $N \times 8$ dimensional matrix, where each row is a data sample. The vector $\mathbf{b}$ is a column vector whose elements are the manually-derived CI values for each of the samples. The vector $\mathbf{x}$ contains the coefficients that define the linear relationship between the feature vectors and the CI values, and can be found by minimizing the standard Euclidean norm of the error,

$$\| \mathbf{Ax} - \mathbf{b} \|_2 .$$

Since our linear system is overdetermined, we use Singular Value Decomposition (SVD) to find the least squares solution of the minimization problem. The CCI for sample n may then be computed as:

$$CCI_n = PLA * x_1 + PFT * x_2 + MCD * x_3 + SDCD * x_4 +$$
$$MHA * x_5 + SDHA * x_6 + MS * x_7 + SDS * x_8. \qquad (2)$$

where the x_ks are the elements of the least squares solution vector, and the remaining variables are the components of the feature vector described in Section 3.7, e.g. PLA is *Percentage of frames when one fly is Looking At the other*.

3.9   Clustering

We also group the samples in terms of their learning and memory response based solely on extracted features. This is accomplished by applying k-means clustering [19] to the *learning* and *memory* feature vectors derived from the

---

[1] To obtain the immediate recall memory response of a male fly, it is transferred into a separate courtship chamber within 2 minutes after its training phase. In the new courtship chamber the male is assayed for 10 minutes with a virgin female unlike the training phase, in which the male courts a fertilized female.

[2] *Sham training* is a training phase 60 minutes in length where the male fly is not paired with a female fly.

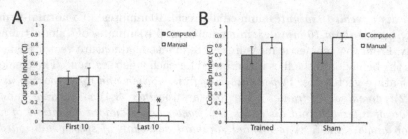

Fig. 6. Comparison of average CI and CCI values for flies with good learning and memory ability. Error bars represent ± SEM. * indicates p < 0.05.

raw 8-dimensional feature vectors. The Euclidean norm is used during cluster analysis to measure the distance between two feature vectors. When clustering samples by learning ability the feature vector calculated from a sample's last 10 minutes of a one-hour training session is subtracted from the feature vector computed from the first 10 minutes. This is done because the difference between a fly's first and last 10-minute behaviors characterizes its ability to learn. This produces a *learning* feature vector for each fly. Similarly, when clustering samples by memory ability the average vector calculated from all of the feature vectors for the sham-trained flies is subtracted from the immediate recall feature vector for a trained fly. This produces a *memory* feature vector for each fly.

4 Results

Video sequences for 22 fly specimens, which were recorded with a Sony DCR-SR47 Handycam fitted with a Carl Zeiss Vario-Tessarare lens, were processed and analyzed.

4.1 Evaluation of the Computational Courtship Index

We evaluated the effectiveness of our video analysis algorithms to produce a Computational Courtship Index (CCI) by comparing the CCI values to traditional CI values. CCI values were generated with a 10-fold cross-validation process. In the first study, we selected 7 different flies that individually showed normal learning and memory by manual CI analysis (Figures 6, A, B white columns). These flies showed a significant decrease in their overall courtship behavior in the last ten minutes of a one-hour training session compared to the first 10 minutes, both in the CI (Figure 6 A, white columns) and CCI (Figure 6 A, gray columns) values. Trained flies showed a significant reduction in courtship behavior as compared to sham-trained flies, indicative of normal immediate-recall memory of training (Figure 6 B, white columns). The CCI values also show a significant decrease for the trained flies as compared to the sham-trained flies (Figure 6 B, gray columns). There is no significant difference between the manually calculated CIs and the CCIs produced via video analysis.

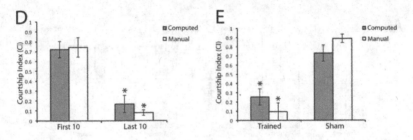

Fig. 7. Comparison of average CI and CCI values for flies with poor memory ability

We next determined whether the CCI could also detect flies with memory deficits. We selected 7 different flies that individually showed a successful learning response to training, but no memory of that training. In both cases, the manual and computed CIs showed a significant difference between the first and last 10 minutes of training (Figure 7 A), indicative of a successful learning response. However, for both the manual and computed CIs, there was no significant change in CIs from trained flies compared to sham controls (Figure 7 B), indicating that these flies have no memory of training. Again, there is no significant difference between the CIs calculated manually and the CCIs calculated via video analysis. Taken together, both sets of data suggest that the CCI is capable of determining successful learning and memory from those flies capable of performing this behavior robustly. Additionally the CCI is able to determine unsuccessful memory in the context of successful learning.

4.2 Automatic Grouping via k-Means Clustering

We also categorized fly specimens by learning and memory capabilities based solely on the feature vectors derived from video analysis with no prior human intervention. We applied k-means clustering [19] to the *learning* and *memory* feature vectors described in Section 3.9. In this study, we increased our total fly specimen size to 22 with varying CIs for both learning and memory. Using the *memory* feature vectors k-means cluster analysis (for k = 2) separated the fly specimens into two groups based on their memory capabilities. We evaluated the effectiveness of the analysis by computing average CI and CCI values for the two groups. The group with good memory is indicated by the higher difference in CI/CCI values (Figure 8 A) between trained flies and sham-trained flies. The poor memory group is indicated by the small difference in CI/CCI values (Figure 8 B) between trained flies and sham-trained flies.

We next performed this same k-means cluster analysis based on the learning capabilities of the flies, followed by a calculation of both the average CI and the CCI for each of these clustered groups. Again, it was clear that these two groups represent flies with good learning (Figure 8 C), and poor learning (Figure 8 D). Significant difference between the CI/CCI values between the first 10 minutes and last 10 minutes of the courtship suppression assay training phase indicates good learning. A small difference indicates poor or no learning.

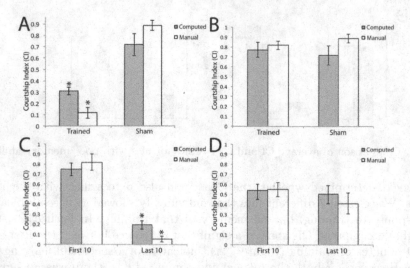

Fig. 8. Automated grouping of fly specimens by memory (A (good) and B (poor)) and learning (C (good) and D (poor)) ability. Error bars represent ± SEM. * indicates p < 0.05.

This analysis does not take into account the genotypes of the flies, only learning and memory capabilities as captured by our feature vectors derived via video analysis. Our data suggest that the automated analysis is capable of successfully creating two groups of flies with differing learning and memory capabilities without human intervention.

5   Conclusion

We have presented a set of algorithms that may be applied to videos of courting fruit flies. The algorithms compute a feature vector for a video segment that captures the behavior of the flies and may be used to categorize the learning and memory capabilities of the specimens. The algorithms include identifying individual flies in the video, tracking their motion, and computing geometric measures. Our results show that we are able to reproduce conventional CI values and automatically differentiate between normal and memory/learning-defective flies using the data derived from our video analysis.

Future work includes improving the performance of the computations via parallelization, automatic image cropping, more robust segmentation, and multivariate non-linear regression analysis for the CCI computation. We anticipate that our analysis will be capable of distinguishing between more subtle differences in fly behavior. Analyzing larger samples of flies and clustering them into several groups may allow for a more detailed analysis of learning and memory in both wild type and mutant flies. Finally, our automated system can also allow for increased throughput of pharmacological screenings in *Drosophila* models of human disease, such as Alzheimer's disease.

Acknowledgements. We would like to thank the members of the Marenda and Saunders labs for their helpful contributions to this paper. This work was supported by grants from the NIH, R01NS057295 (AJS), R21RR026074 (DRM).

References

1. Wu, J., Luo, L.: A protocol for mosaic analysis with a repressible cell marker (marcm) in Drosophila. Nat. Protoc. 1, 2583–2589 (2006)
2. Yu, H., Chen, C., Shi, L., Huang, Y., Lee, T.: Twin-spot marcm to reveal the developmental origin and identity of neurons. Nature Neuroscience 12, 947–953 (2009)
3. Pitman, J., DasGupta, S., Krashes, M., Leung, B., Perrat, P., Waddell, S.: There are many ways to train a fly. Fly 3, 3 (2009)
4. Bastock, M., Manning, A.: The courtship of Drosophila melanogaster. Behaviour 8, 85–111 (1955)
5. Siegel, R., Hall, J.: Conditioned responses in courtship behavior of normal and mutant Drosophila. Proceedings of the National Academy of Sciences 76, 3430 (1979)
6. Siwicki, K., Riccio, P., Ladewski, L., Marcillac, F., Dartevelle, L., Cross, S., Ferveur, J.: The role of cuticular pheromones in courtship conditioning of Drosophila males. Learning & Memory 12, 636 (2005)
7. Kamyshev, N., Iliadi, K., Bragina, J.: Drosophila conditioned courtship: Two ways of testing memory. Learning & Memory 6, 1 (1999)
8. Reza, M.: Automated categorization of Drosophila learning and memory behaviors using video analysis. Master's thesis. Drexel University, Philadelphia, PA (2011)
9. Maggio, E., Cavallaro, A.: Video Tracking: Theory and Practice. Wiley (2011)
10. Isard, M., MacCormick, J.: BraMBLe: A Bayesian multiple-blob tracker. In: Proc. IEEE International Conference on Computer Vision, vol. 2, pp. 34–41 (2001)
11. Khan, Z., Balch, T., Dellaert, F.: Mcmc-based particle filtering for tracking a variable number of interacting targets. IEEE Transactions on Pattern Analysis and Machine Intelligence 27, 1805–1918 (2005)
12. Branson, K., Rabaud, V., Belongie, S.: Three brown mice: See how they run. In: IEEE International Workshop on Performance Evaluation of Tracking and Surveillance, PETS (2003)
13. Isard, M., Blake, A.: Condensation – conditional density propagation for visual tracking. International Journal of Computer Vision 29, 5–28 (1998)
14. Tao, H., Sawhney, H.S., Kumar, R.: A Sampling Algorithm for Tracking Multiple Objects. In: Triggs, B., Zisserman, A., Szeliski, R. (eds.) ICCV-WS 1999. LNCS, vol. 1883, pp. 53–68. Springer, Heidelberg (2000)
15. Comaniciu, D., Ramesh, V., Meer, P.: Kernel-based object tracking. IEEE Transactions on Pattern Analysis and Machine Intelligence 25, 564–577 (2003)
16. Dankert, H., Wang, L., Hoopfer, E., Anderson, D., Perona, P.: Automated monitoring and analysis of social behavior in Drosophila. Nature Methods 6, 297 (2009)
17. Branson, K., Robie, A., Bender, J., Perona, P., Dickinson, M.: High-throughput ethomics in large groups of Drosophila. Nature Methods 6, 451 (2009)
18. Valente, D., Golani, I., Mitra, P.: Analysis of the trajectory of Drosophila melanogaster in a circular open field arena. PloS One 2, e1083 (2007)
19. Everitt, B., Landau, S., Leese, M., Stahl, D.: Cluster Analysis, 5th edn. John Wiley & Sons, Chichester (2011)

Panorama Image Construction Using Multiple-Photos Stitching from Biological Data

Joshua Rosenkranz[1], Yuan Xu[1], Xing Zhang[1], Lijun Yin[1], and William Stein[2]

[1] Department of Computer Science
[2] Department of Biology,
State University of New York at Binghamton

Abstract. This paper presents an image construction tool for biological image visualization and education using image matching and stitching approaches. The image matching technique is based on the algorithm SURF (Speeded-up Robust Feature) [3, 4], a successor to the popular feature detection algorithm SIFT (Scale Invariant Feature Transform) [1, 2]. Unlike a traditional image stitching approach, our tool assumes that biological images are taken on a linear model with similar degrees of overlap and orientation angle towards ground from air. With these aspects in mind, generated panoramas will display less distortion and more raw valuable details. Such a tool will facilitate the scientific research and education through applications of visual information processing in fields of Biology, Astronomy, Geology, etc.

Keywords: image synthesis, image stitching, feature matching.

1   Introduction

In a significant variety of scientific research fields, images are constructed for the purposes of being analyzed; part of the intent of the researcher is to prove or disprove a claim or observation about the image. In order to meet these standards, researchers must obtain high resolution images that portray the raw valuable detail within their cameras observation window. For these reasons, in today's times, it is common to see scientists capture multiple close-ups of a single object or landscape, with the aim of keeping records of the crucial details embedded within their images. A procedure known as image stitching is used as a way to preserve these indispensable features while keeping the object as a whole, without which would render the produced image useless for a scientific researcher [8]. Image stitching is the process of taking multiple images of varying degrees of overlap and combining these images to construct a virtual mosaic that resembles the object or landscape being viewed as intact. Previous methods of producing high resolution panoramas upheld the belief that as long as an image appears first-rate, it will be judged as if the algorithm has succeeded in creating what appears to be the full object or landscape being photographed. But, when viewed by a researcher, the distortion and manipulation of the image that had made this image look clean, has also corrupted the detail of the original image, rendering it as inadequate data in the research world. Unlike the existing work, this implementation is focusing on using strict geometry and known parameters such as the type and lens of

G. Bebis et al. (Eds.): ISVC 2012, Part II, LNCS 7432, pp. 242–252, 2012.

the camera, to produce a picture that better resembles the fine details of the original images that are so critical to researchers in the biological fields. This method adopts the viewpoint that, the center of every image contains the most valuable detail, the photograph taken should be modified as little as possible and we trust the original detail of the photograph.

This research was first started for usage in paleontological fields, specifically, a forest floor in Gilboa, NY, that has been preserved for millions of years. The particular site, which has now become a window to the past for vital information regarding the evolution of plant-life, will be unreachable to researchers due to construction constraints. In order to save this valuable data, an image must be constructed to map the site accurately. Although our current software was implemented considering these motivations, the future of this research can be endless regarding many other research fields that may need accurate image stitching.

In a research setting, photographs taken under the assumption that they will be stitched are generally taken under controlled conditions. We anticipate that these assumptions will allow our method to perform better under research conditions, thus granting our program the ability to sufficiently be applied in the research community more-so than the existing commercial software.

In Section 2, we will describe the image construction method through image matching and stitching implementation. Section 3 will report experimental results using real biological data. A conclusion and future work will be discussed in Section 4.

2   Image Construction by Matching and Stitching

For this image stitching method, a general guideline was always followed. In Figure 1, the general design of the whole project can be observed. Each process will be further explained in more detail following the figure.

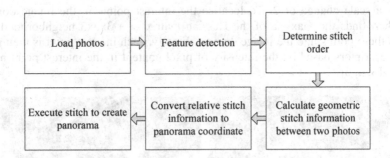

Fig. 1. General framework of the developed image construction tool

2.1   Preprocessing Photos

For this method, we simply loaded each photo and normalized the histogram for pixel intensity values in order to deal with light exposure. Lastly, we store each photo in memory for later use. This section of the program is one of the most time- consuming throughout our program.

2.2   Feature Detection

In this implementation, the feature detection algorithm, SURF, is applied. In most previous existing image stitching methods, SIFT was used for the purposes of feature detection, but in this implementation, SURF is used because it has shown some matching and time improvements over the previous existing algorithm. The main idea of SURF is to detect blob structures within the images [5]. SURF can be broken down into three main processes. First, SURF attempts to detect all possible interest points in the input image. For computational complexity issues, SURF works in the integral image domain [9]. When the image domain is computed, SURF uses a Hessian matrix based computation method along with a second derivative box filter to gather interest points. The detector is based on the Hessian matrix because of its good performance in computation time and accuracy. However, rather than using a different measure for selecting the location and the scale, as was done in the Hessian-Laplace detector [11], this method relies on the determinant of the Hessian for both. Given a point $\mathbf{x} = (x, y)$ in an image I, the Hessian matrix $H(\mathbf{x}, \sigma)$ in $\mathbf{x}$ at scale σ is defined as follows:

$$H(x,\sigma) = \begin{bmatrix} L_{xx}(x,\sigma) & L_{xy}(x,\sigma) \\ L_{xy}(x,\sigma) & L_{yy}(x,\sigma) \end{bmatrix} \tag{1}$$

Lxx($\mathbf{x}$, σ) is the convolution of the Gaussian second order derivative $\frac{\partial^2}{\partial x^2} g(\sigma)$ with the image I in point $\mathbf{x}$, and similarly for Lxy($\mathbf{x}$, σ) and Lyy($\mathbf{x}$, σ). Gaussians are optimal for scale-space analysis, as shown in [12].

Because SURF is implemented in the integral image domain and it utilizes box filters in its design, the scale space that is normally generated with an image pyramid, can now be implemented by simply increasing the filter [10]. The scale space is then divided into octaves based on filter response maps; convolute the input image with a filter of increasing size. The octaves are further reduced into particular scale levels. Scale spaces are constructed with 9x9, 15x15, 21x21, and 27x27 filters, which achieves a scale change of 2. To localize the interest points in the image domain, SURF then finds the maxima of the Hessian matrix in a 3x3x3 neighborhood, and converts these points into the image domain. Finally, each interest point is then given certain descriptors based on the intensity of pixel content in the interest point neighborhood. [3, 4, 6]

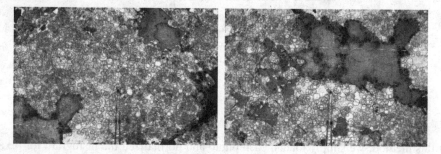

Fig. 2. An example of two images that have some form of similarity. The red circles denote features detected by the SURF algorithm. These images were taken from a devonian fossil forest site for use in biological research. Image generated using the SURF algorithm.

2.3   Finding Correspondence between images

Both images being reviewed at this stage in the process have a large number of possible important features. We use a comparison based on the descriptors of feature points produced by the SURF algorithm to determine the feasible features that exists between both images being reviewed. The comparison method determines whether two descriptors are similar enough by using a specific threshold. Ideally, all of the features that were detected to exist in both images are correct, but this is rarely the case, which required the creation of an outlier removal method. Figure 3 shows an example of detected correspondences for a pair of Devonian Fossil Forest images taken from Upstate New York.

2.4   Outlier Removal

Like all feature detection algorithms, SURF is not perfect, so this implementation removes any poor correspondence pairs that can possibly skew our final data. In this application, a simple K-means clustering algorithm (Lloyd's Algorithm) is used to find the feasible pair key points set [9]. This algorithm is based off the similarity between angles and distances of key points. Figure 3 also displays a pictorial representation of this outlier removal method.

Fig. 3. In this figure, the endpoints of each line represent matching features. Blue lines denote features determined to be useful in the image stitch application process. Red lines denote outliers to be removed from the process. This image was generated using the K-means clustering algorithm in this projects implementation. Please Note: Blue lines have similar angle and length orientation.

2.5   Image Geometry Stitch

2.5.1 Stitch Order

The methodology for panorama creation involves propagating each single image onto a larger panoramic image one-by-one. This order by which the images can be properly added to the panoramic image is called the "stitch order." For generality purposes, special structures such as Best-Bin-First (Lowe, 1999 [1]) and k-d tree to search the nearest neighbor are used for this intent.

2.5.2 General Stitch Information Calculation

Based on the determined stitch order and previously calculated key point pairs from (2.4), this step then computes the stitch information including rotation, translation, etc. to determine the information needed between two images. By doing this, the problem never becomes more complex, considering every computation between images is only done with two smaller base images.

2.5.3 Stitch Process – Two Base Images

In order to find relative translation and rotation with respect to two base images, two pairs of feasible matched key points generated from (2.4) are selected. More specifically, suppose we have two photos, photo A and photo B, both which were determined to share some similar features. Denote $p1(x1,y1)$, $p2(x2,y2)$ as two pairs in photo A, and $p3(x3,y3),p4(x4,y4)$ as two pairs in photo B. $p1$ and $p3$ are a pair of feature points that represent ideally, the same features that exist within photos A and B, at those points. Similarly, $p2$ and $p4$ are considered another pair of similar feature points detected between the two images A and B. Photo A, the first photo, is regarded always as the fixed position photo; no translation or rotation is required. It is essential for photo B to be translated and rotated in order to be stitched properly to the first image, as illustrated in Figure 4.

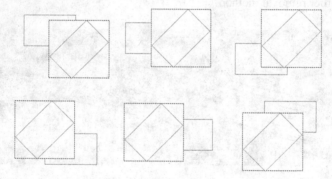

Fig. 4. Illustration of different stitch scenarios that exist between two base input images. The red rectangle represents "photo A", the fixed photo, whereas the blue rectangle represents "photo B", the photo that requires rotation and translation.

The angle of rotation is calculated by the equation,

$$\theta = \arctan(\frac{y_1-y_2}{x_1-x_2}) - \arctan(\frac{y_3-y_4}{x_3-x_4}) \tag{2}$$

According to the information gathered from both the rotation and translation model with respect to the correlation between locations of the two photos, we can then determine the geometric information needed to stitch these two images. Once calculated, the information is stored for use in the next stage.

2.5.4 Stitch Process – Panorama Coordinate Calculation

The stitch information gathered from the two base image stitch process is used to convert all coordinates to their general panoramic coordinates. Note that as specified in Figure 4, that of the two images being considered, the first photo is fixed and the second photo is subject to translation and rotation. With the geometric information gathered and stored in the buffer, this information can be used to stitch a given photo to another photo that has been added to the overall panorama. As denoted earlier, in this case, we regard the panorama as the fixed image and the new image to be added as the image that is subject to translation and rotation. Using this model, information about geometric transformations regarding the panoramic image can be obtained and stored; this information denotes the locations and translation of each image within the panorama. By storing this information rather than just creating the panorama in this step, we make these two processes independent of one another; once this information is gathered, panoramic image creation is created by affine transformation (translation and rotation). Note that this simulation is purely based on geometric calculation. The following can be observed in Figure 5.

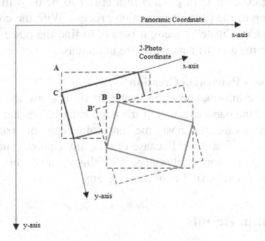

Fig. 5. Illustration of the geometry for calculating the location of an image for the panoramic coordinates system. The red rectangle denotes "Photo A" and the blue rectangle denotes "Photo B" from the two base image stitch process example. The "Panorama Coordinate" represents the coordinate whose origin is at the upper-left corner of the panorama. "A" represents the upper-left corner of "Photo A"; the start point for pixel-by-pixel execution in the final panorama creation. "C" represents the upper-left corner of "Photo A" before rotation. "B" denotes the upper-left corner of "Photo B", and "D" denotes the upper-left corner of "Photo B" before rotation.

As seen in Figure 5, the coordinates of $A(x_A, y_A)$ and $C(x_C, y_C)$ under "panoramic coordinates" are known. From the stitch information provided in the saved buffer, the coordinates of points $B'(x_{B'}, y_{B'})$ and $D'(x_{D'}, y_{D'})$ are identified under the "2-Photo" coordinate, where D and D' are the same position. In order to complete the conversion, we still require the coordinate of $B(x_B, y_B)$ under the "Panorama Coordinate" system. Since the positive y direction of the coordinate system is different from the Cartesian coordinate system, the rotation matrix requires a change. This can be computed in the following revised rotation matrix equation,

$$R(\theta) = \begin{pmatrix} \cos \theta & \sin \theta \\ \sin \theta & -\cos \theta \end{pmatrix} \tag{3}$$

Denote the angle that "Photo A" needs to rotate in order to be stitched to the panorama as β. With this information, the coordinate of point B in "Panoramic Coordinates" can be calculated using the following formula,

$$\begin{pmatrix} x_B \\ y_B \end{pmatrix} = \begin{pmatrix} x..offset..of..\overline{BD} \\ y..offset..of..\overline{BD} \end{pmatrix} + \begin{pmatrix} x_C \\ y_C \end{pmatrix} + R(\beta) \begin{pmatrix} x_{D'} \\ y_{D'} \end{pmatrix} \tag{4}$$

Lastly, we store the coordinate of point B in a buffer to be used in the execution of stitch in the final step of the panorama creation process. With the coordinate of point B in the "Panorama Coordinate" system, it is easy to find the place to insert the new photo into the panorama and the panorama size to increase.

2.5.5 Stitch Process – Panorama Creation

Finally, with all of the information given in the previous steps, the program can add photos one-by-one to the panorama. All of the math calculations before this final step in the process guarantees that the most time consuming stage of pixel-by-pixel stitching will be saved for the final task. Because of this, any changes such as blending of overlapped regions can be done at this step. With this in mind, this stitching process ensures high efficiency and flexibility of the program.

3   Experimental Results

This implementation displayed fast results while still maintaining consistency in the scientific photographs taken. We tested our approach using a large amount of photos (about 1,500 photos) taken from Devonian Fossil Forest and Ancient Biological Plantlife of Upstate New York through the HHMI research and education program. Figure 6 – 9 show several sample results. Figure 10 illustrated the feature correspondence matching from a sample set of photos. The stitched images are inspected by experts from the Biology department of our university, and are compared to the ground-truth images created by manual stitching from the biologists. They have shown the correct match with sufficient details among those data.

Fig. 6. This figure displays a panorama of 13 images taken from a devonian fossil forest site. The original photo-data set is shown in the upper rows. The stitched result is shown in the bottom row.

Fig. 7. This figure displays a panorama of 34 images taken from ancient biological plantlife. The original photo-data set is shown in the upper rows. The stitched result is shown in the bottom row.

Fig. 8. This figure displays a panorama of 53 images taken from ancient biological plantlife. The original photo-data set is shown in the upper rows. The stitched result is shown in the bottom row.

Fig. 9. This figure displays a panorama of 24 images taken from ancient biological plantlife. The original photo-data set is shown in the upper rows. The stitched result is shown in the bottom row.

Fig. 10. Examples of outlier removal in corresponding keypoint matching

4   Conclusion and Future Work

In this paper, we presented an image stitching tool for biological data visualization through the image matching and stitching implementation. This is a preliminary study for biological visual information processing through the undergraduate and graduate education and research program. The future work is to test our results using SIFT algorithm for feature detection to obtain a comprehensive comparison between the two algorithms. We will also add a more advanced blending function to create a smoother stitch between photos. Considering the distortion properties of camera lenses, it is imperative to include distortion correction in this program given a specific camera lens as input. Finally, we will develop an approach for incorporating more advanced camera calibration information for distortion-proof image stitching in order to improve the accuracy, reliability, and realism of stitching results.

Acknowledgements. This work is supported under a HHMI program grant. We would like to thank the Howard Hughes Medical Institute (HHMI) for their support of this work in biomedical and biological research from visual information processing applications. In addition, we would like to thank the HHMI program for supporting graduate and undergraduate interdisciplinary research in computer science and biology.

References

1. Brown, M., Lowe, D.: Automatic Panoramic image stitching using invariant features. IJCV 74, 59–73 (2007)
2. Lowe, D.: Object recognition from local scale-invariant features. In: ICCV (1999)
3. Lowe, D.: Distinctive image features from scale-invariant key points, cascade filtering approach. IJCV 60, 91–110 (2004)
4. Bay, H., Tuytelaars, T., Van Gool, L.: SURF: Speeded Up Robust Features. In: Leonardis, A., Bischof, H., Pinz, A. (eds.) ECCV 2006. LNCS, vol. 3951, pp. 404–417. Springer, Heidelberg (2006)
5. Bay, H., Ess, A., Tuytelaars, T., Van Gool, V.: Speeded-Up Robust Features (SURF), technical report (2005)
6. Koenderink, J.: The structure of images. Biological Cybernetics 50, 363–370 (1984)
7. Mikolajczyk, K., Schmid, C.: Indexing based on scale invariant interest points. In: IEEE International Conference of Computer Vision (ICCV 2001), pp. 525–531 (2001)
8. Mikolajczyk, K., Schmid, C.: A performance evaluation of local descriptors. IEEE Trans. on PAMI 27, 1615–1630 (2005)
9. Viola, P., Jones, M.: Rapid object detection using a boosted cascade of simple features. In: CVPR, vol. 1, pp. 511–518 (2001)
10. Szeliski, R.: Image Alignment and Stitching: A Tutorial. In: Microsoft Research (2006)
11. Lloyd, S.P.: Least Squares Quantization in PCM. IEEE Trans. on Information Theory IT-28(2) (March 1982)
12. Lindeberg, T.: Scale-space theory: A basic tool for analyzing structures at differ-ent scales. Journal of Applied Statistics 21(2), 225–270 (1994) (Supple-ment Advances in Applied Statistics: Statistics and Images)

Function Field Analysis for the Visualization of Flow Similarity in Time-Varying Vector Fields

Harald Obermaier and Kenneth I. Joy

Institute for Data Analysis and Visualization, University of California, Davis, USA

Abstract. Modern time-varying flow visualization techniques that rely on advection are able to convey fluid transport, but cannot provide an accurate insight into local flow behavior over time or locally corresponding patterns in unsteady vector fields. We overcome these limitations of purely Lagrangian approaches by generalizing the concept of function fields to time-varying flows. This representation of unsteady vector-fields as stationary function fields, where every position in space is a vector-valued function supports the application of novel analysis techniques based on function correlation, and allows to answer data analysis questions that remain unanswered with classic time-varying vector field analysis techniques. Our results demonstrate how analysis of time-varying flow fields can benefit from a conversion into function field representations and show the robustness of our presented clustering techniques.

1 Introduction

In areas such as meteorology or aerodynamics complex flow fields are being produced whose analysis and interpretation requires the development of suitable visualization techniques. There are flow analysis problems, e.g., in meteorology, that are, unlike many modern flow analysis approaches, not only concerned with the transport behavior of flow, but with the detection, and analysis of (repeating) local flow behavior. Previously, this important task has rarely played a role during the design process of unsteady flow visualization techniques.

We develop new methods that aim at supporting this task by visualizing flow similarity and allowing analysis and discovery of similar or relevant local behaviors in the flow field. For this matter, we interpret unsteady flow fields as stationary function fields. With this notion we are able to apply function processing and cross-correlation concepts and develop meaningful flow similarity metrics. These novel metrics are used for automatic flow domain clustering and manual flow behavior querying and prove to be valuable in answering relevant flow analysis questions that are hard to answer with existing techniques. Our visualization solution reduces the visual complexity of multi-temporal flow visualization that often suffers from the presence of features with a spatio-temporal character, such as path-lines. For the first time, it is now possible to identify regions that yield similar flow behavior over time, a notion that is of central importance in location based flow analysis such as in meteorology or geometry optimization in aerodynamics and industrial mixing. Furthermore, our novel techniques support interactive querying in vector-valued function fields.

G. Bebis et al. (Eds.): ISVC 2012, Part II, LNCS 7432, pp. 253–264, 2012.

Section 2 summarizes related work and provides a motivation. Section 3 introduces function fields, before novel flow similarity metrics are presented in Section 4. Visualization and analysis techniques are detailed in Section 5, whereas concrete results are presented in Section 6. Section 7 concludes this work.

2   Motivation and Related Work

Most modern unsteady flow analysis techniques focus on the analysis of material transport and the extraction of flow trajectories [1, 2], mimicking physical experiments such as dye advection in a Lagrangian setting. Even in combination with querying [3–6], these analysis techniques neglect a large field of application areas, where recurring patterns [7–9] and local flow behavior is of central importance. In such contexts, a Eulerian perspective of the flow field is more suitable for effective visual analysis.

In meteorology, for example, the behavior of wind directions in town X in a given month may be compared to wind behavior at a different location during the same time period. Alternatively, wind behavior may be studied at the same location for different seasons. In combination, this requires a multi-temporal, local analysis of flow. Physically, such Eulerian flow measurements may be obtained e.g., by means of Doppler Velocimetry [10]. Here, no longer is fluid transport the focus of interest, but characteristics of local flow behavior over time and space.

Changing the representation of an unsteady vector field into a different form (cf. FTLE [11] or scalar function fields [12]) has long known to make certain analysis and visualization tasks feasible or computationally more straight-forward. We therefore choose to represent unsteady flow as vector-valued function fields. This localized representation supports the use of flow pattern recognition, tracking, and querying of similar local flow behaviors. These techniques are related to research [5, 13–15] on time-series, where 2D graphs [16] still dominate the field. For the first time, the presented techniques allow automatic time-varying flow similarity analysis, straight-forward identification of 'dead' mixing regions and interactive behavior querying. Thus, we regard function field representations as suitable solutions to local flow analysis problems. The workflow structure of the remainder of this paper is illustrated in Figure 1.

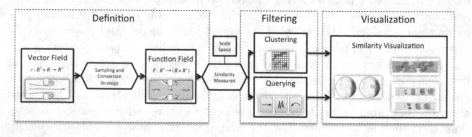

Fig. 1. Workflow of flow similarity evaluation. Definition of similarity measures supports clustering and querying of similar flow behavior in a function field.

3 Function Fields

3.1 Definitions and Concepts

A time-varying 3D flow field $v : \mathbb{R}^3 \times \mathbb{R} \to \mathbb{R}^3$ is a continuous representation of velocity vectors at positions $x \in \mathbb{R}^3$ and time $t \in \mathbb{R}$. An equivalent notion of such a flow field can be obtained by modifying the role of an independent variable:

$$F : \mathbb{R}^3 \to (\mathbb{R} \to \mathbb{R}^3) \tag{1}$$

$$F(x) = v(x, .) \tag{2}$$

F contains a vector-valued function at every position in space and is therefore called a *function field*. An individual vector-valued function at position $x \in \mathbb{R}^3$ in F is denoted by f_x and called a *local flow pattern* in the following. A *normalized* flow pattern corresponds to a normalized v with unit vectors. Note that there may be equivalent flow patterns at different positions in the flow field, for which reason the position index may be dropped. This representation is comparable to feature-space representations used in visualization as it represents a set of (independent) functions. For scalars, such a notation was used by Anderson et al. [12]. A *flow motif* is obtained by restricting the parameter space of a flow pattern to an interval $[t_0, t_1] \subset \mathbb{R}$. We denote such a flow motif as $(f_x, [t_0, t_1])$.

3.2 Function Field Creation

The conversion of an unsteady flow field into its numerical representation as a function field requires finding an appropriate sampling strategy for the placement of local flow functions. We propose the use of three alternative sampling strategies. The first and most simple strategy uses a regular grid [17] to position local flow patterns. This is fast to compute and facilitates flexible user control along with level-of-detail sampling. Since the visual output of the techniques presented in this paper does not directly depend on the density and uniformity of the chosen sampling, we can make use of further techniques. The second strategy places a flow function at every node of the original computational grid of the flow field. This technique has a more optimal sampling resolution, but is only applicable to fields that maintain a static mesh. The third approach adaptively samples the flow field. In this approach we construct a octtree representation of the flow field whose cells are subdivided if they span a region with a large variance in velocity direction at any point in time (see Figure 2a). In all cases, a single local flow pattern f_x represents the flow direction of a constrained region $C \subset \mathbb{R}^3$ (cell) of the flow field. Numerically, a flow pattern f_x is stored as a series of average vector values $[v(x, t_0), \ldots, v(x, t_n)]$ together with standard deviations of velocity $[SD(C, t_0), \ldots, SD(C, t_n)]$ computed from discrete velocity histograms per cell C and time step. Note that standard deviations for strategies whose grid cells correspond to less than a cell in the original resolution evaluate to zero. In the course of this paper we normalize all flow patterns to $f_x(.)/\|f_x(.)\|$. Flow patterns with a standard deviation magnitude above 1 are considered non-representative and dissimilar to all other patterns.

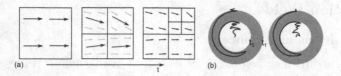

Fig. 2. (a) Cell subdivision in octree sampling is based on time-varying velocity deviation. If velocities within a cell are too dissimilar in a time step, it is subdivided. (b) Illustration of 2D flow radar glyphs as introduced in [17]. Our motif similarity extraction supports the detection of similar flow behavior as highlighted in blue.

4   Flow Similarity

Consisting of a set of independent, local flow patterns, the function field representation is a suitable tool for local, domain centered flow analysis. Through pattern clustering, the identification of dead regions, recurring flow patterns, and distinct flow behaviors may be achieved. Metrics for the efficient, manual and automatic local flow pattern similarity evaluation are presented in the following.

4.1   Visual Similarity

With a direct function visualization technique, similarity evaluation in a function field may be performed visually. An example of such a technique are flow radar glyphs that map flow directions to curves on a radial glyph [17]. While the proposed glyphs (Figure 2b) are a step towards analysis of unsteady flow, there are limits w.r.t. accurate similarity evaluation and expressiveness as the number of time-steps is increased. In this work we devise methods to effectively display and analyze flow behavior similarity in unsteady 3D flow.

4.2   Similarity Metrics

The key to efficient clustering and similarity analysis in function space is the definition of mathematical distance measures d for any two flow motifs f and g that allow efficient and meaningful comparison of flow patterns. In the context of this work we identify the following key characteristics of flow distance measures:

- **Efficiency:** d should be fast to compute to facilitate interactive querying.
- **Robustness:** Influence of small scale noise from the simulation on the value of d is to be reduced.
- **Flexibility:** d should be invariant w.r.t. selected linear transformations such as rotation or scaling of patterns.

The flexibility criterion exists to support not only analysis of direct similarity between flow functions (i.e., $d(f, g) = 0$ iff $f = g$), but also facilitate similarity comparison with respect to abstract flow behaviors (i.e., oscillation along an arbitrary direction, $d(f, g) = 0$, iff $f = Rg$, with linear transformation R).

Thus, similarity of local flow patterns is defined on multiple levels of abstraction. In the following we provide details about similarity measures on two levels of abstraction. In the remainder of this work we assume normalized flow motifs.

Relative Similarity. A low-level similarity comparison only requires direct comparison of functions. Ideally, behavioral similarity that includes linear transformations is computed by maximizing similarity with respect to a linear transformation matrix R. Function similarity or (cross-)correlation is commonly based on application of the inner product over the range of two functions. We therefore adopt a modified version of the discrete maximum norm

$$d(f,g) = min_R(max_t(1- < f(t), Rg(t) > \cdot s(t))), \qquad (3)$$

where Euclidean distances in standard deviations are used to modify the value of the inner product in the form of $s(t) = 1 - |atan(\|SD_f(t) - R \cdot SD_g(t)\|)\frac{2}{\pi})|$. Numerically the arctangent is approximated by a second order polynomial. For computational efficiency, we approximate this similarity by first computing average flow direction of f and g: $\bar{f}$ and $\bar{g}$, which are used to align f and g, such that $\bar{f} = \bar{g}$. Finally, we rotate g around $\bar{g}$ until $d(f,g)$ is minimal. This rotation is performed in 20 degree angles and uses angle bisection to iterate to maxima. Note that simply choosing $R = I$ violates the flexibility criterion, but still allows for simple function comparison. This distance measure identifies flow behaviors that are identical under rotational transformation and returns overall maximal angular deviation between velocities present in flow patterns.

Behavioral Similarity. A more abstract measure compares similarity of abstract flow behaviors. For this matter, a vector-valued function has to be converted into a behavior-sequence description. We exploit the fact that a range of typical flow behaviors can be described mathematically and be used to classify flow motifs. We identify these typical flow behaviors B (see Figure 2b for backflow) as boolean predicates on flow motifs $(f, [t_0, t_n]) = (f(t_0), \ldots, f(t_n))$:

- **P(Constant)** = **true** iff time-variation of flow direction is minimal.
  $f(t) \cdot \bar{f} < \epsilon$ for all $t \in [t_0, t_n]$, $\bar{f}$ average flow direction of the motif.
- **P(Oscillation)** = **true** iff velocity directions oscillate along a direction.
  $|\angle(f(t_i) - f(t_{i-1}), f(t_{i+1}) - f(t_i))| \ll \frac{\pi}{2}$ for subsequent i.
- **P(Backflow)** = **true** iff flow direction is inverted.
  $f(t) \cdot \bar{f} \ll 0$ for at least one $t \in [t_0, t_n]$
- **P(Rotation)** = **true** iff flow directions rotate around a center direction c.
  $\angle(f(t_{i-1}) - c, f(t_i) - c) \cdot \angle(f(t_i) - c, f(t_{i+1}) - c) > 0$
  $\sum \angle(f(t_i) - c, f(t_{i+1}) - c) > 2\pi$, $|\angle(f(t_i) - f(t_{i-1}), f(t_{i+1}) - f(t_i))| < \epsilon$.
- **P(Swirling)** = **true** iff flow performs planar rotation.
  Same as rotation, but $\bar{f} \approx 0$

Note that behaviors based on velocity magnitude, such as pulsation are excluded from this paper and are subject to future research. Every one of these behaviors B may hold for a given flow motif. Thus, a local flow pattern shows a selected behavior B for a length $t_1 - t_0$ if it contains a flow motif $(f, [t_0, t_1])$ with $P(B) = true$. This allows for behavior querying and distance computation. A flow pattern is converted into a sequence of flow behaviors by performing a sliding-window search with varying window length according to user specified minimal and maximal window lengths. This creates flow motifs for the evaluation of behavior predicates. At every point in time, the B that holds for a maximal window length is selected. For equal lengths, priorities are Constant < Oscillation < Backflow < Rotation < Swirling. If no pattern matches, behavior is assumed to be chaotic. Two flow patterns are considered similar if they represent the same sequence of flow behaviors. The distance function then evaluates to the Euclidean distance between the behavior-length vectors of the two functions.

4.3   Improvements

In combination, the presented similarity measures satisfy all key requirements listed in the last section with exception of robustness with respect to noise in the spatial and time dimension. We implement a scale space representation along with a time-shift similarity measure to allow for robust similarity evaluation.

The representation of flow patterns as vector-valued functions allows the application of concepts from scale space theory. Thus, we represent individual flow patterns as one parameter family of low-pass filtered [13] curves. Consequently, we do not use f and g in (3), but the resulting curves after Gaussian filtering with different kernel widths. This reduces the impact of small scale variations in matching and allows robust scale-based matching with a user-specified scale.

The time-varying nature of our problem can cause different locations in the flow field to show very similar flow behavior at slightly different times. Thus, a rigid comparison based on flow pattern entries at corresponding times (static inner-product in (3)) is inferior to time shift matching methods. Consequently, we implement the *Dynamic Time Warping* (DTW) [18] method to provide a time-shift tolerant similarity measure between flow patterns.

5   Function Field Analysis and Visualization

Clustering of function fields based on the presented distance measures can help identify similar and relevant behaviors automatically. However, in flow analysis, interactive exploration of flow properties is often superior to automatic feature detection, making the definition of suitable operators necessary. Thus, we allow the application of feature space querying techniques [6, 19] to function fields. We make use of semantically linked visualizations of the function field along with its domain geometry and a visualization of function space for efficient querying and cluster representation as described in the following.

Function Space Visualization. We use a spherical representation for the visualization of function space that maps flow motifs to radial curves in a sphere [15, 20]. Goal of this visualization is to plot all or a selection of normalized patterns in a common space in a manner that allows a concise overview of the behavior of the set of patterns with respect to time and orientation. Additionally, we facilitate brushing operations such as pattern selection and region of interest selection on the surface of the sphere. The user is enabled to draw on the sphere to mark and filter individual behaviors. The visualization of this potentially vast amount of functions, as well as temporal components require special attention.

As a first step, we create curve geometry on the GPU to allow interactive framerates in the presence of a large number of curves. The resulting curve geometry, projected polygon strips, are color coded by t, and blended together in order to emphasize behaviors that are present in multiple flow patterns. While this gives a good impression of the distribution of orientations throughout the parameter space of the flow field, it still lacks an expressive notion of time, even when the rendered curves are color-coded. Radially scaling these functions according to time allows efficient perception of temporal characteristics, similar to generalized flow radar glyphs. For joint behavior analysis, we offer the option to rotate all flow patterns such that all average flow pattern directions line up, giving a better insight into backflow, rotation and overall constant behavior. We optionally visualize clusters of flow motifs or subsets of function space as separate spherical plots by filtering out flow patterns that do not belong to a cluster.

Function Field Visualization. Clustering, similarity evaluation, and querying results in the selection of (possibly disconnected or overlapping) flow motifs, i.e., spatio-temporal selections in the function field domain. Visualization of these function sets requires the simultaneous display of individual and compound function properties. On an individual function level, we draw a single generalized flow radar glyph at the position of flow patterns. If only a specific time range of the function (i.e. a flow motif) is selected, the corresponding glyph is faded out if the currently displayed time step is outside of the range of the motif.

To highlight the shape and geometry of the current position of a cluster of motifs, we extract a metaball based [21] isovolume for each cluster, thus giving an insight into positions of selected function sets. Metaballs are scaled according to the selected function cell size. We provide the user with a region-of-interest selection tool that allows the selection of a set of local flow patterns in space followed by a specification of the relevant time interval. This effectively selects flow motifs in the function field. Furthermore, the user can query for behaviors B in the flow field by composing a sequence of behaviors.

6 Results

We demonstrate the efficiency and applicability of function field analysis with the help of selected data sets that were generated by flow simulations and converted into function field representation by uniform sampling with up to 10000 flow patterns. Functions are represented at 10 scales, around 30 time steps were processed per simulation.

6.1   Performance

During pre-processing, the field is evaluated and converted into function field representation. The speed of these operations is directly dependent on the field evaluation method, number of time-steps, and sampling strategy/density. These computations took less than a minute for all presented data sets on a 64 bit Intel Corei7 at 2.2 Ghz with 8 GB of memory.

In Figure 3b we present average computation times for automatic clustering of the function field for different numbers of clustered flow observers and two different clustering techniques. The *distance* times denote time spent to compute a full similarity matrix for all patterns. Note that distance computation, as is used during interactive querying as well, is fast even for large numbers of flow patterns, which clustering techniques with quadratic or higher complexity become unfeasable for very large numbers of patterns. In practice this can be avoided by restricting function field analysis locally.

6.2   Clustering and Querying

We perform automatic clustering of flow regions by k-means or QT-clustering [22] with (3). QT-clustering, a method that creates a clustering given a specified maximal intra-cluster distance, of an extraction column data set reveals large sets of similar flow behavior and produces a decomposition into regions with persistent flow behavior. Sphere-plot visualizations of the user selection and the 11 largest clusters are shown in Figure 3a. Clustering reveals regions with distinct flow behaviors as indicated by a range of different cluster variances and oscillation patterns. This shows how the local effects of boundary properties like corners or flow obstacles onto the flow field can efficiently be detected, visualized, and analyzed as indicated by pockets of turbulent flow behind flow obstacles. The mixing column example hints at possible applications in medical areas, where the effect of blood-vessel anomalies on local blood flow have to be studied. As demonstrated in this figure, function field analysis can answer the abstract question *"Where in this data set do similar 'flow events' happen?"* in time-varying flow fields. Furthermore, flow pattern clustering is able to reveal symmetries in the flow field. The shown flow-fields are virtually symmetric along their x-axis (z-axis), as shown by extracted pattern outlines. This represents a novel way of analyzing time-varying flow symmetries, which could not be captured by previous visualization methods. Automatic clustering of flow fields in a spatio-temporal setting has potential application areas in flow field compression or acceleration of trajectory computation. In our given example, we can clearly distinguish different constant and unsteady flow behaviors, i.e., regions where time plays a minor vs. a major role.

Function space visualization and brushing is useful, if overall flow behaviors in the data set are to be visualized or analyzed. We have implemented proof-of-concept techniques for the described brushing operations in function-space. We use interactive sphere brushing for pattern selection on the sphere plot, performing brushing in time-varying space of flow patterns and orientations.

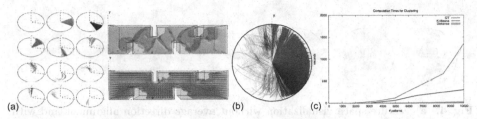

Fig. 3. (a) Result of QT clustering on an extraction column data set. Visualization shows user selection (top left) and the 11 largest clusters produced by QT clustering. The four largest clusters are rendered in the data set (top). The different flow behaviors in these regions is clearly depicted in the spherical plots. The 3D visualization facilitates analysis of similar, low, and high turbulence regions, supporting analysis of mixing properties. A cluster obtained by querying for constant flow is shown at the bottom along with vector splats indicating flow direction. (b) Constant flow selection shown in function space.Comparatively low variance in the green, blue and purple regions indicate large regions with linear, homogeneous flow behavior. (c) Computation times for clustering.

We employ a simple combination of ray-sphere intersection and stencil-buffering to allow selection of regions on the sphere. Selected flow patterns are found by collecting vertices of the projected flow pattern representation on the sphere that lie within the selected region-of-interest. Such brushing queries can reach from simply selecting regions that have backflow behavior at a certain point in time over to querying flow patterns that have comparable turbulence or oscillation properties. Figure 4 shows the effect of function rotation and possible manual selections. In the shown cross-flow data-set, distinct rotating backflow behavior can be identified as the flow passes sharp corners. Virtually all other flow regions are covered by constant flow, as indicated by generalized flow radar glyph visualization and metaball-based selection outlining. The given function space visualization indicates locations, time, and extend of backflow in different regions of the data set and enable direct filtering or highlighting of such flow patterns. Direct manual function space operations together with metaball visualization are able to discover small scale boundary effects on the flow field that are hidden with glyph-based rendering techniques.

A jet stream data set is displayed in Figure 5. Function space display in such a turbulent data set becomes challenging if no flow patterns are pre-filtered by automatic clustering. Automatic and manual clustering highlights relevant flow behaviors and emphasizes a point-symmetry present throughout the complete evolution of the data set.

We recorded motifs by region-of-interest selection in the function field domain close to vortex cores in a simulated Kármán vortex street. Similarity evaluation highlights flow motif sets in other regions in space and time of the data set that behave similarly to the recorded motif set, see Figure 6. The results of such motif recording are promising for two reasons: They can be applied to detect and track flow anomalies in flow fields and to uncover error sources such as faults in the simulation code or mesh.

Fig. 4. (a) Feature space visualization without average direction alignment and with average direction alignment. Note the four distinct flow directions in the non-rotated rendering. (b) Manual selection of flow patterns with constant flow behavior. Metaball outlining gives a look at set outlines, revealing a small region around the bottom left inner corner, were non-linear flow occurs. (c) Manual selection of pattern with backflow behavior. A close-up illustrates, how different flow patterns as glyphs may appear despite having similar flow behaviors. This illustrates, why pure visual comparison of flow glyphs is not sufficient for accurate similarity analysis.

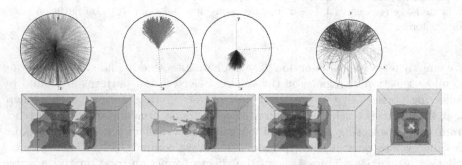

Fig. 5. Analysis of a short sequence of a jet flow data set reveals multiple orthogonal flow behaviors. Automatic clustering produces homogeneous constant backflow and forward (at the front and center) flow clusters and emphasizes point-symmetry of the data set. Note that function space display alone without cluster information is not expressive for this data set. Manual querying in function space allows the extraction of flow details, such as the main jet core shown from the side and front.

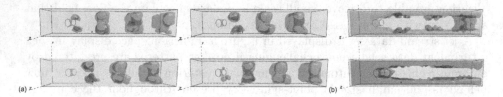

Fig. 6. (a) Flow pattern recording in a Kármán vortex street. Distinct flow behaviors are reliably tracked over time. An interesting property made evident by this visualization is the fact that these features are born on both sides of the obstacle and merge after passing the region with prominent backflow behavior. (b) Results from automatic clustering (top) reveal the spatial frequency of recurring flow behavior. Constant and backflow behavior querying (bottom) show regions unaffected by turbulent vortex core paths.

7 Conclusions and Future Work

In this work we have presented techniques to create and utilize a function field definition for unsteady flow fields. This conversion allowed the definition of effective function space similarity measures that support novel unsteady vector-field analysis and visualization techniques based on flow behavior similarity and clustering. Function field representations and processing techniques allow unsteady flow field analysis that is not possible with transport based Lagrangian approaches. A set of presented querying and visualization techniques facilitates interactive function field analysis in a wide number of potential application areas. A drawback of the presented methods is their limitation to normalized flow patterns, while this is can be compensated by taking spatial variation of velocity vectors into account, it cannot model behaviors based on velocity magnitude. Future work is focused on the creation and evaluation of further interaction techniques, such as interactive flow observer placement and the evaluation of function field techniques in concrete application areas.

Acknowledgements. The authors wish to thank the members of the Institute for Data Analysis and Visualization (IDAV) for valuable discussions and support.

References

1. Weiskopf, D., Erlebacher, G.: Flow visualization overview. In: Handbook of Visualization, pp. 261–278. Elsevier, Amsterdam (2005)
2. McLouglin, T., Laramee, R.S., Peikert, R., Post, F.H., Chen, M.: Over two decades of integration-based geometric flow visualization. Comp. Graph. Forum 29, 1807–1829 (2010)
3. Shi, K., Theisel, H., Hauser, H., Weinkauf, T., Matkovic, K., Hege, H.-C., Seidel, H.-P.: Path line attributes - an information visualization approach to analyzing the dynamic behavior of 3d time-dependent flow fields. In: Topology-Based Methods in Visualization II, Mathematics and Visualization, pp. 75–88. Springer (2009)
4. Xu, L., Shen, H.W.: Flow web: a graph based user interface for 3d flow field exploration. Visualization and Data Analysis 7530, 75300F (2010)
5. Wei, J., Wang, C., Yu, H., Ma, K.: A sketch-based interface for classifying and visualizing vector fields. In: Proc. of PacificVis 2010, pp. 129–136 (2010)
6. Lez, A., Zajic, A., Matkovic, K., Pobitzer, A., Mayer, M., Hauser, H.: Interactive exploration and analysis of pathlines in flow data. In: Proc. Int. Conf. in Central Europe on Comp. Grap., Vis. and Comp. Vision (WSCG 2011), pp. 17–24 (2011)
7. Garcke, H., Preußer, T., Rumpf, M., Telea, A., Weikard, U., van Wijk, J.: A continuous clustering method for vector fields. In: Proc. of the Conf. on Visualization 2000, pp. 351–358. IEEE Computer Society Press, Los Alamitos (2000)
8. Schlemmer, M., Heringer, M., Morr, F., Hotz, I., Bertram, M.H., Garth, C., Kollmann, W., Hamann, B., Hagen, H.: Moment invariants for the analysis of 2d flow fields. IEEE Trans. Vis. Comput. Graph. 13, 1743–1750 (2007)
9. Nagaraj, S., Natarajan, V., Nanjundiah, R.S.: A gradient-based comparison measure for visual analysis of multifield data. Comp. Graph. Forum 30, 1101–1110 (2011)

10. Albrecht, H.-E.: Laser doppler and phase doppler measurement technique, 2nd edn. Springer, New York (2003)
11. Haller, G.: Distinguished material surfaces and coherent structures in three-dimensional fluid flows. Physica D: Nonlinear Phenomena 149, 248–277 (2001)
12. Anderson, J.C., Gosink, L.J., Duchaineau, M.A., Joy, K.I.: Interactive visualization of function fields by range-space segmentation. Comp. Graph. Forum 28, 727–734 (2009)
13. Mokhtarian, F., Mackworth, A.: Scale-based description and recognition of planar curves and two-dimensional objects. IEEE Trans. Pattern Anal. Mach. Intell. 8, 34–43 (1986)
14. Agrawal, R., Lin, K., Sawhney, H.S., Shim, K.: Fast similarity search in the presence of noise, scaling, and translation in time-series databases. In: Proc. Int. Conf. on Very Large Data Bases, pp. 490–501. Morgan Kaufmann Publishers Inc., San Francisco (1995)
15. Grundy, E., Jones, M.W., Laramee, R.S., Wilson, R.P., Shepard, E.L.C.: Visualisation of Sensor Data from Animal Movement. Comp. Graph. Forum 28, 815–822 (2009)
16. Van Wijk, J.J., Van Selow, E.R.: Cluster and calendar based visualization of time series data. In: Proc. IEEE Symposium on Information Visualization, pp. 4–9. IEEE Computer Society, Washington, DC (1999)
17. Hlawatsch, M., Leube, P., Nowak, W., Weiskopf, D.: Flow radar glyphs - static visualization of unsteady flow with uncertainty. IEEE Trans. Vis. Comput. Graph. 17, 1949–1958 (2011)
18. Yi, B., Jagadish, H.V., Faloutsos, C.: Efficient retrieval of similar time sequences under time warping. In: Proc. of the 14th Int. Conf. on Data Engineering, ICDE 1998, pp. 201–208. IEEE Computer Society, Washington, DC (1998)
19. Burger, R., Muigg, P., Doleisch, H., Hauser, H.: Interactive cross-detector analysis of vortical flow data. In: Fifth Int. Conf. on Coordinated and Multiple Views in Exploratory Visualization, CMV 2007, pp. 98–110 (2007)
20. Qu, H., Chan, W., Xu, A., Chung, K., Lau, K., Guo, P.: Visual analysis of the air pollution problem in hong kong. IEEE Trans. Vis. Comput. Graph. 13, 1408–1415 (2007)
21. Blinn, J.F.: A generalization of algebraic surface drawing. ACM Trans. Graph. 1, 235–256 (1982)
22. Heyer, L.J., Kruglyak, S., Yooseph, S.: Exploring expression data: identification and analysis of coexpressed genes. Genome Research 9, 1106–1115 (1999)

A Novel Algorithm for Computing Riemannian Geodesic Distance in Rectangular 2D Grids

Ola Nilsson[1], Martin Reimers[2], Ken Museth[1], and Anders Brun[3],[*]

[1] Department of Science and Technology, Linköping University, Sweden
[2] Department of Informatics, University of Oslo, Norway
[3] Centre for Image Analysis,
Swedish University of Agricultural Sciences, Sweden
anders@cb.uu.se

Abstract. We present a novel way to efficiently compute Riemannian geodesic distance over a two-dimensional domain. It is based on a previously presented method for computation of geodesic distances on surface meshes. Our method is adapted for rectangular grids, equipped with a variable anisotropic metric tensor. Processing and visualization of such tensor fields is common in certain applications, for instance structure tensor fields in image analysis and diffusion tensor fields in medical imaging.

The included benchmark study shows that our method provides significantly better results in anisotropic regions and is faster than current stat-of-the-art solvers. Additionally, our method is straightforward to code; the test implementation is less than 150 lines of C++ code.

1 Introduction

The computation of distances in manifolds is important in both academic and industrial applications, e.g. computational geometry [1], seismology, optics, computer vision [2], computer graphics and image analysis [3,4]. Another cross disciplinary example is the versatile level set method [5] that propagates interfaces embedded in scalar distance fields. Often the computations are performed on rectangular grids. One way to define distance is the *eikonal equation*

$$\|\nabla d(\boldsymbol{x})\|_2 = 1, \ \boldsymbol{x} \in \Omega, \tag{1}$$

which is a nonlinear Hamilton-Jacobi PDE with boundary condition $d|_{\partial\Omega} = 0$. It defines $d(\boldsymbol{x})$ as the Euclidian distance from $\boldsymbol{x}$ to an implicit source, $\partial\Omega$. Substituting the right-hand side in equation (1) with a positive scalar "speed function" that depends on position, $F(\boldsymbol{x})$, gives a nonuniform eikonal equation that is often used in applications. However, this is not the most general case. The nonuniform and anisotropic case is described by adapting the norm $\|\cdot\|$ to an arbitrary and spatially varying metric tensor field $g_{ij}(\boldsymbol{x})$. A metric tensor is a positive definite quadratic form that defines the scalar product between tangent vectors in a point $\langle \boldsymbol{v}, \boldsymbol{u} \rangle_g = \boldsymbol{v}^T g_{ij} \boldsymbol{u}$. This, in turn, interprets the length

---

[*] Corresponding author.

G. Bebis et al. (Eds.): ISVC 2012, Part II, LNCS 7432, pp. 265–274, 2012.
© Springer-Verlag Berlin Heidelberg 2012

of a tangent vector in a point as $\|\boldsymbol{x}\|_g = \sqrt{\langle \boldsymbol{x}, \boldsymbol{x} \rangle_g}$. Thus equation (1) can be generalized to take the metric into account

$$\|\nabla d(\boldsymbol{x})\|_g = 1, \ \boldsymbol{x} \in \Omega. \tag{2}$$

2   Previous Work

Considerable attention has been devoted to the computation of distances in an anisotropic setting. Early work include shading from shape [6,7] and the salience distance transforms [8]. In optics the problem is typically solved using a shooting type of algorithm, like in [9] where a method based on solving ODEs and tracking individual rays of light is used. This type of Lagrangian approach is not efficient when it comes to computing a a continuous approximation over the rays' embedding domain, which is the case for a distance map.

It is well known that a distance function is the so called *viscosity solution*, see e.g. [10,1]. Much of the previous work on anisotropic distances is based on the discretization in [11] and involves a Godunov-type approximation of the gradient. The solution (per grid point) is selected from a set containing solutions to 8 quadratic equations in 2-D. In the original paper, an iterative sweeping update scheme was used to find the global solution in $O(k\,n)$ time where n is the number of grid points in the discretization and k the number of sweeps. This algorithm updates all grid points the same number of times, which is suboptimal. A different iterative update scheme was proposed in [12] based on the same discretization. It keeps an unsorted list holding the expanding solution-front and updates nodes on a need-to basis. Even though the run time is not bounded, the method is fast in practice.

If the distance value at each node only depends on neighboring nodes with smaller values, it is possible to construct an update scheme that finds all distances in the domain in one single pass. This causality property is the basis for the well known *fast marching method* (FMM) [13], see also [14], which approximates isotropic distances on a regular grid in n steps with $O(n \log n)$ time complexity. In [1], this is extended to parametric three dimensional manifolds. Two recent one-pass methods are proposed in [15] and [16]; the former is uses a reaction-diffusion setting and the latter is based on control theory. In [17] a modified version of the FMM scheme was proposed that is supposed to be more accurate in certain cases. Finally, [18] finds the exact distance over a triangle mesh in $O(n^2 \log n)$ time.

3   Basic Scheme for 2-D Grids

The problem of computing distances from a source over a regular grid is conceptually simple. It can be broken down to successive applications of grid point updates. If the distance computed is valid locally, and a monotonic update is enforced, then it can be proved that distances will converge globally[19].

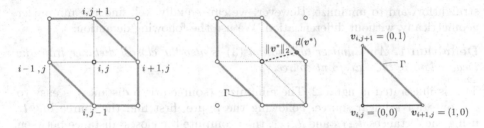

Fig. 1. Left: A generalized setup around grid point i, j with local coordinates $(0, 0)$. Middle: Assuming an implicit triangulation (dotted) we define a boundary Γ (green) for our domain. Local optimality ensures that the distance at origin is minimal, $d(\mathbf{0}) = \min \|\boldsymbol{v}^*\|_2 + d(\boldsymbol{v}^*)$ for $\boldsymbol{v}^* \in \Gamma$. Right: One of the four triangular simplices resulting from splitting the 4-connectivity neighborhood. Unit grid spacing, $h = 1$, is assumed.

A local update can be constructed by placing the update point in a local origin and minimizing the following nonlinear functional

$$d(\mathbf{0}) = \min_{\boldsymbol{v}^* \in \Gamma} \|\boldsymbol{v}^*\|_2 + d(\boldsymbol{v}^*), \tag{3}$$

for any domain with a closed boundary, requiring that the source is outside of the domain. For a grid with an implicit triangulation a possible setup is shown in figure 1. Equation (3) states local optimality and was discretized in [19,20] for triangle meshes. The solution at $\mathbf{0}$ finds a point $\boldsymbol{v}^*$ on the boundary for which the sum $\|\boldsymbol{v}^*\|_2 + d(\boldsymbol{v}^*)$ is minimal.

The update step of this type of method can be summarized as follows, given a grid point (i, j) and its neighborhood (see figure 1):

1. Split the domain into 6 right angled triangular simplices
2. Solve equation (3) for each of the triangles and set $d_{i,j}$ to the minimum

The outer loop then repeats this for all grid points (in some order) until convergence. As the 6 different configurations and their solutions are similar, we only depict one simplex illustrated in figure 1 (right).

The update value is found in two steps. First, interpolate the distance values at $\boldsymbol{v}_{i+1,j}$ and $\boldsymbol{v}_{i,j+1}$ along Γ and denote the interpolant $\tilde{d}(\boldsymbol{v}^*)$. Then compute the minimum of the sum of the edge interpolant and the distance to the edge:

$$d_{i,j} = d(\boldsymbol{v}_{i,j}) = \min_{\boldsymbol{v}^* \in \Gamma} \|\boldsymbol{v}^*\|_2 + \tilde{d}(\boldsymbol{v}^*). \tag{4}$$

As seen from equation (3), the minimization of (4) would yield the exact solution if the interpolation was exact. Finding and differentiating exact interpolants for complex wave fronts is nontrivial and potentially unstable, thus we constrain our search to interpolants of order one and two.

Linear Sources. An early discretization was proposed in 1985 by Gonzalez and Rofman [20]. They deployed a linear interpolant in equation (4), which is straightforward to minimize. However, we can equally well find the minimizer geometrically, without differentiation. We use the following definition:

Definition 1. *A minimizer to equation* (3) *is also the closest distance from the local origin to the wavefront source.*

This is illustrated in figure 2. The minimizer (source-origin distance) is easy to compute for a linear source. Following the figure, first find the source line l_s using the distances $d_{i+1,j}$ and $d_{i,j+1}$, then compute the closest distance between l_s and origin. A similar (linear) approach is taken in the popular fast marching method (FMM) [13], which was derived from a finite difference discretization.

Point Sources. A linear scheme can not accurately handle curved wave fronts. This is especially evident close to a point source where curvature is high and linearization is a poor approximation, see figure 2 (right). Both [17] and [19] recognized the need for a more accurate scheme for point sources, and independently came up with similar solutions.

From the geometry of the problem it is straightforward to find a closed form solution. Placing the source point p_s in figure 2 (right) reduces to a two-circle intersection problem [19], and the minimizer is found as $||p_s||_2$.

A point source approximation will, of course, not accurately find minimizers if the wavefronts are linear functions. Thus care needs to be taken when choosing source type. For this conference paper, we will focus mainly on point sources since they are most commonly found in applications, e.g. when geodesic distance is sought between two points in a manifold.

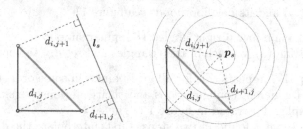

Fig. 2. (Left) The minimizer using linear interpolation can be found as the closest distance from origin to the source line l_s, for a linear wave front. (Right) For a point source the minimizer is the distance between origin and the source, p_s

Accuracy and Algorithms. We have seen that it is possible to find the update as a distance in the plane. For a 2-D grid it is possible to place the source *exactly*. Thus the update values, and hence the solutions, computed above are exact for their respective type of source. For other types of sources this type of scheme introduces a first order approximation [20,19].

For a flat metric with a linear interpolant and a regular grid, a one-pass algorithm for the global solution of (4) can be found using the fact that larger values always depend solely on smaller values. However, when using the quadratic interpolant for point sources, this is no longer true. Additionally, obtuse triangle configurations can (and are likely to) occur when the metric is not flat, as we will see in the next section. A robust and practical algorithm is described in [19] that efficiently addresses these problems. Let S be the source that is initialized to distance 0. Then add the index (or indices) of S to the sorted candidate set C. Push and pop adds and extracts indices of C, and $\min C$ refers to the indices associated with the smallest distance.

> **while** $C \neq \emptyset$ **do**
> $(i, j) \leftarrow \text{pop}(\min C)$
>   **for** $(k, l) \in \text{neighbor}(i, j)$ **do**
> $d'_{k,l} \leftarrow \text{update}(k, l)$
>   **if** $d'_{k,l} < d_{k,l}$ **then**
> $d_{k,l} \leftarrow d'_{k,l}$
> $C \leftarrow \text{push}(k, l)$

From an implementation point of view, this scheme has a complexity similar to the popular FMM. Due to space constraints we refer to [19] for more details.

4 Anisotropic Speed Functions

The update of distance values in section 2 was derived in the Euclidean plane. However, the geometric construction indicates that we may incorporate a metric into equation (4) by adapting the edge lenghts of the triangles. The generalized metric dependent minimization reads

$$d_{i,j} = \min_{\boldsymbol{v}^* \in \Gamma} ||\boldsymbol{v}^*||_g + \tilde{d}(\boldsymbol{v}^*), \tag{5}$$

where $||\cdot||_g$ is measuring the geodesic length of $\boldsymbol{v}^*$ under g. Geodesic distance between two points, $\boldsymbol{a}$ and $\boldsymbol{b}$, on a manifold is defined by a minimization over all curves γ (in the manifold) joining the two points

$$d_g(\boldsymbol{a}, \boldsymbol{b}) = \inf_{\gamma(0)=\boldsymbol{a}, \gamma(1)=\boldsymbol{b}} L(\gamma), \text{ where } L(\gamma) = \int_{t=0}^{t=1} \sqrt{\gamma'(t)^T g_{ij}(\gamma_x(t), \gamma_y(t)) \gamma'(t)} \, dt.$$

If two points are close, like two vertices in a triangle, the space-variant metric can be approximated by a constant metric. Thus,

$$d_g(\boldsymbol{a}, \boldsymbol{b}) \approx \sqrt{(\boldsymbol{b} - \boldsymbol{a})^T (g_{ij}(\boldsymbol{a}) + g_{ij}(\boldsymbol{a}))/2(\boldsymbol{b} - \boldsymbol{a})}, \tag{6}$$

where $\boldsymbol{a}$ and $\boldsymbol{b}$ are the vertices of a triangle edge [1]. We could now insert a second order interpolant into equation (5), together with the metric distance equation (6) to find minimizers. Directly doing so, however, involves finding roots of a

---

[1] A similar scheme was described for isotropic speed functions in [19]. This reference also mentions the possibility of using anisotropic speed functions.

fifth order polynomial; this is feasible but impractical. Since the solution is still only a linear approximation, we propose a simpler and more intuitive strategy.

Using the linear approximation of metric distance is approximately equivalent to linearly "flatten" the simplex under $\|\cdot\|_g$, as shown in figure 3. From this follows that minimizing equation 5 is approximately the same as minimizing equation 4 for a transformed simplex.

A consistent positioning of vertices, in Euclidean space, using the metric is shown in figure 3 (left). Since the transformed simplex is flat, the Euclidian distance function can be used to find distances from the origin to $\boldsymbol{p}_s$, and thus approximately solve equation (5). In the rare cases when the edge lengths fails to satisfy the triangle inequality, we have used Dijkstra update is deployed instead.

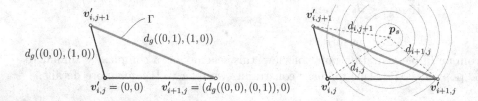

Fig. 3. Left: The simplex is (linearly) flattened using the metric g. For a consistent positioning we first place $\boldsymbol{v}'_{i,j}$ in origin, then $\boldsymbol{v}'_{i+1,j}$ along the positive x-axis. The position of $\boldsymbol{v}'_{i,j+1}$ is then found using inter-vertex distances. Right: The minimizers to equation (5) can be found in a flattened space as the Euclidian distance from origin to $\boldsymbol{p}_s$, respectively. This is treated analogously for line sources.

5   Results and Benchmarks

To test the accuracy of geodesic distance computation, it is desirable to have a set of representative "test manifolds" for which distances are known analytically. In a 2D-manifold, the local geometry is completely characterized by the Gauss curvature, K. From a geometric point of view, up to a scaling of the metric, there are three principal cases to test: $K > 0$, $K = 0$ and $K < 0$, corresponding to local isometry with the sphere, the plane and the hyperbolic plane. For this reason, these manifolds have been chosen as test manifolds. In addition, we also included the cone, which has zero curvature in all points except at the apex. We compare our method against two state-of-the-art anisotropic eikonal solvers. One from Lenglet *et al.* [16] and one by Jeong and Whitaker [12].

The test grid consists of 400×400 grid points with spacing $h = .005$. For the metrics (see appendix A) that only live in the unit disk, we have restrained computations outside a radius of 0.98 to avoid degenerate tensors at the border. We place the sources exactly on grid points and report the mean of the absolute error, that is l_1/n, the max error l_∞, and the time for the computations. The findings are listed in table 1 and figure 4. We have also performed a numerical convergence study that is reported in figure 5. Finally, we investigated the smoothness of the derivatives of the distance function for our and competing works, see figure 6.

Table 1. Comparison of accuracy and speed against competing work. S_p means that point sources are used and S_l corresponds to line sources. Also see figure 4.

Metric	Our method			Jeong and Whitaker			Lenglet et al.		
	l_1/n	l_∞	Time	l_1/n	l_∞	Time	l_1/n	l_∞	Time
Plane, S_p	7.2e-16	6.2e-15	.34s	5.1e-03	9.3e-03	1.4s	5.1e-03	9.3e-03	1.9s
Sphere S_p	6.3e-05	5.8e-04	.82s	6.2e-03	1.7e-02	1.9s	6.3e-03	1.8e-02	3.4s
Poincaré S_p	1.3e-03	2.5e-02	.54s	1.0e-02	7.1e-02	1.6s	1.0e-02	7.1e-02	2s

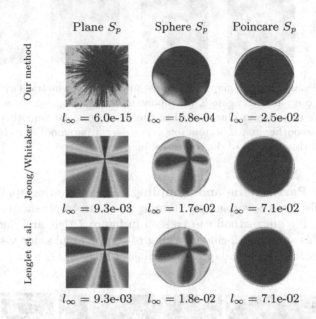

Fig. 4. Error distribution plots. Color shows absolute error (blue to red) and is linearly scaled to fall in $[0, l_\infty]$. For more statistics, see table 1 and figure 5.

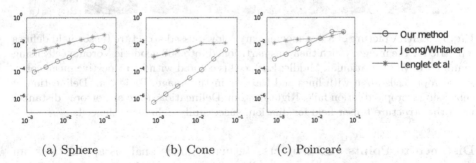

(a) Sphere   (b) Cone   (c) Poincaré

Fig. 5. Convergence under regular refinement using different metrics and the point source interpolant (S_p). Plots show l_1/n as a function of step size h. In the Poincaré example (right), the metric tensor is isotropic and [16,12] behave identically.

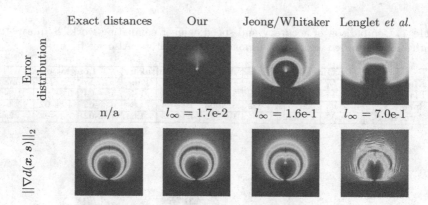

Exact distances   Our   Jeong/Whitaker Lenglet *et al.*

Error distribution

$\|\nabla d(\boldsymbol{x}, \boldsymbol{s})\|_2$

n/a $l_\infty = 1.7\text{e-}2$ $l_\infty = 1.6\text{e-}1$ $l_\infty = 7.0\text{e-}1$

Fig. 6. Visualizing distance map smoothness under a cone metric (with an extreme slope ($s = 5$), compared to figure 5, to emphasize errors). Top row shows color coding of the absolute error (from blue to red) of the distance map. Smoother transition in color indicate smoother maps. Bottom row shows derivative smoothness by color coding of the norm of the gradient of the respective map.

Anisotropic Partitioning and Sampling. One interesting application using anisotropic distances is voronoi partitioning and anisotropic sampling [4,3]. One experiment using our method can be seen in figure 7 (left and middle). Applications for this sampling include packing of tensor glyphs, for visualization of diffusion tensor MRI, and adaptive sampling of images.

Fig. 7. Left: A picture of a fast moving car. A low-passed structure tensor field defines a metric for the image, which then was partitioned into 300 geodesic Voronoi cells using manifold Loyd relaxation[3]. Middle: Each cell is colored with its respective mean color. Anisotropic cells align with lines and features in the image. Right top: Delineation of cells using isoropic dilation fails. Right bottom: Delineation using anisotropic distances from the structure tensor field for dilation, successfully delineates the cell.

Distance to Points and Objects. In many image analysis applications, an important procedure is to compute distances from an object or point. These methods are called distance transforms and they compute all distances from a set of pixels in an image to all other pixels in the image [21,13]. Figure 7 (right) shows one example of an application that benefits from our method, namely anisotropic curve closing for cell nuclei segmentation [22]. In short, our method enables accurate anisotropic versions of morphological operations such as dilation, erosion, skeletonization and watershed segmentation.

6 Conclusion

We have presented a novel metric dependent solver that converges to geodesic distances under grid-refinement. It is simple to implement, less than 150 lines of C++ code, and performs in a predictable manner. It appears to have better convergence properties than two current state-of-the-art algorithms.

Our main motivation for publishing these results is to present a solver that we have found to be easy to implement and adapt. It is a solver that is different from many current approaches, in particular FMM.

A   Test Metrics and Analytic Distances

A spherical shell: $K = 1$, $\{x, y : x^2 + y^2 \leq 0.98^2\}$,

$$g_{ij}(x,y) = \frac{1}{1-x^2-y^2}\begin{bmatrix} 1-y^2 & xy \\ yx & 1-x^2 \end{bmatrix}, \text{and}$$

$$d_S(\mathbf{a},\mathbf{b}) = \cos^{-1}(\mathbf{a}^T\mathbf{b} + \sqrt{(1-\mathbf{a}^T\mathbf{a})(1-\mathbf{b}^T\mathbf{b})}).$$

A flat square: $K = 0$, $(x,y) \in [-1,1] \times [-1,1]$, $g_{ij} = \delta_{ij}$, $d(\mathbf{a},\mathbf{b}) = ||\mathbf{a}-\mathbf{b}||_2$.
The Poincaré disk model: $K = -1$, $\{x, y : x^2 + y^2 \leq 0.98^2\}$,

$$g_{ij}(x,y) = \frac{\delta_{ij}}{(1-x^2-y^2)^2}, \quad d_P(\mathbf{a},\mathbf{b}) = \frac{1}{2}\cosh^{-1}\left(1 + \frac{2||\mathbf{a}-\mathbf{b}||_2^2}{(1-||\mathbf{a}||_2^2)(1-||\mathbf{b}||_2^2)}\right).$$

The cone metric: $K = 0$, a cone $z = s\sqrt{x^2+y^2}$ with slope s, $\{(x,y) \neq (0,0)\}$.

$$g_{ij}(x,y) = \delta_{ij} + \frac{s^2}{x^2+y^2}\begin{bmatrix} x^2 & xy \\ yx & y^2 \end{bmatrix}$$

$$d(\mathbf{a},\mathbf{b}) = \sqrt{[||\mathbf{a}||_2^2 + ||\mathbf{b}||_2^2 - 2||\mathbf{a}||_2||\mathbf{b}||_2\cos(\frac{\angle(\mathbf{a},\mathbf{b})}{\sqrt{(1+s^2)}})](1+s^2)}$$

References

1. Bronstein, A.M., Bronstein, M.M., Kimmel, R.: Weighted distance maps computation on parametric three-dimensional manifolds. Journal of Computational Physics 225, 771–784 (2007)
2. Bruss, A.R.: The eikonal equation: Some results applicable to computer vision. In: Horn, B.K.P., Brooks, M.J. (eds.) Shape from Shading, pp. 69–87. MIT Press, Cambridge (1989)
3. Feng, L., Hotz, I., Hamann, B., Joy, K.: Anisotropic noise samples. IEEE Transactions on Visualization and Computer Graphics 14, 342–354 (2008)
4. Du, Q., Wang, D.: Anisotropic centroidal voronoi tessellations and their applications. SIAM J. Sci. Comput. 26, 737–761 (2005)

5. Osher, S., Sethian, J.A.: Fronts propagating with curvature-dependent speed: Algorithms based on Hamilton-Jacobi formulations. Journal of Computational Physics 79, 12–49 (1988)
6. Verbeek, P.W., Verwer, B.J.: Shading from shape, the eikonal equation solved by grey-weighted distance transform. Pattern Recogn. Lett. 11, 681–690 (1990)
7. Rouy, E., Tourin, A.: A viscosity solutions approach to shape-from-shading. SIAM Journal on Numerical Analysis 29, 867–884 (1992)
8. Rosin, P.L., West, G.A.W.: Salience distance transforms. Graph. Models Image Process. 57, 483–521 (1995)
9. Parazzoli, C.G., Koltenbah, B.E.C., Greegor, R.B., Lam, T.A., Tanielian, M.H.: Eikonal equation for a general anisotropic or chiral medium: application to a negative-graded index-of-refraction lens with an anisotropic material. J. Opt. Soc. Am. B 23, 439–450 (2006)
10. Bardi, M., Capuzzo-Dolcetta, I.: Optimal control and viscosity solutions of Hamilton-Jacobi-Bellman equations. Birkhauser (1997)
11. Tsai, Y.H.R., Cheng, L.T., Osher, S., Zhao, H.K.: Fast sweeping algorithms for a class of hamilton-jacobi equations. SIAM Journal on Numerical Analysis 41, 673–694 (2004)
12. Jeong, W.-K., Fletcher, P.T., Tao, R., Whitaker, R.T.: Interactive visualization of volumetric white matter connectivity in dt-mri using a parallel-hardware hamilton-jacobi solver. IEEE Transactions on Visualization and Computer Graphics (Proceedings of IEEE Visualization 2007), 1480–1487 (2007)
13. Sethian, J.A.: A fast marching level set method for monotonically advancing fronts. Proc. Nat. Acad. Sci. 93, 1591–1595 (1996)
14. Tsitsiklis, J.N.: Efficient algorithms for globally optimal trajectories. IEEE Transactions on Automatic Control 40, 1528–1538 (1995)
15. Konukoglu, E., Sermesant, M., Clatz, O., Peyrat, J.-M., Delingette, H., Ayache, N.: A Recursive Anisotropic Fast Marching Approach to Reaction Diffusion Equation: Application to Tumor Growth Modeling. In: Karssemeijer, N., Lelieveldt, B. (eds.) IPMI 2007. LNCS, vol. 4584, pp. 687–699. Springer, Heidelberg (2007)
16. Lenglet, C., Prados, E., Pons, J.P., Deriche, R., Faugeras, O.: Brain connectivity mapping using Riemannian geometry, control theory and PDEs. SIAM Journal on Imaging Sciences (2008)
17. Novotni, M., Klein, R.: Computing geodesic distances on triangular meshes. In: Proceedings of The 10th International Conference in Central Europe on Computer Graphics, Visualization and Computer Vision, WSCG 2002 (2002)
18. Surazhsky, V., Surazhsky, T., Kirsanov, D., Gortler, S.J., Hoppe, H.: Fast exact and approximate geodesics on meshes. In: SIGGRAPH 2005: ACM SIGGRAPH 2005 Papers, pp. 553–560. ACM Press, New York (2005)
19. Reimers, M.: Topics in Mesh based Modelling. PhD thesis, Univ. of Oslo (2004)
20. Gonzalez, R., Rofman, E.: On deterministic control problems: An approximation procedure for the optimal cost i. the stationary problem. SIAM Journal on Control and Optimization 23, 242–266 (1985)
21. Sonka, M., Hlavac, V., Boyle, R.: Image Processing: Analysis and Machine Vision. Thomson-Engineering (1998)
22. Malm, P., Brun, A.: Closing curves with Riemannian dilation: Application to segmentation in automated cervical cancer screening. In: Proc. of 5th International Symposium on Visual Computing, Las Vegas, Nevada, USA (2009)

Visualization of Taxi Drivers' Income and Mobility Intelligence

Yuan Gao, Panpan Xu, Lu Lu, He Liu, Siyuan Liu, and Huamin Qu

Hong Kong University of Science and Technology

Abstract. Different taxi drivers may use different strategies to choose operating regions and find customers, which is called mobility intelligence. In this paper, we present a visualization system to analyze a large amount of spatial-temporal multi-dimensional trajectory data and identify some key factors that differentiate the top drivers and ordinary drivers according to their income. Two novel encoding schemes, Choice-of-Location graph and Move/Wait Strategy tree, have been proposed to analyze drivers' behaviors when choosing operating locations and drivers' move/wait strategies when their taxis are vacant. We have applied our system to the trajectories of thousands of taxis in a major city and have gained some interesting findings on taxi drivers' mobility intelligence.

1   Introduction

Taxi drivers usually have the flexibility to choose their working hours and operating locations. They use different strategies and skills to look for and deliver customers to their destinations. For example, some taxi drivers prefer to drive along the streets and look for customers while other drivers may just wait at certain locations like train stations. These strategies and skills, called mobility intelligence, greatly affect their income.

In the past, the study of taxi drivers' mobility intelligence has only relied on questionnaires and interviews. Many taxis are now equipped with GPS devices, so their trajectory data and other driving data logs such as vacant or occupied status can be collected over a long period of time, which gives a comprehensive description of the behaviors of taxi drivers. Analyzing this large amount of taxi trajectory data reveals important patterns and behaviors of taxi drivers and shed new light on their mobility intelligence. A mobility intelligence analysis system is highly desirable to taxi companies, taxi drivers, economists, and so on. For example, with such a system, taxi companies can monitor the behavior of their employees, and taxi drivers may analyze their own performances to improve their income.

Some presumptions about taxi drivers' mobility intelligence are widely believed, though not necessarily correct, in the taxi business. These presumptions are about the possible patterns exhibited by the higher income drivers that can be major factors affecting their income. In this study, we will ascertain the presence of any strong correlation between drivers' income and those factors in order

G. Bebis et al. (Eds.): ISVC 2012, Part II, LNCS 7432, pp. 275–284, 2012.

to validate the presumptions. The presumptions about taxi drivers' income and their mobility strategies to be validated in this study are as follows:

1. High-income drivers are more intelligent
2. High-income drivers serve more customers
3. There are correlations between drivers income and operating regions
4. High-income drivers are more aggressive and they actively search for customers
5. High-income drivers change strategies less often when vacant

In this paper we use visualization to investigate taxi drivers' mobility intelligence based on taxi trajectory data. The purpose of this research is to discover how taxi drivers' strategies affect their income and identify the key factors that differentiate top-income drivers from ordinary drivers. We analyze this problem from two aspects: (1) drivers' behaviors when choosing operating locations; (2) drivers' move/wait strategies when their taxis are vacant. We present several novel visualization techniques to analyze the taxi drivers' trajectory data. Our visualization can reveal the preferred locations of the taxi drivers, the correlation between hot locations and drivers' preferred areas, and the move/wait strategies of the drivers. We apply our methods to real-world data and obtain some interesting findings. Our encoding schemes can also be applied to other trajectory-related problems.

2   Related Work

Trajectory Visualization. Space-Time Cube (STC) [1,2] presents a trajectory in both geography and time to visualize temporal and geographical changes but only a limited number of trajectories could be displayed without occlusion. Orellana et al. [3] proposed an approach to analyze movement suspension patterns and generalized sequential patterns to find out place of interests and commonality of trajectories. Geographical positions could be transformed into abstract space to find certain patterns in multivariate trajectory data [4]. The occlusion problem caused by a large amount of trajectories could be solved by density-based clustering [5] or aggregation after dividing temporal and geographical spaces into short time intervals and compartments [6,7]. In the visualization of vessel movements [8], kernel-based methods provide an overview of area usage and speed variations of individual vessels. Financial market data can also be visualized as trajectories for further analysis [9]. The geographical information associated with trajectory data can be visualized by various schemes [10,11].

Trajectory Pattern Analysis. The trajectory patterns are defined as frequent behaviors in space and time by Giannotti et al. [12], and several methods are introduced to extract trajectory patterns. Similarity between trajectories can be measured by the Longest Common Subsequence model [13]. Interesting stops in trajectories can be discovered by a speed-based clustering method [14]. The pendulum of the inhabitants daily mobility could be reflected by traffic records of buses, metros and trajectories of taxis [15]. The taxi drivers' mobility intelligence has been studied by Liu et al. [16] by analyzing five top-income drivers. They found that the top drivers changed their operating regions every several hours to work in better traffic conditions.

Income Analysis. Economists have been looking into the labor supply of taxi drivers for years since taxi drivers can decide their working hours more freely than other occupations. Camerer et al. [17] discovered that the elasticity of taxi drivers' labor supply is negative, which means the taxi drivers quit early when the hourly-wage is high and work longer when the hourly-wage is low. Chou [18] applied statistical research on data from Singapore taxi drivers and the result is similar to that of New York drivers. Farber [19] presented a different model for taxi drivers' labor supply which is the standard neoclassical inter-temporal labor supply model. The result suggests that the drivers stop working mainly because of the 12-hour shift instead of the target income.

3   Choice-of-Location Visualization

3.1   Definitions

Hotness. We define the hotness of a location to be the rank of the number of customers getting on/off taxis at that location. High hotness areas refer to the areas with lots of customers, such as downtown areas. Low hotness areas refer to the locations with few customers such as suburban areas. Figure 1 (a) shows the hotness of the Shanghai city by heatmap.

Trip. A trip is the maximum trajectory during which the taxi's operating status (i.e., vacant or occupied) does not change. An occupied trip starts by picking up a customer and a vacant trip starts when a customer gets off the taxi.

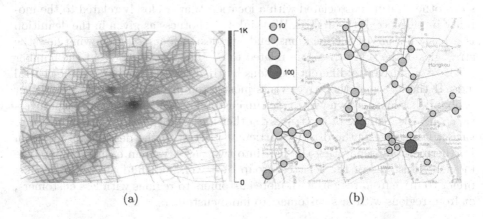

(a)   (b)

Fig. 1. (a) The heatmap of locations according to the average number of customers in Shanghai. (b) The pairwise relation shows the movements between the two regions.

3.2   Visual Encoding Schemes

We propose a Choice-of-Location scheme graph to reveal the intelligence of choosing operating locations by taxi drivers. In our graph we have a horizontal central axis which shows the locations ranked by their hotness. Above and below the axis we display the trip information. Specifically, our scheme graph encodes the following information:

Hotness of the Locations. The locations on the central line are sorted by hotness, which is proportional to the total number of customers getting on/off taxis at those locations. The hotness of the locations increases from left to right. Figure 2 (a) shows the location clusters and hotness used in this design.

Number of Visits. The height of the bars on the central axis encodes the number of visits to each location by the selected driver or the selected group of drivers. The color of the bars also represents the number of visits. The bar is divided into two parts. The upper part represents the number of visits to this area, while the lower part means the number of trips leaving from this area.

Trip Information. We show the trip information by connecting the origin and destination locations on the central axis by Bezier curves. The height of the curves encodes the average length of the trip. The color of the curves represent the number of visits to the destination location, while red means frequent visits and blue means rare visits. All curves above the central axis are trips from suburban areas (low hotness) to downtown areas (high hotness), and all curves below the central axis are trips from downtown areas (high hotness) to suburban areas (low hotness).

3.3   Design Rationale

Instead of encoding the locations directly by their geographical coordinates, we show other features associated with a location that is closely related to the mobility intelligence of the drivers. We think that hotness as given in the definition is a basic feature of the locations, and our design focuses on showing how the strategies of the taxi drivers are related to the hotness. We present the number of visits of a driver to different locations by the height of the bars on the central axis. If the locations the driver visits most often are not hot spots, the driver probably has his/her own way to pick up customers, which is worth further investigation. The distance of the trip is another important feature. A long-distance vacant trip suggests the driver's particular interest in the destination location. Furthermore, by separating the plot into two parts, we can easily compare the moving directions of the selected group of drivers and find out whether they prefer to drive from regions with more customers to regions with less customers, or from regions with less customers to more customers.

4   Strategy Tree Visualization

We would like to investigate drivers' various behaviors starting from the same location where the taxi becomes vacant. Once the taxi becomes vacant, the driver has to make successive decisions about whether to stay around the current location or travel farther away. We use "Move/Wait Strategy" to refer to the decisions that the driver makes after the taxi becomes vacant. A driver's strategies are closely related to his/her income. Our task is to visually present the different strategies the drivers adopt.

4.1   Visual Encoding Scheme

A binary tree is used to visually present the decisions made by the drivers after they have dropped customers off at a location chosen for analysis. Starting from the root, a driver chooses which branches to take at each node. Choosing the upper branch means that the driver drives away from his current location and choosing the lower branch means the driver stays around the current location. The tree can either display the different decisions made by a group of drivers or a driver who has dropped off customers multiple times at the same location. Our Strategy Tree extends the traditional decision tree by encoding the following additional information:

Time. The time is encoded by the x-coordinate of the node.

Distance. The distance of the current location to the original location is encoded by the y-coordinate of the node.

Decisions Made by Majority. The thickness of the edge shows the number of drivers who choose that branch. Therefore we can identify the decisions made by the majority of the drivers.

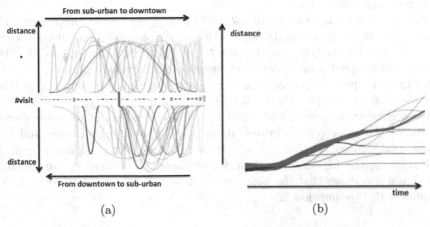

Fig. 2. (a) The Choice-of-Location view. (b) The strategy tree.

4.2   Design Rationale

A tree is a natural way to visually present successive decisions. We use a binary tree to show the gradually deviating paths of the drivers: they choose to either keep moving or stay at each node. Each individual path in the tree can be seen as a line-view of time-variant distance data. An overview of the tree can give a sense about the variations of the decisions made: a large number of branches and an even distribution in thickness of the edges means more variations in decisions. More branches means the drivers change their move/wait strategies more often during the vacant trip. If the drivers do not change strategies, their Strategy Tree has two branches: one for keep moving and one for keep waiting. If the decisions of only one driver is displayed, the depth of the tree can show whether the driver has made good decisions such that he can find a new customer in a short time.

5   Experiments

The data used in this study is the taxi trajectory data of over eight thousand taxi drivers collected in Shanghai, China during a non-continuous eight-month period (92 days in total). Each GPS record contains the taxi ID, the latitude and longitude of the taxi, the date, the time of the day in seconds, the taxi's status (occupied / vacant), and the speed of the taxi. The data is sanitized by removing erroneous trajectories that have impossible speed, immediate location changes or large off-road distance errors. After the data is sanitized, we are able to get the valid trajectory data of more than four thousand taxis.

5.1   Drivers Income

In this section, we analyze the total income of all taxi drivers. Based on the trajectory data, we can compute the income of each trip for a taxi according to a publicly-accessible formula.

We first draw on the income of each driver as a time series curve. From the curve, we select two groups of drivers – 30 high-income drivers and 30 low-income drivers for further investigation (Figure 3 (a)). From the figure, we see that the income of these drivers are quite stable over time in a weekly pattern.

We then use parallel coordinates to show the multiple attributes associated with these 60 drivers (see Figure 3 (b)). It is clear that there exists a strong correlation between income and other attributes such as the total number of customers. We observe large income differences between high-income and low-income taxi drivers. The results effectively indicate the existence of certain intelligence of the high-income drivers and therefore confirm Presumption 1. This figure also shows that the high-income drivers have more customers, which complies with Presumption 2.

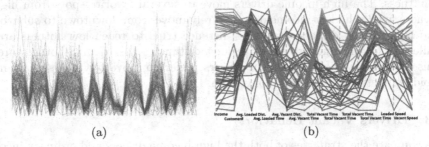

(a)   (b)

Fig. 3. (a) Time series plots showing the drivers' income varies over time with the red lines indicating 30 high-income taxi drivers and dark blue lines indicating 30 low-income taxi drivers. (b) Parallel coordinates showing the multiple attributes related to taxi driver, including income, customer number, average loaded and vacant distance and time, and loaded and vacant speeds.

5.2   Drivers Preferences of Locations

We want to answer the question of whether taxi drivers deliberately choose certain locations to find customers and study the correlation of taxi drivers income with the area of operation. We applied our Choice-of-Location scheme to two groups of drivers (Figure 4). From the figure we observed that the high-income drivers have different preferences of operating locations from the low-income drivers. Instead of choosing one hottest spot, the high-income drivers prefer multiple locations in downtown that have more customers. On the contrary, the low-income drivers demonstrate high interests in the hottest areas. This suggests an interesting possibility: intelligent drivers choose multiple working regions so that they can avoid congestion and competition when they know there are too many drivers waiting and less chance to pick up a customer in a short period of time, or the traffic of the hottest spot is relatively heavier. The upper part and the lower part in Figure 4 show the moving direction of taxi drivers in terms

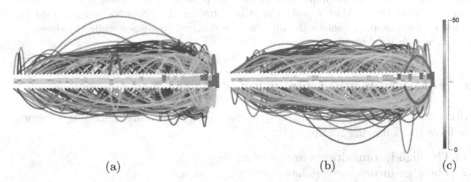

(a)   (b)   (c)

Fig. 4. Drivers' moving trends between areas: (a) The high-income drivers' choices; (b) The low-income drivers' choices; (c) The encoding scheme

of hotness. The high-income drivers move to several favorite spots from higher hotness, while the low-income drivers rarely move from downtown to suburban. The high-income drivers drive long distance trips to reach low hotness areas, while the low-income drivers drive long distance trips to high-hotness areas. This trend shows that intelligent drivers prefer to move towards suburban to avoid competition and traffic congestion.

5.3   Drivers Vacant Behaviors

We compare the strategies of both the high-income drivers and ordinary-income drivers at the same location using our Strategy Tree. The drivers are the same as in the previous experiment. From Figure 5 we found that the high-income drivers are more insistent about their decisions. The majority of the high-income drivers choose to leave at the beginning and will not change their decisions later on. On the contrary, many low-income drivers choose to wait at the beginning. This proved Presumption 4 that the high-income drivers are more aggressive in searching for customers. For the low-income drivers, it is common that the drivers wait at the original location for three minutes and then keep moving to a new location. The change of decision is very often and consequently they spend more time and may have to travel long distances to pick up the next customer. Since the high-income drivers are insistent about their decisions, Presumption 5 is right.

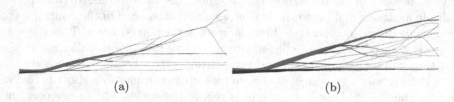

(a)   (b)

Fig. 5. Strategy Tree of top 30 drivers and ordinary 30 drivers. Each node in X-axis represents 3 minutes. Ordinary drivers may keep moving for about 6 to 9 minutes and then stay there. Most top drivers either stay in the origin or search along the street without stop. (a) shows the high-income drivers' strategies while (b) shows the low-income drivers' strategies.

6   Discussion and Conclusion

After studying the taxi drivers' trajectories using our visual encoding schemes, we have the following findings:

1. The high-income drivers are more intelligent.
2. The high-income drivers have more customers.
3. There are correlations between drivers income and operating regions. The high-income drivers have multiple favored regions but the low-income drivers concentrated in small number of regions.

4. The high-income drivers are more aggressive. They actively search for customers.
5. The high-income drivers change their move/wait strategies less often during vacant trip.

In this paper we have presented two novel visual encoding schemes to visualize trajectory data of taxi drivers and gain insight into their mobility intelligence. The Choice-of-Location scheme graph can simultaneously reveal multiple attributes that are critical for analyzing drivers' tactics of choosing operating locations. In addition, with an intuitive layout, it allows easy comparison of two different driver groups. The Move/Wait Strategy tree extends the traditional decision tree by encoding the average waiting time and the travel distance to reveal different decisions made by top-income drivers and ordinary-income drivers when their taxis are vacant. We have demonstrated the effectiveness of our methods by visualizing real taxi trajectory data and obtained some interesting findings about some drivers' preferred operating locations and the major differences between the high-income drivers and the low-income drivers. Compared with traditional methods, our encoding schemes are especially effective at revealing the key features related to mobility intelligence. They can also be applied to other trajectory-related problems, such as the mobility patterns of animals in a forest, tourists' travel paths in a city, and visitors' traces in an exhibition hall.

In the future, we plan to further improve our system by adding more interaction schemes for users to select a group of trips or locations, clustering and bundling similar trips to reduce visual clutter, and encoding some temporal information of trips to reveal temporal patterns of trajectories and the changes drivers' strategies over time.

Acknowledgements. We thank Weiwei Cui, Nan Cao, Conglei Shi, and Jiansu Pu for their support and suggestions. We thank the anonymous reviewers for their constructive and valuable comments. This work was supported in part by grant HK RGC GRF 619309.

References

1. Hägerstrand, T.: What about people in regional science? Papers, Regional Science Association 24 (1970)
2. Hedley, N., Drew, C., Arfin, E., Lee, A.: Hägerstrand revisited: Interactive space-time visualizations of complex spatial data. Informatica: An International Journal of Computing and Informatics 23 (1999)
3. Orellana, D., Bregt, A., Ligtenberg, A., Wachowicz, M.: Exploring visitor movement patterns in natural recreational areas. Tourism Management 33, 672–682 (2012)
4. Crnovrsanin, T., Muelder, C., Correa, C., Ma, K.L.: Proximity-based visualization of movement trace data. In: IEEE Symposium on Visual Analytics Science and Technology, pp. 11–18 (2009)

5. Nanni, M., Pedreschi, D.: Time-focused clustering of trajectories of moving objects. Journal of Intelligent Information System 27, 267–289 (2006)
6. Andrienko, G., Andrienko, N.: Spatio-temporal aggregation for visual analysis of movements. In: IEEE Symposium on Visual Analytics Science and Technology, pp. 51–58 (2008)
7. Rinzivillo, S., Pedreschi, D., Nanni, M., Giannotti, F., Andrienko, N., Andrienko, G.: Visually driven analysis of movement data by progressive clustering. In: Information Visualization, pp. 225–239 (2008)
8. Willems, N., van de Wetering, H., van Wijk, J.J.: Visualization of vessel movements. Computer Graphics Forum 28, 959–966 (2009)
9. Schreck, T., Tekusova, T., Kohlhammer, J., Fellner, D.: Trajectory-based visual analysis of large financial time series data. IEEE Transactions on Visualization and Computer Graphics 9, 30–37 (2007)
10. Fisher, D.: Hotmap: Looking at geographic attention. IEEE Transactions on Visualization Computer Graphics 13, 1184–1191 (2007)
11. Dorling, D., Barford, A., Newman, M.: Worldmapper: the world as you've never seen it before. IEEE Transactions on Visualization and Computer Graphics 12, 757–764 (2006)
12. Giannotti, F., Nanni, M., Pinelli, F., Pedreschi, D.: Trajectory pattern mining. In: International Conference on Knowledge Discovery and Data Mining, pp. 330–339 (2007)
13. Vlachos, M., Kollios, G., Gunopulos, D.: Discovering similar multidimensional trajectories. Proceedings of The International Conference on Data Engineering 18, 673–684 (2002)
14. Palma, A., Bogorny, V., Kuijpers, B., Alvares, L.O.: A clustering-based approach for discovering interesting places in trajectories. In: Proceedings of the 2008 ACM Symposium on Applied Computing, pp. 863–868 (2008)
15. Liu, L., Biderman, A., Ratti, C.: Urban mobility landscape: Real time monitoring of urban mobility patterns and data on maps. Computers in Urban Planning and Urban Management (2009)
16. Liu, L., Andris, C., Biderman, A., Ratti, C.: Uncovering taxi driver's mobility intelligence through his trace. Environment and Urban Systems (2009)
17. Colin, C.: Labor supply of new york city cabdrivers: One day at a time. Quarterly Journal of Economics - Cambridge Massachusetts 112, 307–442 (1997)
18. Chou, Y.K.: Testing alternative models of labor supply: Evidence from taxi drivers in sinapore. The University of Melbourne. Department of Economics - Working Papers Series (2000); All
19. Farber, H.S.: Is tomorrow another day? the labor supply of new york cab drivers. Journal of Political Economy 113, 46–82 (2005)

Frame Cache Management
for Multi-frame Rate Systems

Stefan Hauswiesner, Philipp Grasmug, Denis Kalkofen, and Dieter Schmalstieg

Institute for Computer Graphics and Vision, Graz University of Technology

Abstract. Multi-frame rate systems decouple viewing from rendering in
an asynchronous pipeline. Multiple GPUs can be used as frame sources,
while a primary GPU is responsible for viewing and display update.
Conventionally, the last rendering result is used for display. However,
modern GPUs are equipped with a fairly large amount of memory which
allows frames to be cached in video memory. As long as the data is
static, caching allows for a more sophisticated reference frame selection
that increases the output quality. With a growing frame database, images for most viewpoints can be queried from the cache and the system
converges into a conventional image-based rendering system. However,
multi-frame rate systems use purely virtual image sources. As a consequence, the rendering process can be actively steered by the viewing process, which allows for advanced strategies. Moreover, by picking multiple
reference frames from the cache, we can avoid display artifacts arising
from occlusions.

1 Introduction

Multi-frame rate rendering helps interactive systems to achieve little latency
and high frame rates [1]. It decouples display updates from image generation in
a pipeline with asynchronous communication across multiple GPUs. The display
update stage can guarantee high frame rates and nearly latency-free response to
user interaction, such as moving the viewpoint. At the same time, frame sources
in the backend can produce new high quality images at their own, usually slower
pace. Such systems employ image-based rendering (IBR) to hide the latency of its
rendering nodes. Conventionally, the last rendering result is used as a reference
frame to display the current viewpoint. This is a natural choice, because it
requires little memory and provides acceptable quality. However, modern GPUs
are equipped with a large amount of on-board memory, which can be used to
improve the quality.

Therefore, we introduce frame caching strategies, which allow to reuse already
rendered content. These strategies maximize the utility of available GPU memory by storing frames which are likely to improve future visual quality. Every
frame which is received from a frame source is evaluated for its usefulness (section 4). By using more than a single reference frame for warping, the output
quality can be increased especially at depth discontinuities. We suggest a heuristic for selecting multiple frames from the cache (section 5). Finally, we evaluate
the visual quality (section 6).

G. Bebis et al. (Eds.): ISVC 2012, Part II, LNCS 7432, pp. 285–294, 2012.
© Springer-Verlag Berlin Heidelberg 2012

0-360 degrees of rotation: 361-720 degrees of rotation:

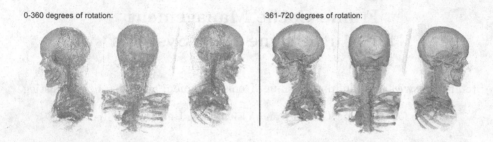

Fig. 1. The effect of caching reference frames when rotating a dataset around the Y axis. The first turn contains visible artifacts, while the second turn offers considerably better visual quality.

2   Related Work

For coping with situations where the computation load exceeds the interactive rendering power of a parallel system, but some computations have to be performed with minimum latency, the concept of multi-frame rate rendering was introduced [1]. Multi-frame rate systems use a number of rendering nodes to independently render parts of the scene. The results are combined without waiting for updates. Thus, the display node (in our case: the *primary GPU*) is not slowed down by the rendering nodes, which enables the display node to perform computations with bounded latency to user input or other events. This approach is especially useful for applications, in which high interactivity has to be guaranteed, e. g., augmented reality, stereoscopic display or interaction with magic lenses etc.

Image warping [2] is a form of image based rendering that allows to extrapolate new views from existing images with per-pixel depth information. Our system uses forward image warping for latency compensation.

The work in [3–5] describes a multi-frame rate architecture, which is able to use the GPU to accelerate image warping by using vertex shader programs for the required scatter operation. However, this work is not suitable for transparency and volumetric datasets and does not employ an image cache. [6] support multi-frame rate volume rendering with transparency, but assume a fixed view point.

Recent implementations of the render cache [7] utilize GPUs for improved performance. In contrast to the render cache notion, our system does not operate on single points. Modern GPUs are efficient at data parallel execution. Our system therefore manages the creation, storage and transfer of frames, or images. The Tapestry system [8] also elaborates on exploiting frame-to-frame coherence by utilizing view-dependent meshes. However, these meshes are not suitable for transparent objects, like volumetric data visualizations.

3 Multi-frame Rate Rendering with Caching

This work is based on a multi-frame rate architecture [9] that is capable of rendering scenes consisting of meshes as well as volumetric objects. Volumetric objects are layered to account for motion parallax withing the object. Rendering can be performed on the GPUs of a single PC, or even on remote machines. The display stage employs image-warping, because the reference frame that is used for display was generally not rendered exactly from the desired viewpoint. It is therefore able to compensate for the latency of the image-generating GPUs.

From the users point of view, our system looks like a simple model viewer. The camera is able to orbit the scene objects and move towards the model or away from it, thus enabling zoom operations. Currently, our system does not support viewpoints inside the scene. This is not a general limitation, but simplifies the frame cache operations.

4 Cache Organization

In our previous work [9], we use the last known rendering result of a frame source GPU for image warping and display. However, the visual quality directly scales with the number and density of available frames along the navigation path [10]. Therefore, our new system stores as many valuable reference frames as possible, but also extends the image cache automatically during idle phases. Note, that reusing rendered images introduces a limitation: the scene's objects are not allowed to deform arbitrarily. Animations can only be performed when they can be described by a single transformation matrix per object.

4.1 Retaining Frames

The maximum number of frames that a reference frame cache can store is limited by the size of GPU memory. A simple round-robin strategy will not create a database with good memory utilization. For example, a reference frame, which shows approximately the same features as another, does not contribute to the display result and therefore wastes memory.

When receiving a new reference frame, its viewpoint distance to the closest neighbor is compared to a threshold. New frames below this threshold do not add enough information to the cache database and are discarded. Above the threshold, the frame is stored if a free slot in the cache is available. If no free slot is available, a trade-off has to be found: from the closest pair of frames, one is replaced. Figure 2(a) illustrates how the frame cache improves over time. Figure 1 shows the result of this strategy.

Finding an appropriate sampling density by setting the mentioned threshold requires knowledge of the dataset and available GPU memory. We use the angle between two viewing directions to determine the distance of two frames. For the datasets shown in this paper, we used a threshold of 4.5°.

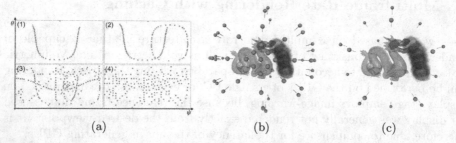

Fig. 2. (a) spherical coordinates θ, ψ of reference frame viewpoints during user navigation. Image (1) shows 65 reference frames and is captured after a complete rotation around the object. (2) shows 109 reference frames after another rotation. (3) shows the situation when the cache is full (200 frames). After a long and diverse navigation session the camera positions are evenly distributed around the object according to our update strategy (4). (b) and (c) show different prerendering strategies with camera positions augmented as spheres.

4.2   Estimating Subsequent Camera Locations

When the system is not in continuous motion, and no user interaction is performed for a period of time, no redraw is necessary. During these idle periods, the rendering nodes can be scheduled to render new viewpoints, without interfering with the display output. The primary GPU stores the new information as described above, which helps to improve future image quality. We call this process *prerendering*.

Generating valuable reference frames for future user interaction requires to predict future navigation. In typical exploration scenarios, the user is only interested in a relatively small range of viewing angles, whereas in navigation tasks all angles may be visible at some point. If the application follows an easily predictable animation, it is also possible to extrapolate the last known movements.

Our system supports these three types of scenarios: for examination, we assume that future viewing angles are close to the current one, and therefore start with a fine sampling around it. The more different the viewing angles, the sparser we sample. See Figure 2 (b) for a visualization of such a prerendering strategy. For navigation applications, camera positions for prerendering are spaced evenly on the bounding sphere of the scene. This can be achieved by placing the reference camera positions at the vertices of a regular polyhedron, which guarantees an equal distance between neighboring positions (see Figure 2 (c)). If the viewpoint follows a certain path, this information can greatly improve image quality. However, it is domain specific, so we provide just a simple motion prediction in addition to the strategies above. This prediction places reference frame viewpoints along the extrapolated path of the last user-induced camera motion with a step size that accounts for the speed of the motion. When a steady rotation around the object is started, this method is capable of producing reference frames just in time for the display process.

4.3   Handling Transfer Function Changes

Multi-frame rate rendering is often employed to improve the performance of viewing volumetric datasets, which are especially common in medical applications. In such applications, the transfer function that maps density values to color values can be changed. This invalidates all frames in the cache, reducing the usefulness of the suggested approach. However, we employ an algorithm that transforms all images in the frame cache to approximate the change of the transfer function.

As described above, every frame in the cache consists of a set of layers. Each layer consists of color, depth and normal values at each pixel. Using this information alone, it is not possible to transform the images to reflect a change of the transfer function. In addition to our previous work [9], we therefore also store representative density values and the layer thickness per pixel, which help to reclassify every pixel using the new transfer function. The representative density value for each ray segment is selected from the sample that has maximum impact on the output color: the sample with the highest opacity weight after classification.

The process of reclassifying a pixel starts by classifying the pixel using its representative density value and the new transfer function. The queried color value is accumulated according to the thickness (or length) of the ray segment. The resulting color represents a homogeneously filled ray segment, which is not necessarily a good approximation. However, there is more information available to improve the quality: the representative density value can be classified by the old transfer function. The resulting color value is then compared to the actual color value of the ray segment. The opacity ratio between these two colors describes how strong the opacity of a ray increased during accumulation. This ratio is multiplied with the color resulting from the new classification to obtain the final approximation.

Results of this reclassification can be seen in figure 3. We evaluated the usefulness of the algorithm by examining the reclassification results between several transfer functions. We observed that the layering criterion that is used to separate the volume dataset into view-dependent layers has strong impact on the visual quality of the result. It controls how well features are distributed between the layers. If more features are mixed in a single layer, the chosen density value is not a good representative. To accommodate for this fact, we modified the layering criterion of [9] to create layer boundaries whenever the density variance of a ray segment becomes higher than a threshold. A threshold of zero variance maximizes the visual quality, because only one density value per layer remains. This requires many layers. Therefore the resulting quality mainly scales with the number of layers. Increasing the number of layers improves the reclassification quality, but decreases overall performance. The reclassification itself takes less than 2 milliseconds per layer on the evaluation system (see below) and is therefore not a limiting factor.

old TF   new TF reclassified   new TF reference

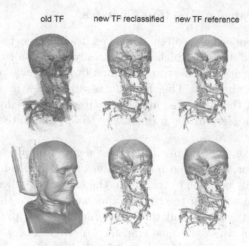

Fig. 3. Reclassification results: the left column shows the initial transfer functions (TF). The right column shows renderings of the new TF. The middle column results are approximated from the left TFs. Ten layers are used for all of the renderings.

5 Reference Frame Selection

When previously unseen parts of the scene become visible, no information is available for these parts. This is called a disocclusion (see the top right of Figure 4(a) for an example). To better deal with disocclusions, we use multiple reference views that are selected from the cache. Warping from multiple reference frames increases the chance that suitable information is found for every pixel.

As a quality criterion for selecting reference frames, the similarity of the reference frame's view to the current view seems obvious. Similarity can be computed as a function of the Euclidean distance of the camera positions, or the viewing angle between them. However, reference frames which are closest to the desired camera position are not necessarily the best. This is especially true for an unstructured viewpoint database, which is created by our system. Figure 4(a) illustrates how warping from the two closest cameras may not contain enough information for a properly generated image. Unfortunately, problems such as the one depicted in figure 4(a) occur frequently.

Diverse viewpoints are less likely to suffer from the same occlusions. Therefore, a suitable way to avoid occlusions is to introduce a penalty factor depending on the distance between two views of the selected set. This factor enforces a certain distance between the selected reference frame viewpoints, while the set stays close to the current view. The quality of a reference frame f_i for the current frame c can therefore be expressed as:

$$q(f_i) = -\alpha * dist(f_i, c) + \beta * dist(f_i, closest(f_i))$$

with $closest(x)$ searching for the closest reference frame to x in the set of selected frames, and α and β representing weighting coefficients. The ratio α/β defines

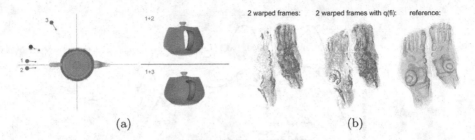

(a)   (b)

Fig. 4. (a) in this example, the two closest frames (green 1 and 2) to the current camera position (red) do not include information behind the handle of the teapot. Selecting more distant, but more diverse views can lead to a better result in such a case. (b) screenshots of a latency compensated rendering after a large viewpoint motion. Two warped frames selected by distance only (left) and by our $q(f_i)$ metric (middle). Right: reference rendering of the VIX feet dataset.

how the distance to the current view and the distance to the nearest neighbor relate. A high α/β favors nearby frames, whereas a low α/β favors diverse views. To find the set of selected frames, we start with an empty set and evaluate f_i for all frames. We pick the frame with the highest f_i and add it to the selection set. Then, f_i is evaluated again for all remaining frames and again the highest is added to the set. This is performed until the desired amount of reference frames are in the set for warping.

When using angles as the distance measurement and $\alpha/\beta = 3$, the example of figure 4(a) ranks the reference frames from best to worst $1 > 3 > 2$, thus resolving the problem. During our evaluations a ratio α/β in the range $[2, 3]$ worked out best. Figure 4(b) illustrates the benefit of our metric.

When warping from more than one reference frame, several fragments are transformed to the same pixel location in most cases. Using a depth test to dissolve fragment collisions was suggested in [11], but the depth test does not necessarily sort out fragments of low quality (inaccurately positioned). Also, the depth test possibly may not produce homogeneously generated surfaces, especially when the source frames contain fragments with similar depth, which is a situation prone to numerical issues. We found that blending all or several fragments of a pixel is not a good option either, because wrongly positioned fragments from lower similarity views interfere with the object's final appearance.

As a remedy, reference frames are sorted by their $q(f_i)$, and then rendered from low to high quality without using any depth buffer test. Layers are not mixed, so the overall depth order stays intact. This way fragments from a potentially better (closer) reference frame overwrite bad fragments. The surfaces appear more homogeneous (smoother), while still being able to fill holes.

5.1 Resolution of the Reference Frames

The above explanations assume that all reference frames have sufficient resolution for warping. This implies that the user's viewpoint lies on a sphere around

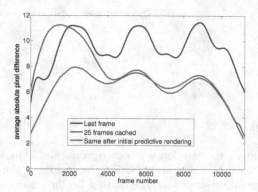

Fig. 5. Pixel error comparison of multi-frame rate rendering from the previous frame or from a reference frame cache (blue and green graphs). After initialization by pre-rendering, also the first seconds have improved quality (red graph).

the object, effectively prohibiting any zooming operations. To remove this limitation, all viewpoints that are used for rendering lie on a sphere with an as-small-as-possible radius. Therefore, the user can only freely select the viewpoint of the display node. The viewpoints of the frame source nodes always lie on the smallest sphere, but copy the selected viewing angle of the user. This ensures that all images have a sufficiently high resolution for future reuse, as long as the viewpoint stays outside this sphere. The radius can be trivially computed from the scene bounding box's extents.

6   Performance and Quality Evaluation

We recorded the viewpoint motion during a user session that contains a rotation around the Y-axis. Viewpoint location, orientation and time-stamps are stored for each frame during recording. When a session is replayed, the current viewpoint is interpolated from the recorded data and passed to our system as input. To obtain ground truth images, we rendered the recorded session using an offline raycaster. We used volumetric objects for evaluation, because they are highly complex even for modern GPUs.

For evaluation we used a single desktop PC equipped with 2 GPUs: a GeForce GTX 480 (Fermi generation) for latency compensation and display, and a GeForce GTX 275 as the frame source. The rendering resolution of all GPUs was set to 580x610. The frame rate of the output GPU was stable between 27 and 30 frames per second, while the frame source GPU performed raycasting at a stable frame rate between 0.3 and 0.5 frames per second.

For evaluation, we replay the recorded user session in real-time. The output images are stored with time stamps for an offline quality comparison. For every image in the evaluation track, the closest image from the reference track is found by comparing timestamps. From these two images, the average absolute pixel difference (AAPD) is computed as a quality metric.

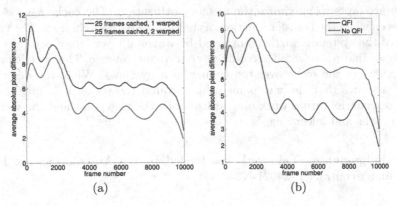

Fig. 6. AAPD comparison of our system using either one or two reference frames (a), and a comparison between our $q(f_i)$ metric and the angular distance

Our first evaluation uses the MANIX head dataset (see figure 1) from the OsiriX database [12]. Figure 5 shows a pixel error comparison of warping from the previous frame or from a reference frame cache, which is dynamically updated during the evaluation. The camera orbits around the dataset for 4 full rounds. Around frame number 2800 the first round is completed, and the frame cache succeeds in improving the output quality. This proves that the reference frame cache is effective when coherence is present.

Figure 5 additionally shows the effect of prerendering. The frame cache has been filled automatically prior to the evaluation run. We used the prerendering strategy that places viewpoints regularly on the bounding sphere. During the first camera orbit, the predictively filled frame cache provides superior visual quality. During further revolutions, the dynamic frame cache without prior rendering catches up and the implementations behave similarly.

The second evaluation uses the VIX feet dataset. We performed a pixel error comparison of multi-frame rate rendering from a frame cache using either one or two reference frames. Two frames help to cover disocclusion artifacts. When rendering from two reference frames, our $q(f_i)$ metric improves the quality when compared to a simple two-nearest-viewpoints selection strategy. Figure 6 shows the measured AAPDs. Rendering from more reference frames does not generally improve the visual quality. When the selected frames were rendered from too distant viewpoints, the quality even declines.

7   Conclusions and Future Work

We have described a system that is suitable for displaying arbitrarily rendered static scenes at high frame rates using a multi-frame rate approach in combination with image warping. Because scenes are static, our system can improve over time by creating and maintaining an image cache, which is updated either by user interaction or prerendering during idle phases. The suggested update strategy attempts to maximize the memory utilization. The cache can be preserved even when the transfer function of the volume dataset changes. The visual quality of the system is further improved by warping layers from several reference frames that are selected according to our quality metric. This metric favors diverse views of the scene over too similar, nearby views. While the described method assumes that the viewpoint orbits around the scene, it is not limited to such a scenario: future work may utilize both the viewpoint and the viewing angle to cache and select images.

Acknowledgements. This work was funded by the Austrian Science Fund (FWF) under contract P-24021-N23.

References

1. Springer, J.P., Beck, S., Weiszig, F., Reiners, D., Froehlich, B.: Multi-frame rate rendering and display. In: Sherman, W.R., Lin, M., Steed, A. (eds.) VR, pp. 195–202. IEEE Computer Society (2007)
2. Mark, W.R., Mcmillan, L., Bishop, G.: Post-rendering 3d warping. In: 1997 Symposium on Interactive 3D Graphics, pp. 7–16 (1997)
3. Smit, F.A., van Liere, R., Fröhlich, B.: An image-warping vr-architecture: design, implementation and applications. In: VRST 2008: Proceedings of the 2008 ACM Symposium on Virtual Reality Software and Technology, pp. 115–122. ACM, New York (2008)
4. Smit, F.A., van Liere, R., Fröhlich, B.: The design and implementation of a vr-architecture for smooth motion. In: VRST 2007: Proceedings of the 2007 ACM Symposium on Virtual Reality Software and Technology, pp. 153–156. ACM, New York (2007)
5. Smit, F.A., van Liere, R, Beck, S., Fröhlich, B.: An image-warping architecture for vr: Low latency versus image quality. In: VR, pp. 27–34. IEEE (2009)
6. Springer, J.P., Lux, C., Reiners, D., Froehlich, B.: Advanced multi-frame rate rendering techniques. In: VR, pp. 177–184. IEEE (2008)
7. Velázquez-Armendáriz, E., Lee, E., Bala, K., Walter, B.: Implementing the render cache and the edge-and-point image on graphics hardware. In: Proceedings of Graphics Interface 2006, GI 2006, Toronto, Ont., Canada, pp. 211–217. Canadian Information Processing Society, Canada (2006)
8. Simmons, M.: Tapestry: An Efficient Mesh-based Display Representation for Interactive Rendering. PhD thesis, EECS Department. University of California, Berkeley (2001)
9. Hauswiesner, S., Kalkofen, D., Schmalstieg, D.: Multi-frame rate volume rendering. In: EGPGV. Eurographics Association, Norrköping (2010)
10. Mark, W.: Post-rendering 3d image warping: Visibility, reconstruction, and performance for depth-image warping. Technical report. Chapel Hill, NC, USA (1999)
11. Chen, S.E., Williams, L.: View interpolation for image synthesis. In: SIGGRAPH 1993: Proceedings of the 20th Annual Conference on Computer Graphics and Interactive Techniques, pp. 279–288. ACM, New York (1993)
12. OsiriX: Dicom sample image sets. OsiriX Imaging Software (2009), http://pubimage.hcuge.ch:8080/

Detecting Periodicity in Serial Data through Visualization*

E.N. Argyriou and A. Symvonis

Department of Mathematics,
School of Applied Mathematical & Physical Sciences,
National Technical University of Athens, Greece
{fargyriou,symvonis}@math.ntua.gr

Abstract. Detecting suspicious or malicious user behavior in large networks is an essential task for administrators which requires significant effort due to the huge amount of log data to be processed. However, several of these activities can be rapidly identified since they usually demonstrate periodic behavior. For instance, periodic activities by specific users accessing the billing system of a financial institution may conceal fraud. Detecting periodicity in user behavior not only offers security to the network, but may prevent future malicious activities. In this paper, we present visualization techniques that aim to detect authorized (or unauthorized) user activities that seem to appear at regular time intervals.

1 Introduction

Due to the continuous increase of the size and complexity of computer networks, monitoring the user or network activity in a continuous basis is a necessary and, at the same time a time-consuming task for maintaining the network security. Traditionally, the network monitoring process is achieved by a combination of log file analysis, traffic analysis and intrusion detection systems. Even though most systems are equipped with mechanisms that produce sufficient log files, processing the huge amount of data requires significant effort, and usually is performed with little or no automated support.

Visualization is essential in cases of large data sets such the ones produced in a network, since it interprets the huge amount of data rows into a more comprehensive visual image. The necessity of the visualization aids is due to the fact that it is more difficult to immediately grasp the essence of something, if it is just described in words. In fact, it is hard for the brain to process text. Pictures or images, on the other hand, can be processed extremely well. They can encode a wealth of information and are therefore, well suited to communicate much larger information of data to human. Thus, by taking advantage of the human perception, the analysis of the visualization and the corresponding decision making becomes easier and more efficient. For this reason, over the last few years much research effort has been focused on seeking for visualizations of the network activity that aim to efficiently detect malicious activities.

* The work of E.N. Argyriou has been co-financed by the European Union (European Social Fund - ESF) and Greek national funds through the Operational Program "Education and Lifelong Learning" of the National Strategic Reference Framework (NSRF) - Research Funding Program: Heracleitus II. Investing in knowledge society through the European Social Fund.

G. Bebis et al. (Eds.): ISVC 2012, Part II, LNCS 7432, pp. 295–304, 2012.

The experience of examining malicious events in a network has revealed that many suspicious attempts appear in regular time basis. In several systems such as the billing system of a company, membership renewals systems, etc, periodic events may conceal fraud. For instance, in a billing system, an employee's monthly activity towards a specific customer account is considered to be suspicious, especially if it occurs before the billing day.

Motivated by the fact that detecting periodicity in serial data helps in rapidly identifying suspicious events, we present a system that visualizes serial data (either static or dynamic) produced by systems similar to the ones mentioned above. The main goal is to identify suspicious activities that may consist fraud. In our approach, each event corresponds to a pair of employee-customer due to the nature of the data sets examined. However, this approach can be generalized to other similar systems where appropriately defined pairs of entities can be identified in the system. The proposed system produces different types of visualizations, such that periodic events that are considered to be suspicious are easily identified. In order to produce aesthetically pleasant visualizations that are eventually easy to read and interpret, we employ standard techniques adopted from graph drawing in conjunction with our visualization techniques. As expected from such a system, it is equipped with supplementary functionalities such as support for storing, reloading and post-processing of data. It provides advanced graphic functionality, including popup menus, printing capabilities, custom zoom, fit-in window, selection, dragging and resizing of objects.

The rest of this paper is structured as follows: Section 2 overviews related work. In Section 3, we sketch our contribution. In Section 4, we describe in detail the proposed system. We conclude in Section 5 with open problems and future work.

2   Previous Work

During the last few years various visualization approaches have been proposed for network monitoring. Mansmann et al. [1] presented a visual analytics tool that visualizes the behavior of hosts or higher level network entities over time. Yin et al. [2] presented a novel approach to the visualization of traffic flows to detect and investigate anomalous traffic between a local network and external domains, whose central aspect is a parallel axes view used to represent the origin and destination of network traffic. Shabtai et al. [3] presented two tools that enable the user to visualize and explore time-oriented security data. Lakkaraju et al. [4] presented NVisionIP, a tool that supports different visualizations in order to provide a snapshot of the activity of a network, which supports filtering and aggregation of the input data based on a number of attributes that are important for security analysis. Vandenberghe [5] presented a data visualization tool that analyzes a security event from a range of visual perspectives using different detection algorithms. Finn and North [6] presented a security visualization tool capable of representing tens of thousands of hosts simultaneously and allows the user to display communication patterns between arbitrary locations.

Regarding visualizations of data produced by intrusion detection systems (or IDS for short), Abdullah et al. [7] presented IDS Rainstorm, a tool that provides high-level overviews of intrusion detection alerts. Tolle and Niggemann [8] propose a system supporting the detection of intrusions and network anomalies by analyzing and visualizing traffic flows in computer networks by means of graph drawing techniques. Oline and Reiners [9] propose several 3-dimensional visualizations, each of which emphasizes on different aspects of IDS alerts. Erbacher et al. [10] presented different techniques for the visual representation, exploration and analysis of IDS related data in order to ease the identification and analysis of network attacks.

Carlis and Konstan [11] presented a spiral visualization technique to highlight a type of data (called it serial periodic), which occurs frequently. According to their approach, serial attributes of the data set are displayed along the spiral axis, while the periodic ones along the radii of the spiral. Weber et al. [12] presented a visualization system for time-series data based on spirals that processes large data sets and detects periodic data patterns. According to their approach, the spiral corresponds to the time axis, while the other attributes of the data are represented by points, colors, lines or bars. Bertini et al. [13] proposed SpiralView, a tool that supports spiral visualizations to monitor network traffic and helps understanding the evolving of network alarms over time. It also provides identification of periodic patterns. An overview on the visualization of time-series data and the available techniques can be found in [14,15,16].

In the context of graph drawing, force-directed methods [17,18,19,20] are quite common when visualizing combinatorial information by means of directed or undirected graphs. In such a framework, a graph is treated as a physical system on which appropriate forces (either attractive or repulsive, or both) are applied. The equilibrium state of the system produces a good configuration or a drawing of the graph. An overview of force-directed methods and their variations can be found in classical graph drawing books [21,22].

3   Our Contribution

Our contribution consists of three different visualization methods that aim to help in identifying periodic activity in time-series data stemming from system involving pairs of entities, e.g., a billing system. The main goal is to contribute in fraud detection. Each visualization results in drawings which can be utilized in order to detect periodicity in the analyzed events. Note that, we measure periodicity by introducing a new metric which is appropriately defined to reveal the frequency in which an event occurs within a time window. Our main visualization method results in drawings consisting of concentric circles whose radius correspond to the periodicity of the activity of each pair of entities of the system. Events that are considered to be suspicious are easily identified since they are dragged towards the center of the circles. Also, a force-directed approach is employed in order to provide a better configuration of the events over the visualization. The system is equipped with supplementary functionalities such as information regarding the entities' activities, examination of certain period of interest or visualization of the series of events.

4   Description of the System

The system's input data sources can be either a log file or a set of records of a database of a system involving pairs of entities, e.g., a billing system. However, it is extensible to other similar data sources. Each pair of entities is associated with a series of *events* involving them (e.g., a phone call between them, a transaction, etc.), which we assume to be sorted by date. In order to produce a visualization, the system preprocesses these series of events, and estimates a proper period of activity for each pair of entities.

4.1 Periodicity Estimation

For each pair of entities, we introduce a metric that estimates a periodicity value with a specific confidence degree. Let ρ be a pair of entities and n_ρ the number of events associated with pair ρ. Assume that e_i^ρ, e_{i+1}^ρ, $i = 1, \ldots, n_\rho - 1$, are two consecutive events, and let $d_{i,i+1}^\rho$ be their time distance (say measured in days). A time-series $T_\rho = (t_1^\rho, \ldots, t_{n_\rho-1}^\rho)$ is generated by assigning to each event e_i^ρ a value t_i^ρ according to the following formula:

$$t_i^\rho = \begin{cases} \sum_{j=1}^{i-1} d_{j,j+1}^\rho & \text{if } 1 \leq i < n_\rho \\ 0 & \text{if } i = 0 \end{cases}$$

For a given period value s, the ideal time-series $D_s = (0, s, 2s, \ldots)$ is defined by the time stamps that occur if the events between the entities of ρ appear in time intervals that equal exactly to s (see Figure 1). For instance, in case of a period value of 30 days, the ideal time series is $D_{30} = (0, 30, 60, \ldots)$.

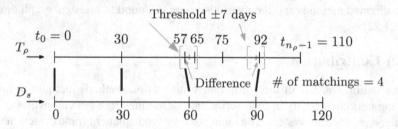

Fig. 1. Line T_ρ corresponds to time-series events of a pair of employee-customer, whereas line T_s' to the ideal time-series for a period of 30 days

Let $t \in T_\rho$ and $\lambda \in D_s$. We say that t and λ *match each other with respect to a threshold value* $\tau \in [1, s/2)$, if it holds that $t \in [\lambda - \tau, \lambda + \tau]$. Let N_ρ^τ be the set of time stamps of ideal time-series D_s, which can be matched with a time-stamp of time-series T_ρ, i.e., $N_\rho^\tau = \{\lambda \in D_s : \exists t \in T_\rho \text{ s.t., } t \text{ and } \lambda \text{ match}\}$. With slight abuse of terminology, we refer to the cardinality of N_ρ^τ as the *number of matchings* of pair ρ. Let also, $diff_\rho^\tau : N_\rho^\tau \to \mathbb{R}$ with:

$$diff_\rho^\tau(\lambda) = \min\{|t - \lambda| : t \in T_\rho, \lambda \in N_\rho^\tau \text{ and } (t, \lambda) \text{ match}\}$$

The confidence level of a pair ρ of entities with periodicity value s and threshold matching value τ is given by the following formula:

$$confidence(\rho, s, \tau) = \frac{\sum_{\lambda \in N_\rho^\tau} 1 - \frac{diff_\rho^\tau(\lambda)}{\tau}}{|N_\rho^i|}$$

Observe that the confidence values belong in $[0, 1]$. Obviously, if the time-series T_ρ is identified with the ideal time-series D_s, then $confidence(\rho, s, \tau) = 1, \forall \tau \in [1, s/2)$. In order to provide a more accurate estimation of the confidence value of a pair ρ for a given period s, with respect to a threshold value τ, one can alternatively compute the confidence value of ρ for all ideal time-series $D_s^i = (i, s + i, \ldots), i = 0, \ldots, s/2$ and, keep the one that maximizes the confidence value.

For a given pair of entities ρ and a prespecified threshold value τ, we measure its confidence for all periodicity values $s \in [1, s_{max}]$, where s_{max} corresponds to the maximum periodicity value defined by the user. Having determined all confidence values of ρ, the *periodicity* of pair ρ (with respect to the specified threshold value τ), equals to the periodicity value that maximizes its confidence value.

Note that for each pair of entities, the system is able to produce a visualization similar to the one of Figure 1, in order to present the series of events associated with the specific pair and the matchings with the ideal time-series. In addition, the system is capable of identifying weekends and feast days of each year, and adapts appropriately the ideal sequences. However, for simplicity reasons, in our description we ignored this functionality.

4.2 Periodicity Visualization

The main visualization of the system is illustrated in Figure 2, where we seek for monthly periodic activity. It consists of a system of concentric circles whose radius correspond to different periodicity values. The nodes of the visualization correspond to pairs of entities. The outermost circle corresponds to a period of 8 days, while the innermost to 31. We only compute periodicity values that are greater than the threshold value, which in the visualization of Figure 2 is set to 7 days. However, this is a value determined by the user. With this configuration, nodes with periodicity of 30 or 31 days are dragged towards the center of the system.

The system is also split in circular sectors that correspond to different number of matchings with the ideal time-series for each period, as discussed above. For a more uniform arrangement of the nodes in the system of circles, nodes whose number of matchings is greater than the median value lie on the upper semicircle, while the remaining ones at the bottom. The maximum number of matchings corresponds to the midpoint of each upper semicircle. In the visualization, we ignore nodes whose time-series had up to two matchings with the ideal time-series since otherwise, we always have a perfect sequence of value two. The gray colored areas of Figure 2 illustrate nodes that appear to have suspicious activity (due to their periodicity values) and need to be further examined. We have also chosen to highlight the entire ring of periods greater than 27, even in cases with few matchings, since this may reveal a suspicious behavior that is about to start.

The system provides the capability to the user to select a node (especially a suspicious one) and draws with the same color or shape all nodes of the system that contain

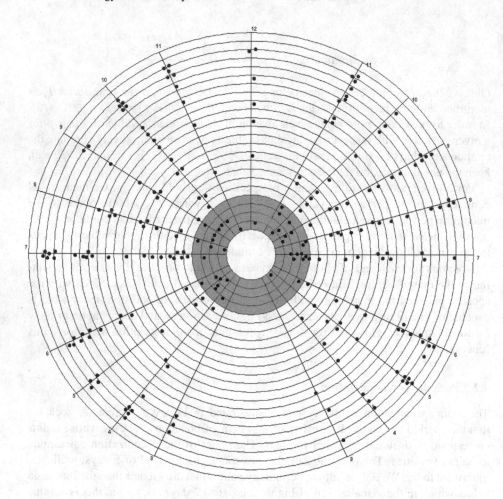

Fig. 2. A concentric circle system in which each radius correspond to a periodicity value. The gray-colored areas are the ones that have to be examined first for suspicious activities.

the same entity with the selected one. In this manner, the user can identify whether the entity appears to have a continuous suspicious activity. Also, the system can draw with different colors or shapes the most suspicious nodes, such that they can be easily distinguished from the remaining ones, as in Figure 3. The system is equipped with popup menus at each node which reveal additional information, such as periodicity, confidence value and so on. The system also provides supplementary functionalities such as support for storing, reloading and post-processing of the visualization. It also supports advanced graphic functionality, including printing capabilities, custom zoom, fit-in window, selection, dragging and resizing of objects.

In order to obtain more legible drawings, we have used the classical force-directed algorithm of Eades [18] in conjunction with our visualization technique. A force-directed algorithm models the vertices of the graph as electrically charged particles that repel each other, and its edges by springs in order to attract adjacent vertices. However,

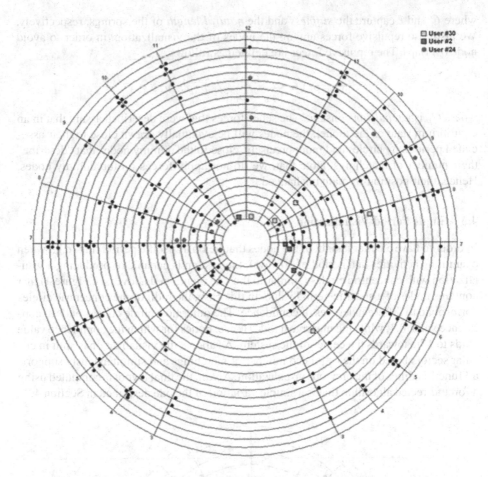

Fig. 3. The top three most dangerous entities are illustrated with different colors and shapes

before we proceed with the detailed description of the algorithm, we introduce some necessary notation. Let $G = (V, E)$ be an undirected graph. Given a drawing $\Gamma(G)$ of G, we denote by $p_u = (x_u, y_u)$ the position of node $u \in V$ on the plane. The unit length vector from p_u to p_v is denoted, by $\overrightarrow{p_u p_v}$, where $u, v \in V$.

In our approach, we add dummy nodes on each circle and along the lines that splits these circles in circular sectors. Each dummy node corresponds to the number of matchings for a given periodicity. Then, we use springs to connect each node with the dummy node of its period circle that corresponds to its number of matchings. The springs follow the logarithmic law instead of the Hooke's law, in order to avoid exerting strong forces on distant nodes. The attractive forces follow the formula:

$$\mathcal{F}_{spring}(p_u, p_v) = C \cdot \log \frac{||p_u - p_v||}{\ell} \cdot \overrightarrow{p_u p_v}, \ (u, v) \in E$$

where C and ℓ capture the *stiffness* and the *natural length* of the springs, respectively. We also, use repulsive forces among the nodes of the visualization, in order to avoid node overlaps. The repulsive forces are defined as follows:

$$\mathcal{F}_{rep}(p_u, p_v) = \frac{C_p}{||p_u - p_v||^2} \cdot \overrightarrow{p_u p_v}, \ u, v \in V$$

where C_p is a *repulsion* constant. The set of forces that were described assure that in an equilibrium state of the model, the nodes will be eventually drawn close to their associated periodicity circles and more precisely, close to the dummy nodes that "describe" their number of matchings. Note that, we do not apply forces on the dummy nodes. Hence, their positions remain unchanged.

4.3   Single Period Visualization

In order to have a better insight of the nodes that lie on a specific period ring (e.g., when examining activities in a period of 30 days), the system is capable of producing a visualization with concentric circles (similar to the one mentioned above; see Figure 4) that contains nodes of a specific period value. In this case, the radii of the concentric circles correspond to different degrees of confidence. The outermost circle corresponds to confidence value 0.1, while the innermost to 1. Hence, nodes for which the confidence value tends to 1 lie towards the center of the system. As above, the visualization is split in circular sectors based on the estimated number of matchings and simultaneously supports all functionalities of the previous visualization. Again, the final layout is computed using a force-directed algorithm that is a simple variation of the one described in Section 4.2.

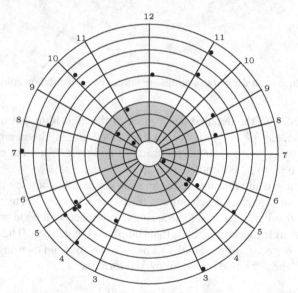

Fig. 4. A concentric circle visualization for a period of 30 days. The radii of each circle correspond to different confidence values. Nodes with confidence value 1 move towards the center of the system.

5 Conclusions and Future Work

In this paper, we presented a system that aims to detect periodic event in time-series data. The system is oriented towards fraud detection in data stemming from billing or other similar business systems. However, it can be extended to support data from other data sources. The presented visualizations help the security managers to identify employees that appear to have suspicious activity towards specific customer accounts. Of course, our work opens several directions for future work:

- One of the main future goals of this system is to be enhanced with several other visualizations methods that reveal periodicity. More sophisticated algorithms adopted from Graph Drawing or Information Visualization need to be incorporated.
- Alternative metrics to measure the confidence degree can be used in order to obtain more accurate periodicity estimations. This may also affect the quality or the type of the produced visualizations.
- Identifying group of users (instead of a particular user) that appear to have similar suspicious behavior is also of interest. Standard clustering techniques adopted from Graph Drawing may be useful for the production of such visualizations.
- Incorporating more functionalities required for a security manager such as statistic analysis of the activity for each entity, plots, bar charts, etc.

Acknowledgements. We would like to thank Vassilis Vassiliou for his useful suggestions and comments related to fraud detection.

References

1. Mansman, F., Meier, L., Keim, D.A.: Visualization of host behavior for network security. In: VizSEC 2007, pp. 187–202. Springer, Heidelberg (2008)
2. Yin, X., Yurcik, W., Treaster, M., Li, Y., Lakkaraju, K.: Visflowconnect: netflow visualizations of link relationships for security situational awareness. In: Proceedings of the 2004 ACM Workshop on Visualization and Data Mining for Computer Security, VizSEC/DMSEC 2004, pp. 26–34. ACM, New York (2004)
3. Shabtai, A., Klimov, D., Shahar, Y., Elovici, Y.: An intelligent, interactive tool for exploration and visualization of time-oriented security data. In: Proceedings of the 3rd International Workshop on Visualization for Computer Security, VizSEC 2006, pp. 15–22. ACM (2006)
4. Lakkaraju, K., Yurcik, W., Lee, A.J.: Nvisionip: netflow visualizations of system state for security situational awareness. In: Proceedings of the 2004 ACM Workshop on Visualization and Data Mining for Computer Security, VizSEC/DMSEC 2004, pp. 65–72. ACM (2004)
5. Vandenberghe, G.: Network Traffic Exploration Application: A Tool to Assess, Visualize, and Analyze Network Security Events. In: Goodall, J.R., Conti, G., Ma, K.-L. (eds.) VizSec 2008. LNCS, vol. 5210, pp. 181–196. Springer, Heidelberg (2008)
6. Fink, G.A., North, C.: Root polar layout of internet address data for security administration. In: Proceedings of the IEEE Workshops on Visualization for Computer Security, VIZSEC 2005, pp. 55–64. IEEE Computer Society (2005)
7. Abdullah, K., Lee, C., Conti, G., Copeland, J.A., Stasko, J.: Ids rainstorm: Visualizing ids alarms. In: Proceedings of the IEEE Workshops on Visualization for Computer Security, VIZSEC 2005, pp. 1–10. IEEE Computer Society (2005)

8. Toelle, J., Niggemann, O.: Supporting intrusion detection by graph clustering and graph drawing. In. In: Proc. of 3rd Int. Workshop on Recent Advances in Intrusion Detection, RAID 2000 (2005)

9. Oline, A., Reiners, D.: Exploring three-dimensional visualization for intrusion detection. In: Proceedings of the IEEE Workshops on Visualization for Computer Security, VIZSEC 2005, pp. 113–120. IEEE Computer Society (2005)

10. Erbacher, R.F., Christensen, K., Sundberg, A.: Designing visualization capabilities for ids challenges. In: Proceedings of the IEEE Workshops on Visualization for Computer Security, VIZSEC 2005, pp. 121–127. IEEE Computer Society (2005)

11. Carlis, J.V., Konstan, J.A.: Interactive visualization of serial periodic data. In: Proceedings of the 11th Annual ACM Symposium on User Interface Software and Technology, UIST 1998, pp. 29–38. ACM (1998)

12. Weber, M., Alexa, M., Müller, W.: Visualizing time-series on spirals. In: Proceedings of the IEEE Symposium on Information Visualization 2001 (INFOVIS 2001), pp. 7–14 (2001)

13. Bertini, E., Hertzog, P., Lalanne, D.: Spiralview: Towards security policies assessment through visual correlation of network resources with evolution of alarms. In: Proceedings of the 2007 IEEE Symposium on Visual Analytics Science and Technology, VAST 2007, pp. 139–146. IEEE Computer Society (2007)

14. Silva, S.F., Catarci, T.: Visualization of linear time-oriented data: A survey. In: Proceedings of the First International Conference on Web Information Systems Engineering (WISE 2000), vol. 1, pp. 310–319. IEEE Computer Society (2000)

15. Müller, W., Schumann, H.: Visualization for modeling and simulation: visualization methods for time-dependent data - an overview. In: Proceedings of the 35th Conference on Winter Simulation: Driving Innovation, WSC 2003, pp. 737–745 (2003)

16. Aigner, W., Bertone, A., Miksch, S., Tominski, C., Schumann, H.: Towards a conceptual framework for visual analytics of time and time-oriented data. In: Proceedings of the 39th Conference on Winter Simulation: 40 Years! The Best is Yet to Come, WSC 2007, pp. 721–729 (2007)

17. Davidson, R., Harel, D.: Drawing graphs nicely using simulated annealing. ACM Transactions on Graphics 15, 301–331 (1996)

18. Eades, P.: A heuristic for graph drawing. Congressus Numerantium 42, 149–160 (1984)

19. Fruchterman, T., Reingold, E.M.: Graph drawing by force-directed placement. Software-Practice and Experience 21, 1129–1164 (1991)

20. Kamada, T., Kawai, S.: An algorithm for drawing general undirected graphs. Information Processing Letters 31, 7–15 (1989)

21. Kaufmann, M., Wagner, D. (eds.): Drawing Graphs. LNCS, vol. 2025. Springer, Heidelberg (2001)

22. Di Battista, G., Eades, P., Tamassia, R., Tollis, I.G.: Graph Drawing: Algorithms for the Visualization of Graphs. Prentice Hall (1999)

Practical Implementation of a Graphics Turing Test

M. Borg[1], S.S. Johansen[1], D.L. Thomsen[1], and M. Kraus[2]

[1] School of Information and Communication Technology,
Aalborg University, Denmark
[2] Department of Architecture, Design and Media Technology,
Aalborg University, Denmark

Abstract. We present a practical implementation of a variation of the
Turing Test for realistic computer graphics. The test determines whether
virtual representations of objects appear as real as genuine objects. Two
experiments were conducted wherein a real object and a similar virtual object is presented to test subjects under specific restrictions. A
criterion for passing the test is presented based on the probability for
the subjects to be unable to recognise a computer generated object as
virtual. The experiments show that the specific setup can be used to determine the quality of virtual reality graphics. Based on the results from
these experiments, future versions of the Graphics Turing Test could ease
the restrictions currently necessary in order to test object telepresence
under more general conditions. Furthermore, the test could be used to
determine the minimum requirements to achieve object telepresence.

1 Introduction

The purpose of the work presented in this paper is to evaluate the realism of
computer graphics through a specific setup which restricts the human visual
system and takes into account the limitations of today's monitors, for example
insufficient black-levels and colour range. In this way, we can create a setup
that allows us to test the realism of a virtual object compared to a genuine
object. Instead of letting the test subjects decide which of two virtual objects
looks more realistic, this setup forces the test subjects to compare a virtual
object to a genuine real object in a similar environment. This results in an
indirect measurement comparable to the just-noticeable-difference between a
real and a virtual scene based on subjective assessments of the realism of the
scenes. The aim is to present a setup which can assist in improving virtual reality
environments.

Based on work by Slater and Usoh [1] and Steuer [2] we define telepresence
as a sense of being physically present in a virtual environment. Presence has
been defined as *"the subjective experience of being in one place or environment,
even when one is physically situated in another"* [3]. Similarly, we define object
telepresence as the subjective experience that a virtual object is situated in the
real world.

G. Bebis et al. (Eds.): ISVC 2012, Part II, LNCS 7432, pp. 305–313, 2012.

In accordance with Mel Slater's statement *"Computer graphics is for ... people"* [4], results are obtained through a study of human vision akin the approach used in the Turing Test suggested by Alan Turing [5]. Rather than, for example, evaluating the pixel colour difference between a synthesized image and a photograph, test subjects assess object telepresence directly. A sufficient number of such subjective measurements leads to an evaluation of object telepresence. This conscious evaluation can complement previous results that measured unconscious reactions to virtual environments, e.g. Slater and Usoh [1]. Our main contribution is to implement a restricted Graphics Turing Test and to propose a practicable criterion for passing the test. We show that it is possible to pass the proposed test without excessive effort. While the setup has been tested with simple and static shapes, it is also applicable to more complex scenes.

2   Related Work

Several previous studies propose a virtual reality test in the spirit of Alan Turing's artificial intelligence test. Instead of testing artificial intelligence, these studies test how a real scene compares to a similar representation on a screen. Meyer et al. [6] tested a setup which required test subjects to view a real physical setup as well as the setup being displayed on a colour television (both through a view camera). The study showed that within the specific setup, the test subjects could not tell which scene was real. In this setup, however, the genuine scene was only indirectly compared to the virtual scene because the test subjects were watching it through a view camera. Similar setups have also been used to compare lightness in virtual and genuine scenes [7] and to identify specific parameters of realistic scenes [8].

Moving closer to virtual reality, Brack et al. [9] propose to test the presence of a virtual object when compared to a similar real object. The study concludes that the display of a virtual object would require fidelity reduction techniques to appear as present as the real object. However, it provides no recommendations as to which restrictions are necessary and to which extent. McGuigan [10] determines that computers are not able to create photo-realistic 3D real-time rendering yet. Similarly, Ventrella et al. [11] argue that computer-generated 3D movement is still not believable. In contrast to this earlier work, we show that under certain restrictions it is possible to render and display virtual, static 3D objects so convincingly that people cannot distinguish them from real objects.

3   Graphics Turing Test

Based on previous work on the Graphics Turing Test, we define a restricted version of the Graphics Turing Test, which can actually be put into practice. Thus, our approach is to define certain restrictions that are sufficient to make a virtual object appear as present as a similar real object. To determine the specific restrictions, we summarize the most important depth cues of the human vision [12]:

- Monocular cues: occlusion, size, position, ocular accommodation, linear perspective and motion parallax.
- Binocular cues: stereopsis and convergence.

The experiment is executed by restricting test subjects from using a number of the above cues. To follow the structure of the Turing Test, the test subjects are shown a real object and a virtual representation. They are told before the experiment that one is real and one is not. After viewing both objects, they are asked to pick the object that in their opinion is the real one.

4   Experimental Setup

To carry out the experiment described above, a rectangular box was built, which is illustrated in Fig. 1. One end of the box can be removed and replaced by a screen displaying a virtual representation of the scene inside the removable end. This removable end of the box has a hole in the top, which is 1.6 cm in diameter and allows for the effect of a spotlight on the object inside the box by placing a halogen lamp above the hole. The lamp was measured to illuminate with approximately 2000 lux just below the top hole. For the sphere scene (see below), the illumination was reduced to approximately 1500 lux.

To compensate for the expected limitations of the monitor, the view needs to be restricted. Since the acuity of the human eye is much higher than it is possible to show on a screen and many monitors suffer from insufficient black-levels, a setup was created which takes the capabilities of the human eye into account. A wooden plate with a 2.1 cm hole in the middle (large enough for the test subject to be able to see the entire object) covers the front of the box. The distance from the front of the box to the screen or the separate end of the box is determined by an equation for the visual acuity which specifies the minimum distance necessary for a human to be unable to separate the pixels [13]. Given a resolution of 1,680 × 1,050 pixels on a 22" display, the minimum distance is 163 cm. In order to accommodate variations of the human acuity (e.g. age of the test subjects), the removable part with the object is placed 200 cm from the front of the box.

The test subjects are asked to look through the small hole using only one eye. This restricts the depth perception by avoiding stereopsis, convergence and motion parallax. If the depth of the scene is small enough relative to the length of the box, the ocular accommodation will approximately be constant over the whole scene and therefore, it will not provide any depth cues [12]. It should be emphasized that — according to the measured illumination — the well lit parts of the objects inside the box were very well visible (even under the imposed viewing restrictions).

To ensure that it is not possible to recognise the screen from the box, the test subjects have to keep a distance of about 10 cm from the hole. This distance makes the eyes of the test subjects adapt to a bright environment because of the emittance of 50 lux from the well lit room. This makes them unable to adapt

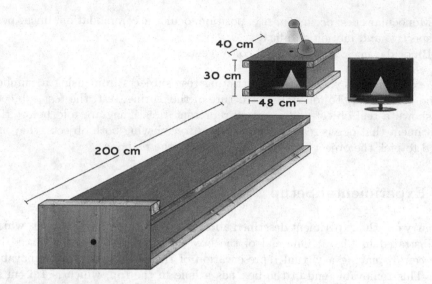

Fig. 1. A representation of the box used for the experiment

to the dark interior of the box, which is covered with a black fabric. Thus, the ability of the test subjects to distinguish the not perfectly black colour of the screen from the also not perfectly black colour of the black fabric is inhibited. To prevent their eyes from adapting further to the dark environment inside of the box, the subjects are only given 10 seconds to look through the hole. [12]

The box is placed in the test environment so that the test subjects are only able to see the front of the box — the rest is hidden from the test subjects such that they cannot estimate the length of the box without looking inside it. Furthermore, this will keep them from knowing whether they are viewing the screen or the end of the box.

5   Virtual Scene

A blue cardboard pyramid and a blue styrofoam sphere were chosen for the experiment because of the materials, which show only limited specular reflection and have simple textures. These surfaces can be displayed on the employed screen (Samsung SyncMaster 2233RZ), while a highly specular surface would be more difficult to display on the screen as it requires a wider range of colours. The pyramid has a height of 12 cm and a 16 cm base (measured diagonally). The sphere has a diameter of 12 cm.

5.1   Screen Usability and Reference Image

A short test was conducted with the purpose of determining the usability of the screen as well as making a reference image for creating the virtual scene.

A photograph of the pyramid inside of the box was captured and adjusted — giving the base more light and a higher contrast, the pyramid a slightly darker top and brighter bottom and the general image warmer colours (see Fig. 2). 24 test subjects were asked to view the photograph on the screen as well as the object in a randomized order. When asked, 19 of the subjects could not tell which object was real (3 of which chose to guess but did so incorrectly). These results indicate that virtual representations of the pyramid and the sphere can seem as present as the real objects within the scope of this experimental setup.

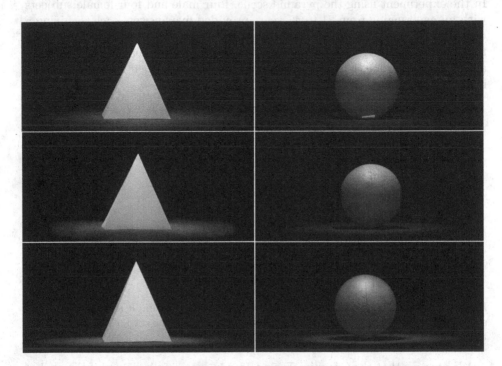

Fig. 2. Top row: The reference images captured. A slow shutter speed reveals the tape in the image of the sphere. This is not visible in the experiment. Middle row: The final reference images which have been adjusted and used for testing the screen. Bottom row: The final scene in Unity as displayed on the screen. Both the middle and the bottom images are displayed in darker colours due to a dynamic contrast function of the employed screen.

5.2   Modeling and Setup

The cardboard pyramid is modeled using a standard square pyramid shape. The edges are chamfered and additional vertices are added to make them crooked. The sphere is a standard sphere shape.

Both textures are created from adjusted photographs of the original object materials. They are applied by making UVW maps of the models. The texture for the planar base of the box is created using multiple layers of noise in yellow, orange and red colours on a black layer. All objects are imported to a game engine as FBX files. The reason for choosing a game engine is the possibility to extend the experiment (see Section 8).

The light in the virtual pyramid scene was set up to simulate the employed halogen lamp. The final result can be seen in Fig. 2.

6   Results

In the experiment using the pyramid scene, four male and four female subjects with an age ranging from 21 to 47 years provided five answers each — giving a total of 40 answers. Of these answers, 22 were correct and 18 were incorrect. For the experiment using the sphere scene, six male and two female subjects with an age ranging from 21 to 28 years were used. They provided 25 correct and 15 incorrect answers.

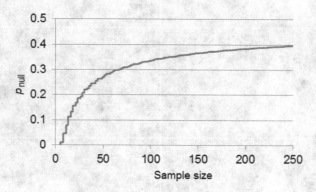

Fig. 3. Graph of the relation between p_{null} and the required sample size that results in a probability of 95 % that the null hypothesis is rejected in the case of complete object telepresence (i.e. subjects are just guessing).

We propose that the Graphics Turing Test is successfully passed if the probability for test subjects to incorrectly identify the virtual object as the real one is estimated as greater than 19 % with a confidence level of 95 %. In other words, we suggest that the null hypothesis of a probability $p_{null} = 0.19$ (or less) for identifying the objects incorrectly, has to be rejected with a significance level of 5 % by the results. This respects true hypothesis testing, where the null hypothesis is either rejected or not. The threshold corresponds to the commonly used threshold of subjects guessing incorrectly at least 25 % of at least 100 trials [14]. With 40 trials and $p_{null} = 0.19$, the corresponding threshold is slightly higher to compensate for lower samples sizes. This threshold may also be understood in terms of Alan Turing's statement *"a considerable proportion of a jury, who should not be expert about machines, must be taken in by the pretence"* [15] and we define "a considerable proportion" to be more than 19 %. Note that significantly higher values of p_{null} would require impractically large sample sizes as illustrated in Fig. 3.

The probability mass function for the number i of incorrectly identified objects under the null hypothesis is:

$$f(i|n, p_{null}) = \frac{n!}{i!\,(n-i)!}\,(p_{null})^i\,(1 - p_{null})^{(n-i)} \tag{1}$$

where n is the number of trials, i is the number of incorrect answers and p_{null} is the probability for the null hypothesis. With this function, a critical number i_c of incorrectly identified objects can be computed such that the probability of incorrectly rejecting the null hypothesis is less than 5 %:

$$i_c(n, p_{null}) = \min\left\{ i \;\middle|\; \sum_{j=i}^{n} f(j|n, p_{null}) < 0.05 \right\}. \tag{2}$$

With 40 trials, the critical number of incorrect answers is 13 (see Fig. 4). Since 18 and 15 incorrect answers have been observed, both tests have been passed.

The results show that a "considerable proportion" of the test subjects are unable to recognise which object is real and which is not. A possible source of error is that each test subject had five trials, i.e. not all trials are independent. Another possible source of error is the test setup which required the test participants to exit the room and wait approximately 20 seconds before being allowed to have a new look. Also, the visual differences between the real and the virtual object in both experiments can be a source of error as the test subjects have a tendency to either guess correctly or incorrectly rather consistently. However, the experiment is not about comparing the real object to a 3D model but to make the test subjects believe that the 3D object is real.

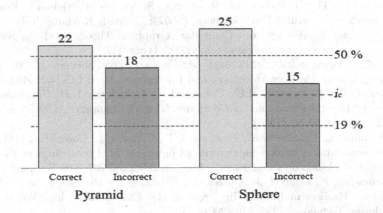

Fig. 4. Results from the two tests with the 50% line representing complete object telepresence (i.e. subjects are just guessing) and the 19% line representing the null hypothesis. i_c represents the required critical number of incorrect answers to reject the null hypothesis.

7   Conclusion

In this work, a restricted Graphics Turing Test has been defined and implemented by the approach of avoiding certain cues for depth perception. A practicable criterion for passing the test has been presented. Our experiments show that within the specific scope, the test can be passed without an excessive number of test subjects. A positive outcome of the proposed test strongly suggests that a limited form of object telepresence has been achieved.

8   Future Work

We are currently planning to extend the test setup by including stereoscopic vision and motion parallax. This would allow us to test less restricted forms of object telepresence. Furthermore, the proposed test could be applied to test the quality of many display technologies and rendering techniques. For example, it is also possible to compare animated renderings with real movements to test visual effects such as motion blur.

References

1. Slater, M., Usoh, M.: Body Centred Interaction in Immersive Virtual Environments. Artificial Life and Virtual Reality 1 (1994)
2. Steuer, J.: Defining virtual reality: dimensions determining telepresence. Journal of Communication 42, 73–93 (1992)
3. Witmer, B., Singer, M.J.: Measuring Presence in Virtual Environments: A Presence Questionnaire. Presence: Teleoperators and Virtual Environments 7, 225–240 (1998)
4. Slater, M.: The Influence of Rendering Styles on Participant Responses in Immersive Virtual Environments (2007); Invited Keynote Talk: Second International Conference on Computer Graphics Theory and Applications, http://www.cs.ucl.ac.uk/research/vr/Projects/VLF/MediaCGTA_keynote_2007/vlf%20pit.pdf (last accessed February 24, 2012)
5. Turing, A.M.: Computer Machinery and Intelligence. Mind LIX, 422–460 (1950)
6. Meyer, G.W., Rushmeier, H.E., Cohen, M.F., Greenberg, D.P., Torrance, K.E.: An Experimental Evaluation of Computer Graphics Imagery. ACM Transactions on Graphics 5, 30–50 (1986)
7. McNamara, A., Chalmers, A., Troscianko, T., Gilchrist, I.: Comparing real & synthetic scenes using human judgements of lightness. In: Proceedings of the Eurographics Workshop on Rendering Techniques 2000, pp. 207–218. Springer (2000)
8. Rademacher, P., Lengyel, J., Cutrell, E., Whitted, T.: Measuring the Perception of Visual Realism in Images. In: Proc. of the 12th Eurographics Workshop on Rendering Techniques, 235–248 (2001)
9. Brack, C.D., Clewlow, J.C., Kessel, I.: Human Factors Issues in the Design of Stereo-rendered Photorealistic Objects: A Stereoscopic Turing Test. In: Proc. SPIE 2010, vol. 7524 (2010)
10. McGuigan, M.D.: Graphics Turing Test (2006), http://arxiv.org/abs/cs/0603132

11. Ventrella, J., El-Nasr, M.S., Aghabeigi, B., Overington, R.: Gestural Turing Test: A Motion-Capture Experience for Exploring Nonverbal Communication. In: Proc. AAMAS International Workshop on Interacting with ECAs as Virtual Characters (2010)
12. Wolfe, J.M., Kluender, K.R., Levi, D.M., Bartoshuk, L.M., Herz, R.S., Klatzky, R.L., Lederman, S.J., Merfeld, D.M.: Sensation & Perception, 2nd edn., pp. 32–33. Sinauer (2009)
13. Clark, R.: Notes on the Resolution and Other Details of the Human Eye (2005) last updated 2009, http://clarkvision.com/imagedetail/eye-resolution.html (last accessed December 9, 2011)
14. McKee, S.P., Klein, S.A., Teller, D.Y.: Statistical Properties of Forced-Choice Psychometric Functions: Implications of Probit Analysis. Perception & Psychophysics 37, 786–298 (1985)
15. Turing, A.M.: Can automatic calculating machines be said to think?, Typescrip of broadcast discussion transmitted on BBC Third Programme, 61th sheet (1952), http://www.turingarchive.org/viewer/?id=460&title=5 (last accessed December 9, 2011)

The Hybrid Algorithm for Procedural Generation of Virtual Scene Components

Tomasz Zawadzki and Dominik Kujawa

Faculty of Electrical Engineering and Telecommunications, University of Zielona Góra
{t.zawadzki,d.kujawa}@weit.uz.zgora.pl

Abstract. The aim of this paper is to present a 3D hybrid shape construction that benefits from discrete and continuous modeling approaches. The proposed technique addresses the problem of automated modeling of virtual scene components such as caves, buildings and clouds. The approach combines two independent methods well known in three-dimensional computer graphics: shape grammar and shape morphing. The modeled structures are characterized by geometrical complexity with inner graph structure more optimized than in classical CSG approach. In this paper, we mainly focus on the description of the algorithm.

Keywords: Algorithms, 3D, procedural modeling, shape grammar, computer graphics.

1   Introduction

Complex structures are necessary elements of visually convincing virtual scenes. Buildings [1], whole urban structures [2-3], terrains [4-5], clouds [6-11], plants [12] or caves [13-17] can be modeled with help of systems based on automated shape construction. The algorithms that enable full automation of the modeling process help to achieve large savings in the digital media production time and budget. We can observe a constant development of new methods, i.e. merging technology and dynamical systems [18-19]. The problem of automated shape modeling constitutes an important area of computer graphics activity and has drawn attention of digital media industry for several years. Digital movies have created constant demand for pleasing visual effects in three-dimensional graphics. Apart from pure entertainment interests, shape modeling has the practical use ranging from CAD engineering applications, through scientific visualization to advanced game programming and Virtual Environments. As far as we consider real-time simulations, it is very hard to easily satisfy the above mentioned demand. Moreover, it is almost not possible to do so without the use of systems based on procedural modeling of geometry. We propose to extend the set of currently available procedural methods with a hybrid of shape grammar and morphing, offering better performance and versatility. The idea of shape metamorphosis commonly known as morphing forms a wide and important area of computer graphics. Generally, morphing is usually defined as continuous (i.e. over time) and smooth process of transformation of one shape into another. So-called key shapes (by analogy

G. Bebis et al. (Eds.): ISVC 2012, Part II, LNCS 7432, pp. 314–323, 2012.

to key-framed animation) may have different topologies and the smoothness of trans-formation does not have to be the case of homeomorphism [23]. Typically the problem of morphing procedure is treated twofold, namely the problem of determination of 'features' of key shapes to be morphed and the interpolation of shapes according to 'trajectories'. The second one results in the intermediate shapes that possess some topological characteristics of both key shapes (the 'beginning' and the 'final' one). The morphic intermediate shapes are used in this paper as alternative to CSG Boolean productions (e.g. the 'beginning' and 'final' shape). There is a number of methods for metamorphosis for 2D shapes represented mostly by polygons [24] but also images [25]. The problem has also been investigated in 3D domain and according to Lazarus [26] they span the methods based on polygonal mesh representation [27-28] and the methods utilizing voxel representations [29]. In the hybrid model presented in this paper, we use the polygonal mesh representation which is based on functional description.

2   Shape Grammar Development – Morphic Rules

Shape grammars was developed by Stiny and Gips [20-22] for supporting the design process. In this paper, we present an innovative method based on hybrid algorithm which is a combination of two independent techniques: continuous morphing and discrete shape grammars. The purpose of this work is to bring morphing into a discrete function and abandon the intermediate states (Fig. 1).

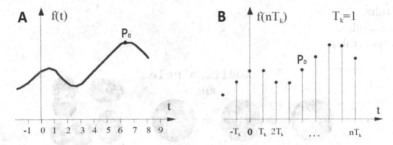

Fig. 1. A – function describing morphing, B – function describing shape grammars, P_0 – point in R^3, t – time

Definition 1. Shape grammars are defined as a four:

$$SG = < V_T, V_M, R, I >$$ (1)

where:
V_T - is a set of terminal shapes.
V_M - is a set of non-terminal shapes.
R - is a finite set of ordered pairs (u, v) (shape rules).
I - is an initial shape (in our case, $I_0 = \{\text{point in } R^3\}$).

In our case, the modeling process start in the initial rule (R_0) by summing the initial shape (I_0 - point P) with a non-terminal shape (V_M - sphere, torus, cuboid, cone, cylinder etc.) randomly selected from a database, which selection is different for caves, architecture and clouds (1). The database of terminal shapes (V_T) is larger than non-terminal shapes, because it is dynamically created during the modeling process. After that, we define the second rule (n-th rule) which can based on terminal and non-terminal shapes as an operation on two shapes (Boolean – sum, difference, union or morphing – new extend) (2). New rules can be applied in several ways based on labeling and connected by joining points existing on the grid (3). Shapes are described by functions or their assemblies, where: $f(P)$, $g(P)$, $k(P)$ are example functions (4).

(1) $(P(x,y,z)) \Rightarrow (P(x,y,z) \cup (random.shape))$

(2) $(random.shape) \Rightarrow (random.shape \cap random.shape)$

(3) $(random.shape)\{x,y,z,-x,-y,-z\}$

(4) $(random.shape) = [f(P) + g(P)] - k(P)$

Classic rules are based on addition and subtraction but the morphic relies only on subtraction rules (Fig. 2). This work is focused on procedural modeling of shapes by extending scene graph with new grammatical rules. The total sum of all rules is given by:

$$R_T = R + R_M \tag{2}$$

where:
R_T – all rules,
R – classic rules,
R_M – morphic rules.

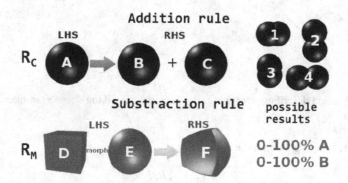

Fig. 2. An example of addition (R - LHS, Left-Hand Side shape) and subtraction rules (RHS, Right-Hand Side shape). 1,2,3,4 and 0-100% A,B – the possibility of shape combination.

Computation (derivation) of shape grammars based on processing the rules from the initial shape to the last rule – example (Fig. 3):

1) $I_0 \Rightarrow R_1 \Rightarrow R_4$ (main rule),
2) $R_2 \cup R_3 \Rightarrow R_4$ (addition rule),
3) $D(morph)E \Rightarrow F$ (morphic rule),
4) $A \cup B \Rightarrow C$ (classic rule)

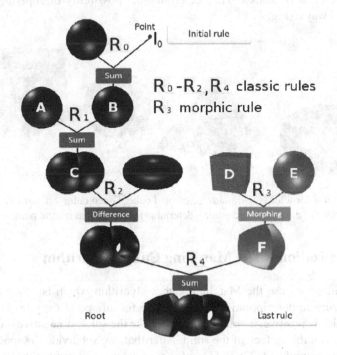

Fig. 3. Example of realization classic and morphic rules (one variant)

The algorithm define shapes and performs operations only on functions or their assemblies based on definitions from implicit objects [30]. We extended Boolean operations by morphing which is defined using linear interpolation:

$$f * g = f(P) * (1 - m) + g(P) * m \qquad (3)$$

where m is morphing parameter.

Our algorithm controls the labeling (L_B) which is responsible for combining two shapes using appropriately selected bonds. Two bonds make a label when the points P overlapping and normal vectors N are opposite:

$$L_B = (B_1, B_2) \Leftrightarrow (B_1.P = B_2.P \wedge B_1.N = -B_2.N) \qquad (4)$$

where:
L_B – labels,
B_1, B_2 – bonds,

P – point in $\mathbf{R}^3$,
N – normal vectors.

In addition the algorithm provides that the bond does not have a direction if $N = (0, 0, 0)$ and then it can create a label with any other bonds. Joint points are fixed and predefined for each class of shapes. They determine the possibility of applying the same rule in several ways. (Fig. 4).

Fig. 4. Joint points for a torus, pyramid, sphere and cube. Green color - determine joint points on the side surface or close to it, red color – determine joint points in central point of the solid.

3   Visualization – The Marching Cubes Algorithm

For visualization, we use the Marching Cubes algorithm which is based on *density function* for surface description. We check value for any point (x, y, z) in 3D space and if this value is positive, a point is located inside the solid, if negative – outside but when is 0 then on the surface of the solid. After that, we subdivide the space by sample blocks, for example 32x32x32 (L_D – *Level of detail*), and inside these cubes (each have 8 vertices), we construct polygons (triangles) that represent the solid surface (Fig. 5).

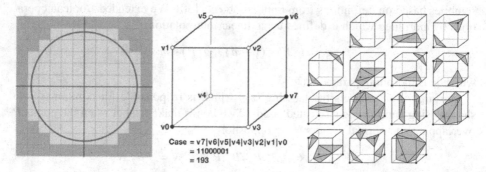

Fig. 5. Left - cross-section distribution of cubes intersected by a solid (green color depicts the selected cubes). Middle - a single voxel with known density values at its eight corners. Right - the 14 fundamental cases in *Marching Cubes*.

4   Results

The results generated by the hybrid algorithm are random but in order to restrict them to a specific class of structures, we have to make some minor changes which affect the selection of appropriate shapes and rules for modeling mode, the combining method, and the range of used operations (Fig. 8).

For all modes, we can adjust :

- the level of detail (LOD), the model grid (L_D, 1 - 32) based on sample density. Fig. 6 shows an architecture example.

For *caves*, the number of iterations is equal to the sum of all rules: classic and morphic. In each rule, the scene base shape is summed with the shape formed by morphing of a sphere and a cube. It is possible to select the direction of the labeling under which new rules will be added (+X/-X, +Y/-Y, +Z/-Z). Labeling in these rules has a random shift of the joints points in order to avoid symmetry. We can control existing parameters in the cave modeling mode:

- amount of classic rules (R),
- labeling (X/-X, Y/-Y, Z/-Z, L_B, 3 direction)
- random shift of the labels (S_{LB}, 0 - 100),
- amount of morphic rules (R_M)
- morphing parameter (M_P, 0..100%)

Clouds modeling is generally based on the summing of many spheres in successive rules in accordance with specified labels. The labels are the directions on the selected coordinate axis. The number of rules is calculated as the maximum of existing value labels. In each rule, all labels are reduced by 1 and from these labels, which have a positive value, new connections are created. In the *n-th* rule, *n* spheres are added to the scene. The radius of the sphere in the *n-th* iteration is interpolated between the initial and the final radius. For shapes connecting points, some offset is introduced by additional randomness. After performing all operations, we can morph our structure with selected shape from database (cylinder, torus etc.). We can model clouds based on the following parameters:

- begin and end sphere radius (20 - 200)
- labeling (X, Y, Z, -X, -Y, -Z, L_B, 6 direction) (amount of classic rules)
- random shift of the labels (R_{SL}, 0 - 100),
- cloud deformation with basic shapes (morphing parameter, M_P, 0..100%).

For *architecture*, we assume that the buildings are multistory. We describe the first and last floor, while intermediate results are generated based on grammatical rules. For each level, we can determine: its height, width and convexity. A floor is a cube that can be glued (sum operation) to additional 4 blocks on the sides when it is raised or cut (difference operation) when it is a concave. After the last rule, the roof is added. It is possible to morph the roof shape from sharp (Pyramid) to a spherical

(Sphere). The construction of buildings does not offer the possibility of labeling control. All labels are selected automatically according to certain principles of building: labels -X.X, -Z/Z are selected from sticking/cutting sides, but label +Y is chosen for building up the next floor and adding a roof. We can adjust architecture modeling by:

- number of floors (1..20) (amount of classic rules, R),
- lower floor: height, width, convexity.
- higher floor: height, width, convexity.
- roof: from spherical to sharp (morphing parameter, M_P, 0..100%).

Fig. 6. Level of detail (L_D), from left L_D=2, L_D=5, L_D=10 of the same object (architecture)

One of very important aspects in the analysis of results on example below (for L_D=5) is estimating execution time based on comparing mesh detail expressed in triangles and model density vs time (Fig. 7):

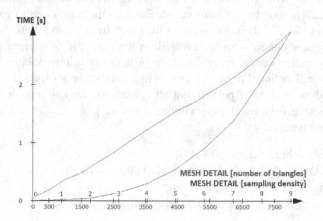

Fig. 7. Time vs mesh detail (L_D) represented by sampling density and number of triangles. All simulations were performed on nVidia GeForce GTX 460M GPU, i7-2630QM CPU and 12 GB RAM.

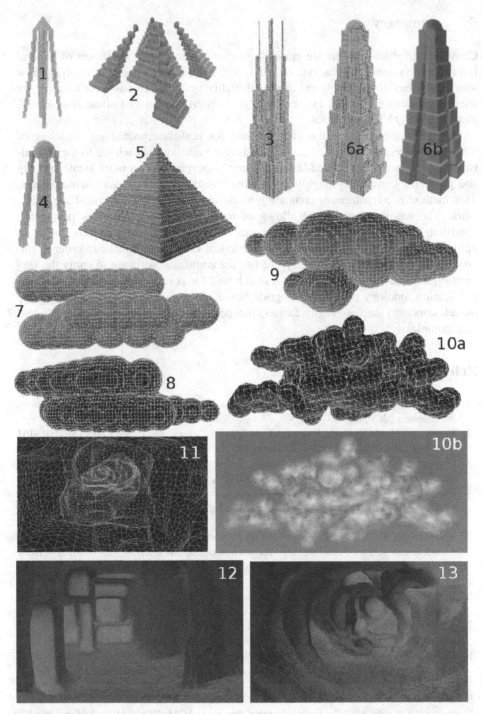

Fig. 8. The results grid obtained for the architecture (1-6), clouds (7-10), caves (11-13). Some example was illustrated by renders: 6b, 10b, 12 (cubical cave), 13 (spherical cave).

5   Summary

Computer graphics systems are currently one of the most important part of general information systems. Nowadays, we can observe a noticeable trend of applying new methods of modeling 3D objects in virtual reality systems. This is forced mainly by the market demand created in areas of digital entertainment and simulation for 3D gaming and 3D VFX industries.

Our paper presents an innovative method for real-time procedural modeling of three dimensional geometry of caves, clouds and buildings. By adding to the formalism of shape grammars, the additional feature – morphing, we obtain greater variety and geometric complexity (highly influencing visual realism) of synthesized objects. This method is advantageous comparing to classical methods based on shape grammars. The morphing parameter allows us to establish the continuous percentage contribution of two input shapes to produce a single output object. This results in optimization of modeling process and reduction of redundant objects. In order to improve realism of modeled geometry, we use the additional software to apply the textures on the final object. Our further research will focus on development analysis and verification topology of output 3D grids based on surface area and volume of the model, concavity and convexity factors, independent model objects, histograms and time complexity.

References

1. Wonka, P., Wimmer, M., Sillion, F., Ribarsky, W.: Instant architecture. ACM Transactions on Graphics 22(3), 669–677 (2003)
2. Parish, Y.I.H., Muller, P.: Procedural modeling of cities. In: Proceedings (SIGGRAPH 2001), pp. 301–308. ACM Press, E. Fiume (2001)
3. Greuter, S., Parker, J., Stewart, N., Leach, G.: Real-time procedural generation of pseudo infinite cities. In: Proceedings (GRAPHITE 2003), pp. 87–95. ACM Press (2003)
4. Peytavie, A., Galin, E., Grosjean, J., Merrilou, S.: Arches: a Framework for Modelling Complex Terrains, Computer Graphics Forum. In: Proceedings EUROGRAPHICS 2009, vol. 28(2), pp. 457–467 (2009)
5. Warszawski, K., Nikiel, S.: A proposition of particle system-based technique for automated terrain surface modeling. In: Proceedings of the 5th International North American Conference on Intelligent Games and Simulation (Game-On-NA 2009), pp. 17–19 (2009) ISBN 978-9077381-49-6
6. Bouthors, A., Neyret, F.: Modelling Clouds Shape. In: Proceedings EUROGRAPHICS (2004)
7. Schpok, J., Simons, J., Ebert, D.S., Hansen, C.: A real-time cloud modeling, rendering, and animation system. In: Symposium on Computer Animation 2003, pp. 160–166 (2003)
8. Dobashi, Y., Kaneda, K., Yamashita, H., Okita, T., Nishita, T.: A simple, efficient method for realistic animation of clouds. In: Proceedings of ACM SIGGRAPH 2000, pp. 19–28 (2000)
9. Ebert, D.S.: Volumetric procedural implicit functions: A cloud is born. In: Whitted, T. (ed.) SIGGRAPH 1997 Technical Sketches Program, ACM SIGGRAPH. Addison Wesley (1997) ISBN 0-89791-896-7

10. Elinas, P., Sturzlinger, W.: Real-time rendering of 3D clouds. Journal of Graphics Tools 5(4), 33–45 (2000)
11. Nishita, T., Nakamae, E., Dobashi, Y.: Display of clouds taking into account multiple aniso-tropic scattering and sky light. In: Rushmeier, H. (ed.) SIGGRAPH 1996 Conference Proceedings, ACM SIGGRAPH, pp. 379–386. Addison Wesley (1996)
12. Prusinkiewicz, P., Lindenmayer, A.: The Algorithmic Beauty of Plants, pp. 101–107. Springer (1991) ISBN 978-0387972978
13. Am Ende, B.A.: 3D Mapping of Underwater Caves. IEEE Computer Graphics Applications 21(2), 14–20 (2001)
14. Boggus, M., Crawfis, R.: Procedural Creation of 3D Solution Cave Models. In: Proceedings of the 20th IASTED International Conference on Modelling and Simulation, pp. 180–186 (2009)
15. Boggus, M., Crawfis, R.: Explicit Generation of 3D Models of Solution Caves for Virtual Environments. In: Proceedings of the 2009 International Conference on Computer Graphics and Virtual Reality, pp. 85–90 (2009)
16. Schuchardt, P., Bowman, D.A.: The Benefits of Immersion for Spatial Understanding of Complex Underground Cave Systems. In: Proceedings of the 2007 ACM Symposium on Virtual Reality Software and Technology (VRST 2007), pp. 121–124 (2007)
17. Johnson, L., Yannakakis, G.N., Togelius, J.: Cellular Automata for Real-time Generation of Infinite Cave Levels. In: Proceedings of the 2010 Workshop on Procedural Content Generation in Games (PC Games 2010), pp. 1–4 (2010)
18. Clempner, J.B., Poznyak, A.S.: Convergence method, properties and computational complexity for Lyapunov games. The International Journal of Applied Mathematics and Computer Science 21(2), 349–361 (2011)
19. Di Trapani, L.J., Inanc, T.: NTGsim: A graphical user interface and a 3D simulator for nonlinear trajectory generation methodology. The International Journal of Applied Mathematics and Computer 20(2), 305–316 (2010)
20. Stiny, G., Gips, J.: Shape grammars and the generative specification of painting and sculpture. In: Information Processing, vol. 71, pp. 1460–1465. North-Holland Publishing Company (1972)
21. Stiny, G.: Pictorial and Formal Aspects of Shape and Shape Grammars. Birkhauser Verlag, Basel (1975)
22. Stiny, G.: Introduction to shape and shape grammars. Environment Planning B 7(3), 343–361 (1980)
23. Martyn, T.: A new approach to morphing 2D affine IFS fractals. Computers & Graphics 28, 249–272 (2004)
24. Alexa, M., Cohen-Or, D., Levin, D.: As rigid as possible polygon morphing. Computers Graphics (SIGGRAPH 2000) 34, 157–164 (2000)
25. Wolberg, G.: Image morphing: a survey. The Visual Computer 14(8-9), 360–372 (1998)
26. Lazarus, F., Verrous, A.: Three-dimensional metamorphosis: a survey. The Visual Computer 14(8-9), 373–389 (1998)
27. Kent, J.R., Carlson, W.E., Parent, R.E.: Shape transformation for polyhedral objects. Computer Graphics (SIGGRAPH 1992) 26, 47–54 (1992)
28. Lee, A.W.F., Dobkin, D., Sweldens, W., Shroeder, P.: Multiresolution mesh morphing. Computer Graphics (SIGGRAPH 1999) 26, 43–46 (1999)
29. Turk, G., O'Brien, J.F.: Shape Transformation using variational implicit functions. Computer Graphics (SIGGRAPH 1999) 33, 335–342 (1999)
30. Velho, L., Gomes, J., Figueiredo, L.H.: Implicit Objects in Computer Graphics. Springer (2002) ISBN: 978-0387984247

Initialization of Model-Based Camera Tracking with Analysis-by-Synthesis

Martin Schumann, Sebastian Kowalczyk, and Stefan Müller

Institute of Computational Visualistics, University of Koblenz, Germany
{martin.schumann,skowalczyk,stefanm}@uni-koblenz.de

Abstract. In applications of augmented reality it is an essential task to retrieve the camera pose for correct overlay with virtual content. This can be realized by using a model-based camera tracking approach that fits a given model of the scene to the images captured by the camera. These systems have to be initialized properly for the pose estimation process of continuous tracking. We present a two-step concept for the global initialization of such model-based tracking systems. With a model database and known GPS coordinates as well as compass orientation, it is possible to determine which part of the scene is visible and to obtain a first rough pose. We also introduce a method to refine the initialization pose to overcome GPS inaccuracies. It has been successfully tested in an urban context.

1 Introduction

In applications of augmented reality the visual perception of the real world is enriched with virtual content. Potential application scenarios for this technique are maintenance tasks, teaching, touristic guides and augmented gaming on mobile systems. The central problem for rendering the virtual overlay correctly onto the image of the real world is to retrieve the position and orientation (pose) of the camera, i.e. the position and viewing direction of the user. The process of continuously estimating this camera pose is called camera tracking. Using a model-based approach the only prerequisite is a given model of the scene that is registered to the camera images by matching 2D image features like points or lines to correspondences in the 3D model data. For estimation of the camera pose from these correspondences there exists a collection of algorithms which can be divided in two groups. Linear methods like the Direct Linear Transform (DLT) [1] are known to calculate a global linear solution for the camera pose in a single step without need of previous initialization. This class of algorithms is fast but the precision is limited because they are very sensitive to false correspondences which influence the robustness of the pose result negatively.

Second, there are non-linear or indirect methods to estimate the camera pose. Most of them are based on classical iterative algorithms of optimization, like the Gauss-Newton or Levenberg-Marquardt-Method [2][3]. The distances of the measured correspondences are iteratively minimized by local linearization until the result converges into a stable minimum or a maximal number of iterations

G. Bebis et al. (Eds.): ISVC 2012, Part II, LNCS 7432, pp. 324–333, 2012.

is reached. These methods are very precise and robust even on noisy correspondences. However, they need an initial input pose for starting the estimation. This pose should be preferably close to the optimal pose. Otherwise the convergence rate may get very slow or the method converges against a local minimum instead of finding the globally optimal pose.

The initialization of the system could be given by the user manually selecting correspondences, adjusting the first pose overlay or localizing the position in a map. Another way is the placing of a marker in the scene to set a known initialization reference frame. However, user input may be erroneous and preliminary preparation of the environment is not always possible or an undesired workload. Therefore an automatic initialization is more appropriate. Several approaches are fusing GPS, inertial sensors and vision-based techniques into hybrid tracking systems. Initialization may be realized by inertial-GPS and planar feature search [4], using key frames in the tracking sequence [5] or by GPS error prediction and sampling as [6] do in an approach similar to ours. Further, there are methods developed to correct mobile GPS output with a database of geo-referenced images [7].

In times of high pervasion with GPS capable smartphones and fast growing model databases available online, possibilities show up to use these information for initialization of urban tracking scenarios. We propose a two-step approach with rough pose initialization from GPS, electronic compass and a model database, combined with a pose refinement process based on analysis-by-synthesis, i.e. comparing rendered images from sampled poses to the camera reference image.

2   Model-Based Initialization

In the first step of the initialization we are given a coordinate by the Global Positioning System (GPS) that tells us the location of the camera and the orientation from an electronic compass which represents the viewing direction of the camera in degrees of [0,359] based on north direction. We want to find the coordinate transform from world GPS position and orientation into the local coordinates of the tracking system. Further, only those models have to be loaded which are currently visible. What we need is a database representation of the world that is indexed to query the models seen by the user.

First, an adequate representation of the GPS position and subsequent coordinate transform has to be defined. For mapping between the model database and the real world we use the Universal Transverse Mercator (UTM) geographic coordinate system which is acknowledged internationally for navigation. The advantage of UTM is a subdivision of the earth surface in separate zones each with orthogonal coordinate system which is desirable for conversion when working with a virtual Cartesian coordinate system in the tracking context. One unit within each zone correlates to one meter in reality. Thus, the conversion from GPS coordinates to the local tracking coordinate system simply consists of a translation and a rotation, where the tracking coordinate system is equally scaled in meters. Thinking of the world as a collection of models, we consider a tiled organization structure. Models related spatially are modeled on one common ground tile, e.g. a city or a street. Using tiles instead of individual models is advantageous because the alignment between all models on one tile is held consistent due to one common origin. Further, handling the topography of the ground and the modeling process itself are eased.

At the start of the initialization position and orientation of the camera are known from GPS and compass. Accompanying information that has to be measured and stored in advance is the GPS coordinates of the tile origin and its orientation angle R_T based on north direction. At initialization this tile origin also becomes the origin in the local tracking coordinate system. The z-axis of the tracking coordinate system is defined as the vector showing north and the tile is aligned accordingly. The model is rotated and translated from the tile origin with the parameters R_M and T_M known from modeling. A difference vector is spanned from the GPS coordinates of the tile origin to the GPS position of the real camera. This becomes the translation vector T_C for the virtual camera. The north angle from the compass can be set as the rotation R_C of the virtual camera (Fig.1).

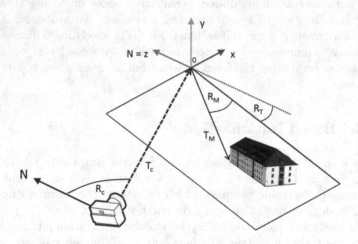

Fig. 1. Transformations of camera, tile and model

Now the planar position and orientation of the model with respect to the camera is known. The height of the camera has not been considered yet. We could retrieve a height value from GPS as well, but again with limited precision. Ideally, for each tile the terrain is represented by a mesh together with the models, e.g. from aerial survey. If terrain information is available in the database, the height of the camera can be retrieved as follows. At the known position of the camera a perpendicular ray is shot from a fixed height down to the terrain. The difference between the length of the ray to the bottom of the bounding box and the intersection with the geometry returns the height value

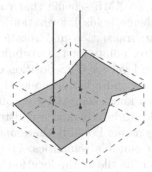 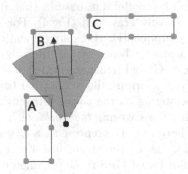

Fig. 2. Retrieving camera height by terrain-ray intersection

Fig. 3. Checking control points for visibility test

of the terrain at that point (Fig.2). To this result the body height of a standard user (e.g. 1.8 meters) has to be added to calculate the viewing height.

After knowing the conversion from world to local tracking coordinates, the next step is to determine visible models. For each model control points are defined equally distributed along the borders of the corresponding building and their GPS coordinates are measured manually. Ideal distances between the control points range from 10 to 20 meters. Thus, at runtime those models can be filtered which are visible under consideration of the current camera position and orientation. This dynamic loading enables efficient handling of arbitrarily large environments. For large databases it would also be possible to calculate these control points automatically from the midpoint GPS position of each building and its dimensions.

Figure 3 shows the scheme for checking model control points against position and orientation of the camera. Around the camera a small and a large radius are set by the minimum and maximum viewing distance. We considered a minimum distance of 0 meters and a maximum of 100 meters. The angle of the field of vision defines the search area where the viewing direction is the bisectors of the angle. This parameter is taken from the camera. Together they characterize a viewing frustum for the camera. A control point is regarded being valid when it lies within the borders of that frustum, i.e. the Euclidean distance of the control point to the camera is in between the previously defined values of minimum and maximum viewing range and the angle between the vector from the camera to the control point and the viewing direction of the camera is smaller than half of the field of vision. In the example there exist three valid control points from two different models (A and B) which have to be loaded. After all control points of the models on the tile have been checked against the position and orientation of the camera, the source paths of the models with at least one valid control point are sent to the model loader of the rendering system. While running the tracking process, this method can be used for dynamically loading models, e.g. when walking along a street.

The model database in this approach is built on an XML scheme that can be easily traversed (Fig.4). For each model the scheme holds information of its unique ID, the source location path for loading, translation and rotation parameter based on the tile origin and a set of control points for the visibility test. Global information that has to be present at tile level are GPS position of the tile origin, tile orientation based on north direction and a list of model IDs belonging to the tile. Up to now, our approach is working exemplarily on one tile. For covering *the world* with tiles it is conceivable storing them in a register where each tile represents a city and is tagged by its name. Possibly there may also be a subdivision like street name or any other suitable semantics. From map-based GPS these parameters can be retrieved so the tile at the location of the user can be queried and the system achieves access to all relevant models through the tile.

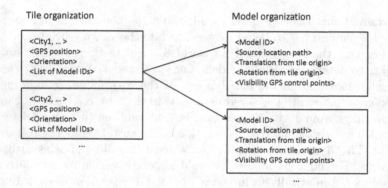

Fig. 4. Data scheme for storing tile and model information

Determining the position by GPS and compass gives us a rough pose estimate which may have an offset of several meters. This is due to common lack of precision in the civil GPS system and the signal might be noisy due to the urban environment. Besides, only the viewing direction around the y-axis has been taken into account, neglecting possible error of the compass. Further, the user is likely to tilt the head up and down. These issues can be solved with the analysis-by-synthesis part of the initialization.

3   Pose Refinement

The first rough camera pose may be too inaccurate for starting a non-linear pose estimation. Thus an additional refinement step seems to be meaningful. This is realized with an approach based on analysis-by-synthesis, originally stemming from speech signal processing where the parameters of a real signal are approximated by varying parameters of a synthetic signal and comparing both. The concept can easily be transferred to the context of camera pose tracking as in

[8] or [9]. The parameters we are looking for are translation and rotation of the real camera. What we can do, is rendering many images of the scene with a virtual camera, changing the known pose parameters of that camera for every new image. These rendered images are then compared to the camera reference image and the one with the highest similarity is selected giving us the parameters wanted.

To generate these poses for rendering, a series of translations and rotations has to be applied to the virtual camera. Based on the first rough pose taken as central point, the sampling is done on a regular grid with 24 meters range, i.e. 12 meters in each direction, which complies to the average standard accuracy of the GPS signal. Within this grid the virtual camera is translated to left/right and front/back in steps of two meters. The camera height is retrieved for each step as described previously. At all of these new poses, four angles in steps of 5 degrees in range of [0,15] are applied for camera tilt to consider possible nod of the users head, which is very likely standing in front of a building. Camera panning can also be modeled by varying the angle around the y-axis if the viewing direction given by the compass data is too imprecise. In our approach we compensated angular errors by applying a Gaussian filter on the camera image. Further, it is assumed that the camera is fixed around the axis of vision. These sampled poses are stored in a list and for each pose a synthetic image is rendered.

For determining the similarity of the rendered images and the real camera image the prominent structures in the images prove useful. In urban context many strong edges of the buildings occur which can be found by edge detectors returning the gradient of the image. Thus, we apply the Sobel filter implementation of the OpenCV library[1] with kernel 3x3 on the camera frame and the rendered image. Calculating the magnitude of the gradient for each pixel we obtain a gradient intensity image. The intensity values I_c and I_r of both images can now be directly compared. The higher both intensities are, the higher the resulting normalized edge intensity value I. To prevent edge pixels from being compared when they do not belong to the same edge, additionally the directions D_c and D_r of the gradient for each pixel in the camera image and the rendering are calculated. A small difference in the gradient direction of both pixels leads to a high contribution of the product $I * D$ by edge intensity and edge direction to the summed image similarity S in the range of [0,1]. If the difference is larger than 8 degrees, D is set to 0 and the pixels compared will not contribute to the image similarity value. The following steps are executed for all pixels of each image pair:

$$I = \frac{I_c}{255.0} * \frac{I_r}{255.0} \qquad (1)$$

$$D = 1.0 - \frac{|D_c - D_r|}{8.0} \qquad (2)$$

$$S+ = I * D \qquad (3)$$

The similarity of two images grows the more pixels with similar gradient magnitude and similar direction can be found. To gain tolerance against small movements of the camera or inaccurate compass data, the reference camera image

---

[1] http://opencv.willowgarage.com/

is blurred by a Gaussian operator before gradient calculation. The image edges
are broadened spatially which makes the pixel comparison less sensitive against
slight pixel offsets. After comparing all renderings with the camera image, the
list of poses is sorted by the corresponding value for S and the pose best ranked
is proposed for initialization. Figure 5 shows the gradient output for a rendered
image and the blurred camera image. The rendered image on the left has been
chosen by the algorithm as best match to the camera image. Figure 6 shows
the difference between the rough pose based on GPS and compass only and the
refinement result.

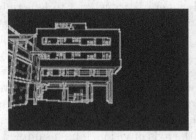 

Fig. 5. Sobel images to compare. The rendered image (left) was returned as best fit to
the blurred camera image (right).

Fig. 6. Left: Rough initialization. Middle: Refinement. Right: Camera image.

4   Results

We tested our initialization approach on a computer with AMD Phenom II
X4 965 (4 x 3,40 GHz) processor and ATI Radeon HD 5850 1 GB graphics
card. The resolution of the pictures taken was 968x648 pixels. To speed up the
image comparison process they were scaled down to 242x162 pixels. For every
image used in the initialization, the GPS coordinate and viewing direction of
the camera was known as ground truth. After determining the first rough pose,
676 new poses were generated for rendering views and comparing them with the
real reference camera image. For the rendering process, the intrinsic parameters
of the real camera were known. During the pose refinement process averagely
up to 200 images could be compared per second. With a maximum of four
models displayed the rate decreased down to 150 compared images per second.

Fig. 7. Results of the initialization after pose refinement. Virtual overlay (left) and camera image (right). Bottom row: Insufficient sampling and an improper model lead to a shift in the pose selected.

The initial camera pose is retrieved within 4-5 seconds, which is not yet fully optimized. Computation time could be significantly reduced by implementing the comparison with a shader-based image filter. Fast GPU-based Sobel and Canny concepts are shown by [10] and [11] in the context of a particle filter tracking framework. If the initialization process runs in realtime, it can even be used for reinitialization in case of tracking loss.

Figure 7 shows some of the initialization results at several locations in our testing scenario. The real images from the camera are shown on the right and the initial pose result as an overlay is shown on the left. We tested the initialization at 21 locations with several buildings and in over 75% of the passes no manual correction of the pose selected by the refinement step was necessary. On the average, the pose was selected in a range of 2-6 meters from the rough estimate, which is quite high enough for successful initialization of a non-linear pose estimator. One has to recall, that the initial pose should be located near to the optimum, but does not need to be entirely exact. Smoothing the reference image previously to applying the Sobel operator also proved sufficient for compensating the deviation of the viewing direction. If stronger differences in the rotation angle occur, alternatively they could be avoided by also considering rotation of the virtual camera around the y-axis when sampling poses. However, this results in more rendered images and a higher computation time.

The degree of detail of the models used also has a major influence on the comparability of the images. The more significant details like windows or doors are modeled the higher the precision of the pose refinement. When the sampling is not dense enough, improper or missing elements lead to a shift in the pose selected, as can be seen in figure 7 bottom row. It should be mentioned, that the pose refinement process is able to select a valid pose only, when the GPS error is not larger than the assumed pose sampling area of 24 meters. This could also be handled by using a larger area for sampling the poses but again reduces the performance.

5   Conclusion

High availability of GPS and compass data delivered by smartphones and online access to large model databases as *Google Earth* move us towards ubiquitous urban tracking scenarios. Regarding this background, it was shown that it is possible to initialize a model-based camera pose tracking system from GPS and compass together with a model database. The visible models are selected dynamically by annotated reference points which are checked against the position and viewing direction of the camera. While the rough pose from GPS may be inaccurate due to noisy signals, the proposed analysis-by-synthesis approach for pose refinement proved precise enough to successfully start continuous non-linear pose estimation.

Higher initialization performance could be reached by integrating an iterative sampling approach for generating the camera poses during refinement. A better search strategy may lower the amount of images to compare. Further, the pose refinement is based on CPU processing so far, but the image comparison could also be realized by a shader implementation. When the tracking is run on a mobile device, the time consuming task of initialization may also be performed by a server that receives the GPS and compass data together with the image and returns the initial pose calculated to the client.

Acknowledgements. This work was supported by grant no. MU 2783/3-1 of the German Research Foundation (DFG). We would also like to thank the image recognition group of Koblenz University for the 3D campus models.

References

1. Hartley, R., Zisserman, A.: Multiple View Geometry in Computer Vision, 2nd edn. Cambridge University Press (2003)
2. Levenberg, K.: A Method for the Solution of Certain Problems in Least Squares. The Quarterly of Applied Mathematics 2 (1944)
3. Marquardt, D.W.: An Algorithm for Least-Square Estimation of Nonlinear Parameters. Journal of the Society for Industrial and Applied Mathematics 11 (1963)
4. Fong, W.T., et al.: Computer Vision Centric Hybrid Tracking for Augmented Reality in Outdoor Urban Environments. In: Proceedings of the 8th International Conference on Virtual Reality Continuum and its Applications in Industry, VR-CAI 2009, pp. 185–190 (2009)
5. Reitmayr, G., Drummond, T.: Going Out: Robust Model-Based Tracking for Outdoor Augmented Reality. In: Proceedings of the 5th IEEE and ACM International Symposium on Mixed and Augmented Reality, ISMAR 2006, pp. 109–118 (2006)
6. Reitmayr, G., Drummond, T.: Initialisation for Visual Tracking in Urban Environments. In: Proceedings of the 2007 6th IEEE and ACM International Symposium on Mixed and Augmented Reality, ISMAR 2007, pp. 1–9 (2007)
7. Marimon, D., et al.: Enhancing Global Positioning by Image Recognition (2011)
8. Wuest, H., Wientapper, F., Stricker, D.: Adaptable Model-Based Tracking Using Analysis-by-Synthesis Techniques. In: Kropatsch, W.G., Kampel, M., Hanbury, A. (eds.) CAIP 2007. LNCS, vol. 4673, pp. 20–27. Springer, Heidelberg (2007)
9. Schumann, M., et al.: Analysis by Synthesis Techniques for Markerless Tracking. In: 6th Workshop on Virtual and Augmented Reality, GI Workgroup VR/AR (2009)
10. Klein, G., Murray, D.: Full-3D Edge Tracking with a Particle Filter. In: Proc. British Machine Vision Conference (BMVC 2006), vol. 3, pp. 1119–1128 (2006)
11. Brown, J.A., Capson, D.W.: A Framework for 3D Model-Based Visual Tracking Using a GPU-Accelerated Particle Filter. IEEE Transactions on Visualization and Computer Graphics 18, 68–80 (2012)

Real-Time Rendering of Teeth
with No Preprocessing

Christian Thode Larsen[1], Jeppe Revall Frisvad[1],
Peter Dahl Ejby Jensen[2], and Jakob Andreas Bærentzen[1]

[1] Technical University of Denmark
[2] 3Shape A/S, Holmens Kanal 7, 1060 Kbh. K, Denmark

Abstract. We present a technique for real-time rendering of teeth with
no need for computational or artistic preprocessing. Teeth constitute a
translucent material consisting of several layers; a highly scattering material (dentine) beneath a semitransparent layer (enamel) with a transparent coating (saliva). In this study we examine how light interacts with
this multilayered structure. In the past, rendering of teeth has mostly
been done using image-based texturing or volumetric scans. We work
with surface scans and have therefore developed a simple way of estimating layer thicknesses. We use scattering properties based on measurements reported in the optics literature, and we compare rendered
results qualitatively to images of ceramic teeth created by denturists.

1   Introduction

It is possible with existing scanning and rendering techniques to create highly
realistic digital versions of real actors. These digital versions can blend in naturally with real characters in feature films [1]. The main problem with this type of
digital (face) acquisition is the cost in artistic preprocessing. In the cited work,
this preprocessing takes up three months of an artist's time, just to refine rendering parameters and do compositing. Once acquired, the digital actor can be
rendered in photorealistic movie production quality, or, as an alternative, rendered in real-time at a lower but still relatively high level of realism. However,
when we impose a real-time constraint, the eye and mouth regions especially
become less realistic [2]. Since mouth and eyes are areas of the face that in particular draw attention, realistic real-time rendering of teeth has been identified
as an important avenue for future work [2].

In this paper, we aim at realistic real-time rendering of teeth that does not require artistic preprocessing, image-based texturing, or precise knowledge about
the internal structure of teeth. We constrain our rendering technique in this way
because it is intended for use in the dental industry. New scanners have been
developed that enable fast in-clinic dental impression scanning [3]. Such scanners capture the surface geometry of teeth and eliminate the need for difficult
and time consuming plaster casts. If rendering of the scanned teeth is realistic, the dentist can talk to customers about options with respect to crowns or
bridges shortly after scanning the oral cavity. Thus these new scanners generate

G. Bebis et al. (Eds.): ISVC 2012, Part II, LNCS 7432, pp. 334–345, 2012.

a demand for a rendering technique for teeth that adheres to the constraints mentioned above.

The appearance model we present for teeth maintains a strong coupling to existing models for subsurface scattering. In particular, we use a model for single scattering in multiple layers by Hanrahan and Krueger [4]. We combine this model with a technique for real-time skin rendering by Hable et al. [5], in order to estimate the multiple scattering component. Finally, we introduce adjustments to better capture the visual traits of the tooth material while keeping frame rates above 20 fps. By following this particular design, the model has maintained a level of generality that should make it useful for rendering any translucent material with a semi-transparent coating (porcelain is another example).

1.1 Related Work

Little work has been published on realistic rendering of teeth. The work most closely related to ours is the layered model presented by Shetty and Bailey [6] and by Shetty [7]. However, artistic preprocessing is needed in order to define the distribution of dentine inside the teeth. This means that the model is only suitable for library tooth geometry and therefore not well-suited for the application we have in mind.

Using a CT scanner, it is possible to capture a volumetric description of a set of teeth. Volume visualisation techniques are typically used to render such a scan [8,9]. This has the purpose of visualising the different layers of the teeth rather than reproducing their appearance. It is, however, possible to extract the surfaces of the different layers from a CT scan and use them for rendering. Kim and Park [10] and Rhienmora et al. [11] did this to develop systems to be used in training situations for dentists. Their systems include real-time visualisation of teeth, but the authors focus on haptic rendering rather than visual realism, and, apparently, they use the standard OpenGL pipeline. To improve visual quality and render rate, Yau and Hsu [12] use surfel models and surfel rendering for their dental training system. However, they do not include any translucency effects.

We have found no previous work where layer surfaces extracted from a CT scan have been used to enhance the rendered appearance of a tooth. Although we focus on surface scans, the rendering technique we present would also apply to a tooth surface extracted from a CT scan. And precise knowledge about the dentine surface beneath the enamel should lead to a higher degree of realism.

Wang et al. [13,14] have developed a dental training system based on surface scans of teeth. They face the same problem as we do, namely that they have to estimate the layered geometry inside a tooth. As far as we can tell, they do this manually. They use the tooth's layers, not to get translucency effects in the graphical rendering, but to get effects in the haptic rendering and to display different colours for different tissue types as the dentist drills into the tooth. Again, shading seems to be done using the standard OpenGL pipeline.

Another approach to rendering teeth is image-based texturing. The best example of the realism that can be achieved with this approach is probably the teeth of the digital actress Emily [1]. As discussed in the introduction, the artistic preprocessing needed for this approach makes it unsuitable for our application. An image-based texturing approach where artistic preprocessing is not needed has been presented by Pighin et al. [15]. Some manual work is still required to specify correspondences between images and a generic face model. It is unlikely that this type of manual work would be acceptable in an in-clinic dental impression scanning context.

Zhang et al. [16] describe a face reconstruction model where a generic skull (including teeth) is automatically morphed to fit a range scanned face. Their model might enable automatic image-based texturing of the teeth in the morphed skull. Unfortunately, the authors leave this investigation for future work.

Image-based texturing has the potential of achieving highly realistic results. The required image data is, however, not always available, as is the case in the application we are working with.

2   Theory

Tooth is a complex composite material that consists of several layers. The innermost layer which is optically significant is dentine. This is a highly scattering material. The dentine layer is beneath an enamel layer which is semi-transparent and possibly covered by a transparent saliva coating. Light scattering in these layers is affected by variations in optical properties, and by variations in the thickness of the enamel and dentine layers. In the following, we describe this subsurface scattering and our simple model for estimating layer thicknesses. Then we suggest a real-time rendering technique based on this theory.

2.1   Subsurface Scattering

The input needed for rendering a scattering material are the optical properties of the material, that is, the phase function, p, the absorption and scattering coefficients, σ_a and σ_s, and the index of refraction, η. These determine the appearance of the material. The Fresnel equations for reflection and the laws of reflection and refraction are used to handle scattering at the surface. These require the index of refraction as input. The remaining properties are used for computing subsurface scattering.

It is common to approximate subsurface scattering by a sum of radiance contributions due to 0th-order, 1st-order and n^{th}-order scattering. These three contributions are determined by finding partial or full solutions to the radiative transfer equation [17], and are commonly referred to as reduced intensity, single scattering, and multiple scattering, respectively. Analytical solutions for these terms only exist in special cases. To find analytical expressions for the reduced intensity and the single scattering terms, we assume that the layers are homogeneous and locally plane-parallel.

The reduced intensity passing through a layer of thickness t (the layer of refractive index η_2 in figure 1) is then [17,4]

$$L^{(0)} = F_t^{12} F_t^{23} e^{-\tau/\mu_0} L_i \; , \tag{1}$$

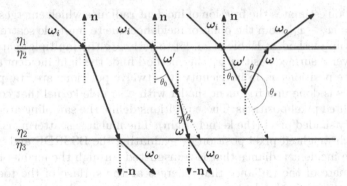

Fig. 1. Illustrating from left to right: reduced intensity, inward single scattering, and outward single scattering

where F_t^{12} and F_t^{23} are the Fresnel transmittances at the interfaces between the layers, $\mu_0 = -\boldsymbol{n} \cdot \boldsymbol{\omega}$ is the cosine of the angle between the inward surface normal and the direction of the refracted light, and, if we let $\sigma_t = \sigma_a + \sigma_s$ denote the extinction coefficient of the layer, $\tau = \sigma_t t$ is its optical thickness.

Single scattering is the contribution from light that scatters exactly once inside the medium before it reemerges. If the light emerges at the plane where it was incident, it is called outward scattered light, whereas it is called inward scattered light if it emerges on the opposing plane (figure 1). Both inward and outward single scattering have analytical solutions for constant, collimated light incident on a homogeneous, plane-parallel medium. Under these assumptions, single scattering only depends on the optical thickness of the layer τ and on the direction of the incident illumination $\boldsymbol{\omega}_i$. Inward scattering (+) and outward scattering (−) are then given by [17,4]:

$$L_+^{(1)} = F_t^{12} F_t^{23} \alpha\, p(\theta_s) \frac{\mu_0}{\mu_0 - \mu}(e^{-\tau/\mu_0} - e^{-\tau/\mu})L_i \ , \qquad (2)$$

$$L_-^{(1)} = F_t^{12} F_t^{21} \alpha\, p(\theta_s) \frac{\mu_0}{\mu_0 + \mu}(1 - e^{-\tau(1/\mu_0+1/\mu)})L_i \ , \qquad (3)$$

where $\alpha = \frac{\sigma_s}{\sigma_t}$ is the scattering albedo, which describes the probability of scattering versus absorption, and p is the previously mentioned phase function. The law of refraction is used to find the forward direction $\boldsymbol{\omega}'$ and the scattering direction $\boldsymbol{\omega}$ inside the medium. These directions lead to the cosine terms $\mu_0 = -\boldsymbol{n} \cdot \boldsymbol{\omega}'$ and $\mu = \boldsymbol{n} \cdot \boldsymbol{\omega}$ and the scattering angle $\theta_s = \cos^{-1}(\boldsymbol{\omega}' \cdot \boldsymbol{\omega})$. If light scatters back into the forward direction, the cosine terms μ and μ_0 are equal. This results in a division by zero in eq. 2. In this special case, the following formula should be used instead:

$$L_+^{(1)} = F_t^{12} F_t^{23} \alpha\, p(\theta_s) \frac{\tau}{\mu} e^{-\tau/\mu} L_i. \qquad (4)$$

Assuming a semi-infinite, homogenous medium with a planar surface, it is possible to derive a dipole approximation for the bidirection scattering-surface reflectance distribution function (BSSRDF) using diffusion theory [18]. The dipole approximation estimates the fraction of incident radiance which emerges at some distance $r = |\boldsymbol{x}_o - \boldsymbol{x}_i|$ from the point of incidence due to multiple scattering. We use the 12-tap method of Hable et al. [5] to estimate the multiple scattering in dentine. Given a surface position $\boldsymbol{x}_o$ this method finds the light incident at twelve other surface positions $\boldsymbol{x}_i$ in its vicinity (the twelve positions are "tapped"). In practice, this is done in a fragment shader with a sample kernel that consists of twelve weighted pixel positions. These positions define the sampling area around the currently shaded pixel (the kernel origin). The multiple scattering is sampled once for each weighted pixel position by evaluating the BSSRDF and multiply-ing with the incident radiance that is transmitted through the enamel layer. To get an estimate of the radiance that emerges at the surface of the tooth after multiple scattering in the dentine, the samples are accumulated and attenuated by transmission back through the enamel layer.

2.2   Optical Properties of Enamel and Dentine

A study of optics literature has provided us with measured intervals for the absorption and scattering coefficients in enamel and dentine. Table 1 has been constructed from values found in several references [19,20,21,22], with emphasis on the work of Fried et al. [21] as it, to our knowledge, presents the most detailed information. In addition, we use a refractive index of 1.33 to model a coating of saliva on the teeth.

Table 1. Approximate optical properties of enamel and dentine at three wavelengths. Scattering and absorption coefficients (σ_a and σ_s) are measured in cm^{-1}.

material	η	σ_a	$\sigma_{s,1053nm}$	$\sigma_{s,632nm}$	$\sigma_{s,543nm}$	g
enamel	1.63	<1	15 ± 5	60 ± 18	105 ± 30	0.96 ± 0.02
dentine	1.49	3.5 ± 0.5	260 ± 78	280 ± 84	280 ± 84	0.93 ± 0.02

The approximate mean thickness of the enamel layer is one millimetre (1 mm), while the average distance to the first scattering event ($1/\sigma_t$) is around 0.2 mm. Around five scattering events before emergence is too little for the multiple scattering approximation to be accurate. Especially considering that the asymmetry parameter g is close to one, which means that the scattering angle most often is very small. In other words, most light passes through the enamel with very little deviation from the forward direction. We have therefore decided to use only single scattering and reduced intensity for the enamel layer.

Similar analysis reveals that the dentine layer should be approximately one millimetre thick for the multiple scattering approximation to be useful. Since the dentine layer always seems to be a few millimetres thick, we have decided to use the multiple scattering approximation for this material.

Since scattering is highly asymmetric in both enamel and dentine (g is close to 1), we use the Henyey-Greenstein phase function [23] for $p(\theta_s)$ in eqs. 3–4.

2.3   Layer Geometry

The saliva layer is included in our model using the method of Jensen et al. [24]. This method is purely based on refractive indices and Fresnel's equations, and thus we need not model the thickness of the saliva layer.

To estimate enamel thickness t, we model it as varying linearly between top y_{max} and bottom y_{min} of the tooth bounding box with height $l_y = (y_{max} - y_{min})$:

$$t = t_{min} + (t_{max} - t_{min})(v_y - y_{min})/l_y \ , \tag{5}$$

where $v = (v_x, v_y, v_z)$ is a tooth vertex position and t_{min} and t_{max} are suitable minimum and maximum enamel thicknesses that are chosen according to the type of tooth (incisor or molar). It is also necessary to choose a reference direction for the interpolation since position of top and bottom depends on whether the tooth is placed in the upper or the lower jaw.

To create the dentine layer, we transform a copy of the tooth surface as follows ($v \mapsto v^{new}$). The geometry is first scaled in the xz-plane which is normal to the tooth height axis (the y-axis). It is then scaled and offset along the height axis to align the two meshes at the bottom of the bounding box. The scaling vector in the xz-plane is determined by

$$s_{xz} = ((l_x - 2t)/l_x, (l_z - 2t)/l_z) \ , \tag{6}$$

where $l = (l_x, l_y, l_z)$ contains the extend of the bounding box along the three coordinate axes. Tooth vertices are displaced in the xz-plane by

$$v_{xz}^{new} = s_{xz} * (v_{xz} - b_{xz}) + b_{xz} \ , \tag{7}$$

where b_{xz} is the mesh barycentre projected to the xz-plane and $*$ denotes element-wise multiplication. To displace the vertices along the y-axis, we use

$$v_y^{new} = y_{min} + (v_y - y_{min})(l_y - t)/l_y \ . \tag{8}$$

Note that this scaling method is not robust if the concavity of the tooth is high, as the dentine vertices then may become ill positioned.

2.4   Translucency Near Edges

The semi-transparency near the edges of teeth is a result of the dentine layer coming to an end. These near-edge regions pose a particular challenge to our method because the remaining enamel is thick enough for more than one scattering event to occur, but too thin for the diffusion approximation. The problem is illustrated in figure 2.

Our model handles this by sampling the enamel surface with the 12-tap kernel, and subsequently mapping each sample point to a corresponding dentine position. This results in an over-contribution of radiance from the dentine layer, which in turn eliminates the semi-transparency near the edges. To account for this, the dentine-contributed radiance should be corrected by an attenuation factor. We prefer to compute this factor using $a = p_t t^2$, where t is the thickness of the tooth in the view direction, and $p_t \in [0; 1]$ controls the level of attenuation.

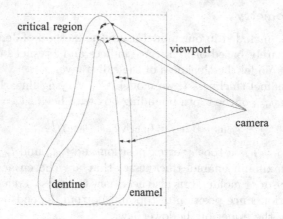

Fig. 2. The near-edge regions of the teeth are critical as they consist entirely of enamel in the view direction. Scattering from the dentine which reaches the edges of the tooth due to a few scattering events is difficult to evaluate in a fragment shader. We use an ad hoc approach to handle this critical region.

2.5   Tooth Roughness

While the surface of enamel appears to be relatively smooth, small variations in the surface micro structure may justify some simulation of roughness. A crude yet simple way to introduce roughness is to raise the Fresnel reflectance to an exponent [24]. This roughness value should be chosen entirely based on preference, and more aggressive values will effectively result in a higher amount of transmitted radiance through the layers. In our renderings, we use a Fresnel reflectance exponent of 1.3 for the enamel layer. This has the effect of making the teeth slightly less reflective and slightly whiter. We have not been able to find any physically based measurements of enamel roughness.

3   Implementation

The shader implementation combines the covered models and techniques using several vertex shader passes that render information about the geometry of the teeth to textures and a main fragment shader pass that shades the teeth. For each tooth, the solution requires as input the enamel surface geometry, its bounding box and barycentre, optical properties, attenuation factor and roughness exponent. We provide the essential components of the shader solution in the following pseudo-algorithm.

1. First pass: Generate texture maps.
   (a) Render enamel front positions and normals, enamel back positions and normals, and enamel thickness in the viewing direction.
   (b) Scale enamel vertex positions to obtain dentine front and back positions. Render dentine back normals, enamel to dentine position map, and dentine thickness in the viewing direction.

2. Second pass: Render tooth.
    (a) Compute multiple scattering in dentine:
        i. Determine pixel sample positions on the enamel surface given the 12-tap sample kernel.
        ii. Interpolate enamel thickness for each sample.
        iii. Estimate incident radiance for each sample by look-up along the normal direction into a prefiltered environment map [25,26].
        iv. Compute radiance transmitted to each mapped dentine sample position using eqs. 1 and 4.
        v. Evaluate the BSSRDF for each sample position [18].
        vi. Sum the contributions and compute the radiance that emerges from the enamel layer using eqs. 1 and 4.
        vii. Attenuate the total radiance contribution.
    (b) Compute outward single scattering:
        i. Look up incident radiance in the (normal) environment map using the reflection of the direction toward the camera.
        ii. Compute refracted directions into enamel and dentine, and use eq. 3 to compute the outward scattered radiance.
    (c) Compute back-lit transmission:
        i. Look up enamel and dentine thicknesses in textures.
        ii. Compute direct transmission using eq. 1.
    (d) Compute reflection in saliva and enamel.
    (e) Tone map total radiance to get colour in rgb.

Since the mouth in many cases has little light inside it, it often makes sense with respect to efficiency to omit back-lit transmission (2.c) and outward single scattering from dentine (2.b, dentine part). For optimization, the diffuse reflectance can be precomputed and stored as a single texture for a number of distances corresponding to the desired texture resolution. The diffuse reflectance texture needs only be generated once for a particular set of parameters. The texture map pass should be optimized using multiple render targets.

4 Results

Figure 3 is renderings of an incisor (~25k triangles) and a full set of teeth in the lower jaw (~200k triangles). Both were rendered at 20–30 frames per second on an NVIDIA GTX 580 without using the optimizations mentioned at the end of the previous section. Our method is not geometry bound, so the frame rate is similar whether we render a single tooth or a full set of teeth.

Figure 4 allows a qualitative comparison of a crown photographed from two different camera angles and a rendered molar tooth. The molar tooth does not have the same geometry as the crown. The crown is quite realistic (as it should be) with its variations in colour. This is a result of the artistic handwork associated with its production. The renderings only exhibit subtle colour variations which are mostly a result of reflections from the environment map. The attenuation of light scattered from the dentine layer produces a semi-transparent effect

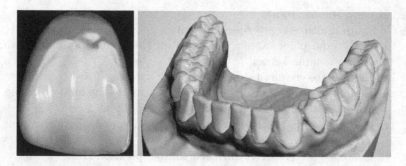

Fig. 3. An incisor (left) and a full set of teeth in the lower jaw (right). Both rendered at 20–30 frames per second.

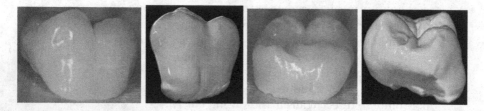

Fig. 4. A molar crown from two different angles compared to renderings of a molar model. The model geometry is not based on the crown.

near the edges, and a slight blueish hue (opalescence) appears due to backscattering in the enamel layer.

In figure 5, we compare two incisor crowns to an incisor model. The incisor comparison is not as convincing as the molar comparison. We believe the reason is a larger deviation between crown geometry and rendered geometry. The dent near the top of the rendered tooth is a damaged area of the tooth. It actually illustrates the softness in illumination that is a result of the 12-tap sampling. The difference in shade between the top and bottom of the rendered tooth is because of variation in the illumination incident from the environment map. Another primary reason for the qualitative differences between crown photos and rendered images is that our appearance model does not include age-related deterioration and discolouring effects (due to smoking, for example).

The techniques that we use for subsurface scattering provide direct control over the material appearance. We use measured optical properties (table 1) as a guide to set reasonable values for absorption and scattering in enamel and dentine. By varying the properties within the allowed margins, we are able to vary how light interacts with the two layers. As illustrated in figure 6, this leads to apparent differences in the appearance of the rendered tooth.

Using our simple interpolation scheme (section 2.3), the enamel thickness is controlled by the parameters t_{min} and t_{max}. As illustrated in figure 7, the enamel thickness directly affects the tooth appearance.

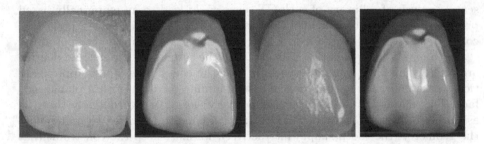

Fig. 5. Two incisor crowns compared to the incisor model rendered with two different sets of parameters

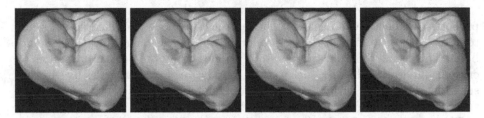

Fig. 6. Variations in colour due to variation in scattering properties. Enamel is changed in the second image, dentine is changed in the third, and both are changed in the fourth.

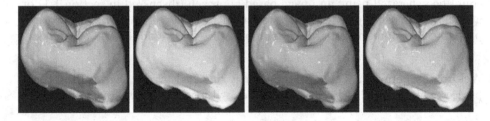

Fig. 7. The effect of enamel thickness. More variation between top and bottom thickness in the second image, more scattering in the third, both in the fourth.

5   Discussion

We have presented what we believe is the first real-time rendering technique for teeth that is based on the optical properties of teeth and does not require artistic or computational preprocessing. In this final section, we discuss options for improvement.

The enamel layer scatters light more than what can be captured by single scattering and too little to fulfil the assumption of being a highly scattering material as is required by the diffusion theory. A good real-time method for handling this type of material is left for future work.

The enamel thickness interpolation scheme could be improved. As mentioned in section 1.1, it is possible to acquire the surface of the dentine layer from a CT scan. Analysis of a library of such CT scans could lead to an empirically based procedure for estimating enamel thicknesses. We believe that such a procedure could greatly enhance the realism of the renderings that our method can produce.

Teeth have a fairly complex geometry, and the surfaces of enamel and dentine are often curved and dented, especially near the edges. The dipole approximation works under the assumption of a semi-infinite medium with a planar surface. This means that sampling must be (very) local, or alternatively that the assumption must be ignored. Since a dentist should be allowed to manipulate the geometry (the tooth shape) interactively, we must keep the constraint that no precomputation is allowed. Development of an interactive method for subsurface scattering that, without use of precomputation, more accurately handles geometry with sharper features is also left for future work.

It would be an advantage to have more detailed measurements of optical properties for dentine and enamel. The references we have identified do not seem to agree on how to measure these properties nor on the range that they should be in. This leaves room for further study on measuring the optical properties of enamel and dentine. Ideally, such a study would categorise properties by subject, such that it would be easier to analyse and render sets of teeth where parameters such as sex, race, and age are visually reflected.

The proposed model does not handle visual effects related to the natural wear and tear of teeth, such as caries or prolonged exposure to coffee, red wine or cigarettes. Such effects can be introduced through texture mapping. Textures could be generated using an image-based approach (discussed in section 1.1) or automatically using procedural or statistical methods. To apply such textures properly, a study of automatic UV mapping of teeth would also be of interest.

Acknowledgement. This work was partly financed by the Digital Prototypes project which is funded by the Danish Council for Technology and Innovation.

References

1. Alexander, O., Rogers, M., Lambeth, W., Chiang, J.Y., Ma, W.C., Wang, C.C., Debevec, P.: The digital Emily project: Achieving a photorealistic digital actor. IEEE Computer Graphics and Applications 30, 20–31 (2010)
2. Alexander, O., Rogers, M., Lambeth, W., Chiang, M., Debevec, P.: The digital Emily project: Photoreal facial modeling and animation. ACM SIGGRAPH 2009 Course Notes (2009)
3. van Noort, R.: The future of dental devices is digital. Dental Materials 28, 3–12 (2012)
4. Hanrahan, P., Krueger, W.: Reflection from layered surfaces due to subsurface scattering. Proceedings of ACM SIGGRAPH 1993, 165–174 (1993)
5. Hable, J., Borshukov, G., Hejl, J.: Fast skin shading. In: Engel, W. (ed.) ShaderX^7, pp. 161–173. Charles River Media (2009)
6. Shetty, S., Bailey, M.: A physical rendering model for human teeth. ACM SIGGRAPH 2010 Posters (2010)

7. Shetty, S.: Layered rendering model for human teeth. Master's thesis. Oregon State University (2011)
8. Botha, C.P., Post, F.H.: New technique for transfer function specification in direct volume rendering using real-time visual feedback. In: Mun, S.K. (ed.) Proceedings of the SPIE International Symposium on Medical Imaging 2002: Visualization, Image-Guided Procedures, and Display, vol. 4681, pp. 349–356 (2002)
9. Kniss, J., Kindlmann, G., Hansen, C.: Multidimensional transfer functions for interactive volume rendering. IEEE Transactions on Visualization and Computer Graphics 8, 270–285 (2002)
10. Kim, L., Park, S.H.: Haptic interaction and volume modeling techniques for realistic dental simulation. The Visual Computer 22, 90–98 (2006)
11. Rhienmora, P., Gajananan, K., Haddawy, P., Dailey, M.N., Suebnukarn, S.: Augmented reality haptics system for dental surgical skills training. In: Proceedings of the 17th Symposium on Virtual Reality Software and Technology (VRST 2010), pp. 97–98 (2010)
12. Yau, H.T., Hsu, C.Y.: Development of a dental training system based on point-based models. Computer-Aided Design and Applications 3, 779–787 (2006)
13. Wang, D.X., Zhang, Y., Wang, Y., Lu, P., Wang, Y.: Development of dental training system with haptic display. In: Proceedings of the 2003 IEEE International Workshop on Robot and Human Interactive Communication, pp. 159–164 (2003)
14. Wang, D., Zhang, Y., Wang, Y., Lü, P., Zhou, R., Zhou, W.: Haptic rendering for dental training system. Science in China Series F: Information Sciences 52, 529–546 (2009)
15. Pighin, F., Szeliski, R., Salesin, D.H.: Modeling and animating realistic faces from images. International Journal of Computer Vision 50, 143–169 (2002)
16. Zhang, Y., Sim, T., Tan, C.L., Sung, E.: Anatomy-based face reconstruction for animation using multi-layer deformation. Journal of Visual Languages and Computing 17, 126–160 (2006)
17. Ishimaru, A.: Wave Propagation and Scattering in Random Media. Academic Press (1978)
18. Jensen, H.W., Marschner, S.R., Levoy, M., Hanrahan, P.: A practical model for subsurface light transport. Proceedings of ACM SIGGRAPH 2001, 511–518 (2001)
19. Spitzer, D., Bosch, J.T.: The absorption and scattering of light in bovine and human dental enamel. Calcified Tissue Research 17, 129–137 (1975)
20. Zijp, J., ten Bosch, J., Groenhuis, R.: HeNe-laser light scattering by human dental enamel. Journal of Dental Research 74, 1891–1898 (1995)
21. Fried, D., Glena, R.E., Featherstone, J.D.B., Seka, W.: Nature of light scattering in dental enamel and dentin at visible and near-infrared wavelengths. Applied Optics 34, 1278–1285 (1995)
22. Zijp, J.R.: Optical properties of dental hard tissues. Doctoral dissertation. University of Groningen (2001)
23. Henyey, L.G., Greenstein, J.L.: Diffuse radiation in the galaxy. Annales d'Astrophysique 3, 117–137 (1940)
24. Jensen, H.W., Legakis, J., Dorsey, J.: Rendering of wet materials. In: Lischinski, D., Larson, G.W. (eds.) Rendering Techniques 1999 (Proceedings of EGWR 1999), pp. 273–282. Springer (1999)
25. Miller, G.S., Hoffman, C.R.: Illumination and reflection maps: Simulated objects in real environments. ACM SIGGRAPH 1984 Course Notes for Advanced Computer Graphics Animation (1984)
26. Greene, N.: Environment mapping and other applications of world projections. IEEE Computer Graphics and Applications 6(11), 21–29 (1986)

An Evaluation of Open Source Physics Engines for Use in Virtual Reality Assembly Simulations

Johannes Hummel[1], Robin Wolff[1], Tobias Stein[2],
Andreas Gerndt[1], and Torsten Kuhlen[3]

[1] German Aerospace Center (DLR), Germany
{Johannes.Hummel,Robin.Wolff,Andreas.Gerndt}@dlr.de
[2] Otto-von-Guericke University, Germany
contact@tobias-stein.info
[3] Virtual Reality Group, RWTH Aachen University, Germany
kuhlen@vr.rwth-aachen.de

Abstract. We present a comparison of five freely available physics engines with specific focus on robotic assembly simulation in virtual reality (VR) environments. The aim was to evaluate the engines with generic settings and minimum parameter tweaking. Our benchmarks consider the minimum collision detection time for a large number of objects, restitution characteristics, as well as constraint reliability and body interpenetration. A further benchmark tests the simulation of a screw and nut mechanism made of rigid-bodies only, without any analytic approximation. Our results show large deviations across the tested engines and reveal benefits and disadvantages that help in selecting the appropriate physics engine for assembly simulations in VR.

1 Introduction

The simulation of the kinematic and dynamic behavior of multi body systems is an important aspect in many interactive simulation and training applications. In recent years, a large number of frameworks have evolved that intend to bundle the algorithms and data structures required for computing the physics effects in a library so that they can be reused in several projects. Such physics engines stem from academic, commercial and hobbyist efforts. Some try to provide a set of common features to support a wide range of applications, such as games and animation, while others specialize on very specific application areas, such as deformable tissue simulation in surgical training applications. Often, this comes with a trade-off between number of supported features, performance and accuracy. The large number of available physics engines makes it difficult for an application developer to select an engine that suits best for a given project.

The work presented in this paper was mainly driven by an ongoing project called VR-OOS (Virtual Reality for On-Orbit Servicing)[1]. Its goal is to develop an interactive application for the planning, training and analysis of on-orbit servicing tasks within a haptic enabled virtual reality environment used to train

G. Bebis et al. (Eds.): ISVC 2012, Part II, LNCS 7432, pp. 346–357, 2012.

astronauts but also robot control. A fundamental feature is the real-time simulation of the realistic dynamic and kinematic behavior of satellite (and robot) components in various on-orbit servicing tasks. These include, for example, the removal of the protection foil (MLI - multi-layer insulation) from the satellite, operating lever switches, removing and inserting module boards using a bayonet handle, and handling connectors and locking mechanisms. As opposed to games, where it is usually sufficient that results look plausible, for this training simulator high accuracy at interactive frame rates is crucial.

The VR-OOS project currently uses Bullet Physics[1] as the physics engine. Bullet was chosen due to its popularity in related projects and its open source license. During preliminary evaluations of the integrated simulation environment, however, there have been observations that the results of the physics engine tend to get unstable in certain scenarios. For example, a switch vibrated during and long after moving the switch, or rigid modules occasionally penetrated their shaft when pushing them in. Such effects are undesired, as they make the simulation environment appear unrealistic and may influence the training effect. Tweaking the engine's parameters did lead to some improvements. However, it influenced stability in other scenarios. Consequently, we had to spend a considerable amount of time tweaking the parameters for every single mechanism for each occurring scenario. As scenarios are modified or combined to simulate a variety of assembly scenarios supporting various on-orbit servicing tasks this reduced the convenient utility of our simulation environment. Hence, we started an investigation to see whether there was an alternative physics engine that would fulfill our needs with generic parameters without scenario based, fine grained tweaking and set up a number of benchmarks specifically designed to test particular aspects of our application scenarios.

2 Related Work

Seugling et al. [2] evaluated three physics engines (Open Dynamics Engine, AGEIA NovodeX, Newton Game Dynamics) and created nine benchmarks to test the energy preservation, constraint handling and collision detection of the engines. Their energy preservation test consisted of two parts, pushing a sphere slightly into a box and measuring the resulting forces, and shooting a sphere against a box and measuring the resulting velocity and direction. They used a fixed restitution coefficient of 1.0, which resembles a perfect rebound of the objects. Constraint handling was tested by monitoring how a constraint set up as pendulum was reacting on increased weight, and whether the engines could handle more than 20 chained constraints combined to a long pendulum. However, it was not explored whether there was a correlation between these two tests. In order to investigate the scalability of collision contacts, they stacked up to 20 boxes on top of each other and measured the impact on computation performance of each engine. In practice, however, for an assembly simulation the physics engine usually has to handle more than 50 and more objects. Thus a

---

[1] Bullet Physics Engine: http://bulletphysics.org

scalability test should have been realized with at least 100 objects to challenge the collision detection of the physics engines.

A similar work by Boeing et al. [3] presented a qualitative evaluation of freely available physics engines. With seven physics engines (Open Dynamics Engine, AGEIA NovodeX, Newton Game Dynamics, Tokamak, True Axis, Bullet Physics and JigLib) they ran several tests measuring the integrator performance, the material properties, the constraint stability and the performance of collision detection. Integrator performance was measured based on accuracy only. The speed was not measured. For measuring the stability of the constraints they used a chain of spheres connected to each other. They mounted both sides of the chain to fixed points. They also varied the number of the spheres in the chain. The behavior of the engines with different weights of the spheres was not tested. Collision detection performance was measured by stacking about 20 boxes, too. Both papers provided valuable information on the generic performance of the then popular physics engines. Nevertheless, we needed an evaluation of up-to-date engines and with more focus on our specific requirements.

3   Benchmarks

Our overall requirements on the physics engine could be divided into five aspects. Firstly, the physics engine had to be able to detect collisions between objects of relatively large numbers (approx. 250 to 500 objects). The detection of collisions is a fundamental part of physics simulation. This had to be performed at interactive frame rates - with today's common display refresh rates of 60Hz, this means within approx. 16ms, which even goes to 1kHz or 1ms respectively when integrating haptic feedback. Secondly, the computed responses of dynamic objects to collisions had to result in realistic behavior. For example, a rigid object colliding with a hard surface should bounce back with high velocity, while when colliding with a soft surface should bounce with little velocity only. Thirdly, constraints, which limit the degree of freedom of movement of an object with regard to another object, should obey their defined limits and not diverge from them. For example, a switch should only rotate around its single rotation axis and not any other axis. Neither should it move along any translational axis when pushed from the side. Furthermore, rigid bodies should not inter-penetrate each other. When a satellite module is pushed into its module shaft, for example, it should only move along the contact plane within the shaft, but not move through the walls or get pushed through the end. Finally, we needed the physics engine to support the realistic simulation of a screw mechanism. This required the stable computation of both collision and friction for complex geometric objects.

This section outlines the design of a number of benchmarks intended to test the five requirements mentioned above. The first three are broadly based on benchmarks in related work and test the performance and accuracy of more generic features. The benchmarks four to six were designed to test particular features that are specific to our on-orbit servicing simulation and training application.

3.1 Collision Computation Performance

The first benchmark was designed to test the performance and scalability of collision detection with a large number of rigid objects. It was closely based on those tests in related work and consisted of a simple physics scene with a specified amount of equal objects (1 to 2500) placed in a three-dimensional grid along x, y, z axes floating above the ground, see Fig. 1. In contrast to related work, we used spheres instead of boxes, as these are the simplest three-dimensional collision shape, and showed us the minimum time required to detect collisions. Each sphere had a radius of 0.5m and a weight of 1kg. Gravity in this benchmark was defined as $9.81 \frac{m}{s^2}$. We measured the time it took to detect the collisions, compute respective collision responses and advance to the next simulation step.

3.2 Accuracy of Collision Response

In order to test the accuracy of collision responses calculated by the physics engines, we measured the height of a bouncing ball with a defined coefficient of restitution. For the benchmark, we created a sphere with the material properties of a tennis ball (static friction coefficient 0.29, dynamic friction coefficient 0.22, coefficient of restitution 0.636) [4,5]. The sphere had a radius of 6.75cm and a weight of 57g. Gravity was set to $9.81 \frac{m}{s^2}$. The sphere was dropped from height H on the floor, see Fig. 2. Over several time steps the heights h' of bounces were measured and compared to the expected heights.

3.3 Fixed Constraint Stability

In order to test the stability of constraints, we created two separate benchmarks. For testing fixed constraints, our benchmark defined a long chain of spheres with radius of 0.5m, see Fig. 3. Each sphere was attached to its neighbor by a fixed constraint between the centers of each sphere at a distance of the double radius so that they touched each other on the surface. The constraints limited the movement around all translation and rotation axes to zero. The outer spheres on the left and right were anchor-points where the chain was connected to horizontally. These were set as static objects, so that their position was not modified by the physics engine during the simulation, but consider collisions and constraints to other objects. When applying gravity of $9.81 \frac{m}{s^2}$, the chain was expected to stay in its initial position. In the benchmark, we measured the deviation from the starting position. The test was repeated with different chain lengths up to 20 (without counting the anchor-points) and with increasing weights up to 1000kg per sphere.

3.4 1-DOF Constraint Stability

One issue in our current simulation was that simulated switches often showed unrealistic behavior. The lever trigger vibrated a lot while being moved and after snapping over to the new resting position. In order to test how other physics

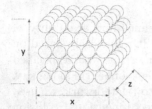

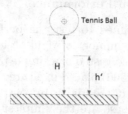

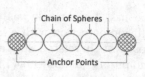

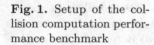

Fig. 1. Setup of the collision computation performance benchmark

Fig. 2. Setup of the collision response accuracy benchmark

Fig. 3. Setup of the fixed constraint stability benchmark

engines would handle such a scenario, the fourth benchmark recreated the switch mechanism. It consisted of two simple rigid bodies, a base and a trigger, see Fig. 4. Both were connected using a hinge constraint between the lower end of the trigger and centered above the base, high enough to avoid both bodies touching each other. A hinge is a 1-DOF constraint allowing rotation along the z axis (pointing out of the image) in the x,y plane. The constraint has been further limited to only rotate between angles of $\alpha=\pm 45°$. In its resting position, the trigger would be at $\alpha=-45°$, pushed down by constant force F_2. A third rigid body, the slider, would move slowly toward the trigger and push it with constant force F_1 to toggle the switch. When the trigger passes $\alpha=0°$, the force F_2 would be inverted to F_2' pushing the trigger to the opposite resting position at $\alpha=+45°$. The expected behavior was that the trigger flips to the opposite side and rests at $+45°$, without bouncing back or penetrating the base or slider bodies. In the benchmark, we measured the translation along each axis and rotation angles around each axis. Gravity was set to zero in this benchmark, as we wanted to simulate a space environment.

3.5   Interpenetration

Another issue in our current simulation was that grasped rigid objects occasionally penetrated other rigid objects. This was especially apparent when interacting with the environment without haptic feedback. Thus, we designed a benchmark resembling the module and shaft scenario. The setup consisted of one dynamic rigid body, the module, surrounded by four static rigid bodies, that held it in place, resembling the shaft. The module had the dimensions of 0.5m x 2m x 1m and a weight of 1kg. The bodies of the shaft were large enough to closely surround the module. The module was slightly taller than the shaft, so that it sticks out at the top, see Fig. 5. A constant force F was applied to another rigid object, the slider, so that it moved towards the module and pushed it. The expected behavior was that the module did not move or rotate. We measured changes in the position and rotation of the module in its center point. Additionally, we repeated the test in several iterations with increased F.

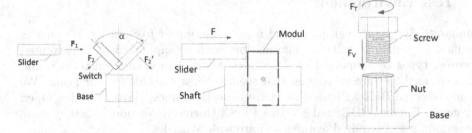

Fig. 4. Setup of the 1-DOF constraint benchmark

Fig. 5. Setup of the rigid body interpenetration benchmark

Fig. 6. Setup of the advanced collision and friction benchmark

3.6 Advanced Collision and Friction

A common scenario in many on-orbit servicing tasks is to screw and unscrew certain components. The simulation of a screw mechanism could be accomplished with a 2-DOF constraint allowing both translation and rotation along one single axis only and using a function that makes the translation dependent on the rotation. This, however, must be scripted in the application layer, as common physics engines do not support a screw mechanism. Additionally, the interactions between screw, thread and other involved rigid bodies have to be scripted too, as the 2-DOF constraint must be attached dynamically to a screw and thread as soon as they collide, but only if their axes are aligned and either screw or thread rotates in the correct direction.

Although it was done for the switch scenario described above, we wanted to avoid scripted behavior and rather solve this problem with rigid bodies and standard constraints only. Simulating the loosening and tightening of a screw using rigid bodies, however, puts a high challenge on a physics engine, as it involves a large number of colliding faces between screw and screw thread. A screw has a complex geometry. The thread is a concave object, which is computationally expensive during collision detection. Therefore, we approximated the geometry using a compound collision shape made up of several convex collision shapes arranged in a spiral. Fig. 6 shows the general setup of the benchmark.

The collision shapes of the screw spiral (bolt) were placed around a cylinder, while the collision shapes of the thread spiral (nut) were placed inside a number of tall collision shapes of 1m, arranged in a circle with radius of 0.5m to form a hollow cylinder as fixed base. We used segments of 30 box collision shapes with 5cm height per turn of the spiral. The static and dynamic friction coefficient was set to 0.001 for each physics engine in the test. Gravity was set to zero.

A constant torque F_T of 1Nm and a vertical force F_V of 1N, were applied every simulation step. The expected behavior was a rotation of the the screw with increasing speed and a continuous subtle movement into the nut without any interruptions and penetrations. The benchmark tested the stability of the computed collision and friction for complex compound objects and could be transferred to other scenarios, such as a bayonet mechanism.

4   Test Environment

Similar to Seugling et al. [2] we did a pre-selection of freely available physics engines, in which we collected information, such as supported platforms, used licenses, types of supported rigid body collision shapes and constraints, and then chose those physics engines that seemed most suitable for our needs. We finally selected Bullet Physics, Newton Game Dynamics, Havok Physics, Open Dynamics Engine (ODE) and NVIDIA PhysX (formerly Novodex), as they share a common set of required features and are widely used.

Bullet is the physics engine in our current simulation environment and was used as reference. It is distributed under the zlib license, runs on all major platforms and supports various optimized collision shapes. Many games and some movies are using Bullet making it one of the most popular open source physics libraries. We used version 2.80 of Bullet, which was released on March 5th, 2012. Newton[2] recently switched to the open source zlib license and provides the common set of features for rigid body simulation. However, instead of using linear complementarity problem or iterative methods, Newton uses a deterministic approach in its solver, promising very accurate results. We used version 2.33, which was released on April 1st, 2011. Havok[3] is a commercial physics engine with its own end-user license agreement (EULA) and promises robust dynamics and constraint solving. It is used in more than 150 game releases. We used version 7.1 of Havok, which was released on December 22th, 2009. ODE[4] is available under the open source LGPL/BSD license. Since its introduction in 2001, ODE has gained great popularity for hobbyist robotics simulation. We used version 0.12 released on February 11, 2012. PhysX[5] is now developed by NVIDIA and distributed as closed source under the terms of its own EULA. Used in approx. 200 game releases, similar as Havok, it is widely adopted by the games industry. We used the stable release 9.12.0213, which was published on March 13, 2012.

In order to reduce the programming overhead and the possibility of errors while implementing the scenarios separately for each engine, as well as to easily iterate several test runs with varying parameters and collect data, we desired a method to decouple the high-level benchmark implementation from the low-level execution in the underlying engine without compromising any features. A few frameworks exist, such as Physics Abstraction Layer (PAL)[6], *Open Physics Abstraction Layer (OPAL)*[7] and *The Gangsta Wrapper (TGW)*[8], that provide an abstraction layer and bindings for multiple physics engines, enabling developers to easily exchange physics engines at run-time without reinitializing the entire simulation. However, they have partly deprecated bindings or limited support for our chosen physics engines. Hence, we implemented our own physics

---

[2] Newton Game Dynamics: http://www.newtondynamics.com
[3] Havok Physics: http://www.havok.com
[4] ODE: http://www.ode.org
[5] PhysX: http://developer.nvidia.com/physx
[6] PAL: http://www.adrianboeing.com/pal
[7] OPAL: http://thatopal.sourceforge.net
[8] TGW: http://gangsta.sourceforge.net

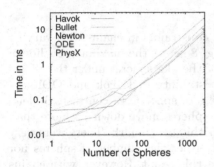

Fig. 7. Mean computation times in ms for collision detection with 1 to 2500 spheres

Table 1. Mean computation times in ms for collision detection with 1 to 2500 spheres

Spheres	Bullet	Havok	Newton	ODE	PhysX
1	0.02	0.05	0.02	0.02	0.05
10	0.06	0.05	0.08	0.03	0.13
25	0.17	0.07	0.14	0.09	0.20
50	0.13	0.15	0.31	0.10	0.20
100	0.26	0.25	0.66	0.26	0.44
250	0.74	0.56	1.69	0.61	0.95
500	1.62	1.17	3.34	1.27	2.26
1000	3.75	2.58	7.13	2.96	4.38
2500	13.17	8.13	21.07	9.39	12.21

abstraction layer framework. It consists of a number of C++ classes that encapsulate repeatedly used interface and utility methods for creating, running, controlling and monitoring a benchmark.

The benchmarks were performed on a workstation with an Intel i7-870 CPU with 2.93GHz, 8GB RAM, Nvidia GeForce GTX 460, running Windows 7 Enterprise 64 Bit (Service Pack 1).

5   Results

5.1   Collision Computation Performance

Fig. 7 shows a graph of the average time for computing collisions with 1 to 2500 spheres, and Table 1 shows the mean values. The computation times had slight deviations with few collision objects and continued to increase almost logarithmically with 50 objects and more. Most plots bend at between 20 and 50 objects, suggesting that these areas were influenced from the engines' internal optimizations. All engines were able to compute collisions of 100 spheres in under 1 ms and all, except Newton, could handle up to 250 spheres in under 1 ms. In overall, all engines performed equally well, with Newton lagging slightly behind.

5.2   Accuracy of Collision Response

In the accuracy of collision response benchmark, we measured the height of a simulated bouncing tennis ball. Table 2 shows the heights produced by the different physics engines compared to the expected height. Seven rebounds were recorded. Apart from Bullet, all physics engines computed bounces very close to the expected value with less than 1cm difference. Havok and PhysX simulated only the first three bounces and then set the rigid body to sleep mode to save computation time, which is a normal optimization method, but makes the simulation appear unrealistic. Bullet appeared to damp the bounces more than other engines, although the restitution coefficients have been set correctly for both the ball and the floor.

5.3   Fixed Constraint Stability

This benchmark tested the stability of fixed constraints in chains with varying numbers of spheres with 1kg and 1000kg. Fig. 8 shows the maximum euclidean distance between actual and start position of the 1kg spheres under the effects of gravity. One can see that the constraint stability of Havok and ODE decreased exponentially with increasing numbers of spheres. In contrast, Bullet's constraints appeared somewhat soft and let spheres move down by more than 30cm. Newton and PhysX performed best in this benchmark. There were very little to no deviations in sphere positions. Neither the number of spheres nor the increased mass had significant impact on this result. However, with chains of 1000kg spheres (not shown in the graph), the constraint stability in Newton decreased proportionally towards that of ODE and Havok. The other physics engines did not appear to be affected by very large masses.

5.4   1-DOF Constraint Stability

Fig. 9 shows the difference in the angle around the z-axis over time, at which the trigger object rotated to the specified limit of the constraint from the moment it has reached its end position at 45°. It was expected that all physics engines leave the trigger at its final position, and thus show a line around 0 degrees. However, only PhysX and Newton approached zero after shortly exceeding the constraint limit with approx. 2°. Newton performed best in this benchmark. The constraint limit was exceeded by 1.87° for 0.68 ms, and converged to a distance of 0.07°. PhysX caused spikes of 1.6° every 0.4 ms. A closer look into the data showed, that these occurred each next calculation step when the angle was exactly 45°. ODE and Bullet both started to vibrate after the trigger bounced back from its end position, suggesting a high default restitution value. Bullet's vibration appeared stable with a frequency of about 18.5 Hz and an amplitude of 0.86°, mostly below the constraint limit. The amplitude of ODE's vibration was about five times the size of that of Bullet and did not appear to have distinct frequencies in it. The plot for Havok shows that the force F_2' was stronger than the constraint limit and pushed the trigger beyond with up to 16°.

5.5   Interpenetration

Table 3 shows the maximum deviation in translation along the x and y axes and rotation around the z axis of the module object, resting in its module shaft, when being pushed at the top end by the slider object along the positive x axis with varying force from 25N to 500N. Looking at the horizontal translation along the x axis only, Bullet appeared to perform best. The module moved very little, less than a millimeter at forces up to 100N. The same applied to PhysX, with only slightly larger values. However, the module appeared to be moved along the y axis with Bullet. Also, the rotation around the z axis in Bullet was with up to 10° significantly larger than that of PhysX. Havok, ODE and Newton showed a similar linear increase of deviation from the resting position with increasing

Table 2. Comparison of computed heights of bounces of a tennis ball dropped from 2.54m as reported by the physics engines (All values in meter.)

# Bounce	Expected	Bullet	PhysX	Havok	ODE	Newton
1	1.027	0.567	1.067	1.007	1.018	1.084
2	0.416	0.145	0.493	0.423	0.384	0.441
3	0.168	0.035	0.237	0.210	0.157	0.195
4	0.068	0.010	0.133	0.132	0.061	0.080
5	0.028	0.003	0	0	0.026	0.029
6	0.011	0.002	0	0	0.011	0.008
7	0.005	0	0	0	0.005	0.001

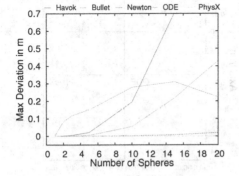

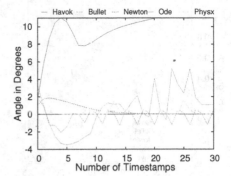

Fig. 8. Maximum deviation in horizontal position in chains of 1kg spheres in the fixed constraint stability benchmark

Fig. 9. Difference of actual angle around the z-axis to the specified limit in the 1-DOF constraint stability benchmark

forces. Both, the x and y translations were approx. 5x, 15x and 20x, respectively, more than that of PhysX. With ODE, however, the module started to get pushed out of the module shaft, with movements of more than 20cm along y. The module rotated slightly with 1° to 2° at forces up to 100N. At 250N and 500N, Bullet showed rotations about double of that with the other engines. However, PhysX appeared very stable and always showed rotations less than 1°.

5.6 Advanced Collision and Friction

In the advanced collision and friction benchmark, we measured the position and orientation of the head of the screw when screwed into the nut. Fig. 10 shows the translation over time along the y axis, which corresponded to the height of the screw. One can see, that Havok did not manage to solve the task. After about 12% of the total length it jumped out of the nut and fell down. Although ODE performed well half way, it also did not manage to complete the task. The screw got stuck in the long compound object resembling the nut and did not move anymore. This behavior was reproduced each time we ran the test with ODE. Bullet and Newton passed the test very well. While Newton showed a more continuous movement, with one interruption at about one third of the height, Bullet showed more occasional interruptions. This was presumably

Table 3. Maximum deviations with different forces applied to the slider for translation in x & y direction (in m) and the rotation on the z axis (in deg)

Force in N	Bullet			PhysX			Havok			ODE			Newton		
	X	Y	α	X	Y	α	X	Y	α	X	Y	α	X	Y	α
25	0.02	0.07	0.42	0.02	0.00	0.029	0.13	0.20	0.29	0.32	0.04	0.18	0.46	0.13	0.35
50	0.07	0.27	0.93	0.04	0.01	0.061	0.25	0.34	0.59	0.67	0.33	0.92	0.89	0.05	0.71
100	0.07	0.68	1.95	0.09	0.02	0.126	0.46	0.45	0.99	1.24	0.31	0.90	1.82	0.38	1.49
250	0.11	2.21	5.13	0.28	0.07	0.379	1.15	1.13	2.48	2.91	2.14	2.57	4.79	1.77	3.72
500	0.45	4.49	10.45	0.63	0.16	0.845	2.29	2.27	4.98	5.78	23.80	5.34	9.85	4.20	7.69

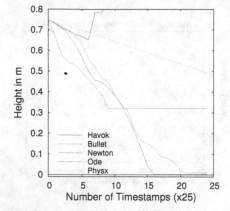

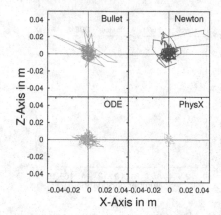

Fig. 10. Height of the head of the screw (in meter)

Fig. 11. Movement of the head of the screw on the x-z plane (in meter)

due to increasing collisions while moving into the nut. Compared to the other engines, PhysX showed a very slow but linear motion. It suggests that the default damping setting was very large. Due to the slow movement, PhysX could not manage to complete the task within our given 1000 timestamps.

Fig. 11 shows the translation in meter of the head of the screw on the x-z plane (looking from top at the screw). The expected behavior was that it would not show any movement to the sides. We did not plot Havok, as it could not complete the task. In the graph of PhysX, one can see the most subtle movement of less than 10cm (remember the screw was 1x0.5m). The graphs of ODE, Bullet and Newton show average translations of similar distances. However, Bullet and Newton showed occasional longer deviations, presumably when collisions occurred.

6   Conclusion and Future Work

As Boeing [3] mentioned, there is no general physics engine that performs best for any given task. Each engine we tested showed strengths and weaknesses. PhysX showed the most stable constraints and little inter-penetrations, presumably due to its high damping default settings. However, the high damping hindered the calculation of realistic bounces and advanced friction, as in the screw benchmark, leading to unrealistic behavior. Havok was highly optimized for speed. But this came at the expense of less accuracy in most of our benchmarks. Bullet performed poorly in the restitution test and tended to vibrate in the 1-DOF constraint stability test, which confirmed our observations within our target simulation environment. However, Bullet showed reliable constraints and performed very well in the advanced collision and friction benchmark simulating screw. Newton's approach of the deterministic solver appeared to produce indeed most accurate results. Apart from the collision computation performance benchmark, Newton performed always very well. ODE was fast and accurate, as long as its constraint and collision handling was not stressed too much. In this case, it reacted unpredictable.

Our benchmarks were designed with assembly simulations in mind. They test the performance of collision computations, preservation of energy, constraint reliability, inter-penetration, and evaluate the computation of both collision and friction for complex compound objects simulating a screw mechanism. The first three tests were mainly standard tests, seen in related work too. We added three novel benchmarks that evaluated the constraint reliability for a switch mechanism and the penetration between rigid bodies when applying a large force. In a more complex benchmark, we evaluated how the physics engines could handle a screw mechanism. Based on our requirements and the results of the benchmarks, we can now say that Newton and PhysX would be valuable candidates to compete with Bullet for integration in our current simulation environment.

As our test environment decouples the benchmark implementation from the actual underlying physics engine, it can be easily extended to evaluate and compare further physics engines as well. As described in section 4, we only selected the five most common engines. Other engines, such as OpenTissue Library, SOFA, Tokamak, Dynamechs, True Axis, or CM-labs Vortex, could be evaluated too. Furthermore, we consider extending our suite of benchmarks to also test the performance of mesh-mesh collisions with concave geometries, as well as handling soft body dynamics.

References

1. Wolff, R., Preusche, C., Gerndt, A.: A modular architecture for an interactive real-time simulation and trainning enviroment for satellite on-orbit servicing. In: Proceedings of IEEE Distributed Simulation and Real-Time Applications, pp. 72–80 (2011)
2. Seugling, A., Rölin, M.: Evaluation of physics engines and implementation of a physics module in a 3d-authoring tool. Master's thesis, Umea University Department of Computing Science Sweden (2006)
3. Boeing, A., Bräunl, T.: Evaluation of real-time physics simulation systems. In: Proceedings of the 5th International Conference on Computer Graphics and Interactive Techniques in Australia and Southeast Asia (GRAPHITE), pp. 281–288 (2007)
4. Federation, I.T.: Itf approved tennis balls, classified surfaces & recognised courts 2011 - a guide to products and test methods (2011)
5. Farkas, N., Ramsier, R.D.: Measurement of coefficient of restitution made easy. Physics Education 41, 73–75 (2006)

A Framework for User Tests in a Virtual Environment

Volker Wittstock, Mario Lorenz, Eckhart Wittstock, and Franziska Pürzel

Chemnitz University of Technology
{volker.wittstock,mario.lorenz,eckhart.wittstock,
franziska.puerzel}@mb.tu-chemnitz.de

Abstract. This paper describes the setup and development of an innovative framework for conducting user tests in a virtual environment. As it is the main purpose to evaluate the user's reactions to user interfaces on mobile devices, an Android-based smartphone and a tablet computer were linked to a Virtual Reality (VR) system. The framework allows user interaction to trigger certain events in the immersive environment and vice versa interaction with the virtual environment can trigger interface behaviour. For hand-free navigation through the virtual scene a Wii Balance Board is used. The framework is the basis for user tests to be conducted within the uTRUSTit (Usable Trust in the Internet of Things) project supported by the EU under the Seventh Framework Programme.

1   Introduction

This paper describes a framework for enabling user tests in a Virtual Environment within the uTRUSTit project. The project aims to provide guidelines and a software toolkit to express the underlying security in the Internet of Things (IoT) in a comprehensible way to users. This effort allows them to make valid judgements on the trustworthiness of IoT-Applications.

Virtual Reality (VR) Technology within the uTRUSTit project (http://www.utrustit.eu) is employed to provide a realistic testing environment at an early stage of development. It will be used to simulate and visualize the previously defined scenarios "Smart Home" and "Smart Office" (Fig. 1) including the interaction with the smart devices and the provided trust feedback. In the "Smart Home" scenario the user interacts with the surroundings in his/her home, like a smart medical cabinet or an electronic door. "Smart Office" is a business scenario where the user fulfils certain tasks in an office building, like holding a presentation or using an elevator.

The VR evaluation takes places at an early stage in the project, when the real prototypes of the smart devices are not yet available. Therefore interaction devices have to be developed that represent and simulate the "real" smart devices and also allow interaction with the VR environment.

We aimed to develop a framework that eases the implementation of VR user tests by providing flexible tools for hand-free navigation in VR and interaction with the virtual objects.

G. Bebis et al. (Eds.): ISVC 2012, Part II, LNCS 7432, pp. 358–367, 2012.

Fig. 1. Bathroom of the "Smart Home" scenario (left) and Entrance hall in the "Smart Office" scenario (right) in a 5-side-CAVE

2 Testing Environment – Requirements and Description

In both scenarios, the interaction devices for the proposed user tests have to fulfil two different purposes:

1. Enabling the user to navigate inside the VR scene
2. Simulating the behaviour of a smart device and its effect on the virtual environment.

Interaction within the user tests should come as naturally as in real-life. Therefore the interaction devices have to be able to display the user interfaces with the developed trust feedback. According to the decisions the user makes while using the devices and user interfaces, certain resulting actions have to be executed in the virtual environment, like the opening of a door. Therefore a robust two-way communication between the device and the virtual environment is needed. Some of the described use cases require the test system to "know" where the interaction device is located in the scene. Therefore it is necessary to implement a method to determine the position of the device and communicate it to the framework.

To be as close as possible to reality, it was decided that the interaction devices have to be a mobile phone or tablet computer. Furthermore it was agreed that navigation should not be done with these devices and that it should be hands-free, so reality is matched as close as possible. Since many users will have no experience with VR-systems, navigation must be easily usable. The implemented navigation also has to be linked with the actual VR-system in a robust way that ensures functionality.

3 System Model

3.1 Using Mobile Devices for Interaction

Android was chosen as the programming platform for the mobile interaction devices because it is easy to use, provides high level access to the device functions and a large variety of usable devices are available. A coupling between mobile devices and VR has been successfully implemented in [1-6]. The tracking issue of the mobile devices was solved by taping them to a wooden board which has a tree target attached for optical tracking.

3.2   Using the Wii Balance Board for Navigation

For hands-free navigation through the virtual scene the Wii Balance Board was chosen. It is a robust, cheap product and easy to access by using existing programming libraries. The Wii Balance Board is a simple board like device with four pressure sensors on each corner. The data is transmitted through a Bluetooth connection to a host device. Originally it was designed to work with the Nintendo Wii Game console but thanks to the work of volunteers, different programming interfaces are available. One of the most complete is the Wiimotelib Library by Brian Peek [7] that was used for our implementation.

Furthermore it has already been employed for related applications. De Hann, Griffith and Post [2] inspected the possibilities of using the Wii Balance Board as a navigation device in VR and described different metaphors for triggering movements. Hilsendeger, Brandauer, Tolksdorf and Fröhlich [3] used the Wii Balance Board for navigation through a previously created virtual BioSphere and found out, that sigmoid functions serve best as transfer functions for rotation and acceleration. They confirmed this by an explorative pre-study. Fikkert, Hoeijmaker, Vet and Nijholt [1] compared the ease of using a Wii Balance Board to a Wiimote handheld control. They tested how long it took the participants to navigate through a maze and found that using the Wiimote was the faster method but moving with the Wii Balance Board was considered to be more intuitive and 'fun'.

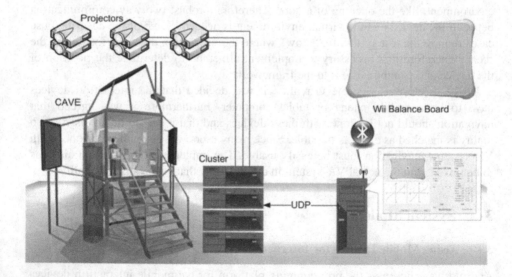

Fig. 2. Wii Balance Board System Setup

In Fig. 2 the robust setup for realizing the navigation is shown. The Wii Balance Board is connected to a computer using the standard Bluetooth connection.

An application has been developed that uses the afore mentioned Wiimote Library for reading the values from the Balance Board and calculating the vectors for translation and rotation of the user's viewpoint. The application sends the calculated vectors to the VR system using a UDP connection, where it is transformed into user movement. For a realistic user impression the so-called 'walk' mode is used, that retains a constant height above ground.

When using the Wii Balance Board as an interaction device for games, the user usually stands on the board and controls e.g. a water skiing game by leaning in the desired direction. The sensors detect the weight the user puts on them. To derive a movement vector from the sensor's values, the balances between back and front and left and right are mapped to values between -1 and 1 for both directional balances as described in [3]. To get as close as possible to natural movements, the back to front balance is used for control of speed and the right to left balance for control of direction. To translate these into movements they are multiplied with different transfer functions, derived from pre-tests. We started with the functions recommended by [3] and adjusted them through testing. We found out that it was more comfortable to use linear functions for turning and piecewise linear functions for moving backward and forward. Because moving backward often made people feel dizzy, we reduced it to a fixed-speed movement. For standing still, these functions contain an interval where the resulting value is zero, so the users can interact with their devices without paying too much attention to his/her weight balance.

The used functions are shown in Fig. 3, the resulting values are sent to the VR system by an UDP connection. The Wii Balance Boards sample rate results in a data stream of about 40 values per second so the use of a connectionless protocol is sufficient. Inside the VR system these values are mapped to maximum values of 2 m/s for moving and 0.3 rotations per second around the vertical axis for turning.

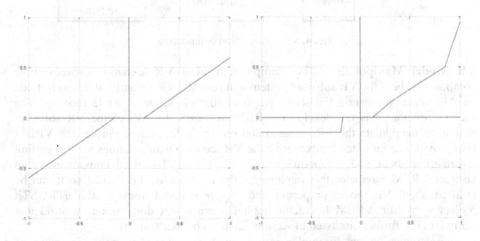

Fig. 3. Transfer functions for rotational (left) and translational (right) movements

3.3   Designing an Android VR System Interaction Framework

Concept. The interaction between the Android device and the VR scenario is bi-directional. Users can manipulate objects in the VR scenario and a change in the VR scenario can trigger a reaction from the user. Therefore the communication between the Android device and the VR scenario has to be designed so that the Android device can send information to the VR scenario (running inside the VR software system) to execute the manipulations. Additionally the VR software system must be able to send information about changes in the VR scenario to the Android device.

As there are many different VR software systems and interaction devices, it is wise to choose a design and implementation that can be adapted with relatively little effort to other VR software systems and interaction devices differing from those chosen in the uTRUSTit project.

The interaction between the user and the VR scenario is event driven. At a random point in time the user will fulfil a certain task by using the Android device to trigger a manipulation in the VR scenario. The reaction of the VR scenario is also not deterministic, it can depend on different factors, like the position of the user in the VR scenario or on previously triggered manipulations. Therefore an event driven software system for the User–VR scenario interaction was designed and implemented.

This approach made it easy to reach a high coherence and a loose coupling of the implemented framework. The Android VR system interaction framework consists of two modules. One, responsible for executing manipulations and for notifying about changes in the VR scenario, is named VR Event Manipulator (VREM). The second module is a framework that allows scripting event driven Android applications for the VR interaction scenarios, named Builder for Event Driven Interaction Scenarios (BEDIS). Both modules are communicating over TCP/IP sockets (Fig. 4).

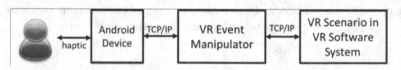

Fig. 4. VR – interaction components

VR Model Manipulation. The manipulation of a VR scenario involves three components. First the VR software system that runs the VR scenario. It has to provide an API for manipulating the scene graph of the VR scenario while running. The second component is the developed VREM that uses the API of the VR software system to manipulate the VR scenario and retrieve information. Further the VREM also provides an interface for executing the VR scenario manipulations and for getting information about node properties of the VR model. The third component is an abstract VREM Interactor that implements the interface of the VREM so it can be notified about VR model properties and trigger manipulations. In the uTRUSTit project a specific VREM Interactor implementation was done as an Android app using BEDIS, further discussed in section Scenario builder (Fig. 5).

There are two reasons for implementing a mediator between the actual interactor of the VR scenario and the VR software system API. First, the VR software system interface only supplies low level access to the VR scenario by setting and getting single node property values. The VREM provides functionality that allows a more comfortable access on a higher level to the VR scenario.

Second the mediator is used to couple the VR software system and the interactor very loosely. So a change of the VR software system would not result in a re-implementation of the interactor. Only the VREM would have to be extended to support the new VR software system.

The VREM module itself consists of two components. The first is the core module, the VR manipulation event system that provides the high-level interaction access to the VR software system. The second is the VREM Player which is the interface for the VREM Interactor (Fig. 5).

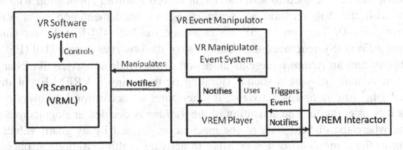

Fig. 5. The design of the VREM and the way of manipulating the VR scenario

As stated in the previous section, the whole User–VR interaction is event driven. The high level VR software system access is realized as an *Event–Action–Condition–System* (Fig. 6). In this system an *Event* is simply a container for *Actions* that are representing a unit.

An *Action* wraps a low level interaction with the VR software system. There are three specific *Action* types. A *ManipulationAction* sets a new value of a VR scenario node property using the functions provided by the VR software system, for instance the toggling of visibility of a node in the VR model, the change of a texture or the transformation of a node. A *StartAction* can start an animation, sound or video in the VR scenario on execution. While these two described *Action* types represent input to the VR scenario, the third *Action* type represents output. A *NotificationEventAction* is bound to a node property of the VR model that notifies its listeners about changes including the new value. For example when an object is made invisible the *NotificationEventAction* passes this information on to the VREM Player.

The execution of an *Action* can further be bound to *Conditions* that must be met. In the case of the VR manipulation event system, a *Condition* represents the state of a property of a node of the VR model.

This system allows defining complex VR scenario interaction. The execution is controlled by the VREM Player. It triggers the executions of an *Event* when the connected VREM Interactor wants it and notifies about changes in the VR scenario that the VREM Interactor wants to know about. The VREM Player is also responsible for the handling of the *Events, Actions* and *Conditions*. These are saved in an XML file (Fig. 6).

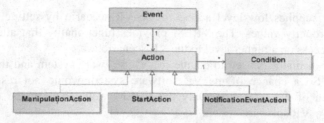

Fig. 6. Class model of the Event-Action-Condition-System of the VR manipulation event system

Scenario Builder. The basic idea behind BEDIS is to design and implement a framework that can be used to script an event driven Android application which can interact with the VR scenario by representing an implementation of a VREM Interactor (Fig. 4). The core of BEDIS is a state machine [9] like *Action–Guard–NotificationEvent*–System that makes heavy use of the observer pattern [10] (Fig. 7).

In this system an *Action* represents an action of the Android device like changing the screen content, playing a sound, vibrating or triggering a VREM-Event that is executed when the *Action* is activated. It determines its activation by observing the *Guards* it is listening to. The activation of the *Actions* is decided analogous to a state machine where the current state of the machine is defined by its guard values. By listening to the same *Guards* it is possible to activate multiple *Actions* at once, like changing the screen content on the Android device and playing a sound on it while also triggering the start of an animation in the VR scenario by sending the according *Event* to the VREM Player.

A *Guard* wraps a Boolean value and notifies its listeners when its state has changed. The state changing of the *Guard* depends on *NotificationEvents* that are sent by *NotificationEventEmitters* the *Guard* is listening to.

Currently *NotificationEventEmitters* are limited to be either GUI controls like buttons and changes of node properties in the VR-Model sent by the VREM Player. But it is possible to bind it to other programmatically accessible applications or devices.

The abstract base class *NotificationEventLinearLayout* is used to integrate GUI controls as *NotificationEventEmitters*. It is derived from the *android.widget.LinearLayout* class and has a member variable named *encapsuledView* of the type *android.view.View*. The only object the *NotificationEventLinearLayout* contains, is this *encapsuledView*. To use a common GUI control like a button as a *NotificationEventEmitter* that leads to the activation of an *Action*, it is necessary to derive a class like *NotificationEventButton* from *NotificationEventLinearLayout* and create an *android.widget.Button* as the *encapsuledView*. The advantage of this approach is that the *NotificationEventButton* can be used just like a normal *android.widget.Button* in the Android Layout Editor. A further advantage is that also non common GUI controls can be integrated this way.

The *VREMP_Connector* is the interface for communicating with the VREM Player and also a *NotificationEventEmitter*. That means it is notifying its listeners about *NotificationEvents* coming from the VR scenario and is sending VREM-*Events* to the VREM Player when a *SendingAction* is activated.

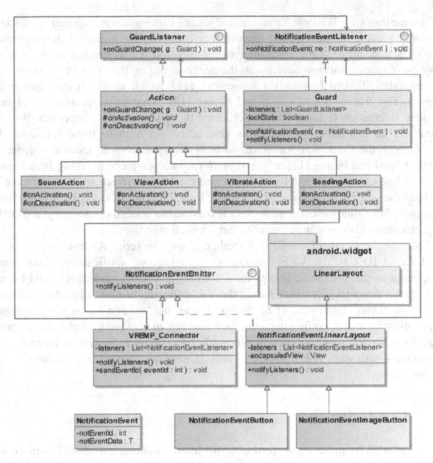

Fig. 7. Class model of BEDIS

4   Discussion

With our described framework we provide an infrastructure for conducting the user tests in the uTRUSTit project. The goal for the user tests was to find out how good our developed guidelines for user interface (UI) design for the interaction with the internet of things are and what adoptions are necessary.

When designing the framework, we paid special attention to the independent reusability of the three modules. The Wii Balance Board can be used for hand-free navigation through every VR scene in walk mode without using the VREM or BEDIS module.

The VREM module provides an interface for bidirectional event driven interaction with a VR scene. The specific interactions with the VR scene are defined in an XML file and are executed when the interaction device sends the *Event* ID and the *Conditions* of the *Actions* are met. Of course this device can be a smart phone or tablet as it is in the uTRUSTit user tests but it is easily possible to map the *Event* ID's on the buttons of a Wiimote or on speech/gesture commands recognized by a Kinect sensor. By using the notification functionality it is possible to track behaviour inside the VR scene, for example when an avatar enters and leaves defined areas. It is also easy to connect various devices at the same time to the VREM module. This distinguishes it from Latoschik's framework [11] which relies on data gloves and speech. The framework of Olwal and Feiner [12] achieves an abstraction of input devices, interaction techniques and application through a dataflow approach. With our event driven approach we reach this abstraction on a more basic level, too but furthermore the *Conditions* enable us to bind manipulations to the state of the VR scene. Olwal and Feiner [12] achieved platform independence by using Java Remote Method Invocation. As the VREM Player and the VREM Interactor communicate via sockets the implementation of the VREM Interactor isn't bound to a programing language. We think the usage of sockets is reasonable instead of using e.g. a CORBA implementation because only primitive data is exchanged.

The novel concept of the BEDIS module allows to script Android apps that can send *Event* ID's to the VREM module and to receive notifications about events happening in the VR scene. In conjunction with the VREM module it would be easy to extract a menu like UI that is part of the VR scene and bring it on an Android smart phone or tablet. Unlike in [1-6] the goal was not to develop a single purpose application for one specific interaction use case but to provide a platform that can be used for other VR scenarios as well. Currently BEDIS has to be seen as an early work in progress with some restrictions, e.g. *ViewActions* currently cannot handle dynamic UI behaviour.

5   Conclusion

In the uTRUSTit project user testing in a virtual environment should help to gain experience and knowledge about the best design and behaviour of user interfaces for trustworthy interaction with the Internet of Things at a very early stage. Two virtual scenarios, with which the users should interact using mobile devices like a smart phone and a tablet, were designed for this purpose. Because of these requirements, common navigation devices, like a flystick could not be used. Instead we decided to use a Wii Balance Board for navigation where the user has to shift his/her weight to navigate through the virtual world. We decided to use mobile devices running on the Android platform because it is easy to use, provides high level access to the devices functions and has successfully been used for VR interaction by others like [2], [4] and [6].

For the interaction between Android and the VR scenario we designed two modules that are loosely coupled. The VREM module provides an event driven access to a VR software system for manipulating the VR model and providing information about node properties. The BEDIS module is a framework that can be used to script an event driven Android application. By coupling such an application with the VREM module the Android-VR interaction is realized. The advantage of our proposed framework for VR based user tests is that it can easily be used for building other scenarios, besides those in the uTRUSTit project.

Concerning further development, we are aiming to improve the VREM and BEDIS module as they currently have some technical restrictions. We are also planning to investigate other interaction devices, like the Kinect for navigating through the VR scenarios and interacting with it.

Acknowledgements. This work was partially funded by the European Union Seventh Framework Program (FP7/2007-2013) under grant 258360 (uTRUSTit; see http://www.utrustit.eu/).

References

1. Neugebauer, R., Wittstock, V., Meyer, A., Glänzel, J., Pätzold, M., Schumann, M.: VR tools for the development of energy-efficient products. CIRP Journal of Manufacturing Science and Technology 4 (2011)
2. Arca, A., Villalar, J.L., Diaz-Nicolas, J.A., Arredondo, M.T.: Haptic Interaction in Virtual Reality Environments through Android-based Handheld Terminals. In: 3rd European Conference on Ambient Intelligence, AmI 2009, pp. 259–263 (2009)
3. Jimenez-Mixco, V., de las Heras, R., Villalar, J.L., Arredondo, M.: A new approach for accessible interaction within smart homes through virtual reality. In: Universal Access in Human-Computer Interaction. Intelligent and Ubiquitous Interaction Environments, pp. 75–81 (2009)
4. Jiménez-Mixco, V., Villalar González, J.L., Arca, A., Cabrera-Umpierrez, M.F., Arredondo, M.T., Manchado, P., Garcia-Robledo, M.: Application of virtual reality technologies in rapid development and assessment of ambient assisted living environments. In: Proceedings of the 1st ACM SIGMM International Workshop on Media Studies and Implementations that Help Improving Access to Disabled Users, pp. 7–12. ACM (2009)
5. Kim, J.-S., Tech, V., Quek, F., Tech, V.: iPhone / iPod Touch as Input Devices for Navigation in Immersive Virtual Environments. IEEE Virtual Reality, 261–262 (2009)
6. Lee, J.Y., Kim, M.S., Seo, D.W., Lee, C.W., Kim, J.S., Lee, S.M.: Dual interactions between multi-display and smartphone for collaborative design and sharing. IEEE Virtual Reality, 221–222 (2011)
7. Peek, B.: Managed library for Nintendo's Wiimote, http://wiimotelib.codeplex.com/
8. Hilsendeger, A., Brandauer, S., Tolksdorf, J., Fröhlich, C.: Navigation in Virtual Reality with the Wii Balance Board. In: 6th Workshop on Virtual and Augmented Reality (2009)
9. Alhir, S.S.: Learning UML. Communication Software Design Graphically. O'Reilly Media, Bejing (2003)
10. Gamma, E., Helm, R., Johnson, R., Vlissides, J.: Design Patterns. Elements of Reusable Object-Oriented Software. Addison Wesley, Amsterdam (1995)
11. Latoschik, M.E.: A user interface framework for multimodal VR interactions. In: Proceedings of the 7th International Conference on Multimodal interfaces - ICMI 2005, vol. 76. ACM Press, New York (2005)
12. Olwal, A., Feiner, S.: Unit: modular development of distributed interaction techniques for highly interactive user interfaces. On Computer Graphics and Interactive Techniques 1(212), 131--138 (2004), http://dl.acm.org/citation.cfm?id=988857 (retrieved from)

Continuous Pain Intensity Estimation
from Facial Expressions

Sebastian Kaltwang, Ognjen Rudovic, and Maja Pantic

Department of Computing, Imperial College London, UK
{sebastian.kaltwang08,o.rudovic,m.pantic}@imperial.ac.uk

Abstract. Automatic pain recognition is an evolving research area with promising applications in health care. In this paper, we propose the first fully automatic approach to continuous pain intensity estimation from facial images. We first learn a set of independent regression functions for continuous pain intensity estimation using different shape (facial landmarks) and appearance (DCT and LBP) features, and then perform their late fusion. We show on the recently published UNBC-MacMaster Shoulder Pain Expression Archive Database that late fusion of the afore-mentioned features leads to better pain intensity estimation compared to feature-specific pain intensity estimation.

1 Introduction

Automatic pain recognition has received increased attention in the recent years mostly because of its applications in health care, ranging from monitoring patients in intensive-care units to assessment of chronic lower back pain [1]. Current research on automatic pain detection is based on automatic analysis of facial expressions, since it has been shown that facial cues are very informative for pain detection [2].

To date, there are only few works that have addressed the problem of automatic pain detection [3,4,5,6,7]. Brahnam et al. [3] used Principal Component Analysis, Linear Discriminant Analysis and Support Vector Machines (SVMs) for binary classification of pain images (i.e., pain vs. no pain). Gholami et al. [4] used intensities from facial images to train a Relevance Vector Machine (RVM) classifier for pain detection. Little-wort et al. [5] proposed a two-layer SVM-based approach for the classification of image sequences in terms of real pain and posed pain. In their approach, the presence of Facial Action Units (AUs) (see [8] for AU description) per frame is detected with a set of AU-specific SVM classifiers based on Gabor features. The outputs of the AU-specific SVMs are then temporally filtered and used as an input to the SVM classifier. The work by Lucey et al. [6] also addresses AU and pain detection based on SVMs. They detect pain either directly using image features or by applying a two-step approach, where first AUs are detected and then this output is fused by Logistical Linear Regression in order to detect pain. In their more recent work in [7], the authors train separate SVM classifiers for three-level pain intensity estimation.

Except from the approach proposed in [7], the rest of the aforementioned methods have been proposed for pain detection only (i.e., pain vs. no pain). In this paper, we propose a three-step approach to continuous pain intensity estimation per video frame

G. Bebis et al. (Eds.): ISVC 2012, Part II, LNCS 7432, pp. 368–377, 2012.

(in contrast to [7], which estimates pain for a whole video sequence only). The outline of the proposed approach is depicted in Fig. 1. In the first step, we extract shape-based features (i.e, locations of characteristic facial points) and appearance-based features (Local Binary Patterns (LBPs) [9] and Discrete Cosine Transform (DCT) [10]) from facial images of subjects displaying different intensities of pain. The pain intensity was annotated by the database creators using sixteen discrete values (0 to 15), with 0 meaning no pain and 15 meaning its peak. In the second step, for each set of features we train separate regression models (in this paper, we employ Relevance Vector Regression (RVR) [11]) for prediction of the pain intensity levels. Note that although the regressor training is performed using discrete outputs (i.e., intensity labels from 0 to 15), during inference the regressors give a continuous estimation of the pain intensity. Finally, the outputs of the regressors trained using different feature sets are combined in two ways: (i) by computing the mean estimate of the regressors, and (ii) by using the outputs of separate regressors as an input to another RVR, which gives a single estimate for the pain intensity. In contrast to the aforementioned methods which deal with pain detection only (i.e., pain vs. no pain), the proposed approach is the first one that performs continuous pain intensity estimation. Furthermore, we show that the proposed feature-fusion scheme outperforms the separately trained RVRs on different feature sets, whereby the combination of appearance features (DCT and LBP) performs best. We also demonstrate the performance of the proposed approach in the task of continuous intensity estimation of the facial AUs.

The rest of the paper is organized as follows. Section 2 describes the employed database. Feature extraction is detailed in Section 3. The regression-based approach to continuous pain intensity estimation is presented in Section 4. Section 5 shows the experiments and discusses the results. Section 6 concludes the paper.

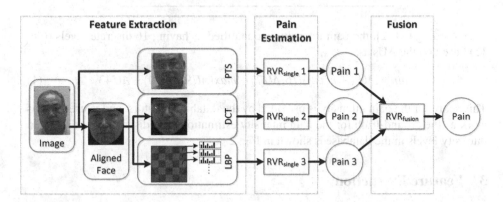

Fig. 1. Overview of the proposed approach to continuous pain intensity estimation. We first extract three feature sets from a face image: facial landmarks (PTS), Discrete Cosine Transform coefficients (DCT) and Local Binary Patterns (LBP). We then use Relevance Vector Regression (RVR) to learn the feature-specific functions, which independently estimate the pain intensity from each feature set. In the final step, we use a second layer RVR to perform the fusion of the pain intensity estimations obtained by the feature-specific functions.

2 Database

We use the publicly available UNBC-MacMaster Shoulder Pain Expression Archive Database [6] for our experiments. It contains face videos of patients suffering from shoulder pain while performing range-of-motion tests of their arms. Two different movements are recorded: (1) the subject moves the arm himself, and (2) the subject's arm is moved by a physiotherapist. Only one of the arms is affected by pain, but movements of the other arm are recorded as well as a control set. 200 sequences of 25 subjects were recorded (in total 48,398 frames). For each frame, AU intensities are provided for the AUs 4, 6, 7, 9, 10, 12, 20, 25, 26 and 27 on a 6 level scale (0-5) and for AU 43 on a 2 level scale (present or not). The number of frames available per AU intensity level is shown in Table 1.

Table 1. Frame distribution over AU intensity levels

Intensity	0	1	2	3	4	5
AU4	47324	202	509	225	74	64
AU6	42841	1776	1663	1327	681	110
AU7	45034	1360	991	608	305	100
AU9	47975	93	151	68	76	35
AU10	47873	171	208	63	61	22
AU12	41511	2145	1799	2158	736	49
AU20	47692	286	282	118	0	20
AU25	45992	766	803	611	138	88
AU26	46306	430	918	265	478	1
AU27	48380	6	3	3	6	0
AU43	45964	2434	-	-	-	-

According to [12], the pain intensity is quantified as having 16 discrete levels (0 to 15) based on the AUs as:

$$Pain = AU4 + max(AU6, AU7) + max(AU9, AU10) + AU43 \qquad (1)$$

This score for the pain intensity is provided by the database creators, and is used in this work as the ground-truth for the pain intensity estimation. The distribution of the pain intensity levels in the database is shown in Fig. 2.

3 Feature Extraction

In the first step of our approach, we perform extraction of three different sets of features. The first set, denoted as Set 1, contains the locations of 66 facial landmark points (PTS) (see Fig.1) that are extracted by the database creators using the Active Appearance Model (AAM) [13]. As a preprocessing step, these points were aligned by applying Procrustes analysis.

The second set, denoted as Set 2, contains features obtained by applying the Discrete Cosine Transform (DCT) [10] to the aligned facial images. Specifically, the faces were

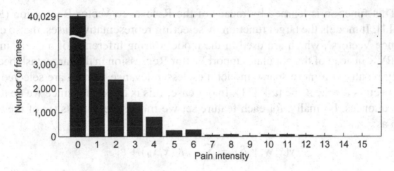

Fig. 2. Frame distribution over pain intensity levels

first aligned to a base shape using the points from the triangulated mesh of the AAM (see [6] for details). Later, the 2-dimensional DCT was applied to the aligned images, and the first 500 coefficients were used as features, which were selected based on the *zig-zag* scheme [14]. We employ the PTS and DCT features in this paper since they have been previously proposed for pain detection (i.e., pain vs. no pain) in [6] .

The third set, denoted as Set 3, contains Local Binary Pattern (LBP) [9] features extracted from the above-mentioned shape-aligned images. We use these features since they have been shown to be effective for facial expression recognition [15]. An aligned face image was divided into patches and LBP histograms were extracted from each patch separately. After an initial parameter search, we chose uniform LBPs with 8 radial points on a radius of 2 pixel. The image was divided into 9x9 equally sized non-overlapping patches with a resolution of 14x13 pixel. The LBP histogram extracted from a patch resulted in a 59-D feature vector. The final LBP feature vector was the 4779-D concatenation of all 81 histograms.

We employ these three sets of features because they contain different types of information. PTS are geometric features, and are robust to illumination changes. However, they cannot accurately capture subtle facial movements (e.g., the eye wrinkles). This can be well described by the appearance features (i.e., DCT and LBP) that are derived from pixel intensities of an image. Compared to PTS, DCT and LBP are much more sensitive to skin color variation, and texture variation due to the illumination changes. Note, however, that DCT and LBP capture different characteristics of the texture changes. Specifically, DCT features describe image appearance on a large scale, which can be seen from a DCT reconstructed image: The overall image structure is still preserved, but sharp edges are lost (see Fig. 1). Conversely, LBP features are local descriptors that model statistics of the gradient orientations within a small pixel neighborhood, i.e. they describe the edges. For the aforementioned reasons, we hope that by fusion of these three types of features we can improve the overall accuracy of the continuous pain intensity estimation, as proposed in this paper.

4   Continuous Pain Intensity Estimation

To perform continuous pain intensity estimation from a single feature set, we learn a regression function that maps the features to the corresponding (discrete) pain intensity levels. This function is learned by means of the Relevance Vector Regression (RVR) model [11]. It models the target function by selecting representative cases, the so called 'Relevance Vectors', which are used in the model during inference of a query image. We use RVR instead of the popular Support Vector Regression in the target task because it usually results in a more sparse model, i.e., less relevance vectors are selected than support vectors for the same task [11]. In our case, this is important since we deal with image sequences. Formally, for each feature set we model the outputs (y) of the target function as:

$$y(x; \mathbf{w}, \gamma, \delta) = \sum_{n=1}^{N} w_n K(x, x_n) + \varepsilon, \tag{2}$$

where x is the input feature vector, $\{x_1, ..., x_N\}$ are N training inputs and $\mathbf{w} = (w_1, ..., w_N)$ are the weight parameters. Here, the sparsity of the model comes from the fact that most of the weights parameters tend to go to zero, thus, the corresponding training samples are not used for inference. As the kernel function, we use the standard Radial Basis Function (RBF) kernel with the length scale parameter γ. The noise on the outputs is modeled as a Gaussian with zero mean and the variance δ.

Once the feature-set-specific target functions are learned, we perform late fusion of their outputs. This fusion is performed in two ways: (i) mean fusion and (ii) RVR fusion. In the mean fusion approach, we calculate the mean of the outputs, obtained by the feature-set-specific target functions $\{y_1, ..., y_L\}$, as $y_f = \frac{1}{L} \sum_{l=1}^{L} y_l$, where y_f is the mean fusion output and L is the number of the feature sets. RVR fusion is performed by learning another RVR model that uses the outputs of the feature-set-specific target functions as an input, i.e., $\hat{y} = (y_1, ..., y_L)$, which are continuous estimates of the pain level intensities, and the (discrete) pain level intensities as outputs. This fusion function is given by

$$y_f(\hat{y}; \mathbf{w}^{\mathbf{f}}, \gamma^f, \delta^f) = \sum_{m=1}^{M} w_m^f K^f(\hat{y}, \hat{y}_m) + \varepsilon^f, \tag{3}$$

where $\{\hat{y}_1, ..., \hat{y}_M\}$ are M training inputs, obtained from the first-layer outputs, and $\mathbf{w}^{\mathbf{f}} = (w_1^f, ..., w_M^f)$ are the weight parameters, γ^f is the length scale of the Radial Basis Function kernel K^f and ε^f is the noise, as defined above. Note that the training samples used to learn the feature-set-specific target functions may differ from the samples used to learn the fusion function.

5   Experiments and Results

We performed two sets of experiments. In the first set of experiments we evaluated the performance of the proposed approach in the task of continuous AU intensity estimation. In the second set, we evaluated the performance in the task of continuous pain intensity estimation. In all our experiments we applied a leave-one-out cross-validation procedure. Specifically, we used facial images of 24 subjects for training and one subject for testing. The feature-specific target functions were trained using the same training data as for the fusion functions. Note that in terms of generalization performance, the performance of the proposed 2-layer approach is expected to be better if the feature-specific target functions and the fusion function are trained using data corresponding to different subjects. However, we found that this strategy results in worse performance than using the same training data to train both layers. This could be due to the limited number of available subjects: if the subjects are split between the first and the second layer, then each of the layers is trained on less subjects, and hence the performance decreases. Note also that AU27 was left out, since only few examples with intensities greater than zero are present in the dataset (see Table 1). We measured the performance of the proposed approach using the mean squared error (MSE) and the Pearson correlation coefficient (CORR). The MSE and CORR were computed on the differences between the predicted pain/AU intensities and the relevant ground truth. Furthermore, MSE and CORR were computed per subject and per sequence, and then correspondingly weighted by the number of frames in each sequence, in order to obtain an average value for each measure.

Table 2 shows the results for the feature-specific target function learned in the task of continuous pain/AU intensity estimation. In the case of pain intensity estimation, the ground truth contains 16 discrete intensity levels, while in the case of AUs there are 6 discrete intensity levels. In addition, we show the results of two methods for pain intensity estimation: Pain (I) is directly estimated from the features as described in Section 4, where Pain (II) is calculated from the estimated AU intensities by using Eq. 1. As can be seen, the results obtained by the latter method are in some cases outperformed by the former method, where the pain intensity is estimated directly from the training data. This is a consequence of the error propagation in the AU estimation, since for some AUs only few positive examples (i.e., with intensity level greater than zero) were available during training. Since the Pain (II) is computed by using a deterministic formula, the inaccuracies in the estimation of each AU are added in the final estimate of the pain intensity. Note also from Table 2 that for AU intensity estimation, LBP features outperform PTS and DCT features. This is because the LBPs are local descriptors and are able to better capture appearance variation caused by changes in AU intensities, since different AUs are located in different regions of a facial image. The accuracy in AU intensity estimation attained by using LBPs directly translates into the accuracy attained by the Pain (II) approach, which outperforms Pain (I) in the case of the LBPs. On the other hand, in the case of DCT features, which capture global changes in appearance, the Pain (I) is more accurate than the Pain (II) approach. This again shows that estimating the pain intensity level from AU intensities is sensitive to the errors in AU intensity estimation. We also observe that in the case of predicting AU20 with LBP features, the MSE can be misleading: the result of 0.103 seems better than the MSE of other AUs, but CORR is only 0.092. This is due to the imbalanced data used for training (see Table 1 and Fig. 2) where the vast majority of the frames have the zero intensity. Overall, LBP features perform best in terms of the MSE measure, while in the case of the CORR measure, the difference is not that apparent, though LBPs are still the best in most cases. On the other hand, DCT features perform best in the task of pain intensity estimation. Overall, appearance features (DCT and LBP) work better than shape features (PTS). However, the poor performance of shape features might be caused by

Table 2. Single feature results for Action Unit (AU) and pain intensity estimation, measured by the mean squared error (MSE) and the Pearson correlation coefficient (CORR). Pain (I) is estimated directly from the features and Pain (II) is calculated from the estimated AU intensities using Eq. 1. The best result for each target and each measure is printed in bold letters.

Measure	MSE			CORR		
Features	PTS	DCT	LBP	PTS	DCT	LBP
AU4	0.341	0.254	**0.204**	.096	**.140**	.133
AU6	0.906	0.592	**0.590**	.385	**.528**	.527
AU7	0.806	0.504	**0.379**	.120	.303	**.342**
AU9	0.119	**0.119**	0.113	**.246**	.224	.190
AU10	0.084	**0.079**	0.097	.171	**.203**	.169
AU12	1.010	0.717	**0.600**	.330	.484	**.548**
AU20	0.505	0.158	**0.103**	.012	**.092**	.092
AU25	0.707	0.579	**0.486**	.130	.104	**.204**
AU26	0.896	0.834	**0.475**	.013	.016	**.111**
AU43	0.300	0.273	**0.176**	.240	.291	**.465**
Pain (I)	2.592	**1.712**	1.812	.363	**.528**	.483
Pain (II)	2.532	1.716	**1.484**	.348	.480	**.518**

registration errors, because the Procrustes alignment cannot cope properly with out-of-plane rotations. A better registration will likely improve the single shape and the fusion results, therefore we would not suggest to rely on appearance features alone.

The results for the mean-fusion approach are shown in Table 3. In most cases, MSE and CORR improves over the results obtained with single features only. This shows that the employed features contain complementary information. Based on the CORR results, the DCT+LBP fusion gives the best results in most cases. This is not surprising, because DCT and LBP, although both being appearance-based features, capture different information: DCT captures global, while LBP captures local appearance variation.

The results for the RVR feature fusion are shown in Table 4. The results are similar to those obtained by the mean fusion in the sense that almost all values improve over the single feature results, as expected. However, the improved performance of DCT+LBP features is even more pronounced in the case of the RVR fusion approach, giving the best CORR results overall. Although we would expect the RVR fusion to perform at least as good as the mean fusion in all tasks, this does not seem to be the case. A reason for this could be the fact that both layers in the proposed approach are trained on the same data (because of the limited training data), which could have led the 2nd-layer RVR to over-fit the data. We plan to address this in our future work.

Fig. 3 shows an example of the pain intensity estimation from one test image sequence. The estimation is based on our best model, i.e., DCT+LBP RVR fusion (the Pain (I) approach). In most cases, the continuous pain intensity estimation is close to the ground-truth. Note, however, the peaks around the frames 95, 120 and 336, which are all caused by the eye blinks. This is a consequence of the fact that the proposed approach is static (i.e., it is trained per frame), and therefore, it cannot differentiate between an eye blink (short time) and eye closure (long time). During the training stage, the model has learned that the closed eyes are related to pain, and that is why the eye blinks result in sudden peaks in the estimated pain intensity, as shown in Fig. 3.

Table 3. Mean feature fusion results for Action Unit (AU) and pain intensity estimation, measured by the mean squared error (MSE) and the Pearson correlation coefficient (CORR). The best results are given in bold letters.

Measure	MSE				CORR			
Features	PTS+DCT	PTS+LBP	DCT+LBP	all	PTS+DCT	PTS+LBP	DCT+LBP	all
AU4	0.224	0.201	0.206	**0.191**	.205	.260	.294	**.295**
AU6	0.543	0.544	0.496	**0.472**	.500	.508	.526	**.543**
AU7	0.479	0.429	**0.361**	0.376	.276	.276	**.376**	.343
AU9	0.087	0.091	0.096	**0.083**	.370	.339	.323	**.382**
AU10	0.064	0.070	0.075	**0.064**	.371	.312	.334	**.370**
AU12	0.656	0.625	0.568	**0.563**	.529	.545	.582	**.588**
AU20	0.177	0.179	**0.103**	0.119	.103	.095	**.133**	.129
AU25	0.474	0.449	0.455	**0.415**	.212	.213	**.264**	.252
AU26	0.622	**0.482**	0.557	0.493	.090	.118	.090	**.120**
AU43	0.232	**0.184**	0.191	0.187	.360	.396	**.462**	.439
Pain (I)	1.469	1.642	1.508	**1.373**	.489	.481	**.554**	.547
Pain (II)	1.928	1.850	**1.368**	1.480	.395	.403	**.529**	.494

Table 4. RVR feature fusion results for Action Unit (AU) and pain intensity estimation, measured by the mean squared error (MSE) and the Pearson correlation coefficient (CORR). The best results are given in bold letters.

Measure	MSE				CORR			
Features	PTS+DCT	PTS+LBP	DCT+LBP	all	PTS+DCT	PTS+LBP	DCT+LBP	all
AU4	0.264	0.248	**0.242**	0.274	.209	.199	**.243**	.177
AU6	0.539	0.550	**0.480**	0.549	.487	.514	**.533**	.502
AU7	0.423	0.428	**0.343**	0.400	.248	.321	**.402**	.314
AU9	0.132	0.233	**0.120**	0.201	.401	.326	**.479**	.414
AU10	0.087	0.074	0.071	**0.070**	.080	.243	**.424**	.294
AU12	0.782	0.713	**0.617**	0.657	.507	.542	**.576**	.545
AU20	0.140	**0.088**	0.109	0.147	.049	.059	**.086**	.049
AU25	0.669	**0.538**	0.572	0.762	.106	.199	**.235**	.090
AU26	0.604	**0.414**	0.490	0.582	.005	.060	**.090**	.015
AU43	0.243	**0.158**	0.179	0.182	.352	.512	**.516**	.437
Pain (I)	1.801	1.567	**1.386**	1.804	.489	.485	**.590**	.502
Pain (II)	1.867	1.899	**1.633**	1.770	.342	.345	**.471**	.369

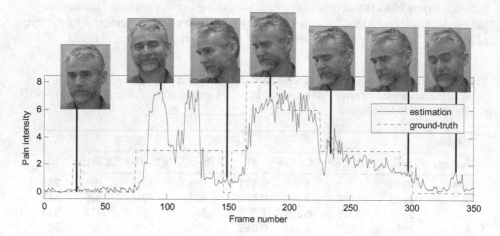

Fig. 3. Example pain estimation sequence for DCT+LBP RVR fusion

6 Conclusion

We have proposed a three-step approach to continuous pain intensity estimation based on Relevance Vector Regression. We have shown that for the task of continuous pain and AU intensity estimation, the proposed approach achieves better results when trained using appearance-based features (either DCT or LBP) than with the shape features (PTS). Also, when used as single input features, LBPs worked best in most cases. Furthermore, we showed that the fusion of DCT and LBP features gives the best performance in the target task. However, we believe that by a proper alignment of the shape-based features (e.g. by using [16]), the overall performance attained by the fusion of these three feature sets should improve. We also showed that direct pain estimation can be more accurate than calculation from the the AUs, which is probably due to the inaccuracies in AU intensity estimation. The approach presented in this paper estimates the AU intensities independently and does not exploit information about their co-occurrences. Furthermore, the current approach is static, and it cannot distinguish between eye blinks and eye closures, which are important cues for pain intensity estimation. These limitations of the proposed approach are the focus of our future research.

Acknowledgments. This work has been supported by EPSRC grant EP/H016988/1: Pain rehabilitation: E/Motion-based automated coaching. The work by Ognjen Rudovic and Maja Pantic is funded in part by the European Research Council under the ERC Starting Grant agreement no. ERC-2007-StG-203143 (MAHNOB).

References

1. Prkachin, K.M., Hughes, E., Schultz, I., Joy, P., Hunt, D.: Real-time assessment of pain behavior during clinical assessment of low back pain patients. Pain 95, 23–30 (2002)
2. Williams, A.C.d.C.: Facial expression of pain: An evolutionary account. Behavioral and Brain Sciences 25, 439–455 (2002)
3. Brahnam, S., Chuang, C.F., Shih, F.Y., Slack, M.R.: Machine recognition and representation of neonatal facial displays of acute pain. Artificial Intelligence in Medicine 36, 211–222 (2006)
4. Gholami, B., Haddad, W.M., Tannenbaum, A.R.: Agitation and pain assessment using digital imaging. In: Int'l Conf. of the Engineering in Medicine and Biology Society, IEEE, pp. 2176–2179 (2009)
5. Littlewort, G.C., Bartlett, M.S., Lee, K.: Automatic coding of facial expressions displayed during posed and genuine pain. Image and Vision Computing 27, 1797–1803 (2009)
6. Lucey, P., Cohn, J., Prkachin, K., Solomon, P., Matthews, I.: Painful data: The UNBC-McMaster shoulder pain expression archive database. In: Int'l Conf. on Automatic Face & Gesture Recognition and Workshops, IEEE, pp. 57–64 (2011)
7. Lucey, P., Cohn, J.F., Prkachin, K.M., Solomon, P.E., Chew, S., Matthews, I.: Painful monitoring: Automatic pain monitoring using the UNBC-McMaster shoulder pain expression archive database. Image and Vision Computing 30, 197–205 (2012)
8. Ekman, P., Friesen, W.V.: Facial action coding system: A technique for the measurement of facial movement (1978)
9. Ojala, T., Pietikainen, M., Maenpaa, T.: Multiresolution gray-scale and rotation invariant texture classification with local binary patterns. IEEE Trans. on Pattern Analysis and Machine Intelligence 24, 971–987 (2002)
10. Ahmed, N., Natarajan, T., Rao, K.R.: Discrete Cosine Transform. IEEE Trans. on Computers C-23, 90–93 (1974)
11. Tipping, M.E.: Sparse Bayesian learning and the relevance vector machine. The Journal of Machine Learning Research 1, 211–244 (2001)
12. Prkachin, K., Solomon, P.: The structure, reliability and validity of pain expression: Evidence from patients with shoulder pain. Pain 139, 267–274 (2008)
13. Cootes, T., Edwards, G., Taylor, C.: Active appearance models. IEEE Trans. on Pattern Analysis and Machine Intelligence 23, 681–685 (2001)
14. Wallace, G.K.: The JPEG still picture compression standard. Communications of the ACM 34, 30–44 (1991)
15. Shan, C., Gong, S., McOwan, P.W.: Facial expression recognition based on Local Binary Patterns: A comprehensive study. Image and Vision Computing 27, 803–816 (2009)
16. Rudovic, O., Pantic, M.: Shape-constrained Gaussian Process Regression for Facial-point-based Head-pose Normalization. In: Int'l Conf. on Computer Vision, IEEE, pp. 1495–1502 (2011)

Local Alignment of Gradient Features
for Face Sketch Recognition

Ann Theja Alex, Vijayan K. Asari, and Alex Mathew

Computer Vision and Wide Area Surveillance Laboratory,
Department of Electrical and Computer Engineering,
University of Dayton, Dayton, Ohio, USA
{alexa1,vasari1,mathewa3}@udayton.edu

Abstract. Automatic recognition of face sketches is a challenging problem. It has application in forensics. An artist drawn sketch based on the descriptions from the witnesses can be used as the test image to recognize a person from the photo database of suspects. In this paper, we propose a novel method for face sketch recognition. We use the edge features of a face sketch and face photo image to create a feature string called 'edge-string'. The edge-strings of the face photo and face sketch are then compared using the Smith-Waterman algorithm for local alignments. The results on CUHK (Chinese University of Hong Kong) student dataset show the effectiveness of the proposed approach in face sketch recognition.

1 Introduction

Face sketch recognition refers to the problem of recognizing a person from an artist drawn sketch. This has wide application in criminal investigation. A sketch drawn based on the descriptions is compared against a large database of face images to retrieve similar images. Since artist drawn sketches are vulnerable to variations due to the artist's rendering style, the problem presented is not a one to one matching problem- the focus is on eliminating unrelated images, and narrowing down the search. The victim or the witnesses are allowed to cross-verify and identify the convict from the list of images retrieved from database.

In face sketch recognition, the only features that are available are the edges that are common in the photo and the sketch. The machine learning concept of using line edge maps for face recognition tasks was first introduced by Takeo Kanade in 1973 and is also mentioned by Brunelli & Poggio [1][2]. The same idea was used by Gao & Leung for face recognition [3]. The approach was to generate edge maps and to use them in conjunction with the Hausdorff distance to recognize faces. Takács used the Sobel edge maps and combined it with Hausdorff distance [4]. The same concept was modified by Gao & Leung. They used Sobel edges to generate line segments and create line edge maps. Gao & Leung proposed another approach for human profile faces [5]. A similar method was proposed by Chen & Gao, where the line edge map is computed first and string

G. Bebis et al. (Eds.): ISVC 2012, Part II, LNCS 7432, pp. 378–387, 2012.

matching is performed for occlusion invariant face recognition [6]. We use an edge feature extraction technique inspired by the above findings.

Once the features are extracted, there should be a way to recognize and classify them. A promising memory representation for visual patterns is to represent them as strings, trees or a set of propositions which are commonly referred to as syntactic pattern recognition methods. Models that treat visual pattern as a symbolic description have been proposed in literatures in psychology [7]. The studies and experiments by Richard A. Chechile et al. provide psychological justification for the significance of syntactic approaches in the area of familiarity analysis for visual patterns [8]. Structural and syntactic method is a high level approach that finds a numeric or non-numeric description of a pattern [9][10]. We use a string representation of the edge features called the 'edge-string' in the proposed method.

There are two categories of methods for face sketch recognition. One set of methods that involve preprocessing to reduce the distance between the sketch space and photo space [11][12][13]. The sketches corresponding to the photo images are first synthesized. Once we have a sketch corresponding to the face photo, the final recognition task is performed in the sketch space using conventional classification methods. The photo images corresponding to the synthesized sketches that show highest similarity are then retrieved from the database for analysis by the witnesses or the victims. The second category involves methods which allow classification across modalities, and does not require sketch synthesis [14]. The proposed method belongs to the second category. The edges in the sketches and the photo images are used as the features for classification and image retrieval.

All sketch synthesis based methods require a very large set of training images for training the system. The images are first divided into patches. The final comparison step required for sketch synthesis has a complexity of $O((nd)^2)$, where n is the number of images and d the number of patches associated with each image. In addition to this, the conventional methods that are used in the final recognition task after sketch synthesis also involves additional complexity. Thus, all sketch synthesis based methods are computationally expensive. As the classifiers used are standard classifiers such as PCA and LDA, the accuracy of all these methods depends on the accurate synthesis of sketches from the training patches. However, for very accurate sketch synthesis a very large training patch set is required. Another drawback of these methods is that they are not scalable. The photos corresponding to the training patches should belong to the same race as the test image subject. If a new test subject from a different race is added to the system, a very large set of training images from the specific race should also be added.

The difficulties posed by conventional methods call for a new approach for face sketch recognition. The major advantage of our method is that it requires only one image for training and hence the proposed approach is scalable. Even if we add a subject of a different gender or a different race to the system, it can easily perform recognition task as long as there is one training image available. Our method does not rely on any preprocessing sketch synthesis step. This greatly reduces the computational complexity. Our algorithm finds application in face image retrieval from a large database such as a police mug shot database. Our technique is a novel approach that uses the inherent edges in the face photo and sketch images to solve the problem of face sketch recognition. To the best of our knowledge this is the first contribution that uses a syntactic approach for face sketch recognition. Experiments are carried out on the CUHK student dataset to prove the effectiveness of the proposed approach for face sketch recognition and sketch based face image retrieval.

2   Proposed Method

We assume that face regions are detected, cropped, registered/aligned, resized and centered prior to using our algorithm. The algorithm is a supervised learning technique- the training input and their respective classes are known *a priori*. The training images are also referred to as reference images. We need only one reference image per class. The reference images are the face photo images in the mugshot database. In training phase, the edges are detected for all reference images. The 'edge-string' representations corresponding to each edge image is created. These 'edge-strings' corresponding to the reference images are referred to as 'reference-strings'.

In the testing phase, the edge image corresponding to the test image (the artist drawn sketch) is first generated. Then, the edge string corresponding to the edge image is computed. The test edge-string (Λ_1) is matched with the reference string (Λ_2) using Smith-Waterman algorithm. The match count is then calculated. As the match count (σ) depends on the length of the two sequences, a second measure called the percentage similarity ρ is used which is computed as given in Eq.1.

$$\rho = \frac{\sigma}{min(\lambda_1, \lambda_2)} \tag{1}$$

Here, λ_1 and λ_2 are the lengths of the strings Λ_1 and Λ_2 respectively. The class corresponding to the maximum ρ value is identified as the class to which the test image belongs. A schematic diagram of the testing phase of the algorithm is shown in fig. 1. The proposed method is described in more detail in the following sections. Edge feature extraction is described in section 2.1 and the string alignment using Smith-Waterman algorithm is described in section 2.2.

2.1   Edge Feature Extraction

The two steps involved in the edge feature extraction process are Canny edge detection and 'edge- string' generation.

Edge Detection: The proposed method uses Canny's algorithm for edge detection. It offers better edge output with good localization and works well even if the image is noisy [15]. Canny suggests using Sobel, Robert's Cross gradient

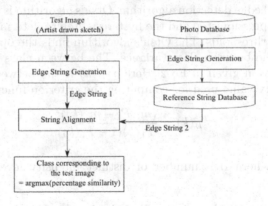

Fig. 1. A schematic diagram of the proposed method

or Prewitt kernels for edge detection. We use Robert's Cross gradient, since it is the simplest kernel. The Gaussian kernel that is used in the Canny edge detection algorithm for smoothing the image contributes to the removal of any noise. The σ value (the standard deviation of the Gaussian kernel) used in the edge detection algorithm is 1.4. In the final hysteresis step, the method uses an adaptive threshold selection mechanism using the Otsu's algorithm to determine the Canny edge detector thresholds [16][17]. The block diagram of the adaptive approach is shown in fig. 2.

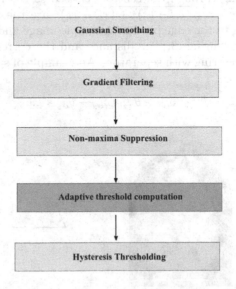

Fig. 2. Modified Canny edge detection algorithm

In the modified edge detection algorithm, Otsu's algorithm is used between the Non-maxima suppression step and the hysteresis step to determine the high (T_h) and the low (T_l) thresholds. The Otsu's algorithm finds the optimum threshold by maximizing the between class variance. The between class variance is given by criterion function given in Eq.2. Here $\sigma^2{}_B(k)$ is the between class variance when the intensity value used for computing the criterion function is k.

$$\sigma^2{}_B(k) = \frac{(m_G P_1 - m_1(k))^2}{P_1(k)(1 - P_1(k))} \tag{2}$$

$m_G = \sum_{i=0}^{L-1} i p_i$, where L is number of distinct intensity levels in the image.

$m_1(k) = \frac{\sum_{i=0}^{k} i p_i}{P_1(k)}$ and $P_1(k) = \sum_{i=0}^{k} p_i$. The criterion function is computed for every intensity value k. The value of k that maximizes Eq.2. is chosen as the high threshold (T_h) for the Canny edge detection. The lower threshold is set as $T_l = 0.5 \times T_h$ as in [17].

'Edge-String' Generation: In this step, 'edge-strings' are generated from the edge image. The edge images are scanned in the raster order. The polar coordinate representations of the pixel in the edge image are concatenated to create the 'edge-string'. The centroid (x_0, y_0) of the edge image is first identified. The distance 'δ' of each edge pixel to the centroid and the angle 'θ' from the horizontal axis fully define the location of a pixel. For computing the angle, the horizontal line passing through the centroid is used as the reference line. $\theta_\alpha = arctan(\frac{y_\alpha - y_0}{x_\alpha - x_0})$ and $\delta_\alpha = \sqrt{(y_\alpha - y_0)^2 + (x_\alpha - x_0)^2}$ are the angle and distance values corresponding to the edge pixel α. A sample image and the corresponding edge image are shown in fig. 3. The distance and the corresponding angle are then represented as a string with separators. An example of such a string is given below.

$$\Lambda_1 = \delta_{\alpha_1} - \theta_{\alpha_1} * \delta_{\alpha_2} - \theta_{\alpha_2} * \delta_{\alpha_3} - \theta_{\alpha_3} * ... \delta_{\alpha_n} - \theta_{\alpha_n} *$$

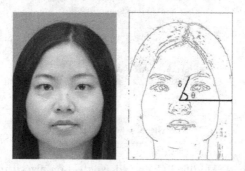

Fig. 3. Sample image and the corresponding edge image with the distance and angle values marked

Here $\delta_{\alpha_i} - \theta_{\alpha_i}$ represents a string primitive corresponding to the edge pixel 'α_i'. The symbol '*' separates the edge pixel representations, and the symbol '-' separates the distance and angle values representing a pixel.

2.2   Local String Alignment

The similarity between the strings is computed using Smith-Waterman algorithm [18]. The algorithm offers an optimal local alignment which finds local similarities. It is also robust to noise. The algorithm was originally proposed for genetic sequence alignments where minor mutations may render similar sequences dissimilar. Similar to the mutations producing dissimilarities in protein sequences, the edge strings of face sketches are prone to local changes due to noise, expression variations and additional artistic strokes. These variations create dissimilarities between the facial image and the sketch of the same person while retaining some locally similar regions. We use a modified version of the Smith-Waterman algorithm in our method. In our modification, when we align an element of the string to any element in the other string, a comparison is carried out. An element in string 'Λ_1' is said to match with an element in string 'Λ_2' only if the distance and angle values of the element match.

Modified Local String Alignment: The steps involved in the modified Smith-Waterman algorithm are described below:

Λ_1 and Λ_2 are the two strings under consideration
λ_1=length(Λ_1) and λ_2=length(Λ_2)
The dynamic programming matrix Ψ of dimension $\lambda_1 X \lambda_2$
$\mu=1$, is the match value, $\gamma=-1$, is the gap penalty,
σ is the match count, η_1 and η_2 the indices

Initialization:
$\psi(0,0) \leftarrow 0$
for $\eta_1 \leftarrow 1$ to λ_1
    do $\Psi(\eta_1,0)=0$
end for
for $\eta_2 \leftarrow 1$ to λ_2
    do $\Psi(0,\eta_1)=0$
end for

Recursion:
for $\eta_1 \leftarrow 1$ to λ_1
    for $\eta_2 \leftarrow 1$ to λ_2

$$\text{do } \Psi(\eta_1,\eta_2)=\text{argmax}\begin{cases} 0 \\ \Psi(\eta_1-1,\eta_2-1)+\mu \\ \Psi(\eta_1-1,\eta_2)+\gamma \\ \Psi(\eta_1,\eta_2-1)+\gamma \end{cases}$$

end for
end for

Backtracking:
$(\eta_1, \eta_2) = \text{argmax}(\Psi)$
while $\Psi(\eta_1, \eta_2) \neq 0$ do

$$\text{if } \Psi(\eta_1, \eta_2) = \begin{cases} (\eta_1 - 1, \eta_2 - 1) \text{ then } \sigma = \sigma + 1 \\ (\eta_1 - 1, \eta_2) \text{ then } \sigma = \sigma + 0 \\ (\eta_1, \eta_2 - 1) \text{ then } \sigma = \sigma + 0 \end{cases}$$

end while

To compare between the global and local alignment performance, we analyzed the recognition rates of both these techniques for a small subset of the CUHK database with 10 images. The combination of local alignment and 'edge-string' provided a 100% recognition while global alignment provides only 80%. Hence, local alignment using Smith-Waterman algorithm is the best choice. The proposed algorithm for string comparison is inherently robust to noise.

Every string alignment algorithm involves a scoring metric. To learn any scoring metric, a large number of training images is required for probabilistic analysis of the data. In face sketch recognition, the number of available images for training is usually limited. Hence the proposed method uses the simplest of all scoring metrics associated with the Smith Waterman algorithm. The gap penalty is -1 and match value is 1 in the matching algorithm.

3   Experiments and Results

For face sketch recognition, most of the time only one sketch image is available as the reference image. The proposed method uses the test sketch image and retrieves related photo images from the database based on the local alignment similarity measure. The retrieved images may be shown to the witnesses or the victims to identify the suspect.

Experiments on the CUHK Student Dataset: In the CUHK student dataset, the probe set consists of 100 face sketches and a gallery set of 100 photo images [11][12][13]. The rank curve in fig. 4, shows the effectiveness of the proposed method in face sketch recognition as compared with other methods. The Geometry method and Eigen face method does not require sketch synthesis just like our method.

Our method performs better than other methods as shown in fig.4. This is because the edges are the only features that are common across the photos and sketches and the methods such as Geometry method and Eigen face method are not capturing these obvious features. The proposed method offers a 100% recognition at rank 19 as shown in Table.1. Thus the correct match to any sketch will be in the best 19 matches provided by the proposed method.

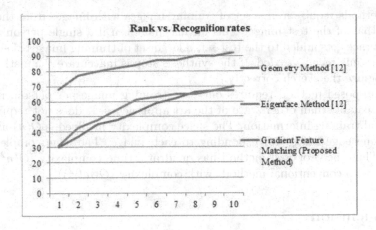

Fig. 4. Rank curve for CUHK student dataset

Table 1. Rank vs. Recognition Rate for the ranks 11 to 19

Rank	11	12	13	14	15	16	17	18	19
%	92	93	95	96	97	98	99	99	100

The reduced set of possibilities is then analyzed by a human being for the final recognition of the convict. 19 images is a small number for a human being to analyze. The primary aim of any sketch recognition algorithm is just to reduce the number of possibilities to a small number as a choice is hard to make from a very large number of possibilities. Our method is capable of achieving this primary objective and hence is useful for sketch recognition applications.

The methods that need sketch synthesis require a large set of training images. There is an additional preprocessing step involved, where the sketches are divided into patches of small size such as 3×3, 5×5, 7×7 or higher sizes. As the patch size increases the accuracy of the sketch synthesis reduces. On the other hand, as the patch size decreases, the space and time complexity of the method increases. The accuracy of the sketch synthesis based methods depends on the number of training patches and consequently, the requirement of a large training set is inevitable. The entire training set and the test set need to be first divided into such small patches. This is a very costly preprocessing step. The space complexity of these methods is $O(nd)$, where n is the number of images and d is the size of the patch. The classification step introduces additional complexity. After preprocessing, the sketch synthesis is performed by matching each patch. The time complexity associated with the synthesis step is $O((nd)^2)$. As the number of patches (d) is a very large quantity the d^2 term further increases the complexity. This makes sketch synthesis based methods highly inefficient in terms of space and time complexities. Furthermore, the approach is not scalable. If the patches belong to a race different from that to which the test image belongs, these methods require a new set of training images with images from the same race as that of the test image. This means that even if a single person from a different race gets added to the test set, a large set of training images of subjects from the same race is needed. If the synthesis step is inaccurate the method will not recognize the sketch correctly.

The proposed method requires only one training image per person, making it more scalable than other state of the art approaches. It does not require any additional training information. The space complexity involved is $O(n)$ as there is only one edge image corresponding to each image. The time complexity is $O(n^2)$. Thus the proposed method has quadratic time complexity ($O(n^2)$) and is superior to conventional methods with complexity ($O(nd)^2$)).

4 Conclusion

We have proposed a solution for face sketch recognition, a very important forensic problem. The proposed algorithm does not require face sketch synthesis. The effectiveness of our approach for face sketch recognition, is verified by the experimental results on the CUHK student dataset. The use of edge features and attributed strings for sketch recognition is a novel contribution of this paper. Our method can be further improved by incorporating learning rules to extract other commonalities between the photo and sketch images. Future work includes modification and testing the proposed algorithm on other face sketch databases such as CUFSF (CUHK Face Sketch FERET Database).

References

1. Kanade, T.: Picture Processing by Computer Complex and Recognition of Human Faces, Technical report, Dept. Information Science, Kyoto Univ. (1973)
2. Brunelli, R., Poggio, T.: Face Recognition: Features versus Templates. IEEE Transactions on Pattern Analysis and Machine Intelligence 15, 1042–1052 (1993)
3. Gao, Y., Leung, M.K.H.: Face recognition using line edge map. IEEE Transactions on Pattern Analysis And Machine Intelligence 24(6), 764–779 (2002)
4. Takács, B.: Comparing face images using the modified Hausdorff distance. Pattern Recognition 31, 1873–1881 (1998)
5. Gao, Y., Leung, M.K.H.: Human face profile recognition using attributed string. Pattern Recognition 35, 353–360 (2002)
6. Chen, W., Gao, Y.: Recognizing partially occluded faces from a single sample per class using string-based matching. In: Proceedings of the European Conference on Computer Vision, vol. 3, pp. 496–509 (2010)
7. van der Helm, P.A., Leeuwenberg, E.L.J.: Accessibility: a criterion for regularity and hierarchy in visual pattern codes. Journal of Mathematical Psychology 35, 151–213 (1991)
8. Chechile, R.A., Anderson, J.E., Krafczek, S.A., Coley, S.L.: A syntactic complexity effect with visual patterns: evidence for the syntactic nature of the memory representation. Journal of Experimental Psychology: Learning, Memory, and Cognition 22, 654–669 (1996)

9. Fu, K.S.: Syntactic pattern recognition and applications. Prentice-Hall, Englewood Cliffs (1982)
10. Bunke, H.: Structural and syntactic pattern recognition. In: Chen, C.H., Pau, L.F., Wang, P.S.P. (eds.) Handbook of Pattern Recognition and Computer Vision, pp. 163–209. World Scientific Publishing Company, Singapore (1994)
11. Tang, X., Wang, X.: Face Photo Recognition Using Sketch. Proceedings of IEEE International Conference on Image Processing 1, 257–260 (2002)
12. Tang, X., Wang, X.: Face Sketch Recognition. IEEE Transactions on Circuits and Systems for Video Technology (CSVT), Special Issue on Image- and Video- Based Biometrics 14(1), 50–57 (2004)
13. Wang, X., Tang, X.: Face Photo-Sketch Synthesis and Recognition. IEEE Transactions on Pattern Analysis and Machine Intelligence 31, 1955–1967 (2009)
14. Zhang, W., Wang, X., Tang, X.: Coupled Information-Theoretic Encoding for Face Photo-Sketch Recognition. Proceedings of IEEE Conference on Computer Vision and Pattern Recognition, 513–520 (2011)
15. Canny, J.: A computational approach to edge detection. IEEE Transactions on Pattern Analysis and Machine Intelligence 8, 679–698 (1986)
16. Otsu, N.: A threshold selection method from gray-level histogram. IEEE Transactions on System, Man and Cybernetics SMC-9(1), 62–66 (1979)
17. Huo, Y., Wei, G., Zhang, Y., Wu, L.: An adaptive threshold for the Canny Operator of edge detection, iccsee. In: International Conference on Image Analysis and Signal Processing, pp. 371–374 (2010)
18. Smith, T.F., Waterman, M.S.: Identification of common molecular subsequences. Journal of Molecular Biology 147, 195–197 (1981)

Towards the Usage of Optical Flow Temporal Features for Facial Expression Classification

Raymond Ptucha and Andreas Savakis

Computing and Information Sciences and Computer Engineering,
Rochester Institute of Technology, Rochester, NY

Abstract. Psychological evidence suggests that the human ability to recognize facial expression improves with the addition of temporal stimuli. While the facial action coding community has largely migrated towards temporal information, the facial expression recognition community has been slow to utilize facial dynamics. This paper contrasts the contributions of static vs. temporal features, including both dense and sparse facial tracking methodologies in combination with sparse representation classification. The temporal methods of facial feature point tracking, motion history images, free form deformation, and SIFT flow are adapted for facial expression classification. Dense optical flow for facial expression recognition is successfully utilized. We show that when used in isolation, the best temporal methods are just as good as static methods. However, when fusing temporal dynamics with static imagery significant increases in facial expression classification are achieved.

1   Introduction

The ability to extract facial semantic information has widespread implications on security, entertainment, and human computer interfaces. Lew [1] argues that future devices will need to engage and embrace all the human subtleties in order to fully convey the true underlying message. Pantic [2] found that humans relied on facial expression more than body gestures or vocal expression in the judgment of behavioral cues.

The study of recognizing the six culture agnostic emotions (fear, sadness, happiness, anger, disgust, and surprise) has made great strides in recent years from constrained frontal posed faces to unconstrained faces in natural conditions [3-5]. Two main approaches include geometric and appearance based modeling. Geometric methods localize facial landmarks (outline of eyes, lips, nose, etc.) using techniques such as Active Appearance Models (AAM) [6]. Appearance based methods work holistically with facial pixels enabling the capture of facial muscle subtleties (such as nose wrinkles or dimple formation). Static methods of facial expression classification work on a single captured image. State of the art classifiers use Gabor wavelets [7], Local Binary Patterns (LBP) [8], Sparse Representations [9] or a combination thereof.

In an effort to more fully describe the varied nature of the human face, facial muscle action descriptors called Action Units (AUs) have been defined in the Facial

G. Bebis et al. (Eds.): ISVC 2012, Part II, LNCS 7432, pp. 388–397, 2012.
© Springer-Verlag Berlin Heidelberg 2012

Action Coding System (FACS) [10]. Rather than attempting to interpret the facial emotion, FACS captures all the possible atomic facial signals which can then be used as features into a reasoning engine. Although AU detection works well with static imagery, recent publications have shown that temporal dynamics can improve AU detection considerably [11].

Psychological studies indicate that temporal evidence is necessary towards the full comprehension of the emotional state of the face of interest [2, 12]. However, the usage of facial dynamics in the facial expression community is quite limited. As evidence of such, in the 2011 Facial Expression Recognition and Analysis Challenge (FERA2011), only four (out of fifteen) entrants utilized facial dynamics. Popular methods are extensions to static methods such as LBP-TOP [8] and LPQ-TOP [13].

The usage and interpretation of facial dynamics [12] by humans is an active area of research. Facial expressions typically contain an onset, apex, offset, and neutral stage. The timing and duration of each stage are critical to the interpretation of the observed behavior. These temporal dynamics have been used to discern between genuine and acted pain [14], telling the truth vs. lying [15], detection of depression [16], synthesizing facial expressions for avatars [17], and much more.

Facial dynamic methods include temporal tracking of geometric landmarks as well as tracking changes in facial appearance. Salient landmarks, such as corners of the eyes and mouth are generally well behaved and track well using optical flow techniques. Less well defined features, e.g. nose wrinkle, can be tracked with dense optical flow techniques such as Motion History Images (MHI), Free Form Deformations (FFD), or SIFT flow. MHI was initially introduced for human movement recognition [18], and was later adopted for facial AU detection [11]. FFD was initially introduced for medical image registration and later adopted for facial AU detection [11]. SIFT flow was introduced for generic image registration [19], and adopted for face alignment by [19] and [20]. To the best of our knowledge, this is the first paper to use motion vectors from MHI, FFD, and SIFT flow for facial expression classification.

This paper contrasts the contribution of static vs. temporal features and specifically investigates the advantages and disadvantages of utilizing temporal predictor methods. The rest of this paper is organized as follows. Section 2 reviews the baseline static image processing technique, Section 3 reviews the temporal processing techniques, Section 4 presents experimental results, and Section 5 summarizes with conclusions.

2   Static Image Processing

While there are many methods of performing facial expression classification using static imagery, we propose to use a baseline multiclass SVM method and a Sparse Representation (SR) method. The multiclass SVM is used because of its ubiquitous usage and excellent performance. The sparse method is used because of its tie in to human biology and excellent performance. While it is not clear how the human brain extracts basic emotional estimation of faces, evidence of sparsity in V1, the first layer of the human visual cortex, has gained considerable attention [22, 23]. Because humans are still better than machines at facial expression classification, borrowing from biology is an interesting endeavor. Although SRs are not common in the facial understanding community, there have been several successful implementations [9, 21, 24]. We summarize the two methods below.

Following face detection, AAM models automatically localize key facial features such as the eyes, mouth, and eyebrow boundaries. The eye and mouth corner points are used to define an affine mapping to a frontal canonical face representation of 60x51 pixels. Image masks extract targeted facial regions in preparation for Gabor wavelet processing. To make the subsequent classification more discriminative, the Gabor processed facial regions are written in lexicographic ordering and mapped onto low dimensional manifold surfaces.

In order to support the extension of the manifold models to new examples, linear techniques such as the Locality Preserving Projections (LPP) [25] solve a linear approximation of the non-linear manifold object. LPP creates an adjacency mapping of the top k neighbors for each feature point x_i by weighting each neighbor by distance to form a nxn adjacency matrix W with entries w_{ij}. W is defined similarly for input X and output Y, such that if neighbors x_i and x_j are close, y_i and y_j are also close. When supervised labels are available, a discriminative embedding can be achieved. Formally, W is initialized to all zeros, and then w_{ij} entries corresponding to the same classes are set to $1/k$, where k is the number of samples per class. Finally, LPP computes eigenvectors of the generalized eigenvector problem:

$$XLX^TU = \lambda XDX^TU \tag{1}$$

where D is a diagonal matrix of the column sums of W, L is the Laplacian matrix, $L=D-W$, and U is the resulting dimensionality reduction projection matrix

For the two static methods, the dimensionality reduced data is passed into both a multiclass linear SVM classification engine as well as a SR classification engine. In the SR framework, a test sample is represented as a sparse linear combination of exemplars from a training dictionary, Φ. Let the input signal be $y \in \mathbf{R}^d$ and the dictionary of examples $\Phi \in \mathbf{R}^{dxn}$. A natural way to represent y from Φ is by solving the linear equations system $\hat{y}= \Phi a$. Finding the sparsest solution in the presence of noise is called Basis Pursuit Denoising (BPDN):

$$\hat{a} = \arg \min \|a\|_1 \quad s.t. \ \|y - \Phi a\|_2 \leq \varepsilon \tag{2}$$

Given the sparse representation coefficients $\hat{a}$ of a test sample, a minimum reconstruction error estimates the class $c*$ of our test sample by comparing the error between our input sparse sample y and a reconstruction of $\hat{y}$ using the coefficients a^c corresponding to each class c, one class at a time:

$$c* = \arg \min_{c=1...z} \|y - \Phi \, a^c\|_2 \tag{3}$$

The class of the test sample is assigned to be the class with the minimum error.

3   Temporal Representation

The communication between humans naturally contains a temporal signature. For example, the rolling of the eyes, or raising of an eyebrow carries significant observable information that is best represented in a temporal fashion. To explore this problem more thoroughly, we adopt both sparse and dense optical flow techniques across the human face and incorporate both into our expression classifiers.

Facial expressions or gestures can occur at any point in time and are variable in length. Thus, we define m sliding temporal windows W^θ_l of duration θ, where θ is the number of frames in an expression sequence, and $l=1..m$. Each of these temporal windows will be used as inputs to our facial dynamic classifiers. Each temporal sliding window W^θ_l produces one of several estimates of facial expression per video segment. These estimates are stored in a vector and converted into a single expression estimate via voting. To combine static and temporal features into a single expression estimate we apply the following formula:

$$\psi = mode(hist(static) \cdot (hist(temporal)/sum(temporal))) \qquad (4)$$

Where *static* and *temporal* are the vectors of predicted votes; *hist()* is a histogram operation; and *mode()* computes the most frequently occurring prediction. Equation (4) weighs the votes of the static model by the confidence values from the temporal model.

3.1   Facial Feature Point Tracking

When accurately placed, facial feature locations such as corners of eyebrows, outline of mouth, etc., can produce accurate expression estimations. Active Shape Models (ASMs) and Active Appearance Models (AAMs), initially introduced by [26] were used for expression estimation in [6, 27]. ASM is applied independently on each frame of our temporal window W^θ_l, localizing 82 facial feature points per frame. Each set of 82 points is transformed to a canonical face representation using a generalized Procrustes analysis. With θ frames per W^θ_l, we get $\theta*82*2=\theta*164$ dimensions per sample.

Each sample is SLPP dimensionality reduced and classified using multiclass SVM. Alternatively, we can pass the average of our 82 points per W^θ_l into SLPP, or convert these points to $[\Delta x, \Delta y]$ or [magnitude,phase] motion vectors before passing to SLPP.

3.2   Motion History Images

Motion History Images (MHI) were initially introduced for human movement recognition [18], and were later adopted for facial AU detection [11]. MHI compresses the motion over each sliding temporal window W^θ_l into a single template. The methods adopted here are similar to [11], except the conversion from MHI image to motion vectors is modified for improved facial expression performance. MHI evaluates the movement between all possible frames f and $f+1$ in W^θ_l, where $f=1..\theta-1$. For each pair of frames $\{f, f+1\}$ in W^θ_l, we first calculate motion detection at the pixel level in a binary fashion:

$$d_f(x,y) = \begin{cases} 1 & |g(x,y,f) - g(x,y,f+1) > \gamma| \\ 0 & otherwise \end{cases} \tag{5}$$

where $g(x,y,f)$ is a Gaussian filtered version of frame f and γ is a noise threshold. These difference frames are morphologically filtered with an opening operation to remove isolated noise. Each of the m sliding windows produces a single MHI^θ_l template:

$$MHI^\theta_l = \frac{1}{\theta - 1} \max_f \{f \, d_f(x,y) \; 0 \le f \le \theta - 1\} \tag{6}$$

In this context, more recent movements are assigned higher weights. The MHI^θ_l templates are converted to motion vectors by replacing each pixel code value with a motion vector that points in the direction of the highest (most recent motion) code value within a 7x7 neighborhood. This neighborhood is constrained such that we may only point to pixels that are monotonically increasing. Further, if there are several pixels of the same value, the average value is used. ΔX, ΔY, magnitude, and phase-magnitude versions of this motion vector image are passed into SLPP dimensionality reduction.

3.3   Free Form Deformations

Free Form Deformations (FFD) are a dense optical flow technique initially introduced by [28] for medical image registration, and were latter adopted for facial AU detection [11]. Given two images, FFD computes a rigid global motion and a non-rigid local motion model representing the movement of each pixel from one frame to the next. The global motion model iteratively solves for a 3x3 affine transformation matrix using convex optimization techniques across all pixels in both images. The local motion model solves for the displacement of a mesh of grid points. Given input pixel $(x,y)_t$ in frame t, we solve for the estimate of its position $(x',y')_{t+1}$ in frame $t+1$:

$$(x',y')_{t+1} = (x,y)_t + F_t(x,y) \tag{7}$$

We solve for $F_t(x,y)$ for each pair of neighboring frames $\{f, f+1\}$ in our temporal window W^θ_l, and pass the set of all $F_t(x,y)$ per W^θ_l as the input to SLPP dimensionality reduction. To solve for each $F_t(x,y)$, FFD solves a gradient descent optimization across a sparse mesh of control points, minimizing sum of square difference of pixels values. Given the deformed sparse mesh of control points, any interpolation method can solve for the dense motion $F_t(x,y)$ at each input pixel location. FFD uses cubic B-splines as the resulting fit is smooth and continuous across mesh vertices. To avoid local minima and make this process more computationally tractable, a hierarchical approach, solving from a low to high resolution mesh is utilized. Each mesh point position is initialized by the lower resolution mesh preceding it. Gaussian blurring with a σ twice the grid spacing at each level in the pyramid ensures robust behavior.

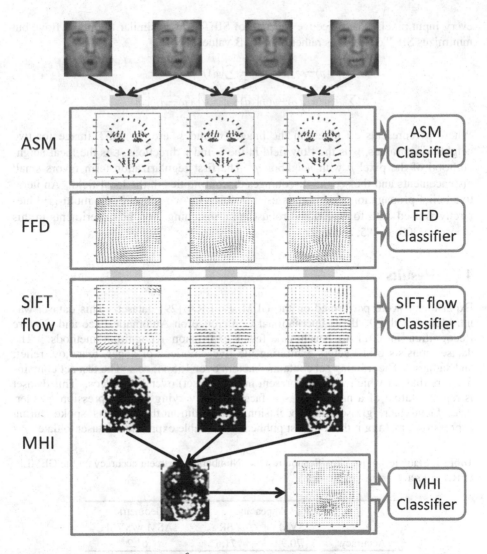

Fig. 1. Sample temporal window W^θ_l, θ=4 frames, from an 'angry' video. From top to bottom we have input 60x51 cropped and affine warped faces, 82 point ASM motion vectors (black arrows have blue tail, red tip), grid of 28x24 dense FFD optical flow vectors, grid of 21x17 dense SIFT flow vectors, and MHI fields. For MHI we have three difference frames, MHI template, and 30x26 dense MHI flow field.

3.4   SIFT Flow

SIFT flow [19] is an image alignment algorithm initially introduced to register two similar images and further adopted for facial registration by [19] [20]. Similar to FFD, SIFT flow produces a dense optical flow field between all neighboring frames {f, $f+1$} in our temporal window W^θ_l. SIFT descriptors are densely computed on every input pixel. The objective function of SIFT flow is similar to optical flow, but minimizes SIFT descriptors rather than RGB values:

$$E(w) = \sum_p \left| s_1(p) - s_2(p+w) \right|_1 + \frac{1}{\sigma^2} \sum_p \left(u^2(p) + v^2(p) \right) + \qquad (8)$$

$$\sum_{(p,q)\in\varepsilon} \min(\alpha \left| u(p) - u(q) \right|_1, d) + \min(\alpha \left| v(p) - v(q) \right|_1, d)$$

Where p represents all pixels in the image, s_1 and s_2 are the SIFT image for two neighboring frames, w is the flow field in the u and v directions, ε is the local neighborhood of the pixel p with neighbor q. The first regularization term favors small displacements and the second discourages discontinuities in the local field. An iterative belief propagation algorithm is used to solve for w. Like FFD, a multi-grid hierarchy is used both for speed and robust point matching. For the experiments in this report, $\sigma=300$, $\alpha=0.5$, and $d=2$.

4   Results

Experiments were performed on the GEMEP-FERA [29] dataset. This dataset was introduced at the 9th IEEE International Conference on Automatic Face and Gesture Recognition in 2011 to benchmark facial expression recognition methods. The dataset consists of 10 actors exhibiting the five emotions of anger, fear, joy, relief, and sadness. The training set contains 7 actors over 155 videos. The test set contains 6 actors (half of which were not present in the test set) over 134 videos. This dataset is representative of a natural dataset- faces are of varying pose, expression, gender, race, facial hair, glasses, and occlusion. In addition the subjects spoke during expressions, making it the toughest publically available expression dataset to date.

Table 1. Static technique classification results. Numbers are percent accuracy on the GEMEP-FERA test set.

	Appearance		Geometric
	SVM	SR	ASM w/SVM
Accuracy	70.9	**71.6**	61.2

With respect to the static models, Table 1 shows similar performance for multiclass linear SVM and the SR techniques, with the SR method being slightly favored. The geometric method's performance is lagging behind the appearance based methods because this dataset has many non-frontal poses and open mouths, both which are difficult to track. The appearance based methods allow for processing of multiple facial regions independently. If we allow the SR method to use multiple facial regions as in [21], the accuracy increases to 74.5%.

Table 2. Temporal technique classification results. Numbers are percent accuracy on the GEMEP-FERA test set.

	ASM	MHI	FFD	SIFT flow
Accuracy	71.6	**73.1**	63.4	44.8

With respect to the temporal models, Table 2 shows the ASM and MHI methods generally outperform the FFD method which generally outperforms the SIFT flow method. Temporal methods almost always utilize the knowledge of a neutral face. In this work, we have no neutral frames, perhaps putting the temporal methods at a disadvantage. For example, if the starting frame of our temporal window W^θ_l is a frown, upward movement of the corner of the mouth may be classified as happiness, when in reality it would be indicative of neutral. For the ASM points, we have the canonical set of facial points from which each frame is warped via Procrustes analysis. Using these canonical set of points as a neutral expression, we can form motion vectors between the raw ASM points after warping and the corresponding points on the canonical face, which we call Δ ASM vectors. The set of all Δ ASM vectors per W^θ_l is used as input features in Table 2.

For the MHI results the $[\Delta X, \Delta Y]$ motion vectors from the single MHI template per window W^θ_l is used. For the FFD and SIFT flow results, the set of m $[\Delta X, \Delta Y]$ motion vectors from each $F_t(x,y)$ per W^θ_l is used as input features. The dense flow field used in MHI, FFD, and SIFT flow was evaluated on a mesh grid of resolution 30x26, 28x24, and 21x17 respectively.

The sliding window size θ affected performance significantly. Quick actions such as eye-blink favor small θ, while longer actions such as smile favor longer values of θ. Window sizes of $\theta = \{2,4,8,12,16,20\}$ were implemented for each method. The results in Table 2 show the optimal window size θ for each method which was 20 frames for ASM and SIFT flow and 16 frames for MHI and FFD.

To gauge how well the above methods compare to state-of-the-art, the median submission in the 2011 FERA challenge reported 70% accuracy. Compared to the FERA challenge results, the SR static or MHI temporal methods score in 5th and 6th place respectively (out of 15 submissions). In an effort to improve the overall performance further, the static and temporal methods can be combined using Eq. (4). For example, if we combine the sparse static and MHI temporal methods, the overall accuracy increases to 79.1%, placing it 2nd in the FERA challenge.

The static methods perform best at joy and angry where characteristic mouth and eyebrow positions induced during these emotions enable accurate classification by humans and machine alike. The temporal methods perform best at relief and sadness where characteristically slow movements enabled accurate classifications. Fear was difficult for all methods, although these confused human viewers most often as well.

5   Conclusions

Inspired by the presence of sparsity in the visual cortex and psychological studies showing the benefit of temporal facial cues, this paper investigates the contributions of sparsity and temporal dynamics in automatic facial expression classifiers. In particular, a static method based upon sparse representations is contrasted with several implementations of temporal expression classifiers. When used in isolation, temporal features can be just as good as static features. However, when static and temporal features are fused together, significant increases in expression recognition performance are achieved.

References

[1] Lew, M., Bakker, E.M., Sebe, N., Huang, T.S.: Human-Computer Intelligent Interaction: A Survey. In: Lew, M., Sebe, N., Huang, T.S., Bakker, E.M. (eds.) HCI 2007. LNCS, vol. 4796, pp. 1–5. Springer, Heidelberg (2007)

[2] Pantic, M., Pentland, A., Nijholt, A., Huang, T.S.: Human computing and machine understanding of human behavior: a survey. In: ICMI 2006 and IJCAI 2007 International Workshops, Artifical Intelligence for Human Computing, Berlin, Germany (2007)

[3] Pantic, M., Rothkrantz, L.U.M.: Automatic analysis of facial expressions: The state of the art. IEEE Transactions on Pattern Analysis and Machine Intelligence 22, 1424–1445 (2000)

[4] Zhihong, Z., Pantic, M., Roisman, G.I., Huang, T.S.: A survey of affect recognition methods: audio, visual, and spontaneous expressions. IEEE Transactions on Pattern Analysis and Machine Intelligence 31, 39–58 (2009)

[5] Shuai-Shi, L., Yan-Tao, T., Dong, L.: New research advances of facial expression recognition. In: Eighth International Conference on Machine Learning and Cybernetics (2009)

[6] Martin, C., Werner, U., Gross, H.M.: A real-time facial expression recognition system based on active appearance models using gray images and edge images. In: 8th IEEE International Conference on Automatic Face & Gesture Recognition (2008)

[7] Buciu, I., Kotropoulos, C., Pitas, I.: ICA and Gabor representation for facial expression recognition. In: Proceedings of International Conference on Image Processing (2003)

[8] Shan, C., Gong, S., McOwan, P.W.: Facial expression recognition based on Local Binary Patterns: A comprehensive study. Image and Vision Computing 27, 803–816 (2009)

[9] Zafeiriou, S., Petrou, M.: Sparse representations for facial expressions recognition via l1 optimization. In: IEEE Computer Society Conference on Computer Vision and Pattern Recognition - Workshops, CVPRW 2010, San Francisco, CA (2010)

[10] Ekman, P., Friesen, W.V.: The Facial Action Coding System. Consulting Psychologists Press, Inc., San Francisco (1978)

[11] Koelstra, S., Pantic, M., Patras, I.: A dynamic texture-based approach to recognition of facial actions and their temporal models. IEEE Transactions on Pattern Analysis and Machine Intelligence 32, 1940–1954 (2010)

[12] Curio, C., Bulthoff, H., Giese, M.: Dynamic Faces: Insights from Experiments and Computation, 1st edn. The MIT Press (2010)

[13] Jiang, B., Valstar, M.F., Pantic, M.: Action Unit Detection Using Sparse Appearance Descriptors in Space-Time Video Volumes. In: Face and Gesture Recognition Face and Gesture Recognition, Santa Barbara, CA (2011)

[14] Lucey, P., Cohn, J., Matthews, I., Prkachin, K., Solomon, P.: Painful Data: The UNBC-McMaster Shoulder Pain Expression Archive Database. In: Face and Gesture Recognition, Santa Barbara, CA (2011)

[15] Bhaskaran, N., Nwogu, I., Frank, M., Govindaraju, V.: Lie to Me: Deceit Detection via Online Behavioral Learning. In: Face and Gesture Recognition, Santa Barbara, CA (2011)

[16] Cohn, J.F., et al.: Detecting depression from facial actions and vocal prosody. In: 3rd International Conference on Affective Computing and Intelligent Interaction and Workshops, ACII 2009 (2009)

[17] Saragih, J., Lucey, S., Cohn, J.: Real-time Avatar Animation from a Single Image. In: Face and Gesture Recognition, Santa Barbara, CA (2011)

[18] Bobick, A.F., Davis, J.W.: The recognition of human movement using temporal templates. IEEE Transactions on Pattern Analysis and Machine Intelligence 23, 257–267 (2001)

[19] Ce, L., Yuen, J., Torralba, A.: SIFT flow: dense correspondence across scenes and its applications. IEEE Transactions on Pattern Analysis and Machine Intelligence 33, 978–994 (2011)

[20] Songfan, Y., Bhanu, B.: Facial expression recognition using emotion avatar image. In: 2011 IEEE International Conference on Automatic Face & Gesture Recognition (2011)

[21] Ptucha, R., Tsagkatakis, G., Savakis, A.: Manifold Based Sparse Representation for Robust Expression Recognition without Neutral Subtraction. In: Presented at the BeFIT 2011 Workshop, International Conference on Computer Vision, Barcelona, Spain (2011)

[22] Olshausen, B.A., Field, D.J.: Sparse coding with an overcomplete basis set: a strategy employed by V1? Vision Research 37, 3311–3325 (1997)

[23] Olshausen, B.A., Field, D.J.: Emergence of simple-cell receptive field properties by learning a sparse code for natural images. Nature 381, 607–609 (1996)

[24] Wright, J., Yang, A.Y., Ganesh, A., Sastry, S.S., Yi, M.: Robust face recognition via sparse representation. IEEE Transactions on Pattern Analysis and Machine Intelligence 31, 210–227 (2009)

[25] He, X., Niyogi, P.: Locality Preserving Projections. In: Advances in Neural Information Processing Systems, Vancouver, Canada, vol. 16 (2003)

[26] Cootes, T.F., Taylor, C.J., Cooper, D.H., Graham, J.: Active shape models - their training and application. Computer Vision and Image Understanding 61, 38–59 (1995)

[27] Kotsia, I., Pitas, I.: Facial expression recognition in image sequences using geometric deformation features and support vector machines. IEEE Transactions on Image Processing 16, 172–187 (2007)

[28] Rueckert, D., Sonoda, L.I., Hayes, C., Hill, D.L.G., Leach, M.O., Hawkes, D.J.: Nonrigid registration using free-form deformations: Application to breast MR images. IEEE Transactions on Medical Imaging 18, 712–721 (1999)

[29] Valstar, M., Jiang, B., Mehu, M., Pantic, M., Scherer, K.R.: The First Facial Expression Recognition and Analysis Challenge. In: Face and Gesture Recognition, Santa Barbara, CA (2011)

Using Detailed Independent 3D Sub-models to Improve Facial Feature Localisation and Pose Estimation

Angela Caunce, Chris Taylor, and Tim Cootes

Imaging Science and Biomedical Engineering, The University of Manchester, UK

Abstract. We show that the results from searching 2D images or a video sequence with a 3D head model can be improved by using detailed sub-models. These parts are initialised with the full model result and are allowed to search independently of that model, and each other, using the same algorithm. The final results for the sub-models can be reported exactly, or optionally fed back into the full model to be constrained by its parameter space. In the case of a video sequence this can then be used in the initialisation of the next frame. We tested various data sets, constrained and unconstrained, including a variety of lighting conditions, poses, and expressions. Our investigation showed that using the sub-models improved on the original full model result on all but one of the data sets.

1   Introduction

Head and facial feature tracking can provide important information in various environments with respect to the activity and attitude of the subject. For example, in a driving scenario, head orientation alone can indicate attentiveness. Feature localisation and subsequent behaviour analysis can give a detailed picture of the driver's state and possible intent. This could identify critical or dangerous situations. Recently, 2D models have been used with great success to localise and analyse features of the face [1-3], however in some unconstrained scenarios with large pose variation this approach may be limited. Multiple 2D models and detectors may be required for different views [4]. There is also the additional problem of view-based occlusion. As a consequence, authors have been experimenting with augmented 2D [5, 6] and 3D [7]. Due to pose invariance, a 3D model requires less training data than its 2D counterpart, and is able to report critical pose information directly without additional calculation. In [8] the authors showed that their 3D method outperformed an established 2D approach on out of plane rotations and in [9] they extended this by integrating some limited facial actions for preliminary behaviour analysis. To do this they used two sparse, largely symmetrical, statistical point models of the whole face. However, the complex interplay of the various parts of the face may not have been fully realised due to global constraints. Also, the ability to deal with some individual quirks, like a crooked smile, is limited by the training set. This is a standard problem with deformable objects containing complex sub-parts. One solution is to define the sub-parts separately and model their inter-relationship. The search is performed by locating the

G. Bebis et al. (Eds.): ISVC 2012, Part II, LNCS 7432, pp. 398–408, 2012.

sub-parts and confirming an acceptable configuration. Methods include: pictorial structures [10]; star models [11]; Hierarchical Deformable Templates [12]; and probabilistic approaches [13, 14]. Martinez [15] used a probabilistic approach and weighted abstract sub-parts of the face based on their involvement in the test expression. However working in 3D has advantages in that many configurations of parts (those derived from the object's pose) are already constrained by the model structure and only the articulation problem remains. Generally, authors working in 3D have taken a bottom up approach, that is finding sub-parts and combining them in some meaningful way. In [16] Blanz and Vetter show that this produces visually pleasing results. Tena et al [17] illustrated that sub-parts improved performance when reconstructing motion capture data.

1.1 Contribution

Our method takes a top down approach, in that we localise the face with a full model using the method in [9], and then refine the result by allowing the parts to search independently afterwards. By extending the method in this way, we have increased its accuracy by improving its versatility. The system can deal with new feature configurations, without requiring additional training data. This may even reduce the amount of data required for training in future. We demonstrate the improvements by reporting results on 7 large datasets with various challenges. These show improvements either overall, or in some localised region of the face. The method has all the advantages of 3D outlined above, and can deal with greater pose variation than demonstrated in other works. We report median average point-to-point errors of less than 15 pixels (inter-ocular distance ~100 pixels) on headings up to 70 degrees and on all pitch angles tested (up to 60 degrees).

2 Search Method

The search method uses two sparse 3D statistical models of the form:

$$x = \bar{x} + Pb \tag{1}$$

Where each example x is represented by a vector of n 3D co-ordinates $(x_1, y_1, z_1, \ldots\ldots x_n, y_n, z_n)$. Each is expressed in (1) as the mean vector, $\bar{x}$, plus a linear combination of the principal components, P, with coefficients, b.

One of the two models is built from 'identity' training data, i.e. data from 923 individuals, with a close to neutral expression, eyes open, and mouth closed. The other is built from a small set of facial actions created from a neutral base.

Unlike other approaches which use a combined model strategy [18, 19], these two models are used in an alternating process to localise the features of the face and to provide a basic representation of some simple behaviours. This is done by substituting the results from the ID model into the actions model, and vice-versa, when matching to the target points. Therefore, at each iteration of the algorithm, both models are fitted in sequence to the same target before moving on to the next iteration. This is represented in the following equations:

$$\mathbf{x}^{k(1)} = \overline{\mathbf{x}}^{k(1)} + \mathbf{P}_{ID}\mathbf{b}_{ID}^{k} \tag{2}$$

$$\overline{\mathbf{x}}^{k(1)} = \overline{\mathbf{x}}_{ID} + \mathbf{P}_{A}\mathbf{b}_{A}^{k-1} \tag{3}$$

$$\mathbf{x}^{k(2)} = \overline{\mathbf{x}}^{k(2)} + \mathbf{P}_{A}\mathbf{b}_{A}^{k} \tag{4}$$

$$\overline{\mathbf{x}}^{k(2)} = \overline{\mathbf{x}}_{ID} + \mathbf{P}_{ID}\mathbf{b}_{ID}^{k} \tag{5}$$

Where $k(1)$ and $k(2)$ refer to the 1st and 2nd fit at each iteration k, (5) is the current identity result and is used as the action model mean, (3) is devised from the current action result and is used as the identity model mean, and $\mathbf{b}_{A}^{0}$ is the zero vector. Notice that the action model mean, $\overline{\mathbf{x}}_{A}$, is not used since this has no meaning in this context. See [9] for further details.

Most of the target points are located using an independent local template matching at each model point. There are 238 points in the model and each can search with a small (5x5) view based texture patch. This patch is extracted from a mean texture generated from 913 subjects. The population variation in texture is not modelled. The content of the patch is updated at every iteration to reflect the current pose, and is compared using normalised correlation to a neighbourhood around the point to find the best match. This has the advantage of providing some robustness to illumination variation over the face and between images.

Those points at approximately 90 degrees to the viewpoint do not use this technique but search along the surface normal for the strongest edge. Once the target points are established the whole model is fitted using the active shape model fitting method in [20] extended to 3D, assuming an orthogonal projection. This is a two-stage process. Firstly, the points are rigidly aligned to minimise the sum of squared distances between matched points, then the shape model parameters ($\mathbf{b}$ in (1)) are updated using a least squares approximation. The global fit has the advantage of minimising the effect of badly matched or occluded points.

The search is conducted at multiple resolutions, starting at the lowest, and is completed at each resolution before moving onto the next.

Initialisation is achieved using the Viola-Jones (V-J) face detector [21]. For the video sequences this occurs once at the start and again only when the search fails during the sequence. Failure is determined by comparing the average match value of the template patches to some threshold. If the match does not fail, the search on the next frame is initialised using the latest result. For still images the V-J detector is used on each example.

3   Sub-models

Fitting the model to the target using global constraints, as outlined above, has the advantage of keeping all of the areas of the model in their expected place, as well as having a neutralising effect on rogue matches, such as those found at occluded points. However, once this process is completed, it may be expected that the final result will also suffer, since un-correlated movements of the individual sub-parts may be lost as noise. Plus, individual quirks, such as a crooked smile, cannot be localised unless they are specifically included in the training set.

To provide the added flexibility necessary we propose extending the method by allowing sub-models to continue searching independently, after the full model search has completed.

The face models described in section 2 have a sparse representation. They are built from 238 vertices of a full 3D head mesh, warped [22] to fit each individual in the training set. However, for the sub-models we chose to use all the vertices but only from parts of the mesh around the eyes and mouth. We thus built two sub-models from the areas shown in Figure 1. As with the full face models, an identity and an actions model were constructed for each sub-part.

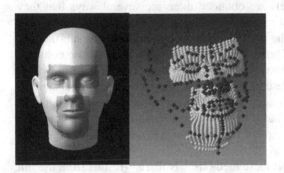
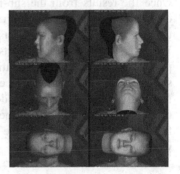

Fig. 1. The two sub-models are built from all the vertices in the eye and mouth areas *(left)*. The relationship between the sparse full face model and the sub-models is shown right.

Fig. 2. The Artificial Driver data set contains extreme poses

4   Sub-model Search

The full model search is conducted over a series of three increasing image/texture resolutions. The search is completed at each resolution before continuing to the next. For the sub-models we use exactly the same search method except that searching always begins at the penultimate resolution. Experiments indicated that results were improved if two resolutions were used but there appeared to be no advantage to using all three.

When the full model search is complete the sub-models are initialised in their respective positions. Since the sub-models have many more points than the full model an initialisation target is constructed for each sub-part from just those points common to both. To do this the common points have a weighting of 1 whereas all other points have weight 0. The fitting method is the same method used in the image search, i.e. the 3D extension of that in [20], whereby an equation of the form in (6) is minimised to find the new model point positions.

$$\min_{b,t}\left\{\left|W\left(T_t\left(\overline{x}+Pb\right)-x_{obs}\right)\right|^2\right\}$$ (6)

T represents the camera transform (in this case orthogonal) and pose with parameters t, W is the diagonal weighting matrix, and x_{obs} is the observed, or target, set of points.

The identity and action sub-models are fitted to this target alternately as described above using equations (2)-(5). After this initialisation, each sub-part is allowed to search the image independently of the other and of the full model.

4.1   Feedback and Reporting Strategies

Once the results from the sub-models are obtained there are several ways that they can be integrated with the full model results for analysis. The most obvious is to use the parts to replace all the points in the full model that are common to the sub-models. We refer to this as 'Exact Parts'. The alternative is to feed back (FB) the common points into the full model by fitting it to the Exact Parts result. This fitting follows the same iterative alternating process described above for the sub-model initialisations. Normally, when analysing a video stream, the result of each frame is used to initialise the search on the next. This means that if the sub-model results are fed back into the full model these will influence subsequent frames. For a series of still images this will have no effect.

Therefore for still images we report these methods: 'NoParts' (results from the full model search); 'NFBExact' (exact parts substituted); and 'FB' (feedback – i.e. full model fitted to parts results). For video data, method 'FB' will affect future results and there is an additional combination: 'FBExact' (feedback affects future results but exact parts reported).

5   Data Sets

For our analysis we used 7 different data sets presenting various challenges:

- XM2VTS [23]: 2344 still images of 295 individuals at 720x576 pixels. The subjects are posed against a fairly uniform backdrop and in general have neutral expressions and near frontal poses.
- BioID [24]: 1520 still images of 23 individuals at 384x286 pixels. The data was acquired in an office setting so has cluttered backgrounds. It shows much more natural poses (although still mainly near frontal) and a variety of expressions.

- Expressions: 401 still images of 103 individuals at 1024x768 pixels. Each person is making some or all expressions selected from: eyes closed; smile; neutral; frown; and surprise. These images are near frontal and posed against a uniform background so the main challenge is from the facial actions.
- 3 video sequences of 3 different drivers. These were taken in real driving conditions. Each sequence is 2000 frames but only a subset of each was used for evaluation: 150, 156, and 136 frames. There is a great deal of lighting variation, within and between frames, and there is a wider variation in poses than in the still images.
- Artificial driver short sequences [8]. This publicly available dataset was devised to assess the ability of search methods to deal with large poses. It comprises of a series of images of 20 synthetic subjects in known poses. The subjects were arranged against a real in-car background (Figure 2). Each sequence starts at zero rotations and runs in a single direction. The sequences are as follows: Heading +/- 90°; Pitch +/- 60°; and Roll +/- 90°.

6 Results

The available annotations for the different data sets do not mark the same features. Due to this, and the differences between the annotations and the model points, only a subset of 12 points was used for point to point error evaluation. The points chosen are located on the better defined features, common to all sets: the ends of the eyebrows; the corners of the eyes; the corners of the mouth; and the top and bottom of the mouth.

On the still image databases the search is initialised using the V-J detector on every image. Since the method relies on initialisation in the area of the face, we assessed the detector's performance by comparing the 12 points of interest to the box it returned. If any manually marked points fell outside the box the detection was considered a failure. It was found that the detector failed on 8% of the BioID data set. These examples were therefore excluded from the analysis.

The generic model can handle a variety of challenges including: occlusion; pose variation; and variable lighting [9]. In many cases the sub-models will make only a small correction. However, Figure 3 illustrates the ability of the sub-parts to correct poorer results from the full model. If this is very inaccurate, sometimes only one part can recover (top row, column 2) and sometimes recovery is not possible (top row, column 3). The middle left image of Figure 3 illustrates the ability of the method to adapt to non-standard configurations, in this case the mouth is not symmetric with the face. Normally this would need to be included in a training set to be handled by a full face model of this kind.

Figure 4 shows the cumulative error proportion plotted against average point-to-point (P2P) distance for each dataset. In most cases this is a percentage of the inter-ocular distance (IOD). This normalises for the fact that the head size can vary across the data set, which is particularly true of the BioID images. For the artificial data the errors are presented as pixels because of the large pose variation. However, the head size is fairly constant on all images and IOD on the frontal faces is ~100 pixels.

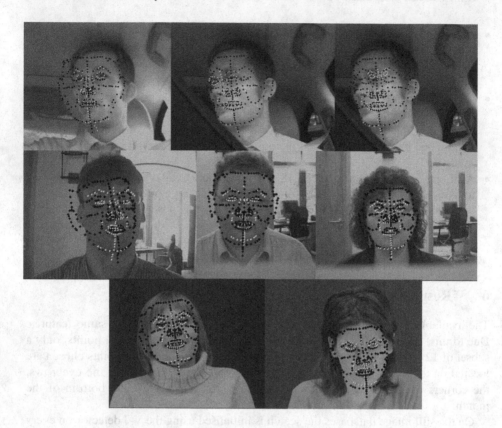

Fig. 3. Examples showing how the sub-models can correct a poor full model result. Top row shows real-world driver video. The second image illustrates the independence of the models: only one has recovered. The third image shows that if the result is very poor the parts cannot always recover. The middle row shows BioID images, including glasses. The first image illustrates how the sub-models can allow adaptation to non-symmetric features (crooked mouth). The bottom row shows Expressions Surprise and Expressions Frown. Black points are the full model result and white points show the sub-models (common points).

Generally, including the parts improves on the original 'NoParts' result. However, examining the cumulative error curve for Video 1 up to the 15% threshold, we observed that performance got worse when the parts were introduced. Extending the curve to show errors beyond 15% revealed that, although the errors were increased the proportion of failures was reduced albeit at larger thresholds (shown in Figure 4).

Examining the results for different parts of the face revealed that the mouth area was much improved even at the lower thresholds (also shown in the figure). This indicates that the mouth model was able to correct many of the failures thus pushing the curve higher. For the other data sets the eyes showed the best performance. This is unsurprising since the corners of the eyes are the most easily located when marking ground truth on the data. It is therefore likely to produce the lowest errors in a model search. The variant producing the best results for the eye corners is therefore shown on the graphs of Figure 4. For the artificial data the results of using a 2D CLM are included for comparison.

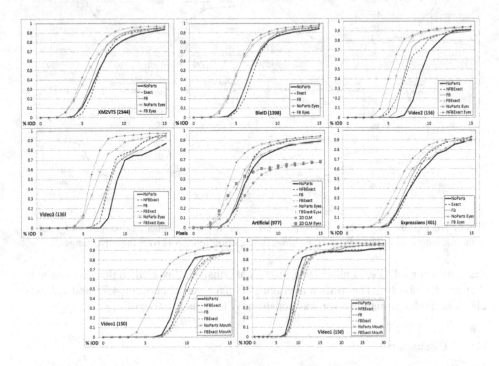

Fig. 4. Cumulative Error Distributions. The proportion of data is plotted against the point-to-point error as % inter-ocular distance (IOD), except for the artificial data which is shown in pixels. The number of images in the data set is given in brackets. The best local result is also shown, which is the eyes for all but Video 1. 2D CLM results are also given for the artificial data and an extended error plot is shown for Video 1.

Figure 5 shows the P2P errors broken down by rotation angle for the artificial images. Also shown are the errors on the estimated pose. Here the pose is reported as a quaternion, which represents a rotation about an axis vector. We examined the error on both the angle and the axis. The latter is calculated as the rotation angle between the known and estimated axes. Since the pose is not affected when using 'Exact' methods only the original and FB methods are shown. There are no axis results for 0 degrees rotation because of the high degree of uncertainty on the rotation axis at that point. As with figure 4, including the sub-parts has had a generally beneficial effect. For heading, the improved P2P errors are below 8 pixels for angles between +/- 50 degrees, and the estimated angle error is 5 degrees or less for the same range. For pitch the P2P error is below 13 pixels as far as +60 degrees and below 8 for -60 degrees. The angle estimation error is no more than 9 degrees for the entire range. Since

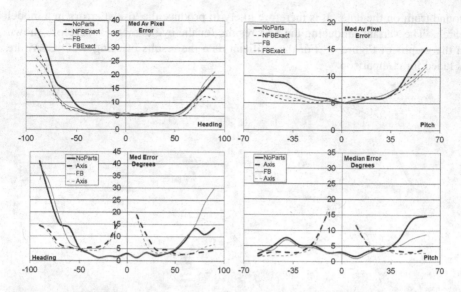

Fig. 5. The top row shows the point-to-point errors for the artificial data, broken down by out of plane rotation angle. The bottom row shows pose estimation errors over the same ranges.

the results for roll (i.e. in plane rotations) were consistent over all angles and methods they are not shown. The median average P2P error varied between 4.75 and 7.7 pixels. The angle estimation error was consistently less than 2.9 degrees.

6.1 Discussion

The easiest data sets in this group are XM2VTS and BioID and, as might be expected, the overall performance (all methods) is relatively better on these sets, closely followed by the expressions and the artificial data. The real-world videos obviously present the greatest challenge and the relative performance reflects this.

If we allow that Video 1 was improved in one area only, the mouth, then there is an overall improvement on every other data set from adding in the subparts. The other two videos and the artificial data show the most obvious improvements which implies that using sub-models has the most beneficial effect when used in the more difficult situations, as might be hoped, since it is a corrective technique. For the still images it seems that the best improvement is when the full model is re-fitted to the parts result (FB). This implies that the ability of the parts to correct the full model and deal with non-standard examples has aided the method but a final regularisation step is still needed. However, in the case of the sequence data this is less clear cut. For Videos 2&3 and the artificial data, 'NoParts' is the worst performer but whether exact or corrected parts should be reported needs further investigation.

When compared to state-of the-art 2D systems such as [1, 3], our method does not perform as well on the common BioID data set. However, in [3] the authors use near frontal faces only and in [1] state that their experimental data set, acquired from the internet, does not have profile or near profile images and that all faces were detected by an off the shelf face detector. This will tend to exclude not only extreme poses but also unusual lighting and some occlusions. We have shown that our method deals with large rotations, a wide variety of data types and conditions and, in addition, can provide an estimate of pose, for those gaze-critical applications.

7   Summary

We have presented an extension to 3D model search which allows refinement of the results using independent 3D sub-parts. From the graphs presented in Figures 4 & 5 it can be seen that, for all data sets but Video 1, including the sub-parts has had a clearly positive effect. Even in the case of Video 1, when breaking down the errors between parts of the face, it can be seen that the mouth area is vastly improved. However, there is uncertainty as to whether reporting the exact parts points or refitting the model gives the best individual result. This requires further study and implies that an iterative approach, alternating between the full and parts models, may yield even more improvements.

Acknowledgements. This project is funded by Toyota Motor Europe who provided the driver videos. We would like to thank Genemation Ltd. for the 3D data markups and head textures.

References

1. Belhumeur, P.N., Jacobs, D.W., Kriegman, D.J., Kumar, N.: Localizing Parts of Faces Using a Consensus of Exemplars. Computer Vision and Pattern Recognition, 545–552 (2011)
2. Cristinacce, D., Cootes, T.: Automatic feature localisation with constrained local models. Pattern Recognition 41, 3054–3067 (2007)
3. Valstar, M., Martinez, B., Binefa, X., Pantic, M.: Facial Point Detection Using Boosted Regression and Graph Models. Computer Vision and Pattern Recognition, 2729–2736 (2010)
4. Pentland, A., Moghaddam, B., Starner, T.: View-Based and Modular Eigenspaces for Face Recognition. Computer Vision and Pattern Recognition, 1–7 (1994)
5. Vogler, C., Li, Z., Kanaujia, A., Goldenstein, S., Metaxas, D.: The Best of Both Worlds: Combining 3D Deformable Models with Active Shape Models. In: International Conference on Computer Vision, PP. 1–7 (2007)
6. Xiao, J., Baker, S., Matthews, I., Kanade, T.: Real-Time Combined 2D+3D Active Appearance Models. In: Conference on Computer Vision and Pattern Recognition, vol. 2, pp. 535–542 (2004)
7. Romdhani, S., Ho, J., Vetter, T., Kriegman, D.J.: Face Recognition Using 3-D Models: Pose and Illumination. Proceedings of the IEEE 94, 1977–1999 (2006)
8. Caunce, A., Cristinacce, D., Taylor, C., Cootes, T.: Locating Facial Features and Pose Estimation Using a 3D Shape Model. In: International Symposium on Visual Computing, Las Vegas, pp. 750–761 (2009)

9. Caunce, A., Taylor, C., Cootes, T.: Adding Facial Actions into 3D Model Search to Analyse Behaviour in an Unconstrained Environment. In: International Symposium on Visual Computing, Las Vegas, pp. 132–142 (2010)

10. Fischler, M.A., Elschlager, R.A.: The Representation and Matching of Pictorial Structures. IEEE Transactions on Computers C-22, 67–92 (1973)

11. Felzenswalb, P.F., Girshick, R.B., McAllester, D.: Cascade Object Detection with Deformable Part Models. Computer Vision and Pattern Recognition, 1–8 (2010)

12. Zhu, L., Chen, Y., Yuille, A.: Learning a Hierarchical Deformable Template for Rapid deformable Object Parsing. IEEE Trans. Pattern Analysis and Machine Intelligence 32, 1029–1043 (2010)

13. Burl, M.C., Weber, M., Perona, P.: A Probabilistic Approach to Object Recognition Using Local Photometry and Global Geometry. In: Burkhardt, H., Neumann, B. (eds.) ECCV 1998. LNCS, vol. 1407, pp. 628–641. Springer, Heidelberg (1998)

14. Hua, G., Wu, Y.: Sequential Mean Field Variational Analysis of Structures De-formable Shapes. Computer Vision and Image Understanding 101, 87–99 (2006)

15. Martinez, A.M.: Recognizing Imprecisely Localized, Partially Occluded, and Expression Variant Faces from a Single Sample per Class. Pattern Analysis and Machine Intelligence 24, 748–763 (2002)

16. Blanz, V., Vetter, T.: A Morphable Model for the Synthesis of 3D Faces. SIGGRAPH, pp. 187–194 (1999)

17. Tena, J.R., Torre, F.D.l., Matthews, I.: Interactive Region-Based Linear 3D Face Models. SIGGRAPH (2011)

18. Amberg, B., Knothe, R., Vetter, T.: Expression Invariant 3D Face Recognition with a Morphable Model. In: International Conference on Automatic Face Gesture Recognition, Amsterdam, pp. 1–6 (2008)

19. Basso, C., Vetter, T.: Registration of Expressions Data Using a 3D Morphable Model. Journal of Multimedia 1, 37–45 (2006)

20. Cootes, T.F., Cooper, D.H., Taylor, C.J., Graham, J.: Active Shape Models - Their Training and Application. Computer Vision and Image Understanding 61, 38–59 (1995)

21. Viola, P., Jones, M.J.: Robust Real-Time Face Detection. International Journal of Computer Vision 57, 137–154 (2004)

22. Bookstein, F.L.: Principal Warps: Thin-Plate Splines and the Decomposition of Deformations. IEEE Transactions on Pattern Analysis and Machine Intelligence 11, 567–585 (1989)

23. Messer, K., Matas, J., Kittler, J., Jonsson, K.: XM2VTSDB: The Extended M2VTS Database. In: International Conference on Audio and Video-based Biometric Person Authentication, Washington DC, USA (1999)

24. Jesorsky, O., Kirchberg, K.J., Frischholz, R.W.: Robust Face Detection Using the Hausdorff Distance. In: International Conference on Audio and Video-based Person Authentication, Halmstaad, Sweden, pp. 90–95 (2001)

Gender Recognition from Face Images with Dyadic Wavelet Transform and Local Binary Pattern

Ihsan Ullah[1], Muhammad Hussain[1], Hatim Aboalsamh[1], Ghulam Muhammad[2], Anwar M. Mirza[2], and George Bebis[1,3]

[1] Department of Computer Science,
[2] Department of Computer Engineering,
College of Computer and Information Sciences,
King Saud University, Riyadh 11543, Saudi Arabia
[3] Department of Computer Science and Engineering,
University of Nevada at Reno

Abstract. Gender recognition from facial images plays an important role in biometric applications. We investigated Dyadic wavelet Transform (DyWT) and Local Binary Pattern (LBP) for gender recognition in this paper. DyWT is a multi-scale image transformation technique that decomposes an image into a number of subbands which separate the features at different scales. On the other hand, LBP is a texture descriptor and represents the local information in a better way. Also, DyWT is a kind of translation invariant wavelet transform that has better potential for detection than DWT (Discrete Wavelet Transform). Employing both DyWT and LBP, we propose a new technique of face representation that performs better for gender recognition. DyWT is based on spline wavelets, we investigated a number of spline wavelets for finding the best spline wavelets for gender recognition. Through a large number of experiments performed on FERET database, we report the best combination of parameters for DyWT and LBP that results in maximum accuracy. The proposed system outperforms the stat-of-the-art gender recognition approaches; it achieves a recognition rate of 99.25% on FERET database.

1 Introduction

Category specific approach for face recognition can perform better but the bottleneck for this approach is categorization i.e. to categorize the facial images into different categories based on visual cues like gender and race. In this paper, we address the problem of face categorization based on gender i.e. gender recognition problem. Gender recognition is important due to other reasons as well; it can increase the performance of a wide range of applications including identity authentication, search engine retrieval accuracy, demographic data collection, human-computer interaction, access control, and surveillance, involving frontal facial images.

Many techniques have been used for extracting discriminative features from facial images, which are given to a binary classifier. The feature extraction step is done through either geometric or appearance based methods. In previous methods

G. Bebis et al. (Eds.): ISVC 2012, Part II, LNCS 7432, pp. 409–419, 2012.

geometric features like distance between eyes, eyes and ears length, face length and width, etc. are considered. Whereas in appearance based methods image as a whole is considered rather than taking features from different parts of a face as local features. To deal with the problem of high dimension, some researchers used Principal Component Analysis (PCA), Linear Discriminant Analysis (LDA). For classification, different techniques like neural network, nearest neighbor method, LDA and other binary classification techniques have been used.

Techniques like Artificial Neural Networks (ANNs) [1-2] and Principal Component Analysis (PCA) [3] were first used for gender classification. A Hybrid technique was proposed by Gutta et. al. [4] consisting of an ensemble of Radial Basis Functions and C4.5 decision trees. Another method proposed in [5] achieved the recognition rate of 96% on FERRET database. SVMs were used by Moghaddam et. al. [6] for gender classification; they reported 3% of misclassification on the color FERET database. Neural Network was exploited by Nakano et. al. [7] for the information extracted from edges of facial images for gender recognition. Lu et. al. [8] used SVM to exploit the range and intensity information of human faces for ethnicity and gender identification. Not only sophisticated classifiers but simple techniques were also used for gender recognition. Yang et. al. [9] improved gender classification using texture normalization. Gaussian Process Classifier is used by Kim et al. [10] in their proposed system for gender recognition.

Several weak classifiers were combined by Baluja and Rowley [11] for pixel value comparisons on low resolution gray scale images in their AdaBoost based gender classifier. They used normalized images of size 20x20 in their test performed on FERET database, which showed an overall recognition rate of 90%. Lu and Shi [12] employed the fusion of left eye, upper face region and nose in their gender classification approach. Their results showed that their fusion of face region approach outperforms the whole face approach. Extending this idea, Alexandre [13] used a fusion approach based on features from multiple scales. They worked on normalized images of resolutions (20 x 20, 36 x 36 and 128 x 128) to extract shape and texture features. For texture features, they used Local Binary Pattern [14] approach.

DyWT decomposes the image features at different scales into different subbands which makes the analysis easy. DWT transform has been used for face description but it does not have better potential for features extraction because of being translation invariant. DyWT is translation invariant and is a better choice for face description. On the other hand, LBP captures local detail in a better way. Employing both DyWT and LBP in a novel way, we present a new face description technique. This approach outperforms the state-of-the-art techniques.

The rest of the paper is organized as follows. Section 2 presents an overview of Dyadic wavelet Transform (DyWT). In Section 3 Spatial Local Binary Pattern is discussed in detail. Gender recognition system based on our methodology is discussed in Section 4. Section 5 presents experimental results and their discussion. In the last Section 6, paper is concluded.

2   Dyadic Wavelet Transform

In this paper Dyadic Wavelet Transform (DyWT) is used for face description. Unlike DWT, it is translation invariant and can capture the micropatterns like edges in a better way. In the following paragraphs, we give an over view of DyWT. Complete detail can be found in [15].

DyWT wavelet transform involves two types of bases functions: scaling and wavelet functions. A scaling function $\phi(t)$ satisfies the following two-scale relation:

$$\phi(t) = \sum_k h[k]\sqrt{2}\phi(2t - k). \tag{2.1}$$

Its Fourier Transform (FT) satisfies the following relation:

$$\dot{\phi}(w) = \frac{1}{\sqrt{2}}\dot{h}\left(\frac{w}{2}\right)\dot{\phi}\left(\frac{w}{2}\right) \tag{2.2}$$

Using the scaling function $\phi(t)$, define a function $\psi(t)$ with the following relation:

$$\psi(t) = \sum_k g[k]\sqrt{2}\emptyset(2t - k).$$

Its Fourier transform is given by

$$\dot{\psi}(t) = \frac{1}{\sqrt{2}}\dot{g}\left(\frac{w}{2}\right)\dot{\emptyset}\left(\frac{w}{2}\right). \tag{2.3}$$

The function $\psi(t)$ is called dyadic wavelet transform if for some $A > 0 \,\& \, B$, if it satisfies the following inequality:

$$A \leq \sum_{-\infty}^{+\infty}\left|\dot{\psi}\left(2^{jw}\right)\right|^2 \leq B.$$

Projection of any L_2 function on dyadic wavelet space requires that the reconstruction condition must be satisfied, which further needs corresponding dual scaling and dual wavelet functions. The dual scaling function $\tilde{\phi}(t)$ is defined by the following two-scale relation:

$$\tilde{\phi}(t) = \sum_k \tilde{h}[k]\sqrt{2}\,\tilde{\emptyset}(2t - k),$$

and the dual wavelet function $\tilde{\psi}(t)$ satisfies the following two scale relation:

$$\tilde{\psi}(t) = \sum_k \tilde{g}[k]\sqrt{2}\,\tilde{\emptyset}(2t - k).$$

The Discrete Fourier Transform (DFT) of the filters $h[k], g[k], \tilde{h}[k], and\, \tilde{g}[k]$ are denoted by $\dot{h}(w), \dot{g}(w), \dot{\tilde{h}}(w), and\, \dot{\tilde{g}}(w)$ respectively. These filters are dyadic wavelet filters if the following condition is satisfied:

$$\dot{\tilde{h}}(w)\dot{h}^*(w) + \dot{\tilde{g}}(w)\dot{g}^*(w) = 2, w \in [-\pi, +\pi]. \tag{2.4}$$

The by symbol (*) denotes the complex conjugation. The above condition is called the reconstruction condition for dyadic wavelet filters.

Theorem 1. (á trousAlgorithm) the reconstruction condition (2.4) is used to obtain the following decomposition formulae

$$a_{j+1}[n] = \sum_k h[k]\ a_j[n + 2^j k], j = 0,1,...., \tag{2.5}$$

$$d_{j+1}[n] = \sum_k g[k]\ a_j[n + 2^j k], j = 0,1,...., \tag{2.6}$$

where $a_0[n]$ is given by $a_0[n] = \int_{-\infty}^{+\infty} f(t)\emptyset(t-n)dt$, and the following reconstruction formula

$$a_j[n] = \frac{1}{2}\sum_k (\tilde{h}[k]\ a_{j+1}[n - 2^j k] + \tilde{g}[k]d_{j+1}[n - 2^j k], j = 0,1,....,. \tag{2.7}$$

Equations (2.5) and (2.6) define the Fast dyadic wavelet transform (FDyWT) and are used for projection of 1-d function onto the space of dyadic wavelets. In case of 2-d function i.e. images, the projection is obtained by applying FDyWT in x-axis (horizontal) and then in y-axis (vertical) direction. Equation (2.7) defines the Inverse dyadic wavelet transform (IDyWT).

Spline dyadic wavelets are dyadic wavelets. A family of spline dyadic wavelets is defined with wavelet filters $h[k]$ and $g[k]$ whose Fourier transforms are given by:

$$\hat{h}(w) = \sqrt{2}\ e^{-i\varepsilon/2}(cos\frac{\omega}{2})^{m+1} \tag{2.8}$$

$$\hat{g}(w) = -i\sqrt{2}e^{-i\frac{w}{2}}sin\frac{w}{2} \tag{2.9}$$

where $m \geq 0$ denotes the degree of the box-spline and

$$\varepsilon = \begin{cases} 1 \text{ if } m \text{ is even} \\ 0 \text{ if } m \text{ is odd} \end{cases}$$

and

$$s = \begin{cases} 1 \text{ if } r \text{ is odd} \\ 0 \text{ if } r \text{ is even.} \end{cases}$$

The degree r is independent of m. Different values of r and m defines a family of spline dyadic wavelets. In this paper we explore this family for face representation for gender recognition.

3   Spatial Local Binary Pattern (SLBP)

LBP descriptor computed using LBP operator introduced by Ojala et al. [17] is one of the widely used texture descriptors that have shown promising results in many applications [14], [18], [19], and [20]. Ahonen et al. [27] used it for face recognition, Lian and Lu [21] and Sun et al. [13] employed it for gender recognition. The initial LBP operator associates a label with each pixel of an image; the label is obtained by converting each pixel value in the 3x3-neighbourhood of a pixel into a binary digit (0 or 1) using the center value as a threshold and concatenating the bits, as shown in Figure 1. Later the operator was extended to general neighborhood sizes, and its rotation invariant and uniform versions were introduced [14].

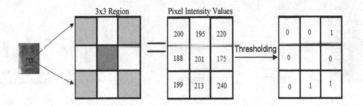

Fig. 1. LBP Operator

The general LBP operator is denoted by $LBP_{P,R}$ and is defined as follows:

$$LBP_{P,R} = \sum_{i=1}^{P-1} 2^i S(p_i - p_c) \tag{3.1}$$

where P is the total number of pixels in the neighborhood and R is its radius, pc is the center pixel and the thresholding operation is defined as follows:

$$S(p_i - p_c) = \begin{cases} 1 & p_i - p_c \geq 0 \\ 0 & p_i - p_c < 0. \end{cases} \tag{3.2}$$

Commonly used neighborhoods are (8, 1), (8, 2), and (16, 2). The histogram of the labels is used as a texture descriptor. The histogram of labeled image $f_l(x, y)$ is defined as:

$$H(i) = \sum_{x,y} I\{f_l(x, y) = i\}, \; i = 0, \dots, n - 1 \tag{3.3}$$

where n is the number of different labels produced by the LBP operator and

$$I\{x\} = \begin{cases} 1, & x \text{ is } true \\ 0, & x \text{ is } false. \end{cases} \tag{3.4}$$

Figure 2 shows the histogram extracted from an image with LBP operator. An LBP histogram in this approach contains information about facial micro-patterns like the distribution of edges, spots and flat areas over the whole image. In case of (8, R) neighborhood, there are 256 unique labels, and the dimension of LBP descriptor is 256. The basic LBP histogram is global and represents the facial patterns but their spatial location information is lost.

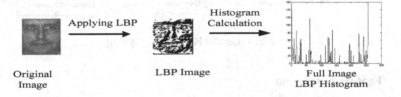

Fig. 2. LBP Histogram Calculation for Full image

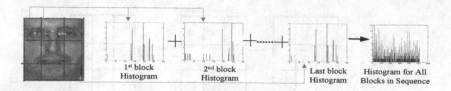

Fig. 3. Full enhanced LBP Histogram generation by Proposed Technique

To overcome this issue, spatially enhanced LBP histogram is calculated. Figure.6 shows the process of computing spatially enhanced LBP histogram. An image is divided into blocks; LBP histogram is calculated from each block and concatenated.

General LBP operator has three parameters: circular neighborhood (P, R), rotation invariance (ri) and uniformity (u2). For a particular application, it is necessary to explore this parameter space to come up with the best combination of these parameters. In this Paper we will explore Uniform version of LBP with P and R as 8 and 1.

4   Gender Recognition

The proposed system for gender recognition follows the general architecture of a recognition system i.e. it consists of four main parts: pre-processing, feature extraction, feature selection and classification. Various existing systems differ in the choice of feature extraction and classification techniques. Preprocessing step involves the normalization of face images. We introduced a new method for feature extraction based on LBP and DyWT. Further we apply feature subset selection method to increase accuracy and to reduce the time complexity. Simplest minimum distance classifiers based on L1, L2, and CS distance classifiers are used.

The block diagram of the recognition system which we used for gender recognition is shown in Figure 4.

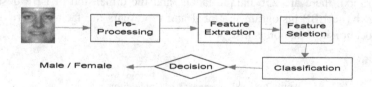

Fig. 4. Gender Recognition system

4.1   Feature Extraction

For feature extraction, we used SLBP and DyWT. DyWT decomposes an image in to a number of sub-bands at different scales. Figure 6 shows an image which is decomposed using DyWT up to scale 2. After decomposition SLBP operation is used to extract features from each sub-band.

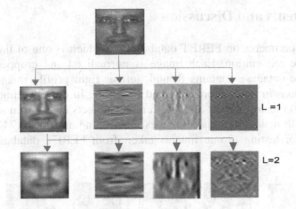

Fig. 5. Example of image from FERET database in to sub-bands

Specifically the following steps are used to extract features from each face:

a) Normalize the image
b) Decompose the image with DyWT up to scale N
c) Apply SLBP on each sub-band
d) Concatenating SLBP histograms for each subband, a multiscale LBP histogram is generated.

These steps have been shown in Figure 6.

DyWT parameters involve scales and filters. These filters are made from the combination of the spline values R and M as mentioned in section 2. SLBP involves many parameters: Neighborhood P, Radius R, Mapping, and block sizes. By experiment we found the best set of parameters which produces maximum result. The dimension of the features becomes big in some cases. To reduce the dimension and to enhance the accuracy, we apply SUN's FSS algorithm [22].

Fig. 6. Proposed Methodology

4.2   Classifier

In our system we preferred to employ minimum distance classifiers for achieving maximum accuracy for gender recognition and keeping the system simple. SLBP and DyWT with FSS can give better or comparable results to many stat-of-art techniques using city block distance (L1), Euclidean Distance (L2), and Chi-Square (CS). The accuracy of a gender recognition system depends on the choice of a suitable metric.

5   Experiments and Discussion

We performed experiments on FERET database [5], which is one of the challenging databases for face recognition. Each image is normalized and cropped to the size 60x48 pixels. The database contains frontal, left or right profile images and could have some variations in pose, expression and lightning. In our experiments, we used 2400 images of 403 male subjects and 403 female subjects taken from sets fa and fb. We used 1204 (746 male+458female) images for training and 1196 (740 male + 456 female) images for testing. Some images taken from FERET database are shown below.

Fig. 7. Samples from FERET Database

In case of SLBP, we tested LBP variation for uniform mapping and no mapping with P and R values of 8 and 1 respectively. Furthermore two types of histogram and calculated: normalized and simple. The images are divided into block sizes of 15x12, 12x12, and 10x12 resolutions. In case of DyWT we tested it for scale 1 to 5 all sub-band images i.e. LL, LH1, HL1...etc. For DyWT not only we tested it for various levels but also applied different spline values. R and M values (R = 1 and 2, M = 0, 1, 2, 3, and 4) are used which specify different filters. Each filter has different impact on different sub-band image. In FSS algorithm SUN, sigma and lambda parameters in the range of 0.1 to 2 are applied with an increase of 0.2 in each interval. This has great impact in the reduction of features and enhancement of result.

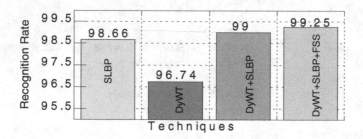

Fig. 8. Best results of proposed techniques

In our experiments the best result by simple SLBP is given with block size of 10x12 having uniform mapping, simple histogram and with L1 minimum distance classifier i.e. 98.66%. This result is shown in Figure.8 which is comparable to many other stat-of-art techniques. The effect of proposed methodology SLBP against different block sizes is shown in Figure.9. In [13] simple LBP with uniform mapping and block size of 16x16 and 32x32 is used which resulted in 93.46% accuracy as shown in Figure.9. It clearly shows that our system outperforms with an increase of 5.20% then the system proposed in [13]. It is noted in the experiments that smaller block sizes increase accuracy but also increases number of features which increases time complexity. Due to this reason we used SUN's algorithm to reduce features and time complexity.

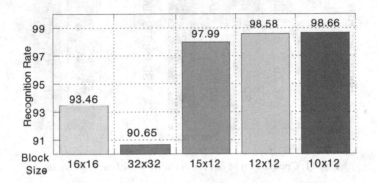

Fig. 9. Effect of Block Sizes for Proposed technique in comparison to [13]

In case of simple DyWT the best result is given by scale 3 sub-band with Low-High frequency (LH). This result is given with filter (R=1, M=1) and L1 minimum distance classifier i.e. 96.74% as shown in Figure.9. The effect of different filters can be seen in Figure 8 with the images resulting in maximum accuracy.

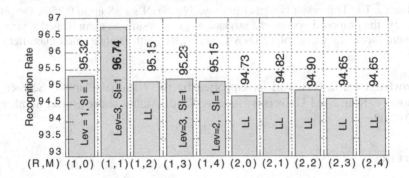

Fig. 10. Effect of different Filters on decomposed images

The maximum result is reported by L1 minimum distance classifier with the resultant matrix achieved from the addition of all the MSLBP histograms as shown in Figure.6. These In the end we compared results with reported results in literature for some of the complex stat-of-art techniques like Local Gabor Binary Pattern with LDA and SVMAC (*LGBP-LDA SVMAC*) [16], Local Gabor Binary Pattern with LDA and SVM (*LGBP-LDA SVM*) [16], and Multi-resolution Decision Fusion method (*MDF*) [13]. These approaches were having recorded maximum result till date in literature.

But our proposed system showed promising and comparable results to those result, as can be seen in Figure.11. It is also clearly noticeable from the figure that proposed system overcomes holistic approach of LBP and PCA.

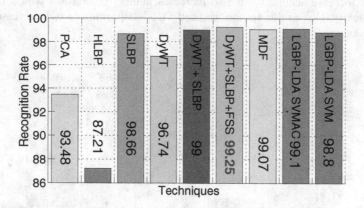

Fig. 11. Comparison of our results with stat-of-art techniques

6   Conclusion

DyWT is multi resolution descriptor which in combination with SLBP and FSS results in great improvement in gender recognition problem. It clearly overcome other global, and multi-resolution techniques i.e. PCA, Gabor etc. The best result is obtained with applying uniform SLBP on DyWT and adding the resultant features from decomposed images with block size of 12 x 12 and P,R values of 8,1 respectively while using simple classifier i.e. L1. This result is further improved by SUN's FSS up to 0.25%. Despite its simplicity, the proposed system can produce as good results as complicated systems. In our future work we will explore DyWT and SLBP with sophisticated classifiers like SVM.

Acknowledgement. This work is supported by the National Plan for Science and Technology, King Saud University, Riyadh, Saudi Arabia under project number 10-INF1044-02.

References

1. Golom, A., Lawrence, D.T., Sejnowski, T.J.: SEXNET: A neural network identifies gender from human faces, Advances in Neural Information Processing Systems, 572–577 (1991)
2. Edelman, B., Valentin, D., Adbi, H.: Sex classification of face areas: how well can a linear neural network predict human performance. Journal of Biological System 6(3), 241–264 (1998)
3. Sun, Z.: Genetic feature subset selection for gender classification: a comparison study. In: Proc. IEEE Conference on Applications of Computer Vision, pp. 165–170 (2009)
4. Gutta, S., Wechsler, H., Phillips, P.: Gender and ethnic classification of face images. In: Third IEEE International Conference on Automatic Face and Gesture Recognition (FG 1998), pp. 194–199 (1998)

5. Phillips, P.J., Hyeonjoon, M., Rizvi, S.A., Rauss, P.J.: The FERET evaluation methodology for face-recognition algorithms. In: IEEE Trans. Pattern Analysis and Machine Intelligence, 22nd edn., pp. 1090–1104 (October 2000)
6. Moghaddam, B., Yang, M.-H.: Gender classification with support vector machines. In: Proc. IEEE International Conference on Automatic Face and Gesture Recognition, pp. 306–311 (March 2000)
7. Nakano, M., Yasukata, F., Fukumi, M.: Age and gender classification from face images using neural networks. Proc. of Signal and Image Processing (2004)
8. Lu, X., Chen, H., Jain, A.K.: Multimodal Facial Gender and Ethnicity Identification. In: Zhang, D., Jain, A.K. (eds.) ICB 2005. LNCS, vol. 3832, pp. 554–561. Springer, Heidelberg (2005)
9. Yang, Z., Li, M., Ai, H.: An experimental study on automatic face gender classification. In: Proc. IEEE Int. Conf. on Pattern Recognition, pp. 1099–1102 (2006)
10. Kim, H.-C., et al.: Appearance based gender classification with Gaussian processes. Pattern Recognition Letters 27(6), 618–626 (2006)
11. Baluja, S., Rowley, H.: Boosting sex identification performance. International Journal of Computer Vision 71(1), 111–119 (2007)
12. Lu, L., Shi, P.: Fusion of multiple facial regions for expression-invariant gender classification. IEICE Electron. Exp. 6(10), 587–593 (2009)
13. Alexandre, L.A.: Gender recognition: A multiscale decision fusion approach. Pattern Recognition Letters 31, 1422–1427 (2010)
14. Ojala, T., Pietkainen, M., Maenpaa, T.: Multiresolution Gray-Scale and Rotation Invariant Texture Classification with Local Binary Patterns. IEEE Trans. Pattern Analysis and Machine Intelligence 24(7), 971–987 (2002)
15. Turghunjan Abdukirim, M., Hussain, K.: The Dyadic Lifting Schemes and the Denoising of Digital Images. International Journal of Wavelets, Multiresolution and Information Processing 6(3), 331–351 (2008)
16. Ojala, T., Pietkainen, M., Harwood, D.: A Comparative Study of Texture Measures with Classification Based on Feature Distributions. Pattern Recognition 29, 51–59 (1996)
17. Zhang, G., Huang, X., Li, S.Z., Wang, Y., Wu, X.: Boosting Local Binary Pattern (LBP)-Based Face Recognition. In: Li, S.Z., Lai, J.-H., Tan, T., Feng, G.-C., Wang, Y. (eds.) SINOBIOMETRICS 2004. LNCS, vol. 3338, pp. 179–186. Springer, Heidelberg (2004)
18. Ojala, T., et al.: Performance evaluation of texture measures with classification based on Kullback discrimination of distributions. In: Proceedings of the 12th IAPR, International Conference on Pattern Recognition (ICPR 1994), vol. 1, pp. 582–585 (1994)
19. Liu, H., Sun, J., Liu, L., Zhang, H.: Feature selection with dynamic mutual information. Journal of Pattern Recognition 42(7) (July 2009)
20. Meng, J., Gao, Y., Wang, X., Lin, T., Zhang, J.: Face Recognition based on Local Binary Patterns with Threshold. IEEE (2010), doi:10.1109/GrC.2010.72
21. Sun, N., Zheng, W., Sun, C., Zou, C.-r., Zhao, L.: Gender Classification Based on Boosting Local Binary Pattern. In: Wang, J., Yi, Z., Żurada, J.M., Lu, B.-L., Yin, H. (eds.) ISNN 2006. LNCS, vol. 3972, pp. 194–201. Springer, Heidelberg (2006)
22. Sun, Y., Todorovic, S., Goodison, S.: Local Learning Based Feature Selection for High Dimensional Data Analysis. IEEE Trans. on Pattern Analysis and Machine Intelligence 32(9), 1610–1626 (2010)
23. Zang, J., Lu, B.L.: A support vector machine classifier with automatic confidence and its application to gender classification. Neurocomputing 74, 1926–1935 (2011)

Architectural Style Classification of Domes

Gayane Shalunts[1,*], Yll Haxhimusa[2], and Robert Sablatnig[1]

[1] Vienna University of Technology
Institute of Computer Aided Automation
Computer Vision Lab
{shal,sab}@caa.tuwien.ac.at
[2] Vienna University of Technology
Institute of Computer Graphics and Algorithms
Pattern Recongition and Image Processing Lab
yll@prip.tuwien.ac.at

Abstract. Domes are architectural structural elements characteristic for ecclesiastical and secular monumental buildings, like churches, basilicas, mosques, capitols and city halls. In the scope of building facade architectural style classification the current paper addresses the problem of architectural style classification of facade domes. Building facade classification by architectural styles is achieved by classification and voting of separate architectural elements, like domes, windows, towers, etc. Typical forms of the structural elements bear the signature of each architectural style. Our approach classifies domes of three architectural styles - Renaissance, Russian and Islamic. We present a three-step approach, which in the first step analyzes the height and width of the dome for the identification of Islamic saucer domes, in the second step detects golden color in YCbCr color space to determine Russian golden onion domes and in the third step performs classification based on dome shapes, using clustering and learning of local features. Thus we combine three features - the relation of dome width and height, color and shape, in a single methodology to achieve high classification rate.

1 Introduction

Architectural styles are phases of development that classify architecture in the sense of historic periods, regions and cultural influences. Architectural elements, like windows, domes, towers, etc, are building components forming the architectural style of buildings. The forms and proportions of the structural elements submit to rules specific for each style. Classification of building facade images into different architectural styles by computer vision technologies will allow indexing of building databases into categories which belong to certain historic periods. Such a categorization narrows the search of building image databases to semantic category portions for the purposes of building recognition [1, 2], Content Based Image Retrieval (CBIR) [3], 3D reconstruction, 3D city-modeling [4]

---

* Supported by the Doctoral College on Computational Perception.

G. Bebis et al. (Eds.): ISVC 2012, Part II, LNCS 7432, pp. 420–429, 2012.
© Springer-Verlag Berlin Heidelberg 2012

and virtual tourism [5]. Real tourism is also a potential field of application for architectural style classification systems on mobile platforms.

The topic of classification of building facade images into architectural styles by computer vision methodologies has recently emerged [6–8]. The authors in [6] classify facade windows of Romanesque, Gothic and Baroque styles. The approach proposed in [7] classifies architectural elements, called tracery, in Gothic class and elements, called pediment and balustrade - in Baroque class. In [8] classification of Flemish Renaissance, Haussmannian and Neoclassical styles is performed on complete facade images.

Our methodology of building facade architectural style classification consists of three major phases - segmentation of building facades by architectural elements, classification of the segmented elements by architectural styles and architectural style voting of the classified elements. The proposed approach allows classification of partly occluded facades by a single typical architectural element, as well as classification of facades, which are a mixture of architectural styles. In the scope of facade architectural style classification by a voting mechanism of structural elements, the current paper focuses on classifying domes into three architectural styles - Renaissance, Russian and Islamic. At the current state our system requires user interaction to deliver the bounding box of the dome as a query image. We propose a three-step approach for the classification. At the first step the relation of the width and height of the dome image is analyzed to identify Islamic saucer domes. The second step checks if the dome has golden color in YCbCr color space to recognize Russian gilded onion domes. At the final third step a shape classification is performed. Each dome class uses certain geometrical rules for construction, so certain gradient directions dominate in each class. We perform clustering and learning of the local features to find out image dominant gradient directions and thus categorize the classes of the three architectural styles. Our system yields an overall classification rate of 90.24%.

The paper is organized as follows: Sect. 2 illustrates dome types and their relation to architectural styles. Sect. 3 explains the chosen method for the architectural style classification of domes. The classification experiments and results are presented in Sect. 4. And in Sect. 5 we conclude the paper.

2   Dome Types and Architectural Styles

A dome is a convex roof. Domes are categorized according to the base shape and the section through the dome center. Hemispherical domes have a circular base with a semicircular section. This dome type is characteristic for grand buildings of Renaissance architecture. The domes of Baroque, Neo-Renaissance and Neoclassical buildings imitate visual forms of Renaissance domes, thus our approach classifies them as Renaissance style. In the further stage of classification other structural elements will be taken into account to categorize the whole building as Renaissance, Baroque, Neo-Renaissance and Neoclassical. The mentioned imitation is displayed on the examples of the domes of St. Charles's Baroque

a) St. Peter's Basilica b) St. Charles's church c) St. Stephan's Basilica d) San-Francisco city hall

e) The saucer dome of Vienna Islamic center

Fig. 1. Hemispherical and saucer domes

church in Vienna (Fig. 1b), St. Stephan's Neo-Renaissance Basilica in Budapest (Fig. 1c) and Neoclassical city hall of San-Francisco (Fig. 1d). The most copied Renaissance domes belong to St. Peter's Basilica in Vatican (Fig. 1a) and St. Paul's Cathedral in London. Saucer domes have a circular base and a segmental, less than a semicircle section. This dome type is typical for Islamic Ottoman mosques, some of which are former Byzantine churches. The saucer dome of Vienna Islamic center is displayed in Fig. 1e. Onion domes have a circular or polygonal base and an onion-shaped section and are a typical feature of Russian Orthodox churches and Islamic mosques. Russian and Islamic onion domes are shown respectively on the examples of one of the domes of St. Nicholas's Basilica in Vienna (Fig. 2a) and the Blue Mosque dome in Yerevan (Fig. 2b). Russian architecture displays its most peculiar feature in the shape and number of the domes [9]. Architectural definitions are taken from the Illustrated Architecture

a) St. Nicholas's Basilica dome   b) Yerevan Blue mosque dome

Fig. 2. Onion domes

Dictionary[1]. Our approach categorizes three architectural styles - Renaissance, Russian, Islamic and seven intra-class types - Hemispherical domes of Renaissance, Baroque, Neo-Renaissance, Neoclassical buildings in Renaissance class, Russian onion domes, Islamic onion and saucer domes. As our goal is architectural style classification, we classify only between the three main architectural classes, but not the intra-class types.

3 Dome Classification: A Three-Step Approach

Classification of dome images by architectural styles is a highly complex task, since dome images are projections of 3D domes in 2D and due to high variability of architectural details and ornaments. The proposed methodology for solving the task is a three-step approach. At the **first step**, the dimensions of the query image are analyzed. Islamic saucer dome is the only dome type for which the width is always greater than the height, because of its shallow geometry. Also Islamic onion domes may have their width greater than the height. In either case the query image is classified as Islamic, when the mentioned condition is true.

At the **second step** we use color as a feature to identify Russian golden onion domes. Though gilded domes are typical for Russian Orthodox churches, they are not the only option. Russian churches also have examples of blue and colorful domes. So by golden color detection at this step we raise the classification rate among Russian gilded onion domes. For golden color detection we choose YCbCr color space, where Y color channel represents luminance, Cb and Cr channels - chrominance. The advantage of YCbCr color space is that it represents luminance (Y) in a single channel and is thus luma-independent. YCbCr color space is widely used for different color detection purposes, including skin detection [10] and facial feature detection [11]. Our goal is to find experimentally the ranges in Cb and Cr channels corresponding to golden color, while discarding Y channel. Thus, we count the mean pixel values of Cb and Cr components of 97 golden patches, which were cut manually from Russian gilded dome images. The reason for considering the mean Cb, Cr values of all image pixels, but not Cb, Cr values of individual pixels is that golden color due to its high reflectivity tends to appear from white to dark brown. So while determining the initial Cb and Cr ranges for golden color the individual pixel values may be false. From each patch a pair of mean Cb and mean Cr components is extracted. Each mean Cb value on Cb line corresponds to a mean Cr value located above it on Cr line (Fig. 3). We sort mean Cb values in ascending order to find out if there is a relation between Cb and Cr components. As seen in Fig. 3, with ascending of Cb values the Cr values tend to decrease. Low Cb and high Cr values correspond to highly saturated golden color. The full range of Cb and Cr values is between 16 and 240. In the mentioned range the initial Cb, Cr ranges found corresponding to golden color are as follows: $58 < Cb < 118$; $126 < Cr < 181$. The final Cb, Cr ranges are determined iteratively, maximizing the true positives and minimizing the false positives while golden color image detection on the training dataset (Sect. 4).

---

[1] Illustrated Architecture Dictionary: http://buffaloah.com/a/DCTNRY/vocab.html

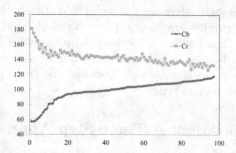

Fig. 3. Cb and Cr mean values of 97 golden dome patches

At the **final third step** classification of domes is done based on their shapes. Different texture features [12, 13], and well established shape descriptors [14, 15] can be used for this purpose. Authors in [16] show that shapes can be represented by local features like peaks and ridges. Since on the shapes of each class certain gradient directions are dominating, we choose local features to describe shapes of domes. We choose a local feature-based approach, as it incorporates texture and gradients into an image descriptor. Different local features can be used, like Harris-Laplacian corner detectors [17, 18], DoG corner detectors [19] or region based detectors [20, 21] and different local image descriptors [19–21]. We choose the bag of visual words (BoW) approach [22] (Fig. 4) for classification. The Scale Invariant Feature Transform (SIFT) [19] is used in the learning phase to extract the information of gradient directions. After performing the difference of Gaussians on different octaves and finding minimas/maximas, i.e. finding interest points, we only perform rejection of interest points with low contrast by setting a low threshold. All interest points that lie on dome edges are kept. Note that we do not follow the original work in [19] at this step, since we do not suppress the response of the filter along the edges. The last phase is finding the interest points and local image descriptors (SIFT image descriptors) and performing

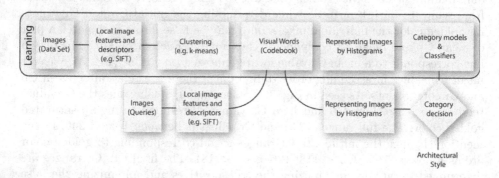

Fig. 4. Learning visual words and classification scheme [6]

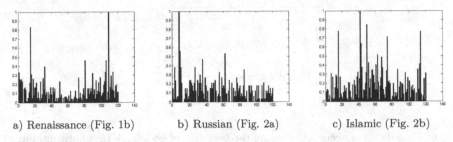

a) Renaissance (Fig. 1b)   b) Russian (Fig. 2a)   c) Islamic (Fig. 2b)

Fig. 5. Histograms of visual words for the images of different dome styles

their normalization. Since the number of local features is usually large, learning of a visual vocabulary (codebook) from the training set is done by clustering. An unsupervised clustering method (k-means) is used to find the visual cluster centers and to create the codebook of separate classes. The classification of a query image follows similar steps (Fig. 4). After extracting local image features and descriptors, the histogram representation is built up by using the codebook learnt on the training stage (Fig. 5).

The sum of all histograms responses for each class (called *integrated response*) is our simple category model. As our category model yields acceptable results (Sect. 4), we refrain from using a (complex) classifier for building a model. This classifier may be used as a weak classifier in a more complex classification line in boosting schemes [23]. The image dome class is determined by finding the maxima of integrated responses of the three classes. For example, for the histogram representation shown in Fig. 5c, the sum of all responses of Russian class is 5.72, Islamic class – 10.76 and Renaissance class – 6.03. Thus the image is classified as Islamic. The histograms shown in Fig. 5 are built using a codebook of 40 cluster centers for each class. Note that for Russian class histogram high responses are located on the bins from 1 to 40, for Islamic class - from 41 to 80 and for Renaissance class - from 81 to 120. The category model based on the maxima of the integrated class responses proves to be effective, as it makes the vote for the right class strong by integration of the high responses and suppresses the false class peaks, which may occur due to irrelevant descriptors located on architectural details. Islamic saucer domes are not included in the training stage, as query saucer dome images pass accurate classification at the first step of our methodology and do not need a codebook for classification.

The steps of our methodology are counted by their priorities. If the image is identified as Islamic at the first step, it does not pass to the further steps for analysis. Otherwise, it is passed to the second step for golden color detection. If the response at this step is positive, the dome is classified as Russian, otherwise it is passed to the third step for shape analysis.

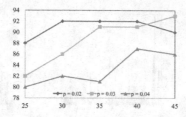

Fig. 6. Classification accuracy. Finding the best size of codebook (k – horizontal axis) and SIFT peak threshold (p).

4   Experiments of Dome Classification and Discussion

To the best knowledge of the authors there is no annotated image database of facade domes. To test and evaluate our method we created a database of 520 images, consisting of images from our own and Flickr[2] image datasets. 100 images of the database make the training dataset, annotated with labels of golden color and architectural styles. Golden labels are needed for identification of Cb and Cr ranges corresponding to golden color, while architectural style labels are used for codebook generation. 20 images out of 100 are labeled as golden. 30 images belong to Russian, 30 - to Islamic and 40 - to Renaissance styles. The resolution range of the images is from 127×191 to 3753×4314 pixels. Our database includes the most famous domes of the world, like those of St. Peter's Basilica in Vatican, Florence Cathedral, St. Paul's Cathedral in London, Pantheon in Paris, United States capitol in Washington, capitol buildings of 24 US states, Taj Mahal in Agra, St. Basil's Cathedral in Moscow, Hagia Sophia in Istanbul, etc.

In order to find the final Cb and Cr ranges corresponding to golden color, we use the fact that with the ascending of Cb values Cr values tend to decrease (Fig. 3). Thus instead of having one Cb and Cr range pair, we devide the final Cb range into Cb subranges and corresponding Cr ranges, so that the number of golden true positives is maximized, while keeping the number of golden false positives minimum. The final Cb subranges and their corresponding Cr ranges are found experimentally and are as follows:

Cb (31-100]; Cr (134-202]	Cb (100-110]; Cr (131-142]
Cb (110-115]; Cr (126-137]	Cb (115-120]; Cr (124-130]

Table 1. Classification accuracy on the training set with different codebook sizes

Peak Threshold (p)	k = 25	k = 30	k = 35	k = 40	k = 45
0,02	88	92	92	92	90
0,03	82	86	91	91	93
0,04	80	82	81	87	86

[2] http://www.flickr.com

Table 2. Confusion matrix and the (accuracy rate) excluding (table a)) and including (table b)) classification step 2

		Russian	Islamic	Renaissance	Σ
	Russian	84 (73.04%)	6	25	115
a)	Islamic	7	115 (92.74%)	2	124
	Renaissance	5	9	167 (92.27%)	181
	Σ	96	130	194	420

		Russian	Islamic	Renaissance	Σ
	Russian	98 (85.22%)	4	13	115
b)	Islamic	7	115 (92.74%)	2	124
	Renaissance	6	9	166 (91.71%)	181
	Σ	111	128	181	420

Golden color detection with the above mentioned subranges on our training dataset of 100 images results in 14 true positives out of 20 golden images and only 1 false positive.

To determine the codebook size best fitting our database (vocabulary size) an experiment with different codebook sizes (k) was performed (Table 1 and Fig. 6). The value of peak threshold for SIFT feature extraction and the value of k for k-means clustering algorithm are searched so that the final classification rate is maximized on the training dataset. Raising the SIFT peak threshold leads to the decrease of the extracted SIFT descriptors. As shown in Fig. 6, experiments with SIFT peak threshold value 0.04 have lower classification rate, since the number of extracted SIFT descriptors, describing the dominating gradients of each dome class, is decreased. Extraction of a bigger number of SIFT descriptors than that with peak threshold value equal to 0.03 tends to extract descriptors located on construction material textures, i.e. we are overfitting. Fig. 6 also shows that the best choice for k-means algorithm k parameter is in the range $25 - 45$. k values smaller than 25 decrease the classification rate, as the number of cluster centers is not enough for the discrimination of visual words of different classes. Whereas values greater than 45 make the image histograms sparser, i.e. we get non-representative visual words. Our final codebook choice for testing the system is the one corresponding to $k = 40$ and peak threshold equal to 0.03.

At first the classification is performed using only the first and the third steps, i.e. without golden color detection. The result is 54 false classified images out of 420 testing images, which yields an average classification rate of 87.14%. The confusion matrix, with true positives, is given in Table 2a). After we run the classification switching on the second step of our approach for Russian golden onion dome detection. To detect Russian golden domes we crop the lower 1/3 of the query image, since we assume that dome basement is on the bottom of the image (Fig. 2a), as we did not find any dome images online taken upside down or under high angle tilt. The image patch to be analyzed is further trimmed from right, left and bottom by 1/8 size of the initial image. The cropping of the query image is done to avoid the segmentation of the sky and clouds and to exclude false positives among Renaissance domes, which have examples with golden decorations on the dome upper part. The query image is considered golden, if more than 70% of pixels of the analyzed patch passes the golden condition, i.e. the pixel Cb and Cr values fall within our defined range of golden color in Cb and Cr channels. The confusion matrix in Table 2b shows that the classification rate of Russian domes is raised by 14 true positives, while resulting in 1 false positive. 60 out of 80 golden domes were detected, which yields 75% rate for our golden color detection module and raises the rate of classification for Russian class from 73.04% to 85.22%. Golden color detection raises the final classification rate of the whole testing database from 87.14% to 90.24%. Since the approach presented in this paper is a part of the larger framework of building facade architectural style classification, the achieved classification rate is acceptable.

5   Conclusion

In the scope of building facade architectural style classification by a voting mechanism of structural elements, the current paper purpose was to classify the architectural style of domes, which is a typical architectural element of religious and secular grand buildings. A three-step approach for dome classification of Renaissance, Russian and Islamic architectural styles was introduced. In our approach the first step determines Islamic saucer domes taking as a feature the relation of the dome width and height, the second step identifies Russian golden onion domes by color analysis in YCbCr color space and in the final step a classification of shapes is performed based on clustering and learning of local features. We proved experimentally that the methodology yielded an acceptable high classification rate. While in the proposed method user interaction is required to deliver the dome bounding box, our current work towards creating a method for automatic segmentation of domes on complete facade images shows nice preliminary results. Future work in the context of architectural style classification of building facades includes semantic segmantation, classification and voting of architectural elements. Applications like virtual tourism, 3D building reconstruction, 3D city-modeling, indexing of cultural heritage buildings will benefit from this work, since it limits the search in the image databases to semantic portions.

References

1. Zheng, Y.T., Zhao, M., Song, Y., Adam, H., Buddemeier, U., Bissacco, A., Brucher, F., Chua, T.S., Neven, H.: Tour the world: building a web-scale landmark recognition engine. In: Proc. of ICCV and PR, pp. 1085–1092 (2009)
2. Zhang, W., Kosecka, J.: Hierarchical building recognition. Image and Vision Computing 25(5), 704–716 (2004)
3. Li, Y., Crandall, D., Huttenlocher, D.: Landmark classification in large-scale image collections. In: Proc. of IEEE 12th ICCV, pp. 1957–1964 (2009)
4. Cornelis, N., Leibe, B., Cornelis, K., Gool, L.V.: 3d urban scene modeling integrating recognition and reconstruction. IJCV 78, 121–141 (2008)
5. Snavely, N., Seitz, S.M., Szeliski, R.: Photo tourism: exploring photo collections in 3d. ACM Transaction on Graphics 25, 835–846 (2006)

6. Shalunts, G., Haxhimusa, Y., Sablatnig, R.: Architectural Style Classification of Building Facade Windows. In: Bebis, G., Boyle, R., Parvin, B., Koracin, D., Wang, S., Kyungnam, K., Benes, B., Moreland, K., Borst, C., DiVerdi, S., Yi-Jen, C., Ming, J. (eds.) ISVC 2011, Part II. LNCS, vol. 6939, pp. 280–289. Springer, Heidelberg (2011)

7. Shalunts, G., Haxhimusa, Y., Sablatnig, R.: Classification of gothic and baroque architectural elements. In: Proc. of the 19th IWSSIP, Vienna, Austria, pp. 330–333 (2012)

8. Mathias, M., Martinovic, A., Weissenberg, J., Haegler, S., Gool, L.V.: Automatic architectural style recognition. In: Proc. of the 4th International Workshop on 3D Virtual Reconstruction and Visualization of Complex Architectures. International Society for Photogrammetry and Remote Sensing, Trento (2011)

9. Rosengarten, A.: A handbook of architectural styles. Chatto and Windus, London (1912)

10. Basilio, J.A.M., Torres, G.A., Pérez, G.S., Medina, L.K.T., Meana, H.M.P.: Explicit image detection using ycbcr space color model as skin detection. In: Proc. of the 2011 American Conference on Applied Mathematics and the 5th WSEAS International Conference on Computer Engineering and Applications, pp. 123–128 (2011)

11. Maglogiannis, I., Vouyioukas, D., Aggelopoulos, C.: Face detection and recognition of natural human emotion using markov random fields. Personal and Ubiquitous Computing 13(1), 95–101 (2009)

12. Ojala, T., Pietikäinen, M., Mäenpää, T.: Multiresolution grayscale and rotation invariant texture classification with local binary patterns. IEEE Trans. on Pattern Analysis and Machine Intelligence 24, 971–987 (2002)

13. Haralick, R.M.: Statistical and structural approaches to texture. Proc. IEEE 67, 786–804 (1979)

14. Zhang, D., Lu, G.: Review of shape representation and description techniques. Pattern Recognition 37, 1–19 (2004)

15. Belongie, S., Malik, J., Puzicha, J.: Shape matching and object recognition using shape contexts. IEEE Trans. on Pattern Analysis and Machine Intelligence 24, 509–522 (2002)

16. Crowley, J.L., Parker, A.C.: A representation for shape based on peaks and ridges in the difference of lowpass transform. IEEE Trans. on Pattern Analysis and Machine Intelligence 6(2), 156–170 (1984)

17. Harris, C., Stephens, M.: A combined corner and edge detector. In: Proc. of the 4th Alvey Vision Conference, pp. 147–151 (1998)

18. Mikolajczyk, K., Schmid, C.: Indexing based on scale invariant interest points. In: Internationl Conference in Computer Vision, pp. 525–531 (2001)

19. Lowe, D.G.: Distinctive image features from scale-invariant keypoints. IJCV 60(2), 91–110 (2004)

20. Matas, J., Chum, O., Urban, M., Pajdla1, T.: Robust wide baseline stereo from maximally stable extremal regions. In: BMVC, pp. 384–393 (2002)

21. Tuytelaars, T., Gool, L.V.: Wide baseline stereo matching based on local, affinely invariant regions. In: BMVC, pp. 412–425 (2000)

22. Csurka, G., Dance, C.R., Fan, L., Willamowski, J., Bray, C.: Visual categorization with bags of keypoints. In: Workshop on Statistical Learning in Computer Vision, ECCV, pp. 1–22 (2004)

23. Freund, Y., Schapire, R.E.: A decision-theoretic generalization of on-line learning and an application to boosting. Journal of Computer and System Sciences 55(1), 119–139 (1997)

Contour Detection by Image Analogies

Slimane Larabi[1] and Neil M. Robertson[2]

[1] University of Sciences and Technology
Houari Boumediene,
Computer Science Department,
BP 32 El Alia, Algiers, Algeria
[2] School of Engineering and Physical Sciences,
Earl Mountbatten Building Gait 2,
Heriot-Watt University,
Edinburgh, EH14 4AS, UK

Abstract. In this paper we deal only contour detection based on image analogy principle which has been used in super resolution images, texture, curves synthesis and interactive editing. Human is able to hand drawn best outlines that may considered as benchmarks for contour detection and image segmentation algorithms. Our goal is to model this expertise and to pass on it at the computer for contour detection. Giving a reference image where outlines are drawn by human, we propose a method based on the learning of this expertise to locate outlines of a query image in the same way that is done for the reference. Experiments are conducted on different data sets and the obtained results are presented and discussed.

Keywords: Image Analogies, Contour Detection, Outline Shape.

1 Introduction

Contour detection is an important task in many computer vision applications such as object recognition, motion, medical image analysis, image enhancement and image compression. There is an huge number of methods in literature devoted to contour detection and many states of the art have been published giving a complete review of proposed techniques [21], [11], [18], [19].

Image analogies constitutes a natural means of specifying filters and image transformations [13] and we can simply supply an appropriate exemplar and say, in effect: "Make it look like this".

Assuming that the transformation between two images A and A' is "learned", image analogies is defined as a method of creating an image filter which allows to recover by analogy from any given different image B the image B' in the same way as A' is related to A [6] [13].

An advantage of image analogies is that they provide a very natural means of specifying image transformations. Rather than selecting from among myriad different filters and their settings, a user can simply supply an appropriate exemplar (along with a corresponding unfiltered source image) and say, in effect:

G. Bebis et al. (Eds.): ISVC 2012, Part II, LNCS 7432, pp. 430–439, 2012.

"Make it look like this". Ideally, image analogies should make it possible to learn very complex and non-linear image filters [13].

Few works have been devoted for the use of image analogies in image processing. A method for supervised segmentation of medical images is proposed by Lackey and Colagrosso [15] applying directly the algorithm of Hertzmann [13]. The method is applied only to find by analogies the same colored regions in medical images as those processed by the expert.

Contrary to image processing, image analogies has been largely used in different applications such as super resolution [10], texture [12], [7], [9], [4], [2], [3], curves synthesis [14], image colorization, texture transfer, image enhancement and artistic filters [17], [20].

However, human can locate easily the contours and results are mainly identical from one person. Our goal is to to model human expertise and to pass on it at the computer for contour detection.

Giving a reference image where outlines are drawn by human, we propose in this paper a method based on the learning of this expertise to locate outlines of a query image in the same way that is done for the reference. In section 2 we present a theoretical foundation as proof that contour may be detected using this technique. Different data sets have been used to validate our approach. The obtained results are presented in section 3.

2   Contour Detection Using Image Analogies

2.1   Position of the Problem

Human is able to draw on natural image accurate contour and therefore to produce best dataset that will serve as a benchmark for comparing different segmentation and boundary detection algorithms [1]. Our aim is then to locate contours such as human does it including low resolution images where objects have small sizes. For such images, the detection in image of outlines becomes a hard task and human must zoom in to hand drawn them accurately. Let L_A be the image of considered scene (see figure 1). We assume that outline shapes are manually located on L_A and highlighted with a specific color. We obtain then a new image H_A which contain a pertinent information (colored contours) that are clearly visible for users and useful for processing. Given a query image L_B, the problem is then how it can be computed the image H_B which will contains outline shapes located and highlighted in the same way as those located in H_A.

2.2   Basic Principle of the Method

Before the describing of the proposed method, we discuss about the human expertise for contour detection. Given a low resolution image, we believe that human takes into account in this process two criteria for outline shapes locating:

- The first one is the neatness of the difference of gray level intensity (or color) between two neighboring sets of pixels.

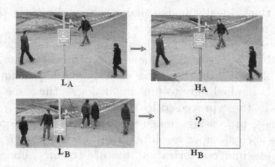

L_A

H_A

L_B

H_B

Fig. 1. Contour detection by analogy: Basic principle

- The second one is the knowledge of outline shape geometry inferred from context or some features such as outlines of dominant parts [8]. Indeed, during the process of outline drawing, human can't localize some parts of the outline due to the high similarity between pixels of background and object part but can avoid this difficulty using the prior knowledge. In this paper we deal only with the first criterion and we present our approach in order that computer locates pixels of contour in similar way to what is done by human.

In hand drawing contours task, we assume that a pixel p is considered belonging to the outline if the two following constraints are verified:

(1)- There is a brightness discontinuity at the considered pixel p for at least one directional line d_i ($i = 0..n - 1$) passing by p (see figure 2).
(2)- If the constraint (1) above is verified for p, it is also verified for some p-neighboring along other directional lines (d_j).

Let (L_B) be a query image of low resolution. The key idea is to classify each pixel q of L_B using the knowledge that may be inferred from (L_A, H_A): Each pixel q will be classified as its best match p^* in L_A, this means that brightness of p^* and its neighboring are the best similar to those of q and its neighboring.

Matching Process between q and p Pixels

Let $N(p)$, $N(q)$ be the ($m \times m$) neighborhood of p and q in L_A, L_B. Our aim is to search $N(p^*)$ from all $N(p)$ in L_A having the same or equivalent circumstances of $N(q)$ pixels where p^* is classified as contour or shape pixel.

To select the best match $N(p^*)$, we will take into account:

- The structure similarity between $N(p^*)$ and $N(q)$. This means that if p is a contour pixel, q will be also a contour pixel and the direction of the outline pixels in $N(p^*)$ and $N(p^*)$ must be very similar.
- The brightness similarity between $N(p^*)$ and $N(q)$ pixels.
For both $N(p)$ and $N(q)$, we will note the line of interest as the line which have the direction $d_i, i = 1..n$ and passes through the central pixel p or q. Depending on the chosen direction, each one of these lines may be an outline shape, background line or constituted by a set of shape and background pixels. In the

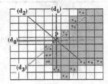

Fig. 2. Properties of contour pixel

comparison process, euclidian distance is used for similarity measure computation given by equation 1 where more importance is given to pixels of interest. To give more importance to pixels of the boundaries, the kernels $K^i(m \times m)$ characterized by high weight associated to pixels of the lines of interest are used.

$$S_m(q,p) = \sum_{u=-m/2}^{u=m/2} \sum_{v=-m/2}^{v=m/2} K^i(u,v) \times (Diff_{u,v})^2 \qquad (1)$$

Where:
- $Diff_{u,v} = N(q)_{(i+u),(j+v)} - N(p)_{(k+u),(l+v)}$
- (i,j), (k,l) are the coordinates of the pixels q, p in the images L_B, L_A
- $N(p)_{i,j}$, $N(q)_{k,l}$ are intensities of pixels (i,j) and (k,l).

For a given q of L_B, the best match p^* of L_A will be chosen so as $S_m(q,p)$ is the minimum from all computed values. If q is a contour pixel, the minimum of $S_m(q,p)$ values is obtained when $N(p)$ has p as a contour pixel and all neighboring contour pixels must have the same direction as those of q. In addition, the difference of luminosity between corresponding pixels must be minimal.

2.3   Study of the Validity of Similarity Measure

The proposed similarity measure must guaranty that any background or shape pixel q can't be classified as contour pixel and each outline pixel q will be correctly classified using the knowledge inferred from (L_A, H_A).

A theoretical study has be done and we present here only the main results. We investigated the cases where pixels are classified correctly using the two constraints cited in subsection 2.2 and the cases where this method fails.

The main result is that if the training pair of images (L_A, H_A) and the query one L_B are taken from the same scene, the location of contour pixels of L_B is done with success. However when L_B is from a different scene, the location of contour pixels can't be done without the loss of many candidates. This fact is due to following constraints that must be verified.

Let:

(1)- G_A, S_A be the average intensity of $N(p)$ background and shape pixels (see figure 3),
(2)- G_B, S_B be the average intensity of $N(q)$ background and shape pixels,
(3)- $d_B = S_B - G_B$, $d_A = S_A - G_A$, $d_G = G_B - G_A$.

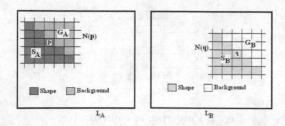

Fig. 3. Basic principle of p, q matching

We demonstrated that:

- if q is a shape or background pixel, it can't be classified as a contour pixel.
- if q is a contour pixel, it will be classified correctly if (see figure 4(a)):

$$(d_A > 0 \text{ and } 2d_G < d_A < 2d_B + 2d_G) \text{ or } (2d_B + 2d_G < d_A < 0) \qquad (2)$$

(Details of this proof aren't included in this paper because the limit of the number of pages). This means that the contour pixel q of a query image will be located if there are in the training images (L_A, H_A) hand drawn contour pixels so as $N(p)$ and $N(q)$ verify the equation 2. Otherwise, it will be classified as shape or background pixel.

To avoid this constraints, we propose the use of a set of artificial patterns (L_A, H_A) instead of real images with hand drawn contours. The values of (G_A, S_A) are chosen so as for any query pixel q, the values of (G_B, S_B) of associated $N(q)$ verifies equation 2 for at least one pattern.

This idea is due to the fact that is we have a pattern (L_A, H_A) as illustrated by figure 4(b), the values of any $S_B > G_B$ where $G_B = (S_A - \varepsilon - G_A)/2$ will verify the constraint (2), because $S_A \in]2d_G, 2d_G + 2d_B]$ and then the pixel q of considered $N(q)$ will be classified correctly. The translation of S_A with the value of $2\delta l$, where δl is the minimal difference intensity between two regions, allows to G_B to be translated by δl. In other side, for a fixed values of G_A, S_A, the value of G_B may decrease from $(S_A - \varepsilon - G_A)/2)$ to $(S_A - \varepsilon - G_A)/2) - \delta l$ because the constraint (2) will be also verified.

To built the patterns, we take firstly $G_A = 0$. The value of S_A must be less than 255 which implies that G_B is less than 128. As example, if we consider $\delta l = 16$ pixels, the values of S_A are:

$32, 64, 96, 128, 160, 192, 224$. The corresponding values of G_B are:

$16, 32, 48, 64, 80, 96, 112$ giving the patterns $P_1, P_2, ..., P_7$.

The value 256 for S_A is excluded because this implies that $G_B = 128$ and then any value of $S_B > G_B$ will allow to have $2d_G + 2d_B$ greater that S_A. Secondly, to obtain other combinations, it is sufficient to take the following values for (G_A, S_A):

$(64, 192), (64, 224), (96, 224), (128, 224), (160, 224), (192, 224), (208, 240)$ which correspond to the following values of G_B: $128, 144, 160, 176, 192, 208, 224$ giving the patterns $P_8, P_9, ..., P_14$. The set of artificial patterns are illustrated by figure 5.

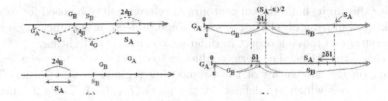

Fig. 4. Basic principle for satisfying all (G_B, S_B) values

Fig. 5. Artificial patterns (L_A, H_A)

3 Results

Experiments have been conducted on different datasets of real images: Caviar [5], PETS [16]. To detect the outline of a query image of a given scene, it is necessary to have a pair of training images (L_A, H_A) where L_A is the image of the scene and H_A is the same image which contain hand drawn contours.

Table 1 illustrates images L_A, H_A from Caviar data set where hand drawn outline shapes are highlighted with red color. The outlines located for some images of the same video are illustrated by the same table. We can see that in the query images only some outlines are located verifying the given constraint. There are many contour pixels not located bacause they not verify the constraints of the equation 2.

Table 1. From the left to right: the H_A image of hand drawn contours, contours computed for two frames of Caviar video

Despite this limit, hand drawn contours as reference may be used for example for tracking of moving object in the same scene. Indeed, it is sufficient to locate only the outline of moving object in different conditions of lighting.

Each one of the artificial patterns enables us to detect a specific level of contour depending on the intensities of the neighboring regions to the border. There are then 14 levels which are defined by the pairs (L_A, H_A). Table 2 illustrates contours located on the same frame of CAVIAR video using some patterns of the set of 14 patterns. We can see that the outline is moving through one region to other.

Table 2. Contours computed using respectively the patterns P_{14}^1, P_9^1, P_5^1

For the BSD500 dataset, we computed the Precision and Recall and we obtained similar results as Arbelaez et al [1] as shown by table 3. For high Recall values, our Precision is better and the difference reaches 20%. This implies that the number of false candidates are less than of Arbelaez et al. However for low Recall values, our Precision values are near from the values of Arbelaez et al [1], the difference is around 3%. This means that the number of missing good candidates are almost the same for both methods.

Table 3. Some BSD dataset results. From left to right: Original image, located contours

We repeated the same experiments for the dataset of Caviar and PETS. High values of Precision and Recall values have been obtained and exceeds those of Arbelaez et al [1] using BSD dataset and the difference reaches respectively 27% and 10%.

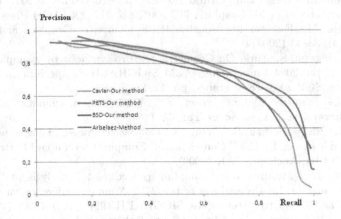

Fig. 6. Obtained ratios (Recall, Precision) for BSD dataset

4   Conclusion

We proposed in this paper a new method for contour detection based on image analogies. In the first part, we presented a theoretical foundation for the use of hand-drawn contours as reference images to be used subsequently for the detection of new contour pixels by analogy in the query image. We found that only pixels that have the same conditions as those of the reference image may be located which is perhaps to be expected in such a data driven technique. This implies that numerous reference images are needed to locate all possible new contour pixels which implies in the hard and time-consuming task of hand drawing reference contours, and thus increasing of the algorithm complexity.

To avoid this constraint and to locate all contour pixels whatever the image query happens to be, we proposed a set of 14 artificial pairs of patterns as reference images of low size and containing the required information to locate contours of different levels of resolution where levels are related to the difference of intensity between neighboring regions.

The proposed method has been applied to different types of images: the "natural" BSD dataset, indoor scenes (CAVIAR) and outdoor scenes (PETS2009). Compared to the reference images, our method demonstrates a very good recall, precision and finds all visible contours of gray level images. In addition, outline shapes of low resolution images such as those in CAVIAR and PETS are well located where the width of human as example is around 10 pixels.

References

1. Arbelaez, P., Maire, M., Fowlkes, C., Malik, J.: Contour Detection and Hierarchical Image Segmentation. IEEE Transactions on Pattern Analysis and Machine Intelligence 33(5), 898–916 (2011)
2. Ashikhmin, M.: Synthesizing Natural Textures. In: Proceedings of 2001 ACM Symposium on Interactive 3D Graphics, I3D 2001, pp. 217–226. ACM Press (2001)
3. Ashikhmin, M.: Fast texture transfer. IEEE Computer Graphics and Applications 23(4), 38–43 (2003)
4. Bhat, P., Ingram, S., Turk, G.: Geometric texture synthesis by example. In: Proceedings of the 2004 Eurographics/ACM SIGGRAPH Symposium on Geometry Processing, SGP 2004, Nice, France, pp. 41–44 (2004)
5. Caviar, EC Funded CAVIAR project/IST 2001 37540, Benchmark Data (2001), http://homepages.inf.ed.ac.uk/rbf/CAVIAR/
6. Cheng, L., Vishwanathan, S., Zhang, X.: Consistent image analogies using semi-supervised learning. In: IEEE Conference on Computer Vision and Pattern Recognition (CVPR), Anchorage, AK (2008)
7. De Bonet, J.S.: Multiresolution sampling procedure for analysis and synthesis of texture images. In: Proceedings of the 24th Annual Conference on Computer Graphics and Interactive Techniques, SIGGRAPH 1997, pp. 361–368 (1997)
8. DeWinter, J., Wagemans, J.: Segmentation of object outlines into parts: A large-scale integrative study. Cognition 99(3), 275–325 (2006)
9. Efros, A.A., Leung, T.K.: Texture Synthesis by Non-Parametric Sampling. In: IEEE International Conference on Computer Vision, vol. 2 (1999)
10. Freeman, W.T., Pasztor, E.C., Carmichael, O.T.: International Journal of Computer Vision 40(1) (2000)
11. Freixenet, J., Munoz, X., Raba, D., Marti, J., Cufi, X.: Yet Another Survey on Image Segmentation: Region and Boundary. Information Integration, ECCV (3), 408–422 (2002)
12. Heeger, D.J., Bergen, J.R.: Pyramid-based texture analysis/synthesis. In: Proceedings of the 22nd Annual Conference on Computer Graphics and Interactive Techniques, SIGGRAPH 1995, New York, NY, USA, pp. 229–238 (1995)
13. Hertzmann, A., Jacobs, C.E., Oliver, N., Curless, B., Salesin, D.H.: Image analogies. In: Proceedings of the 28th Annual ACM Conference on Computer Graphics and Interactive Techniques, SIGGRAPH 2001, New York, NY, USA, pp. 327–340.
14. Hertzmann, A., Oliver, N., Curless, B., Seitz, S.M.: Curve Analogies. In: EGRW 2002 Proceedings of the 13th Eurographics Workshop on Rendering Switzerland, Switzerland (2002)
15. Lackey, J.B., Colagrosso, M.D.: Supervised Segmentation of Visible Human Data with Image Analogies. In: IC-AI, pp. 843–847 (2004)
16. Pets 2009, PETS 2009 Benchmark Data. University of Reading, UK (2009), http://www.cvg.rdg.ac.uk/PETS2009/a.html
17. Sykora, D., Burianek, J., Zara, J.: Unsupervised colorization of black-and-white cartoons. In: Proceedings of the 3rd International Symposium on Non-Photorealistic Animation and Rendering, NPAR 2004, Annecy, France, pp. 121–127 (2004)

18. Suri, J.S., Liu, K., Singh, S., Laxminarayan, S.N., Zeng, X., Reden, L.: Shape recovery algorithms using level sets in 2-D/3-D medical imagery: a state-of-the-art review. IEEE Transactions on Information Technology in Biomedicine 6(1), 8–28 (2002)
19. Papari, G., Petkov, N.: Edge and line oriented contour detection: State of the art. Image Vision Computing Journal 29(2-3), 79–103 (2011)
20. Wang, G., Wong, T., Heng, P.: Deringing cartoons by image analogies. ACM Trans. Graph 25(20), 1360–1379 (2006)
21. Ziou, D., Tabbone, S.: Edge Detection Techniques - An Overview. International Journal of Pattern Recognition and Artificial Intelligence (1998)

Rotation Invariant Texture Recognition Using Discriminant Feature Transform

Nattapong Jundang and Sanun Srisuk

Department of Computer Engineering
Mahanakorn University of Technology,
Bangkok, Thailand, 10530
{jnattapo,sanun}@mut.ac.th

Abstract. This paper presents a new texture representation, the volume Trace transform, based on several Trace transform. The volume Trace transform (VTT) is constructed using multi-trace functional to produce salient features. The VTT is transformed to a distinctive compact representation using our proposed method, the discriminant feature transform (DFT). DFT is a 2-D histogram. The histogram is evaluated by chi-square test statistics. The experimental result was conducted on Brodatz texture database.

1   Introduction

In appearance model, the crucial information used to distinguish objects is its texture pattern. Studying the texture analysis may be useful for many applications including content based image retrieval, medical image, industrial surface inspection, object recognition and remote sensing [1], [2], [3]. The texture analysis has been extensive studied and many techniques have been proposed in the literature [4]. In the real world applications, the local texture descriptor must be invariant to rotation, scaling, gray scale transformation and white noise. Texture is a variation of the gray scale in which it can be represented as a pattern. Similar textures may share some common properties, e.g. frequency, size and structure. Generally, textures can be categorized into statistical and structural patterns. The statistical texture is regarded as the realization of a probability function while the structure texture represented as repetitive deterministic structures. Rotation-invariant texture analysis has been studied extensively and is a major issue in the image processing community [5], [6], [7], [8], [9], [10], [11].

The approaches on rotation invariant texture descriptor are the cooccurrence matrices [12]. Ayala et al. [9] proposed a framework for shape and texture analysis based on newly defined spatial distributions (SSD). The main idea consisted of combining a granulometric analysis of image. The comparison between the geometric covariograms for binary images of the auto-correlation function for gray-scale images and its granulometric transformation was also given. This approach can be regarded as a technique for texture descriptor using the probability distributions of the shape and

G. Bebis et al. (Eds.): ISVC 2012, Part II, LNCS 7432, pp. 440–447, 2012.

texture. Ojala et al. [13] introduced a new texture descriptor based on the local binary pattern (LBP). The LBP is an operator for gray scale and rotation invariant texture descriptor. The difference between the center and the neighbor pixels was used to encode as a binary number. The nonuniform and uniform patterns are used as a local texture descriptor. This method is invariant against the monotonic gray scale transformation and rotation, but sensitive to the noisy images. Some extensions on local binary pattern can be found in [14], [15]. A new HMM, called HMT-3S, for statistical texture characterization in the wavelet domain was introduced by Fan et al. [7]. They studied the wavelet-based texture analysis and synthesis using HMMs. The HMT-3S method was applied to texture analysis including classification, segmentation and texture synthesis. Lazebnik et al. [10] presented a texture representation that is suitable for texture surface recognition using viewpoint and nonrigid deformations. A sparse set of affine Harris and Laplacian regions is discovered in the image. The texture is represented as a texture element having a characteristic elliptic shape and a distinctive appearance pattern. The elliptical shape serves as a discriminative feature for texture recognition. However, some statistical texture images have failed in the classification process since the elliptical shape cannot be constructed. The rotation-invariant texture analysis based on the wavelet and radon transform has been proposed [6], [11]. The radon transform is used to estimate the principal direction of the texture. Then, the texture is rotated back to the 0 degree. The features are extracted from the rotated texture by using the wavelet transform. However, estimating the principal direction is difficult due to the fact that rotation degree of the statistical texture cannot be discovered correctly.

In this paper, we first converted the rotation problem to the circular-shifted one by using the Trace transform. In fact, several Trace transforms are generated from one texture image to produce the volume Trace transform. Finally, the 2-D histogram representation is formed which captures the spatial structure of the VTT. Two histograms are evaluated using chi-square test statistics.

2   Texture Representation

2.1   The Volume Trace Transform

In this section, we introduce a texture representation based on the Trace transform. As stated by Kadyrov and Petrou, 'in order to be able to recognize objects one has the problem of large number of features necessary.' The more classes one has to discriminate, the more feature may be necessary [1]. This is a valuable hint when proposing a texture classification which makes use of robust features from the Trace transform. The important advantage of Trace transform is an alternative image representation that allows one to construct thousands of features from an image. Therefore, the Trace transform may be appropriate for a texture classification system, and that is why we decided to investigate its usefulness in the particular problem. We propose here the volume Trace transform (VTT) which are produced from the n Trace function.

The Trace transform [16], a generalization of the Radon transform, is a tool for image processing which can be used for recognizing objects under transformations, e.g. rotation, translation and scaling. We assume that the texture image is subject to solve a rotation problem. In this approach, we can see that the image remains the same but it is viewed from a linearly distorted coordinate system C_2 with respect to the original coordinate system C_1. If we known the rotation angle ϕ of an image in coordinate system C_2, the line t_1 and t_2, for coordinate system C_1 and C_2 respectively, are equivalent by which t_2 is rotated by $-\phi$. To produce the Trace transform one computes a functional T along tracing lines of an image. Each line is characterized by two parameters, namely its distance ρ from the centre of the axes and the orientation ϕ the normal to the line has with respect to the reference direction. In addition, we define parameter t along the line with its origin at the foot of the normal. The definitions of these three parameters as shown in Fig. 1. With the volume Trace transform, the image is transformed into another "image", which is a 3-D function $g(\phi, \rho, k)$. Consider scanning an image with lines in all directions. Let us denote the set of all these lines with Λ. The Trace transform is a function g defined on Λ with the help of T which is some functional of the image function when it is considered as a function of variable t. T is called "the trace functional".

$$g(\phi, \rho, k) = T_k \left(F(\phi, \rho, k) \right), \tag{1}$$

where $F(\phi, \rho, k)$ stands for the values of the image function along the chosen line. T_k is the kth Trace functional. Parameter t is eliminated after taking the trace functional. The result is therefore a function of parameters ϕ, ρ and k and can be interpreted as another volume image defined on Λ. The resultant Trace transform depends on the functional we used. Different Trace transform can be produced from an image using different functionals T. Let us denote by t_i the sampling points along the tracing line defined by n the number of points along the tracing line. n may be varied depending on the length of the trace line. In our current implementation, we demonstrate only 10 different trace functionals, samples of which are shown in Table 1. Fig. 1 demonstrates the construction of the VTT. When tracing an image with a line in Fig. 1., the intensity values are formed as a vector. It is known in the Trace transform space is computed from a line in the image space. We generate the VTT by using the n Trace functional. The n Trace transform, calculating from n Trace functionals, are stacked as the 'volume' Trace transform. The VTT provides us numerous discriminant features which can be use to recognize the textures more correctly. However, the computational complexity of the VTT makes texture recognition difficulty. We will show our proposed method to convert the VTT to be compact discriminant representation.

Table 1. The Trace functioanls T for the volume Trace transform (VTT)

No.	Teace Functionals	Details
1	$T(f(t)) = \int_0^\infty f(t)\,dt$	The line integral transformation
2	$T(f(t)) = \left[\int_0^\infty |f(t)|^p\,dt\right]^q$	$p = 0.5,\ q = 1/p$
3	$T(f(t)) = \left[\int_0^\infty |f(t)|^p\,dt\right]^q$	$p = 4,\ q = 1/p$
4	$T(f(t)) = \int_0^\infty |f(t)'|\,dt$	The one-dimensional numerical gradient of t. $f(t)' = \left[(t_2 - t_1),(t_3 - t_2),....,(t_n - t_{n-1})\right]$
5	$T(f(t)) = \text{median}_t\left\{f(t),|f(t)|\right\}$	
6	$T(f(t)) = \text{median}_t\left\{f(t),|f(t)'|\right\}$	
7	$T(f(t)) = \left[\int_0^x |F\{f(t)\}|^p\,dt\right]^q$	F denote the discrete Fourier transform computed with the FFT algorithm, $p = 4,\ q = 1/p,\ x = n/2$
8	$T(f(t)) = \int_0^\infty \left|\dfrac{d}{dt}M\{f(t)\}\right|\,dt$	M is a median filter operator, using a local window of length 3, and differentiation means taking the difference of successive samples
9	$T(f(t)) = \left[\int_0^\infty r^p f(t)\,dt\right]$	$p = 2,\ r = |l - c|,\ l = 1,2,...,\ n,\ c = \text{median}_t\{l, f(t)\}$
10	$T(f(t)) = \left[\int_0^\infty r^p f(t)\,dt\right]$	$p = 0.5,\ r = |l - c|,\ l = 1,2,...,\ n,\ c = \text{median}_t\{l, f(t)\}$

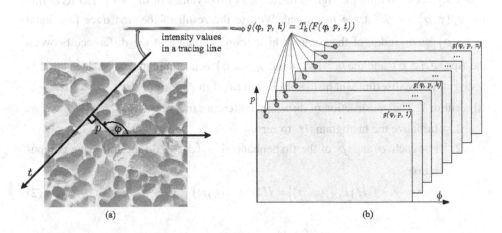

Fig. 1. The construction of volume Trace transform. (a) Tracing line on an image with parameters ϕ, ρ and t. (b) The volume Trace transform with n Trace fucntionals.

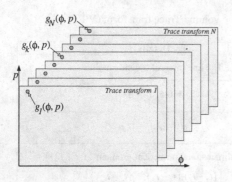

Fig. 2. The construction of the histogram DAFT descriptor. (Left) The volume Trace transform and (Right) the discriminant feature transform.

2.2   Discriminant Feature Transform

The Trace transform provides numerous features for texture classification. We produce several Trace transforms from one texture image to construct the VTT. However, the main problem of the VTT is its computational complexity [17]. Hence, the compact representation must be built in order to reduce its complexity. Usually, the 1-D histogram was constructed from the global and local encoded image [3], [5], [14], [15]. Those 1-D histogram do not preserve the spatial structure of the 2-D encoded images. In this section, we propose a new compact 2-D global descriptor generating from the VTT. Therefore, we first construct the global histogram for each texture image by accumulating the sign of difference of the values of the VTT. Let us define by $g_k(\phi,\rho)$ the kth trace functional. We use the result of the first trace functional $g_1(\phi,\rho)$ as an index of the compact histogram. The sign of the difference between the first $g_1(\phi,\rho)$ and the kth functional $g_k(\phi,\rho)$ is accumulated to produce the 2-D histogram. The discriminant histogram is described in Algorithm 1.

Algorithm 1: The construction of discriminant feature transform

1. Initialize the histogram H to zero.
2. For each ϕ and ρ of the first enhanced $g_1(\phi,\rho)$, do the following computation:

$$H\big(k,g_1(\phi,\rho)\big)=H\big(k,g_1(\phi,\rho)\big)+Z(v),\qquad(2)$$

   with

$$Z(v)=\begin{cases}1 & \text{if } v>0;\\ 0 & \text{if } v=0;\\ -1 & \text{if } v<0;\end{cases}$$

   Where $v=g_1(\phi,p)-g_k(\phi,p), k=2...N.$
3. Repeat step 2 for all ϕ and ρ.

In this paper, N is number of Trace functional. The $H\left(k, g_1(\phi, \rho)\right)$ is a 2-D histogram which characterize the VTT as the compact representation. Based on the robustness of the VTT, we use the VTT value as the index of the histogram which describes the characteristics of the texture images. This 2-D histogram is called the Discriminant Accumulative Feature Transform-DAFT. Please note that, before constructing the DAFT, the VTT is normalized so that their values are in the range [0, 1]. Fig. 2 demonstrates the construction of the DAFT descriptor from the VTT with N Trace functional. The 2-D histogram H is a rich discriminative feature which is transformed from volume image to the 2-D feature space. Hence, the process of classification is now become simple and efficient since the dimensions of the histogram H are fixed irrespective to the resolution of the texture images and the Trace transform.

3   Histogram Matching

The DAFT is a 2-D histogram generating from the VTT. Hence, we measure the similarity between the two histograms by using χ^2 test statistics.

$$R_{i,j} = \frac{1}{2}\sum_{k=1}^{K}\frac{\left[H_i(k) - H_j(k)\right]^2}{H_i(k) + H_j(k)} \qquad (3)$$

4   Experimental Result

We demonstrate the efficiency of our approach with rotation-invariant texture analysis using 72 images from publicly Brodatz texture database [18]. We subsampling all images to the size 154 x 154 for efficiency of the algorithm and used in all of the experiments. These texture images were displayed in Fig. 3. Each texture image was assigned as a class in classification process.

We showed here the powerful of our method which is robust to the rotation of texture image. The volume Trace transform produces from 10 Trace functionals were show in Fig. 4.

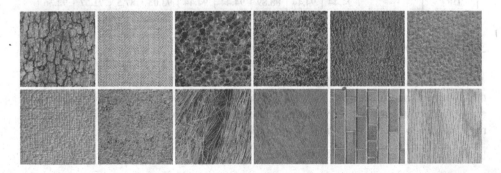

Fig. 3. Samples from the Brodatz texture database [18]

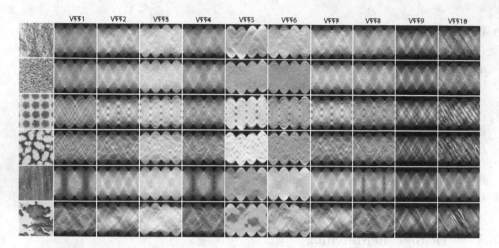

Fig. 4. Example results of the Trace transform produced from 10 trace functional

On the experiment of the rotation problem, the test set was created by rotating the texture image with several degrees of rotations (15°, 30°, 45°, 60°, 75°, 90°, 120° and 150°). Hence, 576 (72 x 8) rotated texture images were created and used them as a test set. Table 2 presented the classification rates using the two methods, LBP and DFT. The average of the maximum classification accuracy was 92.36% for DFT and 84.63% for LBP. Please note that a similar result of the LBP classification rate has been reported in [11].

Table 2. Comparison of the classification accuracies (%) for rotated texture image where the training was done with rotation 0° and test with other degree

Operator	Parameters	Testing Rotation Angle								
		15°	30°	45°	60°	75°	90°	120°	150°	Avg
$LBP_{P,R}^{riu2}$ [5]	P =8, R =1	56.9	60.1	62.4	66.5	63.2	59.8	60.1	58.1	61.23
	P =16, R =2	76.4	77.3	74.9	78.1	77.2	75.6	74.5	78.3	76.54
	P =24, R =3	83.2	84.1	85.3	84.5	85.2	85.7	83.4	85.7	84.63
DFT	-	94.44	94.44	88.89	94.44	94.44	93.05	87.5	91.67	92.36

5   Conclusion

The texture representation, which was produced from the Trace functional, was called the volume Trace transform. The discriminant feature transform produces a compact representation from a VTT. With our proposed method, we achieved 92.36% of accuracy which is outperformed the LBP (84.63%).

References

1. Ahonen, T., Hadid, A., Pietikainen, M.: Face Description with Local Binary Pattern: Application to face recognition. IEEE Trans. PAMI 28(12), 2037–2041 (2006)
2. Ahonen, T., Hadid, A., Pietikäinen, M.: Face Recognition with Local Binary Patterns. In: Pajdla, T., Matas, J(G.) (eds.) ECCV 2004. LNCS, vol. 3021, pp. 469–481. Springer, Heidelberg (2004)
3. Hadid, A., Pietikainen, M., Ahonen, T.: A discriminative feature space for detecting and recognizing face. In: Proc. of IEEE CVPR, pp. 797–804 (2004)
4. Randen, T., Husoy, J.H.: Filtering for Texture Classification: A Comparative Study. IEEE Trans. PAMI 21(4), 291–310 (1999)
5. Ojala, T., Pietikainen, M., Maenpaa, T.: Multiresolution gray-scale and rotation invariant texture classification with local binary patterns. IEEE Trans. PAMI 24(7), 971–987 (2002)
6. Khouzani, K.J., Zadeh, H.S.: Rotation-Invariant Multiresolution Texture Analysis using Radon and Wavelet Transforms. IEEE Trans. IP 14(6), 783–795 (2005)
7. Fan, G., Xia, X.G.: Wavelet-Based Texture Analysis and Synthesis using Hidden Makrov Models. IEEE Trans. On Cir. and Sys.-1: Fund. Therory and Apps. 50(1), 106–120 (2003)
8. Kokkinos, I., Evangelopoulos, G., Maragos, P.: Texture Analysis and Segmentation using Modulation Features, Generative Models, and Weighted Curve Evolution. IEEE Trans. PAMI 31(1), 142–157 (2009)
9. Ayala, G., Domingo, J.: Spatial Size Distributions: Applications to Shape and Texture Analysis. IEEE Trans. PAMI 23(12), 1430–1442 (2001)
10. Lazebnik, S., Schmid, C., Ponce, J.: A Sparse Texture Representation using Local Affine Regions. IEEE Trans. PAMI 27(8), 1265–1278 (2005)
11. Khouzani, K.J., Zadeh, H.S.: Radon transform orientation estimation for rotation invariant texture analysis. IEEE Trans. PAMI 27(6), 1004–1008 (2005)
12. Davis, L.S., Johns, S.A., Aggarwal, J.L.: Texture Analysis using Generalized Cooccurrence Matriced. IEEE Trans. PAMI 1, 251–259 (1979)
13. Ojala, T., Pietikainen, M., Maenpaa, T.: Multiresolution gray-scale and rotation invariant texture classification with local binary patterns. IEEE Trans. PAMI 24(7), 971–987 (2002)
14. Montoya Zegarra, J., Beeck, J., Leite, N., Torres, R., Falcao, A.: Combinning global with local texture information for image retrieval applications. In: Proc. of IEEE ISM, pp. 148–153 (2008)
15. Zhang, B., Shan, S., Chen, X., Gao, W.: Histogram of gabor phase patterns (hgpp): A novel object representation approach for face recognition. IEEE Trans. IP 16(1), 57–68 (2007)
16. Kadyrov, A., Petrou, M.: The trace transform and its application. IEEE Trans. PAMI 23(8), 811–828 (2001)
17. Kadyrov, A., Petrou, M.: A face authentication system using the trace transform. Journal of Pattern Analysis and Applications 8(1-2), 50–61 (2005)
18. Brodatz, P.: Texture: A Photographic Album for Artist and Designers. Dover (1996)

An Unsupervised Evaluation Measure of Image Segmentation: Application to Flower Image Segmentation

Asma Najjar and Ezzeddine Zagrouba

Team of Research SIIVA- Lab. Riadi,
Higher Institute of Computer Science,
University of Tunis Elmanar, Tunisia
najjar.asma@yahoo.fr, E.Zagrouba@fsm.rnu.tn

Abstract. We present a new unsupervised metric for segmentation result evaluation based on Bayes classification error and image global contrast. First, we presented a comparative study between several unsupervised metrics in order to prove their limits. The qualitative study was performed to make a preliminary selection and to discard some measures unsuitable for evaluation of foreground/background segmentation on flower images. For the quantitative study, we proposed a validation protocol based on the vote technique and involving a comparison to the ground truth. Experiments were performed on Oxford flower dataset in order to select the best result between different segmentation results. The obtained result showed that our proposed metric gives the best results.

1 Introduction

Segmentation is one of the most important steps in an image analysis process. It aims to decompose an image in homogeneous regions in order to facilitate the scene interpretation. Although there are many segmentation algorithms in the literature, it remains difficult to decide about their efficiency and the relevance of the given results. This decision is closely related to the domain of application.

There are three approaches that can be used to evaluate a segmentation result. The first one is the analytical approach based on an analysis of segmentation algorithms properties such as complexity, convergence, stability, etc. Some analytical methods was proposed by Liedtke and al. [1], Cho [2] and Voisine [3]. The second approach, known as empirical discrepancy methods [4], requires an available ground-truth and it is considered as a supervised method [5][6]. The ground-truth can be synthetic or made by domain experts. Literature presents supervised metrics for the evaluation of region based segmentation algorithm [7] [8] [9], and others for the evaluation of contour detectors [10] [11] [12]. The last one is the unsupervised approach or also called empirical goodness methods [4] and it does not require a reference image. For example, [13][14][15][16] are unsupervised metric for the evaluation for region based segmentation algorithms, whereas [17] [18] [19] are dedicated for contour detectors evaluation.

In our case, we are interested in the evaluation of the foreground/background segmentation performed on colour flower images. The segmentation is obtained

G. Bebis et al. (Eds.): ISVC 2012, Part II, LNCS 7432, pp. 448–457, 2012.

by, first, transforming the image to the Lab color space, followed by an application of the OTSU thresholding technique on each Lab component, independently. Given those segmentation results and in order to choose the best result, we need to use an evaluation measure.

Our contribution consists on a new unsupervised measure for the evaluation of foreground/background segmentation. This measure is based on Bayes classification error and the image contrast. But, we first establish a comparative study between the most common measures used in literature using quantitative and qualitative methods. This study aims to put out the limits of existent metrics.

The rest of the paper is organized as follows: in section 2, we present a comparative study of several unsupervised measures for segmentation results evaluation. Our proposed measure is described in section 3. Experiments and evaluation results are given in sections 4. Finally, some conclusions and perspectives of this work are given.

2   Comparative Study on Unsupervised Evaluation Metrics of the Segmentation Results

2.1   State of the Art

Because a ground truth is not always available for flower image datasets, we will focus our study on unsupervised measures. In the performed study, we selected the most common measure used in the literature which are those proposed by Weszka and Rosenfeld [13], Levine et Nazif [14], Sahoo et al [15], Cocquerez et Devars [16], Pal et Pal [20], Liu et Yan [21], Zeboudj [22], Rosenberger and Chehdi [23].

Weszka et Rosenfeld [13] designed the discrepancy measure (D_{WR}) given by (1) to evaluate foreground/background segmentation methods. It measures the difference between the gray-level of the original image and the segmented image after thresholding. In (1), $C_{gl}(i,j)$ is the gray-level value of pixel $p(i,j)$ on original image and $L(i,j)$ is the gray-level value of $p(i,j)$ on the image after thresholding. Finally, h and w are the height and the width of the image, respectively.

$$D_{WR} = \sum_{i=1}^{h} \sum_{j=1}^{w} (C_{gl}(i,j) - L(i,j)) \tag{1}$$

To characterize segmentation, Levine and Nazif in [14] had developed a performance vector (PV) including an uniformity measure (U) given by (2) where N is the number of regions, W_j is a weighting factor, Z is a normalization factor, $C_{gl}(p)$ is the gray level value for the pixel p, $\hat{C}_{gl}(R_j)$ is the average gray level value of the region R_j. (PV) includes also a texture measure (R) and a global region contrast measure (GC).

$$U = 1 - \sum_{x=1}^{N} \frac{\sum_{p \in R_j} (C_{gl}(p) - \hat{C}_{gl}(R_j))^2 . w_j}{Z} \tag{2}$$

The equation of the measure R is given by (3) where N_A is the number of regions in area $\mathcal{A}$, S_A and S_I are the size of the aria $\mathcal{A}$ and the image, respectively.

$$R = \frac{N_A . S_I}{N_I . S_A} \tag{3}$$

In (4), we present the formula for the measure (GC). It describes the inter-region disparity. It is the sum of the per-region contrast measures, weighted by a function approximating the human contrast sensitivity curve. The per-region contrast measure is the weighted sum of the differences between the average color of this region and its adjacent regions divided by the sum of their average colors. In (4), p_{ij} is the adjacency value used for weighting the contrast between regions, ν_j is the weight associeted to the region R_j and $\mathcal{N}(R_i)$ is the set of neighbouring regions of R_i.

$$GC = \frac{1}{\sum\limits_{R_i \in \mathcal{A}} \nu_i} . \left(\sum_{R_i \in \mathcal{A}} \nu_i \sum_{R_j \in \mathcal{N}(R_i)} p_{ij} \frac{|\hat{C}_{gl}(R_i) - \hat{C}_{gl}(R_j)|}{\hat{C}_{gl}(R_i) + \hat{C}_{gl}(R_j)} \right) \tag{4}$$

Sahoo et al [15] had proposed a shape measure (SM) defined as the sum of the gradients at each pixel whose feature value exceeds both the segmentation threshold and the average value of its neighbors (5).

$$SM = \frac{1}{C} \sum_{p \in I} \{Sgn[C_{gl}(p) - C_{N(p)}] \, g(p) . Sgn[C_{gl}(p) - T]\} \tag{5}$$

Where $g(p)$ is the gradient at the pixel p, T is the segmentation threashold, $Sgn(.)$ is the sign function, $C_N(p)$ is the average value of the neighbors of the pixel p in the image I, C is a normalization factor and $C_{gl}(p)$ is the gray level value for the pixel p.

Pal et Pal [20] introduced the joint entropy (H_J) and the local entropy (H_L). This measure express the intra-region uniformity and repose on the maximization of the entropy over the cooccurrence matrix.

Liu and Yang [21] had implemented a function (F) which measures the average squared color error of the regions. It penalizes over-segmentation because it is weighted by a quantity that is proportional to the square root of the number of segments. In (6) N presents the number of regions, S_j is the size of a region j.

$$F = \sqrt{N} \sum_{j=1}^{N} \frac{\sum\limits_{p \in R_j} (C_{gl}(p) - \hat{C}_{gl}(R_j))^2}{\sqrt{S_j}} \tag{6}$$

Zeboudj [22] developed a metric (C_z) suitable for too noisy or textured images. It is based on the maximization of the between-region contrast and the minimization of within-region contrast. (7) presents the equation for (C_z) where $C(R_i)$ is the contrast of the region R_i of the image I and it is defined using the region boundary, pixels intensity and pixels neighborhood. Furthermore, S_I and S_{R_i} are the size of the image I and the region R_i, respectively.

$$C_z = \frac{1}{S_I} \sum_{R_i \in I} S_{R_i} . C(R_i) \tag{7}$$

Rosenberger and Chehdi [23] presented a method for estimating the regions homogeneity of the segmented image. The measure (F_{RC}) has two implementations, one designed for non-textured images and the other for textured images. To characterize a region and to identify if it is uniform or textured, they used the co-occurrence matrix and the inter-region and the intra-region disparities.

Cocquerez and Devars [16] use the test of variance homogeneity (G) proposed by Cochran [24].

2.2 Qualitative Comparative Study

We are first going to eliminate some measures that do not verify assumptions made to make the chosen metric suitable to the case of flower photographs segmentation. In fact, we need a measure which is independent of the number, the size and the position of the flower region into the image because we can have more than one flower in the same image and the size of a flower can be insignificant compared to the entire image. Therefore, we analyzed the unsupervised techniques cited above so we can determine which criterion each of them takes into account. Table 1 presents a comparative study between those measures based on the carachteristics of each of them. After this study, we discarded the measures proposed by Zeboudj[22], Rosenberger and Chehdi[23], Liu and Yang[21], Cocquerez and Devars [16] and we only maintain the measure proposed by Sahoo et al [15]. Pal and Pal [20], Levine and Nazif [14], Weszka and Rosenfeld [13]. In fact, we can't use the measure proposed by Zeboudj [22] because it depends on regions size. Those proposed Rosenberger and Chedi in [23] and Lieu and Yang in [21] was discarded because they depend on the number of regions and their size. Cocqurez and Devars [16] evaluation metrics were excluded because it considers the neighborhood.

2.3 Quantitative Comparative Study

Evaluation Methodology. To compare the behaviour of several unsupervised measures, we used a statistical schema based on the voting technique. Indeed, for a given image, we vote for a measure if it selects the same best segmentation result chosen by the supervised measure. We vote also for the second best result if it is close to the first best one. In fact, we can obtain two good segmentation results with close evaluation measures. Then, given an images dataset, we compute a success rate (8), for each measure, as the quotient between the number of obtained votes and the size of the dataset. Finally, we select the unsupervised evaluation measure that behaves in the same way as the supervised one in term of selected segmentation. So, we opt for the measure having the highest success rate.

$$Success\ Rate = \frac{Number\ of\ votes}{Dataset\ size} \tag{8}$$

Obtained Results and Discussion. We compare the given measures using Oxford Flower database provided by Oxford University and used to achieve the

Table 1. Comparative study

Measure	Criteria
D_{WR} [13]	Pixel color, region color
U [14]	Regions variance
R [14]	Regions size
GC [14]	Image contrast
SM [15]	Gradient and neighbourhood of a pixel, segmentation threshold.
G [16]	Regions variance, Regions mean, neighbourhood , Regions size.
H_J and H_L [20]	Intra-region uniformity, entropy.
F [21]	The squared color error of region j, regions size, Regions number.
C_z [22]	Internal and external region contrast, Regions Boundary, Regions size.
F_{RC} [23]	Intra-class homogeneity, inter-class contrast , Regions size, Image type(Textred or uniform).

work of Maria-Elena Nilsback and Andrew Zisserman[25]. The experiments were performed using 846 images classified into 16 classes. We eliminated the class of the cowslips flower because there is not an available ground truth segmentation for this class.

The segmentation results to be evaluated are obtained by, first, transforming each image to the Lab color space, followed by an application of the OTSU thresholding technique on each Lab component, independently.

To be able compute the success rate introduced in section 2.3, we need to apply a supervised metric. We opted for the measure proposed by Shufelt in [26], taking into account the number of true positives (TP), false positives (FP) and false negatives (FN) because it verify assumptions announced in section 2.2 This measure presents the quality percentage given by (9).

$$Quality\ Percentage = \frac{100.TP}{TP + FP + FN} \qquad (9)$$

Fig. 1 shows the success rate for each of the tested measure. We can notice that the best rate is given by the global contrast (54%). This rate stills not enough to adapt this measure as an effective metric for segmentation results evalauation.

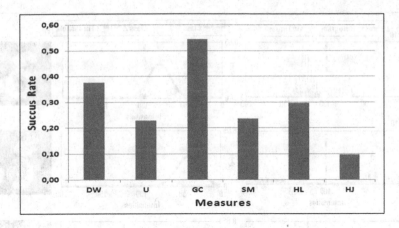

Fig. 1. Succes rate for each of the maintained meausres

3   Proposed Evaluation Metric for Flower Image Segmentation

To improve these results, we used defined a new metric we called (BC) for evaluating the results of unsupervised segmentation based on Bayes classification error (BCE) and global image contrast (GC).

Based on histogram analyses, we can notice that the (BCE) is larger if the separation between the foreground and background is harder, due to the unimodality of the histogram. In the other hand, the (BCE) is smaller if the separation between the two modes is more obvious. If the histogram shows two clearly separated modes, then the (BCE) will null. An example can be found in Fig. 2. Then, we weighted the (BCE) by the inverse of (GC) because it gives good result based on the comparative study in section 2. Equation (10) gives the proposed measure. In (11), μ_b and μ_f denote the average intensity values for background and foreground classes, respectively. Furthermore, to compute the Bayes Classification Error (BCE)(12) we need to determinate the prior probabilities $P(R_f)$ and $P(R_b)$ for the foreground and background, respectively. A prior probability can be approximate by the proportion of pixel in the class. The likelihood probabilities $P(X/R_f)$ and $P(X/R_b)$ can be approximate as Gaussian density of probability function (pdf) with parameters (μ_f, σ_f) for the foreground and (μ_b, σ_b) for the background. With μ refers the average intensity values within a class and σ refers the stander deviation.

$$BC = \frac{BCE}{GC} \tag{10}$$

with:

$$GC = \frac{\mu_o - \mu_b}{\mu_o + \mu_b} \tag{11}$$

and

$$BCE = \int_{R_b} P(X/R_f).P(R_f)\,dX + \int_{R_f} P(X/R_b).P(R_b)\,dX \tag{12}$$

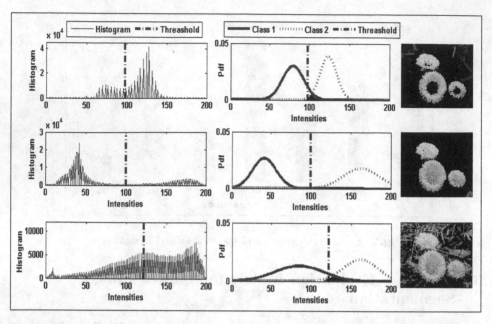

Fig. 2. The impact of the threasholding result on the Bayes Classification Error (BCE). The top row: threasholding of the (a) component. The middle row: threasholding of the (b) component. The bottom row: threasholding of the (L) component.

4   Experimental Results

We performed our experiments on the same dataset we used to establish the quantitative comparative study on unsupervised measure in section 2.3 which is oxford flower collection. Fig. 3 shows the effectiveness of our proposed measure.

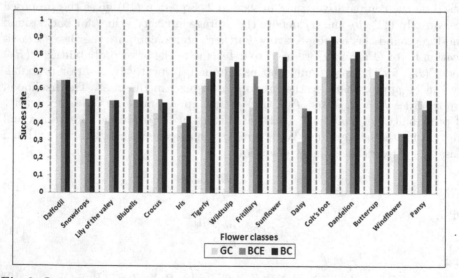

Fig. 3. Quantitative comparison between BC, BCE and GC unsupervised measures

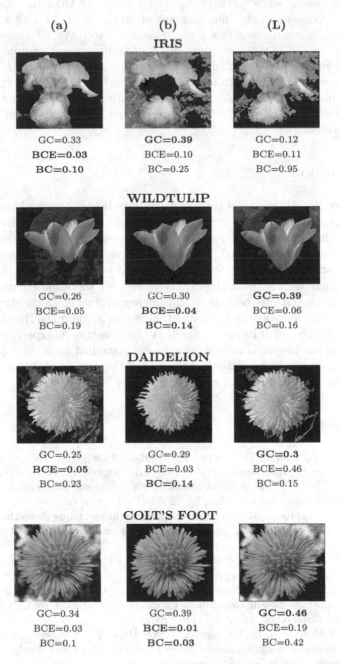

Fig. 4. Example of segmentation results evaluation using the measures GC, BCE and BC. GC is a quantity to maximize. BC and BCE are quantities to minimize. For each measure the elected segmentation as the best one is in bold.

Indeed, the success score given by the (BC) is higher than the one given by (GC) in most cases and the improvement attend 23% in the Colt's foot class. Furthermore, by computing the average success rate over all classes, we can notice that the (BC) measure gives a 7% of improvement comparing to the (GC) measurement and 1% of improvement comparing to the (BCE).

Fig. 4 shows the segmentation results given by (L), (a) and (b) components for several images. It illustrates also the evaluation of those results using the measures (GC), (BCE) and (BC). As we can see, (GC) fails when (BC) and (BCE) succeeds.

5   Conclusion

In this paper, we established a comparative study between several unsupervised evaluation measures for image segmentation evaluation. This study included two phases. The first phase aimed to determine the criteria that the measures take into account in order to discard those that don't verify some made assumptions. The second step was to statistically evaluate the behaviour of the chosen metrics to be able to decide about their effectiveness. So, we proposed an evaluation schema based on the vote technique and the segmentation ground truth. Then, we introduced an unsupervised measure taking into account the Bayes classification error and the global image contrast. We demonstrated, experimentally, the superiority of the proposed metric over the early studied ones. As a perspective of this work, we will investigate the influence of the quality of the segmentation algorithm on the quality of the evaluation.

References

1. Liedtke, C.E., Gahm, T., Kappei, F.: Segmentation of microscopic cell scenes. Analytical and Quantitative Cytologie and Histology 9, 197–211 (1987)
2. Cho, K., Meer, P.: Image segmentation from consensus information. Computer Vision and Image Understanding 68, 72–89 (1997)
3. Voisine, N.: Approche adaptative de coopération hiérarchique de méthodes de segmentation: application aux images multicomposantes. PhD thesis. Université de Rennes I (2002)
4. Zhang, Y.J.: A survey on evaluation methods for image segmentation. Pattern Recognition 29, 1335–1346 (1996)
5. Chabrier, S., Laurent, H., Emile, B., Rosenburger, C., Marche, P.: A comparative study of supervised evaluation criteria for image segmentation. In: European Signal Processing Conference, pp. 1143–1146 (2004)
6. Yang, L., Albregtsen, L., Lonnestad, T., Grottum, P.: A supervised approach to the evaluation of image segmentation methods. In: Computer Analysis of Images and Patterns, pp. 759–765. Springer (1995)
7. Coquin, D., Bolon, P., Chehadeh, Y.: Evaluation quantitative d'images filtrées. In: GRETSI 1997, vol. 2, pp. 1351–1354 (1997)
8. Wilson, D.L., Baddeley, A.J., Owens, R.A.: A new metric for grey-scale image comparison. International Journal of Computer Vision 24, 5–17 (1997)

9. Zamperoni, P., Starovoitov, V.: On measures of dissimilarity between arbitrary gray-scale images. International Journal of Shape Modeling 2, 189–213 (1996)
10. Odet, C., Belaroussi, B., Cattin, H.: Scalable discrepancy measures for segmentation evaluation. In: ICIP 2002, pp. 785–788 (2002)
11. Pratt, W., Faugeras, O.D., Gagalowicz, A.: Visual discrimination of stochastic texture fields. IEEE Transactions on Systems, Man, and Cybernetics 8, 796–804 (1978)
12. Roman-Roldan, R., Gomez-Lopera, J.F., Atae-allah, C., Martinez-Aroza, J., Escamilla, P.L.L.: A measure of quality for evaluating methods of segmentation and edge detection. Pattern Recognition 34, 969–980 (2001)
13. Weszka, J.S., Rosenfeld, A.: Threshold evaluation techniques. IEEE Transactions Systems, Man, and Cybernetics 8, 622–629 (1978)
14. Levine, M.D., Nazif, A.M.: Dynamic measurement of computer generated image segmentations. IEEE Trans. Pattern Anal. Mach. Intell. 7, 155–164 (1985)
15. Sahoo, P.K., Soltani, S., Wong, A.K., Chen, Y.C.: A survey of thresholding techniques. Comput. Vision Graph. Image Process. 41, 233–260 (1988)
16. Cocquerez, J.P., Devars, J.: Détection de contours dans les images aériennes: Nouveaux opérateurs. Traitement du Signal 2, 45–65 (1985)
17. Demigny, D., Kamlé, T.: A discrete expression of canny's criteria for step edge detector performances evaluation. EEE Transactions on Pattern Analysis and Machine Intelligence 19, 1199–1211 (1997)
18. Han, J.H., Kim, T.Y.: Ambiguity distance: an edge evaluation measure using fuzziness of edges. Fuzzy Sets and Systems, 311–324 (2002)
19. Tan, H., Gelfand, S., Delp, E.: A cost minimization approach to edge detection using simulated annealing. IEEE Transactions on Pattern Analysis and Machine Intelligence 14, 3–18 (1992)
20. Pal, N.R., Pal, S.K.: Entropic thresholding. Signal Processing 16, 97–108 (1989)
21. Liu, J., Yang, Y.H.: Multiresolution color image segmentation. IEEE Transactions on Pattern Analysis and Machine Intelligence 16, 689–700 (1994)
22. Zeboudj, R.: Filtrage, Seuillage Automatique, Contraste et Contours: du Pré-Traitement à l'Analyse d'image. PhD thesis. Université de Saint Etienne (1988)
23. Rosenberger, C., Chehdi, K.: Genetic fusion: application to multi-components image segmentation. In: IEEE International Conference on Acoustics, Speech, and Signal Processing, vol. 4, pp. 2223–2226 (2000)
24. Cochran, W.G., Snedecor, G.W.: Méthodes statistiques. Association de Coordination Technique, Agricole, Paris, France (1957)
25. Nilsback, M., Zisserman, A.: Delving deeper into the whorl of flower segmentation. IVC 28, 1049–1062 (2010)
26. Shufelt, J.A.: Performance evaluation and analysis of monocular building extraction from aerial imagery. IEEE Trans. Pattern Anal. Mach. Intell. 21, 311–326 (1999)

Robust Hand Tracking with Hough Forest and Multi-cue Flocks of Features

Hong Liu, Wenhuan Cui, and Runwei Ding

Key Laboratory of Machine Perception and Intelligence,
Shenzhen Graduate School, Peking University, P.R. China
hongliu@pku.edu.cn, cwh@sz.pku.edu.cn, dingrunwei@pkusz.edu.cn

Abstract. Robust hand tracking is highly demanded for many real-world applications relevant to human machine interface. However, current methods achieve no satisfactory robustness in real environments. In this paper a novel hand tracking method was proposed integrating online Hough Forest and Flocks-of-Features tracking. Skin color was integrated in the Hough Forest framework to gain more robustness against drastic hand appearance and pose changes, especially against partial occlusions. Also a novel multi-cue Flocks-of-Features tracking algorithm based on computer graphics was integrated in to enhance the framework's robustness against distractors and background clutter. Additionally, recovery from tracking failure was addressed. Lots of experiments were carried out to evaluate our method, also to compare it with CAMShift, Hough Forest tracker, and the original Flocks-of-Features Tracker, and showed the effectiveness of our method.

1 Introduction

Markerless vision-based hand tracking has been under research for decades, because of its great potential for natural human computer interface, such as robotics, intelligent surveillance, and so on. However, due to the complicated shape and appearance change of human hand in motion, and the change of environment, there is not yet one single method that can give satisfactory robustness.

Hand tracking methods are usually classified as model-based and appearance-based [1]. Hand models are most often used for 3D hand tracking, as in [2] and [3]. On the other hand, appearance-based methods extract image features as appearance. Skin color feature is often chosen for its simplicity and discriminability. It was used in CAMShift tracker for fast face tracking [4], and for simple hand tracking. Contour feature was used to represent the shape of hand in a particle-filter framework to track hand deformation against clutter [5]. Maximally Stable Extremal Regions (MSER) is a new *stable* region feature and was used for hand tracking [6]. Combination of these features can increase a hand tracker's robustness. However, difficult real-world situations, such as occlusion, background clutter, drastic environment change cannot be easily overcome with these methods. Especially a robust hand tracker should be able to recover tracking failure through re-detection. Off-line-trained tree classifier was used in [7] and combined

G. Bebis et al. (Eds.): ISVC 2012, Part II, LNCS 7432, pp. 458–467, 2012.
© Springer-Verlag Berlin Heidelberg 2012

with interpolation for 3D hand tracking. In [8], clustering based on shape context and boosted tree classifier were used for hand detection. However, to build a classifier off-line is a great challenge due to hand's huge space of possible shape and appearance variation. On the other hand, classifier built online can learn more of the appearance and shape space. Online-learning methods are gaining increasing interest in recent years, and have demonstrated great power in rigid object tracking-by-detection [9]. Especially, part-based online-learning methods were used for learning representations for articulated objects [10][11]. In [10] an Implicit Shape Model was proposed to integrate votings from patches of object parts in a *Generalized Hough Transform* manner, and the patches were clustered by their similarity of appearance. In [11] random forest was used to cluster these patches and votings. And in [12] the Hough forest tracking framework is extended in an on-line fashion, with additional steps of backprojection and segmentation, bringing greater adaptivity to drastic pose and shape changes.

This work was built upon the Hough forest tracking-by-detection framework described in [12]. The *stability-plasticity dilemma* [13] of the approach in [12] was tackled by integrating the Hough forest classifier with a modified multi-cue Flocks-of-Features tracker [14]. The modification is based on a computer graphics technique to simulate flock behavior [15]. The resulting framework showed good robustness against difficult situations such as clutter, distraction, occlusion and tracking failure. A system overview of our approach is shown in Fig. 1.

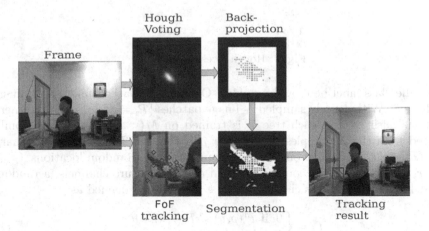

Fig. 1. Overview of our approach

The rest of the paper is organized as follows. Section 2 gives a brief introduction to the Hough forest tracking method, and shows its limitations. Section 3 describes in detail the Flocks-of-Features tracker and its combination with the Hough forest tracking. Experiments and discussions are given in section 4. Finally conclusions and future directions are given in section 5.

2   Hough Forest for Hand Tracking

2.1   Hough Forest: Hough Voting and Random Forest

The Implicit Shape Model (ISM) is a part-based model of appearance and configuration of constituting parts of an object [10]. The Hough forest learns their appearances and configurations by random forest [11]. An illustration of the working scheme of Hough forest detection is given in Fig. 2. Patches around densely sampled pixels and their offset from the hand center are fed to the random forest for training and testing. Details are given in the following.

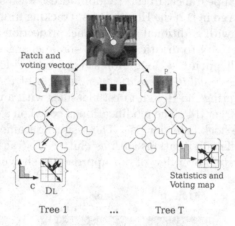

Fig. 2. Hough forest detection

Let the class label be c, and $c \in \{1 = \text{Object}, 0 = \text{Background}\}$, and image patch be $\mathcal{P}$. With densely sampled N image patches $\{\mathcal{P}_i = (\mathcal{I}_i, c_i, \mathbf{d}_i)\}_{i=1}^{N}$, where $\mathbf{d}_i$ is the offset vector, each tree $\mathcal{T}$ is trained on $M(< N)$ samples randomly selected from the N samples. Each node j in the tree $\mathcal{T}$ is assigned a binary test $t(\mathcal{I}) \rightarrow \{1, 2\}$, defined on a patch's appearance $\mathcal{I}_i$. Random locations, (p, q) and (r, s), are chosen for comparison, and of all C feature channels, a random channel $a \in \{1, 2, ..., C\}$ is chosen. The test at node j is defined as:

$$t_j = \begin{cases} 0, & \text{if } I^a(p, q) < I^a(r, s) + \theta \\ 1, & \text{otherwise} \end{cases} \tag{1}$$

where θ is a threshold value chosen based on some optimization criteria. In Hough forest, the patch class statistics and voting vectors are both stored in leaves, forming a voting map $D_L = \{\mathbf{d}_i\}$. In such a way an implicit discriminative *codebook* is constructed at each leaf. All votes from each tree are collected into the Hough space, and the maximum location is determined as the object's center. Notice that this can be replaced by clustering to find multiple centers for multiple objects detection.

2.2 Hough Forest for Tracking: Increasing Stability and Adaptivity

In [12], the Hough forest was formulated in an online learning manner for tracking non-rigid objects. The Grabcut Segmentation algorithm [16] was used to generate ample training samples online. The seeds were generated by backprojection/support of the Hough voting. The Grabcut algorithm builds Gaussian mixture models for foreground and background, and then utilizes Graph-cut algorithm to segment foreground from background. This indicates that the segmentation is based on color similarity and pixel adjacency. The support $S(\mathbf{m}, \rho)$ of a specific point $\mathbf{m}$ in Hough space is the pixel locations that give votings to points within a range of radius ρ to $\mathbf{m}$. Furthermore, a forgetting scheme was used in [12] to reduce the weights of formerly learned samples and increase the online learner's adaptivity, which was applied on both class statistics and voting map in the leaves. Skin color represented by its distribution in Hue channel was used in our work to refine the segmentation, which further increase the accuracy of classification.

3 Hough Forest Tracking with Flocks of Features

3.1 Tracking Flocks of Features

The Flocks-of-Features (FoF) tracker first proposed in [14] exploited a biological phenomenon, the *Flock Behavior*, which states that the members p_i in the flock $\mathcal{F} = \{\{p_i\}_{i=1}^{N_f}\}$ should be neighter too condensed or too sparse:

$$d_{\min} < |p_i - p_j|, \forall i, j \in \{1, 2, ..., N_f\}, \text{and}$$
$$d_{\max} > |p_i - m|, \forall i \in \{1, 2, ..., N_f\}. \tag{2}$$
$$m = \text{median}(\mathcal{F}) \text{ or centroid}(\mathcal{F}).$$

where $|p_i - p_j|$ is the distance between p_i and p_j. The $d_{\min}$ and $d_{\max}$ are maximum and minimum distances which are parameters the flock holds. Unlike [14], in our work techniques in computer graphics for simulating flocking behaviors are considered. We chose the well-known *Boids* algorithm [15], which is widely used to simulate flock behavior in computer graphics. Call the members of the flock as *boids*, the flocking behavior is maintained by the following rules:

Separation boids try to keep a distance away from other boids.
Cohesion   boids try to fly towards the center of mass of neighbours.
Allignment boids try to match velocity with near boids.

Based on the *Boids* algorithm, we proposed a new realization of FoF tracker, which showed better *flock behavior* than that of [14]. The algorithm is summarised in Alg. 1. The key steps of the FoF tracker are KLT optical flow tracking [17] of feature points and a modified Boids algorithm. Major modification is the use of a confidence map to weight the feature points. The intuition beneath is that, for the flock to track a target, members near to the target should have

---

Algorithm 1. The modified Flocks-of-Features tracker

1: **INPUT:** Current search window B, last feature points $\mathcal{F}_{pre}$, image frame I, confidence map W
2: **OUTPUT:** New feature points $\mathcal{F}$
3: **BEGIN:**
4:    $\mathcal{F}_t \leftarrow \{\mathcal{F}_{\text{tracked}}, \mathcal{F}_{\text{lost}}\} \leftarrow \text{KLT}(\mathcal{F}_{pre}, I)$
5:    $\mathcal{F}_{\text{lost}} \leftarrow \text{getNewPoints}(W, B)$
6:    $S_p \leftarrow \sum_{i=1}^{N_f} W(\mathcal{F}_{pre}(i)) * \mathcal{F}_{pre}(i)$
7:    **for** $i = 1 \rightarrow N_f$ **do**
8:        "Perceived center": $s_i \leftarrow \frac{S_p - \mathcal{F}_{pre}(i)}{N_f - 1}$
9:        "Positive Driving Force": $f_p \leftarrow s_i - \mathcal{F}_{pre}(i)$
10:       "Negagive Driving Force": $f_n \leftarrow 0$
11:       **for** $j = 1 \rightarrow N_f$ **do**
12:           **if** $j \neq i$ AND $\|\mathcal{F}_{pre}(i) - \mathcal{F}_{pre}(j)\|_2 \leq d_{\min}$ **then**
13:               $f_n \leftarrow f_n + d$
14:           **end if**
15:       **end for**
16:       $\mathcal{F}_t(i) \leftarrow \mathcal{F}_{pre}(i) + \alpha * f_p + \beta * f_n$
17:   **end for**
18: **END**

---

more importance. In our current implementation, only skin color was used to build the confidence map, which is the backprojection of hue histogram of the hand, similar to the CAMShift tracking [4]. Better confidence map is possible by using other features.

3.2 Integration of FoF Tracker with Hough Forest Framework

Sole tracking-by-detection has the problem of stability-plasticity dilemma, which can be alleviated by integrating different trackers, as shown in [18] and [19]. In the FROST framework [18], trackers occupying different stability-adaptivity scope were combined to compensate each other. Alternatively, in [19] a self-verifying tracker was devised to supervise the on-line *Positive-Negative learning*. These integration methods have shown impressive results on rigid object tracking, however, poor results for tracking human hand were observed in our evaluations. It was observed that the low performance of the online classifier and the base tracker on articulated objects contributed to the poor result.

In this work the Hough forest classifier was chosen to be integrated with FoF tracker, both of which have shown good tracking results on articulated objects. The interactions between relevant modules are shown in Fig. 3 and explained as follows. First, define the *tracking validity* on which the supervision of learning is based. For FoF tracking, its tracking validity can be staightforwardly defined as $V_f = N_t / N_a$, where N_t is the number of feature points that are confidently tracked, and N_a is the number of all feature points. The final tracking result was given by the Hough forest detector, while the update of the Hough forest was extended by the FoF tracker. When the area of the tracked object exceeds

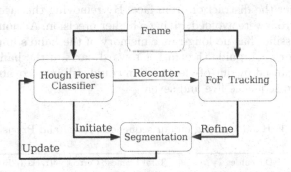

Fig. 3. Integration of Hough Forest classifier and FoF tracker

some threshold, and the validity of FoF tracking V_f is verified, the Grabcut segmentation result was pruned by intersecting with the result of FoF tracking to eliminate highly-possible distracting patches. With one FoF tracker for each cluster, the framework may be extended to track multiple objects.

4   Experiments and Discussions

The proposed approach was tested on seven real-world sequences with about 6225 frames taken by an ordinary RGB camera, whose resolution is 640×480 in pixel. No assumptions of still camera or still body were made. In all sequences illumination change was unconstrained, and face was a constant distractor. An overall performance of the proposed approach is shown in Table. 1. Each of the sequences has specific difficulties designed to test different aspects of tracking robustness, depicted in second column of Table. 1. The performance was measured with *recall* and *precision*, whose definitions were:

$$recall = \frac{\text{Num. of True Positives}}{\text{Num. of True Positives} + \text{Num. of False Negatives}}$$
$$precision = \frac{\text{Num. of True Positives}}{\text{Num. of True Positives} + \text{Num. of False Positives}} \quad (3)$$

where True Positives are tracked targets being real hand, False Positives are wrongly tracked target, and False Negatives are tracking failure report when the hand is in the image.

Most of the recalls in the tests are 100 percent because the hand is always in the scene and tracking failure never happened, so there is no false negatives. In sequence one and four with fast posture change, most false tracking were caused by face' distraction, because the tracker cannot distinguish hand and face when a novel posture suddenly appears over face. On the contrary, sequence seven shows that slow posture change causes no such problems. In sequence five, as re-detection rate was low, a low recall as 79 percent was obtained with many false negatives. This revealed the lack of stability of the Hough forest part in the framework. In sequences five and six, face detection technique was used to counteract the distraction from face. By removing the face' distraction, much tracking errors were avoided, giving a higher precision. Although the online Hough forest classifier has no long-range memory of the hand's appearance, the short-term memory can guide the tracker toward parts of the hand. Therefore, tracking failure caused by out-of-view motion could still be recovered after some time, as shown in sequence five and seven.

Table 1. Hand tracking on our sequences: Recal and Precision

Sequence	Difficulties	Total Frames	Correctly Tracked	Recall	Precision
Seq.1	face distraction	1006	805	100%	80%
Seq.2	background clutter	310	295	100%	95%
Seq.3	large motion	706	693	**100%**	**98%**
Seq.4	occlusion, distraction	527	384	99%	73%
Seq.5	out-of-view motion	413	249	**79%**	87%
Seq.6	3D changing postures	763	697	100%	91%
Seq.7	all above with slow posture change	2500	2420	99 %	96%

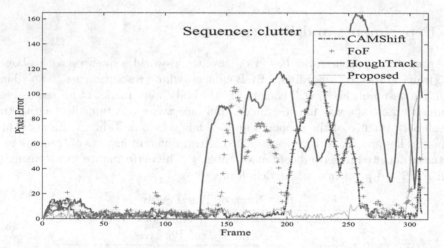

Fig. 4. Tracking error of CAMShift, FoF, HoughTrack and proposed, on sequence 2

The proposed approach was also compared with CAMShift, FoF tracker, HoughTrack on sequence two, because it has much clutter along with face' distraction, a good testbench for all methods. The pixel error between the center of each tracked bounding box and the ground truth as a function of frame number is plotted in Fig. 4. It can be seen that the proposed method gave the longest stable tracking with 300 frames, untill the person was leaving to end the sequence. On the contrary, the HoughTrack behaves as the least stable tracker in the test. After less than 150 frames, it got locked on the background bookshelf and never recovered. This is due to the fact that textured objects are easier to track since they have more stable feature points for Hough forest to learn.

Fig. 5. Robustness against occlusion, frames: 60, 61, 62, 63, 69, 88

Fig. 6. Robustness against face' distraction, frames: 662, 667, 678, 684, 688, 694

Fig. 7. Recovery from tracking failure caused by out-of-view motion, frames: 128, 137, 144, 183, 192, 201

And once the stable appearance of the background was learned, it could hardly be forgotten. The FoF tracker also showed much fluctuation, which were caused by the distraction of face and the bookshelf. It suffered the same problem as that of Hough forest tracker. On the other hand, the CAMShift tracker is based on skin color, thus is robust against environment clutterness, and it kept stable tracking for nearly 200 frames. However, illumination change and other skin color objects can easily attract CAMShift tracker away, as happened around frame 200. But CAMShift resumed tracking when the hand moved over face a later time since it distinguishs no face and hand.

For hand tracking, the robustness against human face' distraction, occlusion, and tracking failure is most challenging and important. Our method displayed good performance against these difficulties mainly because it's a part-based approach with tracking failure recovery. Fig. 5 shows an example of occlusion. Interestingly the track drifts away with another hand after full occlusion, but get recovered when the false track is lost. In Fig. 6, the ISM with FoF tracking alleviates the face' distraction. Fig. 7 shows the recovery of a tracking failure caused by out-of-view motion.

5   Conclusions

This paper proposed a robust hand tracking method based on a framework combining Hough forest detection and Flocks-of-Features tracker. The modified Flocks-of-Features tracker was proposed based on a computer graphics technique. The proposed part-based framework has demonstrated great robustness against partial occlusion, distraction, background clutter, and recovery from tracking failure, yet does not rely on still camera or static body. Experiments also revealed the lack of stability of the Hough forest detection part. Future work would explore shape information for better segmentation, and extend the framework with clustering and multiple-objects tracker for tracking multiple objects.

Acknowledgements. This work is supported by National Natural Science Foundation of China(NSFC, No.60875050, 60675025), National High Technology Research and Development Program of China(863 Program, No.2006AA04Z247), Scientific and Technical Innovation Commission of Shenzhen Municipality (No.JC201005280682A, CXC201104210010A).

References

1. Erol, A., Bebis, G., Nicolescu, M., Boyle, R.D., Twombly, X.: Vision-based hand pose estimation: A review. Computer Vision and Image Understanding 108, 52–73 (2007)
2. Rehg, J., Kanade, T.: Digiteyes: Vision-based hand tracking for human-computer interaction. In: Proceedings of the Workshop on Motion of Non-Rigid and Articulated Bodies, pp. 16–22 (1994)

3. Stenger, B., Thayananthan, A., Torr, P.H.S., Cipolla, R.: Model-based hand tracking using a hierarchical bayesian filter. IEEE Transactions on Pattern Analysis and Machine Intelligence 28, 1372–1384 (2006)

4. Bradski, G.R.: Real-time face and object tracking as a component of a perceptual user interface. In: IEEE Workshop on Applications of Computer Vision, pp. 214–219 (1998)

5. Isard, M., Blake, A.: CONDENSATION - conditional density propagation for visual tracking. International Journal of Computer Vision 29, 5–28 (1998)

6. Donoser, M., Bischof, H.: Real time appearance based hand tracking. In: Proceedings of the International Conference on Pattern Recognition, pp. 1–4 (2008)

7. Tomasi, C., Petrov, S., Sastry, A.: 3D tracking = classification + interpolation. In: Proceedings of the International Conference on Computer Vision, vol. 2, pp. 1441–1448 (2003)

8. Ong, E.J., Bowden, R.: A boosted classifier tree for hand shape detection. In: Proceedings of International Conference on Automatic Face and Gesture Recognition, pp. 889–894 (2004)

9. Avidan, S.: Ensemble tracking. IEEE Transactions on Pattern Analysis and Machine Intelligence 29, 261–271 (2007)

10. Leibe, B., Leonardis, A., Schiele, B.: Combined object categorization and segmentation with an implicit shape model. In: Proceedings of ECCV Workshop on Statistical Learning in Computer Vision, pp. 17–32 (May 2004)

11. Gall, J., Lempitsky, V.: Class-specific hough forests for object detection. In: IEEE Conference on Computer Vision and Pattern Recognition, pp. 1022–1029 (2009)

12. Godec, M., Roth, P.M., Bischof, H.: Hough-based tracking of non-rigid objects. In: International Conference on Computer Vision, pp. 81–88 (2011)

13. Grossberg, S.: Competitive learning: From interactive activation to adaptive resonance. Cognitive Science 11, 23–63 (1987)

14. Kolsch, M., Turk, M.: Fast 2D hand tracking with flocks of features and Multi-Cue integration. In: IEEE Conference on Computer Vision and Pattern Recognition Workshop, pp. 1–8 (2005)

15. Reynolds, C.W.: Flocks, herds and schools: A distributed behavioral model. SIGGRAPH Comput. Graph. 21, 25–34 (1987)

16. Rother, C., Kolmogorov, V., Blake, A.: "grabcut": interactive foreground extraction using iterated graph cuts. ACM Trans. Graph. 23, 309–314 (2006)

17. Lucas, B.D., Kanade, T.: An iterative image registration technique with an application to stereo vision. In: Proceedings of the 1981 DARPA Image Understanding Workshop, pp. 121–130 (1981)

18. Santner, J., Leistner, C., Saffari, A., Pock, T., Bischof, H.: Prost parallel robust online simple tracking. In: IEEE Conference on Computer Vision and Pattern Recognition, pp. 720–730 (2010)

19. Kalal, Z., Matas, J., Mikolajczyk, K.: P-n learning: Bootstrapping binary classifiers by structural constraints. In: IEEE Conference on Computer Vision and Pattern Recognition, pp. 49–56 (2010)

The Impact of Unfocused Vickers Indentation Images on the Segmentation Performance

Michael Gadermayr, Andreas Maier, and Andreas Uhl

Department of Computer Sciences, University of Salzburg, Austria
{mgadermayr,andi,uhl}@cosy.sbg.ac.at

Abstract. Whereas common Vickers indentation segmentation algorithms are precise with high quality images, low quality images often cannot be segmented appropriately. We investigate an approach, where unfocused images are segmented. On the one hand, the segmentation accuracy of low quality images can be improved. On the other hand we aim in reducing the overall runtime of the hardness testing method. We introduce one approach based on single unfocused images and one gradual enhancement approach based on image series.

1 Introduction

In Vickers hardness testing, a pyramidal indenter causes a square indentation in a specimen. A major issue is to measure the diagonal lengths of the indentation. Therefore the square object must be segmented from the background to identify the vertices. Especially images of rough surfaces are likely to be highly noisy or have low contrast. The indentation images which should be segmented, approximately fit the following description: The object has a square geometry and is darker than the background. The diagonals are approximately aligned horizontally and vertically. Figure 1 shows example images and the manually determined vertice positions. Whereas the first image is quite perfect, the others suffer from noise and low contrast, respectively. There are several proposals for automated

Fig. 1. Vickers indentation images - evaluated vertice positions

image segmentation of Vickers indentations. The methods proposed in [1,2] rely on template matching. Others are based on edge detection and Hough transform [3], wavelet analysis [4,5], thresholding [6,7,8] and axis projection [9].

In order to acquire focused images, the Vickers hardness testing facilities rely on autofocus systems. The autofocus system takes pictures, computes the focus metric and moves the camera for one step until the peak of the focus metric

G. Bebis et al. (Eds.): ISVC 2012, Part II, LNCS 7432, pp. 468–478, 2012.
© Springer-Verlag Berlin Heidelberg 2012

(i.e. the focused image) is reached. We investigate if it is possible to compute approximative segmentation results from unfocused images. This could be advantageous, because an unfocused image is earlier available than the focused image as the autofocus takes a significant amount of time. Moreover, a failure of the autofocus might determine the wrong image to be in focus. Furthermore, we introduce a gradual enhancement approach, which is able to utilize free cpu cycles (caused when moving the camera) to incrementally improve the segmentation results.

This paper is structured in to following way: In Sect. 2, optical effects which occur with focused and unfocused images are explained. In Sect. 3, two different strategies are introduced which are based on unfocused images. In Sect. 4, the results are explained and compared with traditional approaches. Section 5 concludes this paper.

2   Focusing in Vickers Hardness Testing

A modern Vickers hardness testing equipment like the emcoTEST DuraScan hardness tester, used in the experiments, includes an inspection unit which is more or less a camera mounted on a microscope. Hardness indentations are analysed and measured with the inspection unit. The size of the indentations is in the millimetre or sub-millimetre range, so the magnification of the microscope is usually between 10x and 100x, and due to the non-transparency of the specimen, the illumination of the specimen takes place through the optics of the microscope.

In an indentation image a high contrast between the indentation and the surface of the specimen is desired. A high contrast facilitates the perception of the indentation when it is measured manually but also simplifies the segmentation when the image is processed automatically. Usually the indentation appears darker than the surrounding because of the groove that is caused by the pyramidal Vickers indenter.

In certain conditions (due to optical effects in high magnification optics) parts of the indentation do appear brighter than the surrounding or the contrast between the indentation and the surrounding is very small or vanishes completely. Such scenarios are challenging for automatic hardness measurement because algorithms often fail to detect the indentation and thus even do not provide approximate numbers for its position and size.

Figure 2a shows a schema of how an indentation image is taken. In a regular configuration the focus of the optical unit is aligned such that the edges and vertices of the indentation are best focused. This corresponds to a focus level that is roughly at the level of the specimen surface. Because the illumination passes through the optics it has the same focus plane. It can be seen from the figure that especially for high magnification optics (with 60x or 100x magnification) the illumination is considerably spread again when it reaches the bottom of the indentation. Due to the spread it includes a substantial amount of light rays that hit the walls of the indentation pyramid in such an angle that they are reflected back into the lens system. These rays act as an illumination for the

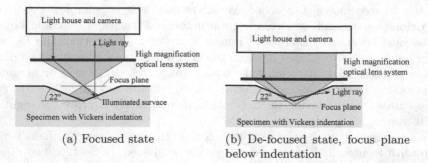

(a) Focused state

(b) De-focused state, focus plane below indentation

Fig. 2. Schema of the optical unit of Vickers hardness testing equipment with different alignments of the focus plane

indentation and are responsible for the reduced or missing contrast with respect to the surrounding.

If the focus of the optical system is shifted down and below the bottom of the indentation, the images are blurred considerably (see Fig. 4) but gain at the same time a substantially increased contrast for the whole or major parts of the indentation. Figure 2b shows such a scenario where the focus plane of the optics and thus the illumination is moved below the deepest point of the indentation. The angle at which the light hits the walls of the indentation is now different and most of the reflected light rays miss the lens system and thus do not illuminate the indentation. The indentation appears darker in many respects and has exceptionally dark areas along the diagonal. The effect increases as the focus plane is lowered but at a certain point the blur becomes so high that the indentation starts to disintegrate in the image and is no longer identifiable.

On such de-focused images the exact measurement of the indentation is no longer possible due to the considerable amount of blur in the image but the increased contrast between indentation and surrounding make de-focused images of this kind a promising candidate for an approximative indentation localisation and size estimation. The result of this first step is then a good starting point for the exact indentation measurement in the focused images.

3   Approaches Based on Unfocused Images

We especially investigate a former 2-stage active contours approach [10] with reference to different kinds of unfocused images: In the first stage, the parameters (position, size and rotation) of a square template are iteratively computed by a gradient descent method. The gradient descent minimizes an energy criterion based on probabilities. The method is not able to exactly segment the indentation, as the indentations slightly vary from a perfect square but a robust localization can be achieved. In the second stage, a region based level set method is initialized with the results of the first stage, to refine the results. To make a more general statement, moreover the 3-stage segmentation method [11] based on approximative template matching [1] is investigated with reference to unfocused indentation images. In Sect. 3.1 a segmentation approach based on single unfocused images and in Sect. 3.2 a gradual enhancement approach based on the approach 2-stage active contours approach is introduced.

3.1 Segmentation of Single Unfocused Images

Our first step is, to find out if it is possible to compute approximative segmentation results from unfocused images. This might be beneficial, as the autofocus algorithm consumes a lot of time, which could be used by an approximative segmentation algorithm based on unfocused images.

We investigate the effect of wrongly focused images on the segmentation algorithms. We have sorted the images according to their focus level (fl). If the focus level is below zero, regions are in focus which are farther away (Fig. 3, dotted line) from the camera than the background (i.e. the indentation might be in focus). If the focus level is above zero, regions are in focus which are nearer to the camera (Fig. 3, dashed line). In our case, no regions are

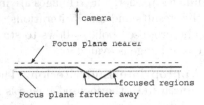

Fig. 3. Cut of Vickers indentation

nearer than the background (i.e. nothing is in focus). The step size between two consecutive focus levels is declared in Sect. 4. Differently focused images of the same indentation are shown in Fig. 4.

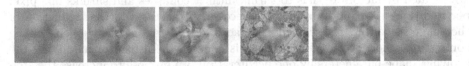

Fig. 4. Different focus settings, reaching from $fl \ll 0$ (left) to $fl \gg 0$ (right)

3.2 Gradual Enhancement Approach

In Sect. 3.1, we considered to segment images of different focus levels. But to identify the focus level, first we have to know the best focus configuration, as the focus levels are defined relatively (focused image: $fl := 0$). Consequently, an approximative segmentation of e.g. the image with $fl = 5$ cannot start before the focused image is known. As we aim in utilizing the free cpu cycles (caused by the autofocus system) for an approximate localization of the indentation, now we consider the following 3 steps based on the gradient descent approach [10]:

1. The focus starting setting is chosen that the focus plane is farther away than any part of the specimen ($fl \ll 0$).
2. Start the proposed first stage gradient descent segmentation algorithm on the unfocused image which is taken with the mentioned focus setting. Approximative results are achieved.

3. Until the end-criterion is reached:
   - Increase the focus level by one step and get the image.
   - Initialize the gradient descent algorithm with the current approximative results and the new image.
   - Run the algorithm with only 5 iterations to enhance the approximative results.
   - New approximative results are achieved.

The first image to segment is highly unfocused. Consequently, an exact segmentation surely cannot be achieved. However, the blurred image can be segmented robustly. Whereas the first image is segmented as proposed in [10], the enhanced images are not. These images are initialized according to the current approximative results and only 5 iterations of the gradient descent approach are applied. The proposed policy allows to start the segmentation even before the final focused image is available.

Appropriate Endcriterion. The intention is, that the results could be enhanced until the focused image is reached. Actually, this is not true. The best results are achieved, when stopping with the image of a focus level below zero. In practice, this is not possible, as the focus levels are defined relatively to the focused image. However, when saving the result history, these results can be recovered.

Speeding up the Initial Segmentation. Whereas the enhancement steps (3) are fast, the initial step (2) takes quite a long time. As the initial contour starts at the boundary of the image (initial radius is about 50 pixels as the images are downscaled by factor 10), has to shrink until it collapses and shrinks one pixel per iteration, about 50 iterations are necessary. Wheras a further reduction of the image size affects the segmentation accuracy, increasing the step size of the contour does not, as far as robustness is concerned. Instead of modifying the evolving shape parameters by one per iteration, we propose to increase the step size (i.e. in one iteration, each parameter is adjusted by the positive or negative step size or stays the same). Increasing the step size to 4, we achieved less accurate results after the initial segmentation step, but after the enhancement steps, the results were exactly the same (the results are shown in Sect. 4.2).

4   Experiments

A database is used with 25 indentations and 40 images (with different focus settings) per indentation. The quality of the images is quite low.

The exact step size between two consecutive focus level images cannot be generally specified, as it depends on the optical zoom of the camera, as shown in Table 1. For example, if an image is 10x magnified, a step size of 10,000 nm is chosen (i.e. while the camera moves, every 10,000 nm a picture is taken). The higher the zoom factor, the smaller the step size must be (because of the different depth of field). For example $fl = 5$ means that the focused plane is five steps nearer to the camera than with the best focus level ($fl = 0$).

Table 1. Focus step size dependent on the zoom factor

zoom factor	step size
10 x	10,000 nm
20 x	5,000 nm
40 x	1,000 nm

Our aim is to detect the four vertices of the approximately square Vickers indentations. In the following analysises, the distances between detected vertices and the ground truth are measured. The ground thruth is determined by taking the mean of the manual measures of four independent experts. In these figures, for each deviation bin (Euclidean distance in pixels) on the x-axis, the number of vertices detected within the deviation is shown on the y-axis.

4.1 Single Unfocused Images

Approximative Stage. First of all, we investigate the effect of unfocused images on the approximative indentation segmentation approach introduced in [10]. Figure 5a shows results of the approximative method. The focus plane is farther away from the camera compared with the best-focus strategy (red line).

The robustness (i.e. few outliers) of the segmentation did not only stay unchanged as expected, but can actually be increased if the unfocused images are used. However, the segmentation accuracy (e.g. the ratio of vertices with a deviation of maximal 20 pixels) slightly decreases. As the curves are crossing, we cannot identify a best configuration just by looking at the results.

In Fig. 5b, a similar effect on the indentation localization stage [1] is shown. Especially focus level −10 seems to be a good choice.

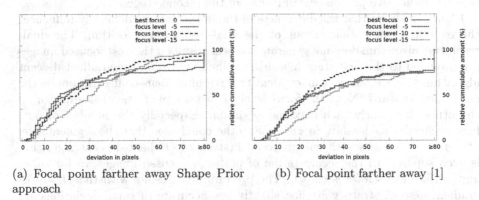

(a) Focal point farther away Shape Prior approach

(b) Focal point farther away [1]

Fig. 5. Single unfocused images: Advantageous settings (2 approaches)

The results in Fig. 6a are achieved with the approximative approach and images where the distance to the focus plane is lower than the distance to any part of the specimen. The segmentation performance with these images definitely decreases. Consequently, we specialize on focus levels shown in Fig. 5a.

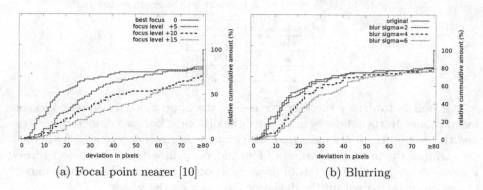

(a) Focal point nearer [10]   (b) Blurring

Fig. 6. Single unfocused images: Disadvantageous settings

As unfocused images look similar to blurred images, we also investigate the impact of differently blurred images on the segmentation performance. Especially we would like to know if a similar enhancement of robustness can be achieved as with appropriate unfocused images. Figure 6b shows the results with different Gaussian filters ($\sigma = 2, 4, 6$). Actually, unlike with unfocused images, the number of outliers cannot be decreased significantly. The probability of a precise segmentation suffers as with the unfocused images.

Precise Stage. Next, we initialized a precise level set segmentation method [10] with the results, gathered from the approximative method with different focus levels to get a knowledge of the impact on the overall performance of the multi-resolution algorithm. However, the level set algorithm still operates on the focused images. The question is, how accurate the first stage results have to be in order to achieve precise overall results of the second stage.

Figure 7a shows that the differences of the initializations definitely influences the overall segmentation output of the dual-resolution algorithm. The dual-resolution algorithm does not generate the best results if the best focused images are provided to the first stage algorithm. When regarding the gradient descent algorithm (Figure 5a) we cannot identify a winning focus-configuration, as the curves are crossing. Now, the focus level $fl = -10$ (first stage) is superior to the others for nearly each maximal deviation. Especially the number of outliers declines considerably. So we come to the conclusion that the segmentation of unfocused images with our proposed first stage gradient descent algorithm is even superior to the segmentation of perfectly focused images, as far as an appropriate focus level is chosen. The precomputed results with the $fl = -10$ gradient descent strategy are just slightly less accurate (if small deviations are regarded) than the best focus strategy, but the number of outliers is minor, which is beneficial. Although the $fl = -15$ strategy has even less outliers, the overall performance decreases, as the accuracy suffers too much.

A similar behavior can be observed with the 3-stage Vickers segmentation algorithm proposed in [11]. The method is based on the approximative template matching [1] and adds 2 enhancement stages. Only in the first stage the

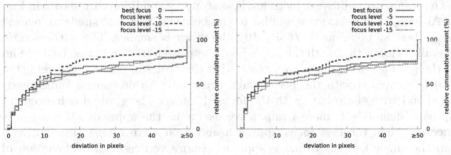

(a) Level set: different initializations   (b) Alternative method: different init. [11]

Fig. 7. Single unfocused images: Effect on precise stages (2 approaches)

unfocused images are segmented. In Fig. 7b, the impact on this enhancement method is shown (based on the different approximative results in Fig. 5b).

As with the level set method, the best overall results can be achieved when the approximative segmentation method is based on the images with the focus level −10.

So far we have investigated the impact of unfocused images on the first approximative stages. As the segmentation performance even increases, next we investigate the impact on the proposed stage 2 (level set) algorithm. We initialize the level set method with the results achieved with the focus level $fl = -10$, as it turned out to be the best choice for our database.

The level set segmentation method is evaluated with different focus levels. In Fig. 8 you can see that the segmentation accuracy definitely suffers, if the images for the second stage algorithm are not focused.

To put it in a nutshell, the overall segmentation performance decreases if the images for the second stage algorithm are not focused. However, the performance even increases, if the approximative first stage algorithm segments appropriate unfocused images (e.g. $fl = -10$).

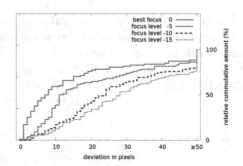

Fig. 8. Single unfocused images: Precise stage also with unfocused images

4.2   Gradual Enhancement Approach

First, we would like to know, if it is advantageous to process until the focused image is reached or if the segmentation should stop earlier, as the results do not necessarily improve until the best focused image is reached. We started with the image of the focus level −20 and iteratively increased the focus level.

The results with different focus levels as stopping conditions are shown in Fig. 9a. Although the behavior is similar to the behavior with one single unfocused image (best stopping level: $fl = -10$), the effect is smaller. The outliers ratio generally is lower than with the single image approach (shown in Fig. 5a). In Fig. 9b, the gradual enhancement approach with the best stopping focus level ($fl = -10$) is compared with the best results achieved with one single (unfocused) image and with the results with the focused image. The gradual enhancement approach definitely is more competitive as far as the approximative stage is concerned than the best focus approach and even more robust (less outliers) than the single unfocused image approach (more vertices with a deviation of ≤ 80 pixels).

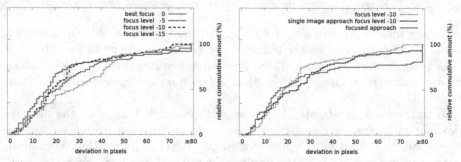

(a) Different stopping levels: first stage   (b) Best stopping level vs. focused approach

Fig. 9. Gradual enhancement approach

The results seem to be more similar compared with the single image approach. However, the impact of the different initialization results, on the level set algorithm is considerable, as shown in Fig. 10a. Especially the number of outliers can considerably be decreased when stopping earlier ($fl = -10$). Consequently, we define the stop level -10 to be the best choice. In Fig. 10b the achieved results of best configurations (gradual enhancement and single unfocused image) are

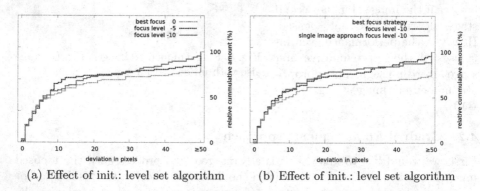

(a) Effect of init.: level set algorithm   (b) Effect of init.: level set algorithm

Fig. 10. Gradual enhancement approach: Effect on the precise stage

compared with the focused image approach. The performance of the methods using unfocused images definitely are higher than the performance of the simple approach dealing with the focused image. The gradual enhancement approach is even slightly more robust (very few outliers) than the single unfocused image approach. In Fig. 11, an indentation is shown which can be segmented with the introduced gradual enhancement approach based on unfocused images, but not with the traditional approach [10].

4.3   Execution Runtimes

We observed the execution runtimes on an Intel Core 2 Duo processor T5500 (1.66 GHz). The approaches are implemented in Java. The traditional approximative gradient descent approach [10] takes about 2.2 s per image. The gradual enhancement approach takes 1.0 s for the initial segmentation and 0.14 s for each enhancement step.

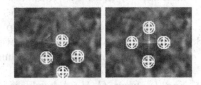

Fig. 11. Segmentation example: Traditional approach (left) and proposed approach (right)

5   Conclusion

The overall segmentation accuracy can be increased with the single unfocused image approach and with the gradual enhancement approach. The accuracy of the approaches is very similar, but significantly better than the traditional approach based on the focused images. With the gradual enhancement approach, the execution runtime potentially can be decreased as the segmentation might start before the focused image is available.

Acknowledgment. This work has been partially supported by the Austrian Federal Ministry for Transport, Innovation and Technology (FFG Bridge 2 project no. 822682).

References

1. Maier, A., Uhl, A.: Robust automatic indentation localisation and size approximation for vickers microindentation hardness indentations. In: Proc. of the 7th Intern. Symposium on Image and Signal Processing, pp. 295–300 (September 2011)
2. Gadermayr, M., Maier, A., Uhl, A.: Algorithms for microindentation measurement in automated Vickers hardness testing. In: Tenth International Conference on Quality Control for Artificial Vision (QCAV 2011). Proceedings of SPIE, vol. 8000, pp. 80000M–1 – 80000M–10. SPIE (June 2011)
3. Ji, Y., Xu, A.: A new method for automatically measurement of vickers hardness using thick line hough transform and least square method. In: Proceedings of the 2nd International Congress on Image and Signal Processing (CISP 2009), pp. 1–4 (2009)

4. Liming, W., Qu, Z., Yaohua, D., Miaoxian, Z.: Automatically analyzing the impress image of vickers hardness test using wavelet. China Mechanics Engineering 15(6) (March 2006)
5. Qu, Z., Guozheng, Y., Yi, Z.: A new method for quickly and automatically analysis of the image of vickers hardness using wavelet theory. Acta Metrologica Sinica 26(3), 245–248 (2005)
6. Macedo, M., Mendes, V.B., Conci, A., Leta, F.R.: Using hough transform as an auxiliary technique for vickers hardness measurement. In: Proceedings of the 13th International Conference on Systems, Signals and Image Processing (IWSSIP 2006), pp. 287–290 (2006)
7. Mendes, V., Leta, F.: Automatic measurement of Brinell and Vickers hardness using computer vision techniques. In: Proceedings of the XVII IMEKO World Congress, Dubrovnik, Croatia, pp. 992–995 (June 2003)
8. Sugimoto, T., Kawaguchi, T.: Development of an automatic Vickers hardness testing system using image processing technology. IEEE Transactions on Industrial Electronics 44(5), 696–702 (1997)
9. Yao, L., Fang, C.-H.: A hardness measuring method based on hough fuzzy vertex detection algorithm. IEEE Trans. on Industrial Electronics 53(3), 963–973 (2006)
10. Gadermayr, M., Uhl, A.: Dual-resolution active contours segmentation of Vickers indentation images. Intern. Conf. on Image and Signal Proc. (June 2012)
11. Gadermayr, M., Maier, A., Uhl, A.: A robust algorithm for automated microindentation measurement in vickers hardness testing. Journal of Electronic Imaging (2012)

GPU-Based Multi-resolution Image Analysis
for Synthesis of Tileable Textures

Gottfried Eibner[1], Anton Fuhrmann[1], and Werner Purgathofer[2]

[1] VRVis Forschungs GmbH
[2] Vienna University of Technology

Abstract. We propose a GPU-based algorithm for texture analysis and synthesis of nearly-regular patterns, in our case scanned textiles or similar manufactured surfaces. The method takes advantage of the highly parallel execution on the GPU to generate correlation maps from captured template images. In an analysis step a lattice encoding the periodicity of the texture is computed. This lattice is used to synthesize the smallest texture tile describing the underlying pattern. Compared to other approaches, our method analyzes and synthesizes a valid lattice model without any user interaction. It is robust against small distortions and fast compared to other, more general approaches.

1 Introduction and Related Work

Many manufactured surfaces contain regular patterns, which could efficiently be used as repetitive textures in a computer graphics. Capturing these patterns by camera or similar means leads in most cases - due to perspective or lens distortion, or due to the geometry of the underlying surface - to images which do not tile correctly. Converting these images into tileable texture maps that correctly represent the input pattern is the goal of this paper. We discuss the problem of finding pattern similarities, extracting a lattice model for repeating patterns and of generating tileable textures from such images. This allows us to render highly detailed surfaces where a base pattern repeats itself at a fixed period. The analysis and synthesis of stochastic or non-regular textures is out of scope of this paper. Input images to our algorithm should contain periodic patterns with at least one full period visible in both dimensions and deformations should be kept as small as possible. Even if those restrictions are not met, the algorithm is in many cases still able to create tileable patterns.

In the field of computer vision Lindeberg [1] introduced the concept of scale space. Lowe [2][3] introduced the idea of image descriptors which led to his SIFT implementation. His work has led to a variety of extensions and new methods to quickly find correspondences between different images even under strong perspective distortions, like ASIFT [4], MSER [5], SURF [6], to name a few of them.

In the area of symmetry detection many researchers like Kiryati and Gofman [7] made some progress, but their methods were very slow. Scognamillo et al. [8] take a different approach, but theirs needs a lot of user interaction, especially in preparing the input image for their detection algorithm. Loy and Eklundh [9] used the concept of scale-invariant features to quickly detect symmetry in a single picture.

G. Bebis et al. (Eds.): ISVC 2012, Part II, LNCS 7432, pp. 479–488, 2012.

Lin et al. [10] use autocorrelation functions, while Hays et al. [11] show that the analysis can be seen as a higher-order correspondence problem and make use of thin plate splines to compensate distortion and synthesize the texture.

Lin and Liu [12] showed an approach to track near-regular textures in motion. Park et al. [13] take advantage of mean-shift belief propagation to efficiently derive a lattice model. Dischler and Zara [14] uses mesh reconstruction to analyze the structure of a pattern.

The work from Liu et al. [15] picks up the essential result from another research area that only *five* possible lattices for 2D textures exist—a fact we can take advantage of for any texture analysis and synthesis. For our approach we will show that under certain circumstances these five possible lattices can be broken down to a single class of lattice, namely the rectangle lattice, which is a prerequisite for most rendering applications. As we will show, this can be done for almost all textures we encountered in our research.

A more detailed view of group theory and lattices can be found in another work of Liu et al. [16].

2   Proposed Method

The proposed algorithm is divided into three stages:

1. *Similarity Analysis*, in which positions of similar regions in the input image are calculated
2. *Lattice Generation*, in which the underlying lattice of the periodic texture is extracted, and
3. *Texture synthesis*, in which the results from the previous steps are used to synthesize a tiling pattern.

2.1   Similarity Analysis

Our algorithm takes a single image as input containing any repeating structure. The algorithm is based on a simple template matching algorithm. It randomly chooses a template which is matched against all pixels of the image. For each pixel a discrete convolution between the template and a corresponding region around the pixel is computed (see equation (1)). This simple approach can be compared to an exhaustive search which would definitely take to much time on a single CPU, but since we take advantage of the parallel execution on the GPU the computation is extremely fast. The per pixel calculation is performed within a fragment shader and each pixel in the new generated image represents the convolution, the sum of squared errors between the given sample region and the same sized region around each pixel. In the fragment shader for a gray scale image the following equation is computed where $i(x, y)$ denotes the intensity of the input image at pixel location x, y; and where x_0, y_0 is the center and w, h is half width and height of the template.

$$f(x,y) = \sum_{\xi=-w}^{w} \sum_{\eta=-h}^{h} (i(x+\xi, y+\eta) - i(x_0+\xi, y_0+\eta))^2 . \qquad (1)$$

By using a color image instead of a gray scale image we obtain better results in the matching algorithm. We use a heuristic metric which is 0 if the colors are visually different, and 1 if the colors are the same. We tried different functions in HSV and RGB space, but generally RGB performed adequately and resulted in faster code on the GPU. The metric is computed with equation (2), where c_i denotes the color's RGB vector, while c_i denotes the color's intensity value.

$$a(c_0, c_1) = 1 - \frac{c_0}{\|c_0\|} \cdot \frac{c_1}{\|c_1\|} * |c_0 - c_1| . \tag{2}$$

Using equation (1) and extending it with (2) we get equation (3) where $i(x,y)$ denotes the RGB vector of the input image at the given position.

$$\tilde{f}(x,y) = \sum_{\xi=-w}^{w} \sum_{\eta=-h}^{h} \left(i(x+\xi, y+\eta) - i(x_0+\xi, y_0+\eta) \right)^2 *$$
$$a\left(i(x+\xi, y+\eta), i(x_0+\xi, y_0+\eta) \right) . \tag{3}$$

Putting equation (3) into equation (4) we obtain a correlation value, where zero means not correlated to the template and one means a full match with the template.

$$p(x,y) = (1 - \alpha * \tilde{f}(x,y))^\gamma \quad , \alpha = \frac{1}{w*h} , \gamma \geq 1 . \tag{4}$$

An exponent greater than one will enhance the response of matching regions and dampen the 'noise'. Choosing a high number for the exponent, will lead to a correlation map with high peaks. A higher exponent makes it easier in the post-process to extract the feature points since regions with high correlations are exaggerated. On the other hand it should be mentioned that by selecting a value that is too large some feature points could be missed.

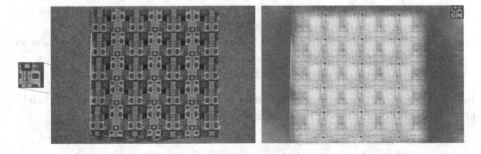

Fig. 1. A randomly chosen region (left) is used as comparison operator on the input image (middle) which leads to the correlation map (right) using equation (4) in a fragment shader

If we perform a naive brute-force search, the computation time grows extremely fast if a bigger sample region is selected and a higher image resolution is chosen. We use a multi-resolution approach (see below) to avoid high computational costs.

As a post-process the correlation map is filtered. In a 6x6 pixel neighborhood only the highest correlation value is kept while all other pixels are set to zero to eliminate nearby matching 'echos' which are in the range of 2-3 pixels. After the filtering, all pixel values greater zero are stored with their correlation value and their image position as potential *feature points*.

2.2 Lattice Generation

All potential feature points from the correlation map are stored with their position and their correlation value. The correlation value represents the matching between the template and the region around the feature point (see figure 1).These correlation values are later used in the lattice model finding process.

The post-processing of the correlation map yields a list S_0 of feature points f_i. Each feature point stores the correlation value of the template and the region around the feature point's location. From these feature points we derive and verify a lattice model. As one can see in figure 2, the feature point candidates contain many false positives. To derive the correct lattice model, we examine various subsets of the feature points and use them to search for a valid lattice model. Each consecutive subset is generated by further raising the threshold level for the correlation values.

$$S_{t_{corr}} = \{f_i \in S_0 \mid value_{corr}(f_i) \geq t_{corr}\} .\qquad(5)$$

In figure 2 four subsets are shown with a threshold of 0.0, 0.125, 0.25, and 0.5. By raising the correlation threshold, candidate feature points vanish until only those feature points remain which span the correct lattice of the texture pattern.

Lets assume we have N feature points extracted from the correlation map and stored in a list. While varying the correlation threshold a new subset from these points is generated and thus another possible lattice model can be found(see figure 2). We use the following method to find the lattice model:

- generate the subset for a given correlation threshold;
- for each feature point in the subset compute two displacement vectors from the two nearest neighbors; these two vectors must be linearly independent;
- cluster all displacement vector pairs with respect to their length and direction so that each cluster contains only vector pairs that go nearly in the same direction;
- take the largest cluster - the one containing the most items - as a starting base.

In this approach we assume that each feature point together with its two closest neighbors builds the correct lattice. To verify this assumption we have to check that

- the lattice model is connected,
- no outliers are taken into account, i.e. regions with a high correlation value but with a weak similarity to the template.

If neither a dominant cluster can be found, i.e. no cluster is found at all, or all clusters have fairly the same amount of vector pairs, nor one of the latter two criteria are met, we raise the correlation threshold until a valid lattice model is found. In figure 2 the lattice finding process is sketched. More and more feature points are removed until the correct lattice model is found. In the first two pictures an unconnected lattice model is generated and thus rejected. In the third picture a lattice model was found which contains some outliers although all vector pairs fall into the same cluster due to their similarity in length and direction. The correct model is not found until the last picture, where all of the false-positive feature points have been removed.

The more feature points we use, the more possibilities exist to find a correct lattice, i.e. we could always use only the three most significant feature points (in respect to their correlation value) and extract at least one single displacement vector pair, but this leads to a suboptimal lattice and to one texture tile only. Therefore, we should make use of as many feature points as possible to verify the overall correctness of the lattice, compensate distortions, and remove outliers. This will be discussed in detail in the section 2.3.

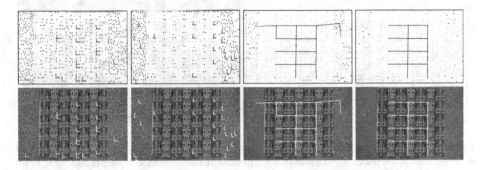

Fig. 2. Extracting a lattice model from left to right: increasing the correlation threshold removes unwanted feature points and leads to a valid lattice model. In the two left-most images a non-valid lattice model is found since it is not connected. In the third image two outliers lie within the lattice and the model is rejected (see top row for outliers). In the rightmost image the correct lattice is found.

Multi-resolution Optimization. We have shown the details of our texture analysis approach in the previous sections. Now we want to improve the execution speed of the method, especially for high resolution images and high resolution sample regions. We use a pyramidal approach by constraining the search space for higher resolution using the results from lower ones. We start with the full resolution input image and successively sample it down to build a level-of-detail (LoD) pyramid. Each individual image of the pyramid is enhanced using a bilateral filter to reduce noise and attenuate color regions. The pyramid is constructed from the highest resolution available to a minimum resolution of at least 256 pixel width or height which ever is reached first.

For a full automatic extraction we decided to use a genetic algorithm approach since our search-space has several dimensions: the correlation threshold (1 dimension), sample size (1 dimension) and location (2 dimensions), and level of detail (1 dimension). So we end up in a 5 dimensional search space. The algorithm starts with the lowest resolution of the LoD-pyramid and a small template region of 16 by 16 pixels. The template region's location is randomly chosen and altered until a valid lattice model is found or a specified amount of attempts is reached. If the maximum amount of attempts is exceeded and no lattice model is found we double the sample region and start again to find a valid lattice model. This is done until a valid lattice model is found or the size of the sample region covers more than a fourth of the LoD image. If we do not succeed in finding a lattice model by then we go to the next higher resolution in the LoD-pyramid and repeat the whole process starting again with a 16 by 16 pixel sample region.

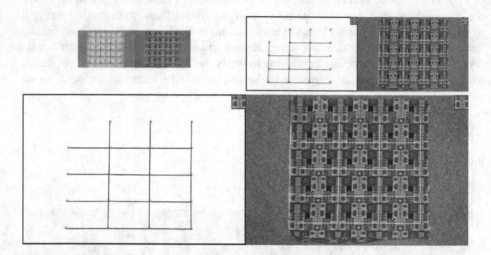

Fig. 3. Constraining the search space through multiple resolutions improves the performance of the algorithm and reduces the number of false-positive feature points at higher resolutions. From top to bottom, left to right. In the first image and lowest resolution all pixels are evaluated and a rough lattice is found. In the next steps only a small area around the feature points is evaluated and all other pixels remain blank.

If at any LoD level a valid lattice is found, we continue with the next higher resolution as follows: With the feature points from the lattice model we generate a stencil map. The fragment shader in combination with the stencil map computes the correlation values given by equation (4) only nearby those feature points. For all other pixels the fragment shader simply returns a value of zero. This saves time at higher resolutions and also reduces the chance of false-positive feature point extraction at higher resolutions since the maximum filter kernel is always kept at a size of 6 by 6 pixel.

Thus we narrow the search-space for higher resolutions and gain a speed-up of more than two orders of magnitude compared to an brute-force matching (see table 1).

The stenciled correlation map generation process is shown in figure 3. The top left image depicts the exhaustive search in the lowest resolution resulting where all pixels are evaluated. This gives a rough estimate of the lattice. The feature points' positions are used in the next higher resolution to constrain the search (top right image). The search is only performed in a small area around each feature point whereas in all other regions no matching is performed leading to blank pixels in the correlation map. As we can see in this example, the direction of the displacement vectors may change from level to level, but this has no influence on the position of feature points or the resulting lattice and texture tiles.

Table 1. Time comparison table for the brute-force and multi-resolution search. We start with a brute-force search at the lowest resolution.

Image res.	Sample size	Time [ms] multi-resolution	brute-force
267 x 178	16 x 16	10 *brute-force*	
534 x 356	32 x 32	3	680
1068 x 712	64 x 64	10	750
2135 x 1424	128 x 128	30	3000

2.3 Texture Synthesis

As soon as we have a valid lattice model and the displacement vectors, we can generate a texture tile. First we have to find two perpendicular displacement vectors. At this stage we know that the vectors are linearly independent (see section 2.2), but to extract a rectangular texture we need them to be perpendicular. We have to find the smallest possible rectangle fitting into the displacement vector model. In figure 4 a lattice with non-perpendicular displacement vectors v_0, v_1 is shown, and three highlighted feature points f_{i_0}, f_{i_1}, and f_{i_2}. Obviously the new displacement vectors should go from f_{i_0} to f_{i_1} and from f_{i_0} to f_{i_2}. Without the loss of generality we take v_0 as fixed and search a vector perpendicular to it. If we look at figure 4 again, one can see that

$$v_1' = \lambda * v_1 + v_0. \tag{6}$$

v_1' is the new second displacement vector we are looking for and since the dot product of perpendicular vectors is always zero, we can write

$$v_1' \cdot v_0 = 0 \quad \Rightarrow \quad (\lambda * v_1 + v_0) \cdot v_0 = 0. \tag{7}$$

By rewriting equation (7) we get

$$\exists \lambda \in \mathbb{Z} \setminus \{0\} \ , \quad \lambda = -\frac{v_0 \cdot v_0}{v_1 \cdot v_0}. \tag{8}$$

With equation (8) a new displacement vector pair and its new lattice are computed. A valid rectangular lattice can only be derived if λ is an integer. Each texture tile is extracted from the original image by taking the rectangle formed by each feature point and the two displacement vectors. If we have enough complete tiles available, we get a stack of valid textures which still may suffer from distortions due to bend and shear of the textile. Figure 5 shows a complete lattice model and the extracted texture tile stack. In the figure the total error of the mean tile and the error of each single tile are shown. At this stage the error in the mean tile is high. To reduce the error we undistort each tile by simply morphing each tile in respect to a randomly choosen one and apply a median filter between all tiles to erase dirt and other artifacts (see figure 6 bottom image where the shadow on the textile is totally removed in the output).

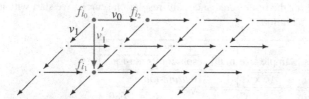

Fig. 4. If the displacement vectors are not perpendicular we have to find a new vector basis that is perpendicular

Fig. 5. A complete valid lattice (*left*) and the extracted texture tile stack on the (*left middle*). The mean tile and the summed error of all tiles (*right middle*) and error tiles for each single tile (*right*). Error coding: black no error, gray to white increasing error.

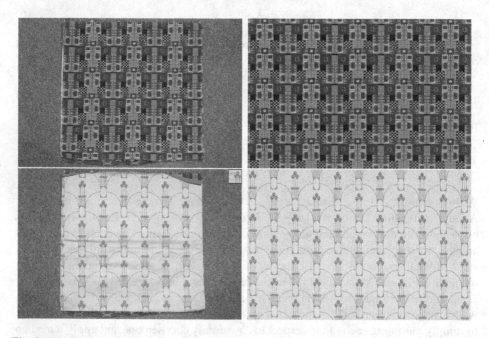

Fig. 6. Real photographs with a repeating pattern (*left*) and the extracted seamless repeating texture tiles (*right*)

3   Results

The overall performance of the algorithm compared to previous approaches is of an order of magnitude faster. The approach of Hays et al. [11] takes some minutes up to half an hour to compute optimal lattice. Our approach takes less than a minute with comparable results.

The main goal of our algorithm was to find a lattice for repeating patterns on woven materials, but it may well be applied to a wide variety of other materials with an inherent priodicity. Its core algorithms are very simple and easy to implement on the GPU. Comparable algorithms can - due to their use of heavy math and their iterative nature - not as easily and efficiently be implemented on the GPU.

Figure 6 shows some results for typical input photos.

4   Conclusion

We have shown a very simple and fast approach to analyze and synthesize textures from images of nearly regular periodic patterns. The overall performance is of an order of magnitude faster than previous approaches with comparable results.

The multi-resolution search step gives a high performance gain since at lower resolutions estimates about the lattice model are made and matching at higher resolutions will be performed only in regions around valid feature points from the found lattice.

Although the idea of template matching is very sophisticated in concept, it is so simple and easy to implement concurrently that it makes a perfect candidate for computation on the GPU.

5   Future Work

Our current work includes a re-implementation of the lattice finding algorithm on the GPU, which should result in even higher performance of our method. We are currently working on a lattice finding algorithm that almost entirely executes on the GPU which will reduce execution times even further.

Tests have shown that the simple template matching works well for extracting a lattice even for severely distorted patterns. The full lattice is currently not automatically found in these cases, since the lattice model is restricted and the displacement vectors variation is constrained. If we constrain the divergence of the displacement vectors only locally, i.e. for nearby feature points, we could use methods describe by Hays [11] or Park [13] to extract the full lattice.

References

1. Lindeberg, T.: Scale-space theory: A basic tool for analysing structures at different scales. Journal of Applied Statistics, 224–270 (1994)
2. Lowe, D.G.: Object recognition from local scale-invariant features. In: International Conference on Computer Vision, pp. 1150–1157 (1999)

3. Lowe, D.G.: Distinctive image features from scale-invariant keypoints. International Journal of Computer Vision 60, 91–110 (2004)
4. Morel, J.M., Yu, G.: On the consistency of the SIFT Method. CMLA 26 (2008) (preprint)
5. Matas, J., Chum, O., Urban, M., Pajdla, T.: Robust wide baseline stereo from. In: British Machine Vision Conference, pp. 384–393 (2002)
6. Bay, H., Tuytelaars, T., Van Gool, L.: SURF: Speeded Up Robust Features. In: Leonardis, A., Bischof, H., Pinz, A. (eds.) ECCV 2006. LNCS, vol. 3951, pp. 404–417. Springer, Heidelberg (2006)
7. Kiryati, N., Gofman, Y.: Detecting symmetry in grey level images: The global optimization approach. In: Proceedings of the 13th International Conference on Pattern Recognition, vol. I, pp. 951–956 (1996)
8. Scognamillo, R., Rhodes, G., Morrone, C., Burr, D.: A feature-based model of symmetry detection. In: Proceedings Biological Sciences the Royal Society, vol. 270, pp. 1727–1733 (2003)
9. Loy, G., Eklundh, J.-O.: Detecting Symmetry and Symmetric Constellations of Features. In: Leonardis, A., Bischof, H., Pinz, A. (eds.) ECCV 2006. LNCS, vol. 3952, pp. 508–521. Springer, Heidelberg (2006)
10. Lin, H.C., Wang, L.L., Yang, S.N.: Extracting periodicity of a regular texture based on autocorrelation functions. Pattern Recogn. Lett. 18, 433–443 (1997)
11. Hays, J., Leordeanu, M., Efros, A.A., Liu, Y.: Discovering Texture Regularity as a Higher-Order Correspondence Problem. In: Leonardis, A., Bischof, H., Pinz, A. (eds.) ECCV 2006. LNCS, vol. 3952, pp. 522–535. Springer, Heidelberg (2006)
12. Lin, W.-C., Liu, Y.: Tracking Dynamic Near-Regular Texture Under Occlusion and Rapid Movements. In: Leonardis, A., Bischof, H., Pinz, A. (eds.) ECCV 2006. LNCS, vol. 3952, pp. 44–55. Springer, Heidelberg (2006)
13. Park, M., Collins, R.T., Liu, Y.: Deformed Lattice Discovery Via Efficient Mean-Shift Belief Propagation. In: Forsyth, D., Torr, P., Zisserman, A. (eds.) ECCV 2008, Part II. LNCS, vol. 5303, pp. 474–485. Springer, Heidelberg (2008)
14. Dischler, J.M., Zara, F.: Real-time structured texture synthesis and editing using image-mesh analogies. The Visual Computer 22, 926–935 (2006)
15. Liu, Y., Liu, a.Y., Tsin, Y.: The promise and the perils of near-regular texture. International Journal of Computer Vision 62, 1–2 (2002)
16. Liu, Y., Collins, R.T., Tsin, Y.: A computational model for periodic pattern perception based on frieze and wallpaper groups. IEEE Transactions on Pattern Analysis and Machine Intelligence 26, 354–371 (2004)

Edge Detection and Smoothing-Filter
of Volumetric Data

Masaki Narita[1], Atsushi Imiya[2], and Hayato Itoh[1]

[1] School of Advanced Integration Science, Chiba University
[2] Institute of Media and Information Technology, Chiba University
Yayoicho 1-33, Inage-ku, Chiba, 263-8522, Japan

Abstract. We develop a higher dimensional version of the Canny edge detection algorithm. The Canny operation detects the zero-crossing of the gradient of the Gaussian-convolved image. The segment edge curve detected by the Canny operation is an approximation of zero-crossing of bilinear form defined by second order derivative of an image. This definition of edge points of segment is dimension independent. This definition also allows us to extend the filtering operation from the Gaussian convolution to general linear and non-linear ones.

1 Introduction

Segmentation [5,6,8] is a fundamental pre-processing in computer vision and image analysis. The Canny operation for planar images is a well-established operation as edge detection for segmentation [8,7]. In this paper, we develop an edge detection method for volumetric and hyper-volumetric data. Since the Canny operation is a planar operation, the edge and segment boundary of a volumetric data is constructed from slices of boundary curves of segments. The boundary constructed by this multi-slice method is widely used as the boundary of volumetric data.

In this paper, we first prove that the boundary detection procedure based on the multi-slice decomposition of volumetric data and rotation in space are generally non-commutative. Then, using the rotation invariant description of the segment-boundary manifold in the space, we derive a rotation invariant boundary detection algorithm for volumetric data. This idea is straight forward to the boundary detection of objects in the higher dimensional Euclidean space. The segment edge-curve detected by the Canny operation is an approximation of zero-crossing of bilinear form defined by second order derivative of an image [7]. This definition of edge-points of segment is dimension independent. This definition also allows us to extend the smoothing pre-operation from the Gaussian convolution to general ones, such as nonlinear diffusion filtering in the higher dimensional space. Therefore, we also clarify relations among pre-smoothing filters from viewpoint of diffusion filtering [11,13].

G. Bebis et al. (Eds.): ISVC 2012, Part II, LNCS 7432, pp. 489–498, 2012.
© Springer-Verlag Berlin Heidelberg 2012

2   Boundary Surfaces

For the function $f(x)$ defined in $\mathbf{R}^n$, we assume the condition $\int_{\mathbf{R}^n} |f(x)|^2 dx < \infty$.

Setting

$$G_2 *_2 f = \int_{\mathbf{R}^2} \frac{1}{\sqrt{2\pi\tau}} \exp\left(-\frac{(x-u)^2 + (y-v)^2}{2\tau}\right) f(u,v) du dv \qquad (1)$$

Canny [8] suggested the following algorithm [8] detects an approximation of the zero-crossing $\frac{\partial^2}{\partial^2 n}(G *_2 f) = 0$ for $n = \frac{\nabla G_2 *_2 f}{|\nabla G_2 *_2 f|}$.

Edge Detection

1. Define the parameters τ^*, T_1 and T_2 such that $T_1 \geq T_2$.
2. Compute $h = G *_2 f$.
3. Mark $\theta(x,y) = \tan^{-1} \frac{h_x}{h_y} = \tan^{-1} \frac{G_x *_2 f}{G_y *_2 f}$ on points as the edge direction.
4. For $|\nabla h|$, select a point $|\nabla h| \geq T_1$ as the starting point of edge tracking.
5. Track peaks using $\theta(x,y)$ of $|\nabla h|$ as for as $|\nabla h| \geq T_2$.

In step 3, the algorithm performs point following using an ordering of points. On two-dimensional manifold, which is the boundary surface of a volumetric object, this point following in the original algorism cannot be achieved, since we cannot order the points on the boundary manifold in the same manner.

On the surface $\phi(x,z) = 0$ for $\phi(x,z) = z - f(x)$, a set of point

$$E(f) = \left\{ x \mid \frac{d^2}{dn^2} |\nabla f| = 0 \right\} \qquad (2)$$

where n is the unit nomal vector on the surface, that is, $n = \nabla f / |\nabla f|$, is the segment boundary.

Since [6,7,9]

$$\frac{d^2}{dn^2} |\nabla f| = \frac{1}{|\nabla f|} (\nabla f)^\top (\nabla \nabla^\top f) \nabla f, \qquad (3)$$

where $\nabla \nabla^\top f$ is the Hessioan of f, we have the relation

$$E(f) = \{ x | \nabla f^\top (\nabla \nabla^\top f) \nabla f = 0 \}. \qquad (4)$$

Furthermore, since

$$\nabla_{Rx} f(Rx) = R^\top \nabla f(x), \quad \nabla_{Rx} \nabla_{Rx}^\top f(Rx) = R^\top (\nabla \nabla^\top f(x)) R, \qquad (5)$$

for $E(f)$, we have the next theorem.

Theorem 1. *The point set $E(f)$ is rotation invariant.*

In medical imaging, the multi-slice expression of volumetric data is traditionally used widely. We examine the relation between the boundary surface of the volumetric data and the boundary curves on a slice of volumetric data. A slice data

$g(a^\perp) = P_{a^\perp,Z} f(x)$ of a volumetric data $f(x)$ on the plane $\Pi_{a,Z} = \{x|a^\top x = Z\}$, where a is the unit normal of the plane, is expressed

$$g(a^\perp, Z) = \int_{\mathbf{R}^n} \delta(a^\top x - Z) f(x) dx \tag{6}$$

where $P_{a^\perp}$ is the orthogonal projection onto the linear subspace $\Pi_a = \Pi_{a,0} = \{x|a^\top x = 0\}$ perpendicular to the vector a and $a^\perp \in \Pi_a$.

For the boundary of g, we have the next theorem.

Theorem 2. *The relation* $E(f) \cap \{x|a^\top x = Z\} = E(g, a, Z)$ *is satisfied iff* $\frac{\partial f}{\partial Z} = 0$ *or* $f(x) = f(a^\perp + Z)$ *for* $a^\perp \in \mathbf{R}^{n-1}$.

(Proof) If $\frac{\partial f}{\partial Z} = 0$ or $f(x) = f(a^\perp)$ for $a^\perp \in \mathbf{R}^{n-1}$, we have the relations $\nabla f = \nabla g$ and $\nabla \nabla^\top f = \nabla \nabla^\top g$ on the plane $a^\top x = Z$. Conversely, if the relations $\nabla f = \nabla g$ and $\nabla \nabla^\top f = \nabla \nabla^\top g$, are satisfied, we have the relation $\frac{\partial f}{\partial Z} = 0$ or $f(x) = f(a^\perp + Z)$ for $a^\perp \in \mathbf{R}^{n-1}$. (Q.E.D.)

From the slices $E(g, a, Z)$, which are perpendicular to the vector a, we can construct three-dimensional manifold as

$$E(f, a) = \bigcup_Z E(g, a, Z). \tag{7}$$

We call the vector a the axis of slices. Setting S^{n-1} to be the unit sphere in $\mathbf{R}^2$, Theorem 2 implies the next theorem for $E(f, a)$.

Theorem 3. *Iff* f *is spherically symmetry, that is,* $f(x) = h(|x|)$ *for an appropriate function* $h(s)$ *defined on* $\mathbf{R}$, *then* $\forall a, \forall b \in S^{n-1}$, *the relation* $E(f, a) = E(f, b)$ *is satisfied.*

If f is spherically symmetry, for all $a \in S^{n-1}$, $E(f, a)$ is independent to a. Conversely if $E(f, a)$ is independent to a. theorem 2 implies f is spherically symmetry. (Q.E.D.)

Theorem 3 derives the following theorem, since for the objects in the higher-dimensional Euclidean space, the condition $a \neq b$ always implies the relation $E(f, a) \neq E(f, b)$.

Theorem 4. *For the functions in the higher-dimensional Euclidean space, which is spherically asymmetry, if* $a \neq b$, *then* $E(f, a) \neq E(f, b)$.

In practice, setting $w_\varepsilon s = w_\varepsilon(-s) \geq 0$ and $w_\varepsilon(s) = 0$ for $|s| \geq \varepsilon$, where a small positive constant ε, the slice date is given

$$g_\varepsilon(a^\perp, Z) = \frac{1}{2\varepsilon} \int_{-\varepsilon}^{\varepsilon} \int_{\mathbf{R}^n} w_\varepsilon(a^\top x - Z) f(x) dx d\varepsilon, \tag{8}$$

that is, we can have a slice data which is smoothed in the normal direction of the slice. Therefore, the smoothed slice g_ε derives a blurred boundary curves on the plane $a^\top x = Z$.

Theorem 2 and eq. (8) imply that for the accurate computation of the boundary of the slice images we are required to compute the boundary surface of the volumetric data. Furthermore, since the boundary surface constructed by using multi-slice decomposition of the volumetric data depend on the axis of the slices, we are required to develop three-dimensional surface extraction algorithm.

Setting $f_{(\tau)}$ to be the smoothed function of f, we extract the zero-crossing

$$E(f, \tau) = \{\boldsymbol{x} | \nabla f_{(\tau)}^{\top} (\nabla \nabla^{\top} f_{(\tau)}) \nabla f_{(\tau)} = 0\}, \tag{9}$$

as the boundary of f. This is a generalisation of the Canny operation to the higher-dimensional function.

3   Filtering Operation

For the edge detection by the Canny operation, images are smoothed by Gaussian filter. This operation removes the small region which yield the artifact segment-boundaries. As an extension the pre-smoothing by the Gaussian filtering, we can use several smooting operations. We summerise relations among the linear diffusion filtering [1,3,2], the non-linear diffusion filtering[11,12], the bilateral filtering[13,14], and the pyramid transform [15].

The generalised function $f(\boldsymbol{x}, \tau)$ in the linear scale space is the solution of the linear diffusion equation [1,2,3,4],

$$\frac{\partial}{\partial \tau} f(\boldsymbol{x}, \tau) = \Delta f(\boldsymbol{x}, \tau), \ f(\boldsymbol{x}, 0) = f(\boldsymbol{x}). \tag{10}$$

The solution $f(\boldsymbol{x}, \tau)$ is given as the Gaussian convolution such that

$$f(\boldsymbol{x}, \tau) = \int_{\mathbf{R}^n} G_n(\boldsymbol{x} - \boldsymbol{y}, \tau) f(\boldsymbol{y}) d\boldsymbol{y} = G_n *_n f, \tag{11}$$

for $G_n(\boldsymbol{x}, \tau) = \frac{1}{(2\pi)^{\frac{n}{2}} \tau} \exp\left(-\frac{|\boldsymbol{x}|^2}{2\tau}\right)$.

The anisotropic linear diffusion equation

$$\frac{\partial}{\partial \tau} f(\boldsymbol{x}, \tau) = \nabla^{\top} \boldsymbol{A} \nabla f(\boldsymbol{x}, \tau), \ f(\boldsymbol{x}, 0) = f(\boldsymbol{x}), \tag{12}$$

for an appropriate symmetry matrix $\boldsymbol{A}$, derive the anisotropic Gaussian convolution

$$f(\boldsymbol{x}, \tau) = \int_{\mathbf{R}^n} G_n((\boldsymbol{x} - \boldsymbol{y})^{\top} \boldsymbol{A} (\boldsymbol{x} - \boldsymbol{y}), \tau) f(\boldsymbol{y}) d\boldsymbol{y}. \tag{13}$$

By assuming the matrix $\boldsymbol{A}$ is diagonal and embedding the n-dimensional manifold $z - f(\boldsymbol{x}) = 0$ in $(n+1)$-dimensional Euclidean space $\mathbf{R}^{n+1}$, as a generalisation of eq. (11) we have the bilateral filtering [13,14],

$$f(\boldsymbol{x}, \tau, \sigma) = \int_{\mathbf{R}^{n+1}} G_n(\boldsymbol{x} - \boldsymbol{y}, \tau) G_1(f(\boldsymbol{x}) - f(\boldsymbol{y}), \sigma) f(\boldsymbol{y}) d\boldsymbol{y} \tag{14}$$

for $z = f(\boldsymbol{x})$.

Next, if we substitute the Hessian operator $\nabla\nabla^\top$, which is a symmetry tensor differential-operator, to $\boldsymbol{A}$ in eq. (12), that is,

$$A\nabla f(\boldsymbol{x},\tau) = (\nabla\nabla^\top)\nabla f(\boldsymbol{x},\tau). \tag{15}$$

we have the bi-Laplacian diffusion equation

$$\frac{\partial}{\partial\tau}f(\boldsymbol{x},\tau) = \Delta^2 f(\boldsymbol{x},\tau), \ \ f(\boldsymbol{x},0) = f(\boldsymbol{x}). \tag{16}$$

An extension of diffusion filtering is to use a non-linear diffusion equation [10,11,12] in the form

$$\frac{\partial}{\partial\tau}f(\boldsymbol{x},\tau) = \nabla^\top F(\nabla f(\boldsymbol{x},\tau)), \tag{17}$$

for an appropriate function F from $\mathbf{R}^n$ to $\mathbf{R}^n$. If $F(\nabla f(\boldsymbol{x},\tau)) = \nabla f(\boldsymbol{x},\tau)$, that is, F is the identity operation, eq. (17) coincides with the linear diffusion filtering. Next, we define a more precise form of F as

$$F(\nabla f(\boldsymbol{x},\tau)) = g(|\nabla f|)\nabla f(\boldsymbol{x},\tau). \tag{18}$$

This class of F means that diffusion is controlled by the gradient map $|\nabla f|$ of f at each τ.

The diffusion functions of the Perona-Malik type [10] and the Weickert type [11] are, for a positive constant λ,

$$g(|\nabla f|) = (1 + |\nabla f|^2/\lambda)^{-1}, \tag{19}$$

and

$$g(|\nabla f|) = \begin{cases} 1 & (|\nabla f| = 0), \\ 1 - \exp\left(-3.31488 \times \left(\frac{|\nabla f|}{\lambda}\right)^{-8}\right) & (|\nabla f| > 0), \end{cases} \tag{20}$$

respectively. These diffusion functions are basically and historically introduced to the edge detection for segmentation. For the Perona-Malik type diffusion function, we can derive the relation $F(\nabla f(\boldsymbol{x},\tau)) = h(|\nabla f|^2)\nabla f(\boldsymbol{x},\tau)$.

The next extension is the combination of diffusion equation and down-sampling operation such that

$$\frac{\partial}{\partial\tau}\bar{f}(\boldsymbol{x},\tau) = \nabla^\top F(\nabla\bar{f}(\boldsymbol{x},\tau)), \ \ \bar{f}(\boldsymbol{x},0) = f(\sigma\boldsymbol{x}), \ \sigma > 0 \tag{21}$$

If $F(\nabla g(\boldsymbol{x},\tau)) = \nabla g$, we have the linear diffusion equation

$$\frac{\partial}{\partial\tau}\bar{f}(\boldsymbol{x},\tau) = \Delta\bar{f}(\boldsymbol{x},\tau), \ \ \bar{f}(\boldsymbol{x},0) = f(\sigma\boldsymbol{x}), \ \sigma > 0 \tag{22}$$

and the convolution transform

$$\bar{f}(\boldsymbol{x},\sigma) = \int_{\mathbf{R}^n} G_n(\sigma\boldsymbol{x} - \boldsymbol{y}, \tau)f(\boldsymbol{y})d\boldsymbol{y}. \tag{23}$$

As a generalisation of the convolution kernel of eq. (23) to the function which satisfies $w(\boldsymbol{x}) = w(-\boldsymbol{x}) > 0$ and $\int_{\mathbf{R}^n} |w(\boldsymbol{x})|^2 d\boldsymbol{x} < \infty$, we have the pyramid transform [15] R of the factor σ

$$Rf(\boldsymbol{x}) = \int_{\mathbf{R}^n} w(\boldsymbol{y}) f(\sigma \boldsymbol{x} - \boldsymbol{y}) d\boldsymbol{y}. \tag{24}$$

The dual transform E of R is

$$Eg(\boldsymbol{x}) = \sigma^n \int_{\mathbf{R}^n} w(\boldsymbol{y}) g\left(\frac{\boldsymbol{x} - \boldsymbol{y}}{\sigma}\right) d\boldsymbol{y}, \tag{25}$$

since these operations satisfy the relation

$$\int_{\mathbf{R}^n} Rf(\boldsymbol{x}) g(\boldsymbol{x}) d\boldsymbol{x} = \int_{\mathbf{R}^n} f(\boldsymbol{x}) Eg(\boldsymbol{x}) d\boldsymbol{x}. \tag{26}$$

Moreover, by combining the pyramid transform and the anisotropic diffusion operation, we have the filtering operation

$$\frac{\partial}{\partial \tau} \bar{f}(\boldsymbol{x}, \tau) = \nabla^\top F(\nabla \bar{f}(\boldsymbol{x}, \tau)), \quad \bar{f}(\boldsymbol{x}, 0) = Rf(\boldsymbol{x}). \tag{27}$$

4 Numerical Method

For the n-dimensional grid points $\mathbf{Z}^n$, setting $\boldsymbol{p} \in \mathbf{Z}^n$, discretisation of diffusion equations derive the following discrete equations

$$\frac{f_{\boldsymbol{p}}^{(k+1)} - f_{\boldsymbol{p}}^{(k)}}{\tau} = \sum_{\boldsymbol{p}' \in N(\boldsymbol{p})} w_{\boldsymbol{p}'\boldsymbol{p}} f_{\boldsymbol{p}'}^{(k)}, \quad \frac{f_{\boldsymbol{p}}^{(k+1)} - f_{\boldsymbol{p}}^{(k)}}{\tau} = \sum_{\boldsymbol{p}' \in N(\boldsymbol{p})} w_{\boldsymbol{p}'\boldsymbol{p}} f_{\boldsymbol{p}'}^{(k+1)}, \tag{28}$$

where $N(\boldsymbol{p})$ is an appropriate neighbourhood of the point $\boldsymbol{p}$ and $w_{\boldsymbol{p}'\boldsymbol{p}}$ is the discrete shift-variant weight function depending on the diffusion process.

Filtering operations for discrete signal and images [13,14] are obteind by selecting an appropriate discrete function $k_{\boldsymbol{p}'\boldsymbol{p}}$. The first equation of eq. (28) derives the discrete filtering operation [13,14]

$$f_{\boldsymbol{p}}^{(k+1)} = \sum_{\boldsymbol{p}' \in N(\boldsymbol{p})} K_{\boldsymbol{p}'\boldsymbol{p}} f_{\boldsymbol{p}'}^{(k)}. \tag{29}$$

If $w_{\boldsymbol{p}'\boldsymbol{p}}$, which defines diffusion process in $\mathbf{Z}^n$, does not depend on the position of the point $\boldsymbol{p}$, eq. (29) becomes

$$f_{\boldsymbol{p}}^{(n+1)} = \sum_{\boldsymbol{p}' \in N(\boldsymbol{p})} K_{\boldsymbol{p}'} f_{\boldsymbol{p}'}^{(n)}. \tag{30}$$

Furthermore, if $N(\boldsymbol{p})$ does not depend on the position of the point $\boldsymbol{p}$, eq. (29) becomes discrete convolution

$$f_{\boldsymbol{p}}^{(k+1)} = \sum_{\boldsymbol{p'} \in \mathbf{R}^n} K_{\boldsymbol{p'}-\boldsymbol{p}} f_{\boldsymbol{p'}}^{(k)}. \tag{31}$$

We can rewrite eq. (29) in the matrix form as

$$\boldsymbol{f}^{(k+1)} = \boldsymbol{A}^{(k)} \boldsymbol{f}^{(k)} + \boldsymbol{c}^{(k)}, \tag{32}$$

where $\boldsymbol{f}^k$ is the vector form of $f_{\boldsymbol{p}}^{(k)}$ constracted by an appropriate ordering of the grid points $\boldsymbol{p}$ and $\boldsymbol{A}^{(k)}$ and $\boldsymbol{c}^{(k)}$ are the matrix and vector depending both on the iteration step and $\boldsymbol{f}^{(k)}$. Furthermore, the second equation of eq. (28) derives the matrix form

$$\boldsymbol{B}^{(k)} \boldsymbol{f}^{(k+1)} = \boldsymbol{f}^{(k)} + \boldsymbol{d}^{(k)}, \tag{33}$$

for the non-singular matrix $\boldsymbol{B}^{(k)}$ and the vector $\boldsymbol{d}^{(k)}$ depending both on the iteration step and $\boldsymbol{f}^{(k)}$.

The discrete version of the pyramid transform of the factor 2 and its dual for three-dimensional digital objects are

$$R f_{kmn} = \sum_{k',m'n'=-1}^{1} w_{k'} w_{m'} w_{n'} f_{2k-k', 2m-m', 2n-n'}, \tag{34}$$

$$E f_{kmn} = 2^3 \sum_{k',m',n'=-2}^{2} w_{k'} w_{m'} w_{n'} f_{\frac{k-k'}{2}, \frac{m-m'}{2}, \frac{n-n'}{2}}, \tag{35}$$

for $w_{\pm 1} = \frac{1}{4}$ and $w_0 = \frac{1}{2}$ and the summation is achieved for $(k-k')/2$, $(m-m')/2$, $(n-n')/2$ are integers.

5   Numerical Examples

For the evaluation of the acceleration of diffusion filtering by the pyramid transform, we evaluate the slices of the boundary surfaces $E_*(f, \tau) \cup \Pi_{a,Z}$ and $E_*(Rf, \sigma^n \tau) \cup \Pi_{a,Z}$ for $* \in \{L, PM, W, BL\}$, where L, PM, and W express the linear diffusion filtering, the Perona-Malic type diffusion filtering, the Weickert type diffusion filtering, and the bi-Laplacian diffusion filtering, respectively. In Fig. 1, $(* := BL)$ expresses the combination of Bi-Laplacian filtering and linear diffusion filtering such that $(f^{(n+1)} = L^5 \text{Bi-Laplacian} f^{(n)})$. The filter L^5 is used to relax over filtering by the bi-Laplacian filtering, since the bi-Laplacian filtering is a high-pass filter.

As $E_*(Rf, \tau)$, we adopt the point set $E_*(Rf, \tau) := \text{Binarisation}[\bar{E}_*(Rf, \tau)]$, for $\bar{E}_*(Rf, \tau) = E[E_*(Rf, \tau)]$ which expresses the expansion and linear interpolation of the point set by the dual transform of the pyramid transform.

For the experiments of $* \in \{L, PM, W\}$, we use discrete pyramid transform of the facter 2 and we set $\tau = k$ in eq. (29). In Figs. 1 and 2, slices of $E_*(f, 2^3 10)$ and $E_*(Rf, 10)$ are shown for $* \in \{L, PM, W\}$, since the pyramid transform reduces the sizes of images by 2^3 for volumetric data in $\mathbf{R}^3$.

Results in top are slices of the boundary surfaces for $\tau = 2^3 \times 10$ and results in bottom are slices of the boundary surfaces for $\tau = 10$ with the pyramid transform. From the left to the right results computed by using the linear diffusion, the Perona-Malick diffusion, and the Weickert diffusion, respectively. For all diffusion filtering, results in top and bottom are very similar. This property of the appearances of the extracted results show that the pyramid transform accelerates the diffusion filtering. by reducing the size of data and removing noises and small regions which derive perturbation for the boundary surface and its slice curves. Our results show the effects of the pyramid-based data reduction for volumetric segmentation by boundary surface detection.

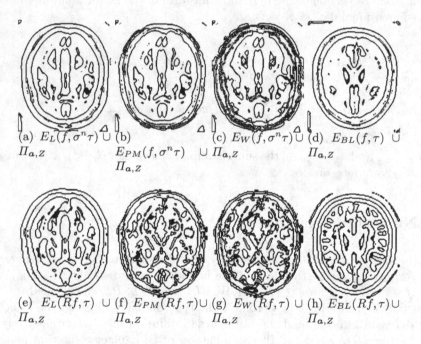

(a) $E_L(f, \sigma^n \tau) \cup$ (b) $E_{PM}(f, \sigma^n \tau) \cup \Pi_{a,z}$ (c) $E_W(f, \sigma^n \tau) \cup$ (d) $E_{BL}(f, \tau) \cup$
$\Pi_{a,z}$ $\Pi_{a,z}$ $\Pi_{a,z}$

(e) $E_L(Rf, \tau) \cup$ (f) $E_{PM}(Rf, \tau) \cup$ (g) $E_W(Rf, \tau) \cup$ (h) $E_{BL}(Rf, \tau) \cup$
$\Pi_{a,z}$ $\Pi_{a,z}$ $\Pi_{a,z}$ $\Pi_{a,z}$

Fig. 1. Computational results I. Results in top are slices of the boundary surfaces for $\tau = 2^3 \times 10$. Results in bottom are slices of the boundary surfaces for $\tau = 10$ with the pyramid transform. From the left to the right results computed by using the linear diffusion, the Perona-Malick diffusion, and the Weickert diffusion, respectively. These results shows the pyramid transform reduces noises and small regions which derive perturbations for the boundary surface and its slice curves. In (h), the down-sampling operation as the pre-processing allows to detect the clear boundaries for the same τ.

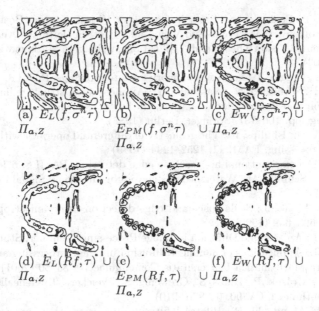

(a) $E_L(f, \sigma^n \tau) \cup \Pi_{a,z}$

(b) $E_{PM}(f, \sigma^n \tau) \cup \Pi_{a,z}$

(c) $E_W(f, \sigma^n \tau) \cup \Pi_{a,z}$

(d) $E_L(Rf, \tau) \cup \Pi_{a,z}$

(e) $E_{PM}(Rf, \tau) \cup \Pi_{a,z}$

(f) $E_W(Rf, \tau) \cup \Pi_{a,z}$

Fig. 2. Computational results II. Results in top are slices of the boundary surfaces for $\tau = 2^3 \times 10$. Results in bottom are slices of the boundary surfaces for $\tau = 10$ with the pyramid transform. From the left to the right results computed by using the linear diffusion, the Perona-Malick diffusion, and the Weickert diffusion, respectively. In the downsampled images, the boundary of teeth are clearly extracted.

6   Conclusions

In this paper, for segmentation of volumetric data we derived a rotation invariant boundary detection algorithm for volumetric data. Furthermore, we shows the acceleration property of the pyramid transform as the pre-processing to diffusion filtering of volumetric data. For the application the bilateral filtering to the boundary extraction, we are required to ecaluate the ration between two parameters $O(\log \frac{\sigma}{\tau}) = \alpha$ which controls the diffusion process in the spatial directions and gray-values in the neighbourhood of each point.

This research was supported by "Computational anatomy for computer-aided diagnosis and therapy: Frontiers of medical image sciences" funded by the Grant-in-Aid for Scientific Research on Innovative Areas, MEXT, Japan and the Grants-in-Aid for Scientific Research funded by Japan Society of the Promotion of Sciences Japan.

References

1. Iijima, T.: Pattern Recognition, Corona-sha, Tokyo (1974) (in Japanese)
2. Witkin, A.P.: Scale space filtering. In: Proc. of 8th IJCAI, pp. 1019–1022 (1993)

3. Lindeberg, T.: Scale-Space Theory in Computer Vision. Kluwer, Boston (1994)
4. ter Haar Romeny, B.M.: Front-End Vision and Multi-Scale Image Analysis Multi-scale Computer Vision Theory and Applications, written in Mathematica. Springer, Berlin (2003)
5. Yuille, A.L., Poggio, T.: Scale space theory for zero crossings. PAMI 8, 15–25 (1986)
6. Enomoto, H., Yonezaki, N., Watanabe, Y.: Application of structure lines to surface construction and 3-dimensional analysis. In: Fu, K.-S., Kunii, T.L. (eds.) Picture Engineering, pp. 106–137. Springer, Berlin (1982)
7. Krueger, W.M., Phillips, K.: The geometry of differential operator with application to image processing. PAMI 11, 1252–1264 (1989)
8. Canny, J.: A computational approach to edge detection. PAMI 8, 679–698 (1986)
9. Najman, L., Schmitt, M.: Watershed of a continuous function. Signal Processing 38, 99–112 (1994)
10. Perona, P., Malik, J.: Scale space and edge detection using anisotropic diffusion. RAMI 12, 629–639 (1990)
11. Weickert, J.: Anisotropic Diffusion in Image Processing, Teubner, Stuttgart (1998)
12. Weickert, J.: Applications of nonlinear diffusion in image processing and computer vision. Acta Mathematica Universitatis Comenianae 70, 33–50 (2001)
13. Pizarro, L., Mrázek, P., Didas, S., Grewenig, S., Weickert, J.: Genealised nonlocal image smoothing. IJCV 90, 62–87 (2010)
14. Tomasi, C., Manduchi, R.: Bilateral filtering dor gray colour images. In: Proc., ICCV 1998, pp. 839–846 (1998)
15. Burt, P.J., Andelson, E.H.: The Laplacian pyramid as a compact image coding. IEEE Trans. Communications 31, 532–540 (1983)

Human Body Orientation Estimation in Multiview Scenarios

Lili Chen[1], Giorgio Panin[2], and Alois Knoll[1]

[1] Department of Informatics,
Technische Universität München,
85748 Garching bei München, Germany
[2] German Aerospace Center (DLR),
Institute of Robotics and Mechatronics,
82230 Wessling, Germany

Abstract. Estimation of human body orientation is an important cue to study and understand human behaviour, for different tasks such as video surveillance or human-robot interaction. In this paper, we propose an approach to simultaneously estimate the body orientation of multiple people in multi-view scenarios, which combines a 3D human body shape and appearance model with a 2D template matching approach. In particular, the 3D model is composed of a generic shape made up of elliptic cylinders, and a 3D colored point cloud (appearance model), obtained by back-projecting pixels from foreground images onto the geometric surfaces. In order to match the reconstructed appearance to target images in arbitrary poses, the appearance is re-projected onto each of the different views, by generating multiple templates that are pixel-wise, robustly matched to the respective foreground images. The effectiveness of the proposed approach is demonstrated through experiments in indoor sequences with manually-labeled ground truth, using a calibrated multi-camera setup.

1 Introduction

In the context of automatic video surveillance, people detection and tracking [1–5] are probably the most important tasks, which received a significant amount of attention in the area of research and development. However, an intelligent surveillance system should not only be able to locate and track people, but also understand and recognize their behaviors. To this aim, a representative cue that is often neglected is body orientation, which provides hints about what the person is going to do, and where the person is probably looking at. The aim of this paper is an approach to robustly estimate body orientation of multiple people, through multiple calibrated views mounted on the ceiling from different viewing angles.

To our knowledge, very few works have been conducted to visually estimate the orientation of human body. Some have addressed this issue by considering the person dynamics, assuming that the orientation is simply given by the walking direction [6, 7]. In [6], estimation is based on the motion of a tracked person and the

G. Bebis et al. (Eds.): ISVC 2012, Part II, LNCS 7432, pp. 499–508, 2012.

size of its bounding ellipse, but it fails in the case of people who are not moving, or slowly walking. [7] also couples human body orientation with motion direction, where the coupling is given by a dynamical model. This work assumes a loose coupling at low speed, whereby the absence of an explicit observation model for the orientation results in similar problems. Some other works [8, 9] consider this task as a classification problem. [8] use HoG descriptors to classify the orientation of pedestrians, recovering an estimate based on 2D low-resolution images. However, they group the orientation of a person very roughly, covering a 45-degrees range for in-plane rotations only. In [9], body pose classification is performed by using multi-level HoG features, and a sparse representation technique, at each frame of the sequence. However, for each human region they end up with a very high dimensional feature vector, which has the drawback of computational complexity, and their experimental scenarios are limited to a single view.

In order to address the shortcomings of motion-based or classification-based techniques, in this paper we propose a methodology combining a 3D human body/appearance model with 2D template-based matching. The primary goal is to robustly and accurately determine the body orientation of a variable number of people, randomly walking in an overlapping, multi-camera environment, which is able to deal with still-standing or slowly moving people, while covering a full 360 degrees range. More precisely, the 3D human appearance model is represented by a colored point cloud, obtained by back-projecting pixels from foreground images onto the surface of a simple geometric model, composed of three cylinders with elliptical section. In order to match the reconstructed model to new images under arbitrary poses, we reproject the 3D colored point cloud onto each view at different locations and orientations, thereby generating multiple 2D templates, which are robustly matched to the underlying foreground image through pixel-level measurements. For this purpose, a recently developed multi-camera, multi-people tracking system based on silhouette edges [10] yields ground plane locations, that are used as a position reference when estimating the body orientation. Subsequently, orientation is estimated by a full search, only restricted to a neighborhood of the previous orientation estimate.

By resuming, the effectiveness of our system is due to three main aspects. One is the reconstruction of a detailed 3D appearance, that makes a pixel-level matching more precise with respect to global or local statistics, such as color histograms. The second is the use of a robust multi-target detector and tracker, since a good location reference is necessary for an accurate orientation estimation. And the third is the integration of calibrated multi-camera views, both for position and orientation estimation.

The remainder of the paper is organized as follows: Sec. 2 describes the overall system architecture, from hardware setup to software framework. Our 3D appearance modeling is then followed by in Sec. 3. The proposed orientation estimation approach is described in Sec. 4. Sec. 5 describes and discusses the experimental results, and Sec. 6 concludes the paper, proposing future development roads.

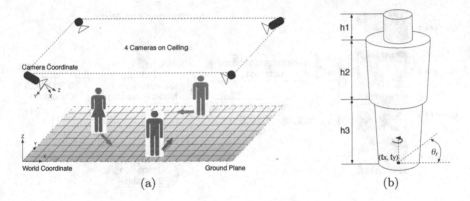

Fig. 1. (a) Hardware setup. (b) 3D human body model.

2 System Overview

In this section, our system for human body orientation estimation is described, starting from the hardware and software setup, followed by details on the specific components that are involved in the implementation.

Our hardware setup is depicted in Fig. 1(a). Four *uEye* USB cameras, with a resolution of 752×480, are mounted overhead on the corners of the ceiling, each of them observing the same 3D scene synchronously from different viewpoints, thus providing an informative measurement set. Furthermore, all cameras are connected to one multi-core PC. A necessary step to get accurate 3D information, is the recovery of intrinsic and extrinsic camera parameters, that we perform with the Matlab Calibration Toolbox [1], with respect to a common *world* coordinate system located on the floor.

Furthermore, Fig. 2 outlines the flow chart of our approach. After acquiring a frame from the four cameras, an edge-based detection using hierarchical grids [10] is performed, to obtain the location of each person in the scene. Meanwhile, in order to avoid collecting background pixels into the appearance model, a fast GPU-based foreground segmentation is utilized. Subsequently, the 3D appearance is reconstructed by applying pixel colors to the predefined body geometry, which makes our approach independent of the camera viewpoint. Also, the appearance model can be rendered onto 2D image plane in arbitrary poses, thereby generating multiple 2D templates, which then can be matched to the segmented foreground image through pixel-level measurements, leading to the orientation estimation result.

3 Appearance Model

If we take into account the flattened shape of a human body along the depth dimension, we can approximate the overall geometry by three cylinders with

---

[1] http://www.vision.caltech.edu/bouguetj/calib_doc/

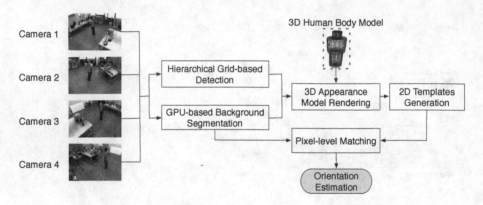

Fig. 2. Flow chart of the proposed approach

elliptical section, as illustrated in Fig. 1(b), enclosing the head, the torso and the legs.

The body pose is given by the person location (t_x, t_y) and orientation θ_r, about an axis perpendicular to the ground plane. In our system, the location is independently provided by a multi-camera, multi-people detector and tracker [10], therefore our aim is to estimate only θ_r.

To this aim, a 3D appearance model is reconstructed by back-projecting pixels from each 2D image onto the respective surfaces, at the estimated location. During this phase, in order to obtain the reference orientation, the person is assumed to be more or less at the center of the observation space, and facing a certain direction. By knowing the full pose, each pixel can be back-projected onto a 3D point on the model surface.

In order to avoid collecting also background pixels, we also perform a background subtraction, with a GPU-based method proposed by Griesser et al. [11], which performs an iterative solution on a 3×3 neighborhood at each pixel, also incorporating darkness compensation. Fig.3(b) shows an example of this algorithm.

The result is a large set of N model points x_n (Fig.3), including position, color values, and local surface normals from the underlying 3D surface. This constitutes a sparse appearance model, that we denote by

$$M \equiv \{(x_1, v_1, n_1), \cdots, (x_N, v_N, n_N)\}. \tag{1}$$

4   Orientation Estimation

Once the sparse point cloud is available, pixel-level measurements can be obtained by reprojecting the appearance model onto the image planes, through a (3×4) projection matrix onto camera $k = 1, \ldots, 4$

$$y = K_k T_{kw} T_{wo} x_o \tag{2}$$

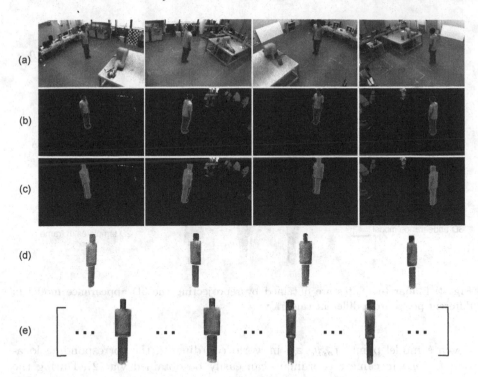

Fig. 3. Appearance model reconstruction. (a) Original input frames from 4 views. (b) Corresponding foreground images. (c) Detected target, with geometry model superimposed onto foreground images. (d) Back-projected partial 3D cloud, at each view. (e) Final 3D appearance model, covering 360°, with some key-poses shown.

where x_o is a local model point, in homogeneous coordinates, y is the corresponding image pixel, K is the intrinsic camera projection, known from off-line calibration

$$K = \begin{bmatrix} f_x & 0 & p_x & 0 \\ 0 & f_y & p_y & 0 \\ 0 & 0 & 1 & 0 \end{bmatrix} \tag{3}$$

with f_x, f_y the focal lengths, (p_x, p_y) the principal point, and T_{kw}, is the camera-to-world constant transform, also known by calibration. Finally, T_{wo} is a homogeneous (4×4) transformation matrix that represents the person location

$$T_{wo} = \begin{bmatrix} R & t \\ 0 & 1 \end{bmatrix} \tag{4}$$

where $t = [X, Y, Z]^T$ is the 3-dimensional translation, and the (3×3) rotation matrix R is expressed in terms of XYZ Euler angles (of which only γ is updated by the orientation estimation):

$$\theta_r = [\alpha, \beta, \gamma]^T \tag{5}$$
$$R(\theta_r) = R_x(\alpha)R_y(\beta)R_z(\gamma)$$

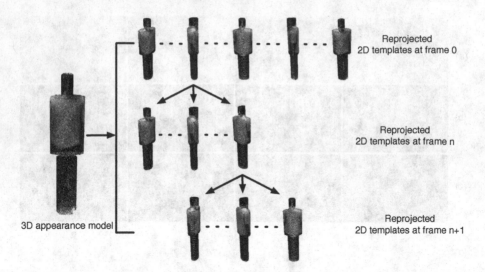

Fig. 4. Planar templates are obtained by reprojecting the 3D appearance model in different poses, from different camera views

Given a model point (x_o, y_o, z_o) in world coordinates, the corresponding location (x_s, y_s) in camera coordinate can easily be obtained via (2). During the reprojection, it is worth noting that, due to the rotations of the body, the visible part of the sparse 3D point cloud should change, only a roughly $180°$ slice of the point cloud is visible in any particular frame, that corresponds with the visible portion of the human body in each video frame.

Thus, it is necessary to test visibility of each point: since the shape is almost everywhere convex, a point p is visible from camera c if the angle between its normal and the camera projection ray through p is less than $90°$, that is

$$V_c \cdot n_c < 0 \tag{6}$$

where V_c is the viewing vector (i.e. the position of p in camera coordinates) and n_c is the respective normal direction.

The point projection (2) and visibility test (6) are done at each pose hypothesis θ_r. Fig. 4 provides an example of re-projected templates on one camera view.

Subsequently, template matching simply amounts to evaluate a likelihood function, by comparing color values of the templates with the underlying foreground pixels. For a predicted pose $\hat{T}_t$ at time t, our similarity measure between the N_k color pixels u from the reprojected template $h_k(\hat{T})$ onto camera k, and the corresponding pixels v of the foreground image, is defined as

$$D = \sum_{k=1}^{4} \frac{1}{N_k} \sum_{u \in h_k} \sqrt{\sum_{c \in (r,g,b)} (u_c - v_c)^2} \tag{7}$$

which is the sum of absolute pixel-wise difference over (r, g, b) channels.

In this formula, the inner sum is an isotropic L_1-norm that, compared to classical L_2-norm, it is more robust to outliers, such as non-Gaussian noise or erroneous colors sampled from the background. Therefore, the template likelihood corresponds to a Laplacian distribution

$$P(z|s) = \frac{1}{2\sigma} \exp\left(-\frac{D}{\sigma}\right) \tag{8}$$

where the pose is described by $s = (t_{ref}, \theta_r)$, and σ is the precision parameter of this distribution.

During orientation estimation, we employ a simple but effective prediction mechanism for computational effiency. As during normal walking, it is unlikely for a person to turn more than 90 degrees over one frame of the sequence; therefore, after the initial detection and modeling phase, reprojection and matching are performed only on the orientations which are in a fixed range around the former estimation, thus saving computation while reducing estimation error. Fig. 4 illustrates this strategy across frames.

5   Experimental Results

We evaluate our orientation estimation algorithm through two video sequences, showing multiple people that move and turn freely, as well as interacting and occasionally occluding each other in some views. The sequences have been simultaneously recorded from all cameras, as described in Section 2, with a resolution of (752×480) and a frame rate of 25 fps.

The implementation is done in C++ on a desktop PC with Intel Core-2 Duo CPU $(1.86GHz)$, 1GB RAM and an Nvidia GeForce 8600 GT graphic card. Before estimating orientations, we run the hierarchical grid-based detector [10] over the sequences, producing quite reliable target locations on ground plane, which are provided as position reference t_{ref}.

Furthermore, during the initialization phase of our system, the 3D appearance model of each target is automatically reconstructed according to the detected 2D location and known orientation, using the technique described in Section 3. During likelihood evaluation, for computational effiency, 12 discrete orientations covering $360°$ are utilized during the reprojection of the appearance model onto the image planes, while at subsequent frames the model is reprojected only within $90°$ around the previous estimate.

The first sequence involves two targets, which interact once by shaking hands, then randomly walk around the room. The second sequence involves three targets that are strongly occluded in some views, and occasionally remain static for a few frames.

Results of both experiments are shown in Fig. 5, where the orientation of each target is indicated by a colored line, pointing to the estimated direction. The sihouette of the geometry model is also superimposed onto each target, to illustrate the tracked location that is used as reference.

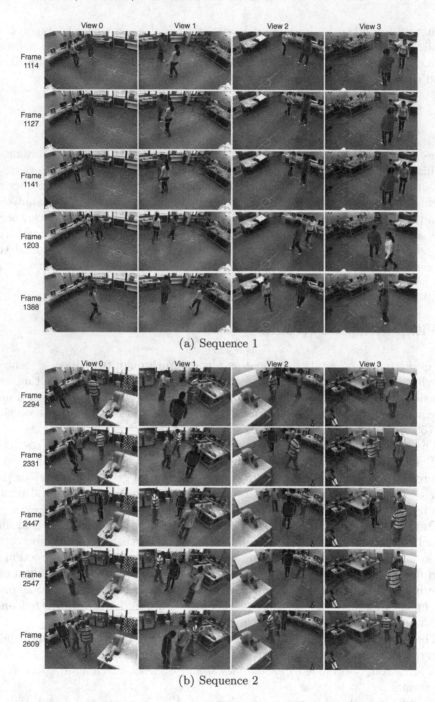

(a) Sequence 1

(b) Sequence 2

Fig. 5. Performance of orientation estimation on four camera views

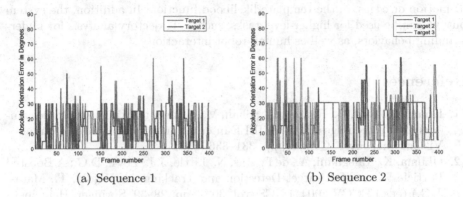

(a) Sequence 1   (b) Sequence 2

Fig. 6. Ground truth evaluation

In particular, as can see from Fig. 5(a), during hand shaking (between frame 1127 and 1141), the two orientations are both correctly estimated. Moreover, in Fig. 5(b), we emphasize the challenges due to mutual occlusions, from one or more views. At frame 2609, although people keep very close to each other, our estimation results are still satisfactory. Around frame 2447, all three targets are almost static throughout several frames, however our algorithm successfully estimates their orientations.

In order to evaluate more precisely the performances of our approach, we also manually labeled ground-truth data for each frame of the sequences, by rendering the 3D body model and visually matching it with the target area, in all views where the person can be seen. The most challenging clip, covering 400 frames from both sequences, was selected for ground truth evaluation, and results are shown in Fig. 6, that shows the absolute error between estimated orientation and ground truth for each person.

We can see that orientation errors are most of the time below 30 degrees. As noticed that in our framework, the likelihood is evaluated on a discrete state-space with an interval of 30 degrees, while ground truth data are labeled with an interval of 5 degrees.

6   Conclusion

In this paper, we presented a robust algorithm for estimating the body orientation of multiple people simultaneously in a calibrated multi-camera environment. Our method uses a generic 3D human body shape model together with a distinctive appearance model, and employs multiple 2D templates for matching with a robust likelihood function. Experiments over real-world sequences have been performed and also evaluated against ground-truth data.

The proposed methodology can be easily applied to different camera setups and different indoor environments. Future work involves increasing processing speed, robustness and versatility, for example including additional features, such as motion or edges, in the template likelihood function. In addition, the system output can be used for higher-level tasks, such as trajectory analysis for understanding behaviors, as well as human-robot interaction.

References

1. Khan, S., Javed, O., Rasheed, Z., Shah, M.: Human tracking in multiple cameras. In: Proceedings of the 8th IEEE International Conference on Computer Vision, Vancouver, Canada, pp. 331–336 (2001)
2. Okuma, K., Taleghani, A., de Freitas, N., Little, J.J., Lowe, D.G.: A Boosted Particle Filter: Multitarget Detection and Tracking. In: Pajdla, T., Matas, J(G.) (eds.) ECCV 2004. LNCS, vol. 3021, pp. 28–39. Springer, Heidelberg (2004)
3. Wu, B., Nevatia, R.: Detection and tracking of multiple, partially occluded humans by bayesian combination of edgelet based part detectors. International Journal of Computer Vision 75, 247–266 (2007)
4. Leibe, B., Schindler, K., Van Gool, L.: Coupled detection and trajectory estimation for multi-object tracking. In: International Conference on Computer Vision (2007)
5. Andriluka, M., Roth, S., Schiele, B.: Monocular 3d pose estimation and tracking by detection. In: IEEE Conference on Computer Vision and Pattern Recognition, CVPR (2010)
6. Lee, M.W., Nevatia, R.: Body part detection for human pose estimation and tracking. In: IEEE Workshop on Motion and Video Computing (2007)
7. Yao, J., Odobez, J.: Multi-camera 3d person tracking with particle filter in a surveillance environment. In: 8th European Signal Processing Conference, EUSIPCO (2008)
8. Gandhi, T., Trivedi, M.: Image based estimation of pedestrian orientation for improving path prediction. In: IEEE IV Symposium, pp. 506–511 (2008)
9. Chen, C., Heili, A., Odobez, J.: Combined estimation of location and body pose in surveillance video. In: IEEE Conf. on Advanced Video and Signal Based Surveillance, AVSS (2011)
10. Chen, L., Panin, G., Knoll, A.: Multi-camera people tracking with hierarchical likelihood grids. In: Proceedings of the 6th International Conference on Computer Vision Theory and Applications, pp. 474–483 (2011)
11. Griesser, A., Roeck, D.S., Neubeck, A., Van Gool, L.: Gpu-based foreground-background segmentation using an extended colinearity criterion. In: Proc. of Vison, Modeling, and Visualization (VMV), pp. 319–326 (2005)

Characterization of Similar Areas
of Two 2D Point Clouds

Sébastien Mavromatis, Christophe Palmann, and Jean Sequeira

Projet SimGraph - LSIS UMR CNRS 7296 - Aix-Marseille Université
Polytech Marseille case 925 - 163, av. de Luminy - 13288 Marseille Cedex 9
{mavromatis,palmann,sequeira}@univ-amu.fr

Abstract. We here present a new approach to characterize similar areas
of two 2D point clouds, which is a major issue in Pattern Recognition
and Image Analysis.

To do so, we define a similarity measure that takes into account several criteria such as invariance by rotation, outlier elimination, and one-
dimensional structure enhancement. We use this similarity measure to
associate locations from one cloud to the other, to use this result in the
frame of a registration process between these two point clouds.

Our main contributions are the integration of various one-dimensional
structure representations into a unified formalism, and the design of a
robust estimator to evaluate the common information related to these
structures.

Finally, we show how to use this approach to register images of
different modalities.

1 Introduction

Image analysis and pattern recognition often use 2D point clouds as processed
representations of images (e.g. these points being the gradient highest value locations, or Laplacian zero-crossing ones,). This binary representation, when using
images of different modalities (e.g. multispectral and radar ones for "Remote
Sensing" applications), can be a useful bridge between these two images in order
to register them.

Several algorithms have been designed to match point clouds but none of them
takes advantage of underlying one-dimensional structures, which is a "knowledge
element" that we have to take into account in most practical cases (these structures being the disrupting lines of the physical support ground, human body,).
The approach we propose here integrates this knowledge element in the matching
process.

We first give a short overview of the existing methods to match point clouds,
and we show why they cannot help us to characterize the similarity of two small
point clouds that represent the neighborhood of two given locations. We next
design a primitive model that can take into account all the possible configurations
of points related to one-dimensional structures, and we propose a method that

G. Bebis et al. (Eds.): ISVC 2012, Part II, LNCS 7432, pp. 509–516, 2012.

analyzes the relations between the primitives of these two neighborhoods, up to a rotation.

The most important outcome of this evaluation is the characterization of landmark pairs that can be used to globally register the two point clouds, without any a priori knowledge on the registration transformation.

We illustrate our approach with several examples related to image registration in the field of Remote Sensing, when images are of different modalities (multispectral, radar).

2   Existing Methods to Match Point Clouds

Most of these methods suppose that one point cloud is derived from the other by a rigid transformation (translation and then rotation). Some approaches do not suppose the points of a cloud to be associated with geometrical structures. The most important ones are the ICP (for Iterative Closest Point) method and the RANSAC (for RANdom SAmple Consensus) method [1]. These approaches are not efficient enough when the noise (i.e. additional points in the two clouds) reaches a given level, and this is practically often the case. In order to register such point clouds, we need to integrate knowledge related to common information: in most cases, relevant points in both point clouds are related to one-dimensional structures. For example, in the field of Remote Sensing, although the extracted information is different in radar and multispectral images, important changes (i.e. edges) in these images are both associated with the same physical elements (coastline, roadside, forest edge), as clearly shown in figure 1. Many methods have been proposed to extract one-dimensional structures. Let us mention the "chain codes" introduced by Li et al in 1995 [2], the polylines used by Zhao and Chen en 2004 [3] and the snakes developed by Kass et al in 1988 [4].

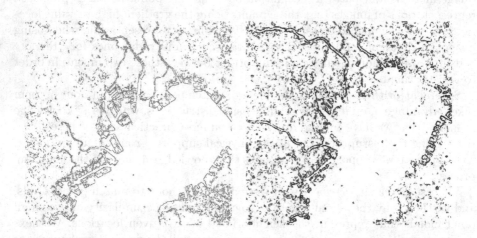

Fig. 1. 2D point clouds obtained by processing two different modality images

These approaches are interesting in specific cases (connected arcs, or coherent contours,) but they cannot be efficient in the case studied here because of the variety of contour representations, as illustrated in figure 2, and also because parts of contours (and not connected ones) are all we have.

Primitives should represent all the situations in figure 2 in a unified form.

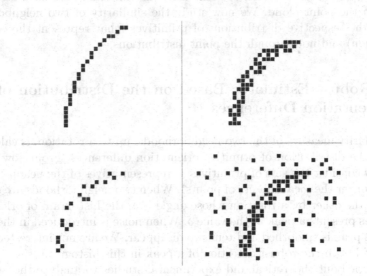

Fig. 2. Different local representations of a one-dimensional structure

We do not describe a contour but only a location (x, y) and an orientation (θ) when, around this location, the points are numerous enough in the *theta* orientation. We obtain these items of the primitive model by using a set of predefined masks with given radius R_m and thickness E, as shown in figure 3.

Fig. 3. A few masks within the set $(R_m = 7, E = 3)$

Finally, once we have obtained a first set of primitives, we select a relevant subset in which all the elements are not closer than a D_{min} distance. We obtain this result by sorting the initial set by decreasing values of primitive scores (number of points covered by the mask that produces the maximum value) and, going through this ordered set, from the highest to the lowest value, we only keep those primitives that are farther than all the previous selected ones. The result we obtain is a set of primitives that show the local orientation of one-dimensional structures, which are uniformly distributed where such structures appear in the point cloud. We now study the similarity of two neighborhoods through the respective distributions of primitives (that represent the essential information) and not through the point distributions.

3   A Robust Estimator Based on the Distribution of Orientation Differences

The similarity measure of the two neighborhoods, up to a rotation, is calculated through the distribution of primitive orientation differences. From now on, we shall only consider the sets of primitives as representatives of the neighborhoods (and no longer the local subsets of points). When two neighborhoods are derived one from the other by a rotation whose angle is δ, the histogram of orientation differences presents a peak at the value δ. When noise is introduced in the initial data, this peak is smoothed and tends to disappear. We are not interested in the value of δ but in the robust detection of a peak in this histogram.

We carried out theoretical and experimental studies to analyze the behavior of this histogram when we introduce noise. We especially studied the probability of having false positive detections and this led us to define a robust estimator that is not based on peaks, but on modes, and that also takes in consideration the relations between the main modes. These studies show that the sensitivity of the detection depends on the amount of noise, on the number of orientation values, and also on the number of possible orientations.

We define k as being the number of primitives (we suppose this number is the same for the two neighborhoods), and $k = k_s + k_b$ (k_s being the number of common primitives   i.e. primitives from one neighborhood we find in the second one with a θ rotation and k_b being the number of added primitives - i.e. noise). We define n as being the number of possible orientations (i.e. the discrete representation of the orientation space). And we define the noise as $\epsilon = k_b/k$.

We can represent all these elements as in figure 4.

Then, at the δ position, the maximum H is calculated by:

$$H = k_s + N \tag{1}$$

or

$$H = (1 - \epsilon).k + N \tag{2}$$

where $N = k^2/n$ is the mean value of the noise.

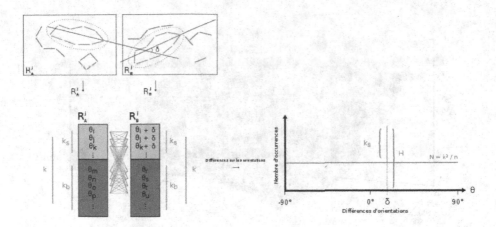

Fig. 4. A dashed line surrounds the areas containing common primitives: histogram of orientation differences

Figure 5 shows the relative importance of these parameters in the emergence of a peak in the histogram. First, we showed that the peak emergence does not depend on the δ angle. Thus we can use any angle (in the following figures, we have $\delta = 10°$). Values of the parameters for this simulation are given in the legend of figure 5.

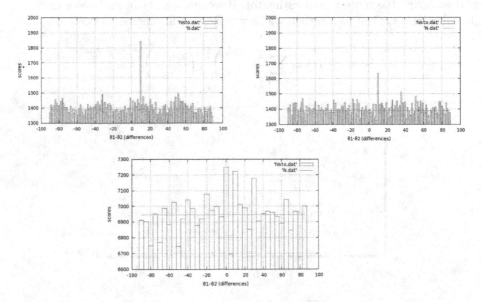

Fig. 5. Left: $k = 500, n = 180(1°), \epsilon = 0\%$. Maximum has been clearly detected at $10°$. Right: $k = 500, n = 180(1°), \epsilon = 50\%$. Maximum has been detected at $10°$. Bottom: $k = 500, n = 36(5°), \epsilon = 50\%$. We obtain a wrong detection at $0°$.

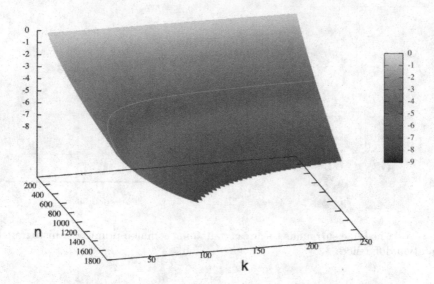

Fig. 6. Probability of having a false positive detection (logarithmic scale) with $\epsilon = 80\%$. The green curve provides the solution of the equation $P(\epsilon, k, n) = P_0$ with $(P_0 = 10^{-4})$.

We also studied the probability of having a false positive detection (as it was the case as in figure 7) in order to evaluate the robustness of the detection and to define the more robust estimator. If we suppose that we have a uniform distribution of orientations, then we have a Bernouilli scheme, hence:

$$P\left(score \geq H\right) = \sum_{s=H}^{k^2} \frac{k^2}{s} \cdot \left(\frac{1}{n}\right)^{s} \cdot \left(\frac{n-1}{n}\right)^{(k^2-s)} \tag{3}$$

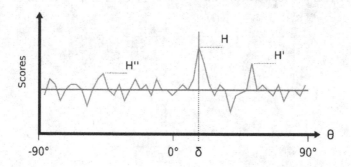

Fig. 7. Main peak H, and secondary ones

This expression can be interpreted as being the risk that a peak related to noise is higher than the peak whose height is H. Then because $k^2 > 50$ and $n > 10$, this law can be replaced by a Poisson law. And we can use a the Chernoff bound to express an upper bound as:

$$P\left(\epsilon, k, n\right) = e^{\left((1-\epsilon).k+\frac{k^2}{n}\right).\log\left(\frac{k^2}{n.\left((1-\epsilon).k+\frac{k^2}{n}\right)}\right)+(1-\epsilon).k} \tag{4}$$

This probability can be visualized as in figure 6:

We define the estimator: $\alpha = \frac{|H-\bar{H}|}{H}$ with $\bar{H} = \frac{H'+H''}{2}$

4   Application to the Identification of Similar Regions within Images of Different Modalities - Conclusion

In the example below, we use a radar image (ERS2), and a multispectral image (Landsat 7), whose resolutions are similar (about 30m) and they represent the Tokyo bay. We used classical specific procedures (as the Touzi operator [5] on radar images, PCA on multispectral ones, with Deriche operator [6]) to obtain representative point clouds.

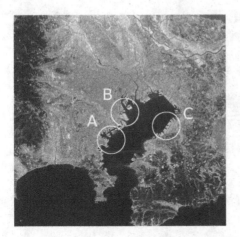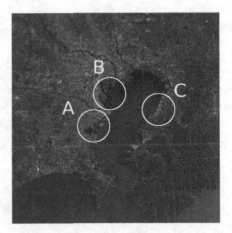

Fig. 8. Pairs of locations whose neighborhood similarity is high

We have presented an approach that enables the definition of a robust estimator which evaluates the similarity of areas of two point clouds. After studying and experimenting this estimator, we have used it on actual data to obtain a quite interesting result.

References

1. Fischler, M., Bolles, R.: Random Sample Consensus. A Paradigm for Model Fitting with Applications to Image Analysis and Automated Cartography. Communications of the ACM 24(6), 381–395 (1981)
2. Li, H., Manjunath, B., Mitra, S.: A contour-based approach to multisensor image registration. IEEE Transactions on Image Processing 4(3), 320–334 (1995)
3. Zhao, Y., Chen, Y.Q.: Connected Equi-Length Line Segments for Curve and Structure Matching. International Journal of Pattern Recognition and Artificial Intelligence 18(6), 1019–1037 (2004)
4. Kass, M., Witkin, A., Terzopoulos, D.: Snakes: Active contour models. International Journal of Computer Vision 1(234), 321–331 (1988)
5. Touzi, R., Lopes, A., Bousquet, P.: A statistical and geometrical edge detector for SAR images. IEEE Transactions on Geoscience and Remote Sensing 30(5), 1054–1060 (1988)
6. Deriche, R.: Using Canny's criteria to derive a recursively implemented optimal edge detector. International Journal of Computer Vision 1(2), 167–187 (1987)

Building an Effective Visual Codebook: Is K-Means Clustering Useful?

Aaron Chavez and David Gustafson

Department of Computer Science
Kansas State University
Manhattan, KS 66506
{mchav,dag}@ksu.edu

Abstract. In this paper we examine the effectiveness of using k-means clustering to build a codebook for image classification. Our findings suggest that k-means clustering does not result in a better codebook than a random sampling of training features. An alternative strategy that uses k-means to cluster features, then ultimately selects a feature from each cluster, also fails to significantly outperform random sampling. We present data in support of our claims, gathered using the training set for the PASCAL Visual Object Classes Challenge. Additionally, we provide a theoretical justification for our findings.

1 Introduction

Many state-of-the-art image classifiers share a similar basic framework. The first step consists of finding a dense sampling of low-level features from the images in the training set. Features are then matched to a codebook in a "bag of features" approach. A spatial pyramid may be employed to further organize the bag of features. As a result, each image produces a vector consisting of feature counts. Finally, a support vector machine develops a classifier based on these vectors.

We consider one step in this algorithm: the process of building a codebook of representative features. A codebook is necessary to reduce an intractable number of distinct features into a finite number of groups. In this way, we can determine two features to be roughly equivalent if they match the same representative in the codebook.

A good codebook satisfies two opposing criteria. On the one hand, a codebook will ideally consist of a small number of highly distinct features. On the other hand, a codebook must be large enough that a representative feature exists for any feature we might encounter in a set of images. If two features are visually distinct, they should not match the same feature or features in the codebook.

Common wisdom in the field suggests the use of k-means clustering on training features to build a codebook. We examine the validity of this approach by testing a classifier on training data from the PASCAL Visual Object Classes Challenge. Our data indicates that k-means is not a useful tool in building a visual codebook.

G. Bebis et al. (Eds.): ISVC 2012, Part II, LNCS 7432, pp. 517–525, 2012.

2   Related Work

2.1   K-Means Clustering

K-means clustering takes a large number of points in n-dimensional space and divides them into a smaller number of clusters, maximizing the distance between each cluster. Currently, most visual codebooks are built using some form of k-means clustering [11]. K-means clustering is not intrinsically a visual algorithm, but it is relatively straightforward to implement.

2.2   Alternative Codebook Generation Algorithms

While K-means is common, other attempts do exist to build an effective codebook from a set of training features. In [7], codebooks are created using a mean-shift approach that aims to produce evenly spaced clusters. Evidence is presented that this approach is more successful than simple k-means clustering for producing a visual codebook. It is argued that the centroids produced by a k-means algorithm are not evenly distributed about the feature space.

Alternately, one can adopt a hierarchical classification mechanism. Work in [12] documents an approach that classifies each feature based on a hierarchical series of smaller decisions. Work in [11] has a similar high-level goal. Using randomized clustering forests, a feature can be classified according to which leaf it matches in multiple decision trees.

2.3   Universal Codebooks

While codebooks are almost always constructed from scratch for a particular classification task, some research suggests that a universal codebook could be comparably effective on any set of natural images [5]. Perhaps contrastingly, other recent work indicates that codebooks specifically tailored to each object category are more discriminative [9].

2.4   PASCAL Challenge

The PASCAL Visual Object Classes Challenge (VOC) is an annual image classification competition in which participants must classify images among twenty different object categories. The image sets used in the challenge are particularly difficult, showing objects in different poses, scales, backgrounds, lighting conditions, and occlusions.

There are two primary competitions each year, a classification competition and a detection competition. In the classification competition, an image must simply be classified based on the object(s) found. The object does not need to be localized in the image. In the detection competition, objects must be detected and localized in each particular image. [13]

While strategies for the competition differ, most successful algorithms share many of the same basic components. These components are outlined in the following section. Matching a dense sampling of low-level features to a codebook is very common. [13]

2.5   Basic Image Classifiers

Codebooks are but one key component in a basic image classifier. Some type of dense sampling of one or more low-level features is necessary. The most commonly used feature is the SIFT feature [10], which introduced the concept of histograms of gradients (HoG). Other features based on HoG also exist, such as Pyramid HoG [1] and deformable parts [4].

The features are then grouped based on a bag-of-features approach. Such an approach takes a large number of features at various points and scales in the image, but does not attempt to rigorously organize them spatially. Intuitively, one would think that image classification requires some sort of geometric knowledge, but this approach (which largely ignores geometry) has been successful [3]. It is at this step in the process where the codebook is needed, to reduce an intractable feature space into a manageable set of representative features.

Spatial Pyramid Matching is a simple addition to a bag-of-features approach that restores some of the spatial information that has been eliminated [8]. The approach is to use two different image segmentations: one into quadrants (2x2), and one vertical segmentation into thirds (1x3). With these segmentations, we can make a distinction between, say, a feature in the upper-left quadrant and a feature in the lower-right quadrant, even if those two features match the same codebook entry. The final step will take advantage of this distinction.

For each image, a bag-of-features approach produces a vector, which represents the number of times each feature is found in that image. A support vector machine takes these somewhat arbitrary vectors of data and separates them into two or more classes [2]. The effectiveness of a support vector machine is largely dependent on the kernel chosen. The kernel must be appropriate given the underlying structure of the input data. The chi-square kernel is a popular choice for a bag-of-features visual classifier, as is the basic linear kernel. [6]

3   Research Question

The de facto argument in favor of k-means clustering is "Why not?" The algorithm is simple. It can be constructed in a way that, while time-consuming, is not prohibitively time-consuming. Furthermore, building the codebook is a separate step from any other in the process. The size of the final codebook and the size of the training set are the only parameters that actually influence how long it takes to to perform k-means.

The conceptual advantage of k-means is that it "spreads out" the features in the codebook. Consider a codebook with a pair of features that are nearly identical. Including both of these features would not be discriminative and thus not useful. It may even be detrimental. K-means prevents such pairs of features from appearing in the codebook, as two very similar features will likely end up in the same cluster during the k-means process.

On the other hand, the k-means algorithm was not specifically designed with visual codebooks in mind. The algorithm can be used with any set of vectors, not just vectors that correspond to visual features. Furthermore, the resulting codebook after k-means clustering contains vectors that do not directly correspond to any of the original visual features. They are instead an "average" of many features. It may be possible that an exact feature vector found in a real image is a better codebook candidate than an "average" vector that never appeared in its exact form in the training data.

Even if an "average" vector is a poor candidate for a codebook feature, we might still build a codebook by selecting one representative feature from each k-means cluster. It would be logical to select the feature closest to the centroid of the cluster.

Then, there are three approaches. There is a baseline approach (a random selection of training features), a basic k-means approach, and a composite approach that uses the k-means clusters to choose a set of actual training features. We will compare the results of each in a basic image classifier.

4   Methodology

Our goal is to evaluate the performance of different types of codebooks in a basic image classifier. Therefore, we need to build a basic image classifier. Once constructed, we can test our classifier with various codebooks and compare them against one another. For testing we use the training data provided for use in VOC2011, the most recent PASCAL challenge, as well as the training data for VOC2007, a common test set used for benchmarking.

A basic classifier must contain several components. First, the classifier must collect SIFT features from every image using a dense sampling strategy. Then, it builds a codebook and matches each image feature to one or more codebook features. From these matches, a vector for each image is constructed, counting the occurrence of each codebook feature. At this step, spatial pyramid information is also incorporated. Finally, the resulting vectors are classified by a linear SVM.

Classifiers in the PASCAL Challenge are not required to run in real time, and in fact are given several weeks to calculate results. While we will not attempt to consider real-time constraints, our classifier must be efficient enough that numerous tests can be performed. Toward this end, some of the steps are implemented to run on a GPU. A GPU can compute much faster than a similar-cost array of CPUs, if the program can be written to accommodate the GPU's massive parallelization.

4.1   Dense Feature Sampling

Many open-source programs exist to extract SIFT data from an image. Provided that the software allows us to extract an arbitrary number of SIFT points from arbitrary locations and scales, it is straightforward to build a program that can employ any sampling strategy. Our implementation uses the open-source software SiftGPU [14].

4.2 Codebook Construction

Constructing a codebook generally consists of gathering a set of image features and performing k-means clustering on them to generate a smaller set of codebook features. Because k-means clustering is time-consuming and easily parallelizable, a GPU implementation is useful. Our implementation uses a GPU algorithm for k-means clustering, written in the CUDA programming language.

We also wish to evaluate the performance of a k-means codebook against a baseline. Our baseline codebook consists of a random sampling of features from the training images. This is straightforward to implement and does not require the GPU.

4.3 Codebook Matching

Matching image features to one or more codebook features is a straightforward problem that requires only some sort of distance measure. We use simple Euclidean distance in our implementation.

This step has the potential to be the most time-consuming step in the process. The classifier must compare every feature in the training and test data against every feature of the codebook. Thus, a parallelized implementation on the GPU is imperative. Our implementation uses a GPU algorithm again written in CUDA. The core operation, finding the distance between one feature and another, is also necessary for the k-means algorithm. Thus, the GPU algorithms for k-means clustering and codebook matching are similar.

4.4 Spatial Pyramid Matching

SPM is easily incorporated into our algorithm. In the dense sampling phase, we simply tag each feature with two tags, indicating which 2x2 and 1x3 segments to which it belongs. Then, in the codebook matching phase, we use those tags to group features according to each segmentation.

4.5 Support Vector Machine

The final step requires us to construct an SVM that can classify every image vector into different categories. Many SVM choices are possible. For simplicity and speed, we use a basic linear, hard-margin SVM for the tests on larger codebooks (*1000* to *16000*). Our implementation uses the SVM-Light [15] software with some modifications for integration purposes.

Tests on smaller codebooks (*31* to *500*) can be run more quickly. Therefore we use a chi-square kernel with error parameter optimization ($C=2^{-5}$ to 2^5), for a configuration that is more similar to the current state-of-the-art.

5   Testing

We can now compare the quality of each codebook construction strategy by observing the quality of the resulting image classifiers. If a codebook construction strategy is superior, it should produce superior classification accuracy.

As mentioned previously, all testing is performed on training data provided for the PASCAL VOC Challenge. Training data for the VOC is known to be difficult to classify. Any classifier that hopes to be successful must be robust with respect to all of the basic visual invariances: scale, position, rotation, affine transformation, illumination, occlusion, and background.

All test runs generate a complete set of results consistent with the VOC classification competition. Thus, the entire test set is classified according to all 20 object categories, and an average precision is found for each category. Our evaluation metric is the mean average precision across all 20 object categories. Because this mean represents the average of 20 different classification tasks, it is robust and statistically meaningful.

Another possibility that must be considered is that a codebook construction strategy may be more or less effective depending on the size of the codebook. Therefore, we test each strategy with codebooks of varying sizes in logarithmic increments of 2.

Our initial testing considers larger codebooks, ranging in size from *1000* to *16000* elements. Subsequent testing is also presented that considers smaller codebooks, ranging in size from *31* to *500* elements.

6   Results

Figure 1 shows the results of each of the three codebook construction strategies, using codebooks of varying size. Note that the smaller codebooks (*31* to *500*) show higher mean average precision only because the optimized chi-square kernel is used for those tests, while the linear kernel is used for the larger codebooks (*1000* to *16000*).

6.1   Larger Codebooks, Linear Kernel

We observe first that the pure k-means clustering method is strictly inferior to both other methods for larger codebooks, except for *size=2000*. In this case, it has a marginally better mean average precision than the codebook built with no k-means. The no k-means method and the representative k-means method are almost equal in all cases.

For larger codebooks, the inferior results from the basic k-means method appear to confirm the theory that codebook features need to consist of actual visual features found in real-life visual data. An "average" feature would weakly match many actual features, as opposed to an actual feature, which would strongly match a smaller number of actual features. Intuitively, the latter seems preferable, and these results indicate the same.

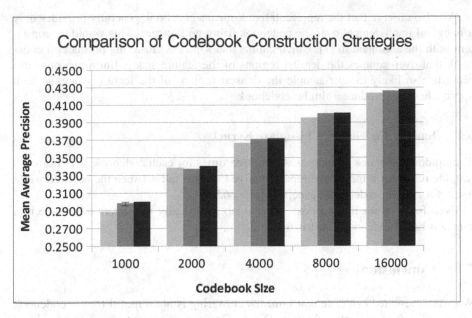

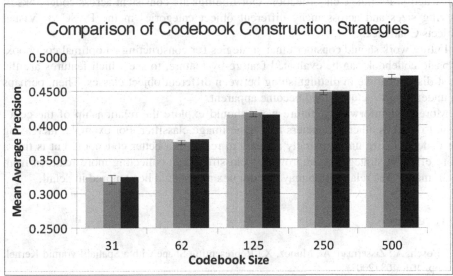

Fig. 1. Mean average precision for the three different codebook types, tested with codebooks of varying sizes. *Top*: larger codebooks, tested with a hard-margin linear kernel on the PASCAL2011 dataset. *Bottom*: smaller codebooks, tested with an optimized chi-square kernel on the PASCAL2007 dataset. For each size, from left to right: 1. Codebook consisting of k-means centroids, 2. Codebook consisting of a random sampling of training features, 3. Codebook consisting of the closest training feature to each k-means centroid.

We also observe that the representative k-means codebook generally provides only a marginal improvement over the codebook using no k-means. This would be consistent with the argument in [7], which claims that k-means clustering produces a codebook that over-samples the densest regions of the feature space. But a random sampling is also likely to over-sample the densest regions of the feature space, so both approaches may produce a similar codebook.

6.2   Smaller Codebooks, Chi-square Kernel

The random codebook strategy was run three times for each codebook size. For each test, the results of k-means and representative k-means fall between the best and worst result for random codebooks, except for *size=62*.

Even for very small codebooks, there is no statistically significant advantage to a k-means based codebook construction strategy.

7   Conclusion

We have presented evidence that k-means clustering is not a useful tool in codebook construction for image classification. Our findings are consistent across codebooks of varying sizes and across many different object categories in the PASCAL Visual Objects Challenge.

Future work should consider other strategies for constructing an optimal codebook. A basic codebook can be evaluated feature-by-feature, to see which features are the most discriminative in distinguishing between different object classes. Then, perhaps an underlying structure would become apparent.

More straightforwardly, future work should explore the relationship of the codebook to various other parameters in a basic image classifier. For example, increasing the codebook size unsurprisingly appears to result in a better codebook, but is this a more efficient strategy for improving a classifier than collecting more features from each image? The relationship may be complex and should be explored in detail.

References

1. Bosch, A., Zisserman, A., Munoz, X.: Representing Shape with a Spatial Pyramid Kernel, pp. 401–408 (2007)
2. Cortes, C., Vapnik, V.: Support-Vector Networks. Machine Learning 20, 273–297 (1995)
3. Csurka, G., Dance, C., Fan, L., et al.: Visual Categorization with Bags of Keypoints (2004)
4. Felzenszwalb, P., McAllester, D., Ramanan, D.: A Discriminatively Trained, Multiscale, Deformable Part Model. In: IEEE Conference on Computer Vision and Pattern Recognition, CVPR 2008, pp. 1–8 (2008)
5. Hou, J., Feng, Z.-S., Yang, Y., Qi, N.-M.: Towards a Universal and Limited Visual Vocabulary. In: Bebis, G., Boyle, R., Parvin, B., Koracin, D., Wang, S., Kyungnam, K., Benes, B., Moreland, K., Borst, C., DiVerdi, S., Yi-Jen, C., Ming, J., et al. (eds.) ISVC 2011, Part II. LNCS, vol. 6939, pp. 398–407. Springer, Heidelberg (2011)

6. Zhang, J., Marszalek, M., Lazebnik, S., et al.: Local Features and Kernels for Classification of Texture and Object Categories: A Comprehensive Study. In: Conference on Computer Vision and Pattern Recognition Workshop, CVPRW 2006, pp. 13–13 (2006)
7. Jurie, F., Triggs, B.: Creating Efficient Codebooks for Visual Recognition. In: Tenth IEEE International Conference on Computer Vision, ICCV 2005, vol. 1, pp. 604–610 (2005)
8. Lazebnik, S., Schmid, C., Ponce, J.: Beyond Bags of Features: Spatial Pyramid Matching for Recognizing Natural Scene Categories. In: 2006 IEEE Computer Society Conference on Computer Vision and Pattern Recognition, vol. 2, pp. 2169–2178 (2006)
9. Yang, L., Jin, R., Sukthankar, R., et al.: Discriminative Visual Codebook Generation with Classifier Training for Object Category Recognition. In: IEEE Conference on Computer Vision and Pattern Recognition, CVPR 2008, pp. 1–8 (2008)
10. Lowe, D.G.: Object Recognition from Local Scale-Invariant Features. In: The Proceedings of the Seventh IEEE International Conference on Computer Vision, vol. 2, pp. 1150–1157 (1999)
11. Moosmann, F., Triggs, B., Jurie, F.: Fast Discriminative Visual Codebooks using Randomized Clustering Forests. Advances in Neural Information Processing Systems 19 (2007)
12. Nister, D., Stewenius, H.: Scalable Recognition with a Vocabulary Tree. In: IEEE Computer Society Conference on Computer Vision and Pattern Recognition, vol. 2, pp. 2161–2168 (2006)
13. PASCAL VOC Homepage, http://pascallin.ecs.soton.ac.uk/challenges/VOC/
14. SiftGPU: A GPU Implementation of Scale Invariant Feature Transform (SIFT), http://cs.unc.edu/~ccwu/siftgpu/
15. SVM-Light: Support Vector Machine, http://svmlight.joachims.org/

Wide Field of View Kinect Undistortion for Social Navigation Implementation

Razali Tomari[1], Yoshinori Kobayashi[1,2], and Yoshinori Kuno[1]

[1] Graduate School of Science & Engineering, Saitama University
255 Shimo-Okubo, Sakura-Ku, Saitama 338-8570, Japan
[2] Japan Science Technology Agency, PRESTO
4-1-8 Honcho Kawaguchi, Saitama 332-0012, Japan
{mdrazali,yosiniro,kuno}@cv.ics.saitama-u.ac.jp

Abstract. In planning navigation schemes for social robots, distinguishing between humans and other obstacles is crucial for obtaining a safe and comfortable motion. A Kinect camera is capable of fulfilling such a task but unfortunately can only deliver a limited field of view (FOV). Recently a lens that is capable of improving the Kinect's FOV has become commercially available from Nyko. However, this lens causes a distortion in the RGB-D data, including the depth values. To address this issue, we propose a two-staged undistortion strategy. Initially, pixel locations in both RGB and depth images are corrected using an inverse radial distortion model. Next, the depth data is post-filtered using 3D point cloud analysis to diminish the noise as a result of the undistorting process and remove the ground/ceiling information. Finally, the depth values are rectified using a neural network filter based on laser-assisted training. Experimental results demonstrate the feasibility of the proposed approach for fixing distorted RGB-D data.

Keywords: Kinect, Fish Eye Lens, Undistortion, Neural Network.

1 Introduction

Social navigation planning can be described as a way for a robot to perform indirect social interaction with humans during navigation. Extending a conventional planner for such purposes requires a sensing system that is capable of not only perceiving obstacles located in the surrounding, but also of distinguishing between human and non-human obstacles. One versatile way to meet this requirement is to employ a vision system, since in comparison to other sensors a vast amount of data can be manipulated to offer hints for human detection and path generation. Utilizing vision for navigation has been extensively researched over the past decade, a comprehensive survey of which can be found in [1], covering land, aerial and underwater robot navigation in almost all possible situations.

In vision navigation, obstacle locations are usually obtained from a depth map, captured either from a stereo setup, range camera or structured light camera. Recently a commercial, low-cost RGB-D Kinect camera was launched by Microsoft. A camera

G. Bebis et al. (Eds.): ISVC 2012, Part II, LNCS 7432, pp. 526–535, 2012.

such as this can provide reliable depth readings in real-time under varying indoor lighting conditions. The practicality of this camera makes it a popular choice for implementation-based robotics, such as 3D SLAM [2], 3D reconstruction [3], and visual tracking [4]. Though this camera is useful, it has a small FOV and hence may constitute a bottleneck for navigation. Basically, the camera's point of view can be broadened by attaching a fish-eye lens system, but this comes at a price because the resulting image is highly distorted in a form of convex non-rectilinear appearance. This means that the straight lines in the image may appear curved and similarly square shapes may not resemble polygons. Such characteristics can be corrected by obtaining the lens projection error through a calibration procedure.

Over the past few decades, computer vision researchers have made substantial progress on improving distorted images via both structured and unstructured models. The former requires the pre-defined error model to be known in advance, while the latter is able to adapt the raw data directly without requiring prior knowledge. Some of the well-known structured model equations are based on the polynomial expansion [5] that is proven to correct both the radial and tangential distortion. According to [6], in a situation where an error can be modeled precisely as in [7], a structured model is reliable. However, when the error is complex and hard to model from a simple curve fitting, an unstructured model (e.g., one usually based on an artificial neural network (ANN) framework) is more appropriate. Such a system has been reported as comparatively effective at correcting a distorted image [8] and also robustly calibrated zoom-lens camera that contains noise in measurement [9]. L.N. Smith, et al. [10] combined ANN and a statistical regression approach and concluded that the combination can lead to a more practical and straightforward lens correction system. While previous studies demonstrate impressive findings, to our knowledge none of them explicitly concerned undistorting Kinect RGB-D data resulting from utilizing a fish eye lens for FOV amplification. However, a wide angle of view is clearly necessary for navigation, as accurate data is essential for the robot to determine the precise size, shape, and location of obstacles.

In this paper, we present a method for recovering the distorted Kinect RGB-D data utilizing a combination of structured and unstructured models. The former is used to rectify RGB-D point location using intrinsic camera attributes. On the other hand, the latter is employed to rectify the depth values through using a neural network architecture trained with laser ground truth data.

2 System Overview

A Kinect sensor consists of an IR projector, an IR camera, and an RGB camera that can reliably acquire RGB-D information in real time by utilizing its onboard processor. The sensor projects known infrared patterns from its surroundings and determines the distance information by deciphering the attributes of these patterns. However, this sensor has a small FOV (57 degrees for the IR camera and 63 degrees for the RGB camera), and hence cannot provide sufficient information to satisfy the navigation requirement. Recently, a lens that is capable of extending the Kinect's FOV has been made commercially available [11], as shown in **Fig.1** (a). Basically the lens in question consists of three fish-eye lenses for amplifying data from the cameras and the projector. **Fig.1** (b) and (c) demonstrate the difference between images taken at the same position (1.3 meters from the ground plane) using the Kinect camera with and without the lens attached. It illustrates how the lens is capable of perceiving valuable information in the proximity, such as the full-size robot on the right side, the monitor on the left side, and a wider view of the ground plane. Such information is extremely useful for the navigation planner to ensure that collisions can be avoided and a smooth trajectory can be generated.

(a)   (b)   (c)

Fig. 1. (a) Kinect sensor with Nyko-Zoom attached, (b) Kinect's images after FOV amplification using the lens, (c) Original images

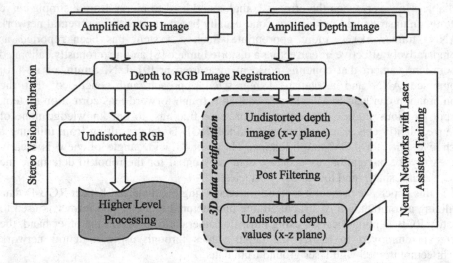

Fig. 2. Method pipeline for undistorting RGB-D data as a result of Kinect's FOV amplification

However, as can be seen in **Fig. 1** (b), the amplified images (RGB and depth) are significantly distorted. Objects located very close to the center lens are imaged accurately. However, the farther an object is located from the center point, the more distortion occurs. For accurate navigation, information about the precise dimensions and location of objects is crucial; therefore, we propose a method for recovering the distorted data. This method is outlined in **Fig. 2**. It begins by registering the depth image information into the RGB image. This process is necessary since the focal points of the cameras are located on different axes. Next, the undistortion procedure for the RGB and depth images is initiated. For the RGB image, the camera's intrinsic parameter is sufficient for correcting the distortion. However, for the depth image, 3D correction is needed to rectify the location and value of the pixels. We therefore utilize the camera to satisfy the former requirement and propose a neural network for fine-tuning the pixel values. In the following sections, we proceed to describe the particulars of each part of this process.

2.1   Depth to RGB Image Registration

To allow for reasoning about RGB pixel placement in 3D world coordinates, we make use of depth data supplied by the camera. Since the focal points of both cameras are located on different axes, we perform a calibration beforehand to tailor the depth image coordinates with the RGB image coordinates. More information about the calibration process can be found in [12]. In short, the process estimates the intrinsic parameters for both RGB and IR cameras, and determines the transformation matrix that maps pixels between both images using equation (1) where K is the intrinsic matrix and H is the extrinsic matrix.

$$X_{RGB} = K_{RGB}^{-1} HK_{Depth} X_{Depth},\qquad(1)$$

For determining the matrices, we capture a pair of 80 chessboard images from both infrared and RGB cameras with the fish-eye lens attached and perform a calibration procedure identical to the one during the stereo calibration step. The parameters are considered optimal when the calibration error is within one pixel. Using these parameters, the depth image points are mapped onto the RGB image, the outcome of which is given in **Fig. 3.** It demonstrates that the depth image dimensions after the calibration (**Fig. 3** (b)) become smaller than the original dimensions (**Fig. 3** (a)). Once both images are calibrated, we can easily determine the RGB pixel locations in the depth image and vice-versa using a linear model. Additionally, any transformation to the RGB image can also be applied to the depth image directly.

(a)   (b)

Fig. 3. Result of depth to RGB image registration: (a) Before calibration, (b) After calibration

2.1.1   RGB and Depth Images Undistortion

As seen in **Fig.3**, the fish-eye lens delivers a significant distortion of the information on both images. Correction can be achieved through a nonlinear inverse distortion model using intrinsic and distortion coefficients from both cameras. A common equation for performing such an operation is given below, where x and y are the corrected pixel locations, x_w and y_w are the original pixel locations in world coordinates, $r^2 = x_w^2 + y_w^2$, and K and P are a series of distortion coefficients for the radial and tangential respectively [8].

$$\begin{bmatrix} x \\ y \end{bmatrix} = \begin{bmatrix} 1 + K_1 r^2 + K_2 r^4 + K_3 r^6 \\ 1 + K_1 r^2 + K_2 r^4 + K_3 r^6 \end{bmatrix} \begin{bmatrix} x_w \\ y_w \end{bmatrix} + \underbrace{\begin{bmatrix} 2P_1 y_w + P_2 (r^2 + 2x_w{}^2) \\ P_1 (r^2 + 2y_w{}^2) + 2P_2 x_w \end{bmatrix}}_{\text{Tangential distortion}}$$ (2)

$$\underbrace{\qquad\qquad\qquad}_{\text{Radial distortion}}$$

For the RGB image, equation (2) can be applied directly and the resulting image can be viewed in **Fig. 4** (a). It indicates that in the corrected image, the black areas in the corners have been eliminated and more natural object shapes have been produced.

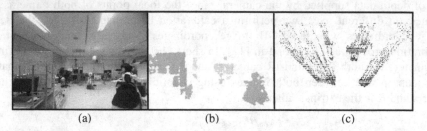

(a)   (b)   (c)

Fig. 4. (a) Undistorted RGB image, (b) Post-filtered and undistorted depth image, (c) Plan view plane (x-z) of the depth image in real world coordinates

As for the depth image, before applying the correction model, a conversion to 8 bits/pixel format is needed since the original format is 11 bits/pixel. This transformation is achieved through adaptive normalization strategies using equation (3). This equation normalizes the image, while retaining depth information on distant points farther than 4 meters away. Once normalized, the floor and ceiling points are removed by eliminating any pixels on the y-axis that are shorter than 20 cm or higher than 180 cm. Next, the undistortion model is applied, and the resulting image can be viewed in **Fig. 4** (b). When the depth values of the image are projected in real world coordinates, as in **Fig. 4** (c), we can see that the plan view map is highly contaminated with noise. This situation occurs because unknown point values are estimated from the surrounding pixels during the undistortion procedure. Such interpolation results in errors and deterioration when the neighboring values are obviously different. In the next section we describe a method to abolish such noise, and we investigate the quality of the projected depth values.

$$G(x, y) = \begin{cases} (d_{x,y} - 490)/510 \, x200 & \text{if } 490 \leq d_{x,y} \leq 1000 \\ 200 + (d_{x,y} - 1000) & \text{if } 1000 < d_{x,y} < 1050 \\ 255 & \text{elsewhere} \end{cases}$$ (3)

2.1.2   Post-Filtering and Depth Values Rectification

To diminish the noise, we further analyzed the plane 3D point cloud. As shown in **Fig. 5.** (a), the noise was scattered randomly behind each object with low density. Since 3D filtering is time-consuming, we chose a 2D approach. For each point location, if, when viewed from the x-z plane, the y-axis density is lower than the predefined threshold value, we can assume the location belongs to noise and remove the whole point on that axis. Comparison between **Figs. 5** (b) and **4** (c) indicates that through this method we managed to remove the noise while retaining the object shapes. The empty space for each object can be wiped out through a morphological operation, and the results can be seen in **Fig. 5**. (c)

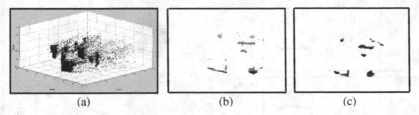

(a)   (b)   (c)

Fig. 5. (a) 3D view of projected depth values in world coordinates; plan view image after post processing (b), and morphological operation (c)

From the process above, the location of the points in both images (RGB (**Fig. 6** (a)) and depth (**Fig. 6** (b))) are rectified, but the depth values remain incorrect. If these values are projected into world coordinates, the obtained distance is incorrect by a significant margin. As shown in **Fig. 6** (c), the distance information is noticeably different from the laser-scanning values (ground truth). The effect is significantly more accentuated when the FOV angles become farther from the center (i.e., 90°).

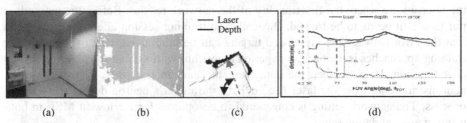

(a)   (b)   (c)   (d)

Fig. 6. Distortion correction of (a) RGB image, and (b) depth image. (c) Projected depth values in world coordinates overlaid with a laser scan. (d) Error in distance, measured between the camera's projection and the laser data.

To overcome such problems, a nonlinear filter is introduced. The filter is constructed by dynamically reducing distance errors between the camera and the laser in every FOV angle. Since the camera only contains valid information in the FOV between 50° - 140°, we mainly used the laser information captured within that range. **Fig. 6** (d) illustrates an example of information attained during the sampling process.

The samples were mainly taken from wall surfaces since we wanted both sensors to perceive information from the obstacles. The samples were taken from several distances ranging from 1m to 5m, in order to portray how the errors evolved broadly. **Fig. 7** (a) to (c) displays some of the readings taken from the positions of 1m, 3m and 5m respectively, while the overall error patterns consisting of 3330 measurements can be seen in **Fig. 7** (d). We can observe from these figures that, as the distance increased the error rate of each FOV angle underwent a change from a positive rate to a negative rate in a nonlinear fashion. Since the error form is relatively complex, we choose to use unstructured methods based on an ANN framework as the basis for constructing the filter prototype.

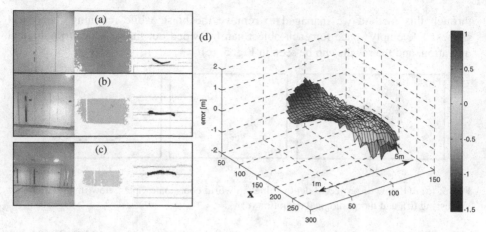

Fig. 7. Sample set measurements from 1m (a), 3m (b), and 5m(c) respectively. (d) Error pattern between camera points and laser points from varying distances in the range of 1m to 5m.

ANN is a mathematical approximation of a biological brain and has been identified as a useful framework for precise modeling of nonlinear responses. It comprises a number of neurons connected together to form a network. The weights linked between the neurons are where the functionality of the network resides. Before the network is put to use, it needs to be trained. Basically the training session alters the weights so that the error between the inputs and targets can be minimized. One of the common training approaches is the backpropagation algorithm with a mean square error (MSE) cost function. Here we feed the data from the camera (i.e., θ_{Fov}, d_d]) to the input neurons and data from the laser (i.e., d_L]) to the target's neuron during the training process. The network setting is considered to be optimal for the lowest MSE in both training and validation sets.

3   Experimental Results

In this section, we evaluate the performance of the proposed RGB-D data correction in real environments. The system was tested in both the lab and in hallway situations. We began by assessing the ANN performance.

Using a set of 3330 samples illustrated in **Fig. 7(d)** for training data set (TDs), and a further 2141 samples for validation data set (VDs), the optimal ANN structure was configured. We tuned it across a number of hidden nodes (HN) selection, with two different activation functions, specifically the sigmoid(S) and tangent sigmoid (T). **Table 1** summarizes our results. It can be seen that the lowest obtained MSE in VDs was 120×10^{-5}, and that it was achieved with an ANN configuration of seven (2L-7S-S) and ten (2L-10S-S) hidden nodes when using the sigmoid activation function, and eight hidden nodes when using the tangent sigmoid activation function (2L-8T-T). When cross-checked between the TDs, the setting of 2L-7S-S gave the lowest MSE (9.8×10^{-5}), and hence it was selected as the optimal network. The residual error of the TDs and VDs, before and after filtering using this network, is shown in **Fig. 8** (*left* and *right* respectively). We can observe from these figures that the network is capable of reducing the variation of readings between the camera and the laser from ±1.5m to +0.2 m. Within the reduced range, 86% of the data is lower than ±0.1m, while the rest is scattered randomly from ±0.1m to ±0.2m. Mainly the error becomes significant at FOV angles between 50° to 55° and 135° to 140° when inferring information from the range of 4m to 5m. Such precision is sufficient for our needs, since we will be applying the system to our autonomous wheelchair system.

Table 1. Kinect's intrinsic and extrinsic parameters with the fish-eye lens attached

Sigmoid Activation function (MSE x10⁻⁵)										
HN	2	3	4	5	6	7	8	9	10	15	20
TDs	22.73	11.84	11.34	10.47	11.41	**9.80**	9.06	7.64	**9.81**	7.07	6.56
VDs	140	130	130	130	130	**120**	130	130	**120**	150	190
Tangent Sigmoid Activation function (MSE x10⁻⁵)										
HN	2	3	4	5	6	7	8	9	10	15	20
TDs	42.54	13.32	11.96	10.39	10.58	10.13	**9.35**	8.72	8.71	7.72	6.88
VDs	160	130	130	130	130	130	**120**	130	130	130	130

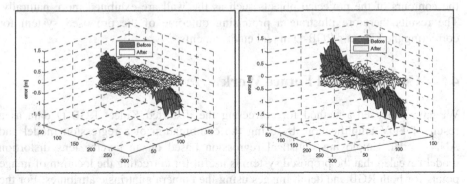

Fig. 8. Residual errors read between laser values and camera values of the training (*left*) and validation set (*right*) before and after the depth values rectification process

We evaluated the performance of the proposed method for rectifying projected depth values in both lab and hallway environments. For the lab setting, the camera was set in a static position while a person moved randomly in the scene. As for the hallway setting, the camera was attached to a wheelchair that navigated the environment at a constant speed. **Fig. 9** shows a sample selection of images captured during the process. The images on the left depict the situation before ANN correction, while those on the right provide the after-effect outcome. As can be seen, before applying the ANN, the measured distances to objects both near and far are significantly different, and the projected wall shapes in the hallway resemble a curve.

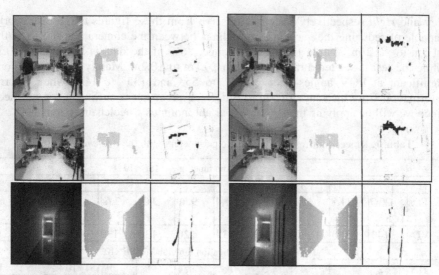

Fig. 9. Examples of depth projection output in lab and hallway environments before (*left*) and after (*right*) depth value rectification using ANN

When the ANN filter is initiated, more accurate distance information is acquired, and the contours of the projected objects such as the wall are exhibited more naturally. The results therefore illustrate a promising outcome of the proposed system for correcting the Kinect's RGB-D in real environments.

4 Conclusion and Future Work

We have proposed a method for correcting the distorted Kinect RGB-D data as a result of FOV amplification, by using the combination of a regression model and ANN model. Employing statistical regression based on an inverse lens distortion model revealed that the proposed system is useful for correcting the location of image points for both RGB and depth images using the camera's intrinsic attributes. For the depth image, apart from correcting point locations, point values also needed to be rectified to suit our navigation requirements. We have shown that the ANN frameworks can effectively correct the depth image values when supplied with sufficient global error patterns. The trained network is optimal in terms of low MSE for both training and validation sets (9.8×10^{-5} and 120×10^{-5} respectively) and is acceptably complex when using seven hidden nodes with sigmoid activation. We have tested our proposed system in both lab and hallway environments, and the results are encouraging. From our preliminary findings, we ascertained that the wide FOV Kinect can be reliably employed in navigation implementation when appropriate distortion correction is applied to the amplified data.

Acknowledgment. This work was supported in part by JST PRESTO, A-Step and KAKENHI (22243037, 24700157).

References

1. Bonin, F., Ortiz, A., Oliver, B.: Visual Navigation for Mobile Robots: A Survey. Journal of Intelligent and Robotic Systems 53(3), 263–296 (2008)
2. Engelhard, N., Endres, F., Hess, J., Sturm, J., Burgard, W.: Real-time 3D Visual SLAM with a Hand-Held RGB-D Camera. In: Proc. of the RGB-D Workshop on 3D Perception in Robotics at the European Robotics Forum (2011)
3. Tran, J.: Low-Cost 3D Scene Reconstruction for Response Robots in Real Time. In: Proc. of IEEE Intl. Symp. on Safety, Security, and Rescue Robotics, pp. 161–166 (2011)
4. Tomari, R., Kobayashi, Y., Kuno, Y.: Multi-view Head Detection and Tracking with Long Range Capability for Social Navigation Planning. In: Bebis, G., Boyle, R., Parvin, B., Koracin, D., Wang, S., Kyungnam, K., Benes, B., Moreland, K., Borst, C., DiVerdi, S., Yi-Jen, C., Ming, J. (eds.) ISVC 2011, Part II. LNCS, vol. 6939, pp. 418–427. Springer, Heidelberg (2011)
5. Brown, D.: Decentering Distortion of Lenses. Photogrammetric Eng. 7, 444–462 (1966)
6. Chang, C., Su, C.: A Comparison of a Statistical Regression and Neural Network Methods in Modeling Measurement Errors for Computer Vision Inspection Systems. Computers & Industrial Engineering 28(3), 593–603 (1995)
7. Shah, S., Aggarwal, J.K.: Mobile Robot Navigation and Scene Modeling Using Stereo Fish-Eye Lens System. Machine Vision and Application, 159–173 (1996)
8. de Villers, J., Nicolls, F.: Application of Neural Networks to Inverse Lens Distortion Modeling. In: Proc. of 21st Annual Symposium of the Pattern Recognition Society of South Africa, vol. 1, pp. 63–68 (2010)
9. Ahmed, M., Hemayed, E., Farag, A.: Neurocalibration: A Neural Network that Can Tell Camera Calibration Parameters. IEEE Trans. PAMI 79, 384–390 (1999)
10. Smith, L.N., Smith, M.L.: Automatic Machine Vision Calibration Using Statistical and Neural Network Methods. Image and Vision Computing 23, 887–899 (2005)
11. http://www.nyko.com
12. http://www.ros.org/wiki/kinect_calibration/technical

Automatic Human Body Parts Detection in a 2D Anthropometric System

Tomáš Kohlschütter and Pavel Herout

Department of Computer Science and Engineering, University of West Bohemia, Plzen, Czech Republic

Abstract. The paper describes the methodology of computer-based measurement for taking anthropometric dimensions. We focus on a 2D anthropometry system which is being designed for the purposes of a clothing company. The system consists of computer software, a digital camera and a background board with calibration dots. Only a few steps are needed to calibrate such a system and prepare it for the measurements. The measured person is captured, his silhouette is extracted and body dimensions are computed. Our research is intended to enrich existing techniques for automatic detection of anatomical landmarks (and body parts such as waist, chest, etc.) on the extracted silhouette. This step is very important to create a complete system without any need of user interaction and it must correspond to the body parts definitions given by clothing standards and tailors. Several potential sources of problems are discussed and some possible solutions are proposed.

1   Introduction

This Ph.D. work concerns how to fulfill the requirements of an important clothing manufacturer whose task is to measure hundreds of people in their everyday clothes to design and create made-to-measure clothing for them. The traditional method of taking human dimensions using a measuring tape can be very time-consuming. Thus, there is a need to create a portable contactless computer-controlled workplace to speed up this measurement process.

In the past few years, there has been active research on replacing traditional anthropometry measurements using a measuring tape with computer-based 2D ([6], [8], [10]) or 3D techniques [9]. Based on literature and our basic experiments, we found that the creation of an image-based 2D anthropometric measurement system will be the best way. Performance of such measurement was already discussed in [8], but we cannot use it directly. Requirements given by the project submitter were that the system must be portable, people cannot be required to get undressed, and the system must be very easy and fully automatic. The objective of this paper is to summarize everything needed for designing a complete, fully automatic 2D anthropometric system and propose a new technique for automatic detection of spatial human body parts that would be useful in the clothing industry.

G. Bebis et al. (Eds.): ISVC 2012, Part II, LNCS 7432, pp. 536–544, 2012.

2   Measuring System

The person to be measured stands in front of a background board with the arms away from the body. A camera is positioned in front of the person, three meters from the board. The frontal image is captured and the person is instructed to turn left to capture the lateral image. This allows modeling spatial dimensions such as waist or chest circumference. The technique from other papers use two different cameras placed in front of and on the side of the measured object. We use one camera only because it is cheaper and satisfies the requirement of portability better – fewer devices, only one background and only one illumination.

The selection of the proper camera is a subject for discussion. It must be relatively cheap and provide a good quality image. The resolution of the camera should not be low but it is not required to be very high. 1280×960 should provide good accuracy (this was verified by experiments) and comfortable performance of image processing. The main requirements are the ability to focus at the distance of three meters, good color contrast, the ability to deal with different conditions of illumination (i.e. setting of exposure value, shutter, etc.) and low noise in the captured image. We tested several web cameras and their main drawback is their inability to focus at the requested distance, regardless of the price of the web camera. The basic experiments can be done using a digital camera which satisfies all the above conditions and is easily available to common users. This allows the end software to be designed for home usage, for example, in web shop applications.

The green backboard covers the whole area behind the object and its dimension is 150 cm × 220 cm. Nine red calibration dots are placed on the backboard and five dots are placed on the floor. This allows automatic camera calibration and fixing camera misplacement. Footprints are printed on the floor to establish the position where the person to be measured should stand.

Several difficulties can arise during measurement. It is very important to deal with them, because they can influence the results' accuracy a lot. The most important potential sources of these problems are:

- camera calibration
- object extraction and landmarks detection
- body part modeling
- person's posture
- person's clothing

Since the whole system should be portable, we did not install the experimental workplace in a special room but in a corridor where the camera looks through a doorway. This nicely simulates difficult circumstances for measuring in real situations and allows testing with several illumination conditions, as light can be directed from several directions.

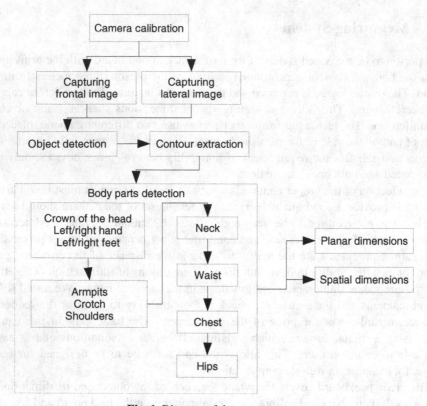

Fig. 1. Diagram of the measurement

3   Camera Calibration

The camera's calibration is one of the most important parts of the whole measurement process. The results' accuracy depends mainly on calibration correctness.

The camera is calibrated applying the Tsai non-coplanar calibration method. Fourteen calibration dots are automatically detected by the Hough transformation [12]. The calibration process detects internal and external camera parameters which allow converting image coordinates to world coordinates. Secondly, it provides the possibility to fix lens distortion. Some calibration error can arise from the calibration process.

For proper measurements, the floor (or a base point) must be identified in the scene. Since the base position is exactly specified (between the footprints), the point image coordinates can be automatically recovered from the known world coordinates.

Calibration becomes invalid when the camera setup is changed. This is not optimal for automatic measurements with a portable solution; thus, there is still room for improvements. It would be useful to perform a correction when the camera is moved during measurement. This could theoretically be easily possible, as calibration dots are still visible during measurement; thus camera misplacement can be easily detected. On the other hand, red dot detection can be more difficult at this moment, because it can interfere with the measured object.

4   Object Detection

The captured image is converted into HSV (Hue-Saturation-Value) color space, which provides a much better view of individual color than RGB (Red-Green-Blue) space. Because the R, G, and B components of an object's color in a digital image are all correlated with the amount of light hitting the object, and therefore with each other, image descriptions in terms of those components make object discrimination difficult. Especially the hue component allows us to detect the particular color regardless of the amount of incident light. A threshold is set and the object is extracted from the background. This technique still needs some improvement, because it is very difficult to extract the areas that blend with the background, e.g. due to bad lightning conditions.

Fig. 2. Object extraction and detection of its contours

Then contours are retrieved from the image using the algorithm [11] and the largest contour is selected to be the person's contour. The rest is rejected. To make the contour as smooth as possible, the polygon approximation is performed. We currently ignore the area behind the knees because we are not interested in any measurement there.

5   Automatic Body Parts Localization

Automatic landmarks and body part localization is the main topic towards which this work should be directed in future research. The current literature talks mainly about manual landmarks localization (e.g. in [2]) and complete automatic detection seems to be done in 3D body scanners [9]. The concept of the automatic landmarks detection in 2D images has been proposed in [5] but this method is very limited and it does not reflect the real position of particular body parts. It is able to detect only basic landmarks such as the crown of the head, armpits or crotch. However, it could serve as the base for creating a new technique.

Another method of automatically extracting human body features has been proposed in [7]. It is based on representing the body contour using Freeman's chain code and feature points are identified by evaluating the difference between the coding sequences. It states that the recognition rate was 100%. However, it does not seem to work correctly when the measured object is dressed, as the clothes can cause unexpected changes in the contour direction. Therefore, a better method needs to be developed.

Our proposal tries to localize body parts so they correspond with the definitions stated in the standards for body and garment dimensions [1] and given by an experienced tailor as much as possible. We are required to follow them to satisfy the project submitter. Whole idea lies in dividing the human body into several segments, specifying a body part level (or an area) and localizing the exact body part emplacement within this area.

The algorithm works in several phases. The first two phases detect the basic points as defined in [5]. This is based on finding the global extremum points followed by finding the local extremum points. The first phase localizes the crown of the head as the contour point with a minimum vertical coordinate, both hands as the contour points with minimum and maximum horizontal coordinates, and both feet as the contour points with maximum vertical coordinates on each side of the body horizontally divided by a center line. The second phase localizes both armpits and the crotch as the contour points with minimum vertical coordinates between the corresponding hands and feet. Both shoulders are located as the contour points with the same horizontal coordinate as the armpits and minimum vertical coordinate. Since it is possible that more points with the same minimum/maximum coordinate can exist, the original algorithm had to be slightly modified to maintain body symmetry.

The following phase iterates over the body height and tries to localize spatial body parts. To achieve this, the human body height is divided into one hundred equal parts, as was introduced by Kollmann at the beginning of the twentieth century [4]. His decimal standard divides the body height into ten segments and each of these is subdivided into ten subunits. Certain body levels can be estimated then. Currently, it is important to correctly arrange the area (level) of the neck, chest, waist and hips. The other body parts are the subject for future research. From the experiments, we deduced the following: Although Kollmann's standard specifies the head height as 13 subunits, this is not always the truth and it is better to begin the neck area at the 10th subunit. The lower boundary can be defined by shoulder points detected in the 2nd phase. The chest will be searched from five subunits above the armpits and one subunit below the armpits. The waist area is distributed from the 40th to 50th subunit and the hip area is localized between the detected waist and the crotch. The back length can be identified using some of these landmarks.

5.1 Circumferences Detection

The spatial body parts are modeled as an ellipse where its major axis is taken from the frontal image and its minor axis is taken from the lateral image. The paper [6] models some body parts, e.g. neck, as an ellipse and some parts, e.g. chest, as a combination of a rectangle and an ellipse. However, in our experiments, the chest approximation with the ellipse gives better results. Hence, the circumferences are currently represented as

$$C = \pi \cdot \left[\frac{3}{2} \cdot (a + b) - \sqrt{a \cdot b} \right]$$

The *neck girth* is defined as the circumference of the neck, taken over the cervicale at the back and the top of the collarbone at the front. However, there can be a problem with the detection of these landmarks, especially when the person is wearing a shirt with a collar. Therefore, after a consultation with the tailor, we decided that we can specify the neck girth as the place with the minimum circumference around the neck at the neck level, measured horizontally in the front and diagonally on the side. The experiments show that this works as expected when the lateral neck height is set to three subunits with the ellipse major axis in the middle. We can specify the neck circumference as

$$N = \min C \ in \ (S_{10}, shoulders)$$

where S_i denotes i-th subunit according to Kollmann's standard.

The *waist girth* is defined as the minimum horizontal circumference around the body at waist level. This definition corresponds with the definition given by the tailor. However, one can see that the waist is not parallel to the floor in the lateral image. Thus we can specify the waist circumference as

$$W = \min C \ in \ (S_{40}, S_{50}) \ .$$

When the waist is identified, we can move to identification of the chest. The *chest circumference* is defined as the maximum horizontal girth at chest level measured under the armpits, over the shoulder blades, and across the nipples with the subject breathing normally; parallel to the floor [9]. First, we approximate the body trunk with two bisectors (left and right side of the trunk as shown in Figure 3) starting at the waist contour points and passing through the armpit points. This ensures that the chest is bounded above the armpits where the ellipse major axis is taken as the horizontal distance between these two lines. Then we specify its circumference as

$$Ch = \max C \ in \ (armpits - 5 \cdot S, armpits + S) \ .$$

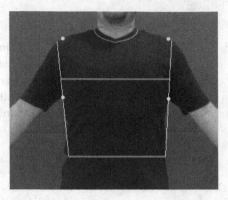

Fig. 3. Approximation of body trunk for the chest detection

The *hip girth* is defined as the maximum horizontal circumference around the body at hip height. This definition corresponds with the definition given by the tailor and it is always parallel to the floor. Thus we can specify the hip circumference as

$$H = \min C \; in \; (W, crotch) \; .$$

6   Experiments

The effectiveness of the proposed method was tested on eight males and two females. Four of the males were tested repeatedly at different times, with different illumination and in different clothes. The laboratory software has been implemented in C++ with the help of the OpenCV image processing library. The purpose of the experiments was not only to compare traditional tape measurements with camera-based measurements, because this was already verified by previous work [8]. They were mainly focused on detection of body parts on dressed persons. The accuracy of the experiments was verified by the experienced tailor. Detection of the neck was not fully accurate in female cases. Although the horizontal position was detected correctly, the interfering hair made the detected part larger. Another two inaccurate cases were caused by very loose clothes which put the waist a little bit higher. These problems were solved by instructing the persons how to behave (i.e. tucking in the shirt or wearing the hair up) during the measurements. Then the recognition rate was fully sufficient and acceptable according to the tailor.

The results of the experiments indicate that we were successful in detection of neck, chest, waist and hips. The only two conditions are the color difference between background and clothes colors and wearing clothes which do not cover or optically increase a person's shape (e.g. thick coat, etc.). Upcoming research will focus on more experiments on females including the proper detection of the under-chest circumference, which is another dimension required in the clothing industry.

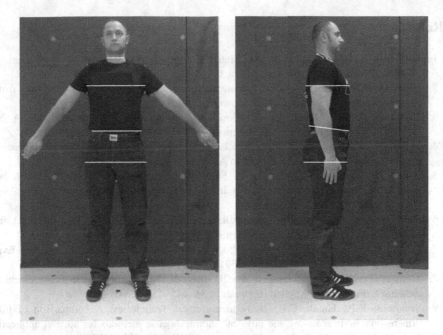

Fig. 4. Successful automatic detection of neck, chest, waist and hips

7   Conclusion

This paper described the methodology of anthropometric measurement by computer. We focused on summarization of everything needed for designing an automatic 2D anthropometric system. We went through camera calibration, separation of the object from the background and simple body parts modeling. A new method for automatic detection of neck, chest, waist and hips circumferences has been proposed. It allows body part localization according to the standards of the clothing industry. Ten persons were tested to verify the effectiveness of the proposed method. All requested body parts were successfully detected when the person was dressed in fitted clothes, i.e. clothes which are not too loose, and hair was not hanging down along the body. The same technique can be used for other body parts, e.g. thigh circumference, etc.

Several problems appeared during system design and we must cope with them in future research. It will be focused mainly on more anatomical landmarks localization, better modeling of body parts and finding a possible solution when a person wears unsuitable clothes. We would like to improve the accuracy of the measurement as much as possible and complete the fully automatic and portable system for clothes sizing.

Acknowledgment. The work was supported by UWB grant SGS-2010-028 Advanced Computer and Information Systems.

References

1. American Standards for Testing and Materials (ASTM), Standard terminology relating to body dimensions for apparel sizing (Vol. 07-02, Designation: D5219-09) (2009)
2. BenAbdelkader, C., Yacoob, Y.: Statistical Estimation of Human Anthropometry from a Single Uncalibrated Image. Computational Forensics (2008)
3. Criminisi, A., Zisserman, A., van Gool, L.J., Bramble, S.K., Compton, D.: A New Approach to Obtain Height Measurements from Video (1998)
4. Drillis, R., Contini, R., Bluestein, M.: Body segment parameters: A survey of measurement techniques. Selected Articles from Artificial Limbs, 329–351 (1964)
5. Hilton, A., Beresford, D., Gentils, T., Smith, R., Sun, W., Illingworth, J.: Whole-body modelling of people from multiview images to populate virtual worlds. The Visual Compute 16(7), 411–436 (2000)
6. Hung, P.C.-Y., Witana, C.P., Goonetilleke, R.S.: Anthropometric Measurements from Photographic Images. Computing Systems 29, 764–769 (2004)
7. Lin, Y.L., Wang, M.J.: Automated body feature extraction from 2D images. Expert Systems with Applications 38, 2585–2591 (2011)
8. Meunier, P., Yin, S.: Performance of a 2D image-based anthropometric measurement and clothing sizing system. Applied Ergonomics 31, 445–451 (2000)
9. Simmons, K.P., Istook, C.L.: Body measurement techniques: A comparison of three-dimensional body scanning and physical anthropometric methods for apparel application. Journal of Fashion Marketing and Management 7(3), 306–332 (2003)
10. Stancic, I., Supuk, T., Cecic, M.: Computer vision system for human anthropometric parameters estimation. WSEAS Transactions on Systems 8, 430–439 (2009)
11. Suzuki, S., Abe, K.: Topological Structural Analysis of Digitized Binary Images by Border Following. CVGIP 30(1), 32–46 (1985)
12. Yuen, H.K., Princen, J., Illingworth, J., Kittler, J.: Comparative study of Hough transform methods for circle finding. Image Vision Comput. 8(1), 71–77 (1990)

Implementation and Analysis of JPEG2000 System on a Chip

John M. McNichols, Eric J. Balster, William F. Turri, and Kerry L. Hill

Dept. of Electrical and Computer Engineering, University of Dayton & The Air Force Research Laboratory

Abstract. This paper presents a novel implementation of the JPEG2000 standard as a system on a chip (SoC). While most of the research in this field centers on acceleration of the EBCOT Tier I encoder, this work focuses on an embedded solution for EBCOT Tier II. Specifically, this paper proposes using an embedded softcore processor to perform Tier II processing as the back end of an encoding pipeline. The Altera NIOS II processor is chosen for the implementation and is coupled with existing embedded processing modules to realize a fully embedded JPEG2000 encoder. The design is synthesized on a Stratix IV FPGA and is shown to out perform other comparable SoC implementations by 39% in computation time.

1   Introduction

One of the most recent image compression schemes, JPEG2000, offers a wide range of features and flexibility over the existing JPEG standard [10]. A block diagram of the JPEG2000 encoder is shown in Fig. 1. The encoder consists of two main parts: the discrete wavelet transform (DWT) and the embedded block coding with optimal trunctation (EBCOT) coder. The wavelet transform takes an image in the spatial domain and transforms it to the wavelet domain. The wavelet domain consists of a frequency representation with the addition of spatial information as well. Once the wavelet transform is completed, the coefficients are scalar quantized if lossy compression is chosen. The quantized wavelet coefficients are then entropy encoded using EBCOT, a two-tier coding algorithm which first divides each wavelet subband into code blocks (either 32x32 or 64x64). EBCOT is composed of Tier I and Tier II encoders. Tier I produces independent embedded bitstreams for each code block using a context-based arithemtic encoder (MQ coder), the context for which is generated by the bit-plane coder. Tier II then

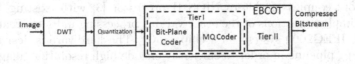

Fig. 1. Block Diagram of JPEG2000 Encoder

G. Bebis et al. (Eds.): ISVC 2012, Part II, LNCS 7432, pp. 545–556, 2012.

reorders the individual compressed bitstreams and applies rate-distortion slope optimization to form the final JPEG2000 bitstream.

While JPEG2000 offers a number of improvements and additional features over JPEG and other image encoding standards, these benefits come with much greater computational cost. JPEG2000 is approximately 4 times more computationally expensive that the original JPEG [14]. Due to these high costs, it becomes impractical to utilize JPEG2000 in applications which require real-time processing of high resolution images, such as wide area imagery or medical imagery. To solve this problem, developers continue to turn to embedded solutions to yield the throughput necessary to meet frame rates for high resolution imagery [15]. Embedded solutions leverage the inherent parallelism of the EBCOT block coders to achieve larger increases in throughput over typical software implementations. Not only do embedded hardware solutions offer dramatic increases in throughput over their software counterparts, but they also free host processors to handle other critical tasks.

Most embedded JPEG2000 solutions focus on performance increases in the EBCOT Tier I, either through novel architectures or simply by leveraging the parallelism of multiple block coders. Research focuses on EBCOT Tier I improvement because this is the most computationally expensive module of JPEG2000, as shown in [2,7].

In [8,11], architectures for the MQ coder are proposed which consume two context-data (CxD) pairs per clock cycle. [2] takes a different approach, increasing performance by using column-based operation combined with pixel and group-of-column skipping techniques. A number of implementations focus on very large scale integration (VLSI) architectures for JPEG2000. Some, such as [15], focus on high-speed VLSI implementations by utilizing pass-parallel EBCOT implementations. Others, such as [12], attempt to reduce the on-chip memory requirements for EBCOT Tier I and Tier II while also improving performance.

However, most of these implementations fail to mention the final piece of JPEG2000 which is the formation of the full bitstream, EBCOT Tier II. Generally it is not mentioned at all and assumed to be left for the host processor to handle in software. [12,13] propose an architecture for EBCOT Tier II which is focused on reducing memory requirements for bit-stream buffering but do not offer high performance. Instead, we propose the use of a softcore co-processor to serve as a Tier II processing module situated at the back end of an encoding pipeline to realize a fully embedded encoder. The softcore co-processor offers more flexibility than [12,13], due to the soft nature of the processor, while also offering adequate performance to meet the demands of high-resolution image compression.

This work couples an Altera NIOS II processor [3] with existing, efficient, hardware implementations of the other various processing units to create a fully embedded JPEG2000 system on a chip (SoC). The hardware is designed as a tile encoding pipeline in order to efficiently encode high resolution imagery. The NIOS II processor interfaces with a FIFO containing the independent code block bitstreams produced by a variable number of hardware block coders in order to create the final bitstream. The NIOS II processor handles all of the Tier II processing as well as transfering data back to the host processor.

While similar to [7] in the use of an Altera NIOS II processor, the proposed implementation utilizes the processor as a separate processing module as opposed to a system scheduler/device arbiter. This avoids the scheduling overhead associated with such an implementation while also preventing the other processing modules from becoming limited by the throughput of the NIOS II system. Additionally, the simplicity of using the NIOS II as a separate processing unit yields a system which is much easier to debug and test, since difficulty debugging in [7] prevented the system from actually being implemented on an FPGA.

The rest of this paper is organized as follows. Section 2 gives a brief overview of the FPGA-based processing architectures used for the front end of the processing pipeline. Section 3 details the selected target platform and the selection of the coprocessor before giving a detailed description of the implementation of the SoC. Section 4 analyzes the performance of the implemented system and discusses the impact of increasing numbers of parallel block coders before comparing the results to other SoC implementations. Finally, Section 5 concludes the paper.

2   JPEG2000 Hardware Modules

2.1   Discrete Wavelet Transform/Quantization

The proposed system implementation uses a lossy CDF 9/7 wavelet transform [1]. The lossy wavelet transform is chosen as it offers additional compression gain over the lossless implementation while still maintaining comparable image quality at lower compression ratios. This implementation is an integer-based approach, utilizing the CDF 9/7 wavelet filter to transform integer input pixels into scaled fixed-point wavelet coefficients. These scaled fixed-point coefficients are then quatized back into integers prior to compression by EBCOT. This DWT module is capable of running at 100 MHz and takes approximately 7 ns to produce the wavelet coefficient for one pixel.

2.2   EBCOT Tier I

EBCOT Tier I is comprised of two main processing modules: the bit plane coder (BPC) and the MQ coder [10]. The BPC for the proposed implementation is a generic implementation, conforming to the standard, and operates at a clock speed of 100 MHz. While most implementations of the BPC aim to maximize throughput, this design instead focuses on reducing resource utilization and does not make use of any of the optimization techniques proposed in [7,8,11,15]. Instead, minimal resource usage is achieved by consolidating the number of memory devices necessary to store code block state data. The MQ coder follows the same design principle as the BPC, with a focus on minimizing hardware resource usage. The MQ coder runs at a clock rate of 200 MHz. The design goal of both the BPC and MQ coder is to maximize the number of Tier I coders which can fit on a device. By achieving high clock rates through resource optimized designs, a Stratix IV device with over 90% resource utilization is capable of yielding throughput in excess of 180 MBytes per second when multiple parallel Tier I coders are coupled with three DWT targets.

3   Proposed System Implementation

3.1   Target Platform

The target platform for the JPEG2000 SoC is a Stratix IV PCIe x4 development board from GiDEL [9]. The platform selected features a Stratix IV E 530 FPGA with a 512 MB DDR2 memory bank and two DDR2 SODIMMS with up to 4 GB each. The platform has two additional ports for expansion daughter cards offered from GiDEL. The high performance offered from a PCIe based platform coupled with the flexibility and size of an Altera Stratix IV FPGA provides an ideal platform capable of meeting the demands of a JPEG2000 SoC [9].

3.2   Selection of Softcore Coprocessor

The soft-core coprocessor chosen is the NIOS II processing core from Altera [3]. The NIOS II processor features a 32-bit reduced instruction set computing (RISC) architecture that is highly configurable, capable of supporting up to 256 custom instructions and clock speeds near 300 MHz on a Stratix IV device. The NIOS II processor system consists of a NIOS II processor core coupled with on-chip peripherals (DMA controllers, timers, custom HW interfaces), on-chip memory as well as interfaces to off-chip memory. All of the various peripherals and memories are managed through the Avalon switching fabric which serves as an arbiter between the various masters and slaves within a system. The Avalon switching fabric allows multiple data/instruction masters to communicate directly with multiple slaves devices simultaneously, assuming no two masters are attempting to communicate with the same slave. The NIOS II processing core is chosen given is high degree of flexibility and ability to support custom peripherals as well as built-in support for embedded C/C++ development using the NIOS II software development suite [3,6].

3.3   Hardware Implementation

In order to realize a full JPEG2000 SoC, the EBCOT Tier II processing module must reside in hardware. Most research on JPEG2000 neglects to mention the implementation of Tier II, presumably leaving it to be handled by the host processor. This paper proposes a novel solution to this gap by leveraging an embedded soft-core coprocessor to serve as an embedded EBCOT Tier II processing

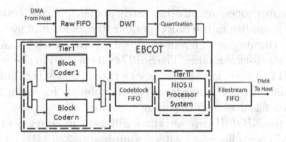

Fig. 2. Data Flow of Proposed JPEG2000 SoC

module. While similar in nature to the proposed architecture in [7], which utilizes the NIOS II system as an arbiter between different hardware modules, this architecture treats the coprocessor as a seperate processing module.

The proposed JPEG2000 SoC is implemented on the target platform by integrating the NIOS II processing unit with existing embedded JPEG2000 processing modules. The existing design features a pipelined DWT architecture coupled with a variable number of parallel EBCOT Tier I block coders. The details of the specific architecture are given in Section 2. The coprocessor serves as the final stage of the pipeline, taking the codeblock streams from the block coders and forming the final JPEG2000 filestream. This eliminates the scheduling overhead associated with arbitration between the various processing modules as in [7]. The dataflow of the proposed implementation can be seen in Fig. 2.

The EBCOT Tier II coprocessor is implemented as the final stage in a tile pipeline. Before the host sends the raw image to the target device, the image is padded and divided into tiles. Tiles are then sent to the target via DMA over the PCIe bus and processed sequentially. Each stage of the pipeline begins once a single tile has been received from the previous stage. Tiling the image reduces the memory requirements for each stage in the pipeline since each stage will need enough memory to store a single tile. Additionally, tile processing enables the use of distributed architectures where a single host leverages multiple target devices to process the tiles of a single image in parallel.

Each tile is handled as a separate entity, with the pipeline retaining no knowledge of the surrounding tiles. The NIOS II processing system creates the independent tile-stream for each tile it receives from the pipeline before placing it in a FIFO for communication back to the host. The host system receives the tile-streams from the device and applies the main headers to form a valid JPEG2000 filestream. While the NIOS II could easily be configured to add the main headers, they are left for the host processor in order to maintain architectural flexibility in the event that more target devices are added to the system.

3.4   NIOS II System Implementation

This NIOS II processor operates within the entity created by Altera's System on a Programmable Chip (SoPC) builder [4]. This tool allows for seamless integration between the soft-core processor and other hardware peripherals through the Avalon switching fabric. In addition, the SoPC builder allows for integration of multiple processing blocks running at different clock rates, with SoPC handling all of the arbitration between clock domains. The SoPC system used in the proposed design features a NIOS II fast core, running at a clock rate of 290 MHz. This NIOS II core is the fastest of the three offerings from Altera, offering high clock speeds and a number of addition features over the other two cores.

Coupled with the NIOS II core are three different memory controllers. One is a DDRII SDRAM controller from Altera, running at 225 MHz, which interfaces directly with one of the two DDRII SODIMMs available on the target platform (Bank B or C).The SDRAM is used as a buffer space for incoming compressed codeblocks while also serving as a buffer to hold the full tile-stream produced by Tier II prior to output. Thanks to the large amount of memory made available through the SODIMM (up to 4 GB), the system is able to buffer a number of tiles if the Tier II processing falls behind the rest of the pipeline.

Additionally, there are also two separate on-chip memory controllers, one which controls a 50 kByte bank and the other which controls a 30 kByte bank. The 50 kByte bank is used to hold the executable code and data sections in addition to the program stack and heap during execution. The other 30 kByte bank is configured as "Tightly Coupled Data" memory. The details of this implementation are elaborated on in Section 3.5. Both of the on-chip memory banks are configured to run on the same clock as the NIOS II core.

Custom interfaces are designed to communicate with the codeblock and filestream FIFOs as seen in Fig. 2. These interfaces are coupled with dedicated DMA controllers. Each controller masters one of the FIFO interfaces as well as the SDRAM controller, allowing the DMA controller to perform block reads and writes to and from the FIFOs as a background task. The processor simply schedules a memory transfer when necessary and receives an interrupt from the DMA controller upon completion of the transfer. This frees the processor to complete other tasks while the transfer is pending.

Additionally, the SoPC system has a number of parallel I/O (PIO) ports which provide direct communication with the hardware. The PIOs are used for a variety of different functions such as receiving interrupts and feedback from the encoding pipeline while also providing feedback to the host device.

3.5 Optimizations

A number of optimizations are made to the system in order to increase throughput. As the implementation couples a coprocessor with existing, optimized, processing modules, these optimizations are focused on the NIOS II processing core. Fig. 3 shows the impact of these optimizations on the processing time for a single tile. High resolution images are compressed and the Tier II processing time is then divided by the number of tiles to yield an average Tier II processing time.

Typically, floating point operations are much more computationally expensive than integer operations. This expense is compounded when using a RISC architecture such as the NIOS II processor. While most of the Tier II algorithm

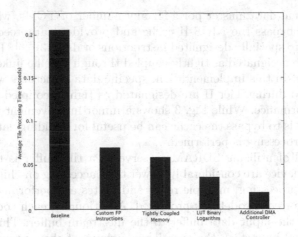

Fig. 3. Impact of Optimizations on Average Processing Time for Single Tile

is performed using integer calculations, a logarithm is required to calculate the length of a codeword segment. Codeword segments are used to signal the number of bytes contributed to a packet by a code block. The number of bits required for to store the codeword is given by

$$bits = Lblock + \lfloor log_2(P) \rfloor, \tag{1}$$

where *Lblock* is the state variable for the current code block and P is the number of coding passes contributed for the current code block [10]. As this calculation is necessary for each code block, the computational cost is high. Software profiling reveals that the calculation of Equation 1 takes over 70% of the total Tier II processing time on the NIOS II processor.

Two approaches are taken to reduce the computation cost of this calculation. The NIOS II processing core supports the addition of custom instructions, such as user created HW implementations of specific operations. Additionally, the tools include a number of premade instructions, including custom floating point (FP) instructions [3,5]. The custom FP hardware is enabled on the NIOS II core, and significant improvement is seen in system throughput, yielding a 66% decrease in processing time. However, this calculation still takes over 35% of the total Tier II time. Since Tier II uses the floored result of the logarithm, the standard library call can be replaced with a custom implementation using a lookup table (LUT) approach. Fig. 3 shows that using the LUT implementation yields better performance than the custom FP hardware, and is used in the final implementation in favor of the custom instructions. The use of a LUT implementation reduces the calculation time to 3% of the total Tier II time, down from over 70%.

As mentioned in Section 3.4, the 30 kByte bank of on-chip memory is configured as "Tightly Coupled" memory (TCM). The NIOS II core can be configured to have additional data master ports for any number of TCMs, which must be on-chip. TCMs bypass the NIOS II cache and provide guaranteed low-latency memory access to specially designated instructions or data [3], [6]. These instructions or data are designated as tightly coupled through specific linker commands at compile time. In this implementation, specific data structures which are frequently accessed during Tier II are designated as tightly coupled to guarantee consistent performance. While Fig. 3 shows a minor improvement in processing time, using TCMs to bypass the cache can be useful for avoiding data corruption while parallel processing is performed.

The downfall of utilizing SDRAM to serve as a tile buffer is that memory accesses to the device are considerably slower than accessing on-chip RAM. During the Tier II processing, mutliple reads and writes are performed within this memory space for each codeblock processed. Additionally, each codeblock must be copied from the input data buffer to the filestream buffer. This copy alone has a detrimental impact on the overall throughput of the system. To address this issue, a third DMA controller is added to the system which masters only the SDRAM controller, allowing the system to schedule non-blocking memory copies within the SDRAM address range. The addition of this DMA controller hides the latency associated with large memory copies, allowing the processor to continue processing the next set of instructions while the transfer completes. Introduction of the third DMA controller results in a 48% decrease in processing time as shown in Fig. 3 when compared to the previous implementation without the additional DMA controller. Care is taken to ensure that all pending memory copies have completed prior to writing out the completed filestream.

The final optimization made to the NIOS II system is to ensure that the system acts as a pipeline in order to maximize throughput of the system. Initially, the system is designed without the codeblock FIFO seen in Fig. 2. Instead, the NIOS II system is directly coupled with the Tier I output, reading code blocks as they become available. While a simplified approach, the downfall is that the Tier I processor must wait for the code block to be read before proceeding to the next code block. In order to eliminate this idle time, the code block FIFO in Fig. 2 is added. Tier I simply writes to the FIFO, which is large enough to buffer multiple tiles, since this FIFO resides in SDRAM. Tier II then reads entire tiles as they become available, instead of reading each code block. Pipelining has two distinct impacts on the system throughput. First, the Tier I encoder is now free to process code blocks as fast as possible, therefore increasing the system throughput. Additionally, the interrupt latency associated with posting read requests is reduced since there is only one read request per tile, instead of one request per code block.

4   Analysis of Results

4.1   Performance and Analysis

The performance of the JPEG2000 SoC is measured using three 2048x2560 ISO test images "Cafe", "Woman", and "Bike". For all tests, each image is compressed and the respective processing times of all three images are averaged together. First, the overall performance of the system is measured with varying numbers of parallel block coders in order to determine the optimum number of block coders and the corresponding throughput of the system. The number of parallel block coders is increased from 1 to 20. The impact of an increased number of parallel block coders on the image processing time as well as the Tier I and Tier II times is shown in Fig. 4.

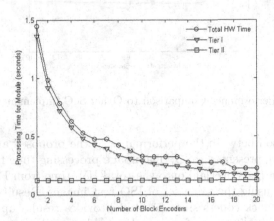

Fig. 4. Performance Profile of Encoding Pipeline with Parallel Block Coders

The total HW time in Fig. 4 shows that, as expected, additional block coders have a large impact on the average image processing time. This is especially true when the block coder count is increased from 1 to 6, resulting in a 67% drop in average processing time. Processing time continues to decrease as the block coder count is increased beyond six with a minimal processing time of 0.22 seconds achieved when 18 block coders are present. Negligible change in performance is seen with block coder counts beyond 18.

The overall throughput of the system is limited by the Tier II processing time. Fig. 4 shows the time spent performing Tier I and Tier II processing, as well as the total time, for a variable number of block coders. Fig. 4 shows that the Tier II processing remains constant at 0.121 sec while the Tier I time decreases as more block coders are added, as expected. As the Tier I processing time approaches the Tier II time, the total processing time begins to flatten out, with little change beyond 18 block coders. Additional block coders have little impact on the total processing time since it has no impact on the Tier II processing time, which is the limiting factor when using 18 or more block coders.

Fig. 4 shows that the proposed architecture scales well with an increasing number of parallel block cloders with the limiting factor being the Tier II processing time on the NIOS. This compares favorably with the SoC architecture presented in [7], which does not scale as well as the proposed architecture. Fig. 5

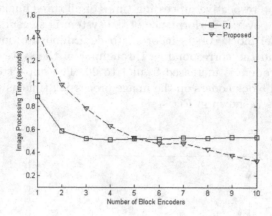

Fig. 5. Performance Comparison to Other SoC Implementations

shows a comparison between the performance of the proposed architecture and th3+e architecture presented in [7]. The image processing time from the fastest implementation of [7] is overlaid onto the total HW time from Fig. 4 for one to ten block coders, using the same set of ISO test images. Results are compared from one to ten block coders since [7] only provides results up to ten coders. It is clear that while [7] outperforms at lower block coder counts, these gains are erased once the count is increased beyond five encoders. At this point the [7] architecture has plateaued while the proposed architecture continues to improve. With 10 parallel block coders, the proposed implementation outperforms [7] by 39%. When the system is scaled to 18 block coders, the proposed design outperforms [7] by 58%. By allowing the other processing modules to operate outside the contexts of the SoPC system, the proposed architecture is able to take full advantage of multiple parallel block coders without the limitations necessarily imposed by the Avalon switching fabric. While extremely effective at integrating multiple different peripherals into a single system, the scheduling and arbitration overhead associated with the NIOS II processing system impose restrictions on the system throughput. By creating a pipelined architecture which utilizes the NIOS II processing core as a separate unit we are able to leverage the flexibility of the NIOS II system while still maintaining the speed of a pipelined encoder.

4.2 Hardware Synthesis Results

The proposed implementation is synthesized on a Stratix IV FPGA using Altera Quartus 10.1. For the purposes of comparing the proposed design to [7], the design is synthesized with 4 parallel block coders. With this encoder count, [7] slightly outperforms the proposed system, but these gains are quickly erased with additional block coders (Fig. 5). The hardware costs of the proposed system and [7] are shown in Table 1. Costs for both the 4 and 18 block encoder implementations are shown. The hardware costs for the proposed system are split into two categories: the DWT and Tier I modules, and the NIOS II system which performs Tier II. The results are presented in this manner to provide an accurate comparison to [7], which only simulates the NIOS II system, so the hardware costs only reflect the DWT and Tier I modules.

Table 1. Hardware Resource Comparison for a System with 4 Block Coders

System	BlockCoders	LCs	Memory	Clock(MHz)
[7]DWT/TierI	4	15,268	622,976	50
DWT/TierI	4	13,123	637,952	100/200
DWT/TierI	18	43,690	1,417,472	100/200
TierII(NIOS)	N/A	10,996	923,008	290

Table 1 shows that the hardware resource costs of the implemented NIOS II system are minimal compared to the other processing modules, whose costs increase as more block coders are added. However, the NIOS II system does have a high memory cost. This is due to the 80 kBytes (50 kByte and 30 kByte banks) of on-chip RAM used along with the instruction and data caches built into the processing core. These large memory modules are necessary to run more complex code requiring larger stack and heap regions in memory. However, due to the flexibility of the NIOS II system, these costs could be shifted off-chip by utilizing more of the SDRAM. However, since the target platform utilizes a Stratix IV FPGA [9] with a large amount of on-chip memory, this is not an issue for the proposed design.

Table 1 shows that the proposed DWT and Tier I designs are comparable in cost to [7] with 4 parallel block coders. The main difference is that the proposed design capable of higher clock speeds, with the DWT and Tier I running at 100 MHz and 200 MHz respectively. This results in a DWT capable of processing one pixel every 7 ns as opposed to 20 ns per pixel in [7]. The higher performance of the DWT prevents the parallel block coders from becoming starved as they do in [7], which yields increased performance up to 18 block coders as opposed to [7], which peaks at 4 block coders due to starvation of the block coders. Instead, the proposed designed is limited by the throughput of the NIOS II system.

5   Conclusion

This paper proposed a fully embedded JPEG2000 SoC which utilized an Altera NIOS II processor as the embedded EBCOT Tier II processing module. The proposed system was synthesized on a Stratix IV FPGA and yielded a 39% performance increase over other JPEG2000 SoC implementations with the same number of parallel block coders. While [7] offers a more flexibile and reconfigurable design, the pipelined architecture of the proposed design allows for a design capable of scaling to higher numbers of parallel block coders with comparable increases in system throughput. While limited by the performance of the

NIOS II performance, future implementations could mitigate this with optimizations to the Tier II algorithm. Additional NIOS II processing cores could also be added to the system to share the processing load, assuming the availability of adequate hardware resources.

In addition to a high performance and scalable design, the proposed system also demonstrates the feasability of utilizing an embedded softcore processor as a dedicated processing unit within a pipeline. Ease of reconfiguration and support for a variety of peripherals allowed for seamless integration of the NIOS II system into an existing encoding pipeline. The proposed design also demonstrates techniques for optimizing the performance of software running on the NIOS II through the use of custom instructions and additional peripherals.

References

1. Balster, E.J., Fortener, B.T., Turri, W.T.: Integer Computation of Lossy JPEG2000 Compression. IEEE Transactions on Image Processing 20(8) (August 2011)
2. Chen, K.-F., Lian, C.-J., Chen, H.-H., Chen, L.-G.: Analysis and Architecture Design of EBCOT for JPEG-2000. In: Proc. ISCAS, vol. 2 (May 2001)
3. Altera Corporation. NIOS II Processor Reference Handbook (December 2010)
4. Altera Corporation. SOPC Builder User Guide (December 2010)
5. Altera Corporation. NIOS II Custom Instruction User Guide (January 2011)
6. Altera Corporation. NIOS II Software Developer's Handbook (February 2011)
7. Dyer, M., Nooshabadi, S., Taubman, D.: Design and Analysis of System on a Chip Encoder for JPEG2000. IEEE Transactions on Circuits and Systems for Video Technology 19(2) (February 2009)
8. Dyer, M., Taubman, D., Nooshabadi, S.: Improved Throughput Arithmetic Coder for JPEG2000. In: Proc. International Conference on Image Processing, ICIP (October 2004)
9. GiDEL. ProceIV Data Book (May 2011)
10. ISO/IEC 1.29.15444-1. JPEG 2000 Part I Final Committee Version 1.0 (September 2004)
11. Kumar, N.R., Xiang, W., Wang, Y.: An FPGA-Based Fast Two-Symbol Processing Architecture for JPEG2000 Arithmetic Coding. In: Proc. IEEE ICASSP (March 2010)
12. Liu, L., Chen, N., Meng, H., Zhang, L., Chen, H.: A VLSI Architecture of JPEG2000 Encoder. IEEE Journal of Solid-State Circuits 39(11) (November 2004)
13. Liu, L., Wang, Z., Chen, N., Zhang, L.: VLSI Architecture of EBCOT Tier-2 Encoder for JPEG2000. In: IEEE Workshop on Signal Processing Systems (2005)
14. Santa-Cruz, D., Grosbois, R., Ebrahimi, T.: JPEG 2000 performance evaluation and assessment. Signal Processing: Image Communication 17 (2002)
15. Sarawadekar, K., Banerjee, S.: An Efficient Pass-Parallel Architecture for Embedded Block Coder in JPEG 2000. IEEE Transactions on Circuits and Systems for Video Technology 21 (June 2011)

Perceiving Ribs in Single-View Wireframe Sketches of Polyhedral Shapes

P. Company[1], P.A.C. Varley[1], R. Plumed[2], and R. Martin[3]

[1] Institute of New Imaging Technology, Universitat Jaume I, Spain
[2] Department of Mechanical Engineering and Construction, Universitat Jaume I, Spain
[3] School of Computer Science & Informatics, Cardiff University, UK

Abstract. As part of a strategy for creating 3D models of engineering objects from sketched input, we attempt to identify *design features*, geometrical structures within objects with a functional meaning. Our input is a 2D B-Rep derived from a single view sketch of a polyhedral shape. In this paper, we show how to use suitable cues to identify algorithmically two additive engineering design features, angular and linear ribs.

1   Introduction

Our aim is to find *design features*, geometrical structures within objects with a functional meaning. In a previous paper [1], we presented a catalogue of common design features, listing for each type of feature any known successful algorithms for detecting the implied presence of such features in sketches. We noted that, as a result of previous interest in *machining features*, there are many known algorithms for detecting subtractive features or "intrusions", but far fewer algorithms for detecting additive features or "protrusions" [2, 3]. Our motivation in this paper is to fill this gap.

We briefly review what is known about both human and artificial perception of features, and we then introduce a new approach aimed at finding one specific design feature, the rib, through indirect cues. We distinguish between angular ribs (or simply *ribs*) and linear ribs (or *rails*). The new contribution here is the search for perceptual cues linked to features embedded in the sketched shapes.

Because of their function, ribs are theoretically important in computer-aided design as the ideal size of a rib for a particular size of part depends on strength calculations. Once a feature has been flagged as a rib, a CAD/CAE package should know that its dimensions are to be fixed not by geometric constraints but by the physical properties of the material. It is thus important to find and classify ribs to distinguish them from more "geometric" features.

Our intended approach differs from other feature recognition performed on inaccurate models in that our input is a 2D graph-like line-drawing, derived from a single view sketch of a polyhedral shape. This introduces one intrinsic limitation: as sketches contain only pictorial information, depth information comes only from the interpretation of perceptual cues. The second limitation, considering only polyhedral shapes, is a simplification that we hope we will relax in the future.

G. Bebis et al. (Eds.): ISVC 2012, Part II, LNCS 7432, pp. 557–567, 2012.

Our goal is to understand the function of an object from a 2D sketch. Finding features is an important step in this process, but it is not the whole story. We envisage a subsequent step in which, given the possible presence of various features, some form of statistical processing is used to combine compatible features and reject contradictory features (although at present we leave open the problem of which particular statistical methodology is to be used for this). Thus, the feature detection algorithms we present here return a statistical likelihood: they do not answer the question "is this a rib?" but the question "how likely is this to be a rib?"

Section 2 describes the current state of the art of perception of design features. Section 3 presents an algorithm for finding ribs and dividers, and Section 4 presents an algorithm for finding rails and slots. Section 5 presents some test results, and Section 6 presents our conclusions.

2   State of the Art in Perception of Features

Although the "rule of concave creases" explains quite well how human perception cuts objects into fragments, and its most famous formulation [4] has been used with reasonable success in areas such as segmentation of 3D meshes [5], it is not enough by itself to produce an algorithmic equivalent for artificial perception of objects from 2D sketches or line drawings.

Artificial perception of fragments of a scene dates back to the "block-world" introduced by Guzman [6]. Early approaches to breaking down objects to perceive their fragments as elements of a CSG tree were proposed by Bergengruen [7] and Wang and Grinstein [8], among others. A survey of contributions to geometrical sketch reconstruction up to 2005 [9] shows that approaches producing CSG output used multiple view input and were less common than those producing B-Rep output. It is noteworthy that nearly all contributions were concerned with geometry and shape, not with functionality. Most sketch based modelling (SBM) systems which provide CSG output use multiple view input, and drive the decomposition by geometric considerations (finding basic geometrical shapes such as prisms and wedges), not design features [10].

Since then, the emphasis has started to change. The most useful recent contributions, by Wen et al. [11] and Rivers et al. [12], deal with multiple view input and are interactive. An interesting aspect of the work by Wen et al. [11] is that they identify features, by searching in multiple view drawings for what they call "semantic information".

Another field, automatic feature recognition (AFR), has evolved in parallel. A recent survey of this field can be found in [2]. Two main branches exist. The first, useless for SBM, is the input of a CSG model containing CAD features which must be converted to Computer-Aided Manufacturing (CAM) features. But the second branch is aimed at the closer goal of finding CAM features in B-Rep 3D models. AFR methods follow three main approaches: graph-based, volumetric decomposition and hint-based. Hint based AFR (where an AFR "hint" is effectively the same thing as a "clue" or "cue" in human perception and SBM) was introduced by Vandenbrande and

Requicha [13], and extended by Han and Requicha [14]. Note, however, that input for AFR is 3D: its goal is to algorithmically extract manufacturing features from CAD 3D models.

Since the goal of SBM is producing 3D models from sketches, while the goal of AFR is finding features in 3D models, a strategy of reconstructing first and then finding features looks obvious. However, this coupling has not yet proved successful, and the problem is still open. Here, we consider the alternative strategy: first find features, then reconstruct. We hope that detecting features first may, for example, help us to constrain the reconstruction (e.g. knowing depth relationships in the feature may help to fix other depth relationships). In order to search for design features in sketches, we must replace AFR's 3D input by 2D input, by suitably adapting the hints.

Detecting features before reconstruction was first applied for single-view sketch-based modelling purposes by Varley [15]. His input was natural single-view line-drawings. He looked for underslots and valleys by trying to match the region around each T-junction in the drawing with some templates. If a match was found, it was given a certain figure of merit. Such template matching remains the best choice for finding features in natural drawings, when no further information is available.

As far as we know, hint-based AFR strategies have never been applied to single-view 2D wireframes. Our hypothesis is that, since more information is available in wireframes (with hidden lines drawn), particularly after faces have been found, than in natural drawings (without hidden lines and faces), AFR strategies could be adapted to 2D input to find features in wireframes.

3 Finding Ribs and Dividers in Single-View Wireframes

Mechanical engineers have no difficulty in detecting ribs embedded in objects: they can imagine how the object will work to resist the loads it is designed for. They can thus decide whether a fragment is there simply to reinforce the whole object without significantly increasing its weight. However, detecting such functionality is a complex ability, hard to replicate, not only for machines but even for novice engineering students. Fortunately, some peculiarities of its shape are easy to detect. The most individual geometrical aspect of ribs is that they are thin walls. The cue linked to a thin wall is that the border faces are narrow and rectangular. We note that a rectangle in 3D usually appears as a parallelogram in an axonometric 2D image, so a parallelogram in 2D is an indirect cue of a rectangle in 3D.

As it is joined to the whole object, most of a rib's border faces are embedded in the object (border faces of webs are fully embedded). However, usually one border face of a rib is not embedded, so the first task when detecting ribs is to look for thin border faces, quadrilateral faces with two quite short opposed edges and two longer edges.

Secondly, as ribs are "walls", they are extruded shapes. The border face should be a parallelogram, and the two faces connected to the longer edges of the border face should be equal. If the input drawing is a wireframe, both faces are known, so this condition helps to disambiguate candidate ribs.

Third, as the rib "interrupts" the faces where it is embedded (at least if objects are polyhedral), edges connected to the vertices of the rib's lateral faces but which do not belong to the rib (we call them "external edges") should be collinear with the corresponding external edges on the opposite side of the rib. This collinearity of external edges is not considered as a requisite, but is an indirect cue which reinforces the assumption than the fragment is a rib. Hence, if this condition is met, the figure of merit which measures the likeliness that a candidate rib should be considered as an actual rib should be increased.

Finally, since the algorithm detects thin walls, and dividers are also thin walls, the algorithm also detects dividers. We must distinguish ribs from dividers, since dividers are purely geometric: they partition space into separate regions or limit movement with respect to another part. (However, multiple dividers may also represent cooling fins, making them another example of design features whose measurements must be derived from physical modelling and not merely from geometric constraints.) Hence, dividers are likely to have more than one border face, and their lateral faces are likely to be quadrilateral. These differences are used in the algorithm to discriminate between ribs and dividers: ribs with multiple border faces are re-labelled as dividers. Parallelogram lateral faces increase the likelihood of the feature being a divider, and decrease that of it being a rib.

The procedure for finding ribs requires 2D B-Reps. Although further improvements are still required, conversion from wireframe sketches into 2D line-drawings is a well known stage in SBM (readers should look for "recognition" in [16]), but conversion from natural into wireframe line-drawings is not so easy [15], [17]. Faces must also have been detected in advance (the process of finding faces in 2D B-Reps of polyhedral objects has been explained elsewhere [18]). Hence, we assume a graph-like wireframe input containing a list of vertices, a list of edges and a list of faces. Vertices are defined by their x and y coordinates. Edges are defined by the two vertices they connect. Faces are defined as closed loops of edges. The proposed procedure for detection of ribs can then be summarised as follows:

Algorithm 1: Pseudocode for the detection of ribs:

```
01:  Begin Thin Wall detect
02:  for each face f
03:      if f is quadrilateral (i.e. a four sided face)
04:          if f does not belong to any rib or divider and is SLENDER
05:              Mark f as a border face
06:              Mark faces connected to the long edges of f as lateral1 and 2
07:              if lateral 1 and 2 are SIMILAR
08:                  return Thin-Wall (border, lateral1, lateral2)
09:              end if
10:          end if
11:      end if
12:  end for
13:  return not-Thin-Wall
14:  End Thin Wall detect
```

```
01:  Begin External Are Collinear
02:  flag_collinear ←true
03:  for each vertex v1 of the lateral face 1
04:      Find the similar vertex v2 of lateral face 2
05:      Find the one, and only one, external edge e1 of vertex v1
06:      Find the one, and only one, external edge e2 of vertex v2
07:      if e1 and e2 are not-collinear
08:          flag_collinear ←false
09:      end if
10:  end for
11:  return flag_collinear
12:  End External Are Collinear
```

```
01:  Begin Ribs detection
02:  do while (Thin Wall detect)
03:      if (lateral1 and lateral2 belong to a previous rib R)
04:          Add the new border to the previous rib R
05:          Re-label rib R as a divider
06:      else
07:          Define a new rib R (border, lateral1, lateral2)
08:          Assign an average figure of merit to the new rib R
09:      end if
10:      if External Are Collinear
11:          Increase the figure of merit of R
12:      end if
13:      if laterals are PARALLELOGRAMS
14:          if R is divider
15:              Increase the figure of merit of the divider R
16:          else
17:              Decrease the figure of merit of the rib R
18:          end if
19:      end if
20:  end while
21:  End Ribs detection
```

The terms in capitals in the pseudocode are functions not listed in the pseudocode, and are explained here. A face is considered *slender* if it is quadrilateral, contains two opposite short edges, and two opposite long edges. Two lateral faces are considered *similar* if they achieve three conditions: the faces must have the same number of vertices, vertices can be paired in pairs whose distance is always equal, and the line segments defined by the paired vertices must be parallel. A face is a *parallelogram* if it is quadrilateral and both pairs of opposed edges (arbitrarily 0-2 and 1-3) have equal length and are parallel.

The terms underlined in the pseudocode depend on tuning parameters. Comparisons for *short, long, equal* and *parallel-collinear* need thresholds, to deal with imperfections inherent in the sketching process.

4   Finding Rails and Slots in Single-View Wireframes

Some linear ribs are present to provide structural reinforcement, but other linear ribs act in pairs with slots in other parts, and their function is to produce sliding joints between adjacent parts in an assembly (Figure 1 left). We call these rails. The shape of rails is the complement of the slots in the paired part. Hence, the procedure for finding rails is similar to that for finding slots.

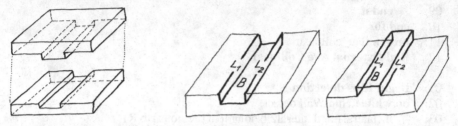

Fig. 1. Rail and slot (left), and some of their 2D cues (right)

Firstly, slots are long and narrow holes machined in flat surfaces. A machined slot produces a result resembling an incomplete prism, where the border/bottom face and the lateral faces of the slot are all quadrilateral (Figure 1 right). Secondly, lateral faces should have the same shape. Thirdly, as the rail/slot interrupts the faces in which it is embedded, the figure of merit which measures the likeliness of candidate rails should be increased if external edges of lateral faces are collinear.

Since most common rails and slots are wider than deeper, the figure of merit is increased if laterals are slender (quadrilateral and narrow).

The proposed procedure can be summarised as follows:

Algorithm 2: Pseudocode for the detection of slots and rails:

```
01:    Begin Candidate-Rail detect
02:    for each face f
03:        if f is SEMIPARALLELOGRAM and does not belong to another rail
04:            Mark f as a border/bottom face
05:            Mark faces connected to the opposed long edges of f as lateral1 and 2
06:            if lateral 1 and 2 are SEMIPARALLELOGRAM
07:                if lateral 1 and 2 are SIMILAR
08:                    return Candidate Rail (border, lateral1, lateral2)
09:                end if
10:            end if
11:            Mark faces connected to the opposed short edges of f as lateral1 and 2
12:            if lateral 1 and 2 are SEMIPARALLELOGRAM
13:                if lateral 1 and 2 are SIMILAR
14:                    return Candidate Rail (border, lateral1, lateral2)
15:                end if
16:            end if
17:        end if
18:    end for
```

```
19:    return no-Candidate-Rail
20:    End Candidate-Rail detect

01:    Begin Rails detection
02:    do while (Candidate Rail detect)
03:        if (lateral1 and lateral2 belong to a previous rail R)
04:            Add the new border to the previous rail R
05:            Re-label rail R as a blind rail
06:        else
07:            Define a new rail R (border, lateral1, lateral2)
08:            Assign an average figure of merit to the rail R
09:            if External Are Collinear
10:                Increase the figure of merit of R
11:            end if
12:            if laterals are SLENDER and share their long sides with the
               border/bottom face
13:                Increase the figure of merit of R
14:            end if
15:        end if
16:    end while
17:    End Rails detection
```

A comparison of the rail pseudocode with the rib pseudocode shows the similarities and differences between them. For instance, in the rail pseudocode the border/bottom need only be a semiparallelogram (there is no need for it to be slender). However, the figure of merit increases if laterals are slender.

The function *semiparallelogram* returns true if faces are quadrilateral and their long opposed sides are parallel to each other. This is used instead of *parallelogram*, since non-parallel short edges are quite common for slots which start or end in inclined faces (Figure 3 centre, Figure 4 centre).

5   Results

After several tests, we tuned the ribs algorithm to the following thresholds: (T1) lines shorter than 20% of the minimum side of the bounding box of the sketch are considered short; (T2) lines more than twice as long as the longer of the two short sides of each quadrilateral (long>2*max(short1, short2)) are considered long for this quadrilateral; (T3) lines are collinear if their orientation differs by less than 9°. The thresholds for similar faces are: (T4) vertices may be paired whose distances differ by less than 20%, and (T5) orientations of segments joined by paired vertices must differ by less than 15°. Both deviations are measured from their respective median values.

The initial average figure of merit is 0.51 (T6). If external edges are collinear, the figure of merit increases by 10% (T7) in all cases. If lateral faces are parallelograms, the figure of merit increases by 25% (T8) for candidate dividers and decreases by 40% (T9) for candidate ribs. If lateral faces are slender, the figure of merit for rails and slots increases by 25% (T10). Candidate features with a figure of merit equal or higher than average are considered true features.

Tuned in this way, the algorithm succeeds in identifying ribs in our sketches.

Parameters T1 and T2 are used to detect border faces, encoding geometrical differences linked to design intent. Changing these thresholds would result in "fat" ribs being accepted, or "thin" ribs being rejected (as far as we know, the limiting thickness for a rib to be still considered as a rib has never been fixed).

Parameters T3, T4 and T5 allow for the imperfections inherent in freehand sketches. Their effect is illustrated in figures 5 and 6.

Parameter T6 is set to 50% plus a small margin to defend against round-offs. Parameter T9 must be greater than T7, to prevent walls with parallelogram lateral faces from being accepted as better than average ribs.

Our results are highly insensitive to changes in parameters T6 to T10.

Figure 2 illustrates how some of the original strokes (left) are first vectorised into wireframes, whose vertices are merged to get graph-like drawings. Then the algorithm detects the ribs and dividers (right).

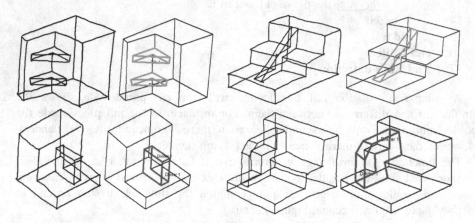

Fig. 2. Four examples where sketches containing ribs or dividers (left), are vectorised into their corresponding line-drawings with the features highlighted (right)

We obtained similar results for rails and slots (Figure 3). Sequential execution of both algorithms generally gives good results too, as shown in Figure 4 left.

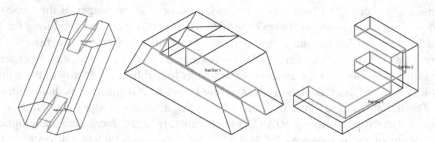

Fig. 3. Sketches containing rails or slots

The hollow rib in the example in Figure 4 centre illustrates a major problem: since the input is a 2D B-Rep, the current algorithm does not distinguish between true ribs and "negative ribs". A possible solution would be labelling the graph-like line-drawing, since labelling edges as convex and concave helps to detect hollow ribs [19, 20]. Before running line labelling, 2D hidden edges detection must be used [21], since line labelling only makes sense for natural drawings.

Finally, ribs at the ends of objects are not yet detectable (Figure 4 right). A possible solution for this case would be detecting whether one of the lateral faces is a "border" face, i.e. a face containing one or more border edges, and accepting such faces as candidate lateral faces (with a figure of merit increasing with the increasing number of border edges). In this case, instruction 7 of the algorithm 1 should be modified as: **if** (lateral 1 and 2 are SIMILAR) or (lateral 1 or lateral 2 is border).

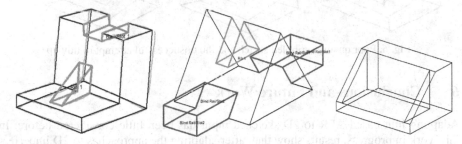

Fig. 4. Sequential detection of ribs and rails (left and centre), and corner ribs (right)

During our tests, false negatives only happened for poor quality sketches (such as the corner highlighted in Figure 5 left) which resulted in distorted graphs. Relaxing tuning parameters would reduce false negatives, but would also produce some false positives.

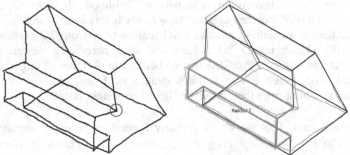

Fig. 5. Poor quality scribbled corner results in a rail difficult to detect

The alternative of tidying up the sketch before detecting features has proved useful to some extent. We used a modified version of the batch beautification described in [22], which is adequate when distortions are local and not too strong (as around the highlighted vertex in Figure 5), but is insufficient for high distortions (such as those in Figure 6).

However, such tidying should not be necessary: our approach is robust, in that, as was said in the introduction, the question we aim to answer is "how likely is this to be a rib?" Hence, we believe that we shall be able to relax the acceptance criteria in order to detect such dubious representations as "weak" features with a low figure of merit. In the end, we want our algorithm to detect what humans can perceive, and we want our algorithm to doubt where humans doubt.

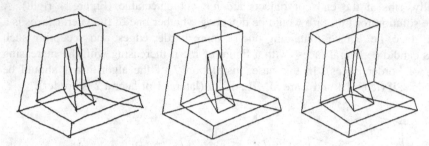

Fig. 6. Poor quality scribbled sketch and the unsuccessful attempt to tidy up

6   Conclusions and Future Work

Adapting hint-based AFR to 2D sketched input has been little considered before. In this work in progress, results show that, after adapting the approaches to 2D imperfect input and tuning the algorithms, they identify many common ribs, dividers, rails and slots from suitable indirect cues.

Our goal is to understand the function of an object from a 2D sketch. Finding features is the first step. We envisage a subsequent step in which, given the possible presence of various features, some form of statistical processing is used to combine compatible features and reject contradictory features. Thus, the feature detection algorithms we present here return a statistical likelihood: they do not answer the question "is this a rib?" but the question "how likely is this to be a rib?"

We hope then to be able to produce CSG feature trees from 2D sketches, by first converting the sketch into a 2D B-Rep, and then perceiving design features to explicitly recreate a mind's eye CSG tree embedded in sketches of engineering parts. Towards this end, when design features cannot be found directly in the sketch, indirect cues resulting from their presence should be sought instead.

Acknowledgements. This work was partially funded by financial support from the Ramon y Cajal Scholarship Programme and by the "Pla de Promoció de la Investigació de la Universitat Jaume I", project P1 1B2010-01.

References

1. Plumed, R., Varley, P.A.C., Company, P.: Features and Design Intent in Engineering Sketches, 3IA2012. Intelligent Computer Graphics (in press, 2012)
2. Babic, B., Nesic, N., Milzkovic, Z.: A review of automated feature recognition with rule-based pattern recognition. Computers in Industry 59(4), 321–337 (2008)

3. Babic, B., Nesic, N., Milzkovic, Z.: Automatic feature recognition using artificial neural networks to integrate design and manufacturing: Review of automatic feature recognition systems. Artificial Intelligence for Engineering Design, Analysis and Manufacturing 25, 289–304 (2011)
4. Biederman, I.: Recognition-by-Components: A Theory of Human Image Understanding. Psychological Review 94(2), 115–147 (1987)
5. Shamir, A.: A Survery on mesh segmentation techniques. Computer Graphics Forum 27(6), 1539–1556 (2008)
6. Guzmán, A.: Decomposition of a visual scene into three-dimensional bodies. In: AFIPS American Federation of Information Proc. Fall Joint Computer Conf., vol. 33, pp. 291–304 (1968)
7. Bergengruen, O.: About 3D-reconstruction from technical drawings. In: Int. Workshop on Industrial Applications of Machine Intelligence and Vision, pp. 46–49 (1989)
8. Wang, W., Grinstein, G.: A Survey of 3D Solid Reconstruction from 2D Projection Line Drawings. Computer Graphics Forum 12(2), 137–158 (1993)
9. Company, P., Piquer, A., Contero, M., Naya, F.: A Survey on Geometrical Reconstruction as a Core Technology to Sketch-Based Modeling. Computers & Graphics 29(6), 892–904 (2005)
10. Zhou, X., Qiu, Y., Hua, G., Wang, H., Raun, X.: A feasible approach to the integration of CAD and CAPP. Computer-Aided Design 39, 324–338 (2007)
11. Wen, Y., Zhang, H., Sun, J., Paul, J.C.: A new method for identifying and validating features from 2D sectional views. Computer Aided Design 43(6), 677–686 (2011)
12. Rivers, A., Durand, F., Igarashi, T.: 3D modeling with silhouettes. ACM Transactions on Graphics 29(4), art. no. 109 (2010)
13. Vandenbrande, Requicha: IEEE Transactions on Pattern Analysis and Machine Intelligence 15(12), 1269–1285 (1993)
14. Han, J.H., Requicha, A.: Integration of feature based design and feature recognition. Computer-Aided Design 29(5), 393–403 (1997)
15. Varley, P.A.C.: Automatic Creation of Boundary-Representation Models from Single Line Drawings. PhD Thesis. Department of Computer Science. University of Wales (2003)
16. Johnson, G., Gross, M.D., Hong, J., Do, E.Y.L.: Computational Support for Sketching in Design: A Review. Foundations and Trends in Human–Computer Interaction 2(1), 1–93 (2009)
17. Kyratzi, S., Sapidis, N.: Extracting a polyhedron from a single-view sketch: Topological construction of a wireframe sketch with minimal hidden elements. Computers and Graphics 33(3), 270–279 (2009)
18. Varley, P.A.C., Company, P.: A new algorithm for finding faces in wireframes. Computer-Aided Design 42(4), 279–309 (2010)
19. Varley, P.A.C., Martin, R.R., Suzuki, H.: Making the Most of Using Depth Reasoning to Label Line Drawings of Engineering Objects. In: Elber, G., Patrikalakis, N., Brunet, P. (eds.) Proc. 9th ACM Symposium on Solid Modeling and Applications, pp. 191–202 (2004)
20. Martin, R.R., Suzuki, H., Varley, P.A.C.: Labelling Engineering Line Drawings Using Depth Reasoning. J. of Computing and Information Science in Engineering 5(2), 158–167 (2005)
21. Conesa, J.: Reconstrucción geométrica de sólidos utilizando técnicas de optimización. PhD dissertation, Universidad de Cartagena, Spain (2001)
22. Company, P., Contero, M., Conesa, J., Piquer, A.: An optimisation-based reconstruction engine for 3D modelling by sketching. Computers & Graphics 28(6), 955–979 (2004)

A Design Framework for an Integrated Sensor Orientation Simulator

Supannee Tanathong and Impyeong Lee

Laboratory for Sensor and Modeling,
Department of Geoinformatics,
The University of Seoul,
Dongdaemun-gu, Seoul, Korea
littlebearproject@yahoo.co.uk,
iplee@uos.ac.kr

Abstract. Integrated sensor orientation (ISO) is an alternative approach in determining the exterior orientation parameters of the camera at the time of exposure. The technique combines the advantages of the direct geocoding and the conventional aerial triangulation. It exploits directly measured orientation parameters as constraints for a bundle adjustment process. This technique is more timely efficient than the traditional adjustment in which ground control points can be eliminated. Also, its accuracy is superior to that of direct geocoding and shows the potential in large scale applications. The accuracy of ISO depends heavily on the accuracy of the directed measured EOs and the distribution and number of tie-points. Various combinations of input produce varying results of ISO. This paper thus presents a design framework of an ISO simulator that can assist operators to investigate real ISO systems with controllable parameters.

1 Introduction

A fundamental photogrammetric problem is to determine the exterior orientation parameters of the camera at the time of exposure and the 3D coordinates of object points [1]. In photogrammetric history, aerial triangulation has long been a conventional technique to solve the orientation of images by exploiting the knowledge of ground control points and corresponding tie-points on a block of images [2]. Although recently tie-points can be obtained less-timely and automatically through various sophisticated image matching techniques invented in the computer vision field, the acquisition of ground control points through ground surveying is still a very time-consuming stage and also not feasible for inaccessible areas or dangerous locations.

With the advancement of micro-electromechanical systems (MEMS) and the improvement in sensor technology, the direct measurement of exterior orientation parameters can be achieved through the integrated GPS/inertial systems. This technique is known as direct georeferencing or direct geocoding. The integration of GPS and INS

G. Bebis et al. (Eds.): ISVC 2012, Part II, LNCS 7432, pp. 568–577, 2012.

sensors makes it possible to eliminate the time-consuming ground control survey that can enormously reduce cost and time [3-5]. However, some researches show that the accuracy of the direct geocoding is not superior to the results of the conventional aerial triangulation but two to three times lower [5-6]. Although a high accuracy can be achieved by employing a high quality but expensive IMU, the y-parallaxes are still troublesome [7]. Typically, a y-parallax above 20 μm causes a difficulty for stereo compilation [5],[7]. This problem is more visible in large scale photography.

Integrated sensor orientation (ISO) is an alternative approach that combines the advantages from both direct geocoding and aerial triangulation [8]. This technique exploits a number of tie-points and the direct observed positions and attitudes as constraints for a bundle block adjustment. The ground control points can be eliminated or used for a quality control. This system helps reduce the y-parallax in the stereo models [5],[7] and that shows more potential for large scale mapping.

The accuracy of ISO depends heavily on the conditions of input parameters. The influence of the direct measured EOs towards the result of ISO comes from the accuracy of the observed data while tie-points form a huge combination. The related tie-point factors can be the distribution of tie-points over image area and the number of tie-points which are resulted from the image matching techniques used. Various combinations of input parameters influence the results of ISO in different ways. Thus this paper presents a design framework for an integrated sensor orientation simulator that can assist operators in deducing the results of the real ISO system through the simulator with adjustable input parameters.

The remainder of this paper is organized as follows: Section 2 details the design of the proposed ISO simulator. Section 3 presents the results of the system. The conclusion is discussed in the last section.

2   A Design Framework for an ISO Simulator

Since the integrated sensor orientation adjusts the accuracy of the directly measured EOs through the conventional aerial triangulation, the simulator thus involves two main processes: (1) ISO parameter generator process and (2) aerial triangulation process. The first process generates the parameters required to perform a bundle adjustment; mainly exterior and interior orientation parameters, tie-points and corresponding ground points for accuracy evaluation. The second process is the actual implementation of the aerial triangulation with a bundle block model. For performance analysis, the simulator offers an accuracy evaluation process to help operators to investigate the results. Fig. 1 illustrates the high-level overview of the simulator which presents the input and output of the system and the main processes. In Fig. 2, the tasks of the simulator are listed out and the flows between the functions are shown. The remainder of this section discusses the input of the simulator, and the aforementioned two main processes.

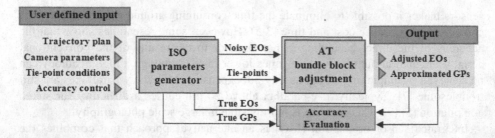

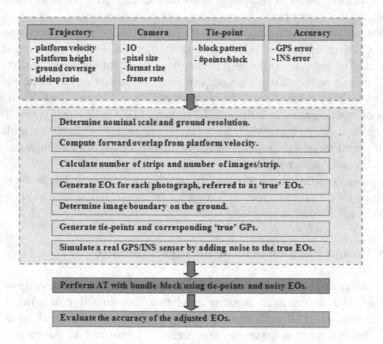

Fig. 1. The high-level overview of the ISO simulator presents two main processes of the system. The input parameters are passed-in by operators to the ISO parameter generator and the output is returned as results from the aerial triangulation process.

Fig. 2. This flowchart lists all tasks performed in the ISO simulator process

2.1   Input of the Simulator

The simulator uses the input passed-in by operators for its computation. The user-defined parameters can be categorized into four groups as follows.

Trajectory Plan. This set of parameters involves the velocity and altitude of the platform. The simulator assumes that the aircraft moves with a constant velocity. Therefore, the number of images per strip is unchanging throughout the trajectory. It also requires the dimension of the ground coverage or project area such that the system can use to calculate the number of photographs per strip and the total number of strips from the given the sidelap ratio.

Camera Parameters. This involves the interior orientation parameters, pixel size and detector dimension. Also, the camera frame rate is required, in conjunction with the aircraft velocity, to compute the distance between each photograph.

Tie-Point Conditions. In the photogrammetric history, tie-points are generated manually by skilled operators. In the recent few decades, feature extraction and matching techniques developed in the computer vision field have played a vital role in automatic tie-point generation in a more timely-efficient manner. One part of this simulator operates as an automatic image matching in generating a set of tie-points for each stereo pair of images. It requires a tie-point pattern such that each image area is divided into separate non-overlapping blocks. A maximum number of tie-points per block are also demanded by the system.

Accuracy Control. Every measurement contains error [9]. Therefore, the system requires the knowledge of uncertainties from the involving components in order to imitate the characteristics of real systems. This set of parameters includes the accuracy of GPS and INS sensors that shows the reliability of the positions and orientations of photographs.

2.2   ISO Parameter Generator Process

This process uses those user-defined input parameters to compute and generate the values required for its internal computation and output for the next process. The resulting values can be categorized into four sets: (a) auxiliary parameters which are non-influential output but used for internal computation, (b) EO parameters for all images, (c) a set of tie-points, and (d) ground points that are corresponding to the generated tie-points. The section below discusses how these outputs are calculated. Since tie-points and ground-points are generated simultaneously, the method to compute them will be discussed together.

Determining Auxiliary Parameters. This stage involves a geometric computation of fundamental properties in planning aerial photographic missions. The variables to be determined are nominal scale, nominal altitude, image ground resolution, area of image coverage on the ground and the endlap ratio. These parameters are computed through basic formulas; consult [10] for more details. Using the user-defined project area, the image dimension, sidelap ratio and the resulting endlap ratio, the number of strips and photographs per strips can be computed. Finally, the total number of images in this project can be derived. The auxiliary parameters computed in this step are illustrated as Fig. 3.

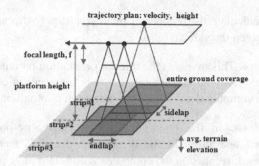

Fig. 3. This figure shows fundamental parameters in planning aerial photogrammetric missions

Generating EOs. This stage is to generate the exterior orientation parameters for each image defined in the project. This set of EO parameters is referred to as 'true EO' and will be used for generating tie-points and their corresponding ground points and for further evaluation process. The other outcome from this process is the set of EOs which have been padded with noises according to the uncertainty of the GPS/INS sensors. This latter set of EOs simulates the direct measured values acquired through the integrated sensors and they are referred to as 'noisy' or 'directly measured' EOs.

It first starts by defining the EO parameters of the first photograph locating at the top-left corner of the project area (the origin is at the bottom-left corner), see Fig. 4. Then EOs of the next photographs are generated by adding a displacement to the first image plus some noises that are computed as random numbers within the range of standard deviation of the platform movement. The trajectory path flows from the top-left corner of the project area to the right-most side such that the Y coordinates of the principal points are nearly constant. Once reaching the other end side of the project area, the trajectory path flows in the opposite direction (right to left) and the Y coordinates of the new set of photographs are deducted by a factor of sidelap. The procedure of generating the true EOs for all photographs in the project area is shown as Fig. 5. Since it is inevitable to avoid errors not only from the devices themselves but environmental factors as well, to imitate the real situation of data acquisition of the GPS/INS sensors, the generated true EOs are added with random noises that bounded within the standard deviation of positioning errors for GPS and orientation errors from INS.

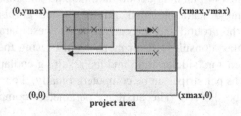

Fig. 4. The trajectory path starts from the first photograph at the top-left corner to the right end before moving in the opposite direction

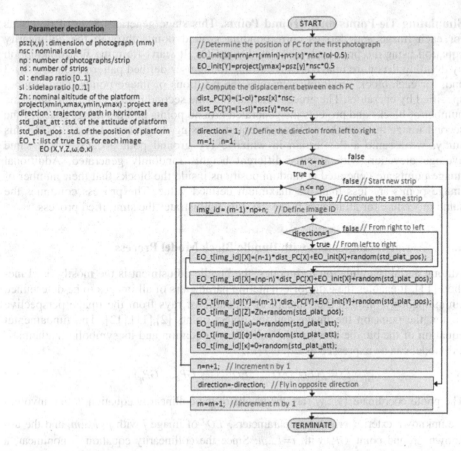

Fig. 5. The procedure in generating exterior orientation parameters for all images in the projecting area

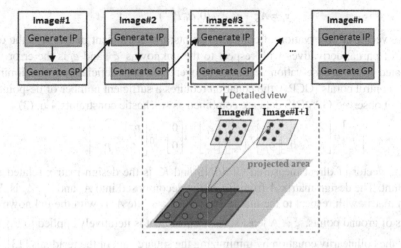

Fig. 6. This picture illustrates the process of generating tie-points and ground points

Simulating Tie-Points and Ground Points. This stage generates a set of tie-points for each image and derives corresponding ground points through the collinearity equation using the previously simulated true EOs. It starts from the first photograph by dividing the entire image area according to the user-defined pattern such as 3-by-3 and, for each block, randomly generates the positions of image points for a number specified by operators. The process then divides the second image's area into the same number of blocks and projects the generated ground points into the image space of the second image through the collinearity equation using its generated true EOs. In this study, we assume a flat terrain in which every ground point locates at the same average elevation with slightly different heights randomly generated. Additional image points are generated at random positions inside the blocks that their number of image points are less than the maximum defined value. The process continues the same procedure for all images in the list. Fig. 6 illustrates the simplified process.

2.3   Aerial Triangulation with Bundle Block Model Process

Among aerial triangulation techniques, the bundle adjustment is the mostly used method [11]. It enables the exterior orientation parameters of all images to be determined simultaneously through the concept of a bundle of rays from the image perspective center, the point on the image and the object points [2],[11],[12]. The fundamental equation of the bundle block is the collinearity equation and its symbolic mathematical model can be expressed as Eq. (1) [2].

$$(x_i^j, y_i^j) = f(EO^1, EO^2, ..., EO^m, GP_1, GP_2, ..., GP_n).$$ (1)

The photo coordinate (x_i^j, y_i^j) are the results of the collinearity equation, f, that involves the unknown exterior orientation parameters, EO^j of image j with $j=1,..,m$, and the unknown ground point GP_i with $i=1,..,n$. Since the collinearity equation is nonlinear, a first-order Taylor series expansion is applied to linearize Eq. (1) [9]. After linearization, the Gauss-Markov model for estimating the unknowns is presented as Eq. (2).

$$y_i = A\xi + e_i, e \sim (0, \sigma_0^2 P_i^{-1}).$$ (2)

y_i is the vector of observation of conjugate points; the coefficient matrix A is the design matrix of partial derivatives with respect to the unknowns ξ, and e_i is the error vector associated with the observation vector y_i. To overcome rank deficiency from eliminating ground control points (GCPs), the equation requires a sufficient number of tie-points and the direct observed GPS/INS data are combined as stochastic constraints, Eq. (3).

$$\begin{bmatrix} y_i \\ y_e \end{bmatrix} = \begin{bmatrix} A_e & A_p \\ K_e & 0 \end{bmatrix} \begin{bmatrix} \xi_e \\ \xi_p \end{bmatrix} + \begin{bmatrix} e_i \\ e_e \end{bmatrix}, \begin{bmatrix} e_i \\ e_e \end{bmatrix} \sim (\begin{bmatrix} 0 \\ 0 \end{bmatrix}, \sigma_0^2 \begin{bmatrix} P_i^{-1} & 0 \\ 0 & P_e^{-1} \end{bmatrix}).$$ (3)

y_e is the vector of direct measured GPS/INS and K_e is the design matrix related to the constraint. The design matrix A from Eq. (2) is decomposed into A_e and A_p. A_e is the Jacobian matrix with respect to the unknown EOs, ξ_e, while A_p is with the unknown coordinates of ground points, ξ_p. A least square adjustment is iteratively applied to Eq. (3) to satisfy the collinearity equation by minimizing the square sum of the residuals [13].

3   Experimental Results

This section presents an example result of the ISO simulator discussed in the previous section. The simulator is programmed in MATLAB. The user-defined parameters input to control the system are listed in Table 1 and the results are presented as Fig. 7.

Table 1. Experimental parameters

Trajectory	Camera	Tie-point	Accuracy
velocity: 36 km/h	pixel size: 3.45μm	pattern: 3×3	GPS error: 0.3 m
altitude: 200 m	image size: 2456×2058 pix	# point: 1 pt/block	INS error: 0.1 deg
sidelap: 30%	focal length: 17mm		
	frame rate: 0.5 s		

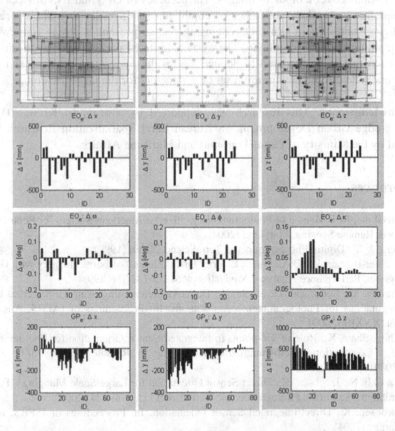

Fig. 7. The results obtained from the simulator. The top row presents the layout of ground coverage for each image and where ground points are located. The second and the third row show the discrepancy between the approximated EOs from the simulator and the generated true EOs. The bottom row presents the discrepancy between the adjusted ground points and the generated true ground points.

With the project area defined as 180×200 m^2, the input trajectory parameters and frame rate produces 3 strips with 8 images per strip. The tie-point simulator generates totally 274 tie-points and 74 corresponding ground points. The bundle block adjustment process consumed 0.32 seconds for processing on a 2.93GHz computer with 4GB-RAM. The resulting EO parameters in the object space are illustrated as Fig. 7 and the average residual in the image space is 17 microns.

4   Concluding Remarks

Among the existing approaches, the integrated sensor orientation is a very promising technique in solving the orientation problem of image as it combines the advantages of both direct geocoding and aerial triangulation. The technique exploits the direct observed position and attitude parameters through integrated GPS/INS sensors and a set of tie-points over a block of images. The accuracy of GPS and INS devices and the tie-point factors produce a number of combinations. In order to assist operators to investigate the results of real ISO systems, this paper presents a design framework for implementing an ISO simulator. The system imitates real systems with adjustable input such that it can be used to deduce the results of real applications and for the operators to determine the optimum set of parameters for the target outcomes.

Acknowledgement. This research was supported by a grant (07KLSGC03) from Cutting-edge Urban Development - a Korean Land Spatialization Research Project funded by the Ministry of Land, Transport and Maritime Affairs.

References

1. McGlone, J.C.: Manual of Photogrammetry, 5th edn. American Society for Photogrammetry & Remote Sensing, Maryland (2004)
2. Schenk, T.: Digital Photogrammetry. TerraScience, Ohio (1999)
3. Grejner-Brzezinska, D.A.: Direct Exterior Orientation of Airborne Imagery with GPS/INS System: Performance Analysis. Navigation 46(4), 261–270 (1999)
4. Cramer, M., Stallmann, D., Haala, N.: Direct Georeferecning using GPS/Inertial Exterior Orientations for Photogrammetric Applications. In: Proceedings of IAPRS, Amsterdam, vol. XXXIII (2000)
5. Khoshelham, K.: Role of Tie-points in Integrated Sensor Orientation for Photogrammetric Map Compilation. Photogrammetric Engineering & Remote Sensing 75(3), 305–311 (2009)
6. Yastikli, N., Jacobsen, K.: Direct Sensor Orientation for Large Scale Mapping – Potential, Problems, Solutions. Photogrammetric Record 20(111), 274–284 (2005)
7. Jacobsen, K.: Direct/Integrated Sensor Orientation. In: Proceedings of ISPRS, Istanbul, Turkey (2004)
8. Ip, A.: Performance Analysis of Integrated Sensor Orientation. Photogrammetric Engineering & Remote Sensing 73(1), 89–97 (2007)

9. Ghilani, C.D., Wolf, P.R.: Adjustment Computations: Spatial Data Analysis. John Wiley & Sons, New Jersey (2006)
10. Falkner, E., Morgan, D.: Aerial Mapping Methods and Applications, 2nd edn. CRC Press, Boca Raton (2002)
11. Mikhail, E.M., Bethel, J.S., McGlone, J.C.: Introduction to Modern Photogrammetry. John Wiley & Sons, New York (2001)
12. Wolf, P.R., Dewitt, B.: Elements of Photogrammetry with Applications in GIS. McGraw-Hill, Boston (1999)
13. Schenk, T.: From Point-based to Feature-based Aerial Triangulation. ISPRS Journal of Photogrammetry & Remote Sensing 58, 315–329 (2004)

Automatic Improvement of Graph Based Image Segmentation

Huyen Vu and Roland Olsson

Østfold University College,
Halden, Norway
roland.olsson@hiof.no

Abstract. Automatic Design of Algorithms through Evolution (ADATE) is a system for fully automatic programming that has the ability to either generate algorithms from scratch or improve existing ones.

In this paper, we employ ADATE to improve a standard image processing algorithm, namely graph based segmentation (GBS), which has emerged as one of the very most popular methods for image segmentation, that is partitioning an image into regions.

The key contribution of the paper is to show that a proven and well-known computer vision code is easy to improve through automatic programming. This may presage a change to the entire field of computer vision where automatic programming becomes a routine way of improving standard as well as state-of-the art image processing and pattern analysis algorithms.

GBS was mostly chosen as case study to investigate how useful the ADATE automatic programming system may be in computer vision. Numerous other algorithms in the field could have been chosen instead.

Keywords: Image processing, Graph based segmentation, Machine learning, Automatic programming, ADATE.

1 Introduction

Automatic Design of Algorithms through Evolution (ADATE) [10] has the ability to either generate algorithms from scratch or improve existing ones.

In this paper, we use ADATE to automatically improve a standard image processing algorithm, namely graph based segmentation (GBS), which has emerged as one of the very most popular methods for segmentation, that is partitioning an image into regions. The new and automatically generated GBS algorithm that we present in this paper gives higher quality segmentations, which we show by running it on a set of 50 test images for which the so-called F-score is increased from 67% to 77%.

The key contribution of the paper is to show that a proven and well-known computer vision code is easy to improve through automatic programming. This may presage a change to the entire field of computer vision where automatic

G. Bebis et al. (Eds.): ISVC 2012, Part II, LNCS 7432, pp. 578–587, 2012.

programming becomes a routine way of improving standard as well as state-of-the art image processing and pattern analysis algorithms.

GBS was mostly chosen as case study to investigate how useful the ADATE automatic programming system may be in computer vision and numerous other algorithms in the field could have been chosen instead. However, a key advantage of the GBS algorithm is its outstanding efficiency, which means that it can be used to process many million pixels per second an a standard CPU. This speeds up automatic programming since the overall run time of ADATE is proportional to the run time of the generated algorithms.

A recent comparison [2] of the segmentation accuracy for eleven different segmentation algorithms, including GBS, showed that four algorithms were worse than GBS, that two are about as accurate as GBS and that four are clearly better. In practice, the reason for choosing GBS instead of the higher quality segmentation algorithms in [2] is its superior speed. In this paper, we improve the accuracy of GBS while retaining its fundamental efficiency advantage.

There are numerous practical applications of image segmentation, for example locating objects in satellite images, preprocessing for face recognition [5], fingerprint recognition [9], auto-braking systems for vehicles, medical imaging [6] and object tracking in the new MPEG-7 video compression standard.

2   Graph Based Segmentation

The algorithm that we chose to be improved by ADATE is as mentioned above graph based segmentation (GBS), which is a minimum spanning tree based algorithm introduced by Felzenszwalb in [7]. GBS is widely used in practice since it is both very efficient and easy to implement. Another reason for choosing GBS is that there are potential alternative region merging heuristics that can be found by automatic programming.

.GBS represents an image as an undirected graph with nodes corresponding to pixels and edges corresponding to pairs of neighbouring pixels. The weight of each edge is the dissimilarity between neighbouring pixels connected by that edge, e.g., the difference in intensity, color, motion, location or some other local attribute. In general, the elements in a component are similar, and elements in different components are dissimilar. In other words, edges between two nodes in the same component should have relatively low weights, and edges between nodes in different components should have higher weights.

Initially, there is exactly one component for each node. Let maxWeight(C) denote the maximum weight in a minimum spanning tree for a component C. GBS sorts all edges in a non-decreasing order and loops through them one by one in that order.

Assume that an edge with weight w that connects components C_1 and C_2 is being considered. Let us also define a function τ that is inversely proportional to the number of nodes in a component as follows.

$$\tau(C) = \frac{k}{|C|}$$

GBS merges the components C_1 and C_2 if and only if

$$w < \min(\mathrm{maxWeight}(C_1) + \tau(C_1), \mathrm{maxWeight}(C_2) + \tau(C_2))$$

The constant k is a scale parameter that is chosen before GBS starts and that controls the size of the segments. Increasing k will result in more merges and larger segments.

3   Automatic Programming

Automatic programming, also known as program induction, is an emerging technology where either a reasonably small program or one or more parts of a potentially huge software system are automatically synthesized.

ADATE [10] is one of the leading systems for induction of programs and is able to induce recursive functional programs with automatic invention of recursive help functions as needed. It could solve standard algorithmic problems such as sorting, deletion from binary search trees, generating all permutations of a given list and so on, two decades ago and has been steadily improved since then. During the last five years, ADATE has been used to improve the nodes in a recurrent neural network [3], to generate algorithms for autonomous cars [4] and to improve state-of-the-art algorithms for combinatorial optimization [8].

In addition to algorithmically sophisticated program transformations that generate new expressions, ADATE contains abstraction (ABSTR) that introduces new functions, and case/lambda-distribution/lifting (CASE-DIST) that manipulates the scope of variables and functions. Both of the latter two transformations preserve semantics and are not as combinatorially difficult as the expression synthesis transformations.

The main challenge when designing a program evolver such as ADATE is to reduce overall run time, which depends on the size of the final program that is to be produced and the average execution time required to run a program on all training inputs.

4   Experiments

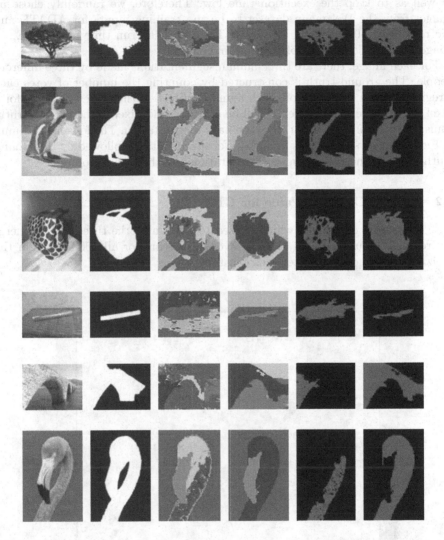

Fig. 1. From left to right: The original images, the ground truth images, segmentations with the original algorithm, segmentations with the improved algorithm, the best segment for each image using the original algorithm, and the best segment using the improved algorithm

4.1   Training and Test Images

We chose the 100 single object gray level images from the Weizmann Institute's dataset [1]. The reason for choosing this dataset is that there for each image is only one object depicted as the foreground object. This means that we can avoid ambiguities in determining which object that is the target for segmentation.

We need to use a suitable number of training inputs to alleviate overfitting as well as to keep the execution time low. Therefore, we randomly chose 50 images from the Weizmann dataset to be the training inputs for ADATE and the remaining 50 images as a test set. To reduce the run time of ADATE, all images were scaled to 25% of their original size.

For each image, there are three human segmentations made by three different people. The ground truth is constructed by counting the number of votes each foreground pixel received from each human subject. If there are two votes for a pixel to be a foreground pixel, it will be a foreground pixel in the ground truth. Thus, the resulting ground truth is a binary segmentation. The left-most column in Figure 1 shows six images from the Weizmann dataset followed by the ground truths, that is the voted human segmentations, in the second column.

4.2   Chosen Constant Value for GBS

GBS contains a constant k which is used to determine the threshold for merging regions. Figure 2 shows how the average F-score for all 100 images in the Weizmann dataset depends on this constant k.

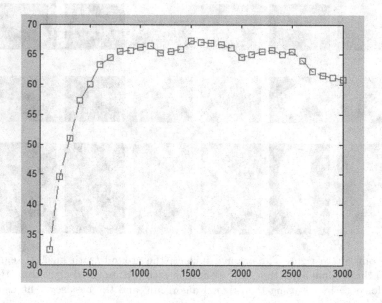

Fig. 2. Constant values and corresponding F-score values

The maximum F-score is rather flat for k values between 800 and 1800 with the best score for $k = 1500$.

Instead of choosing the optimal constant value for the algorithm, we used $k = 1000$ to show that ADATE does not rely on careful a priori optimization of k.

5 Results

5.1 The Best Automatically Synthesized Program

After running ADATE for 24 hours on 192 INTEL i7-2600 cores, we obtained an improved program, called AGBS10, which is shown in Figure 3. All of the code in that program has been subjected to evolution.

ADATE generates programs written in a minimalistic subset of Standard ML, but the code in Figure 3 has been rewritten in more normal functional programming syntax to be easier to understand. Note that the program contains abstract data types with arrays and therefore is not purely functional.

The function f goes through the edges in the list SortedEdges which contains the edges in order of non-decreasing weights, that is the smallest weights first. The parameter Universe is an abstract object that contains the data necessary for union-find operations and also an array that associates a threshold with each component. Initially, each pixel is its own component and given an initial threshold of 1000.

```
fun f( Universe, SortedEdges ) =
  case SortedEdges of
    [] => Universe
  | edge( A, B, W ) :: RestEdges =>
  let
    val ( ComponentA, ThresholdA ) = find( A, Universe )
    val ( ComponentB, ThresholdB ) = find( B, Universe )
    val ( Comp, Thres ) =
      if  W < 4.7 then
        ( ComponentB, ThresholdB )
      else if ThresholdB < ThresholdA orelse
              compSize ComponentA < 57.9
      then
        ( ComponentB, ThresholdA )
      else
        ( ComponentA, ThresholdA )
    val NewThres =
      ( compSize Comp + Thres / tanh Thres ) / ThresholdA
  in
    f(
      updateThresholdValues(
        ComponentA,
        NewThres,
        union( Universe, ComponentA, Comp )
      ),
      RestEdges
    )
  end
```

Fig. 3. The best program, called AGBS310, that was generated by ADATE

Note that it is the algorithm itself that has been automatically improved and that this of course goes far beyond mere parameter optimization.

The code for f is only allowed to access and change Universe through three predefined functions, namely find, union and updateThresholdValues. The find function takes a node as its first argument, finds the component that the node currently belongs to and returns the root node of that component along with the current threshold value for the component. The union function merges two components and returns the resulting new universe. The role of updateThresholdValues is simply to update the array of threshold values in a universe with a new value, called NewThres, that is computed in f. The help function compSize used in Figure 3 returns the number of nodes in a component.

If the list of sorted edges is empty, f terminates and returns Universe. Otherwise, the list is split into its first element and the tail of the list, called RestEdges. The first element in the list is an edge that goes between nodes denoted by A and B with a weight W on the edge.

The totally novel and automatically invented code in Figure 3 is the one that computes first the pair (Comp, Thres) and then the new threshold NewThres for component A. Note that Comp may be the same as component A, which means that the union operation in the tail-recursive call to f leaves the universe unchanged. Thus, ADATE has generated code that performs lots of union operations that have no effect, but this would of course be very easy to rewrite to more efficient code.

If W < 4.7, components A and B will be merged as seen in the code for f. Since we use 8-bit grey scale images, the intensity values range from 0 to 255. The weight on an edge is the difference in intensity between the two pixels that it connects. Therefore, this test can be rewritten to $W \leq 4$ and obviously means that the difference between the pixels is small and that it may make sense to merge. Since the list of edges is sorted in ascending order, all such similar neighbouring pixels will be merged early on during the execution of f.

The rest of the automatically generated code in Figure 3 looks deceptively simple, but actually gives rise to complex behaviour that is not easy to fully understand. The key to analyzing it seems to be the calculations of threshold values using component size and the hyberbolic tangent function. We will provide an experimental analysis below but refrain from a theoretical analysis of the automatically invented calculations of thresholds.

5.2   Comparison of F-Score Results

Table 1 contains the F-scores for GBS and AGBS10 for the 50 training and the 50 test images.

As we can see, with the program generated by ADATE, we got an average F-score of 0.7298 instead of 0.6564 for the training data and 0.7706 instead of 0.6666 for the test data. The reason that the result on the training images actually did not improve as much as for the test images is statistical fluctuations, but it is difficult to compute a confidence interval for our results and this does not seem to be standard practice in papers on image processing.

Table 1. Sum and average F-score for training and test images

Algorithm	The 50 training images		The 50 test images		All 100 images	
	Sum	Average	Sum	Average	Sum	Average
GBS	32.8217	0.6564	33.3286	0.6666	66.1503	0.6615
AGBS10	36.4910	0.7298	38.5289	0.7706	75.0202	0.7502

Note that evolution of programs in ADATE may specialize the programs to the training experience. Therefore, AGBS310 can only be expected to be superior as seen in Table 1 for images of roughly the same size as our 50 training images. The constant values in AGBS310 would need to be changed in order to segment bigger images.

5.3   Experimental Analysis of the Improved Program

Recall that a weight not greater than four leads to merging in AGBS10. This behaviour can be found in most of the dataset images where the intensity difference between neighbouring components is small and avoids over-segmentation.

We see this in the tree images in Figure 1, where there are a smaller number of segments for AGBS10. The best segment found also fits with the ground truth better than the the best segment found by GBS.

The F-score for GBS is in this case 0.83732 while it is 0.92731 for AGBS10. This can be explained when we look at the left side of the tree. Although the sunlight makes the intensity of the branch of the tree a little bit different from the leaves, they should be considered to belong to the whole tree. AGBS10 solves this problem and the branch is merged with the whole tree as it should be.

A significant improvement can be found in the penguin images in Figure 1. In the best segment for GBS, only a small part of the penguin was correctly segmented while its shadow wrongly was segmented as a part of the penguin. Also, the white part of the penguin was segmented incorrectly as a different segment.

In contrast, AGBS10 segments the penguin image well and separates the shadow of the penguin from the penguin itself. In addition, the white part of the penguin was merged into the penguin as it should be. The F-score for this image is 0.4138 with GBS but with AGBS10 it is actually 0.7787!

Thus, GBS sometimes merges regions with similar intensity levels together in the wrong way. For example, it puts the shadow of the penguin in the same segment as the penguin and in the image with a pen on a table it wrongly merges many pixels at the border to the pen with the pen itself.

AGBS 10 avoids these pitfalls by considering both the threshold value and the component size to determine whether to merge components.

6   Conclusions and Future Work

The F-score is a generally accepted way to measure the quality of segmentations. For a set of 50 test images, GBS has an F-score of 67%. The best algorithm generated by ADATE, that is AGBS10, has 77%.

The F-score of GBS was computed after optimizing the constant that is used in its merging threshold calculations. Our GBS implementation was validated against a public domain C code and found to give exactly the same results.

The paper has made two key contributions.

1. A substantial improvement in the segmentation quality for graph based image segmentation [7].
2. An indication that ADATE may be able to improve computer vision algorithms in general.

ADATE has also been used to improve a totally different image segmentation method, namely PCNN [3]. This means ADATE now has been used to automatically improve two different image processing algorithms, that is, GBS and PCNN. These are the only computer vision algorithms that ADATE has been used for so far. ADATE could easily generate better algorithms in both cases. Of course, this opens the possibility that there may be many other computer vision methods that ADATE can improve automatically.

There are at least two challenges to overcome when using ADATE to improve an algorithm.

1. The algorithm must be reimplemented in ADATE-ML, which requires very good skills in functional programming.
2. The run time of the ADATE system may be quite long since it needs to evaluate millions or billions of different algorithm candidates. However, the generated algorithms typically run as fast as the original ones.

The key issue for the future of automatic programming in computer vision is the last one, that is the long computing time required for evolution of better algorithms. This time can be kept within reasonable bounds by focusing on small images or by employing higher performance computer clusters when running ADATE.

References

1. Alpert, S., Galun, M., Basri, R., Brandt, A.: Image segmentation by probabilistic bottom-up aggregation and cue integration. In: Proceedings of the IEEE Conference on Computer Vision and Pattern Recognition (2007)
2. Arbelaez, P., Maire, M., Fowlkes, C., Malik, J.: Contour Detection and Hierarchical Image Segmentation. IEEE Transactions on Pattern Analysis and Machine Intelligence 33, 898–916 (2011)
3. Berg, H., Olsson, R., Lindblad, T., Chilo, J.: Automatic design of pulse coupled neurons for image segmentation. Neurocomputing (2008)

4. Berg, H., Olsson, R., Rusås, P.-O., Jakobsen, M.: Automated synthesis of control algorithms from first principles. In: The 2009 IEEE/RSJ International Conference on Intelligent Robots and Systems, IROS 2009 (2009)
5. Ngan, K.N., Chai, D.: Face segmentation using skin-color map in videophone applications. IEEE Transactions on Circuits and Systems for Video Technology 9, 551–564 (1999)
6. Xu, C., Pham, D.L., Prince, J.L.: Current Methods in Medical Image Segmentation. Annual Review of Bimedical Engineering 2 (2000)
7. Felzenszwalb, P.F., Huttenlocher, D.P.: Efficient Graph-Based Image Segmentation. In: International Journal of Computer Vision (2004)
8. Løkketangen, A., Olsson, R.: Generating meta-heuristic optimization code using ADATE. Journal of Heuristics (2010)
9. Mehtre, B.M., Chatterjee, B.: Segmentation of fingerprint images – A composite method. Pattern Recognition 22, 381–385 (1989)
10. Olsson, R.: Inductive functional programming using incremental program transformation. Artificial Intelligence 74(1) (1995)

Analysis of Deformation of Mining Chains Based on Motion Tracking

Marcin Michalak, Karolina Nurzyńska,
Andrzej Pytlik, and Krzysztof Pacześniowski

Central Mining Institute, Plac Gwarków 1, 44-166 Katowice, Poland
{Marcin.Michalak,Karolina.Nurzynska,Andrzej.Pytlik,
Krzysztof.Paczesniowski}@gig.eu

Abstract. This paper presents the possibilities of using a compact digital camera, which is amateur class of equipment, to a simple analysis of the deformation of mining chains during the tests performed at percussive, dynamic load. When the mining chain works during underground coal-bed exploitation, in the scraper conveyor, which is one of the elements of a longwall system, there could be observed frequent chains fractures due to its dynamic percussive load. It often occurs as a result of an emergency chain locking in the troughs of the chain conveyor. For the purpose of analysis special software was created. Thus it was possible to obtain at low cost, with the assumed accuracy of deformation measurement, deformations as a function of time, without installing on the monitored elements acceleration sensors (e.g. inductive, potentiometric), which were usually destroyed during the tests, mostly due to their low resistance to shock.

1 Introduction

The problem of breaking of the mining chains is very important in the coal mining. When the chain link breaks it usually generates some sparkles that may initialize the fire. From this point of view the analysis of the chain resistance at the dynamic load is very interesting. One of the method of analyzing the chain is to measure the influence of the load on the chain length. This problem may be resolved with the usage of optical flow analysis methods. One of the most popular algorithm of optical flow is the Horn and Schunck [7]. Similar group of algorithms are variational methods [4,9,10]. Other optical flow methods are correlation–based [3], frequency–based [6] and phase–based [5].

This paper presents two important aspects of optical flow analysis: the usage of non–professional camera as the motion capture tool and the algorithm of elimination of irregular lighting. The paper is organized as follows: in the beginning the background of the problem and some previous works are described. Then the method of optical flow is presented together with the simple way of irregular lighting problem elimination. Afterward, the program for motion tracking is presented and results of analysis of link mining chains. The final part of the paper contains some conclusions and plans for the future.

G. Bebis et al. (Eds.): ISVC 2012, Part II, LNCS 7432, pp. 588–596, 2012.

2   Background and Previous Works

In the previous works implemented in the Central Mining Institute, comparative study of dynamic resistance of the chains at the dynamic load was conducted. The method of testing developed in this paper is to determine the resistance of the chain to the dynamic, percussive load, defined as the maximum energy value dissipated by the chain without its damage (rupture). In cooperation of the Central Mining Institute and THIELE GmbH & Co.KG it was designed and constructed a test stand for trial for breaking the chain at the dynamic load and the first research attempts have already been made.

During the test of chains at the test stand specified performance characteristics of the chain work with a dynamic load was determined. It was presented as charts of the load and deformation as a function of time.

In Figure 1 a diagram of the test stand is presented. It consists of two parts:

- a fixed base - equipped with a support plate on which a force sensor with the upper bracket, mounting the section of the test chain is placed,
- movable loading element - consisting of the top and the bottom plates, connected by bearing rods.

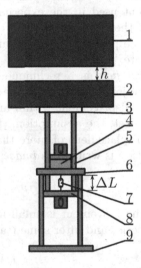

Fig. 1. Schematic of the test stand for the trials of breaking the chain at the dynamic load: 1 – impact mass m_1, 2 – traverse with a mass m_2, 3 – upper plate, 4 – force sensor, 5 – support bars, 6 – carrier plate base, 7 – chain, 8 – bottom plate, 9 – base

In the support plate of the base the holes are made to ensure free movement of the bearing rods of the movable loading element. On the top plate of the loading element there is a traverse weighing $m_2 = 6600$ kg, which statically loads the test section of the chain.

The test consists in free fall of the impact mass $m_1 = 20\,000$ kg on the traverse statically loading the chain in the test stand, from different free-fall height h. The elastic collision of the impact mass and the traverse induces dynamic loading force, which is then transmitted through the bearing rods on the lower section of the chain, which induces dynamic tensile force in the chain.

During the tests, measuring the force as a function of time is performed. Before and after the test, measurements of chain length (7 active cells) are carried out, on the basis of which its total elongation is determined. On one section of the chain one measurement is performed every time. Tests were performed at variable values of drop height up to the chain break. Some attempts have been filmed at 1200 frames per second. Chain loading force was recorded at a measurement frequency rate of 9600 Hz, and images at 1200 frames / second. For the force measurement instrumentation amplifier type HBM's MGCplus (class 0.03) was used, where a tensometric force sensor (class 1) was connected.

3   Object Tracking

The problem of object tracking is very popular in modern literature. There have been designed plenty of solutions which deal with this problem successfully. In the case of presented research, the aim was to track objects marked on the equipment used in the experiments and also enable tracking of characteristic parts of the equipment, when it was impossible to stick a marker. In order to ensure the possibility of accurate motion tracking the acquired films have high frame rate.

Optical flow is a technique which enables recognition of apparent motion on the adjoining frames in the film. It assumes that after short time, Δt, the pixel, p, displacement within an image can be described by coefficient change, $\Delta x, \Delta y$. Yet, the assumption, that the movement is small, and rather does not excess displacement of more than one pixel per frame should be fulfilled. This assumption is written as *image constraint equation*:

$$I(x, t, y) = I(x + \Delta x, y + \Delta y, t + \Delta t). \tag{1}$$

When the movement is small it is possible to approximate this equation with Taylor series and after some transformation the velocity formula is given:

$$I_x V_x + I_y V_y + I_t = 0, \tag{2}$$

where V_x, V_y are the components of object velocity and I_x, I_y, I_t are derivatives of the image $I(x, y, t)$ in corresponding directions. This equation has two unknowns and therefore needs additional constraints to be solvable.

In the literature one of the most famous definition of additional constraints are given by Horn and Schunck [7], who suggest using a global constraint of smoothness to solve the aperture problem. Unfortunately, this approach is very sensitive to noise. On the other hand, Lucas and Kanade [8] based their solution on the assumption that the local neighborhood of the pixel is constant. That results in considering all pixels from neighborhood to resolve the ambiguity of the equation. Due to the much higher number of equations describing the optical flow for a pixel a least squares criterion is applied to choose the solution.

In particular, equations in Lucas-Kanade approach can be written in matrix form $Av = B$ where

$$
A = \begin{bmatrix} I_x(p_1) & I_y(p_1) \\ I_x(p_2) & I_y(p_2) \\ \dots \\ I_x(p_n) & I_y(p_n) \end{bmatrix}, v = \begin{bmatrix} V_x \\ V_y \end{bmatrix}, B = \begin{bmatrix} -I_t(p_1) \\ -I_t(p_2) \\ \dots \\ -I_t(p_n) \end{bmatrix}, \tag{3}
$$

and n corresponds to the pixel neighborhood size. Finally, in order to find values describing the velocity following formula is applied:

$$
\begin{bmatrix} V_x \\ V_y \end{bmatrix} = \begin{bmatrix} \sum_i I_x(p_i)^2 & \sum_i I_x(p_i)I_y(p_i) \\ \sum_i I_x(p_i)I_y(p_i) & \sum_i I_y(p_i)^2 \end{bmatrix}^{-1} \begin{bmatrix} -\sum_i I_x(p_i)I_t(p_i) \\ -\sum_i I_y(p_i)I_t(p_i) \\ \dots \\ -I_t(p_n) \end{bmatrix}. \tag{4}
$$

4   Irregular Lighting Normalization

The measurement setup with high-speed camera uses also halogen lamps to improve scene illumination. Unfortunately, the lamps are connected to an alternating current which causes the light to fluctuate. This change is invisible for a human eye, but when there are taken several images in one light cycle, the change of lighting is noticeable in the video sequence.

Since the optical flow algorithm is applied for object tracking, it implies a constant lighting of the scene. Therefore, it is important to remove or at least diminish the changes of illumination. In presented system, the statistical approach is introduced. It assumes that the average illumination for each frame should be constant and have similar standard deviation, as on all images the same objects are presented. The other difficulty is the fact that the correct average intensity level for image as well as the standard deviation are unknown. Therefore, it is suggested to update these values for each frame.

In presented approach for each frame $I(x, y)$ the average and standard deviation value are calculated following the formula:

$$
\mu = \frac{1}{NM} \sum_{i=0}^{N} \sum_{j=0}^{M} I(i, j), \tag{5}
$$

$$
\sigma = \frac{1}{NM} \sum_{i=0}^{N} \sum_{j=0}^{M} [\mu - I(i, j)]^2, \tag{6}
$$

where N and M are image resolution. Next, the image is normalized according to the equation:

$$
I_{norm}(x, y) = \frac{I(x, y) - \mu}{\sigma}. \tag{7}
$$

Finally, updated average image value and standard deviation value are calculated as:

$$\mu_{total} = \frac{1}{F} \sum_{i=0}^{F} \mu_F, \tag{8}$$

$$\sigma_{total} = \frac{1}{F} \sum_{i=0}^{F} \sigma_F, \tag{9}$$

where F is a number of all frames from the beginning of the movie till current frame, and the image is transformed to reflect this parameters in a following manner:

$$I(x,y)_{out} = I(x,y)_{norm} \cdot \sigma_{total} + \mu_{total}. \tag{10}$$

Fig. 2 depicts the variation of average intensity level in consecutive frames before and after normalization. Similarly on the plot the changes in standard deviation value are presented. The proposed algorithm for varying lighting proved to diminish considerably the influence of non-ideal lamps.

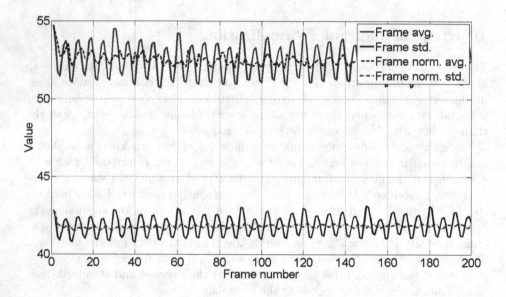

Fig. 2. Luminance fluctuation in consecutive frames of the movie

5   Tracking Software

The software created in the Central Mining Institute implements the algorithm of motion tracking described in detail in the previous part of the paper.

Most important parameters of the algorithm like the width of the marker neighborhood, the ray depth and the number of iteration per each step of the algorithm, may be defined by the user. Their default values are determined by several experiments of tracking objects.

The elimination of the luminance influence may be performed in two ways:

- in the gray scale model: normalization is performed on the gray scaled frames.
- in the HSV model: normalization is performed only on the V component of the HSV color model.

User may define the non-limited list of markers but at most one of them may be defined as the "reference marker". The results of tracking may be generated at every moment of the analysis, It means that user may obtain several "snapshots" from many time-stamps. The "reference marker" plays a very important role. It makes it possible to eliminate some camera movements that change each marker position. If the user place the "reference marker" on the part of the scene that should not move, its possible movement (introduced artificially by the movement of the camera) can be eliminated easily. It can be done on the tab "Results" with clicking the button "Refer series to the reference marker".

The result of the analysis is the set of time series which are listed below:

1. placement – it shows how do the coordinate (X, Y) of the marker change in the following frames,
2. speed – it presents the change of the marker velocity,
3. acceleration – it presents the change of the marker velocity in time,
4. $X(t)$ – decomposition of the first time series: only first coordinate is shown in time,
5. $Y(t)$ – as above, but in the reference to the second coordinate.

Each result is presented in the chart. Charts may be saved as graphics or text files and then postprocessed in the other way.

6   Experiments and Results

6.1   Object of the Analysis

Research on chain slings was performed on sections of chains of size $d \times p$ (defined by the product of the nominal values in [mm] of a diameter of the links and the chain pitch) which is of 34×126 in the THD class [2] (class C according to Polish standard [1]) and in TSD class (no Polish equivalent of the class, class TSD is higher than the class THD) without corrosion protection and with corrosion protection. The producer (company THIELE GmbH & Co.KG) used the substance Tectyl 506 or hot-dip galvanizing as the corrosion protection of the chains. Each section of the chain consisted of nine links, 7 of which were active, with their total measuring length $L = 882$ mm, and the other two extreme links were placed in two brackets (upper and lower) of the test stand.

TSD class chains, at the material strength of about 1500 MPa, during quasi-static tear tests reach accordingly minimum breaking force of the value of 1800 kN, while the THD class chains, with the material strength of about 1200 MPa, in identical conditions reach lower tear force of a minimum value of 1450 kN. Figure 1 presents the test stand for testing chains breaking at the dynamic load. The test stand is located in Łaziska Górne.

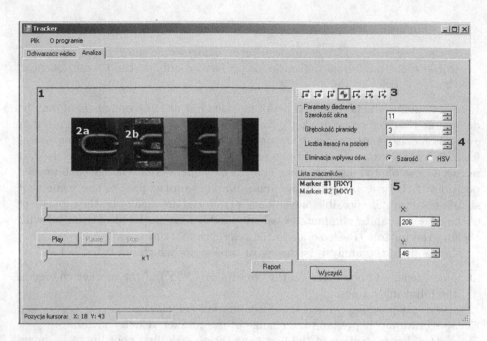

Fig. 3. Movie file with defined markers and tracking parameters: 1 – movie window, 2a,b – markers, 3 – toolbox of markers types, 4 – tracking parameters, 5 – list of defined markers

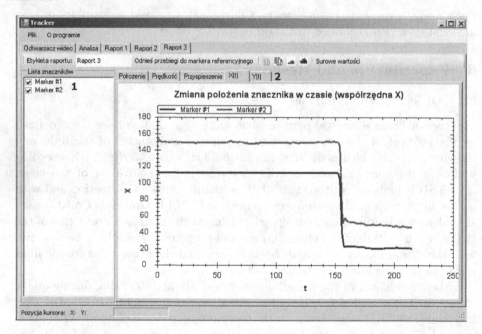

Fig. 4. Results of a single tracking: 1 – marker list (checked markers values are presented in the chart), 2 – different charts

6.2   Results

Fig. 5 shows the chart of loading the THD class chain without the corrosion protection as the function of time. The chain did not break.

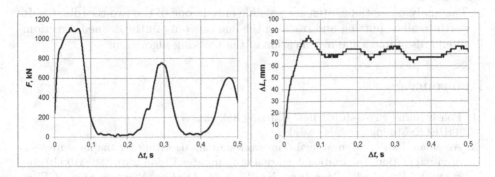

Fig. 5. Charts of loading the THD class chain without the corrosion protection (left) and the deformation (right) as the function of time

Fig. 6 also shows the chart of loading the THD class chain without the corrosion protection as the function of time but in this case the chain was broken. The moment of breaking of the chain is marked with black dot on the right figure.

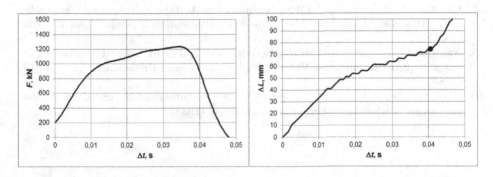

Fig. 6. Charts of loading the THD class chain without the corrosion protection (left) and the deformation (right) as the function of time

7   Conclusions and Further Works

Past and future experience related to research on chains, conducted based on presented methodology on dynamic resistance of chains, can provide manufacturers and designers of conveyors and coal ploughs as well as users, with a lot of important information and tips that may be useful for the proper selection of the chain to the specific conditions of work. By using the analysis of chain deformation, using simple tools such as a compact camera and software for optical image analysis, it is possible to determine the characteristics of deformation as a function of load time. The Central Mining Institute is going to continue this research and to gain further experience in the field of understanding characteristics of chains work at the dynamic, percussive load and various chain destruction phenomena.

The analysed motions often are possible only in one of the two coordinates. It is worth to allow putting and tracking the markers only in the defined direction. It should also force some modification of the tracking algorithm.

References

1. Polish Norm: PN-G-46701:1997 - Mining link chains
2. THIELE: Mining products catalog
3. Anandan, P.: A computational framework and an algorithm for the measurement of visual motion. International Journal of Computer Vision 2(3), 283–310 (1989)
4. Brox, T., Bruhn, A., Papenberg, N., Weickert, J.: High Accuracy Optical Flow Estimation Based on a Theory for Warping. In: Pajdla, T., Matas, J(G.) (eds.) ECCV 2004. LNCS, vol. 3024, pp. 25–36. Springer, Heidelberg (2004)
5. Fleet, D.J., Jepson, A.D.: Computation of component image velocity from local phase information. International Journal of Computer Vision 5(1), 77–104 (1990)
6. Heeger, D.: Optical flow using spatio-temporal filters. International Journal of Computer Vision 1(4), 279–302 (1988)
7. Horn, B.K.P., Schunck, B.G.: Determining optical flow. Artificial Intelligence 17, 185–204 (1981)
8. Lucas, B.D., Kanade, T.: An iterative image registration technique with an application to stereo vision. In: DARPA Image Understanding Workshop, pp. 121–130 (1981)
9. Nagel, H.H.: Displacement vectors derived from 2nd order intensity variations in image sequences. Computer Vision, Graphics and Image Processing 21(1), 85–117 (1983)
10. Uras, S., Girosi, F., Verri, A., Torre, V.: A computational approach to motion perception. Biological Cybernetics 60(2), 79–87 (1988)

A Spatial-Based Approach for Groups of Objects

Lu Cao[1], Yoshinori Kobayashi[1,2], and Yoshinori Kuno[1]

[1] Department of Information and Computer Science, Saitama University
255 Shimo-okubo, Sakura-ku, Saitama 338-8570, Japan
{caolu,yosinori,kuno}@cv.ics.saitama-u.ac.jp
[2] Japan Science and Technology Agency (JST), PRESTO, 4-1-8 Honcho, Kawaguchi,
Saitama 332-0012, Japan

Abstract. We introduce a spatial-based feature approach to locating and recognizing objects in cases where several identical or similar objects are accumulated together. With respect to cognition, humans specify such cases with a group-based reference system, which can be considered as an extension of conventional notions of reference systems. On the other hand, a spatial-based feature is straightforward and distinctive, making it more suitable for object recognition tasks. We evaluate this approach by testing it on eight diverse object categories, and thereby provided comprehensive results. The performance exceeds the state of art by high accuracy, less attempts and fast running time.

1 Introduction

We are interested in utilizing spatial expressions to locate objects. Describing spatial relations between a target and its surrounding objects is an operational way to better understand the context of a scene. This approach can be more accurate and efficient for locating objects than traditional training and testing low-level feature methods. For example, instead of describing a pencil independently as a long, wooden object with a sharpened point, one could say **"The pencil is in front of the book"** for simplicity.

However, talking about space can be notoriously difficult. In a table-top query task, the speaker must select expressions carefully and the interlocutors must be aware of the speaker's perspective or frame of reference. To locate an object, one has to retrieve the cognitive map in one's mind, find suitably relevant reference objects around the object in question, express these accordingly and then check whether the interlocutors understand. The cognitive process is highly complex. When attempting to duplicate this process in a robotic system, two problems arise: 1) how to interpret spatial relations to the robotic system; 2) how to teach the system spatial expressions. With the help of some sophisticated prior research [1-7], we have already designed a computational projective spatial-based feature model [8-10] able to utilize instructions from human users to locate objects. Moreover, it has achieved considerable success in accomplishing various tasks.

G. Bebis et al. (Eds.): ISVC 2012, Part II, LNCS 7432, pp. 597–608, 2012.

Generally, humans discuss and manipulate space through deploying three distinct entities: a reference object (relatum), a target object (referent), and reference frames. In the simplest case, a single object serves as the relatum, and can be detected in a straightforward manner by the speaker and the robotic system. In this paper, however, we will discuss a more complex case, namely one involving a group of objects, a situation that occurs frequently in real-world environments but which has been largely neglected in the literature. In addressing this issue, we focus on two particular questions: 1) since humans consider a group to constitute an integrated object, how can we enable a system to treat a group of objects likewise; and 2) what refrence frames should be taken in these cases? Note that here the conception of "group" refers to either several identical or similar objects.

In this study, beyond a single relatum scheme[8-10], to mimic the human ability to recognize an object we conducted experiments involving groups of objects from which we sought to distinguish other unknown objects via simple and straightforward input instructions, such as **"The pocket calculator is in front of the computer monitors."** The paper is organized as follows: after reviewing the related work in the next section, we introduce the benchmark of spatial representation and then describe the group-based reference system scheme in Section 3, cover the utilized datasets and experimental results in Section 4, and finally draw some conclusions in Section 5.

2   Related Work

There is a significant body of work on spatial representation in linguistics and philosophy (e.g., [11-17]) which has developed since the 1980s and 1990s. The most seminal work is [14], which explicitly analyzed human language for spatial descriptions and proposed an affirmative family of reference frames as a solution. Willem J. M. Levelt in [17] analyzed problems which may arise when spatial expressions are used for scenes, such as **"The book is in front of the computer monitor"** and restricted limitations to avoid ambiguous meanings in different reference systems. More principles are specified in [18]. In contrast to other studies, Schmidt pointed out how the reference frames are determined from the perspective of relatums. These findings are the benchmark for our approach. We have modeled two kinds of reference frames and identify them exclusively.

In computer vision, spatial-based classifications do not have a long history and little has been done in the area of object recognition. In terms of considering the salient advantages of spatial-based features, which are not affected by subtle changes among objects (e.g., differences in size or color), there has been some tentative work since the 1990s. Inchoate tentative approaches represented objects as points. However, the limitations of this approach are obvious: 1) points are only able to reveal local information of an object, and 2) an object can be represented by any of its points. For these two reasons, the bounding region method has attracted more attention, and is becoming widely used for localization regions, such as in [19].

One particularly applicable way to specify spatial relations is to use angular deviation. In a series of rational works, R. Moratz and T. Tenbrink [5-7] developed a computational framework by using projection cues on a plane. Our approach is inspired by theirs. Instead of constructing an automatic system moving forward towards a target, in our application domain we prefer to locate the referent across a sliding window by classifying spatial relationships within an image, providing input instructions to assist the enterprise.

3 Grouping Objects as Relatum

3.1 What Is a Spatial-Based Feature?

Since spatial expressions partition the space into loose regions such as *near*, *back* and *left* [4], space can be viewed as a type of feature alongside other features that characterize objects, such as *color*, or *shape*. In the simplest case, we take a detected object as a relatum, and locate another undetected target object in relation to it by using spatial expressions. We emphasize that the spatial-based feature cannot be utilized independently, but rather must be understood as providing complementary information for other kinds of features. We also note that at least one object should be detected in an image.

3.2 Brief Introduction to Relative Reference Frames

Different spatial configurations make use of distinctive references of frames. Previous research by Levinson in [14] has proposed that when referring the position of one object to another, humans mainly apply three kinds of reference systems: *intrinsic*, *relative*, and *absolute*. In this paper, we only concentrate on the **relative** case. Such a case depends on a position of a further entity (relatum) as an origin from which to refer to a referent, as in the example formulation "**the travel cup is to the left of the box.**" Except for the relatum and referent objects, another crucial component is needed for a locative description: the **viewpoint**. A viewpoint is the point of view from which the relatum is viewed, and which can impose an orientation on it. Different viewpoints can result in different spatial layout configurations. In particular, the speaker's location can serve as the point of view. The listener's location might also serve as the point of view but is used less frequently [18]. For more classification, the ambiguity can be avoided by explicitly stating the viewpoint, for example, "**From my viewpoint, the travel cup is to the left of the box.**" Specifically, if a relatum lacks an inherent front, as is the case with cups for example, or its inherent front is not normally used for establishing the reference system, it can be determined that the reference system is relative.

3.3 Group-Based Reference System

The situation described earlier proves useful for scenes wherein 1-3 different objects serve as relatums. This study seeks to take this notion further, because we argue that it can also be applied in the case of groups of objects. When there are multiple identical or similar objects accumulated together in a scene, humans consider them to constitute a group. In this case, the group-relatum can be viewed as a whole bounding area, which functions as a rectangular sliding window across the group, and thereby suggests that it could be viewed as a type of **relative** reference system as well.

Cognitively, humans first determine whether the relatum has inherent parts. Therefore, conflict can occur if two criteria contradict each other. A sentence such as **"The pocket calculator is in front of the computer monitors"** is ambiguous because it can be interpreted in several ways, namely that the calculator is located in front of a group of monitors from the viewpoint of the speaker, with respect to the orientation of the monitors themselves, or with respect to the inherent orientation of the monitors. Herskovits [13] claims that a situation depends on whether it is seen from the inside or the outside. In our approach, we discuss these two situations, which are called **internal use** and **external use**, respectively. **Internal** refers to those cases in which the relatums can be seen from the inside, namely, the inner spatial relations of the group. On the contrary, **external** refers to those cases in which the relatums can be seen from outside the group.

Internal Use

Speakers describe the position of an object in a group using the spatial relations between the object and the group as a whole, such as: **"the second from the left of those objects."** Additionally, humans also noticeably employ unary **superlatives** to describe group of objects. For example, they may use the terms *leftmost*, *middle* and *rightmost*, which are binary-level spatial relations based on projections of bounding boxes on the x and y axes. Therefore, we define them based on the Manhattan distance between bounding boxes in Eq. (1).

$$D_{manhattan} = |x - x_i| + |y - y_i| \tag{1}$$

where C (x, y) is the centroid of the bounding box of the total group, and $O(x_i, y_i)$ is the centroid of the bounding box of each object in the group. Then, the superlatives can be defined as:

$$\text{Leftmost: } x_i \ll x \text{ and arg max } \{D_{manhattan}\} \tag{2}$$

$$\text{Middle: } x_i \approx x \text{ and arg}\{D_{manhattan}\} \approx 0$$

$$\text{Rightmost: } x_i \gg x \text{ and arg max } \{D_{manhattan}\}$$

3.4   External Use

Speakers can also specify the position of one object which does not belong to the group using the whole group as a relatum. This is defined likewise in the way of a conventional relative reference system presentation. By and large, humans tend to use three canonical axes: up-down, left-right, and front-back. Moratz in [5] has shared detailed spatial configurations in his work, and our method is inspired by his. The baseline work is described in [8-10], where we proposed a computational spatial-based approach based upon angle deviation. The origin is often omitted and the default origin is usually the user (speaker), or sometimes the listener (robotic system). In our implementation, the up-down axis is ignored so that the system is reduced to a two-dimensional plane. We assume that the origin is the speaker. If the speaker and the robot are looking in the same direction, then the viewpoint can be omitted. Following Hernandez [2], a vertical divides the reference plane into *left* and *right* parts while a horizontal manages the *front* and *back*. This is illustrated in Fig. 1, where one can see the spatial configuration is viewed from a frontal viewpoint.

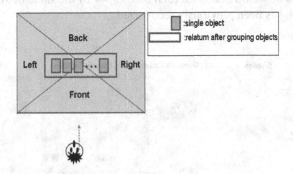

Fig. 1. Group-based Reference System Configuration

We formulate the angle θ below. Fig.2 illustrates the included-angle, which is between the reference direction and the straight line from the relatum to the referent.

referent *front* relatum: $\qquad 0 \leq \theta \leq \dfrac{\pi}{4}$ or $\dfrac{7\pi}{4} \leq \theta \leq 2\pi$

referent *back* relatum: $\qquad \dfrac{3\pi}{4} \leq \theta \leq \dfrac{5\pi}{4}$

referent *left* relatum: $\qquad \dfrac{\pi}{4} \leq \theta \leq \dfrac{3\pi}{4}$

referent *right* relatum: $\qquad \dfrac{3\pi}{4} < \theta < \dfrac{7\pi}{4}$

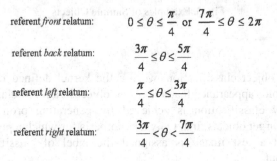
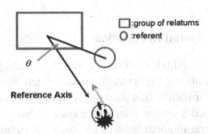

Fig. 2. θ in Group-based Reference System.

4   Experimental Results

4.1   Dataset

Since there is no standard dataset suitable for our experimental purposes, a dataset was collected by the authors. It is built upon ground truth bounding boxes that contain multiple objects per image. There are 380 images, which can be further divided into 8 categories, 2-4 subcategories in each, namely bottles, glasses cases, nozzle bottles, toy cars, food keepers, cups, shoes, and digital cameras. Fig. 3 illustrates the sample objects. All the images have a uniform background, and were taken at medium resolution in 400×300 pixels. In each image, 3-4 of the same or similar objects form a group which has been previously recognized.

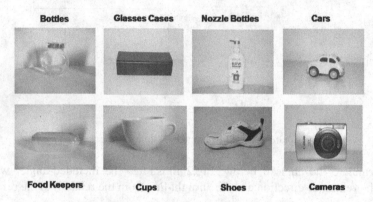

Fig. 3. Examples of Sample Objects

4.2   Presets

For autonomous object classification, we use the kernel defined in [20] combining both shape and color appearance features for an object in a SVM classifier. In the test phase, multi-way classification is achieved by generating promising hypothesis locations for the target objects first. Only a hypothesis for which a positive confidence is measured and a test image is assigned the label of classifier with highest probabilistic output.

4.3   Group-Based Spatial Recognition

We implemented the method outlined above into the spatial recognition model. Human users give hints to the system about which object is the target one by providing position information, and then the system locates the object and responds to the users. We developed a simple interaction system that can understand some simple English words and grammatical structures, such as **bring**, **how many, what** and **which**. To adopt this system to group use, the input toolbox was further developed to understand *noun*-s(es) rules. We restricted user input commands to the following set format: **Bring me xxxx / It is OOOO**. We annotated the undetected objects with the label "unknown" and related them to a list in which the names of categories are recorded first. For the sake of simplicity, we assumed the user and the system shared the same viewpoint. At run-time, if nothing is detected in the previous phase, the system does nothing except terminate. On the other hand, in the process of spatial recognition, it makes three attempts before advocating failure if any target object is unable to be detected by using spatial relations.

We employed two performance measures to evaluate our method's effectiveness.

4.3.1   Internal Use Results

Some representative results are shown in Fig. 4. The basic dialog script was as follows:

> User: Can you see xxx-s/es? (e.g.: car)
> Robot: Yes, I can.
> User: How many xxx-s/es (e.g.: car) can you see?
> Robot: I can see □□□ (e.g.: three).
> User: Bring me the OOO (e.g.: leftmost) one.
> Robot: OK. I can see it. Is this the one?
> User: Yes.

Precision-recall plots of 3 superlatives are shown in Fig. 5 (a-c). Fig. 5 (d) illustrates that the method outperforms in all 8 categories: on average, the accuracy is about 87%.

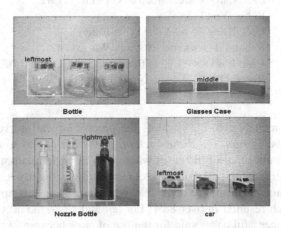

Fig. 4. Examples of detection results. The referent is marked in yellow.

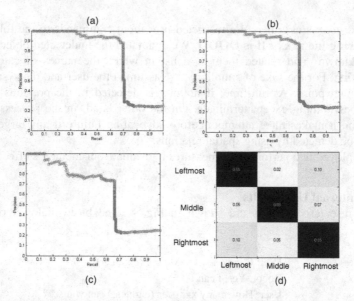

Fig. 5. (a-c): Precision-recall curves for superlatives: leftmost, middle and rightmost. (d): Confusion matrix for superlatives.

4.3.2   External Use Results

Only one undetected object is present in the scene. With a user's instructions, the system located the target object successfully even if it did not know what the item in question was. The basic dialog script was as follows:

> User: Can you see a group of xxx-s? (e.g.: car)
> Robot: Yes, I can.
> User: Bring me ΔΔΔ (e.g.: coin-battery).
> Robot: I do not know what it is. Where is it?
> User: It is OOO (e.g.: in front of) the xxx-s (e.g.: car).
> Robot: OK. I can see it. Is this the one?
> User: Yes.

Results are shown in Fig. 6. Fig. 7 plots the precision-recall curves of 4 kinds of spatial expressions: *in front of*, *at the back of*, *to the left of* and *to the right of*. The performance resembles that of the cases where only a single relatum was referred to, which were investigated in [10], but the accuracy rate was slightly lower. The method achieves an average performance of 85.5%, with the best results appearing with the use of *front*, which resulted in over 88%. For *left* and *right*, the accuracy is closer to 87%. The least successful cases gain for the use of *back* because of partial occlusion. This is shown in Fig. 8. In such cases, the accuracy decreases significantly, especially when two objects are of a similar size. The confusion is raised when detecting spatial relations, and is understandable given that the occlusion greatly affects the bounding area generation in the automatic detection approach, and thereby makes the angle deviation incorrect. Similarly, the main confusion pairs are those that are diametrically opposed (i.e., off by 180 degrees), for example, *front* vs. *back*.

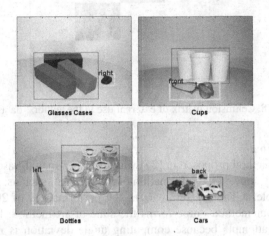

Fig. 6. Examples of detection results in the case of external use. The referent is marked in yellow.

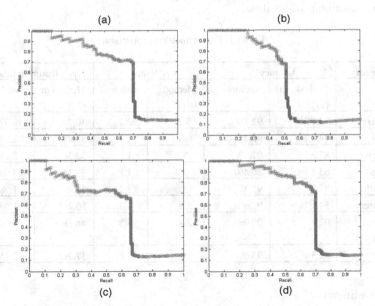

Fig. 7. Precision-recall curves for front, back, left and right

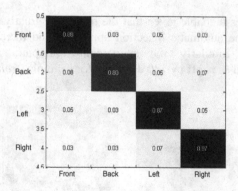

Fig. 8. Confusion matrix of external use in the authors' dataset

4.3.3   Comparision

We compared the method proposed in [21] without using spatial-based feature. In this case, we only imply appearance information of the referent, such as: shape, color, and texture. For example, a Table 1 summarizes the performances of 200 images in the author's dataset. Our method outperforms the state-of-art method. It fasts with high accuracy and less attempts because computing angle deviation is much easier than training a complex descriptor. The benefit of spatial-based feature is significant: rather than the details of what the object looks like, one can discrimitively locate the target by describing where it is.

Table 1. Performance comparison

Category	Accuacy		Attempt		Running time	
	Method in [21]	Authors	Method in [21]	Authors	Method in [21]	Authors
Scissors	72.1%	92.1%	5.2	2.1	30.7s	2.37s
Keys	74.2%	93.4%	6.8	1.7	45.4s	2.35s
Pencils	87%	91.3%	3	1.2	32.3s	2.18s
Clips	63.1%	86.4%	7	1.8	39.1s	2.09s
Candles	83.7%	95.3%	3	1.2	26.4s	1.98s
Lighters	54.1%	92.6%	5.8	1.3	50.2	1.98s
Glue Sticks	92%	98.4%	2.4	1.5	48.3s	2.04s
Combs	52.5%	93.6%	3.6	1.8	37.2s	2.15s

4.4   Summary

Overall, the proposed approach proved efficient and accurate in practice for all the images tested in both cases: the average accuracy was about 86.25% and the response time was 0.23 seconds per round. Compared to the singular relatum cases (where the average accuracy was 83.56%), the accuracy increased due to the wide area of a group of relatums. Moreover, there was a great improvement when the size of the group and referent were significantly different, resulting in accuracy as high as 89.7%.

5   Conclusion

In this paper, we proposed an approach to locate and recognize objects based upon spatial-based features in a group-based reference system. The reference system is an extension of a relative system. Although the performance depends to a great extent on low-level object recognition methods and the novel feature cannot yet be utilized independently, the simplicity and distinctiveness makes it both capable and distinguishable as complementary information for low-level features. We tested the performance of this approach on the authors' datasets, which consisted of 200 images with a uniform background subdivided into 8 categories. The results clearly demonstrated the ability of this approach to accurately localize and detect objects.

However, we would like to develop this approach further in future work. Three points require consideration here: 1) to advance further, the current object localization and detection algorithm needs to be improved so as to be able to perform well when faced with partial occlusion and changes in viewpoint; 2) for complex cases, multiple undetected object recognition tasks will have to be considered; and finally, 3) Further experiments using more complex scenes would be beneficial, and this is something we shall pursue next in our future work.

Acknowledgements. This work was supported by KAKENHI (23300065).

References

1. Eschenbach, C., Tschander, L., Habel, C., Kulik, L.: Lexical Specifications of Paths. In: Habel, C., Brauer, W., Freksa, C., Wender, K.F. (eds.) Spatial Cognition 2000. LNCS (LNAI), vol. 1849, p. 127. Springer, Heidelberg (2000)
2. Hernandez, D.: Qualitative representation of spatial knowledge. LNAI. Springer, Berlin
3. Dobnik, S., Pulman, S., Newman, P., Harrison, A.: Teaching a Robot Spatial Expressions. In: Second ACL-SIGSEM (2005)
4. Skubic, M., Perzanowski, D., Schultz, A., Adams, W.: Using Spatial Language in a Human-Robot Dialog. In: Proc.of IEEE Intl. Conf. on Robotics and Automation (2002)
5. Moratz, R., Tenbrink, T., Bateman, J.A., Fischer, K.: Spatial Knowledge Representation for Human-Robot Interaction. In: Freksa, C., Brauer, W., Habel, C., Wender, K.F. (eds.) Spatial Cognition III. LNCS (LNAI), vol. 2685, pp. 263–286. Springer, Heidelberg (2003)
6. Moratz, R., Tenbrink, T.: Spatial reference in Linguistic Human-Robot Interaction: Iterative, Empirically Supported Development of a Model of Projective Relations. Spatial Cognition and Computation 6(1), 63–106 (2006)
7. Moratz, R., Fischer, K., Tenbrink, T.: Cognitive Modelling of Spatial Reference for Human-Robot Interaction. International Journal on Artificial Intelligence Tools 10(4) (2001)
8. Cao, L., Kobayashi, Y., Kuno, Y.: Spatial Resolution for Robot to Detect Objects. In: IROS (2010)

9. Cao, L., Kobayashi, Y., Kuno, Y.: Object Recognition for Service Robots: Where is the Fronts? In: IEEE ICMA (2011)
10. Cao, L., Kobayashi, Y., Kuno, Y.: Spatial-Based Feature for Locating Objects. In: Huang, D.-S., Ma, J., Jo, K.-H., Gromiha, M.M. (eds.) ICIC 2012. LNCS, vol. 7390, pp. 128–137. Springer, Heidelberg (2012)
11. Jackendoff, R.: Languages of the Mind. The MIT Press (1992)
12. Bloom, P., Peterson, M., Nadel, L., Garrett, M. (eds.): Language and Space. MIT Press (1999)
13. Herskovits, A.: Language and Spatial Cognition: An Inderdisciplinary Study of the Prepositions in English. In: Natural Language Processing. Cambridge University Press (1986)
14. Levinson, S.C.: Frames of reference and Molyneux's question: Crosslinguistic evidence. In: Language and Space, pp. 109–170. MIT Press (1999)
15. Levine, D., Warach, J., Farah, M.: Two visual systems in mental imagery: Dissociation of 'what' and 'where' in imagery disorders due to bilateral posterior cerebral lesions. Neurology 35, 1010–1018 (1985)
16. Eschenbach, C., Tschander, L., Habel, C., Kulik, L.: Lexical Specifications of Paths. In: Habel, C., Brauer, W., Freksa, C., Wender, K.F. (eds.) Spatial Cognition II. LNCS (LNAI), vol. 1849, pp. 127–144. Springer, Heidelberg (2000)
17. Levelt, W.: Some perceptual limitations on talking about space. In: Limits in Perception, pp. 323–358. VNU Science Press (1984)
18. Schmidt, G.: Various views on spatial prepositions. AI Magazine 9, 95–105 (1988)
19. Clementini, E., Billen, R.: Modeling and Computing Ternary Projective Relations between Regions, The House is North of the River: Relative Localization of Extended Objects. IEEE Transactions on Knowledge and Data Engineering 18(6), 799–814 (2006)
20. Das, D., Kobayashi, Y., Kuno, Y.: Multiple Object Category Detection and Localization Using Generative and Discriminative Models. IEICE Transactionon Information and System 92, 2112–2121 (2009)
21. Kuno, Y., Sakata, K., Kobayashi, Y.: Object recognition in service robot: Conducting verbal interaction on color and spatial relationship. In: Proc. IEEE 12th ICCV Workshops, pp. 2025–2031 (2009)

Adaptive Exemplar-Based Particle Filter
for 2D Human Pose Estimation

Chi-Min Oh[1], Yong-Cheol Lee[1], Ki-Tae Bae[2], and Chil-Woo Lee[1]

[1] Chonnam National University, Gwangju, Korea
[2] Korean-German Institute of Technology, Seoul, Korea
{sapeyes,budlbaram}@image.chonnam.ac.kr, ktbae@kgit.ac.kr,
leecw@chonnam.ac.kr

Abstract. This paper proposes how to utilize pose exemplars in the prediction step of particle filter for efficient human pose estimation. The prediction of particle filter is only dependent on the previous posterior distribution. If observation data and reference dataset are used in prediction, the prediction range can be more compact and precise. We use adaptive exemplars for prediction. To do so the similarity between pose exemplar and the pose silhouette observation are measured. Based on the similarity of exemplars corresponding number of particles are predicted either from exemplars and previous posterior distribution. After pose estimation with the likelihoods of predicted particles, the finally estimated pose is used for updating adaptive exemplar dataset for improving performance of next prediction. Therefore, the proposed method efficiently estimates the pose of articulated full body as resultant images represent.

1 Introduction

Human pose estimation has been a difficult issue due to dynamic and nonlinear motion of articulated joints of human body. The pose of human body has been used for applications in human-computer interaction (HCI) field such as image quality improvement of human area, image tagging of human body, cloth analysis, pose and action recognition, gesture analysis and etc.

For general usage of human pose information, using monocular camera, 2D body model-based pose estimation has been interestingly investigated. M. A. Fischler et al. [1] first introduced pictorial structures (PS), 2D deformable model which represents face or human body with the connected rectangular parts where spring-like connections were used to combine parts. Like an energy minimization process Fischler found an optimal position of deformable object with reasonable costs of spring tension and local appearance difference of each part.

After thirty two years, P.F. Felzenszwalb et al. [2] reintroduced Fischler's PS with their sampling-based object pose estimation which is more robust for noisy measurement than Fischler's method. They matched each part based on some samples drawn from the distribution of its parent location without any additional information. X. Zhang [3] proposed an improved sampling method, data-driven Markov chain Monte Carlo, for utilizing part candidates. For avoiding exhaustive search of optimal part

G. Bebis et al. (Eds.): ISVC 2012, Part II, LNCS 7432, pp. 609–615, 2012.

location, before searching, they tested some part candidates to build a better proposal probability. Therefore they achieved a tremendous speed-up and higher efficiency.

In this paper, we propose evolutionary pose exemplars-based particle filter. Rather than using part candidates, exemplar poses are compared to observation image to establish proposal distribution of entire pose directly. From proposal distribution we randomly sample pose candidates as particles and finally choose one from particles with best likelihood which is calculated by chamfer distance. This determined pose is validated whether it could be an additional pose exemplar in the existing set. Then we adaptively update exemplars for pose estimation of human full-body.

The content of this paper is organized as follows. Section 2 describes our PS model and particle filter framework with likelihood estimation. Section 3 describes how to build adaptive exemplars and the proposal distribution. The estimation results are shown in Section 4. The conclusion and future works are explained in Section 5.

2   Full-Body Pose Estimation

Our full body pose is modeled by PS and its pose estimation from observation is fulfilled by adaptive exemplar-based particle filter.

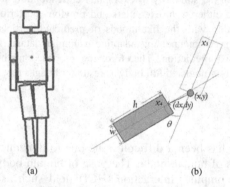

(a)   (b)

Fig. 1. Our full-body pose model is defined by pictorial structures (a) where each part is located with a distance and orientation from its parent part as shown in (b)

2.1   Full-Body Pose Model

We use a full-body pose model designed by PS. The ten parts (torso, head, two arms with upper and lower parts, two legs with upper and lower parts) are connected as a tree graph. The root node is torso part. The PS is shown in Fig. 1(a).

From the root node, torso part, other parts are connected to parent parts. Fig. 1 (b) shows that relation. the parameters of body part in our full-body body model is $x_i=\{(x,y),(dx,dy), \theta ,(w,h)\}$: joint location (x,y) determined by parent part, spring-like displacement (dx,dy), orientation θ and rectangular part size (w,h). Among the parameters, the joint location (x,y) cannot change by part itself and only can change by parent part. This constraint makes the connections between parts based on parent part.

Every parts can move around joint location with (dx,dy) displacement vector. This is also called spring-like connection with joint location and displacement vector which is determined by in Gaussian distribution with the mean (x,y) and a variance of part movement in body. In addition to the displacement, the part can rotate with the orientation parameter θ based on the rotation origin (x,y).

2.2   Particle Filter for Pose Estimation

Most tracking systems assume that current posterior distribution must be much related to and dependent on the previous posterior distribution. Particle filter estimates a proper posterior distribution by updating the previous posterior distribution to current time step. The proposal distribution predicts the current posterior distribution from previous posterior distribution with discrete and weighted particles. In Eq. (1), the posterior distribution $p(x_{t-1}|y_{t-1})$ at previous time step represents itself with discrete and weighted particles, where $x^{(i)}_{t-1}$ is ith particle, $w^{(i)}$ is the weight of ith particle, and N is the total number of particles.

$$p(x_{t-1} \mid y_{t-1}) \approx \{(x^{(i)}_{t-1}, w^{(i)}_{t-1})\}^{N}_{i=1} \tag{1}$$

Particle filter has two steps: prediction and update. In the prediction step, the previous posterior distribution is marginalized to eliminate x_{t-1} and to be updated to x_t based on transition model $p(x_t|x_{t-1})$, Markov chain model.

$$p(x_t \mid y_{t-1}) = \int p(x_t \mid x_{t-1}) p(x_{t-1} \mid y_{1:t-1}) dx_{t-1} \tag{2}$$

In the update step, the posterior distribution is reformulated to adjust to the current observation y_t. Based on Baye's rule, posterior distribution $p(x_t|y_t)$ is represented with the likelihood $p(y_t|x_t)$ and prior distribution $p(x_t|y_{t-1})$.

$$p(x_t \mid y_{1:t}) = \frac{p(y_t \mid x_t) p(x_t \mid y_{1:t-1})}{p(y_t \mid y_{1:t-1})} \tag{3}$$

In addition to the prediction step, the particles for x_t prediction are drawn from the proposal distribution $q(x_t|x_{t-1},y_t)$ and in the update step the weights of particles are determined by below:

$$w^{(i)}_t = w^{(i)}_{t-1} \frac{p(y_t \mid x^{(i)}_t) p(x^{(i)}_t \mid x^{(i)}_{t-1})}{q(x^{(i)}_t \mid x^{(i)}_{t-1}, y_t)} \tag{4}$$

In the process of weighting particles, likelihoods of particles are measured. Our likelihood $p(y_t|x^{(i)}_t)$ is an edge likelihood.

$$p(y_t \mid x^{(i)}_t) = p(I_E \mid x^{(i)}_t) \tag{5}$$

$p(I_E| x^{(i)}_t) = exp(-d(I_E, x^{(i)}_t))$ is the likelihood of chamfer distance d as in Fig. 2.

Fig. 2. Chamfer distance [4] is calculated on a distance map of edge image where PS pose model is overlaid. Chamfer distance measures how badly PS model is overlaid on the map.

3   Adaptive Exemplar-Based Particle Filter

For automatic initialization of efficient particle filter tracking, we use adaptive exemplar-based particle filter. To build the set of exemplars, we have manually made a thousand PS poses and chosen 91 exemplar poses based on motion energy $E_{i,j}$. If given PS pose has no similar exemplar poses with moderate motion energy (sigma), that PS pose will be selected as additional exemplar pose. Below Fig 3 shows this process.

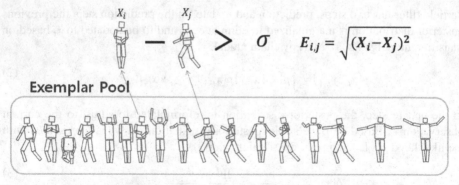

Fig. 3. Exemplar pool is built based on the motion energy. If input exemplar pose has no similar one in exemplar pool, that pose will be included in our system.

Each exemplar is represented by a linear feature vector, silhouette image and PS pose $E_i=\{f_i, sil_i, ps_i\}$. To calculate the feature vector of an exemplar, we estimate all chamfer distances between the given exemplar and all exemplars using Eq. (6) as $f=(d_1,...d_{91})^T$ and the diffusion distance [5] is used for similarity measurement between an exemplar's feature vector and input feature vector.

$$d_i = cd(sil_k, ps_i) \qquad (6)$$

To become adaptive exemplars, after pose estimation of particle filter at every time step, if the estimated pose is newer one by comparing exemplar pool, we try to add it as an exemplar into the pool. Otherwise we gradually update similar exemplars with estimated pose to have similar part size.

In particle filtering process, to use exemplars in prediction step, it is better to modify proposal distribution where all particles are predicted from. To be predicted either from previous posterior distribution or exemplar pool, below proposal distribution is defined.

$$x_t^{(i)} \sim q(x_t \mid x_{t-1}, y_t) = \alpha p(x_t \mid x_{t-1}) + (1-\alpha) p(x_t \mid y_t) \tag{7}$$

$$p(x_t \mid y_t) = \frac{1}{N_{exemplars}} \sum_{E_i} p(x_t \mid E_i) P(E_i \mid y_i) \tag{8}$$

A particle $x_t^{(i)}$ is drawn from the proposal distribution. In the proposal distribution Markov chain model $p(x_t|x_{t-1})$ reuses of the particles in previous posterior distribution and exemplar similarity model $p(x_t|y_t)$ makes the particle predicted similarly from the exemplar which is selected from pool if it is similar to the silhouette image y_i.

Finally the predicted particles can be made either from previous particle set or exemplar pool. This shows in Fig. 4. Based on Markov chain model, particles can be predicted from in a range of Markov chain pdf. Also some particles can be predicted from exemplars based on its similarity with input image ($I_t = y_t$).

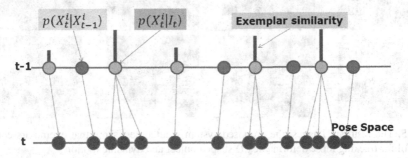

Fig. 4. In the prediction step of particle filtering, our proposal distribution is used to predict particles either from previous particle (dark) set or exemplars (bright) considering exemplar similarity

4   Results

Using the proposed method, adaptive exemplar-based particle filter, we tested our video data. The estimated poses are shown in Fig. 5. The resultant images show that our method is working well with such dynamically changing poses because this system depends both on previous particle set and exemplar pool to cover whole pose space efficiently. One additional benefit as stated in [6], if we use the proposal distribution with observation data, the start of system is automatically initialized and also the system can have a tendency to recover a tracking error that usually happens when prediction area with previous particle set is out of the user motion range but it can be recovered when we use exemplar pool in prediction together with previous particle set.

Fig. 5. The resultant images, the proposed system's outcomes, show that exemplars could be helpful for tracking those deformable 2D shaped object and dynamic motion

5   Conclusion

We have proposed how to utilize exemplars in the prediction step of particle filter. Comparing related works, our method is not dependent on part candidates and their several probability distributions for tracking each child parts. We only use a proposal distribution for predicting full body pose. If we try to predict part-by-part, the previous time information can be ignored, which still gives useful information. If we consider full body pose, the posterior distribution at previous time can be more easily reused. Also that exemplar pool helps prediction, automatic initialization and error recovery. As results shows, the proposed method works well with stated benefits.

In the future, we will investigate more detail with other datasets and try to make a sophisticated evaluation dataset.

Acknowledgements. This research was supported by the MKE(The Ministry of Knowledge Economy), Korea, under both of the ITRC(Information Technology Research Center) support program supervised by the NIPA(National IT Industry Promotion Agency) (NIPA-2012-H0301-12-3005), and the Human Resources Development Program for Convergence Robot Specialists support program supervised by the NIPA (NIPA-2012-H1502-12-1002).

References

1. Fischler, M.A., Elschlager, R.A.: The Representation and Matching of Pictorial Structures. IEEE Trans. on Computer 22(1), 67–92 (1973)
2. Felzenszwalb, P., Huttenlocher, D.: Pictorial Structures for Object Recognition. Int. J. Computer Vision 61(1), 55–79 (2005)
3. Zhang, X., Li, C., Tong, X., Hu, W., Maybank, S., Zhang, Y.: Efficient Human Pose Estimation via Parsing a Tree Structure Based Human Model. In: Int. Conf. Computer Vision, pp. 1349–1356 (2009)
4. Barrow, H.G., Tenenbaum, J.M., Bolles, R.C., Wolf, H.C.: Parametric Correspondence and Chamfer Matching: Two New Technique for Image Matching. In: Int. Conf. Artificial Intelligence, pp. 1175–1177 (1997)
5. Ling, H., Okada, K.: Diffusion Distance for Histogram Comparison. Computer Vision and Pattern Recognition VPR 1, 246–253 (2006)
6. Okuma, K., Taleghani, A., de Freitas, N., Little, J.J., Lowe, D.G.: A Boosted Particle Filter: Multitarget Detection and Tracking. In: Pajdla, T., Matas, J(G.) (eds.) ECCV 2004. LNCS, vol. 3021, pp. 28–39. Springer, Heidelberg (2004)

Estimation of Camera Extrinsic Parameters of Indoor Omni-Directional Images Acquired by a Rotating Line Camera

Sojung Oh and Impyeong Lee

Laboratory for Sensor and Modeling, Department of Geoinformatics,
The University of Seoul, 163 Seoulsiripdaero, Dongdaemun-gu,
Seoul, 130-743, Korea
{osojung0201,iplee}@uos.ac.kr

Abstract. To use omni-directional images obtained by a rotating line camera for indoor services, we should know the position and attitude of the camera at the acquisition time to register the images with respect to an indoor coordinate system. In this study, we thus develop a method for the estimation of the extrinsic orientation parameters of an omni-directional image. First, we derive a collinearity equation for the omni-directional image by geometrically modeling the rotating line camera. We then estimate the extrinsic orientation parameters (EOP) through the collinearity equations with indoor control points which are stochastic constraints. The experimental results indicate that the extrinsic orientation parameters are estimated with the precision of ±1.4 mm and ±0.05° for the position and attitude, respectively. The residuals are within ±3.11 and ±9.20 pixels in horizontal and vertical directions. Using the proposed method for estimating EOP of indoor omni-directional images, we can generate sophisticated indoor 3D models and offer precise indoor services to users based on the models.

1 Introduction

According to the growth of applications developed using omni-directional images such as Street View of Google, the study on acquiring spatial information from omni-directional images is gaining particular interest. The omni-directional image is an image with a 360° field of view at the point where the image is captured. The services using omni-directional images offer usefully the surrounding information to users. For example, a user may find it difficult to figure the right direction on an unfamiliar road with a map, but if one uses the omni-directional image-based services, it will be more convenient as the omni-directional images cover the information of whole directions.

Omni-directional images are classified into two categories according to acquisition methods. One is the single view omni-directional images and the other is the multiple view omni-directional images [1]. The former is taken by a camera with mirrors. This method generates many distortions which are difficult to correct and could not acquire the images with high and uniformed resolution. Therefore, most omni-directional image-based services have been generated using the method of the latter one.

G. Bebis et al. (Eds.): ISVC 2012, Part II, LNCS 7432, pp. 616–625, 2012.

The latter is generally produced by stitching images captured by a number of cameras at the same time. It is possible to acquire images with high and uniformed resolution. However, the duplicated area, what is called stitching area, gains many distortions because the distance from the principle point is further away [2]. It is possible to create omni-directional images with less distortion by using many cameras such that the field of view of each individual camera is narrower. In this case, it is difficult to perfectly correct distortions because it is impossible to know the distance between image points and their corresponding object points.

In order to overcome this problem, Oh et al. (2011) have proposed a rotating line camera which can acquire the omni-directional image using only one camera with small field of view (FOV). It creates omni-directional images by stitching line images that are pictured by rotating the line camera in order [4]. This method produces less distortion because it stitches line images with very small FOV. However, it has a weak point that the rotation axis and objects should not move while rotating the camera. Therefore a rotating line camera is appropriate to the acquisition of indoor omni-directional images. Furthermore, it can be effectively applied to indoor spatial images information services such as indoor 3D modeling or Store View of Google.

Many studies on a rotating line camera have been accomplished through the central projection line camera. Wei et al. (2002) studied the design of a rotating line camera and Tang et al. (2001) derived the geometry for estimating 3D object coordinates. Shum et al. (1999) and Li et al. (2004) calculated the depth map using the ratio of the rotation radius against the distance between object points and the rotation origin, the rotation angle on capturing and the camera tilted angle from the rotation axis. These studies however have some limitation. It is difficult to acquire whole view according to vertical direction when the distance between the camera and object is very close.

In addition, Huang et al. (2006) accomplished the calibration which estimated the extrinsic orientation parameters (EOP), the rotation radius, the camera tilted angle, the principal point and the focal length using the indoor object control points acquired in relative coordinate system. In case of indoor scene, since the absolute coordinate system has not been defined yet, it is hard to acquire the object control points (OCP). To be able to manage indoor database, EOP of indoor images must be estimated.

In this paper, we propose a methodology for estimating EOP of indoor omni-directional image acquired by a rotating line camera. We first derive the collinearity equations of omni-directional image acquired by a rotating line camera which has a fish-eye lens with four types of projection. Second, we define the object coordinate system and obtain OCP in indoor scene. Third, we establish the mathematical model for estimating EOP of omni-directional images acquired by a rotating line camera.

2   The Proposed Method for EOP Estimation

We derive the methodology for estimating EOP of a single omni-directional image. First, we define a rotating line camera and coordinate systems, and we derive the collinearity equations of omni-directional images acquired by a rotating line camera. Second, we present a convenient method to obtain OCP. Finally, we introduce a mathematical model to estimate EOP using the collinearity equations and OCP.

2.1   Definition of a Rotating Line Camera and Coordinate Systems

A rotating line camera is a system that acquires images as a line camera with $1 \times n$ pixels is rotating the rotational radius (r) away from the rotation center. It finally generates an omni-directional image with $m \times n$ pixels by stitching the captured line images captured in order. Fig. 1 is a rotating line camera.

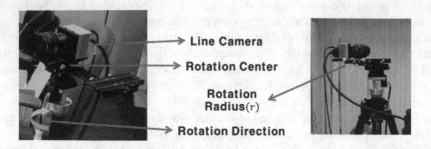

Fig. 1. A Rotating Line Camera

Fig. 2 is the 2D geometry of a rotating line camera. P is an object point, which can be projected onto an image point P' through a rotating line camera. f is the focal length of the line camera. α is the rotation angle on picturing. β is the tilted angle of the line camera from the rotation axis. $^{O}O_{R}$ is the origin of the rotation coordinate system (RCS) expressed in the object coordinate system (OCS).

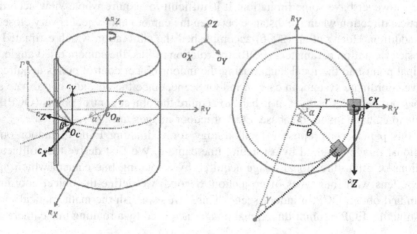

Fig. 2. The Geometry of a Rotating Line Camera

In acquiring an indoor omni-directional image through a rotating line camera, it needs to define three coordinate systems. One is the object coordinate system (OCS) in which the object points locate. Another is the rotation coordinate system (RCS) defined by rotating a camera. The other is the camera coordinate system (CCS) in capturing a line image. The origin and attitude of OCS can be defined by a corner and three edges meeting at the corner. The Z axis of OCS (^{O}Z) can be defined by the edge connecting the corner with the ceiling. The X and Y axis of OCS $(^OX, {}^OY)$ can be defined by the other edges. The origin of RCS can be defined as the center of rotation. The X axis of RCS (^RX) can be defined as the line connecting the center of rotation with the camera position that starts the rotation. The Y axis of RCS (^RY) can be defined by rotating ^RX counter-clockwise by 90 °. The Z axis of RCS (^RZ) can be defined as the cross product of ^RX and ^RY. The Y axis of CCS (^CY) is the same as ^RZ. The Z axis of CCS (^CZ) is the optical axis of the line camera. The X axis of CCS (^CX) can be defined by rotating ^CY counter-clockwise by 90 °.

2.2   The Collinearity Equations for an Omni-Directional Image

The collinearity equations can be derived from the condition that lies on a line with the object point and the image point. We can derive the collinearity equations by separating into three stages. The first stage is the transformation of the object points, second is the projection into the image points and the final step is the transformation of the image points.

First, the object points in OCS have to be transformed into CCS. For this, after being transformed into RCS using Eq. (1), the object point has to be transformed into CCS using Eq. (2). ^RP is the object point in RCS, ^OO_R is the origin of RCS in OCS and ^RR_O is the rotation matrix transforming from OCS into RCS which can be calculated using the attitude of RCS. ^CP is the object point in CCS, ^RO_C is the origin of CCS in RCS and ^CR_R is the rotation matrix transforming from RCS into CCS which can be calculated using the attitude of CCS. ^RO_C can be determined through Eq. (3) and ^CR_R can be determined through Eq. (4). α is calculated using Eq. (5) as shown in Fig. 3. θ can be determined through Eq. (6), ε can be determined through Eq. (7).

$$^RP = {}^RR_O(^OP - {}^OO_R). \tag{1}$$

$$^CP = {}^CR_R(^RP - {}^RO_C). \tag{2}$$

$$^RO_C = \begin{bmatrix} r\cos\alpha \\ r\sin\alpha \\ 0 \end{bmatrix}. \tag{3}$$

$$^CR_R = \begin{bmatrix} 0 & 1 & 0 \\ 0 & 0 & 1 \\ 1 & 0 & 0 \end{bmatrix} \begin{bmatrix} \cos-(\alpha-\beta) & -\sin-(\alpha-\beta) & 0 \\ \sin-(\alpha-\beta) & \cos-(\alpha-\beta) & 0 \\ 0 & 0 & 1 \end{bmatrix}. \tag{4}$$

$$\alpha = \theta + \varepsilon. \tag{5}$$

$$\theta = \begin{cases} atan2(^RY_1, {}^RX_1) & atan2(^RY_1, {}^RX_1) \le 0 \\ atan2(^RY_1, {}^RX_1) - 2\pi & atan2(^RY_1, {}^RX_1) > 0 \end{cases}. \tag{6}$$

$$\varepsilon = \pi - (\pi - \beta) - \sin^{-1}\left(\frac{r\sin(\pi-\beta)}{\sqrt{^RX_1^2 + {}^RY_1^2}}\right) = \beta - \sin^{-1}\left(\frac{r\sin(\pi-\beta)}{\sqrt{^RX_1^2 + {}^RY_1^2}}\right). \tag{7}$$

In the second stage, the object points are projected into the image points. There are four types of projection models for fisheye lens [9]. Initially, we have no idea which project model best suits our system. Therefore, we have to perform all projection models and select the best fit model that has smallest residuals. The fisheye-lens projection model is different with the central projection, Fig. (3) and Eq. (8). r_p is the distance between the principal point and an image point, f is the focal length of the line camera and ρ is the angle between the light ray and the ray axis as showing Eq. (9).

$$r_p = \begin{cases} 2f \tan\left(\frac{\rho}{2}\right) & \text{Pr1} \\ f\rho & \text{Pr2} \\ 2f \sin\left(\frac{\rho}{2}\right) & \text{Pr3} \\ f \sin\rho & \text{Pr4} \end{cases}. \tag{8}$$

$$\rho = \tan^{-1}\left(\frac{{}^cY_1}{{}^cZ_1}\right). \tag{9}$$

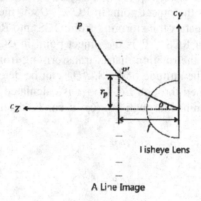

Fig. 3. The projection geometry of the fisheye lens

Third, the image points in CCS must be transformed into image coordinate system (ICS) through Eq. (10). The column coordinate (i) of image points related with the rotation angle can be determined by dividing $\alpha - \beta$ into the individual pixel rotation angle (μ_α) which can be calculated by dividing $360°$ into the number of column (m). The row coordinate (j) of image points related with the projection can be determined by dividing r_p into the size of the row direction of the pixel (μ_j) and adding $\frac{n}{2}$ in which n is the number of pixels existing on a line camera.

$$\begin{bmatrix} i \\ j \end{bmatrix} = \begin{bmatrix} \dfrac{\alpha - \beta}{\mu_\alpha} \\ -\dfrac{r_p}{\mu_j} + \dfrac{n}{2} \end{bmatrix}. \tag{10}$$

2.3   Definition of OCS and Acquisition of OCP in OCS

The georeferencing which estimates EOP of outdoor images uses the control points which could be obtained by GPS or total stations in the absolute coordinate system. However, for indoor images, it is difficult to obtain OCP in OCS which is the absolute coordinate system because OCS could not be defined. Therefore, in order to obtain OCP, we define OCS and transform OCP acquired in the relative coordinate system into OCS.

First, the origin of OCS is defined as a corner and the attitude of OCS is defined as three edges encountering at the corner which are the same as the three axes of OCS. Fig. 4 shows the definition of OCS.

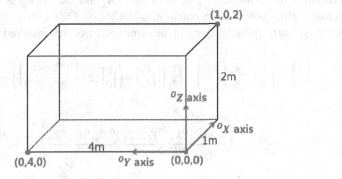

Fig. 4. The definition of OCS

Second, OCPs acquired in the measurement coordinate system (MCS) which is the relative coordinate system have to be transformed into OCS. For this, it is essential to estimate the transformation parameters from MCS to OCS. The transformation parameters are the origin and the attitude of MCS in OCS. Estimating that needs at least six observation equations which can be established through three corner points coordinates in MCS which can be defined by the distance between the corners and the origin of MCS. After estimating the transformation parameters using eq. (11), the OCPs can be transformed from MCS into OCS using eq. (12). (X_a, Y_a, Z_a) is the OCP coordinates in OCS, (X_b, Y_b, Z_b) is the OCP coordinates in MCS, R is the rotational matrix from OCS to MCS which is calculated through the attitude of MCS and t is the translation matrix which is the same as the origin of MCS.

$$\begin{bmatrix} X_b \\ Y_b \\ Z_b \end{bmatrix} = R \begin{bmatrix} X_a \\ Y_a \\ Z_a \end{bmatrix} + t . \tag{11}$$

$$\begin{bmatrix} X_a \\ Y_a \\ Z_a \end{bmatrix} = R^T \left(\begin{bmatrix} X_b \\ Y_b \\ Z_b \end{bmatrix} - t \right) . \tag{12}$$

2.4   Mathematical Model for EOP Estimation

EOP mean the position and attitude of camera. In a rotating line camera, the origin and attitude of RCS has been known in order to calculate the position and attitude of camera. Therefore, we define EOP as the origin and attitude of RCS.

In order to estimate EOP, we propose the observation equations with stochastic constraints presented as eq. (13). The observation equations are classified into two categories. One is the observation equation based on the collinearity equations. A collinearity equations such as eq. (14) represents the relation between an image point (IP) and an OCP. The other is the observation equation of OCPs. In eq. (13), y is the observed image point, z_o is the observed OCP, A_{eo} and A_o is the design matrix differentiating the collinearity equations on EOP and OCPs, K_o is the design matrix differentiating the observation equation of OCPs on OCPs, e_i is the error vector of the observed image points and e_o is the error vector of the observed OCPs.

$$\begin{bmatrix} y \\ z_o \end{bmatrix} = \begin{bmatrix} A_{eo} & A_o \\ 0 & K_o \end{bmatrix} \begin{bmatrix} \xi_{eo} \\ \xi_o \end{bmatrix} + \begin{bmatrix} e_i \\ e_o \end{bmatrix}, \begin{bmatrix} e_i \\ e_o \end{bmatrix} \sim \left(\begin{bmatrix} 0 \\ 0 \end{bmatrix}, \sigma_0^2 \begin{bmatrix} P_i^{-1} & 0 \\ 0 & P_o^{-1} \end{bmatrix} \right). \tag{13}$$

$$\underset{IP}{\underline{(i,j)}} = f \left(\underset{EO(\xi_{eo})}{\underline{\omega, \varphi, \kappa, {}^oX_R, {}^oY_R, {}^oZ_R}}, \underset{OCP(\xi_o)}{\underline{{}^oX_n, {}^oY_n, {}^oZ_n}} \right). \tag{14}$$

3   Experimental Results

In this experiment, we obtained the omni-directional images through a rotating line camera and OCP through the total station. We then estimated EOP by applying the four projection models and selected the best projection model which had the smallest residuals.

3.1   Acquisition of Omni-Directional Images and OCPs

In this experiment, we obtained two omni-directional images through a rotating line camera. The line camera used in this experiment can capture image with size 1×4000 pixel, the focal length is 15mm, and β is 90°. One image, referred to as inner image, was acquired by setting the rotational radius to 0cm. The other image, referred to as outer image, was acquired by setting the rotational radius to 18cm. The size of images is 8000×4000 pixels, the size of pixel is $10\mu m \times 10\mu m$. Fig. (5) is the inner image and Fig. (6) is the outer image. In the figures, the yellow points are OCPs.

Fig. 5. Inner image acquired by a rotating line camera

Fig. 6. Outer image acquired by a rotating line camera

We obtained the 56 OCPs through the total station. First, we defined the origin of OCS as the corner that is the left down from the door in the images. Second, we defined the corner's coordinates without the origin of OCS by calculating the distance between the corners and the origin of OCS. Third, we estimated the origin and attitude of MCS which was defined by the total station. The estimated origin and attitude are listed in Table 1. Finally, we transformed OCP from MCS into OCS.

Table 1. The estimated origin and attitude of MCS

	The origin of MCS			The attitude of MCS		
	X	Y	Z	ω	φ	κ
Estimates	2.3038m	-0.5015m	-0.1721m	0.07°	-0.05°	-247.40°
Precision	±0.0024m	±0.0025m	±0.0023m	±0.05°	±0.05°	±0.03°

3.2   The Results of Estimated EOP

We estimated EOP of the inner and outer images using OCPs applying the four projection models. According to the experimental results, the first projection model (Pr1) had the smallest residuals. Therefore, we selected Pr1 as the best projection model for a rotating line camera in this study. The results listed in Table 2. $\hat{\sigma}_0^2$ is the estimated variance component, STD of $\tilde{e}_i$ is the standard deviation of residuals according to horizontal direction and STD of $\tilde{e}_j$ is the standard deviation of residuals according to vertical direction.

Table 2. The estimated variance component, the standard deviation of residuals according to horizontal direction and the standard deviation of residuals according to vertical direction

	Inner Image			Outer Image		
	$\hat{\sigma}_0^2$	STD of $\tilde{e}_i$ (pixels)	STD of $\tilde{e}_j$ (pixels)	$\hat{\sigma}_0^2$	STD of $\tilde{e}_i$ (pixels)	STD of $\tilde{e}_j$ (pixels)
Pr1	2.1959	2.5816	9.9399	1.7654	3.6395	8.4508
Pr2	4.3921	4.1938	13.9031	2.8265	3.9535	10.9143
Pr3	6.9675	5.3729	17.4683	4.9214	4.8475	14.5179
Pr4	20.8509	8.6791	30.2995	17.6789	7.8879	27.8397

The estimated EOP and its precision of inner and outer images are listed in Table 3. The inner and outer omni-directional images were not acquired at the same time. Therefore, the two images' origins of RCS were the same, the two images' ω and φ among attitude of RCS were similar and the two images' κ among attitude of RCS were different.

Table 3. The estimates and the precision of EOP applying Pr1

	Inner Image		Outer Image	
	Estimates	Precision	Estimates	Precision
$^{O}X_R$	2.5945 m	±0.0014 m	2.5908 m	±0.0012 m
$^{O}Y_R$	1.8621 m	±0.0021 m	1.8597 m	±0.0019 m
$^{O}Z_R$	1.6698 m	±0.0011 m	1.6725 m	±0.0010 m
ω	0.1149 °	±0.0442 °	0.7990 °	±0.0497 °
φ	1.0475 °	±0.0654 °	0.7574 °	±0.0502 °
κ	15.2598 °	±0.0454 °	-21.7092 °	±0.0404 °

4   Conclusions

In this paper, we could estimate EOP of a single omni-directional image acquired by a rotating line camera. We first defined a rotating line camera and the coordinate systems. Second, we derived the collinearity equations. Third, we estimated the transformation parameters from MCS to OCS and obtained OCPs in OCS by transforming.

Finally, we established a mathematical model for estimating EOP and estimated EOP by selecting the best projection model which had the least standard deviation of residuals.

According to the results, we selected the Pr1 that estimated the origin of RCS with ±1.4mm precision and the attitude of RCS with ±0.05° precision. In addition, the standard deviation of residuals according to i-direction and j-direction was ±2.56 pixel and ±9.85 pixel, respectively.

In the future, once we achieve on the calibration estimating the lens parameters which are the principal point and the distortion coefficients and the system parameters (the rotational radius and the titled camera angle), we can accomplish the estimation of EOP more precisely. This enables the 3D coordinates of the indoor objects and the 3D indoor model to be approximated more accurately.

Acknowledgment. This work was jointly supported by the supporting project to educate GIS experts and Seoul R&BD Program (JP100056).

References

1. Huang, F., Wei, S., Klette, R.: Rotating Line Cameras: Model and Calibration. IMA Preprint Series #2104. Institute for mathematics and ITS applications, University of Minnesota (2006)
2. Kang, S.B., Weiss, R.: Characterization of Errors in Compositing Panoramic Images. In: IEEE Computer Society Conference on Computer Vision and Pattern Recognition, San Juan, Puerto Rico, pp. 103–109 (1997)
3. Oh, S., Lee, I.: Estimation of 3D Object Points from Omni-directional Images Acquired by A Rotating Line Camera. In: The 32nd Asian Conference on Remote Sensing, Taipei, Taiwan (2011)
4. Torii, A., Sugimoto, A., Imiya, A.: Mathematics of a Multiple Omni-Directional System. In: The 2003 Conference on Computer Vision and Pattern Recognition Workshop, Washington, DC, USA, pp. 1063–6919 (2003)
5. Wei, S., Huang, F., Klette, R.: The Design of a Stereo Panorama Camera for Scenes of Dynamic Range. In: 16th International Conference on Pattern Recognition (ICPR 2002), Quebec, Canada, pp. 635–638 (2002)
6. Tang, W.K., Hung, Y.S.: Concentric Panorama Geometry. In: 8th IEEE Conference on Mechatrinics and Machine Vision in Practice, Hong Kong, China, pp. 507–512 (2001)
7. Shum, H., Szeliski, R.: Stereo Reconstruction from Multiperspective Panoramas. In: 7th International Conference on Computer Vision ICCV 1999, Kerkyra, Greece, pp. 14–21 (1999)
8. Huang, F., Wei, S., Klette, R.: Technical Report - Rotating Line Cameras: Model and Calibration, IMA Preprint Series #2104, Institute for mathematics and ITS applications, University of Minnesota (2006)
9. Juho, K., Sami, B.: A Generic Camera Calibration Method for Fish-Eye Lenses. In: The 17th International Conference on Pattern Recognition, Cambridge, UK, pp. 10–13 (2004)

Spatter Tracking in Laser Machining

Timo Viitanen[1], Jari Kolehmainen[1], Robert Piché[1], and Yasuhiro Okamoto[2]

[1] Tampere University of Technology
[2] Okayama University

Abstract. In laser drilling, an assist gas is often used to remove material from the drilling point. In order to design assist gas nozzles to minimize spatter formation, measurements of spatter trajectories are required.

We apply computer vision methods to measure the 3D trajectories of spatter particles in a laser cutting event using a stereo camera configuration. We also propose a novel method for calibration of a weak perspective camera that is effective in our application.

The proposed method is evaluated with both computer-generated video and video taken from actual laser drilling events. The method performs well on different workpiece materials.

1 Introduction

In laser drilling, a laser beam is focused on a workpiece to melt or vaporize the material, forming a hole. Often a jet of high-speed assist gas is applied to blow molten material away from the hole. A recent advance in the field is the use of a Laval nozzle whose shape accelerates the gas to supersonic speed [1].

A common defect associated with laser drilling is that some material ejected from the hole lands back on the workpiece and resolidifies, forming spatter. An example of spatter is shown in Figure 1a. This is especially undesirable in applications that require high precision. There is therefore interest in improving the design of an assist gas nozzle to minimize spatter formation. For example, the material could be expelled away from the workpiece, or caught at a shield. To guide the designer, the motion of spatter particles can be predicted with CFD simulation. However, measurements are needed to calibrate the simulation models and to verify the models' fidelity.

The objective of this study is to analyze video of laser drilling events taken with a stereo configuration of two high-speed cameras. An example video frame is shown in Figure 1b. This study is structured as follows. First we discuss calibration of such a camera configuration, and propose a method for calibration of a weak perspective camera that is effective for this application. We then apply a sequential Monte Carlo (SMC) filter to track spatter particles in the video. We pair tracks using the epipolar constraint and perform 3D reconstruction using an unscented Kalman filter (UKF). Finally, we evaluate the proposed method with synthetic video, and video taken from 30 laser drilling events. Results reported in [2] were computed using this method.

G. Bebis et al. (Eds.): ISVC 2012, Part II, LNCS 7432, pp. 626–635, 2012.

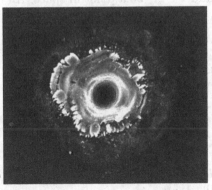

(a) Spatter formation surrounding a laser drilled hole.

(b) A typical input video frame.

Fig. 1. Laser drilling

2   Camera Calibration

2.1   Overview

A main obstacle in this study was obtaining accurate calibration. We initially attempted calibration with Zhang's method [3]. However, the results obtained this way suffered from a large reprojection error. We found that Zhang's method is unsuitable for an application such as ours, where observations are noisy and cameras are close to affine, and developed a method based on the weak perspective camera model.

Like Zhang's method, our method requires images of a calibration object. A sample calibration image is shown in Figure 2d. We first calculate homography matrices, which relate the model plane to each image. Using these matrices, we estimate the intrinsic parameters that are common to all images. We then solve the extrinsic parameters that describe the calibration object's position and orientation in each image. The following sections describe these steps in more detail. [3]

2.2   Weak Perspective Camera Model

The affine camera is an abstract model of a camera whose focal length is very large compared to the object being viewed. The model is accurate for e.g. microscopes. An affine camera constrained to have zero skew is called a *weak perspective camera*. Compared to the projective camera, which has 5 intrinsic and 6 extrinsic parameters, a weak perspective camera has 2 and 5 respectively, making it more analytically tractable.

The cameras used in our study have a focal length of tens of centimeters, and our calibration object is 1 mm wide, making them practically indistinguishable from affine cameras.

Let $x \in \mathbb{R}^2$ be a point in the image plane and $z \in \mathbb{R}^3$ be a point in space. An affine camera model is given by

$$x = Mz + t. \tag{1}$$

where M is a 2×3 matrix and t is a translation vector. Using homogenous coordinates, the affine camera model can be written as

$$\begin{bmatrix} x \\ 1 \end{bmatrix} = P \begin{bmatrix} z \\ 1 \end{bmatrix}, \tag{2}$$

$$P = \begin{bmatrix} M_{11} & M_{12} & M_{13} & t_1 \\ M_{21} & M_{22} & M_{23} & t_2 \\ 0 & 0 & 0 & 1 \end{bmatrix}. \tag{3}$$

The matrix M can be decomposed as $M = A_{22}R_{1:2,1:3}$, where A_{22} is an affine calibration matrix and $R_{1:2,1:3}$ consists of the first two rows of a rotation matrix. The calibration matrix is of the form

$$A_{22} = \begin{bmatrix} \alpha & 0 \\ \gamma & \beta \end{bmatrix} = \begin{bmatrix} \alpha & 0 \\ 0 & \beta \end{bmatrix}. \tag{4}$$

where α and β represent focal length and aspect ratio, and γ the skew of the camera, which we constrain to be 0.

2.3   Homography

An affine mapping from a 3D plane to an image plane can be represented by an affine homography matrix $H \in \mathbb{R}^{3 \times 3}$ whose last row is $[0, 0, 1]$. Without loss of generality we can choose a coordinate system such that the image plane lies in xy-plane, so that z equals zero. We can then write the homography in terms of the projection matrix:

$$H = \begin{bmatrix} M_{11} & M_{12} & t_1 \\ M_{21} & M_{22} & t_2 \\ 0 & 0 & 1 \end{bmatrix}. \tag{5}$$

2.4   Solving for Focal Length and Aspect Ratio

Given a set of homographies $\{H^{(k)}\}$, the objective of affine calibration is to find the calibration matrix A, rotation matrices $\{R^{(k)}\}$ and translation vectors $\{t^{(k)}\}$ that fit the given homographies.

First of all we denote $A_{ii} = \alpha_i$. Now from the first two colums of the homography H we obtain two equations

$$h_{1:2}^{(i)} = \alpha_i r_{1:2}^{(i)}, \text{ for } i \in \{1, 2\}. \tag{6}$$

Adding the missing rotation components, Equation (6) becomes

$$[h_{1:2}^{(i)}, \alpha_i r_3^{(i)}]^\mathrm{T} = \alpha_i r^{(i)}. \tag{7}$$

Knowing that $\|r^{(i)}\| = 1$, and that $r^{(1)} \cdot r^{(2)} = 0$ we obtain three equations

$$\left(r_3^{(i)}\right)^2 = \frac{\|h_{1:2}^{(i)}\|^2 - \alpha_i^2}{\alpha_i^2}, \text{ for } i \in \{1, 2\}, \text{ and} \tag{8}$$

$$h_{1:2}^{(1)} \cdot h_{1:2}^{(2)} = -\alpha_1 \alpha_2 r_3^{(1)} r_3^{(2)}. \tag{9}$$

Eliminating rotation elements $r_3^{(i)}$ yields

$$\left(h_{1:2}^{(1)} \cdot h_{1:2}^{(2)}\right)^2 - \|h_{1:2}^{(1)}\|^2 \|h_{1:2}^{(2)}\|^2 = -\|h_{1:2}^{(2)}\|^2 \alpha_1^2 - \|h_{1:2}^{(1)}\|^2 \alpha_2^2 + (\alpha_1 \alpha_2)^2. \tag{10}$$

Denoting the expression on the left of (10) as b_k, where k denotes the index of homography used, and $a_k = -[\|\|h_{1:2}^{(2)}\|^2, h_{1:2}^{(1)}\|^2]^\mathrm{T}$, we obtain

$$b_k = a_k[\alpha_1^2, \alpha_2^2]^\mathrm{T} + (\alpha_1 \alpha_2)^2. \tag{11}$$

Taking the differences of any two equations with different indices yields

$$b_{k_1} - b_{k_2} = (a_{k_1} - a_{k_2}) [\alpha_1^2, \alpha_2^2]^\mathrm{T}. \tag{12}$$

For a set of n images, (12) determines an overdetermined set of $O(n^2)$ linear equations to be solved in the least-squares sense.

2.5   Solving for Rotation and Translation

Due to the affine nature of the problem, translation is equal to the last column of homography (if it is scaled so that $H_{33} = 1$), that is,

$$t = h^{(3)}. \tag{13}$$

We can obtain the top left 2×2 block of the rotation matrix as

$$r_{1:2}^{(i)} = \alpha_i^{-1} h_{1:2}^{(i)}. \tag{14}$$

Now we need to find a rotation matrix that fits the 2×2 block. The last component of a column can be obtained from the unity constraint as

$$r_3^{(i)} = \pm\sqrt{1 - \min(1, \|r_{1:2}^{(i)}\|^2)}, \tag{15}$$

where the minimum is added to ensure numerical stability. Signs are chosen so that they preserve orthogonality and right-handedness. The last row is obtained using the well-known relationship [3]

$$r^{(3)} = r^{(1)} \times r^{(2)}, \tag{16}$$

After invoking Equation (15), the full rotation matrix is adjusted using SVD-decomposition to ensure its orthogonality. This procedure also eliminates mirroring.

The parameters obtained in this way could be used as an initial guess for further optimization that minimizes the reprojection error, as in [3], but this was not done in this study.

2.6   Evaluation

We compared our method to Zhang's using simulated data. In each trial, we performed camera calibration with ten synthetic views of a calibration object, taken at equal distance from random directions, perturbed with Gaussian noise. We varied the distance and the mean deviation of the noise, ran 100 trials for every parameter pair, and compared the reprojection RMSE statistic for each method. The results are shown in Figure 2.

With zero noise, Zhang's method outperforms ours at any distance, since it solves the original camera matrix, which is not exactly affine. Also, near the calibration object, the affine camera is a poor approximation and this causes large errors for our method. Further away, Zhang's method becomes highly susceptible to small amounts of noise, while our method remains robust. In our application the noise std. dev. is of order 10^{-3} in normalized coordinates and the distance is approximately 10^3. As seen in Figure 2, these conditions are very suitable for our method and unsuitable for Zhang's.

3   Spatter Object Tracking

3.1   SMC Filter

Sequential Monte Carlo (SMC) filtering is a method used to estimate the state of a discrete-time nonlinear dynamic system. It is applied on similar problems for instance in [4]. A dynamic system has a state sequence x_k and a measurement sequence y_k. The state's evolution is specified by a probabilistic model given by[1]

$$x_k|x_{k-1} \sim p(x_k|x_{k-1}). \tag{17}$$

The state x_k is connected to the measurement y_k by the observation model given by

$$y_k|x_k \sim p(y_k|x_k) \propto L(y_k|x_k), \tag{18}$$

where $L(y|x_k)$ is called likelihood. Likelihood needs only be known up to a multiplicative constant. The solution of the problem is the posterior density $p(x_k|y_{1:k})$. In this study, the state is the position of a spatter object. We used the SIFT method to detect candidate spatter particles in the video [5].

---

[1] For the sake of notational simplicity, the distinction between random variables and their realized values (e.g. using bold font or upper case) is not made.

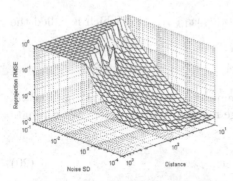
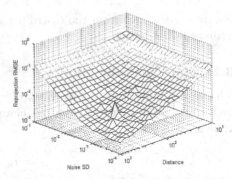

(a) Error analysis of Zhang's method. The method is at its best close to the calibration object, and quickly becomes susceptible to noise as distance grows. The RMSE is capped at 1 for clarity, but it may rise up to 10^6.

(b) Error analysis of the affine method. At close distance the affine approximation is unusable. The method grows more accurate as distance grows. With very high noise, the results are less erratic than in Zhang's but still useless.

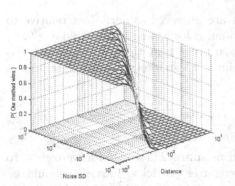

(c) Comparison of the affine method to Zhang's. The Z-axis represents the proportion of trial runs where our method had the lower RMSE.

(d) A typical calibration image.

Fig. 2. Evaluation of the proposed affine calibration method with synthetic data. The calibration object is 10 units wide. The noise is added to normalized image coordinates that range between $-1..1$.

A SMC filter solves the problem by approximating the posterior density with set of N samples $\{x_k^{(i)}\}$ and associated weights $\{w_k^{(i)}\}$ by Equation (19)

$$p(x_k|y_{1:k}) \approx \sum_{i \leq N} w_k^{(i)} \delta(x_k - x_k^{(i)}).$$

(19)

Samples are drawn from the distribution $\pi(x_k|x_{1:k-1}, y_{1:k})$, which is called the proposal distribution. [6]

3.2   State Evolution Model

The state evolution model in this study embodies the assumption that the velocity of a spatter particle does not change much in one time step. We estimate velocity as

$$v_k = x_k - x_{k-1}.$$

(20)

which yields the following state model,

$$x_k = x_{k-1} + v_{k-1} + \varepsilon_k = 2x_{k-1} - x_{k-2} + \varepsilon_k,$$

(21)

where ε_k is a normally distributed error term.

3.3   Observation Model

In a laser drilling video, spatter objects are observed as dark dots relative to background. The user specifies a background color b_k and a threshold $\theta \geq 0$.

Let $I_k(x)$ be the intensity at the image point x at time k. Likelihood was computed in a neighborhood U_x of the point x. Mean intensity in a neighborhood U_x is given by Equation (22)

$$\overline{I}_k(x) = \frac{\int_{U_x} I(x')dx'}{\int_{U_x} dx'}.$$

(22)

In the discrete case integrals are treated as sums. Three criteria were used to formulate a likelihood function that recognizes particles. Particles should be distinct from background color, and the mean intensity and texture of a particle should remain similar in consecutive frames. The likelihood used in this study is given by

$$L(I|x_{1:k}, I_{k-1}) = \left(\overline{I}(x_k) - b_k\right)^2 H\left(\left(\overline{I}(x_k) - b_k\right)^2 - \theta\right) \times$$

$$\exp\left(-\frac{1}{2}\frac{\left(\overline{I}(x_k) - \overline{I}_{k-1}(x_{k-1})\right)^2}{\sigma_I \max(I)^2}\right) \times$$

$$\int_{U_0} \exp\left(-\frac{1}{2}\frac{\left(I(x' + x_k) - I_{k-1}(x' + x_{k-1})\right)^2}{\sigma_I \max(I)^2}\right) dx' \quad (23)$$

where σ_I denotes the assumed intensity covariance of points in consecutive frames and $H(\cdot)$ is the Heaviside function. In this study a square neighborhood of 5×5 pixels was used. An ellipse was also tested, but it gave no noticeable advantage over the square.

3.4   Proposal Distribution

In SMC filtering it is common to use bootstrap filtering, where the state model is the proposal distribution for the sampling. However in a spatter tracking application particles are small and bootstrap filtering leads to sampling from regions that are of no interest. To make tracking more efficient, we used data dependent sampling, where part of the likelihood is merged to the proposal distribution [4]. This lead to a proposal distribution given by

$$\pi(x_k|x_{1:k-1}, I_{1:k}) \propto p_{x_k|x_{k-1}}(x_k|x_{k-1}) \exp\left(-\frac{1}{2}\frac{\left(\overline{I}_k(x_k) - \overline{I}_{k-1}(x_{k-1})\right)^2}{\sigma_I \max(I_k)^2}\right). \tag{24}$$

Sampling from this proposal distribution can be achieved using Metropolis-Hastings algorithm [7] with the state model $p_{x_k|x_{k-1}}(x_k|x_{k-1})$ as a target distribution. This caused most samples to end up inside a spatter particle, allowing the use of very small sample sizes. We achieved good performance with a short random walk of tens of samples.

4   3D Reconstruction

4.1   Spatter Object Matching

After tracking spatter objects in the separate videos from the two cameras, we match them to form pairs. Image points $x, x' \in \mathbb{P}^2$ from different views must satisfy the equation

$$x^T \mathbf{F} x' = 0, \tag{25}$$

where $\mathbf{F}$ is the fundamental matrix between the two views. The fundamental matrix can be computed from 7 point correspondences between two images. [8]

We matched each particle to the particle that minimizes the mean of $x^T \mathbf{F} x'$ over the shared lifetime of the pair. Since the fundamental matrix constrains the particles only to a line, not a point, this may result in spurious matches. To reduce the number of spurious matches, we reject very short tracks, as well as tracks that have a high reprojection error after reconstruction.

4.2   3D Path Reconstruction

In this study, 3D reconstruction was done by an unscented Kalman filter (UKF) [9]. The UKF solves problems similar to the SMC filter. Here the 2D locations of

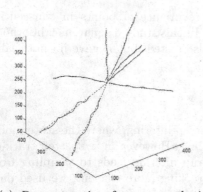
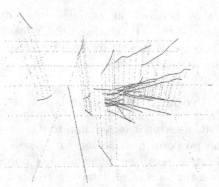

(a) Reconstruction from a synthetic test video with Matlab prototype. Dashed line: ground truth, Cont. line: reconstructed.

(b) Reconstruction from laser drilling data with a threaded C++ implementation.

Fig. 3. Evaluation

a particle were chosen as measurements and the state evolution model was similar to the one used with the SMC filter. An advantage of UKF over the extended Kalman filter (EKF) is that it doesn not require the computation of Jacobian matrices of these models. The standard constant velocity motion model[10] was used as the state evolution model.

Given camera matrices P_x and P_y of both cameras, the 3D position $z \in \mathbb{R}^3$ is connected to homogenous image coordinates x and y by

$$\begin{bmatrix} x \\ y \end{bmatrix} \propto \begin{bmatrix} P_x \\ P_y \end{bmatrix} z. \tag{26}$$

With a perspective camera, the observation model would be nonlinear due to the use of homogenous coordinates.

5 Evaluation of Method

The proposed method was used to track particles in computer generated video and in actual laser drilling data. In synthetic video with constant lighting and low noise, the method successfully tracked eight separate particles with an RMSE of 1.5% of trajectory length, which is less than the particle radius. Typical results from an eight particle test video are shown in Figure 3a.

The method was also tested on data measured from 30 actual laser drilling events. The used laser was a Nd:YAG-laser, with a pulse energy of 0.2 J. Each video recorded a single laser pulse. The assist gas was nitrogen, with cylinder pressure 0.6 MPa. Frame rate was 1 Mfps and exposure time was 0.5 µs.

Tests carried out with real data gave natural looking results. Typical results of spatter tracking can be seen in Figure 3b. Fuirther experimental results using the proposed tracking method are reported in [2].

6   Conclusions

The proposed method for spatter measurements produces plausible trajectory reconstructions from real data recorded by the high speed video cameras and obtains good accuracy in our simulated test cases. There is currently no other way to obtain detailed knowledge such as individual particle velocities or particle sizes for statistical analysis from a cutting event. The results will be used to calibrate and verify CFD simulation in assist gas nozzle design.

The method could also be used for spatter tracking in laser cutting of materials such as steel, aluminium and plastics. The main limitation is the use of oxygen as an assist gas, as it makes the observation of particles difficult.

Practical problems with the method were mostly associated with difficulty of obtaining accurate camera calibration. These problems were resolved by developing the weak-perspective calibration method described earlier. The weak perspective method may be useful for other applications, for example some Augmented Reality research uses Zhang's method to locate fiducial markers, for instance in [11]. Since markers may be relatively far from the camera, our method may be more robust.

References

1. Okamoto, Y., Uno, Y., Suzuki, H.: Effect of nozzle shape on micro-cutting performance of thin metal sheet by pulsed Nd: YAG laser. International Journal of Automation Technology 4 (2010)
2. Kolehmainen, J.T., Okamoto, Y., Yamamoto, H., Okada, A., Viitanen, T.T.: Measurement of the spatter velocity in fine laser cutting. JSPE Spring Meeting (2012)
3. Zhang, Z.: A flexible new technique for camera calibration. IEEE Trans. Pattern Anal. Mach. Intell. 22, 1330–1334 (2000)
4. Smal, I., Niessen, W., Meijering, E.: Advanced particle filtering for multiple object tracking in dynamic fluorescence microscopy images. Biomedical Imaging: From Nano to Macro 4, 1048–1051 (2007)
5. Lowe, D.G.: Object recognition from local scale-invariant features. Computer Vision 7, 1150–1157 (1999)
6. Doucet, A., Godsill, S., Andrieu, C.: On sequential Monte Carlo sampling methods for Bayesian filtering. Statistics and Computing 10, 197–208 (2000)
7. Hastings, W.: Monte Carlo sampling methods using Markov chains and their applications. Biometrika 57, 97–109 (1970)
8. Luong, Q.T., Faugeras, O.D.: The fundamental matrix: Theory, algorithms, and stability analysis. International Journal of Computer Vision 17, 43–75 (1996)
9. Julier, S.J., Uhlmann, J.K.: A new extension of the Kalman filter to nonlinear systems. In: Int. Symp. Aerospace/Defense Sensing, Simul. and Controls, vol. 3 (1997)
10. Bar-Shalom, Y., Li, X.R., Kirubarajan, T.: Estimation with Applications to Tracking and Navigation. John Wiley & Sons, Inc. (2002)
11. Atcheson, B., Heide, F., Heidrich, W.: CALTag: High precision fiducial markers for camera calibration. Vision, Modeling, and Visualization, 41–48 (2010)

Car License Plate Detection under Large Variations Using Covariance and HOG Descriptors

Jongmin Yoon, Bongnam Kang, and Daijin Kim

Department of Computer Science and Engineering
Pohang University of Science and Technology (POSTECH)
{albedo039,bnkang,dkim}@postech.ac.kr

Abstract. This paper presents a novel method that can detect license plates which have large variations including perspective distortion, size variation, blurring. Spatial combinations of covariance descriptors in different positions are used with feed-forward network to extract plate-like region and HOG descriptor is used with LDA for validation. From this method, we could achieve high detection rate 94% while maintaining low FPPW(2.5^{-6}) in road view image.

1 Introduction

License plate detection is very important topic in intelligent transportation systems and surveillance systems. One of its application is privacy protection in road view images. In recent years, many internet portal sites such as Google and Daum (a major portal site in Korea) provide web-based street view services (called 'road view'). The problem of such services is that people's privacy can be violated because the service exposes their life to the public regardless of their intention. To prevent such a problem, most web sites which provide such a service remove or blur all human faces and license plates before releasing images to the public. However, finding and removing them manually is very tedious and time-consuming. Fortunately, thanks to great advances in face detectors, most faces can be automatically detected without any human labor. But, in case of car license plates, an appropriate algorithm for detecting them in road view images has not been proposed yet.

The difficulty of detecting license plates in road view images is that they contain many variations including size or scale variations, perspective distortions, and illumination changes. Moreover, the shape of a license plate is not fixed because the order of digits written it can vary. In addition, images tend to be very large in road view service (in our case 8000x4000) to covers all directions at the specific view position. Therefore false alarm rate must be extremely low to reduce the number of false detections in large scale image.

To overcome such difficulties of detecting car license plates in the road view image, we propose a new method based on neural network classifier using covariance and HOG descriptors.

G. Bebis et al. (Eds.): ISVC 2012, Part II, LNCS 7432, pp. 636–647, 2012.

2   Previous Work

Algorithms for detecting license plate can be divided into two categories according to their approaches.

First approach is based on heuristic search which uses knowledge gotten from human's experiences. For example, we can find some distinctive characteristics of license plates based on our experiences such as high edge density or four border lines in a license plate region. From this point of view, Vahid et al.[1] proposed a filter which outputs high response on an area which has high edge density. It is very simple and fast. Furthermore, it doesn't need many training samples. But a significant drawback of this method is that it causes so many false alarms on the patches of high cornerness because it only measures edge density to detect license-plate-like regions. For this reason, it is not suitable for final decision of license plate area, but for detecting license-plate-like candidates. For another example, Tran et al.[2] proposed a method which uses Hough transform to find four boundary lines of license plate. However, it operates only in ideal environments where all disturbances are suppressed.

Second approach is based on learning which uses knowledge gotten from training samples. From this point of view, Wing et al.[3] proposed a two stage classifier in which first stage is Adaboost and second stage is SIFT-SVM. They used Haar-like features which are based on brightness difference between rectangle regions. But a problem is that distinguishing brightness differences are rare in license plate regions. Moreover, the number of key points for SIFT descriptor should be larger than a certain level; This assumption is not satisfied when an image is small. Faith[4] proposed a method that uses a covariance descriptor in a neural network (NN) framework. It shows outstanding detection rates which has large variations and distortions such as rotation, blurring. However, this method is still not enough to detect license plates in road view image, because extremely large image involves so many false alarms in road view image. Therefore we need to develop a better method which can overcome such disadvantages of existing methods to achieve our goal.

3   Methods

3.1   Overview

From a given input image, integral images[1] of each feature are generated for the preparation of computing covariance descriptors. A detector consists of four cascaded classifiers. Classifiers of stage 1-3 are feed-forward neural networks, which have two hidden layers with an input and an output layer; the classifier of stage 4 is a hyperplane computed by LDA. From stage 1-3, covariance descriptors[2] are computed in each subregion of the given search window. Next, these descriptors are concatenated in a feature vector. This computed feature vectors is put into the neural network. In stage 4, HOG descriptor[3] was used instead of covariance descriptors for final validation and LDA was used as a classifier. To process large size image, the input image is divided into several pieces whose sizes are affordable to be processed because the original image is very large(in our case 8000 x 4000). Then, each image piece is skewed in -30, 0', +30' degrees so that skewed (or rotated by view position) license plates can be detected without making different detectors for them. These skewed images are entered into the detectors sequentially. We made two different detectors according to the aspect ratio (AR) of the license plate to be detected. These two different detectors are run on the each skewed image piece to detect license plates of different AR.

3.2   Features

Covariance Descriptor. Tuzel et al.[4] proposed region covariance for object detection. The correlation coefficient of two random variables X and Y is given by

$$\rho_{X,Y} = \frac{\text{cov}(X,Y)}{\text{var}(X)\text{var}(Y)} = \frac{\text{cov}(X,Y)}{\sigma_x^2 \sigma_y^2}, \tag{1}$$

where cov is the covariance of two random variables, that is,

$$\text{cov}(X,Y) = \mathbf{E}[(X - \mu_X)(Y - \mu_Y)], \tag{2}$$

and μ is the sample mean, and σ^2 is the sample variance. The seven image statistics that we used are: x, y, $|I_x|$, $|I_y|$, $|I_{xx}|$, $|I_{yy}|$, $|I_{xy}|$. To improve the calculation efficiency of covariance matrices, a technique which employs integral image [5] can be applied [6]. By expanding the mean from equation (2), the covariance equation can be written as

$$\text{cov}(X,Y) = \frac{1}{n-1}[\sum_k X_k Y_k - \frac{1}{n}\sum_k X_k \sum_k Y_k]. \tag{3}$$

Hence, to compute covariance quickly in a given rectangular region, the sum of each dimension, e.g., $\sum_k X_k$ and $\sum_k Y_k$, e.g., $\sum_k X_k Y_k$, must be computed using the integral image. In our method , we used a 32-dimensional covariance feature in cascade stage 1; this feature consists of correlation coefficients between image statistics and means of RGB color components. The correlations between x, y are not used because they are always 0. Means of RGB color components are computed by calculating mean of each color component within the box contracted to 75% of the original size.

Although a single covariance descriptor itself has good discriminant power, combining multiple covariances computed from several subregions (Fig.1) shows a better result, To take advantage of the combined features while maintaining its detection speed, we used a cascaded detector which consists of three stage classifiers. 32 dimensional single covariance feature used in the first stage is generated by computing the covariance elements in the entire area of the scanning window and vectorizing them. In the second stage, the feature is generated by concatenating the entire regional feature with the features extracted in the four subregions located in the upper, lower, left, and right sides of the scanning window. These features extracted from four subregions are used to discriminate

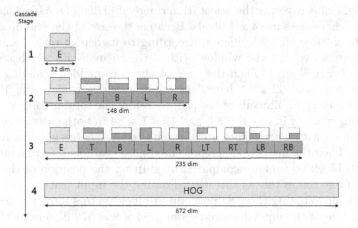

Fig. 1. Features in each cascade stage

horizontal or vertical lines of a license plate: patches that have sharp edges in horizontal or vertial direction can be classified as license plates. In the third stage, the feature is generated by concatenating the entire regional feature with the features extracted in the four sub-regions located in the left-top, right-top, left-bottom, right-bottom corners. These features extracted from four corners are used to discriminate corner shapes of a license plate: patches that have sharp edges in both x and y directions at the four corner of the scanning window can be classified as license plates.

HOG Descriptor. Dalal et al[7] proposed Histogram of Oriented Gradient (HOG) for detection of human. Their method uses a dense grid of histograms of oriented gradients, computed over blocks of various sizes. Each block consists of a number of cells. Blocks can overlap. For each pixel $I(x,y)$, the gradient magnitude $m(x, y)$ and orientation $\theta(x, y)$ is computed from

$$dx = I(x + 1, y) - I(x - 1, y) \tag{4}$$

$$dy = I(x, y + 1) - I(x, y - 1) \tag{5}$$

$$m(x, y) = \sqrt{dx^2 + dy^2} \tag{6}$$

$$\theta(x, y) = tan^{-1}\left(\frac{dy}{dx}\right) \tag{7}$$

A local 1-D orientation histogram of gradients is formed from the gradient orientations of sample points within a region. The histogram divides the range of gradient angles into a predefined number of bins. The gradient magnitudes vote into the orientation histogram. In our method, the orientation histogram of each cell has eight bins covering the orientation range of [0,360]. One block consists of 2 x 2 cells whose sizes are 4 x 4 pixels. Because the size of the scanning window is either 60 x 27 or 50 x 32(different according to its aspect ratio), and the size of cell is 4 x 4, we pad the window with zeros to make its width and height multiples of 4 (cell size). Then the sizes of the zero-padded scanning windows become 60 x 28 and 52 x 32; hence the windows contain 15x7 and 13x8 cells. Blocks can overlap at intervals of one cell; therefore the total number of blocks in the scanning window is $14 \times 6 = 84$, and $12 \times 7 = 84$ in both cases. The gradient values of pixels within each block are accumulated into corresponding bins of the histogram. Then the histogram is represented as vector form. In this manner, all normalized block vectors are computed by shifting the position of the block at intervals of one cell in the x,y direction. Finally, the input feature vector is generated by concatenating these 84 normalized block vectors into one vector. As a result, the size of the input feature becomes $84 \times 8 = 672$ dimensions. We used this HOG descriptor in stage 4 for verification of the detection result. Because we already used various combinations of covariance features from stage 1 to stage 3, most of the negative samples which reach stage 4 are hard to classify using these covariance features. Therefore, a different type of feature must be used in stage 4 to classify those difficult samples; that is why we use HOG feature in stage 4. HOG descriptor has more discriminant power than covariance features because it has more dimensionality and contains more spatial information.

3.3   Detection

In stage 1-3, we used ordinary an feed-forward network with the back-propagation learning method. The network that we used in this method has two hidden layers, and one input and one output layer. These four layers are fully interconnected. The number of nodes in the input layer differs according to the size of the input feature vector in each cascade stage. We used 32, 148, and 235 nodes in the input layer of cascade stages 1-3, respectively. The number of nodes in two hidden layers is 25, and 20 respectively and the number of nodes in output layer is one. Output node of the neural network outputs a confidence value between -1(poor) and 1 (high). We trained the neural network so that it outputs 1 when the input vector is positive and -1 otherwise. Hence, if the input feature vector is similar to the trained license plate samples, the neural network outputs a value near 1. but, if the input feature vector is similar to the trained background samples, it outputs value near -1. In stage 4, we used linear classifier instead of neural network, because a simple linear classifier computed by LDA outperforms the neural network in stage 4.

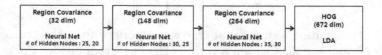

Fig. 2. Cascade structure of the proposed detector

Table 1. Algorithm for detection

Definitions
  get_cov(x, y, w, h) : Computing covariance in (x, y, w, h)
  get_hog(x, y, w, h) : Computing IIOG in (x, y, w, h)
  NN_classify$(cascade\_idx, v)$: Classifying vector v
                                    in $cascade\_idx$ by neural net
  LDA_classify(v) : Classifying vector v by LDA
Outputs
  D : Detected bounding box

Algorithm
//Inserting all possible positions and scales to be scanned.
$P \leftarrow \{x_1, y_1, w_1, h_1), (x_2, y_2, w_2, h_2), \cdots, (x_n, y_n, w_n, h_n)\}$
for $cascade\_idx = 0, \cdots, 3$
  $D \leftarrow \{\emptyset\}$
  for each $(x, y, w, h) \in P$
    if $cascade\_idx == 0$ then)
        $v \leftarrow$ getCov(x, y, w, h)
        if NN_classify$(0, v)$ then $D = D \cup (x, y, w, h)$
    else if $cascade\_idx == 1$ then
        $v_1 \leftarrow$ get_cov(x, y, w, h)
        $v_2 \leftarrow$ get_cov$(x, y, w, h/2)$
        $v_3 \leftarrow$ get_cov$(x, y + h/2, w, h/2)$
        $v_4 \leftarrow$ get_cov$(x, y, w/2, h)$
        $v_5 \leftarrow$ get_cov$(x + w/2, y, w/2, h)$
        $v \leftarrow (v_1\ v_2\ v_3\ v_4\ v_5)$
        if NN_classify$(1, v)$ then $D = D \cup (x, y, w, h)$
    else if $cascade\_idx == 2$ then
        $v_1 \leftarrow$ get_cov(x, y, w, h)
        $v_2 \leftarrow$ get_cov$(x, y, w, h/2)$
        $v_3 \leftarrow$ get_cov$(x, y + h/2, w, h/2)$
        $v_4 \leftarrow$ get_cov$(x, y, w/2, h)$
        $v_5 \leftarrow$ get_cov$(x + w/2, y, w/2, h)$
        $v_6 \leftarrow$ get_cov$(x, y, w/2, h/2)$
        $v_7 \leftarrow$ get_cov$(x + w/2, y, w/2, h/2)$
        $v_8 \leftarrow$ get_cov$(x, y + h/2, w/2, h/2)$
        $v_9 \leftarrow$ get_cov$(x + w/2, y + h/2, w/2, h/2)$
        $v \leftarrow (v_1\ v_2\ v_3\ v_4\ v_5\ v_6\ v_7\ v_8\ v_9)$
        if NN_classify$(2, v)$ then $D = D \cup (x, y, w, h)$
    else if $cascade\_idx == 3$ then
        $v \leftarrow$ get_hog(x, y, w, h)
        if LDA_classify(v) then $D = D \cup (x, y, w, h)$
    end if
  end for
  $P \leftarrow D$
end for

Generally, in most object detectors, the image is resized to different scales to construct an image pyramid, whereas the window size is fixed. But if we use such a method, we must generate 39 integral images every time, including 11 integral images of each image statistics and 28 integral images of multiplied image statistics; this makes the detector too slow. To avoid this problem, we changed the size of window by a power of 1.125 instead of resizing image. Every decision whether the given patch includes license plate is made by the neural network in stages 1-3, and by the LDA in stage 4.

3.4 Generating Training Samples

Because of the great number of variations caused by perspective distortion, view point change, and different combinations of digits, Covering the whole positive space with just the given small training set is difficult. Therefore, we generated more samples using the given pre-training data artificially.

Table 2. Algorithm for generating training samples

Inputs
$image_k$: Image of k'th pre-training data
  NP : Number of pre-training data
  NT : Number of training images to be generated per
               one pre-training data
  $L_k.p_{1,...,4}$: 4 marked corners of k'th pre-training data
Outputs
  trn_images : Images for training
  trn_points : Points for training
Definitions
  $warp(I, S_1, S_2)$: perspective warping function of image I
              from S_1 to S_2
  gamma_cdf : C.D.F of horizontally flipped gamma distribution
            where $K = 4, \theta = 0.15$

Algorithm
//Normalizing scales and positions
for $i = 1, \ldots, NP$
  $c = \overline{L_i.p_j}|_{j=1,\ldots,4}$
  $A_i.p_j = \frac{1}{\|L_i.p_1 - L_i.p_2\|}(L_i.p_j - c)|_{j=1,\ldots,4}$
end for
//Randomly selecting NT samples that are not much different
for $i = 1, \ldots, NP$
  $W = \{A_i\}$
  while $n(W) < NT$
    $r \leftarrow rand[1, n]$
    if $\sum_{j=1}^4 \|A_r.p_j - A_i.p_j\| < T_{ubound}$ **then**
      $R = \{x | x \in W, \sum_{j=1}^4 \|A_r.p_j - X.p_j\| < T_{lbound}\}$
      if $n(R) == 0$ **then**
        $W = W \cup A_r$
  end while
  //Warping to target shape by perspective transform
  for each $x \in W$
    $scale \leftarrow$ gamma_cdf$(rand[0, 1])$
    $x'.p_j = x.p_j \cdot \|L_i \cdot p_1 - L_i \cdot p_2\| \cdot scale$
    $lt\_xy = \{\min_j(x'.p_j.x), \min_j(x'.p_j.y)\}$
    $LO.p_j = lt\_xy + x'.p_j|_{j=1,\ldots,4}$
    $trn\_images \leftarrow trn\_images \cup warp(image, L_i, LO)$
    $trn\_points \leftarrow trn\_points \cup LO$
  end for
end for

We focused on the fact that the possible distortions which can appear on license plates are limited. For example, two horizontal lines the upper and lower boundaries of a license plate are parallel in most cases, as are the side boundaries. Moreover, the lengths of two horizontal and vertical lines are almost the same and the angle between horizontal and vertical lines is within a certain range. However, taking such constraints into consideration by a number of if-else statements is not a sophisticated way to deal with the problem. Instead of that, we designed a novel method (Table.2) for generating training samples. Our method is based on expansion of the knowledge obtained from pre-training data. The points originally marked on the image already have shape information about the license pate. We can use them to generate more training samples so that these generated samples can fill the gaps between positive training samples. Some of marked points are randomly selected from the pre-training data, then source license plate is projected onto those shapes by perspective projection The scale of the training sample is randomly selected according to gamma distribution because window size is changed whereas image size is fixed in our approach. This means that the probability of selecting the original size is highest whereas the probability of selecting bigger size is relatively low.

4   Experimental Results

Test set consisted of 55 images, which included five cars on average. We marked the four corner points of each license plate manually to establish the ground truth. The condition for checking whether the detection result match to the ground truth is as follows

$$\|\mathbf{c}_r - \mathbf{c}_g\| < \|\mathbf{lt}_g - \mathbf{rb}_g\| \tag{8}$$

$$0.7 < \frac{\|\mathbf{lt}_r - \mathbf{rb}_r\|}{\|\mathbf{lt}_g - \mathbf{rb}_g\|} < 1.3 \tag{9}$$

where $\mathbf{c}$ means the center coordinate of the rectangle, and small letters g and r means ground truth and result respectively; $\mathbf{lt}$ means left top corner point and $\mathbf{rb}$ means right bottom corner point of the rectangle. Fig.3 shows ROC curve in each stage. To check whether the NN is suitable as a classifier, we compared its ROC curve with LDA which is one of the simplest linear classifiers. NN showed better performance than LDA from stage 1 to stage 3, but in stage 4, LDA showed better performance than the NN (Fig. 6). This result suggests that LDA has better generalization ability than the NN, provided that the dimensionality of the input feature is sufficiently large, So we used the neural network as the classifier from stage 1 to stage 3 and LDA as the classifier in stage 4.

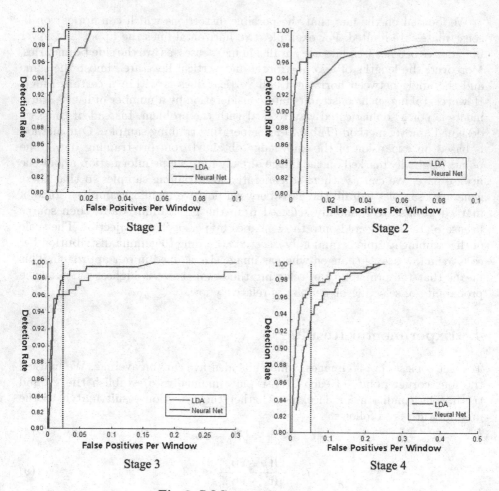

Fig. 3. ROC curves in stage 1~4

Fig.4 shows final detection rate of our cascaded detector. Remaining false alarms could be removed by postprocessing which rejects detected boxes whose intersection count is less than a certain threshold(We set it to 3); After this postprocessing, FPPW could be reduced to 8.3×10^{-7} with just a little drop of detection rate to 93.5%. Most missed samples (Fig. 7a) occurred either when the size of the sample was less than the size of the minimum scanning window or when its skew angle was larger than 45°. Other misses occurred when the aspect ratio of horizontal length to vertical length was too large or too small, or when the color of the license plate were not often found in training sample.

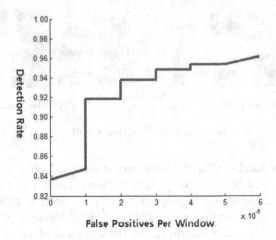

Fig. 4. Performance of the final cascaded detector

Fig.5 shows examples of detection result. Most car license plates which have skew or perspective distortion could be detected successfully. To identify each car, character recognition method could be applied. But it was not, because original research purpose of our case is just blurring car license plates in a road view image for privacy protection.

Fig. 5. Detection results

<table>
<tr><td>(a) Miss samples</td><td>(b) False alarms</td></tr>
</table>

Fig. 6. Wrong detection results

Most false alarm patches (Fig. 6b) look very similar to license plates. Also, the sample can be misidentified if its edge distribution is similar to that of a license plate, or if the sample shows high edge density in both x and y directions. This is because our detector uses covariance and HOG features, which are based on gradient representation. Generally, the digit region of license plate shows high edge density in both x and y directions; hence similar patches whose edge density are also high in x,y directions were detected. For this reason, signboards or patches in which characters are written, whose edge density is high in both directions, are detected.

5   Conclusion

This paper presented a novel method to detect license plates that is based on neural network classifier and LDA using covariance and HOG features. Although The proposed features for license plate detection shows high detection rate even when the simple linear classifier is used, we could achieve higher performance by applying neural network to these proposed features in stage 1-3. Our contributions can be summarized as follows

1. discriminability of the covariance feature was increased by spatial combination of different blocks.
2. A new way of generating training samples was proposed so that they can reflect all possible variations which can occur in real environment.
3. Two heterogeneous features (covariance and HOG feature) were adopted for license plate detection to make up disadvantage of opposite feature.

The proposed method achieved a 94% detection rate while maintaining $2.5 \cdot 10^{-6}$ FPPW in road view images.

Acknowledgement. This work was partially supported by the MKE (The Ministry of Knowledge Economy), Korea, under the Core Technology Development for Breakthrough of Robot Vision Research support program supervised by the NIPA (National IT Industry Promotion Agency)(NIPA-2012-H1502-12-1002). Also, this work was supported by the IT R&D program of MKE/KEIT. [10040246, Development of Robot Vision SoC/Module for acquiring 3D depth information and recognizing objects/faces].

References

1. Vahid, A., Alireza, A.: A Fast Algorithm for License Plate Detection. LNCS, pp. 468–477 (2007)
2. Tran, D.D., Dung, A.D., Tran, L.H.D.: Combining Hough Transform and Contour Alogirhtm for detecting Vehicle's License-Plates. In: Proceedings of 2004 International Symposium on Intelligent Multimedia, Video and Speech Processing
3. Wing, T.H., Hao, W.L., Yong, H.T.: Two-stage License Plate Detection using Gentle AdaBoost and SIF-SVM. In: 2009 First Asian Conference on Intelligent Information and Database Systems, pp. 110–114 (2009)
4. Faith, P., Tekin, K.: Region-based license plate detection. In: Proceedings of the IEEE International Conferenceon Video and Signal Based Surveillance 2006, p. 107 (2006)
5. Viola, J.: Robust Real-time Object Detection. International Journal of Computer Vision, 1–3 (2001)
6. Tuzel, O., Porikli, F., Meer, P.: Region Covariance: A Fast Descriptor for Detection and Classification. In: Leonardis, A., Bischof, H., Pinz, A. (eds.) ECCV 2006. LNCS, vol. 3952, pp. 589–600. Springer, Heidelberg (2006)
7. Navneet, D., Bill, T.: Histograms of Oriented Gradients for Human Detection. Compter Vision and Pattern Recognition (2005)
8. Wenjing, J., Huaifeng, Z., Xiangjian, H.: Region-based license plate detection. Journal of Network and Computer Application, 1324–1333 (2007)

Fast Intra Mode Decision Using the Angle of the Pixel Differences along the Horizontal and Vertical Direction for H.264/AVC

Taeho Kim and Jechang Jeong

Department of Electronics and Computer Engineering,
Hanyang University
{crewx,jjeong}@hanyang.ac.kr

Abstract. In this paper, we proposed a fast intra prediction algorithm for H.264/AVC. Although the H.264/AVC achieved superior coding performance compared with previous video coding standards, the coding computational complexity is also considerable. To reduce the complexity, we select proper candidate intra prediction modes using the angle of the pixel differences along the horizontal and vertical direction to be involved in the RDO, and it results in around 77% of encoding time reduction with negligible coding performance loss. Thus, the proposed algorithm can be utilized usefully for real time high quality high quality video application.

1   Introduction

The H.264/AVC video coding standard [1] has been adopted in common video applications such as mobile service, HDTV, and etc., since it has superior coding performance compared to previous standards [2-3]. This standard achieves coding performance by up to 50% of bit-rate save while preserving visual quality in contrast to previous standard such as MPEG 2 [4-5]. The high performance is achieved due to adoption of various states-of-the-art coding tools such as extended intra/inter coding mode, quarter pixel inter prediction and rate-distortion-optimization (RDO) technique. Particularly the RDO technique adopted in H.264/AVC leads to maximization of coding performance improvement; however, it yields to increase coding computational complexity. Consequently, the encoding time of H.264/AVC is quite high. The RDO technique selects the most suitable prediction mode among the all possible inter/intra prediction modes, and the criterion is the RD cost. Therefore, we can reduce computational complexity by reducing possible candidate prediction modes to be involved in the RDO. Nowadays, this approach has been thoroughly studied.

To reduce candidate prediction modes, the algorithm proposed by Pan et al. [6] exploited edge direction using the Sobel edge operation and its histogram. The detected edge direction employed to select relevant intra prediction modes. Wang et al. [7] implemented an algorithm finding dominant edge strength through 2x2 filters of 5 sets. As another algorithm, Tsai et al. [8] proposed a new approach that detects

G. Bebis et al. (Eds.): ISVC 2012, Part II, LNCS 7432, pp. 648–656, 2012.

reliable edge directions using intensity gradient technique based pixel and sub-block. Zeng et al. [9] proposed the hierarchical intra mode selection method, and it utilized sum of absolute Hadamard transform difference (SAHTD) to select candidate intra prediction modes. It reduces intra modes as well as candidate block size. As the similar approach, Huang et al. [10] proposed two-stage approach to reduce candidate prediction modes and block mode, so that the computational complexity can be reduced considerably. After that, Chen et al. [11] developed a hybrid function to select intra prediction modes, and it considers not only edge information among adjacent pixels but also the prediction value of relevant directions.

In this paper, we focused on selecting candidate intra prediction modes to reduce computational complexity while preserving the coding performance. As a measure to select prediction modes, we utilized the angle of the pixel differences along the horizontal and vertical directions. The experimental results demonstrated that, in term of coding performance and encoding time reduction, the proposed algorithm is more competitive than the state-of-the-art algorithm.

2   Intra Prediction of H.264/AVC

To increase compress rate, the intra prediction of H.264/AVC is conducted by using spatial correlation among adjacent block. In other words, the predicted pixel is determined by previously reconstructed left, upper left, upper, upper right block. H.264/AVC main profile supports two intra prediction block modes for luminance component: intra 4x4 and intra 16x16, which are denoted as I4MB and I16MB.

Fig. 1 and Fig. 2 are illustrated prediction modes in I4MB and I16MB, respectively. Eight directional and one non-directional prediction modes are offered for I4MB, and it is suitable to code texture region [12]. Besides above-mentioned prediction modes, an additional prediction mode called most probable mode (MPM) is provided for I4MB. The MPM is determined as follows:

$$MPM = min(mode\ A, mode\ B), \tag{1}$$

where the mode A presents the prediction mode number determined by RDO in left block, and B presents the best mode in top block. The min denotes the minimum function, thus the MPM is the mode with lower mode number in the left and top block.

On one hand, three directional and one non-directional prediction modes are offered for I16MB, and it is well suitable to code smooth and homogeneous region because the bite expected to code is smaller than I4MB. On the other hand, three directional and one non-directional prediction mode are offered for chrominance component, and its prediction unit size is of only 8x8. The prediction modes are similar as those of I16MB.

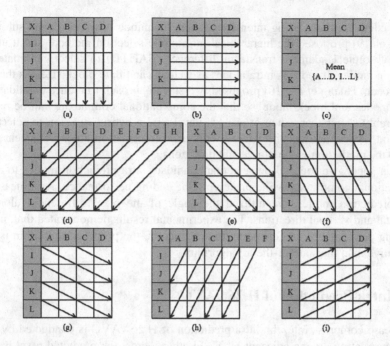

Fig. 1. Nine intra prediction mode and mode number for luminance 4x4 block : (a) vertical, (b) horizontal, (c) DC, (d) diagonal down-left, (e) diagonal down-right, (f) vertical-right, (g) horizontal-down, (h) vertical-left, and (i) horizontal-up

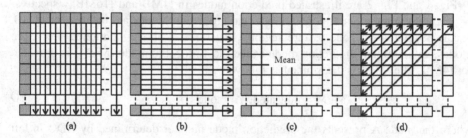

Fig. 2. Four intra prediction modes for luminance 16x16 block: (a) vertical, (b) horizontal, (c) DC, (d) plane

In the reference software JM13.2 [13], the RDO technique examines all prediction modes exhaustively in above mentioned prediction modes, and then the mode with minimum RD cost will be selected as the best mode. The RD cost function is

$$RD \text{ cost} = D + \lambda \times R, \tag{2}$$

where D denotes distortions between the current block and reconstructed block, λ denotes Lagrangian coefficient according to quantization parameter(QP), and R denotes total bit used to code the relevant prediction mode.

Although the RDO technique can find the most suitable prediction mode among all possible modes, the computational complexity is also increased, which makes it difficult to use in real-time application. The number of calculating RD cost is

$$N_{RD\ cost} = N_{chroma} \times (N_{I16MB} + N_{I4MB}),\qquad(2)$$

where N_x denotes the number of calculation relative block type. In more details, $N_{chroma} = 4$, $N_{I16MB} = 4$, and $N_{I4MB} = 16 \times 9$, and it means that $N_{RD\ cost}$ is 592. $N_{RD\ cost}$ of the H.264/AVC high profile is larger than those of main profile, since it offers additional intra 8x8 block mode (I8MB) [2]. The prediction modes of I8MB are the same as I4MB and its block size is of 8x8.

3   Proposed Fast Intra Prediction Algorithm

The aim of the proposed algorithm is to reduce encoding time through minimizing coding complexity; so that we reduced complexity by joining only designated prediction modes to RDO technique. In order to designate candidate prediction modes, we assumed that the angle of the pixel differences along the horizontal and vertical direction (APD) can correspond to the direction of prediction mode.

3.1   The Angle of the Pixel Differences along the Horizontal and Vertical Direction

To compute APD, we have to first calculate the pixel difference along the horizontal and vertical as follows:

$$PD_{Hor.} = \left[\sum_{j=0}^{m} \sum_{i=1}^{n} |img(j,i) - img(j,i-1)| \right] \div [m \times (n-1)],\qquad(4)$$

$$PD_{Ver.} = \left[\sum_{j=1}^{m} \sum_{i=0}^{n} |img(j-1,i) - img(j,i)| \right] \div [(m-1) \times n],\qquad(5)$$

where the PD_X denotes pixel difference along the horizontal and vertical direction, respectively. The m and n denotes column and row in block, and the img(j, i) is original image intensity. The APD Φ is computed as follows:

$$\Phi = \frac{180}{\pi} \arctan\left(PD_{Hor.} \Big/ PD_{Ver.} \right).\qquad(6)$$

$180/\pi$ is to conversion from radian to degree; therefore, the unit of Φ is degree (°). The range of Φ is from 0 to 90°, since the absolute operator in (5) and (6) restricts that the value of PD_X becomes negative value.

3.2   Fast Intra Mode Decision Algorithm

To verify that APD can correspond to the prediction mode, we plot the histogram of percent of selecting each prediction mode along the APD. Fig 3-4 shows an example of this histogram for "Mobile" (CIF) sequence. Those of other sequences show the similar trend to "Mobile" sequence.

In Fig.4 and 5, the lower the APD is, the higher is percent of the selecting horizontal mode (mode 1). On the contrary the higher the APD is the more probable is the selection of the vertical mode (mode 0). On the other hand, at the intermediate APD, the percent of the selecting the mode 0 and 1 shows a falling trend and the probability of selecting directional prediction mode similar to APD shows an upward trend. Consequently, we estimated best prediction mode using several prediction modes based on APD as Table 1.

In case of I4MB, DC and MPM are always selected as the candidate modes, because the number of considerable blocks are selecting best mode as the MPM and DC to maintain the coding performance [9, 10]. In addition, this algorithm can be also applied to I8MB. For I16MB, one mode and DC are selected the candidate prediction modes.

Table 1. Candidate prediction modes for intra 4x4 block based on APD

APD (Φ)	modes for intra 4x4	APD (Φ)	modes for intra 16x16
[0, 7.5]	1, 2, MPM	[0, 30]	0, 2
(7.5, 37.5]	1, 2, 6, 8, MPM	(30, 60]	2, 3
(37.5, 52.5]	0, 1, 2, 3, 4, 6, 8, MPM	(60, 90]	1, 2
(52.5, 67.5]	0, 1, 2, 3, 4, 5, 7, MPM		
(67.5, 82.5]	0, 2, 5, 7, MPM		
(82.5, 90]	0, 2, MPM		

For chrominance component, among the four chrominance prediction modes, the mode with minimum sum of absolute Hardarard transform differences (SATD) is selected the best mode. This algorithm is suggested in JM reference software.

4   Experimental Results and Discussion

The proposed algorithm was implemented on JM13.2. Various MPEG standard sequences with CIF and QCIF are tested under Table 2. The test platform used is Intel Core i7 3.47GHz processor, 3GB of DDR3 RAM and Microsoft Windows 7 x86 for experiments.

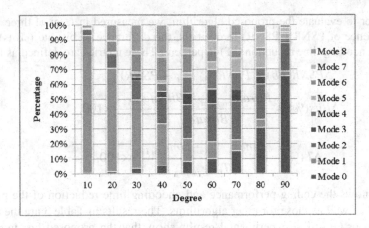

Fig. 3. The percent of selecting prediction mode along APD in intra 4x4 block

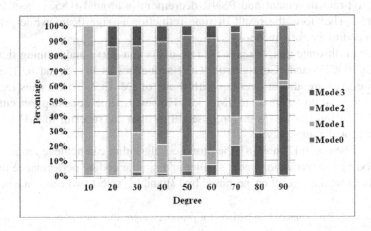

Fig. 4. The percent of selecting prediction mode along APD in intra 16x16 block

Table 2. Encoding conditions

Encode Parameters	Values
Profile	main
GOP Structure	III…
Number of frames	100
R-D optimization	Enabled
QP	24, 28, 32, 36
Entropy coding	CABAC

In order to evaluate the proposed algorithm, we measured in term of three metrics: the difference of PSNR (ΔPSNR) in decibels, the difference of bit-rate (ΔBits) in percent and encoding time reduction (ΔT) in percent. These metrics are defined as follows:

$$\Delta PSNR(db) = PSNRY_{proposed} - PSNRY_{JM13.2},\qquad(7)$$

$$\Delta Bits(\%) = \frac{Bitrate_{proposed} - Bitrate_{JM13.2}}{Bitrate_{JM13.2}} \times 100,\qquad(8)$$

$$\Delta T(\%) = \frac{Time_{proposed} - Time_{JM13.2}}{Time_{JM13.2}} \times 100.\qquad(9)$$

Table 3 shows the coding performance and encoding time reduction of the proposed algorithm as well as those for other algorithms. The results in Table 3 are the average value of four QPs. The experimental results show that the proposed fast intra mode decision using APD can achieve time reduction around 77% compared with JM 13.2 while the bit-rate increment and PSNR decrement is around 0.852% and 0.055dB, respectively. Therefore, the result of time reduction is significant while preserving negligible coding performance loss.

In order to illustrate that, we plot the rate-distortion curve and running time curve of "Stefan" (QCIF) and "Coastguard" (CIF) sequences as Fig. 5 and 6. These rate-distortion curves are almost over-lap with those of JM 13.2. That means the coding performance is almost similar to JM 13.2. However, time consumption curves are under JM 13.2 and this means the proposed algorithm are superior to JM13.2 in term of computational complexity.

In comparison with [7] and [8], the proposed algorithm can achieve around 15% of more encoding time reduction under smaller and similar coding performance loss. This happens because the algorithms proposed by [7] and [8] select two candidate modes for

Table 3. Performance comparison for proposed algorithm and other algorithms

		Wang et al. [7]			Tsai et al. [8]			Chen et al. [11]			Proposed		
		ΔBits (%)	ΔPSNR (dB)	ΔT (%)	ΔBits (%)	ΔPSNR (dB)	ΔT (%)	ΔBits (%)	ΔPSNR (dB)	ΔT (%)	ΔBits (%)	ΔPSNR (dB)	ΔT (%)
QCIF	Carphone	5.453	-0.009	-64.050	3.006	-0.023	-61.232	6.901	-0.027	-82.042	1.798	-0.037	-77.872
	Claire	4.766	0.001	-65.343	6.168	0.034	-64.889	13.274	-0.053	-82.161	2.313	-0.050	-78.838
	Stefan	3.114	-0.059	-62.106	0.814	-0.097	-60.448	4.138	-0.114	-81.816	0.478	-0.107	-78.521
	Table	3.237	-0.005	-60.912	2.202	-0.017	-59.304	5.630	-0.023	-81.976	0.803	-0.032	-77.928
CIF	Bus	4.649	0.006	-62.939	1.607	-0.036	-58.228	4.178	-0.038	-79.896	0.278	-0.055	-75.597
	Costguard	3.853	0.027	-62.907	0.591	-0.035	-61.419	4.339	-0.013	-79.970	0.120	-0.041	-77.851
	Mobile	2.173	-0.040	-59.854	1.090	-0.051	-53.173	3.547	-0.072	-79.538	0.532	-0.065	-74.589
	Football	1.899	-0.029	-61.515	1.491	-0.032	-54.571	3.389	-0.051	-79.852	0.494	-0.054	-74.840
Average		3.643	-0.014	-62.453	2.121	-0.032	-59.158	5.675	-0.049	-80.906	0.852	-0.055	-77.004

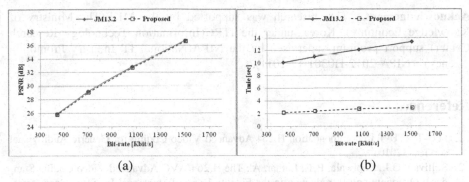

Fig. 5. Rate distortion and time consumption curve for "Stefan" (QCIF) sequence: (a) rate distortion curve, (b) time consumption curve

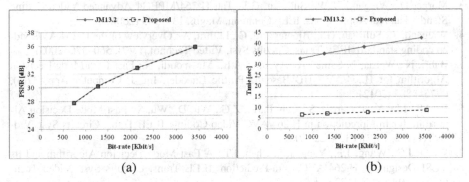

Fig. 6. Rate distortion and time consumption curve for "Coastguard" (CIF) sequence: (a) rate distortion curve, (b) time consumption curve

chrominance component. In other words, those of [7] and [8] employ more candidate prediction modes to determine best prediction mode. Nevertheless, the proposed algorithm using one candidate mode for chrominance component shows further coding performance and encoding time reduction. On the one hand, the proposed algorithm requires more need of encoding time compared with [11]; however, that is still competitive since the proposed algorithm shows better coding performance.

5   Conclusion

In this paper, we proposed a new fast intra prediction algorithm for H.264/AVC. It reduces candidate prediction modes to be involved in the RDO using APD. The extensive simulation results demonstrated that the proposed algorithm shows significant encoding time reduction without negligible coding performance loss. Thus, the proposed algorithm is suitable for real-time application with high quality video.

Acknowledgement. "This research was supported by the MKE(The Ministry of Knowledge Economy), Korea, under the ITRC(Information Technology Research Center) support program supervised by the NIPA(National IT Industry Promotion Agency)" (NIPA-2012-H0301-12-1011).

References

1. ITU-T: ITU-T Recommendation H.264 Advanced Video Coding for Generic Audiovisual services (2010)
2. Sullivan, G.J., Popiwala, P.T., Luthar, A.: The H.264/AVC Advanced Video Coding Standard: Overview and Introduction to the Fidelity Range Extensions. In: SPIE Conf. Apps. of Digital Image Processing XXVII, Special Session on Advances in the New Emerging Standard H.264/AVC, pp. 454–474 (2004)
3. Marpe, D., Wiegand, T.W., Sullivan, G.J.: The H.264/MPEG4 Advanced Video Coding Standard and its Applications. IEEE Commun. Magn., 134–143 (2006)
4. Wiegand, T., Sullivan, G.J., Bjontegaard, G., Luthra, A.: Overview of the H.264/AVC video coding standard. IEEE Trans. Circuits Syst. Video Technol. 13(7), 560–576 (2003)
5. Bahri, N., Werda, I., Samet, A., Ayed, M.A.B., Masmoudi, N.: Fast Intra Mode Decision Algorithm for H264/AVC HD Baseline Profile Encoder. Int. J. Computer Appl. 37(6), 0975–8887 (2012)
6. Pan, F., Lin, X., Rahardja, S., Lim, P., Li, Z.G., Wu, D., Wu, S.: Fast Mode Decision Algorithm for Intraprediction in H.264/AVC Video Coding. IEEE Trans. Circuits Syst. Video Technol. 15(7), 813–822 (2005)
7. Wang, J.C., Wang, J.F., Yang, J.F., Chen, J.T.: A Fast Mode Decision Algorithm and Its VLSI Design for H.264/AVC Intra-Prediction. IEEE Trans. Circuits Syst. Video Technol. 17(10), 1414–1422 (2007)
8. Tsai, A.C., Paul, A., Wang, J.C., Wang, J.F.: Intensity Gradient Technique for Efficient Intra-Prediction in H.264/AVC. IEEE Trans. Circuits Syst. Video Technol. 18(5), 694–698 (2008)
9. Zeng, H., Ma, K.K., Cai, C.: Hierarchical Intra Mode Decision for H.264/AVC. IEEE Trans. Circuits Syst. Video Technol. 20(6), 907–912 (2010)
10. Huang, Y.H., Ou, T.S., Chen, H.H.: Fast Decision of Block Size, Prediction Mode, and Intra Block for H.264 Intra Prediction. IEEE Trans. Circuits Syst. Video Technol. 20(8), 1122–1132 (2010)
11. Chen, C., Chen, J., Ouyang, K., Xia, T., Zhou, J.: A hybrid fast mode decision method for H.264/AVC intra prediction. Int. J. Multimed. Tools Appl. (2011), doi:10.1007/s11042-011-0862-6
12. Wang, P., Huang, H., Tan, Z.: A fast two-step block type decision algorithm for intra prediction in H.264/AVC high profile. Int. J. Multimed. Tools Appl. (2011), doi:10.1007/s11042-011-0807-0
13. H.264/AVC reference software JM13.2 (2011), http://iphome.hhi.de/suehring/tml/download/old_jm/

Interpolation of Reference Images in Sparse Dictionary for Global Image Registration

Hayato Itoh[1], Shuang Lu[1], Tomoya Sakai[2], and Atsushi Imiya[3]

[1] School of Advanced Integration Science, Chiba University
Yayoicho 1-33, Inage-ku, Chiba, 263-8522, Japan
[2] Department of Computer and Information Sciences, Nagasaki University
Bunkyo-cho 1-14, Nagasaki, 852-8521, Japan
[3] Institute of Media and Information Technology, Chiba University
Yayoicho 1-33, Inage-ku, Chiba, 263-8522, Japan

Abstract. In this paper, we introduce an efficient global image registration for medical images. We use the nearest neighbour method for searching appropriate reference image. The NNS based image registration uses pre-computed references in the image-dictionary as the targets of image registration and search the best matched reference using NNS. For speeding up, we apply the random projection to image for reduction of the image size. For the reduction of the number of data in the dictionary, we use interpolation of references images in the image dictionary.

1 Introduction

In this paper, we develop a fast global-image-registration algorithm using an efficient random projection and interpolation in the sparse dictionary of reference images[1,2,3]. Image registration method generally classified into local image registration and global image registration.

Image registration overlays two or more template images, which are same sense observed at different times, from different viewpoints, and or by different sensors, on a reference image. In this paper, we assume that transformation between template and reference images is an affine. Therefore, in image registration we are required to compute an affine transform between two images[1]. An affine transform is decomposed into rotation, scaling, shearing, and translation. Therefore, image registration estimates these geometric transformations that transform all points or most points of the template images to points of the reference image. To estimate these spatial transformations, several methods are developed[4,5,6].

For global alignment images, the linear transformation $x' = Ax + b$ which minimises the criterion

$$D(f, g) = \sqrt{\int_{\Omega} |f(x) - g(x')|^2 dx}, \qquad (1)$$

where Ω is the region of interest of images, for a reference image $f(x)$ and the template image $g(x)$, is used to relate two images.

G. Bebis et al. (Eds.): ISVC 2012, Part II, LNCS 7432, pp. 657–667, 2012.
© Springer-Verlag Berlin Heidelberg 2012

For sampled affine-transforms $\{A_i\}_{i=1}^m$ and translation-vectors $\{b_j\}_{j=1}^m$, we generate transformed reference images such that $g^{ij} = g(A_i x + b_j)$, $i,j = 1, 2, \ldots, m$. In this paper, we call the set of $\{A_i\}_{i=1}^m$, $\{b_j\}_{j=1}^m$ and g^{ij} dictionary. Assuming that

$$A_i x + b_j \neq A_k x + b_l, \quad g^{ij} \neq g^{kl}, \quad i \neq k, \quad j \neq l, \tag{2}$$

we find the A_i and b_j which minimise

$$D(f, g^{ij}) = \sqrt{\int_\Omega |f(x) - g^{ij}(x)|^2 dx} \tag{3}$$

by using the nearest neighbour search (NNS). Figure 1 shows the procedure of global registration strategy using this method. Fig. 1(1) shows the scheme of global image registration. Fig. 1(2) shows the preprocessing of proposed method and Fig. 1(3) shows the registration process of proposed method.

The simplest solution to the NNS problem is to compute the distance from the query point to every other point in the database, preserving track of the "best so far". This algorithm has a computational cost $\mathcal{O}(Nd)$ where N and d are the cardinality of a set of points in a metric space and the dimensionality of the metric space, respectively.

For NNS, dimension of data affects the computational cost and accuracy of search. It is known as the curse of dimensionality. To avoid the curse of dimensionality, we develop a fast algorithm using an efficient random projection[3], which reduces both time and domain complexities of random projection[7]. If the number of data in the data dictionary is small, we call the data dictionary a sparse image-dictionary. Interpolation of reference in the image-dictionary yields new references from $\{g^{ij}\}_{i,j=1}^m$. New references are closer than data in sparse dictionary to query. Therefore we can start the NNS with sparse dictionary. Since N of sparse dictionary is small, we can reduce the computational cost and size of memory for the NNS using interpolation. For interpolation in the image dictionary, we use adaptive Voronoi tessellation[8] in the data dictionary of reference images generated by eq. (1).

In section 2, we briefly summarise an efficient random projection [3]. And in section 3, we introduce interpolation in the data manifold. In section 4, we show numerical examples and analyse complexity of our algorithm.

2   Efficient Random Projection

Random projection is a simple technique for projecting a set of data points from a high-dimensional space to a randomly chosen low-dimensional linear subspace approximately preserving the distance of data points[7]. Conventional random projection is not efficient and efficient random projection (ERP) is shown to be of benefit for global image registration[2].

Setting $w = (w_1, \ldots, w_k)^\top$ to be an independent stochastic vector such that $E_{\mathbb{R}^k}[w] = 0$ in k-dimensional Euclidean space $\mathbb{R}^k$, we have the relation

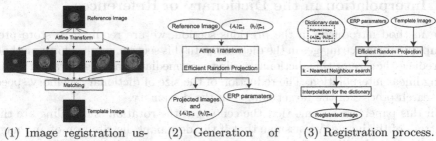

(1) Image registration using pre-generated dictionary.

(2) Generation of data in the dictionary.

(3) Registration process.

Fig. 1. Flow charts of the propsed method

$E_{\mathbb{R}^{k \times k}}[\boldsymbol{w}\boldsymbol{w}^\top] = \gamma^2 \boldsymbol{I}$. Furthermore, setting the $(i-1)$-time shifting of $\boldsymbol{w}$ to be $\boldsymbol{c}_i = (w_i, \ldots, w_k, w_1, \ldots, w_{i-1})^\top$, we define the matrix

$$C = \begin{bmatrix} \boldsymbol{c}_i^\top \\ \vdots \\ \boldsymbol{c}_k^\top \end{bmatrix} = \begin{bmatrix} w_1 & w_2 & \cdots & w_{k-1} & w_k \\ w_2 & w_3 & \cdots & w_k & w_1 \\ \vdots & \vdots & \ddots & \vdots & \vdots \\ w_k & w_1 & \cdots & w_{k-2} & w_{k-1} \end{bmatrix}. \tag{4}$$

Since

$$E_{\mathbb{R}}(\boldsymbol{c}_i^\top \boldsymbol{c}_j) = \begin{cases} \gamma^2 & (i = j) \\ 0 & i \neq j \end{cases} \tag{5}$$

we have the relation

$$E_{\mathbb{R}}[|\boldsymbol{y}|_2^2] = E_{\mathbb{R}}[\sum_{i=1}^k (\boldsymbol{c}_i^\top \boldsymbol{x})^2] = \sum_{i=1}^k \boldsymbol{x}^\top E_{\mathbb{R}^{k \times k}}[\boldsymbol{c}_i \boldsymbol{c}_i^\top] \boldsymbol{x} = \sum_{i=1}^k \boldsymbol{x}^\top \gamma^2 \boldsymbol{I} \boldsymbol{x} = k\gamma^2 |\boldsymbol{x}|^2. \tag{6}$$

We set $\gamma = 1/\sqrt{k}$ in eq. (6).

Setting $\boldsymbol{s} = (s_1, \ldots, s_k)^\top$ to be an independent stochastic vector such that $E_{\mathbb{R}^{k \times k}}[\boldsymbol{s}\boldsymbol{s}^\top] = \sigma^2 \boldsymbol{I}$, the vector *zeta* which have dense spectrum is computed as $\zeta = \boldsymbol{s} \odot \boldsymbol{x} = [\boldsymbol{S}]\boldsymbol{x}$, from the sparse vector $\boldsymbol{x}$, where $[\boldsymbol{S}]$ is the diagonal matrix whose diagonals are $\{s_i\}_{i=1}^k$ for $k \leq n$. The expectation of the norm is $E_{\mathbb{R}^k}[|\boldsymbol{y}|_2^2] = k\gamma^2\sigma^2|\boldsymbol{x}|_2^2$. To preserve $E[|\boldsymbol{y}|_2^2] = |\boldsymbol{x}|_2^2$, we set $\gamma = 1/\sqrt{k}$ and $\sigma = 1$. Since $\boldsymbol{C}\boldsymbol{\eta}$ is achieved by a cyclic convolution between $\boldsymbol{c}_i$ and $\boldsymbol{\eta}$, we can compute $\boldsymbol{C}[\boldsymbol{S}]\boldsymbol{x}$ using the fast Fourier transform (FFT). These algebraic property derive the next theorem.

Theorem 1. *The vector $\boldsymbol{x}$ is projected to the vector $\boldsymbol{y}$ using $\mathcal{O}(d)$ memory area and $\mathcal{O}(nd \log d)$ calculation time.*

3   Interpolation in the Dictionary of References

The method introduced in the previous section, we are required to store pre-computed reference images in the dictionary. In this section, we develop a method to reduce the size of this dictionary using intermediate image generation [9,10] using linear interpolation. The reduction of the size of dictionary memory speed up search process of the target images in the dictionary.

In this paper, we assume that the centers of the rotation and scaling are the centroid of the image. We assume that if the dominant distribution of an image is approximately ellipse, distance between the image and the rotated image, and between the image and scaling image are proportional to θ and τ, respectively. For

$$f_1(\boldsymbol{x}) = arg \min_{f \in D} \left\{ \sqrt{\int_\Omega |g(\boldsymbol{x}) - f(\boldsymbol{x})|^2 d\boldsymbol{x}} \right\}, \tag{7}$$

$$f_2(\boldsymbol{x}) = arg \min_{f \in D \setminus \{f_1\}} \left\{ \sqrt{\int_\Omega |g(\boldsymbol{x}) - f(\boldsymbol{x})|^2 d\boldsymbol{x}} \right\}, \tag{8}$$

in the dictionary D and the affine transform $\boldsymbol{A}_1$ and $\boldsymbol{A}_2$, we assume that

$$g(\boldsymbol{x}) = (1 - \alpha)f_1(\boldsymbol{x}) + \alpha f_2(\boldsymbol{x}), \quad f_i(\boldsymbol{x}) = f(\boldsymbol{A}_i \boldsymbol{x}). \tag{9}$$

From eq. (9), we have the relation

$$g(\boldsymbol{x}) - f_1(\boldsymbol{x}) = \alpha(f_2(\boldsymbol{x}) - f_1(\boldsymbol{x})), \tag{10}$$

for each point. Next by computing square average of both side of eq. (10), we have the relation

$$\int_\Omega |g(\boldsymbol{x}) - f_1(\boldsymbol{x})|^2 d\boldsymbol{x} = \alpha^2 \int_\Omega |f_2(\boldsymbol{x}) - f_1(\boldsymbol{x})|^2 d\boldsymbol{x}, \tag{11}$$

for images. Therefore, we have the relation

$$\alpha = \sqrt{\frac{\int_\Omega |g(\boldsymbol{x}) - f_1(\boldsymbol{x})|^2 d\boldsymbol{x}}{\int_\Omega |f_2(\boldsymbol{x}) - f_1(\boldsymbol{x})|^2 d\boldsymbol{x}}}. \tag{12}$$

In this paper, we deal with the scaling and rotation of images, that is, we assume that

$$\boldsymbol{A} = \tau \boldsymbol{R}(\theta), \quad \boldsymbol{R}(\theta) = \begin{pmatrix} \cos\theta, & -\sin\theta \\ \sin\theta, & \cos\theta \end{pmatrix}, \tag{13}$$

for $\tau > 0$ and $-\pi \leq \theta < \pi$. Therefore, equations (9) and (12) become as

$$g(\tau \boldsymbol{R}(\phi)\boldsymbol{x}) = (1 - \alpha)f(\tau_1 \boldsymbol{R}(\theta_1)\boldsymbol{x}) + \alpha f(\tau_2 \boldsymbol{R}(\theta_2)\boldsymbol{x}), \tag{14}$$

and

$$\alpha = \sqrt{\frac{\int_\Omega |g(\tau \boldsymbol{R}(\phi)\boldsymbol{x}) - f(\tau_1 \boldsymbol{R}(\theta_1)\boldsymbol{x})|^2 d\boldsymbol{x}}{\int_\Omega |f(\tau_1 \boldsymbol{R}(\theta_1)\boldsymbol{x}) - f(\tau_2 \boldsymbol{R}(\theta_2)\boldsymbol{x})|^2 d\boldsymbol{x}}} \tag{15}$$

for $\theta_1 < \phi < \theta_2$ and $\tau_1 < \tau < \tau_2$.

We use the following Algorithm 1 to generate a template which approximates the reference.

---

input : Dictionary data D, Template image g
output: Registered image g_{12}

Find f_1, f_2 using k-NNS in D;
Find $g(\tau R(\phi)x)$ applying eq. (14);
for $i \leftarrow 1$ **to** n **do**
  if $|g(\tau R(\phi)x) - f_1| < |g(\tau R(\phi)x) - f_2|$ **then** $f_1 := g(\tau R(\phi)x)$ and
  $f_2 := f_2$, and apply eq. (14);
  else $f_1 := f_1$ and $f_2 := g(\tau R(\phi)x)$, and apply eq. (14);
end

Algorithm 1. Interpolation for dictionary.

4   Numerical Examples

In this section, we use following expressions. For image f, $f(\theta)$ express the θ degree rotated image of f. For image f, $f(\tau)$ express the τ times scaled image of f. For image f, $f(\theta, \tau)$ express the θ degree rotated and τ times scaled image. We call $D(\cdot, \cdot)$ in eq. (3) a distance of two images. A dictionary $d_\theta(\theta_0, \theta_{\text{step}}, \theta_N)$ is consist of $g^i = g(\theta_i)$, $\theta_i = \theta_0 + (i - 1)\theta_{\text{step}}$. A dictionary $d_\tau(\tau_0, \tau_{\text{step}}, \tau_N)$ is consist of $g^i = g(\tau_i)$, $\tau_i = \tau_0 + (i - 1)\tau_{\text{step}}$.

We first evaluate the validity of eq. (11). We evaluate the dependancy of metric $D(\cdot, \cdot)$ on rotation angle θ and scaling parameter τ using anisotropic Gaussian. P0 in Fig. 2 shows an anisotropic Gaussian. For the rotation angle $0^{\text{deg}} < \theta < 15^{\text{deg}}$, we computed $D(P0, P0(\theta))$. The curve in Fig. 2(2) shows the dependency of the distance $D(P0, P0(\theta))$. The curve shows that the distance linearly increases for the rotation angle θ. For the scaling parameter $0.8^\times < \tau < 1.2^\times$, we computed $D(P0, P0(\tau))$. For $0.8 < \tau < 1.0^\times$, the curve in Fig. 2(3) shows that the distance linearly decreases for the scaling parameter. For $1.0 < \tau < 1.2^\times$, the curve shows that the distance linearly increases for the scaling parameter. If the transformation are rotation and scaling, Fig. 2(2) and Fig. 2(3) show validity of eq. (11) for interpolation.

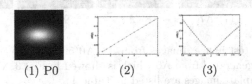

(1) P0   (2)   (3)

Fig. 2. Phantom image and dependancy of metric on parameters. (1) is the phantom data generated from 2-dimensional Gaussian function. (2) shows the square norm between the phantom and rotated phantom as y axis, and rotation parameter θ as x axis. (3) shows the square norm between the phantom and scaled phantom as y axis, and scaling parameter. τ as x axis.

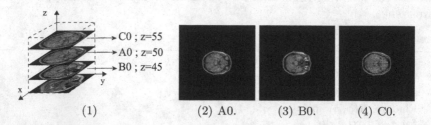

(1) | (2) A0. | (3) B0. | (4) C0.

Fig. 3. Extracted slice images from 3-dimensional volume data. The volume data is MRI simulation data of human brain[11,12]. This volume data has $181 \times 217 \times 181$ image resolution. Slice image $A0$, $B0$, and $C0$ are extracted from $z = 50, z = 45, z = 55$ plane of the volume data. The extracted slices are embedded to 543×543 pixels.(2),(3),(4) are embedded slice images.

Table 1. Parameters for template generation. The parameters $\sharp(\theta, \tau)$ in the table are used for the generation of template A$\sharp$, B$\sharp$ and C$\sharp$.

$\sharp$	0	1	2	3	4	5	6	7	8	9	10
θ	0.00	50.0	30.0	6.00	-18.0	-40.0	0.00	0.00	0.00	0.00	0.00
τ	1.00	1.00	1.00	1.00	1.00	1.00	0.650	0.85	1.05	1.25	1.45

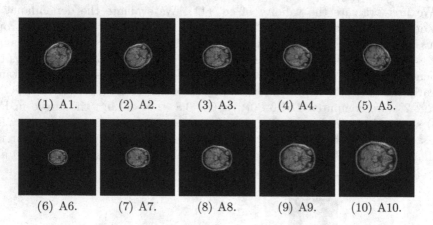

(1) A1. | (2) A2. | (3) A3. | (4) A4. | (5) A5.

(6) A6. | (7) A7. | (8) A8. | (9) A9. | (10) A10.

Fig. 4. Examples of test image. Images from A1(A0(50,1.00)) to A10(A0(0.00,1.45)) are template images generated from A0 which is showed in Fig. 3. The parameters for the generation of template images are listed in Table 1.

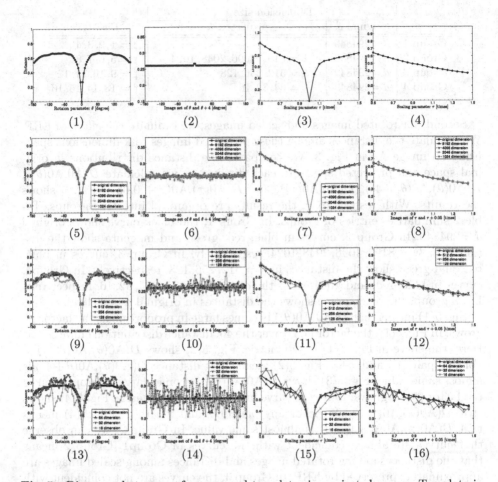

Fig. 5. Distances between references and templates in projected spaces. Template is A0. Reference is A0(θ, τ). Distances are measured in projected k dimension Euclidean space. With respect to k, the results are organized into four groups. Table 2 summarises entries of groups. In Group 1, the black curve shows original $d = 294849$. In Group 2, curves in blue, red, green and magenta show the distances in $k = 8192, 4096, 2048, 1024$, respectively. In Group 3, curves in blue, red, and green show the distances in $k = 512, 256, 128$, respectively. In Group4, curves in blue, red, and green show the distances in $k = 64, 32, 16$, respectively. In all group, the black curve shows the distances in original $d = 294849$. In each dimension, we calculate $D(\text{A0,A0}(\theta))$, $D(\text{A0}(\theta),\text{A0}(\theta+4))$, $D(\text{A0,A0}(\tau))$, and $D(\text{A0}(\tau),\text{A0}(\tau + 0.05))$.

Table 2. Grouping of results in Fig.5. Graphs in Fig.5 are organized into four group with respect to k.

	original d	projected k	Fig.5($\sharp$)
		Dimension size	
Group 1	$d =294849$	-	$\sharp = 1, 2, 3, 4$
Group 2	$d =294849$	$k = 8192, 4096, 2048, 1024$	$\sharp = 5, 6, 7, 8$
Group 3	$d =294849$	$k = 512, 256, 128$	$\sharp = 9, 10, 11, 12$
Group 4	$d =294849$	$k = 64, 32, 16$	$\sharp = 13, 14, 15, 16$

Second, for rotated images and scaled images, we evaluate the effect of ERP. We calculate the distances among these projected images to k dimensions space using an image A0 in Fig. 3. We compared the distance distributions in original space and projected space. In each dimension, we calculate $D(A0,A0(\theta))$, $D(A0(\theta),A0(\theta + 4))$, $D(A0,A0(\tau))$, and $D(A0(\tau),A0(\tau + 0.05))$. Fig. 5 shows the results. With respect to k, the results are organized into four groups. Table 2 summarises entries of groups. In Group 1, the black curve shows original $d =294849$. In Group 2, curves in blue, red, green and magenta show the distances in $k = 8192, 4096, 2048, 1024$, respectively. In Group 3, curves in blue, red, and green show the distances in $k = 512, 256, 128$, respectively. In Group4, curves in blue, red, and green show the distances in $k = 64, 32, 16$, respectively. In all group, the black curve shows the distances in original $d = 294849$.

Fig. 5(1) means that $D(A0,A0(\theta))$ becomes large in proportion as the factor $|\theta|$ grows. In Fig. 5(1), the factor $|\theta|$ go over the 40, distance distribution curve is flat. Here, the curve in Fig. 5(1) is symmetric. Fig. 5(2) shows $D(A0(\theta),A0(\theta + 4))$. For any pair of θ and $\theta + 4$, Fig. 5(2) means that distance $D(A0(\theta),A0(\theta+4))$ is almost same value. Fig. 5(3) shows $D(A0,A0(\tau))$ becomes large in proportion as the factor $|\tau-1|$ grows. Here, the curve in Fig. 5(3) is symmetric. Fig. 5(4) shows that $D(A0(\tau),A0(\tau + 0.05))$. For any pair of τ and $\tau + 0.05$, Fig. 5(4) means that $D(A0(\tau),A0(\tau + 0.05))$ is almost same value. In Group 2, we can observe that curves are almost coincident with the curves in Group1. Group 2 means that the distances among rotated images and distances among scaled images are approximately preserved by ERP. In Group 3, the curves are not coincident with the curves in Group 1. In Group3, the distances are not preserved. In Group 4, the difference of curves from the curves of Group 1 is larger than Group 3.

Fig. 6(1) shows the mean distance errors of distaces $D(A0,A(\theta))$, $D(A0,A(\tau))$, and $D(A0(\theta),A0(\tau))$. The curves in blue, red, and green show rotated images, scaled images, and rotated and scaled images, respectively. From Figs. 5 and 6(1), in higher dimension than 1024 dimensions, the distances among the rotation images and scaled images are approximately preserved. This $k = 1024$ dimensions is approximately 0.34% size of original data dimension $d = 294849$.

Next, we evaluated mean registration errors in the Algorithm 1. From the result, we show the available step size of θ and τ for the interpolation in the dictionary. For rotation transform, Fig. 6(2) shows the mean registration errors in each iteration step, respectively. For scale transform, Fig. 6(3) shows the mean registration errors in each iteration step, respectively. Table 3 summarises color

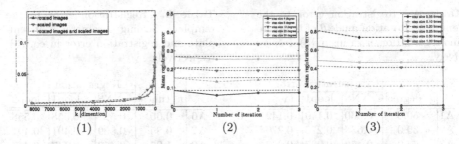

(1)   (2)   (3)

Fig. 6. Mean distance errors. (1),(2) and (3) illustrate the dependencees of the distance errors to dimensions of projected space, to the iteration of θ and to the itaration of τ.

Table 3. Curve's color for each dictionary in Figures 6(2) and 6(3). An angle θ_{size} expresses the angle step size of dictionry in Fig. 6(2). Scale τ_r expresses the scaling step of dictionary in Fig. 6(3).

	Color of curve					
---	blue	green	red	cyan	megenda	black
θ_{setp} [degree]	4	8	12	20	30	-
τ_{setp} [times]	0.050	0.10	0.20	0.25	0.50	1.0

expression of curves in Figures.6(2) and 6(3). Fig. 6(2) and Fig. 6(3) show that the iteration converge in 3 times iteration with the given sparse dictionary. In these step sizes θ_{size} and τ_{size}, interpolation in the dictionary is available.

Finally, we compare the registration error of proposed method with NNS method. Here, we adopted the image A0, B0, and C0 in Fig. 3. We adopted image A0 as a reference image, and transformed images of A0, B0, and C0 as templates. Fig. 4 shows the examples of transformed image and Table 1 summarises the parameters of transformations. Table 4 summarises the registration errors for rotation images by NNS and Algorithm 1. Table 5 summarises the registration errors for scaled images by NNS and Algorithm 1. In Tables 4 and 5, registration errors by NNS are large. However, as the iteration progress registration errors become small by proposed method.

These experiments firstly show that efficient random projection preserves the distance among images, if the target dimension of projection is appropriate. Secondly, we can generate new reference in the image dictionary using linear interpolation, if the sparsity of dictionary is appropriate. These properties lead to the conclusion that we can reduce the dimension d of images in the dictionary by the efficient random projection. Furhtermore, we can reduce the number N of the data in dictionary by interpolation of references in the dictionary.

For example, we show the comparison of computational cost between NNS method and proposed method. Here, we assume that image registration for rotation transfromation. Table 6 summarises the comparison of computational cost between NNS method and proposed method. In this case, we can reduce the

Table 4. Registration error to the template. Rotation angle θ of template and registration error in eq. (3), respectively.

	θ [degree]	NNS	loop1	loop2
A1	50.0	0.660	0.149	0.149
A2	30.0	0.633	0.270	0.270
A3	6.00	0.452	0.267	0.267
A4	-18.0	0.452	0.265	0.265
A5	-40.0	0.633	0.253	0.253
B1	50.0	0.722	0.447	0.447
B2	30.0	0.721	0.464	0.464
B3	6.00	0.605	0.470	0.470
B4	-18.0	0.661	0.466	0.466
B5	-40.0	0.724	0.468	0.468
C1	50.0	0.641	0.333	0.333
C2	30.0	0.653	0.350	0.350
C3	6.00	0.385	0.359	0.359
C4	-18.0	0.644	0.359	0.359
C5	-40.0	0.643	0.368	0.368

Table 5. Registration error to the template. Scaling parameter τ of template and registration error in eq. (3), respectively.

	τ [times]	NNS	loop1	loop2
A6	0.65	0.581	0.558	0.558
A7	0.85	0.509	0.491	0.491
A8	1.05	0.452	0.447	0.447
A8	1.25	0.406	0.396	0.396
A10	1.45	0.368	0.358	0.358
B6	0.65	0.691	0.686	0.686
B7	0.85	0.646	0.644	0.644
B8	1.05	0.610	0.615	0.615
B9	1.25	0.582	0.589	0.589
B10	1.45	0.560	0.566	0.566
C6	0.65	0.595	0.572	0.572
C7	0.85	0.532	0.513	0.513
C8	1.05	0.486	0.480	0.480
C9	1.25	0.451	0.445	0.445
C10	1.45	0.425	0.4204	0.420

Table 6. Example of computational cost. In Table 6, d expresses the dimension of the original space. k expresses the dimension of projected space by ERP. N expresses the number of data in dictionary. Computational cost is $\mathcal{O}(Nd)$ on search stage of NNS. In proposed method, computational cost is $\mathcal{O}(Nk)$ on the search stage.

	Data dimension size		Number of data	Computational cost	
	original d	projected k	N	Nd	Nk
NNS method	294849	-	360	1.0×10^8	-
proposed method	294849	1024	30	-	3.0×10^4

original dimension to 1024 dimensions. Furthermore, we can use the dictionary $d_\theta(-176, 12, 180)$ instead of the dictionary $d_\theta(-176, 1, 180)$. In this exaple, we can reduce the computational cost to approximatery 0.03% of NNS by proposed method.

5   Conclusions

In this paper, using an efficient random projection and interpolation in the dictionary of reference, we developed an efficient algorithm that establishes global image registration. We introduced to use an interpolation in dictionary of reference to reduce computational cost of preprocessing and the dictionary size of the NNS. We show that our method can perform the global registration with high accuracy, using small number of dictionary compare with previous method, in the low dimension in which the distance of dictionary was preserved by projecting efficient random projection from original dimension.

Acknowledgement. This research was supported by "Computational anatomy for computer-aided diagnosis and therapy: Frontiers of medical image sciences" funded by the Grant-in-Aid for Scientific Research on Innovative Areas, MEXT, Japan, the Grants-in-Aid for Scientific Research funded by Japan Society of the Promotion of Sciences and the Grant-in-Aid for Young Scientists (A), NEXT, Japan.

References

1. Healy, D.M., Rohge, G.K.: Fast global image registration using random projections. In: Proc. Biomedical Imaging: From Nano to Macro., pp. 476–479 (2007)
2. Itoh, H., Lu, S., Sakai, T., Imiya, A.: Global image registration by fast random projection. In: Proc. International Symposium on Visual Computing, pp. 23–32 (2011)
3. Sakai, T., Imiya, A.: Practical algorithms of spectral clustering: Toward large-scale vision-based motion analysis. In: Machine Learning for Vision-Based Motion Analysis, pp. 3–26. Springer (2011)
4. Zitová, B., Flusser, J.: Image registration methods: a survey. Image and Vision Computing 21, pp. 977–1000 (2003)
5. Modersitzki, J.: Numerical Methods for Image Registration. Oxford university press (2004)
6. Goshtasby, A.: Image Registration: Principles, Tools and Methods. Springer (2012)
7. Vempala, S.S.: The Random Projection Method, vol. 65. American Mathematical Society (2004)
8. Hjelle, Ø., Dæhlen, M.: Triangulations and Applications. Springer (2006)
9. Hartley, R.I., Zisserman, A.: Multiple View Geometry in Computer Vision. Cambridge University Press (2004)
10. Mahajan, D., Huang, F.C., Matusik, W., Ramamoorthi, R., Belhumeur, P.: Moving gradients: A path-based method for plausible image interpolation. ACM Transactions on Graphics 28, 42:1–42:11 (2009)
11. Cocosco, C., Kollokian, V., Kwan, R.S., Evans, A.: Brainweb. online interface to a 3D MRI simulated brain database. NeuroImage 5, 425 (1997)
12. Brainweb, http://www.bic.mni.mcgill.ca/brainweb/

Customizable Time-Oriented Visualizations

Mohammad Amin Kuhail, Kostas Pandazo, and Soren Lauesen

IT University of Copenhagen, Denmark

Abstract. Most commercial visualization tools support an easy and quick creation of conventional time-oriented visualizations such as line charts, but customization is limited. In contrast, some academic visualization tools and programming languages support the creation of some customizable time-oriented visualizations but it is time consuming and hard. To combine *efficiency*, the effort required to develop a visualization, and *customizability*, the ability to tailor a visualization, we developed time-oriented building blocks that address the specifics of time (e.g. linear vs. cyclic or point-based vs. interval-based) and consist of inner customizable parts (e.g. ticks). A combination of the time-oriented and other primitive graphical building blocks allowed the creation of several customizable advanced time-oriented visualizations. The appearance and behavior of the blocks are specified using spreadsheet-like formulas. We compared our approach with other popular visualization tools. Evaluation showed that our approach rates well in customizability.

1   Introduction

Time is a common and important dimension in many application areas such as healthcare, history and management. Several advanced visualizations, such as LifeLines [1] and Spiral Graph [2] were invented to help practitioners explore their temporal data and find interesting patterns. However, most of these visualizations were created by traditional programming languages and that makes it hard for designers to customize them to match real-life working environment.

Visualization tools allow easier visualization construction than traditional programming languages. However, customizability may be hard to obtain and some time-oriented visualizations that show the time cyclically (e.g. Cyclic Plot) are challenging or impossible to create.

Commercial visualization tools support users to create conventional time-oriented visualizations such as line and plot charts. Nevertheless, customizability is limited to some pre-defined settings.

As an alternative, we created customizable time oriented blocks that address the requirements of time-oriented visualizations, for instance, the level of details (e.g. week numbers vs. months), and whether time is shown linearly or cyclically. The time-oriented blocks consist of inner parts that can be customizable. We created several customizable advanced time-oriented visualizations by combining the time-oriented and other graphical building blocks and specifying their appearance and behavior using spreadsheet-like formulas. The formulas can refer to data fields, properties and functions of building blocks.

G. Bebis et al. (Eds.): ISVC 2012, Part II, LNCS 7432, pp. 668–677, 2012.

To evaluate our approach, we created a time-oriented visualization using our approach and two popular tools. Evaluation showed that our approach rates well in customizability.

2   Related Work

The related work can be divided into two categories: Modeling time and approaches to developing time-oriented visualizations.

2.1   Modeling Time

Depending on the application of use, there are many aspects that need to be considered when modeling the time. Below we summarize the works of Frank [3], Ozul et. al. [4], Bettini [5] and Wilinkson [6].

Domain. Time domains can be categorized into continuous and discrete. In a continuous time domain, there is a point between any two points in time. However, in a discrete domain, time-oriented data can be only mapped to specific discrete time ticks.

Scope. The scope of time-oriented data can be point-based or interval-based. In a point-based scope, no information is given about a region between two points in time. In an interval-based scope, the data are given in intervals, for instance, the start time of the interval, as well as the duration, or the start and end time.

Arrangement. Time can be arranged linearly or cyclically. In linear arrangement, time proceeds from past to future. However, in cyclic arrangement, the domain consists of recurring time values, for instance, holidays or seasons that reoccur yearly. As noted by the survey book [7], strictly cyclic arrangement is rare. However, a combination of cyclic and linear arrangement is more common.

Time Granularities. Time can be expressed at different levels of detail, granularities, such as years, months, weeks, etc. A single or multiple granularities can be used. A detailed discussion can be found on page 52-54 [7].

2.2   Approaches to Developing Time-Oriented Visualizations

Academic Visualization Toolkits. Some toolkits such as Flare [8], Protovis, D3 [9] and Prefuse [10] support time-oriented visualizations by means of linear time scales, and the visualization specifications are program-like (e.g. sequential steps, declared variables, etc.). While this approach is rather flexible, creating a simple thing like a time axis can be unnecessarily tedious. A more complex visualization such a focus + context line chart visualization requires programming, for instance, sequential statements, event handling, and variable declarations. Furthermore, none of these tools support advanced cyclical time-oriented visualizations such as the Spiral Graph.

uVis [11], a formula-based visualization tool, uses spreadsheet-like formulas to specify the visualization. The designer does not need to write for loops, declare variables, etc. However, only a few examples (e.g. custom pie chart, heat map) are available, and it is hard to know whether this approach can support the creation and customization of novel time-oriented visualizations efficiently.

Commercial Visualization Tools. Some commercial tools such as Spotfire [12] and Tableau [13] support users to create conventional time-oriented visualizations such as line and plot charts. Further, users can compare several variables, and filter them as needed. Nevertheless, these tools fall short if users want to customize individual time primitives according to data. For instance, it is not possible to customize time tick marks or gridlines that represent holidays.

Visualization Libraries. Some visualization libraries such as Google Visualization [14], Infragistics [15], and SMILE widgets [16] provide complex widgets, building blocks such as line charts and time lines, suited for one time-oriented visualization. Advanced customization is not possible or requires modification of the widget. Assuming the widget is open-source, it can be programmatically modified. However, this might be hard and tedious.

3   Solution

The main goal of our solution is to allow the creation of customizable time-oriented visualizations at a low effort. Thus, our solution aims to combine expressiveness, efficiency, and customizability. To address *expressiveness*, we created time-oriented blocks based on the design aspects of time discussed in section 2.1. *Efficiency* is supported by providing a number of building blocks that are commonly used in time-oriented visualizations (e.g. time scale, time axis) rather than leaving such a tedious task to the designer. Finally, the time-oriented building blocks and the inner parts (e.g. ticks, gridlines) they contain can be *customized* by specifying a succinct number of visual properties.

The process of creating a visualization consists of finding the right building blocks and *gluing* them together according to data and logic. We chose uVis formulas to specify the building blocks because they are declarative, sequence-free, and do not require variable declaration, for loops, etc. Combining the time-oriented and other general-purpose building blocks, and specifying their properties with uVis formulas, we were able to create and customize a number of novel time-oriented visualizations such as LifeLines, CircleView [17], Time Spiral, etc.

3.1   General-Purpose Building Blocks

We provide geometric primitive building blocks that show data according to position, color, size and shape. Box and Triangle are examples. To receive input from end-users, we use standard .NET input building blocks such as TextBox and Button.

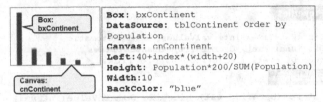

| Box: bxContinent |
| DataSource: tblContinent Order by Population |
| Canvas: cnContinent |
| Left: 40+index*(width+20) |
| Height: Population*200/SUM(Population) |
| Width: 10 |
| BackColor: "blue" |

Fig. 1. Creating a bar chart with uVis formulas

3.2   uVis Formulas

uVis formulas resemble spreadsheet formulas, but can refer to relational data table fields and other block properties. Moreover, they can specify the appearance and behavior of a bundle of instances that share a common record set.

Fig. 1 shows an example of a bar chart created with uVis. The chart visualizes the populations of the world continents. The bars are created by means of a Box building block. The data come from table tblContinent. The DataSource represents the data intended for visualization. The DataSource formula retrieves a record set from table tblContinent ordered by population. This causes uVis to create as many Box instances as rows in the record set. Moreover, the fields of tblContinent are accessible now by the formulas. The Left formula positions the instances 20 pixels far from each other's right. To make the height depend on the population, the Height formula refers to field Population in tblContinent.

3.3   Time-Oriented Building Blocks

The time-oriented building blocks show time and can map other graphical building blocks to time. All the time-oriented blocks have common properties (Fig. 2) that define and customize their appearance and address the time design aspects discussed in section 2.1. Below we explain some of the common properties.

IntervalValues are the start and end times of intervals. More than two interval values means more than an interval is shown (e.g. multi-interval time line). For instance, 01-02-2012, 28-02-2012, 01-03-2012 represents two intervals.

Granularities is the time granularities (year, month, week, etc) that the building block shows. It is of a Dictionary type (collection) so designers can combine multiple granularities of different steps. The steps are the time spans between each time granule. We support the granularities shown in Fig. 3. As an example, to assign a yearly granularity with step 2 and a monthly granularity with step 6 to the first interval (interval with index 0), the designer can define the IntervalValues as (0,Granularity("Y",2),(0,Granularity("M",6).

The **Ticks** is a table that has information about the time tick marks to show. The fields of the Ticks table are shown in Fig. 2. Setting **Automatic** true means the block chooses a suitable granularity to allow maximum coverage for the time interval in the available space to show the time.

```
┌─────────────────────────────────────────────────────────────────┐
│ Properties                                                        │
├─────────────────────────────────────────────────────────────────┤
│ List<DateTime> IntervalValues, bool Automatic                     │
│ Dictionary<int, Granularity> Granularities,                       │
│ bool Discrete (false), DataTable Ticks                            │
├─────────────────────────────────────────────────────────────────┤
│ Functions                                                         │
├─────────────────────────────────────────────────────────────────┤
│ int Pos(DateTime val) DateTime DomainValue(int val)               │
└─────────────────────────────────────────────────────────────────┘

┌─────────────────────────────────────────────────────────────────┐
│ Tick Fields                                                       │
├─────────────────────────────────────────────────────────────────┤
│ DateTime Date, Granularity Granularity, String Format,            │
│ enum Type, int Position                                           │
└─────────────────────────────────────────────────────────────────┘
```

Fig. 2. The common properties and functions of the time-oriented building blocks

```
┌─────────────────────────────────────────────────────────────────┐
│ Granularities                                                     │
├─────────────────────────────────────────────────────────────────┤
│ Yearly:100,50,10,5,2,1 Monthly:6,3,2,1 Weekly:3,2,1,              │
│ Daily: 15,10,5,2,1 Hourly: 12,6,3,1 Minutely:30,20,15,10,5,1      │
│ Secondly:30,20,15,10,5,1                                          │
└─────────────────────────────────────────────────────────────────┘
```

Fig. 3. The time granularities we support and their default steps

The algorithm is inspired by the work of [18]. However, If Automatic is false, the time-oriented block show time based on the Granularities the designer chooses.

Pos translates a DateTime to the corresponding position in pixel. This function can be used to map time-oriented data to the time scales. By default, the block shows continuous time. Hence, data can be mapped to any point in the interval covered by the time-oriented block, but setting **Discrete** true lets the block only map the data to the available Ticks.

The time-oriented building blocks are robust. If the designer does not specify the properties as intended, a default value is assumed, and a warning message is given to the designer. For instance, the IntervalValues are supposed to be defined in an ascending manner, but if the designer specifies the start and end in a descending manner, the start and end's values will be set equal and a warning is generated.

All the time-oriented building blocks comprise inner blocks and have a few properties that can automatically specify properties of the inner blocks. The designer can customize the properties of the individual inner blocks. However, if the designer changes the time-oriented block property specifications after the customization of inner block, the related basic block properties will be affected, and customization might be lost for these properties. Below we present the time-oriented blocks.

TimeLine. TimeLine is a linear time scale. It can show time at multiple granularities and can map other building blocks to it. Depending on the Granularities the designer defines, TimeLine generates a number of inner blocks of type Label that show the tick marks, and positions them, but the designer can change that. Moreover, it generates an inner block of type Line that shows the gridlines. Finally, if the Automatic property is set true, the end-user can drag the time in or out. This lets

`TimeLine` show more or less time, and it adjusts the granularities automatically. Let us look at an example of using the `TimeLine`.

Example. Designer Lise wanted to make a time-oriented visualization for doctor Ibra. The data model and the visualization are shown in Fig.4. Ibra wants to see the medicine prescriptions of a patient mapped to a time line. The time line has to cover the period of time from the admission of the patient till today at yearly and monthly granularities. Moreover, the time line should show the patient holidays in a different color.

Lise chose a `TimeLine` building block because it can show a period of time linearly and allow the mapping of other building blocks. The `DataSource` of the `Time-Line` refers to a particular patient (`TimeLine` specifications in Fig. 5). This created an instance of `TimeLine` and made the fields in the patient row accessible. The `IntervalValues` refers to field `admDate` (admission date), and `Date`. Hence, `TimeLine` covers the intended period of time. Lise assigned the yearly, monthly, and daily granularities (Fig. 5).

Based on these specifications, the `TimeLine` block generated three inner `Label` building blocks representing yearly (`TL.TickMark_0`), monthly (`TL.TickMark_1`), and daily (`TL.TickMark_2`) tick marks, and a `Line` building block representing the grid lines.

The inner building blocks were specified automatically by the `TimeLine` block. For instance, `DataSource` of each inner building block was specified using the `Ticks` table of the `TimeLine` block. Lise customized the daily tick marks (`TL.TickMark_2`) so they show the patient holidays in yellow. She first added an additional property, *designer property*, where she tests using `Contains` function if the `Label` instance shows a date that is within holiday intervals. `Parent-<Holiday` accesses all the patient holiday rows. Then using if formula, she sets `BackColor` yellow if the condition is met.

Finally, Lise chose a `Box` building block to represent the medicine prescriptions. She mapped the `Box` instances to the time line using the `Pos` function.

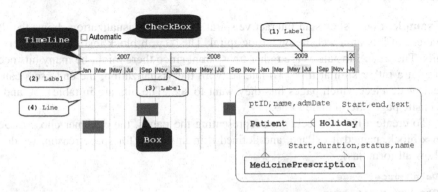

Fig. 4. A time-oriented visualization showing the medicine prescriptions of a patient mapped to a time line. The data model of the visualization is shown on the right-hand side.

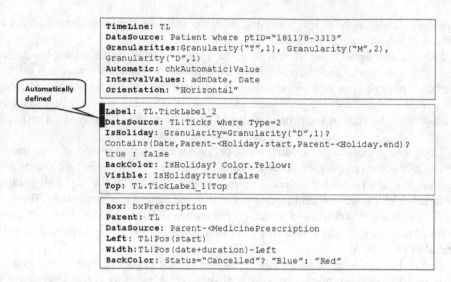

Fig. 5. The specifications of the building blocks that make up the visualization in Fig. 4

TimeAxis. Time axes are commonly used in line charts. Similar to the `TimeLine` block, `TimeAxis` generate customizable inner blocks (`Line`, `Label`, etc.) that draw the ticks. We provide `LinearTimeAxis` and `CyclicalTimeAxis` blocks. Both can be shown horizontally or vertically. The `LinearTimeAxis` supports conventional linear time-oriented visualizations such as line charts. The `CyclicalTimeAxis` is a cyclic axis designed to support the Cycle Plot visualization [19] but can be customized. The `CyclicGranularity` property allows showing cyclic time at several granularities such as `DW` (Day of week), etc. Fig. 5 shows an example of a day-of-week cycle plot. The designer used `Pos` function to position the `Spline` instances to the axis.

Spiral. The `Spiral` block that shows cyclic time where each cycle represents an equal period of time such as 1 day or 2 months, etc.

Example. Fig. 6 shows an interactive spiral graph. The visualization shows the daily pattern of hits on a website on a time spiral. The `Glyph` block instances represent the hits. The `Glyph` instances are rather transparent since there might be many hits occurring at a rather similar time. To interact with the visualization, the end-users can select or deselect which pages hits they want to see .The hits are in table `Hit` and the website pages are in table `Page`.

To create a list of checkboxes representing the pages, the designer chose `Check-Box` block, named it cxPage, and defined it in this way (for space reason, we do not show all formulas).

```
DataSource: Page                                              (1)
Checked: init true                                            (2)
```

The `DataSource` formula created as many `cxPage` instances as `Page` records. `Init` means the initial state of a property; in this case it means the checkboxes are initially checked.

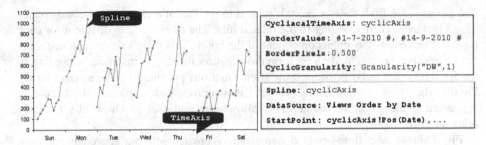

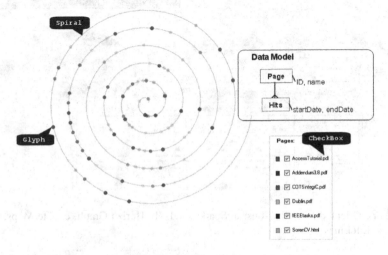

Fig. 6. This cyclic plot shows a week-of-day analysis to the views of a website in a period of time. The visualization uses The CyclicalTimeAxis block. The Spline block instances represent the views.

Fig. 7. An interactive visualization created with uVis

To show time cyclically, the designer defined a Spiral block in this way.

```
Spiral: spiralGraph                                           (3)
IntervalValues: MIN(Hit.StartDate]), MAX(Hit.EndDate)         (4)
CyclicalGranularity: Granularity("D",1)                       (5)
```

The IntervalValues represents the time interval the spiral is covering. The MAX and MIN retrieves the maximum and minimum dates to cover the entire range of hits. The CyclicalGranularity means each cycle in the spiral represents a day. Finally, to create glyphs representing the hits, the designer defined a Glyph block in this way.

```
Parent: cxPage                                                (6)
DataSource: Parent-<Hit                                       (7)
BackColor: PageID=1? Color(30,"red"):PageID=2? Color(30,"blue"):  (8)
Top: spiralGraph!vPos(Date)                                   (9)
Left: spiralGraph!hPos(Date)                                  (10)
Visible: Parent!Checked                                       (11)
```

The DataSource finds the Hit records related to the Parent's Page records, and creates Glyph instances corresponding to the related hits. The BackColor is different for each page, 30 represents the alpha component of the color. This makes the Glyph instances rather transparent. Top and Left formulas position the Glyph instances according to the hit date using polar position translation functions provided by the Spiral block. Finally, the Glyph instances are visible if their corresponding parent (the checkbox representing the page) is visible. This allows the end-user to show hits for only relevant pages.

Fig. 7 shows other time-oriented visualizations created with our approach.

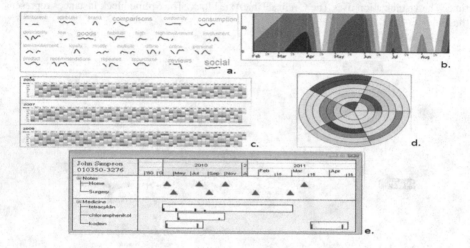

Fig. 8. Time-Oriented Visualizations: a. SparkClouds, b. Horizn Graph, c. Tile Maps, d. CircleView, e. LifeLines.

4   Conclusion and Future Work

In this paper, we showed that it is feasible to create customizable advanced time-oriented visualizations. The approach we propose is that designers create visualizations by combining building blocks and specifying their appearance and behavior using spreadsheet-like formulas.

Since they consist of customizable inner building blocks, the various time-oriented building blocks speeded up the time-oriented visualization construction without compromising customizability. Together with uVis formulas, we were able to create and customize several advanced time-oriented visualizations (e.g. LifeLines, Spiral Graph, etc.). However, similar to spreadsheet applications, the more advanced the visualization is, the longer the formulas are.

Our next step is to conduct a usability test with designers without advanced IT skills and evaluate to what extent they can build and customize time-oriented visualizations.

References

1. Plaisant, C., Mushlin, R., Snyder, A., Li, J., Heller, D., Shneiderman, B., Colorado, K.P.: Lifelines: Using visualization to enhance navigation and analysis of patient records. In: Proceedings of the 1998 American Medical Informatics Association Annual Fall Symposium, pp. 76–80 (1998)
2. Weber, M., Alexa, M., Muller, W.: Visualizing time-series on spirals. In: IEEE Symposium on Information Visualization, INFOVIS 2001, pp. 7–13 (2001), doi:10.1109/INFVIS.2001.963273
3. Frank, A.: Different types of times in gis. Spatial and Temporal Reasoning in Geographic Information Systems, 40–62 (1998)
4. Ozsu, M.T., Goralwal, I.A., Szafron, D.: A framework for temporal data models: Exploiting object-oriented technology. In: Proceedings of the Tools-23: Technology of Object-Oriented Languages and Systems. IEEE Computer Society, Washington, DC (1997)
5. Bettini, C., Jajodia, S.G., Wang, S.X.: Time Granularities in Databases, Data Mining and Temporal Reasoning, 1st edn. Springer-Verlag New York, Inc., Secaucus (2000)
6. Wilkinson, L.: The grammar of graphics. Springer-Verlag New York, Inc., New York (1999)
7. Aigner, W., Miksch, S., Schumann, H., Tominski, C.: Visualization of Time-Oriented Data. Springer, Heidelberg (2011)
8. Flare, http://flare.prefuse.org/ (accessed October 2011)
9. Bostock, M., Ogievetsky, V., Heer, J.: D3: Data driven documents. IEEE Trans. Visualization & Comp. Graphics, (Proc. InfoVis) (2011), http://vis.stanford.edu/papers/d3
10. Heer, J., Card, S.K., Landay, J.A.: Prefuse: a toolkit for interactive information visualization. In: Proceedings of the SIGCHI Conference on Human Factors in Computing Systems, CHI 2005, pp. 421–430. ACM, New York (2005)
11. Lauesen S.: uvis, http://www.itu.dk/~slauesen/SEHR/UnifiedDataVisualization.pdf/ (accessed February 2012)
12. Spottfire: Tibco spotfire, http://spotfiretibco.com (accessed October 2011)
13. Tableau, http://www.tableausoftware.com/ (accessed October 2011)
14. Google chart tools, http://code.google.com/apis/chart/ (accessed November 2011)
15. Infragistics, http://www.infragistics.com/dotnet/netadvantage/silverlight/data-visualization.aspx/ (accessed February 2012)
16. Smile widgets - timeline, http://www.simile-widgets.org/timeline/ (accessed October 2011)
17. Keim, D.A., Schneidewind, J., Sips, M.: Circleview: a new approach for visualizing time-related multidimensional data sets. In: AVI, pp. 179–182 (2004)
18. Talbot, J., Lin, S., Hanrahan, P.: An Extension of Wilkinson's Algorithm for Positioning Tick Labels on Axes. IEEE Trans. Visualization & Comp. Graphics (Proc. InfoVis) (2010)
19. Cleveland, W.S.: Visualizing Data. Hobart Press, NJ (1993) (Summit)

A Visual Cross-Database Comparison
of Metabolic Networks

Markus Rohrschneider, Peter F. Stadler, and Gerik Scheuermann

Leipzig University, Department of Computer Science, Germany

Abstract. Bioinformatics research in general and the exploration of
metabolic networks in particular rely on processing data from different
sources. Visualization in this context supports the exploration process
and helps to evaluate the data quality of the used sources.

In this work, we extend our existing metabolic network visualization
toolbox and hereby address the fundamental task of comparing metabolic
networks from two major bioinformatics resources for the purpose of data
validation and verification. This is done on different levels of granularity
by providing an overview on retrieval rates of chemical compounds and
reactions per pathway on the one hand, as well as giving a detailed insight
into the differences in the biochemical reaction networks on the other.

1 Introduction

During the last decade a wealth of high throughput sequence data of both
genomes and proteomes have become available for a wide variety of organisms.
For a small number of model organisms, on the other hand, detailed information is available on their metabolic chemical reactions and the enzymes that
catalyze them. Combining these sources of knowledge allows the inference of
metabolic networks, see e.g. [1]. Databases, including KEGG Pathway [2], Bio-
Cyc [3], WikiPathways [4], Reactome [5] as well as a wide variety of species-
specific resources such as LeishCyc [6] provide metabolic network data with a
varying degree of manual curation.

Both the computational inference of metabolic networks and the manual curation process is subject to errors, however. Systematic misannotations and ambiguities in assignments of enzyme functions [7,8], for instance have been identified
as sources of errors. The most common type of error is associated with "over-
prediction" of molecular function. Further problems arise from the incomplete
modeling of the chemical reactions themselves, which typically are treated as
annotation texts rather then data in their own right [9] and can lead to stoichiometric inconsistencies [10] that can hamper the analysis of metabolic data. The
notorious incompleteness of genome annotations even in well-studied organisms
such as *E. coli*, yeast, or human, furthermore translates into an incompleteness of metabolic pathway maps. Enzymes also may change their function and
substrate specificity over the course of evolution, fundamentally limiting the accuracy of functional annotations that are rooted in sequence similarities. Taken

G. Bebis et al. (Eds.): ISVC 2012, Part II, LNCS 7432, pp. 678–687, 2012.
© Springer-Verlag Berlin Heidelberg 2012

together, thus even well-curated metabolic network data cannot be assumed to be complete and entirely accurate.

The direct comparative analysis of the chemical reaction networks describing the metabolism can be used to identify and expose potential weaknesses and errors in the representation of biochemical networks. Our approach is inspired by a set-theoretic approach to comparing chemical (and in particular metabolic) networks [11], originally proposed as a means of identifying metabolic innovations.

In this contribution we focus in particular on the KEGG Pathway Database. KEGG pathways are relatively large sub-systems of the metabolic network that combine multiple biological processes from different organisms in a way that matches biological intuition.We use this network as template graph and perform a multi-scale comparison to the metabolic network provided by BioCyc. Firstly, we describe how compounds and reactions are matched between the two networks and how this information is used to infere relationships between pathways defined by KEGG vs. the ontology defined in BioCyc. The results can be viewed by the user from a global point of view, i.e. a quantification of the node matching quality for each pathway of the KEGG network, and in more detail by interactively expanding the pathways of interest to reveal the respective reaction networks. Brushing techniques for highlighting portions of the detailed network are used to draw the users attention to inconsistencies among the two resources or ambiguities. We combine several popular information visualization methods to navigate the presented network, such as semantic zoom, hierarchical exploration by node expansion, and focus & context techniques.

2   Related Works

The comparison of large and complex networks requires suitable visualization and navigation techniques. A key issue with this specific task is to preserve the mental map. Especially in the field of life sciences exist certain standards and drawing conventions to visualize those networks. There currently exist a variety of different software products, graph drawing algorithms and visualization techniques addressing these special requirements. A good overview on problems and current research in the context of visualizing biological networks is given in [12]. The article also addresses the specific task of comparing two similar metabolic networks from different species and focuses on finding a suitable consensus layout (backbone structure). A generic layout algorithm for biological networks takes specific drawing conventions into account [13]. By defining a set of constraints for the algorithm, certain aesthetic criteria can be met, e.g. non-overlapping clusters or compartments, given bounding boxes for subgraphs, layout of cycles or the direction of edges.

Popular tools for exploring, analyzing and navigating biological networks are *Cytoscape* [14] and *VisANT* [15]. Both provide a variety of generic layout algorithms. Additionally, advanced filtering techniques for the reduction of graph complexity and flexible mapping of data attributes to visual properties are implemented. Various plug-ins exist for statistical analysis and simulation.

Some important publications on the specific topic of metabolic network comparison can be found in [16] and [17]. Albrecht et al. focus on finding a suitable layout for the union graph, which is constructed from the two individual graphs to be compared. The union graph layout is used to layout common subsets, while still preserving the differences. The algorithm is designed in a hierarchic manner. An overview graph construction is provided by laying out the backbone first, and the detailed graph representing changes is constructed and laid out afterwards. While the work in [17] addresses generic biological networks, a comparison of similar metabolic pathway graphs is given in [16]. Identical parts of the network are identified to define constraints for a common layout. The proposed method is applied to data obtained from the BioPath System and the KEGG Pathway database.

The two bioinformatics resources we use in this work provide very limited capabilities for comparing different metabolic networks. The KEGG web interface allows the user to project metabolic pathways specific to one of approximately 1400 organism onto the reference pathway diagrams. The respective enzymes are highlighted. The web interface of BioCyc [3] is more flexible in that respect. Pathways are dynamically rendered with a user-specified level of detail. A cross-species comparison is possible for two organisms. MetaCyc also provides the software package *Pathway Tools* [18] for navigating the data, metabolic pathway prediction, analysis and visualization.

3   Data Resources, Graph Structure and Layout

Both databases provide a semi-structured flat-file dump of metabolic network data, either as KGML files (KEGG) or attribute-value files (BioCyc). The reconstruction of the chemical reaction networks from KGML files is a straightforward task and is explained in more detail in [19]. Each KGML file represents a metabolic pathway as defined by KEGG and contains the connectivity information of reaction and compound nodes, as well as layout information for each node. This allows a very similar depiction of the pathway maps as provided by KEGG. It helps preserving the mental map as these drawings constitute a de-facto standard among biologists. After the construction of bipartite graphs representing the pathways, we add parent nodes for each pathway and insert inter-pathway edges connecting two identical compounds in different pathways as defined by the *maplink* elements in the KGML file. These inter-pathway edges are propagated to the higher pathway level and will be visible in the abstracted network overview part of the visualization. The hierarchy introduced by the pathway nodes is non-overlapping due to the duplication of compound nodes present in more than one pathway. Within a pathway, all compound and reaction nodes are unified facilitating the mapping process from one network to the other. We use this non-overlapping pathway clustering for navigating the network by expansion of pathways of interest as suggested in [20] with one major difference. Instead of computing a new layout for the individual reaction networks, we use the node positions given in the KEGG diagram for the sake of mental map preservation and route the edges according to the algorithm given in the aforementioned work.

In the case of BioCyc, we only use the MetaCyc branch of the database collection containing multi-organism metabolic pathways. It relates most closely to the reference pathways from KEGG as these also represent the union of the reaction sets realized in different organisms. Unlike the KEGG Pathway graph, we construct a large unified bipartite graph from a reactions data file. As for the pathway ontology, we add nodes for each occurring pathway and evaluate the super-pathway and sub-pathway relations to reconstruct the ontology represented as a directed acyclic graph. Each pathway references several reactions, creating an overlapping hierarchical clustering of the reaction set. The annotations, i.e. synonym lists for chemical compounds and enzyme commission numbers (EC) for reactions are stored and serve as the basis for the mapping process.

4   Graph Matching

We use the metabolic network constructed from the KEGG Pathway database as template graph, which is considerably smaller than the BioCyc graph, but not necessarily a subset. The mapping we describe here is uni-directional from KEGG to BioCyc, so we identify three cases:

1. Unique match: a node in the KEGG network can be mapped to exactly one node in the BioCyc network.
2. Ambiguous match: a KEGG node will be mapped to more than one BioCyc node.
3. "No hit": the KEGG compound or reaction cannot be found in the BioCyc collection.

The actual mapping process is done in two steps: (1) matching compound nodes, and (2) match reaction nodes based on the compound mapping.

Given two graphs, finding a graph isomorphism or inclusion relations is NP-hard. The problem described here is not of a graph-theoretical nature in the traditional sense, but rather a lexicographical one. With every chemical entity comes a set of annotations from the respective database. For mapping the chemical compounds found in the KEGG network to nodes in BioCyc, we use a list of synonymous chemical names associated with the compound. These two sets of lists will be cross-referenced to identify matches. For every compound node in the template network (KEGG), we hold a – possibly empty – list of matching candidates in the BioCyc network.

In the second step, we do not have to rely on string comparison operations. Instead, we take advantage of the unique adjacency of a reaction node. In general, a reaction can be defined by the sets of chemical compounds it consumes – substrates – and the set of compounds it produces – products. Given this signature, we can robustly identify nodes in the bipartite BioCyc network that fulfill a certain neighborhood configuration. For each reaction node in the KEGG graph, we determine the set of adjacent compounds and identify the set of reaction nodes in the BioCyc graph that have the matched compounds as neighbors. In case of ambiguous compound matches, we have to repeat the search for every combination of potential compound candidates, resulting in a potentially larger set of reaction matches. Reactions can still be robustly identified even if one or more compounds in its neighborhood could not be matched at all. We refer the reader to section 6 for a detailed discussion on the different mapping scenarios. For multiple hits on reaction nodes, we use the EC nomenclature given with the reactions to refine the search result.

As a measure for the overall quality of the mapping serves a simple match score s_p for each pathway p in KEGG, which accumulates the match score s_v for each node $v \in V_p$:

$$s_v = \begin{cases} 0 & , \mid m(v) \mid = 0 \\ \frac{1}{\mid m(v) \mid} & , \mid m(v) \mid > 0 \end{cases} \text{ and } s_p = \frac{\sum_v s_v}{\mid V_p \mid}$$

It takes ambiguous matches $m(v)$ into account and penalizes a large number of candidates for a specific node.

5   Visual Comparison and Exploration

After the mapping process has been completed, the overview on the metabolic network comparison is presented to the user (see Fig. 1). We start with a completely collapsed network with only the pathway nodes visible. Several properties of the underlying networks will be visualized: relative pathway size, match score s_p as the "filling level" and the number of entities without match in relation to the network size. The latter is depicted by the saturation of the red color in the upper portion of the pathway node.

For a selected pathway, the user may investigate the relations to the Bio-Cyc pathway ontology (Fig. 3). On demand, the portion of the directed acyclic graph representing the nesting relations of pathways together with the respective chemical reaction nodes is overlaid onto the current view. The selected KEGG pathway node remains highlighted. The subgraph of the DAG contains at least all the matched reactions belonging to the selected KEGG pathway. Reactions present in the current subset of the ontology but not part of the selected KEGG pathway are not displayed to avoid clutter.

Once the user selects a pathway for further investigation by expanding one or more pathway nodes, the detailed chemical network as found in KEGG Pathway is revealed (see Fig. 2). The three types of matches are indicated by different node colors. Nodes unable to map appear in white (compounds) and red (reactions). A successfully mapped compound node appears in a color on a scale from yellow to green depicting the number of matched BioCyc compounds (green for exact match). If a reaction node can still be mapped to one or more BioCyc compounds even though in its neighborhood is at least one unmapped node, the reaction

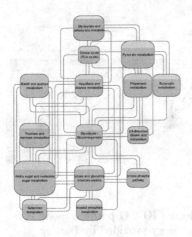

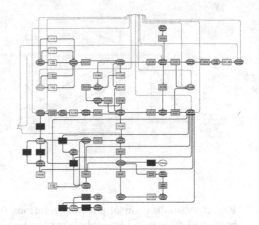

Fig. 1. Overview on the mapping quality of a metabolic network constructed from KEGG Pathway. 15 Pathways associated with the carbohydrate metabolism are shown. The node size depicts the number of reactions and compounds of the respective pathway. The nodes' filling level reflects the match score s_p for the respective pathway, which is closely related to the ratio of matched nodes and total node number, but also penalizes ambiguous matches.

Fig. 2. Match results on the detailed reaction network of the Pentose-Phosphate-Pathway using the coloring scheme described in the text. A reaction node drawn in blue indicates that only a subset of adjacent compound nodes could be found in the BioCyc graph. In those cases, the reaction matches are much less reliable, but could very often verified using the EC number of the associated enzyme.

node appears blue. The saturation channel is used to indicate the ambiguity of the mapping. We make the distinction between reaction nodes with a completely mapped neighborhood vs. an incompletely mapped neighborhood. In the latter case, fewer adjacent nodes are taken into account when searching for a reaction match in the BioCyc graph. This results in a higher degree of freedom and, therefore, to a less reliable match.

The user can finally verify the highlighted discrepancies in the two networks by displaying the context information in the BioCyc graph. For a selected reaction node, the mapped reaction(s) in BioCyc are displayed with the respective substrate and product compounds (Fig. 4, l.h.s.). Vice versa, for a selected compound in KEGG, the corresponding match(es) with their adjacent reaction nodes are overlaid over the current graph (Fig. 4, r.h.s.). The aforementioned node coloring scheme is applied on the BioCyc nodes as well. In addition to the selected node's neighborhood, the partial pathway ontology containing the displayed reaction nodes is presented.

Displaying the KEGG template graph and the reaction graph context in Bio-Cyc simultaneously has another advantage besides mental map preservation. This choice of design allows the user to edit the KEGG pathway graph to correct flaws in the network. The graph editing capability of the software tool

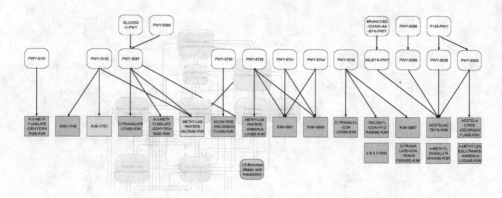

Fig. 3. Visualization of pathway relations from a selected KEGG pathway node (C5-branched dibasic acid metabolism) to the pathway ontology provided by BioCyc

allows manipulating the graph's topology as well as assigning attributes to graph elements.

6   Results

We constructed two different KEGG networks as template graph and ran the comparison on both graphs. The first network contains the complete set of available pathways (136). The second is a compilation of 15 pathways related to the carbohydrate metabolism, for which 380 out of 449 compounds (84.6%) and 519 out of 616 reactions (84.3%) could be identified in the BioCyc graph. For the complete metabolic network, 2872 out of 4817 compounds (60%) and 2169 out of 3129 reactions (69.3%) were successfully mapped. The retrieval rates in the large network were considerably smaller than in the network representing the carbohydrate metabolism. This may have different reasons. Firstly, pathways related to the carbohydrate metabolism are well studied and understood. We can assume, that those pathways contain fewer errors. Secondly, among the more 'exotic' pathways in KEGG, there may be some reactions not present in the BioCyc collection at all.

We could yet make another discovery on the reaction level. In some cases of ambiguous mappings, a perfect match would indeed be a mapping error! As the context graph in Fig. 5 shows, the KEGG compound 'D-Glucose' is mapped to two Glucose nodes in BioCyc: 'D-Glucose' and 'GLC'. Referring to the annotation, the KEGG compound is more precisely *alpha-D-Glucose*, and GLC is *beta-D-Glucose*, a stereo-isomer of the former. Since no annotation is given for the BioCyc node 'D-Glucose', we can only assume, that it represents indeed *alpha-D-Glucose* as specified in KEGG. In that case, the BioCyc pathway contains more – and more accurate – information, since both isomers are metabolized. In addition, the two reaction nodes mapped to the highlighted KEGG reaction are again a more precise description than provided by KEGG. The annotation

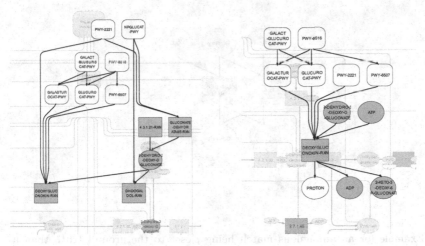

Fig. 4. Details on demand: For the selected compound 2-Dehydro-3-deoxy-D-gluconate (l.h.s.) and the chemical reaction identified by the enzyme 2.7.1.45 (r.h.s.) of the Pentose-Phosphate-PW, the direct neighbors in the BioCyc graph are displayed and the corresponding subset of the pathway ontology

for the KEGG reaction reveals two EC names, which indicate a complex reaction catalyzed by more than one enzyme. The two matches in BioCyc support this assumption as the reaction 'RXN-11334' corresponds to the enzyme number 1.1.99.35, and the reaction 'GLUCOSE-DEHYDROGENASE-ACCEPTOR-RXN' is identified with the EC number 1.1.99.10. Both EC numbers match the annotation of the respective reaction in KEGG.

7   Conclusion and Future Work

In this work we have presented an extension to our graph visualization software capable of comparing metabolic networks from different bioinformatics resources. The network constructed from the KEGG Pathway database served as template graph and was used for navigating and exploring the data exploiting its 2-layer hierarchical structure. The reaction and compound nodes of this network were mapped to the metabolic network provided by the BioCyc database collection for the purpose of validation and verification of the KEGG data. The mapping quality was summarized using a simple, but yet meaningful score and presented to the user as an overview over the matched pathway graph. Discrepancies could be located and investigated in more detail taking advantage of the implemented focus & context technique described in [20] for navigation and exploration.

Some differences in the datasets are plausible as explained in the results section, others are indeed incorrect entries or annotations. The proposed method helps to identify those problems, however, the evaluation of those discrepancies is the task of the user and certainly requires some background knowledge about the biochemical processes in question.

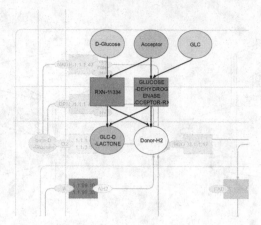

Fig. 5. Example for an ambiguous match being closer to the ground truth than a unique match. Both types of D-Glucose are present in the network. The reaction node in KEGG represents a complex reaction consisting of several alternatives for metabolizing the different stereo-isomers of D-Glucose.

A very useful feature only shortly mentioned in this work is the graph editing capability of the software tool. By displaying portions of the BioCyc graph relevant to the selected elements in the KEGG graph on top of the current graph representation, the user can manually refine the metabolic network and save the changes.

There are, however, a few issues to be addressed. The mapping of compound nodes relies on string comparison of the provided annotations and may miss matches due to syntactic errors in the synonym lists. Although the results suggest the matching process to be rather robust, manual refinement of the synonym lists may be necessary in some cases. On the other hand, identifying reaction nodes by their adjacency relations works very well even if some of the adjacent nodes could not be matched. On the other hand, perfect matches on reaction nodes with under-defined neighborhood can be exploited to match the missing compounds.

As stated in the previous section, the matching scores for 'exotic pathways' were very low suggesting that large parts of these pathways are missing in the BioCyc collection. In the future, we intent to investigate this issue further in close cooperation with bioinformatics experts and biologists.

References

1. Karp, P., Ouzounis, C., Moore-Kochlacs, C., Goldovsky, L., Kaipa, P., Ahrén, D., Tsoka, S., Darzentas, N., Kunin, V., Lóper-Bigas, N.: 33, 6083–6089 (2005)
2. Kanehisa, M., Araki, M., Goto, S., Hattori, M., Hirakawa, M., Itoh, M., Katayama, T., Kawashima, S., Okuda, S., Tokimatsu, T., Yamanishi, Y.: KEGG for linking genomes to life and the environment. Nucleic Acids Research 36, D480–D484 (2008)

3. Caspi, R., Altman, T., Dale, J., Dreher, K., Fulcher, C., Gilham, F., Kaipa, P., Karthikeyan, A., Kothari, A., Krummenacker, M., Latendresse, M., Mueller, L., Paley, S., Popescu, L., Pujar, A., Shearer, A., Zhang, P., Karp, P.: Metacyc: The metacyc database of metabolic pathways and enzymes and the biocyc collection of pathway/genome databases. Nucleic Acids Research 38, D473–D479 (2010)
4. Pico, A., Kelder, T., van Iersel, M., Hanspers, K., Conklin, B., Evelo, C.: WikiPathways: Pathway Editing for the People. PLoS Biology 6, 1403–1407 (2008)
5. Haw, R., Croft, D., Yung, C., Ndegwa, N., D'Eustachio, P., Hermjakob, H., Stein, L.: The Reactome BioMart. In: Database 2011 (2011)
6. Doyle, M., MacRae, J., Souza, D.D., Saunders, E., Likić, M.M.V.: LeishCyc: a biochemical pathways database for Leishmania major. BMC Syst. Biol. 3, 57 (2009)
7. Green, M., Karp, P.: Genome annotation errors in pathway databases due to semantic ambiguity in partial EC numbers. Nucleic Acids Res. 33, 4035–4039 (2005)
8. Schnoes, A., Brown, S., Dodevski, I., Babbitt, P.: Annotation error in public databases: Misannotation of molecular function in enzyme superfamilies. PLoS Comput. Biol. 5, e1000605 (2009)
9. Ott, M., Vriend, G.: Correcting ligands, metabolites, and pathways. BMC Bioinformatics 7, 517 (2006)
10. Gevorgyan, A., Poolman, M., Fell, D.: Detection of stoichiometric inconsistencies in biomolecular models. Bioinformatics 24, 2245–2251 (2008)
11. Forst, V., Flamm, C., Hofacker, I., Stadler, P.: Algebraic comparison of metabolic networks, phylogenetic inference, and metabolic innovation. BMC Bioinformatics 7, 67 (2006) [epub]
12. Albrecht, M., Kerren, A., Klein, K., Kohlbacher, O., Mutzel, P., Paul, W., Schreiber, F., Wybrow, M.: On Open Problems in Biological Network Visualization. In: Eppstein, D., Gansner, E.R. (eds.) GD 2009. LNCS, vol. 5849, pp. 256–267. Springer, Heidelberg (2010)
13. Schreiber, F., Dwyer, T., Marriott, K., Wybrow, M.: A generic algorithm for layout of biological networks. BMC Bioinformatics 10, 375 (2009)
14. Smoot, M.E., Ono, K., Ruscheinski, J., Wang, P., Ideker, T.: Cytoscape 2.8: new features for data integration and network visualization. Bioinformatics 27, 431–432 (2011)
15. Hu, Z., Mellor, J., Wu, J., DeLisi, C.: VisANT: an online visualization and analysis tool for biological interaction data. BMC Bioinf. 5, e17 (2004)
16. Schreiber, F.: Visual comparison of metabolic pathways. J. Vis. Lang. Comput. 14, 327–340 (2003)
17. Albrecht, M., Estrella-Balderrama, A., Geyer, M., Gutwenger, C., Klein, K., Kohlbacher, O., Schulz, M.: 08191 working group summary – visually comparing a set of graphs. In: Borgatti, S.P., Kobourov, S., Kohlbacher, O., Mutzel, P. (eds.) Graph Drawing with Applications to Bioinformatics and Social Sciences. Number 08191 in Dagstuhl Seminar Proceedings, Dagstuhl, Germany, Schloss Dagstuhl - Leibniz-Zentrum fuer Informatik, Germany (2008)
18. Karp, P.D., Paley, S., Romero, P.: The pathway tools software. Bioinformatics 18, S225–S232 (2002)
19. Klukas, C., Schreiber, F.: Dynamic exploration and editing of kegg pathway diagrams. Bioinformatics 23, 344–350 (2007)
20. Rohrschneider, M., Heine, C., Reichenbach, A., Kerren, A., Scheuermann, G.: A Novel Grid-Based Visualization Approach for Metabolic Networks with Advanced Focus&Context View. In: Eppstein, D., Gansner, E.R. (eds.) GD 2009. LNCS, vol. 5849, pp. 268–279. Springer, Heidelberg (2010)

Visual Rating for Given Deployments of Graphical User Interface Elements Using Shadows Algorithm

Daniel Skiera[1], Mark Hoenig[1], Juergen Hoetzel[1],
Slawomir Nikiel[2], and Pawel Dabrowski[1,3]

[1] Bosch Thermotechnik GmbH Thermotechnology
Lollar, Germany
[2] Institute of Computer Engineering and Electronics
University of Zielona Gora, Poland
[3] Institute of Control and Computation Engineering
University of Zielona Gora, Poland

Abstract. Good information design is very important for human computer interfaces since it improves productivity and enhances human understanding. This paper proposes a novel algorithm for estimation of deployment of the visual elements of the GUI. The interface layout can be either created manually or be the output of computer simulation locating uniform rectangular blocks on the layout. Those 'black-box' blocks can be replaced by the system dependent objects representations, e.g. visual metaphors of the heating system. The proposed "shadows" algorithm can be used in visual rating of various GUI models. The paper discusses the theoretical background, the properties of the proposed algorithm together with the sample prototype application.

1 Introduction

Graphical User Interface may influence the user understanding of visualized information systems. Efficient design of complex GUI systems requires a multidisciplinary approach. Issues related to this area of information systems include the perception of people [1], knowledge of how information must be displayed to enhance human acceptance and comprehension [2], [3] and must also consider the capabilities and limitations of the hardware and software of the human-computer interface [4], [5]. Particularly, the software part of GUI design process of complex systems needs some automated tools used in creation processes. Generally the information design, crucial for human-computer interaction (HCI), blends the results of visual design research, knowledge concerning human perception, knowledge about the hardware and software capabilities of the interfaces and artificial intelligence algorithms. Visual information is prepared manually or alternatively, may be an output of numerical simulations. During the design process the individual elements are assigned visual metaphors following the logical flow of information. The next step is to organize and lay out individual elements

G. Bebis et al. (Eds.): ISVC 2012, Part II, LNCS 7432, pp. 688–695, 2012.

of GUI clearly and meaningfully. Good screen layout presentation and depicted background structure will encourage quick and correct information processing, the fastest possible execution of tasks and functions, and enhanced user acceptance [6], [7]. However, it is hard to estimate numerically the information related to aesthetics. There are some attempts to provide useful metrics [8]. This paper will present rationale and reasoning that explains why the proposed method is useful in numerical estimation of deployment of visual elements.

2   Layout Alignment

A number of papers addressed GUI design from the users perspective [9], [10], [11], [12]. One of the designers rules suggests that fewer screen alignments result in a more visually appealing screen layout. It is known as the rule of minimum design [13]. Proper aligning of GUI elements will also make eye movement through the screen much more obvious and reduce the distance it must travel. Good layout is obtained by creating screen balance. But how can we check whether the created pattern is consistent, predictable, and distinct? One of the solutions comes from the visual aesthetics. Symmetries and even spacing are closely related to the human sense of aesthetics, hence the visual complexity measure related to the estimation of the even deployment layout elements described in the following sections.

3   Rating of Even Deployment

The visual rating of given deployments of GUI elements boils down to even deployment of these elements, because the even deployment is treated as one of the major features of aesthetic deployment [14]. The visual rating can be used for classifying various deployments or for selecting the best deployment. The rating can also be treated as aesthetic criteria and can be used for the construction of a cost (objective) function [12]. Fig. 1 shows two deployments of GUI elements in example of heating system elements on the displays of smartphones.

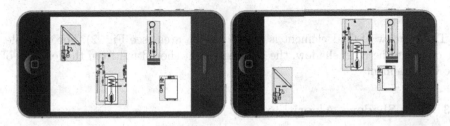

Fig. 1. Above two smartphones with GUI elements are shown. The first (*left*) shows an even deployment of heating system elements. The second (*right*) shows an uneven deployment of the same elements.

3.1   Preconditions

For the Shadows Algorithm some limitations were set:

- the GUI elements are treated as uniform rectangles (see Fig. 1, the rectangles are shown as gray blocks);
- the GUI elements may not overlap;
- the edges of the GUI elements should not touch each other;
- the returned results by the Shadows Algorithm can be compared if the examples of various deployments consider the same GUI elements (with the same dimensions);

This paper does not discuss the connections between the GUI elements.

3.2   The Definition of Shadow

The GUI elements are treated as black blocks, each GUI element has four shadows (left, top, right, bottom) as is shown in the Fig. 2.

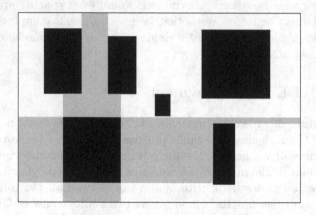

Fig. 2. An example of given deployment of GUI elements (*the black blocks*). For the left-bottom GUI element its shadows are shown(*the gray areas*).

The shadow of GUI element is treated as an area (see Fig. 2). For example, for creating the right shadow, the source of light should fall from the left side of the GUI element.

3.3   The Shadows Algorithm

The Shadows Algorithm is based on the determination of four mean length of shadow (left, top, right and bottom) for each GUI element. The mean length of shadow is determined from the suitable side of GUI element and the shadow (for example, the suitable side for the right shadow is the right side of the GUI element etc.).

In the next step of the algorithm the shadows are divided by the suitable sides of the GUI elements. In this way the four (left, top, right and bottom) mean length of shadows for each GUI element is determined. If the mean length of shadows are more similar, the deployment is more even. Fig. 3 shows an example of the mean length of shadow for one right shadow of the GUI element.

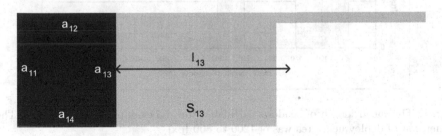

Fig. 3. The mean length of shadow l_{13} is calculated from the shadow (the gray area) S_{13} and the right side a_{13} of the GUI element

Eq. 1 calculates the mean length of shadow from the shadow and the suitable side of GUI element,

$$l_{ij} = \frac{S_{ij}}{a_{ij}}, \tag{1}$$

where:
i – the number of GUI element, $i = 1, 2 \ldots$,
j – the number of side of GUI element, $j = 1, 2, 3, 4$ (e.g. the number 1 means the left side, the number 2 means the top side etc.),
l_{ij} – the mean length of shadow for the i GUI element and the j side,
S_{ij} – the shadow (the gray area, see Fig. 3) for the i GUI element and the j side,
a_{ij} – the side of the i GUI element and the j number of side.

Fig. 4 presents a histogram with all mean length of shadows calculated for the example shown in Fig. 2.

The main task for this algorithm is the determination of the visual rating. The first idea was, that the rating is calculated as standard deviation of the mean length of shadows. Experiments have shown, that the treatment returns good results if the GUI elements are grouped, despite the fact that the deployment is not even. This happens, because if the GUI elements are grouped the short mean length of shadows dominate the longer mean length of shadows and the standard deviation is small. An example of group GUI elements is shown in Fig. 5. The standard deviation for the example of Fig. 5 is equal to 89 [px] and the standard deviation for example shown in Fig. 2 is equal to 143 [px] (the dimension of deployment area was of 1200 to 800 [px]). The fact should mean, that the deployment of GUI elements in Fig. 5 is better than in Fig. 2 (because 89 < 143), but it is not.

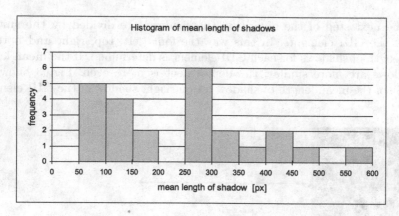

Fig. 4. The mean length of shadows calculated for the example given in Fig. 2. The dimension of deployment area was of 1200 to 800 [px]

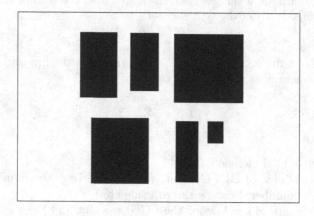

Fig. 5. Deployment of GUI elements. The GUI elements are grouped in the center of the deployment area

In the chart in Fig. 6 the histogram with mean length of shadows for the example in Fig. 5 is shown. The histogram in Fig. 6 shows that the short mean length of shadows dominate.

In order to achieve better results, returned by the Shadows Algorithm the short mean length of shadows have to be "punished" and the long mean length of shadows should be favored. This can be realized by using the sum of inverses of the mean length of shadows. The question is, how severe the short mean length of shadows should be "punished"? The "punish" value can be established by raising the inverses of the mean length of shadows to the appropriate power. The research was conducted for powers equal to 0.25, 0.5, ..., 2. Further research of powers has been interrupted, because the returned results began to impair,

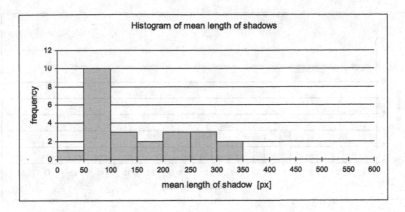

Fig. 6. The mean length of shadows calculated for the example given in Fig. 5. The dimension of deployment area was of 1200 to 800 [px]

or did not change. The best returned results were observed if the visual rating was calculated as the sum of all inverses of the mean length of shadows with the power equal to 0.5,

$$V_r = \sum_{i=1}^{n} \sum_{j=1}^{4} \sqrt{\frac{1}{l_{ij}}}, \tag{2}$$

where:
V_r - the visual rating,
n - the number of GUI elements, $n \in N \setminus 0$.

Finally the visual rating of the given GUI elements deployment is calculated from the Eq. 3.

$$V_r = \sum_{i=1}^{n} \sum_{j=1}^{4} \sqrt{\frac{a_{ij}}{S_{ij}}}. \tag{3}$$

If the value returned by the Shadows Algorithm is smaller, the deployment of GUI elements is more even. The visual rating for the example in Fig. 2 is equal to 0.15 and for the example in Fig. 5 is equal to 0.22. This means that the deployment in Fig. 2 is rated better by the Shadows Algorithm (because $0.15 < 0.22$).

3.4 Discussion

To validate the results, returned by the Shadows Algorithm, a survey was conducted among the users of GUI interfaces. The users received fourteen pictures of deployment examples of GUI layout examples on the cards. The cards are shuffled and the task was to order the deployments from the best to the worst appealing. Fourteen people were interviewed.

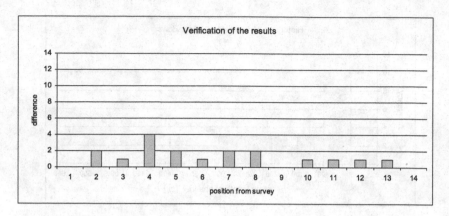

Fig. 7. Differences between the position from the survey and the position from the
Shadows Algorithm

Fig. 7 shows the difference between the order returned by the Shadows Algorithm and the survey.

For better understanding of the chart in Fig. 7 an example is presented:
The survey classified an example of deployment on the position 9 (the x axis)
and the Shadows Algorithm classified the same example of deployment also on
the position 9, because the difference between the position from the survey and
the position from the Shadows Algorithm (the y axis) is equal to 0. The survey
classified an example of deployment on the position 5 and the Shadows Algorithm
classified the same example of deployment on the position 3 $(5-2)$ or 7 $(5+2)$,
because the difference is equal to 2.

As it can be seen from the chart in Fig. 7, the differences between the order
returned from the Shadows Algorithm and the survey for many distributions are
similar. This means, that the Shadows Algorithm is capable of visual rating of
given deployments of GUI elements.

4 Conclusion

Good design is very important, as far as graphical user interfaces are considered.
The paper proposes the novel algorithm for visual rating of deployment of the
GUI elements in the given area. Graphical layout is treated as a set of rectangular
'black-box' blocks. The paper discusses an alternative metric of visual design,
based on the Shadows Algorithm. The method works well with several GUI
design scenarios, and provides results that are close to those obtained with the
survey conducted among experts in he field. The authors plan to develop further
the method in order to support more complex systems.

Acknowledgment. This study is supported by Bosch Thermotechnik GmbH,
and the authors would like to thank the company for the cooperation and assis-
tance rendered.

References

1. Winograd, T., Flores, F.: Understanding computers and cognition. Ablex Publishing, Norwood (1986)
2. Miller, G.: The magical number seven, plus or minus two - some limits on our capacity for processing information. Psychological Review 101, 343–352 (1955)
3. Bodker, S.: Through the interface: a human activity approach to user interface design. Lawrence Erlbaum Associates, Hillsdale (1991)
4. Wood, L.: User Interface Design: Bridging the Gap from User Requirements to Design. CRC Press (1997)
5. Teo, L., Byrne, J., Ngo, D.: A method for determining the properties of multi-screen interfaces. International Journal of Applied Mathematics and Computer Science 10, 413–427 (2000)
6. Gromke, G.: Digital asset management - der effektive umgang mit mediendaten. In: Proceedings of Intl. Conf. EVA 2007, Berlin, pp. 161–166 (2007)
7. Hoffmann, P., Lawo, M., Kalkbrenner, G.: Zur aesthetik interaktiver medien - hypervideo im spannungsfeld zwischen usability und design. In: Proceedings of Intl. Conf. EVA 2007, Berlin, pp. 117–123 (2011)
8. Gamberini, L., Spagnolli, A., Prontu, L., Furlan, S., Martino, F., Solaz, B., Alcaniz, M., Lozano, J.: How natural is a natural interface? an evaluation procedure based on action breakdowns. Personal and Ubiquitous Computing 5 (2011)
9. Hartswood, M., Procter, R.: Design guidelines for dealing with breakdowns and repairs in collaborative work settings. International Journal on Human Computer Studies 53, 91–120 (2000)
10. Keppel, G., Wickens, T.: Design and analysis: a researchers handbook. Pearson/Prentice Hall, Upper Saddle River (2004)
11. Galitz, W.: The Essential Guide to User Interface Design. Wiley (2007)
12. Nikiel, S., Dabrowski, P.: Deployment algorithm using simulated annealing. In: 16th International Conference on Methods and Models in Automation and Robotics, MMAR 2011, Miedzyzdroje, Poland, pp. 111–115 (2011)
13. Sirlin, D.: Subtractive design. Game Developer, 23–28 (2009)
14. Deussen, O.: Aesthetic placement of points using generalized lloyd relaxation. In: Computational Aesthetics 2009, Victoria, British Columbia, Canada, pp. 123–128 (2009)

Hierarchical Visualization of BGP Routing Changes Using Entropy Measures*

Stavros Papadopoulos[1,2], Konstantinos Moustakas[2,3], and Dimitrios Tzovaras[2]

[1] Department of Electrical and Electronic Engineering, Imperial College London,
SW7 2AZ, London, UK
{s.papadopoulos11}@imperial.ac.uk
[2] Information Technologies Institute, Centre for Research and Technology Hellas,
P.O. Box 361, 57001 Thermi-Thessaloniki, Greece,
{spap,moustak,tzovaras}@iti.gr
[3] Electrical and Computer Engineering Department, University of Patras,
Patras, Greece
moustakas@ece.upatras.gr

Abstract. This paper presents a novel framework for optimizing the visual analysis of network related information, and in particular of Border Gateway Protocol (BGP) updates, using information theoretic measures of both the underlying data and the visual information. More precisely, a hierarchical visualization scheme is proposed using a graph metaphor that is optimized, with respect to information theoretic metrics of several visual mapping parameters. Experimental demonstration in state-of-the-art BGP events, illustrate the flexibility of the proposed framework and the significant analytics effect of the proposed optimization scheme.

1 Introduction

Today's rapidly expanding Internet, provides data delivery and communication service to millions of end users. The backbone of the internet infrastructure is the Border Gateway Protocol (BGP), where autonomous systems provide dynamic updates and form the internet routing tables. The distributed nature of BGP and the lack of verification of the validity of the announcements causes Internet routing to be vulnerable to attacks, while the situation will become even more complex after IPv6 will be widely adopted. Therefore, advancements in analytics and in particular visual analytics are expected to prove highly useful in this domain.

1.1 BGP Basics

Internet consists of a large number of networks called autonomous systems (AS). Each AS is assigned a number. The basic components of BGP [3] are the

---

* This work has been partially supported by the European Commission through project FP7-ICT-257495-VIS-SENSE funded by the 7th framework program. The opinions expressed in this paper are those of the authors and do not necessarily reflect the views of the European Commission.

G. Bebis et al. (Eds.): ISVC 2012, Part II, LNCS 7432, pp. 696–705, 2012.

announcements, which have the form {<prefix> : <AS-path>}. An example of an announcement is: <91.194.12.0/23>: <3549 1 2 3> which consists of the prefix: <91.194.12.0/23> describing 512 IP addresses and the AS-path: <3549 1 2 3>. The last AS in the path AS-3 is the one that made the announcement and it is the owner of these 512 IP addresses. Furthermore, the path describes the ASes that one should traverse through to send data to AS-3.

The process of the announcement is as follows: An AS makes an announcement to all its neighbors. Then, they add their AS number at the AS-path and propagate the announcement to the next neighbor.

Since BGP announcements propagate routing information, each AS knows different paths to the same destination. The AS-paths visible from an AS depend on the his location on the network. The AS router from which the data are collected is called monitoring point.

The data in this paper are collected from the RIPE [5] BGP monitoring project. This project collects BGP announcements from various monitoring points around the globe and makes the data available to the public.

1.2   Related Work

One of the most important aspects of BGP behavior visualization is to capture and present its dynamic characteristics that are evolving over time. Moreover, detection of variations in Internet traffic can be also caused by router misconfigurations or malicious attacks.

BGPlay [7] allows Internet Service Providers to monitor the reachability of a specified prefix from the perspective of a given border router while incorporating animation to highlight routing changes. Pier et al. [10] use the idea of topographic maps to enhance BGPlay visualizations and provide an intuitive way to show the ranking of different ASes. This is archived, by positioning the ASes to different areas of the topographic map according to their ranking. Wong, et al. [8] describe in their proposed system, TAMP, statistical methods to aggregate BGP data in order to visualize and diagnose BGP anomalies. The LinkRank visualization [1] provides a high level view of Internet routing changes. Autonomous systems are displayed as nodes while connectivity is illustrated using links. More attributes such as size, color, and width are used to encode auxiliary information. BGP-eye [9] clusters the BGP announcements into events that are capable of better representing an anomaly. Afterwards, these events are correlated across multiple monitoring points to ascertain the extent of the anomaly.

1.3   Motivation-Contribution

Most of the aforementioned approaches provide a high level view of interactions between autonomous systems due to their highly dynamic nature. Moreover, since the overlaying logic is of higher importance than the visualization itself, simpler visualization techniques such as node link graphs have been used. The proposed approach aims to tackle this problem by providing a hierarchical framework for the visualization of the BGP routing changes, allowing therefore the use of high level visualizations in the first steps of the analysis, eliminating therefore visual clutter, while the analyst is capable to see in more detail specific parts of the graph that are of particular interest. Moreover, the present framework makes the first steps in proposing an information theoretic metric for quantifying and subsequently optimizing the generated visualizations. As a result many of the up to know manually and empirically selected parameters and thresholds can be automatically and optimally estimated using the proposed approach.

2   Visualization of the AS-Graph

The present visualization system is closely related with the work of Dan Massey et al[1]. The fundamental objective of the visualization system is to visualize the routing changes. Firstly, from the AS-paths of the BGP announcements the AS-Graph is created. The notation $G(V, E)$ is used to represent the AS-Graph. Afterwards, from all the announcements until a point in time a snapshot is taken that describes the state of the network. For example, from the announcements until time T_i, AS-1 owns N_i^1 IP addresses, and the edge that connects AS-1 to AS-3 is used by N_i^2 IP addresses. Afterwards, at time $T_k > T_i$ the above values change and become N_k^1, N_k^2. It is this difference of IPs $N_k^1 - N_i^1$ and $N_k^2 - N_i^2$ that constitutes the weights of the edges and the vertices. Moreover, the magnitude of the corresponding weight is mapped to the width of the edge or the radius of the vertex respectively. Furthermore, positive values are represented with green color and negative with red. The visualization system is shown in Fig. 5 and 6.

3   Clustering Algorithm

This Section presents the use of hierarchical clustering to visualize the entire AS-Graph. The AS-Graph of the year 2005 has over 20,000 ASes and 28,000 edges, which makes it very difficult to visualize. The method used by [1] was filtering. While this method is practical, part of the information is lost. The effort of this paper is to try to preserve all the information but present it to the user in a more abstract form. Afterwards, the user can use operations, such as zooming to see the actual data. Basically, the hierarchical clustering produces a hierarchy of sparse graphs were in each level, a vertex represents a cluster of ASes. This Section is based on the work of [4]. As in [4] the position of the graph nodes needs to be calculated first, which is achieved using a force-directed algorithm.

To compute the clusters of the first level of the hierarchy, the set of candidate pairs $S_0 = \{v_i, v_j | v_i \neq v_j \ and \ v_i, v_j \in V_0\}$ is constructed for the graph of the zero level $G_0(V_0, E_0) \equiv G(V, E)$. The superscript is used to represent the level of the hierarchy. To find the candidate pairs as in [4], the pairs from the graph that are connected with an edge are firstly added to the set S_0. Moreover, pairs from a proximity graph that are connected with an edge, are also added to the set. The proximity graph used is the Urquhart graph, which is easily constructed by removing the longest edge from each triangle in the Delaunay triangulation. Furthermore Urquhart graph is a good approximation to the relative neighborhood graph [12].

After constructing the set of candidate pairs S_0, the maximal number of disjoint pairs is constructed from it. To achieve this, an iteration is made over all the vertices and for every unmatched vertex i, the set of candidate pairs containing it are searched. Afterwards, the vertex i is matched with the vertex that maximizes the weighted sum of the following measures:

1. Geometric proximity: $\frac{1}{\|p_i - p_j\|}$
2. Similarity of neighborhood: $\frac{|N_i^* \cap N_j^*|}{|N_i^* \cup N_j^*|}$
3. Degree: $\frac{1}{\deg_i * \deg_j}$

where p_i is the position of vertex i, N_i^* is the set of all the vertices that have a coomon edge with i and $\deg_i$ is the degree of vertex i. The measures are normalized to the range $[0, 1]$. In the present implementation the weights used for the measures are set to 0.6, 0.2 and 0.2 respectively.

Afterwards, the maximum number of disjoint pairs, plus the vertices that are unmatched at the end of the process, define the graph of the upper level. For example, to construct the level 1 graph $G_1(V_1, E_1)$ the following procedure is followed:

- To construct the set of vertices V_1: a vertex is created for each pair of the maximum number of disjoint pairs and for each unmatched vertex.
- To construct the set of edges E_1: an edge is added between two vertices, if in the AS-Graph $G(V, E)$ there exists an edge from an AS of the first vertex to an AS of the second.

This procedure is repeated for all the levels of the hierarchy. The position of a vertex is calculated by finding the average position all of the AS nodes that it is comprised from the AS-Graph $G(V, E)$.

Since what is visualized are the edge and AS weights that directly reflect the BGP announcements, the corresponding weights of the vertices and edges of every level of the hierarchy must be calculated:

- The weight of a vertex is the sum of the weights of the ASes it is comprised.
- The weight of an edge between two vertices is the sum of the weights of the edges, that exist between all the ASes of the first vertex and all the ASes of the second.

Furthermore, the clusters must be constructed in such a way that the information content of the graph is as high as possible. Since every announcement ends to a monitoring point, it is deliberately left out from the calculation of the clusters. Also, the large ASes usually have a big portion of the BGP announcements passing through them. So the user can set the number of ASes with the highest ranking from CAIDA [6], that should be left out of the clustering calculation. Fig. 1 shows the results of the clustering algorithm applied on the AS-graph.

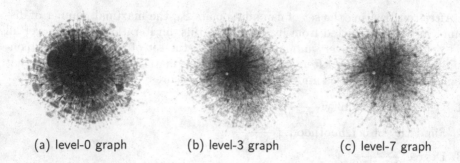

(a) level-0 graph   (b) level-3 graph   (c) level-7 graph

Fig. 1. Cluster Hierarchy of the AS-Graph

4   Quantification

In this Section the input and the output signal are quantified using entropy measures. The need of quantification arises from the need of a metric that could measure the quality of a visualization system and be able to perform comparisons and optimizations.

4.1   Entropy of the Input Signal

The input signal is the weight of the edges and the vertices that are the outcome of the BGP announcements. Two entropy measures are calculated, one with respect to each edge and vertex of the graph and one with respect to all the vertices and edges weights.

The first entropy measure represents the entropy of an edge or a vertex. In general the entropy of an edge is:

$$H_e^{in} = -\sum_{i=1}^{N} \frac{n_i}{n_{total}} \log\left(\frac{n_i}{n_{total}}\right) \qquad (1)$$

where N is the number of different weight occurrences, n_i is the number of occurrences of the i_{th} weight and $n_{total} = \sum_{i=1}^{N} n_i$ the total number of weight occurrences. The superscript "in" represents the input signal. The same procedure is repeated in order to find the entropy of each vertex H_v^{in}.

The second entropy measure, represents the entropy of all the weight occurrences, with respect to the entire graph. In general the entropy of all the edge weights of the entire graph is:

$$H_{ew}^{in} = -\sum_{i=1}^{M} \frac{m_i}{m_{total}} \log\left(\frac{m_i}{m_{total}}\right) \qquad (2)$$

again M is the number of different weight occurrences, m_i is the number of occurrences of the i_{th} weight and $m_{total} = \sum_{i=1}^{M} m_i$ the total number of weight occurrences, but with respect to the edge weights of the entire graph. The same procedure is repeated in order to find the entropy H_{vw}^{in} of all the vertices weights.

4.2 Entropy of the Output Signal

In general a visualization system uses a mapping function $F : \Re^n \to V^m$, where $\Re^n$ is the input signal in the space of real numbers and has n features, and V^m is the visualization (scatterplots, graphs, glyphs etc) that has m features. In the proposed scheme, the output signal is the visualized graph. Because of display capacity and visual clutter, a mapping of all the information of the input signal to the output, is usually not possible. As explained in Section 2, the edge weights are mapped to the width and the color of the edges and the AS weights are mapped to the radius and the color of the vertices. So the output space V^m has four features ($m = 4$). Moreover, the width of edges and the radius of the vertices must have some bounds to reduce the visual clutter and enhance the readability of the graph. The bounds of the present implementation are defined to $[1, 10]$ pixels for the edge width and $[3, 30]$ pixels for the vertex radius. A simple experimentally selected mapping function F is used:

$$\left\{ \begin{array}{l} F_e(W_e) = sqrt\,(W_e)/S_e \\ F_v(W_v) = sqrt\,(W_v)/S_v \end{array} \right\} \tag{3}$$

where W_e represents the edge weight and W_v the vertex weight. S_e and S_v are experimentally selected constants. The value of F_e represents the edge width in which the edge weight W_e is mapped to, and the value of F_v represents the vertex radius in which the vertex weight W_v is mapped to. Positive values are mapped to green color and negative to red. Furthermore:

$$sqrt\,(W) = \left\{ \begin{array}{c} -\sqrt{-W} \\ \sqrt{W} \end{array} \right\} \begin{array}{l} ,if\ W < 0 \\ ,if\ W \geq 0 \end{array} \tag{4}$$

Like it is done for the input signal, two entropy measures are calculated, the entropy of each edge and vertex and the entropy of all the weight occurrences of the entire graph. To find the entropy of the output signal equations (1) and (2) are used, but the difference is that the weights are firstly mapped to the above ranges through the mapping function.

4.3 Entropy Based BGP Clustering

As explained in Section 3, in the clustering algorithm the monitoring point as well as some big ASes are left out of the cluster calculation. In this Section this implementation selection is justified.

Fig. 2 depicts the first 20 edges with the biggest entropy calculated with the use of equation (1). It is apparent that 14 out of 20 edges are connected to the monitoring point AS-3549. The rest are connected to large ASes as one can see from the CAIDA ranking [6]. For example AS-3356 is number 1, and AS-1239 is number 6. Most of the depicted ASes in Fig. 2 are within the first 30 in the

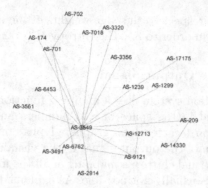

Fig. 2. The first 20 edges with the biggest entropy H_e^{in}

CAIDA ranking. This provides a strong indication that the choice to left out these ASes from the cluster calculation in order to have smaller information loss is correct.

5   Edge Width Mapping Optimization

In this Section, a method is presented that helps to find the optimum mapping function F so as to maximize the entropy H_{ew}^{out} of the output signal. The reason for wanting to maximize the entropy of the output signal is because of the need to have the smaller information loss as the information traverses through the visualization pipeline [2] and present to the user as much information as possible. Without loss of generality only the case of edge width mapping will be examined. As explained in Section 4.2, to map the edge weights to the edge color and width, the mapping function F of equation (3) is used. This function is experimentally selected and does not take into account the particularities of the input signal, as for example its distribution. In other words, an other mapping function $F : \Re^n \to V^m$ needs to be found such that the entropy of the output signal is maximum:

$$\max_{x_i} \left\{ H_{ew}^{out} \left(\, F \left(x_{-10}, x_{-9}, ...x_0, ...x_9, x_{10} \right) \, \right) \right\} \tag{5}$$

where x_i for $i \in [-10, 10]$, is a range of edge weights that is mapped to width i and H_{ew}^{out} is the output entropy of the edges weights of the entire graph.

Furthermore, for perceptual reasons, there is a need for this mapping function to map positive weights to positive widths and negative to negative. Fig. 3 shows the mapping problem.

To maximize the objective function (5) an exhaustive search is used over all the x_i, since more sophisticated search approaches are out of the scope of this paper. The same procedure is followed for the optimization of the vertex weights mapping function. The results in the visualization are shown in Fig. 5 and 6.

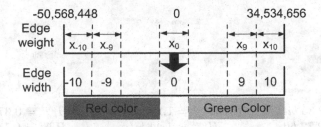

Fig. 3. The mapping problem from the edge weight to the edge width

6   Event Analysis and Results

In this Section the visualization tool developed is used to examine an event that took place on Oct 21, 2005. Furthermore, the results of the entropy analysis are presented and the visualization approach of [1] is compared with the present one.

On Oct 21, 2005, as it is shown in [1], AS-3356 had some internal problems and lost a big portion of the paths leading in and out of it. Fig. 4 shows an example of the visualization approach of [1].
As it is shown in Fig. 4, AS-3356 loses paths and the ASes around it gain paths. In Fig. 4 a simple filtering is used to reduce the graph size. On the contrary Fig. 5 and 6 show the visualization approach of the present work.

Fig. 5(a) shows the view of the event from the level 14 of the cluster hierarchy. The vertex in the center loses a lot of paths and IPs. After zooming this suspicious vertex (Fig. 6(a)), it is apparent that AS-3356 is the one that loses these paths. The difference between the tool developed here and the tool of [1] is that the present one does not need to filter the data and is able to visualize all the BGP announcements that took place in the event in a hierarchical manner.

The visualization entropy of Fig. 5(b),5(c) and 6(b),6(c) is artificially lowered to show the correlation between entropy and visualization. As it is shown

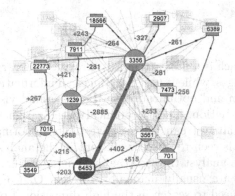

Fig. 4. The tool of [1] Link-Rank

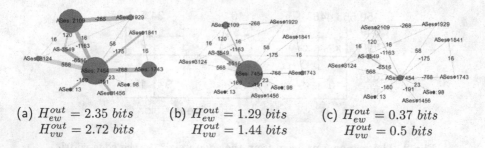

(a) $H_{ew}^{out} = 2.35$ *bits*
$H_{vw}^{out} = 2.72$ *bits*

(b) $H_{ew}^{out} = 1.29$ *bits*
$H_{vw}^{out} = 1.44$ *bits*

(c) $H_{ew}^{out} = 0.37$ *bits*
$H_{vw}^{out} = 0.5$ *bits*

Fig. 5. Abstract view of level 14. The vertex in the center loses a lot of paths and IPs. Subfigure (a) represents the entropy optimized version. The monitoring point is AS-3549.

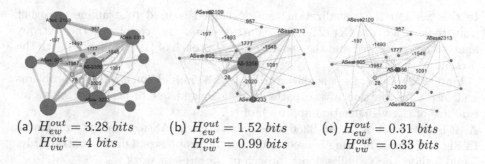

(a) $H_{ew}^{out} = 3.28$ *bits*
$H_{vw}^{out} = 4$ *bits*

(b) $H_{ew}^{out} = 1.52$ *bits*
$H_{vw}^{out} = 0.99$ *bits*

(c) $H_{ew}^{out} = 0.31$ *bits*
$H_{vw}^{out} = 0.33$ *bits*

Fig. 6. Abstract view of level 14 after zooming to the suspicious vertex. Subfigure (a) represents the entropy optimized version. The monitoring point is AS-3549.

in these figures, the lower the entropy, the lower the information content of the visualization. For example in Fig. 5(c) the user can not discriminate between many of the edges as well as the vertices weights. On the contrary, in the optimized version the differences are more apparent and insights about the data are more easily obtained by the user.

7   Conclusions

The proposed approach is a first attempt to provide a hierarchical visualization of BGP routing changes, while it also introduces an information theoretic metric for the quantification and optimization of the generated visualization and the underlying raw information. The hierarchical visualization provides high level overview, while the analyst can further investigate abnormalities in more detail. The framework is seen to be very efficient in the analysis of BGP routing changes and aids the analyst through the automated estimation of parameters and thresholds that are usually manually selected. One of the limitations of the approach is the need to search in the high-dimensional parameter space for a global maximum. However, due to the monotonicity property that entropy functions usually demonstrate with respect to each individual dimension, this problem can be solved in a straightforward manner. From the experiments of this work it is obvious that a small change in the entropy is not perceivable by the user. As a result, the maximum entropy regarding the user perception is considered to be a neighborhood of values around the absolute maximum of (5). The proposed framework can be used to optimize every visualization approach as long as the mapping function F and the entropy are well defined.

References

1. Lad, M., Massey, D., Zhang, L.: Visualizing Internet Routing Changes. IEEE Transactions on Visualization and Computer Graphics 12(6), 1450–1460 (2006)
2. Chen, M., Janicke, H.: An information-theoritic Framework for Visualization. IEEE Transactions on Visualization and Computer Graphics 16(6) (2010)
3. Rekhter, Y., Li, T.: A border Gateway Protocol (BGP-4). Request for Comment (RFC): 1771 (1995)
4. Gansner, E.R., Koren, Y., North, S.C.: Topological Fisheye View for Visualizing Large Graphs. IEEE Transactions on Visualization and Computer Graphics 11(4) (2005)
5. RIPE NCC, Routing Information Service project (RIS), http://www.ripe.net/
6. CAIDA, The Cooperative Association for Internet Data Analysis, http://www.caida.org
7. Colitti, L., Di Battista, G., Mariani, F., Patrignani, M., Pizzonia, M.: Visualizing interdomain routing with BGPlay. Journal of Graph Algorithms and Applications 9, 117–148 (2005)
8. Wong, T., Jacobson, V., Alaettinoglu, C.: Internet routing anomaly detection and visualization. In: Proceedings of International Conference on Dependable Systems and Networks, DSN 2005, pp. 172–181 (2005)
9. Teoh, S.T., Ranjan, S., Nucci, A., Chuah, C.-N.: BGP eye: a new visualization tool for real-time detection and analysis of BGP anomalies. In: Proceedings of the 3rd International Workshop on Visualization for Computer Security (VizSEC 2006), pp. 81–90. ACM, New York (2006)
10. Cortese, P.F., Battista, G.D., Moneta, A., Patrignani, M., Pizzonia, M.: Topographic Visualization of Prefix Propagation in the Internet. IEEE Transactions on Visualization and Computer Graphics 12(5), 725–732 (2006)
11. Press, W.H., Teukolsky, S.A., Vetterling, W.T., Flannery, B.P.: Numerical Recipes in C. Cambridge University Press (1992)
12. Jaromczyk, J.W., Kowaluk, M.: A note on relative neighborhood graphs. In: Proceeding SCG 1987 Proceedings of the Third Annual Symposium on Computational Geometry, pp. 233–241 (1987)

InShape: In-Situ Shape-Based Interactive Multiple-View Exploration of Diffusion MRI Visualizations

Haipeng Cai[1], Jian Chen[2], Alexander P. Auchus[3],
Stephen Correia[4], and David H. Laidlaw[5]

[1] School of Computing, University of Southern Mississippi
hcai@eagles.usm.edu
[2] Computer Science and Electrical Engineering Department,
University of Maryland Baltimore County
jichen@umbc.edu
[3] Department of Neurology, University of Mississippi Medical Center
aauchus@umc.edu
[4] Department of Psychiatry and Human Behavior, Brown University
stephen_correia@brown.edu
[5] Department of Computer Science, Brown University
dhl@cs.brown.edu

Abstract. We present InShape, an in-situ shape-based multiple-view selection interface for interactive exploration of dense tube-based diffusion magnetic resonance imaging (DMRI) visualizations. An optimal experience in such exploration demands concentration on the tract of interest (TOI). InShape facilitates such workflow by leveraging three design principles: (1) shape-enabled precise selection; (2) in-the-flow multi-views for comparison; (3) sculpture-based removal. Results of a pilot study suggested that users have the best interaction experience when the widget shapes match the targeted selection shape. We also found that widget design without losing the flow of operations facilitates focused control. Finally, quick sculpture can help reach the target selection fibers quickly. The contributions of this work are the design principles, together with discussions of usability considerations in interactive exploration in dense 3D DMRI environments.

1 Introduction

The three-dimensional (3D) tractography of diffusion magnetic resonance imaging (DMRI) usually produces a set of integral curves or fiber tracts. When the fibers are constructed from a large DMRI volume, the display can become so cluttered as to impede insights into the data. Tasks such as this require selection and comparative visualizations. Various appoaches exist for selection with DMRI visualizations, from intuitive three-dimensional interaction (e.g., box selection in BrainApp [1]) and pen-based input (e.g., brush-based interface in CINCH [2]) to the use of two-dimensional (2D) embedding [3] and projection [4].

We found at least three difficulties with the current selections. The regular box shape was not precise enough for tube selection when the tubes are dense with high curvature. A second problem is that medical doctors need to reach those tracts of interest (TOI) to

G. Bebis et al. (Eds.): ISVC 2012, Part II, LNCS 7432, pp. 706–715, 2012.

examine the dataset more closely. Sometimes, thoses tracts are in the center of the brain and it would be more intuitive to remove the unnecessary fibers than merely to select the TOIs. Existing 2D methods have addressed this occlusion problem and permited constrained selection. But they suffer from the lack of intuitive domain interpretation. The 2D embedding is often not anatomically meaningful, and thus users must switch between the 3D and 2D views to make selections.

Multiple-view interfaces enable comparative visualizations in medical imaging studies, since investigation of a dense and unfamiliar dataset can be enhanced by comparing it to other better-studied datasets [5]. Our neurologist collaborators also frequently use comparison to study abnormal or normal cases, for example, in comparing brain development of an agenesis patient (ACC) with undeveloped corpus callosum (CC). ACC is a rare developmental anomaly; the patient in this case displayed a complete absence of CC. The doctors postulate that the number of fibers in other areas of the diseased brain (e.g., transverse pontine fibers) may increase to compensate for the loss of CC, and they are interested in knowing where and in what ways the two brain regions differ.

With these points in mind, we have designed InShape, an interactive DMRI visualization tool to allow multiple datasets comparison and precise selection. We call the present tool InShape to stress the in-situ, shape-driven, and shape sculpture design of its DMRI visualization interactions (Fig. 1). In this work, both boxes and spheres are used for selection. We hypothesize that when widget shape matches the target selection shape, more precise selection will be possible. To allow comparison of multiple datasets, we design the multiple-view interface to minimize the cost of switching attention between views. We also hypothesize that concentrating on a specific TOI is crucial in examining a DMRI model. Our third hypothesis is that sculpture-based method will complement to the conventional pure-selection based method. The main contributions

Fig. 1. The InShape interface uses widgets in two shapes: the conventional boxes and our spheres. The sphere shape can improve the user experience by allowing precise selection of regions of high curvature (a and b). In this example, two NOR boxes are colored in blue (c and d). Here selected fibers are within the white-colored boxes (e and f) and spheres (a and b) using boolean logics (a OR b OR e OR f NOR c NOR d). The corpus callosum (yellow) is selected in both views.

of this paper are the above three design principles, the InShape interaction tool for DMRI visualizations, and the pilot study results on the effectiveness of such methods.

2 Related Work

2.1 TOI Section in MRI Visualizations

One class of TOI selection tools uses direct operations in 3D, including those using box-shaped selectors and pen-based inputs. Sherbondy et al. design box widgets to define TOIs on which dynamic queries are constructed to locate specific structures in neural pathways [6]. Blaas et al. propose a geometry-based approach for selection that uses multiple convex objects to yield reproducible bundles [7]. Their multi-box method is efficient in selecting major or regular bundles but less so for bundles consisting of jagged fibers because of the imperfect tractography often seen in practice. Brushes and strokes have also been applied to 3D selection. CINCH allows selection of 3D pathways using pen strokes [2]; in this pen-based interface, TOIs are defined by arbitrary marks drawn by a trackball. Our design facilitates desktop interaction using a mouse as input hopefully to achieve the same performance as those 3D input techniques.

Another class combines 3D and 2D views of a same dataset [4,3] in interactive visualizations. Both Chen et al. [3] and Jianu et al. [4] use a compound interface that integrates a 2D embedding to assist in 3D exploration; here the user can select embedded points on a 2D plane in order to specify the corresponding 3D fiber tracts. Recently, Nowinski et al. present techniques for controlling system states, such as tract labeling, coloring, counting, smoothing, and thresholding [8].

2.2 Multiple-View Visualizations

A common approach to concurrent exploration of multiple datasets is to use multiple views. It is useful to overlay different visual representations in one viewport [9] in order to enhance and clarify visualizations [10]. Multiple views [11] have been used, especially in information visualization, to support simultaneous explorations of multiple models. One powerful approach to coordinating views is brushing-and-linking [12], an exploratory data-analysis technique for displaying a set of data items in multiple views. Items selected by the user in one view are automatically highlighted in other views. For example, the well-known tool Polaris uses brushing and linking with a rubber-band lassoing tool [13].

Keefe et al. [14] employ a multi-view strategy for biomechanical data visualization, making a view of small multiples, a 3D inspection view and parallel coordinates view work together to support a motion sequence data analysis. VisTrail [15] presents the visualization of a set of different instances of time-varying data in multiple views, each view showing one instance. The workflow framework supports multiple visualizations generated from different pipelines as well.

Both 2D and 3D views can be displayed [4,3] for exploratory data analysis. The 2D view can also complement the 3D view by providing contexts for navigation in complex geometries or for interpretation of the actions performed in the 3D view. With a visualization technique called multiple-scale small multiples, Chen et al. use multiple views for dynamic displays of bat-flight kinematics models [16] that give users multiple levels of details and scales of the dataset. In our design, we focus on reducing the cost of switching context between views in the widget interface design.

2.3 Removal-Based Method

Removal-based methods have previously been employed for 3D explorations. Interactively sculpting 3D objects by moving voxel-based tools within a model is used as a modeling technique [17]. This reverse approach has also been applied to interactive volume editing, called volume sculpting [18]. Multiresolution volume sculpting is also made possible in [19]. In this work, we expand the literature to apply the sculpture-based method to DMRI visualizations.

3   Methods

3.1   Shape-Driven Selection

In clinical tasks using DMRI visualizations, neurologists usually want to make accurate investigation of certain fiber bundles. But due to the irregular shapes of DMRI fibers, they often find exact TOI selection very difficult. In a brain DMRI fiber model, some bundles, like those in the stem region, fit a box-shaped selection very well, but efficient and precise selection of others, like those in the genu or splenium part of the corpus callosum, is best done by a sphere-shaped selector due to their curvilinear nature (Fig. 2).

In contrast to the box-shaped selectors in many existing TOI interface, spheres are designed in InShape because their curvilinear shape can make them a better fit in selecting curvilinear fibers than the regular shape of boxes. For example, selecting the CC genu fibers would require at least two selection boxes and the rectangular corners of the boxes would inevitably include undesirable fibers. In comparison, a single properly

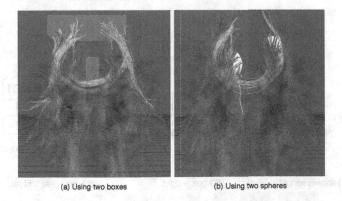

(a) Using two boxes   (b) Using two spheres

Fig. 2. Genu selection. The two spheres in (b) can select the tracts more precisely compared to the boxes in (a).

placed sphere can perform the selection precisely and quickly. In addition, boxes are more visually demanding during interaction because of their many operational faces and thus less desirable to medical doctors than spheres that operate in one dimension only.

Nevertheless, InShape retains box-shaped selectors in order to select approximately regular fibers such as those in the brain-stem regions. Emphatically, InShape features shape-driven TOI selection by which proper selector type can be chosen according to the shape of target TOIs. They can be used together in tasks in which target TOIs include both regular and curvilinear fibers (such as that shown in Fig. 1).

3.2 *In-the-Flow* Multiple-View Visualization

Neurologists often want to examine multiple DMRI models side-by-side so as, for instance, to recognize brain anomalies in a suspect model by comparing it to a normal brain model or to observe disease development over time using images of a same person. In response, multiple views are used in InShape to support multiple comparisons. A known difficulty in multiple-view design is the context switching, necessitated when users look at different portions of the views: this attention shift can derail them from the current workflow. The context-switching problem can be worsened in dense DMRI datasets where a slight modification can cause a complete change in the selection results.

A guiding principle in our design is that all interaction should be performed within the context of the model. One approach we use is to design the hover-over widgets (Fig. 3) for making selections in place. Hovering over a sphere or a box widget activates the hover-over widgets. An entire set of operations can be involved, such as resizing widgets, alternating selector behavior, changing associative logic among selectors, and cloning or deleting the selector (Fig. 3 (c)).

A sphere can be resized through congruent scaling by left-dragging, moved on the viewing plane by right-dragging around the sphere center, and moved along the depth dimension by right-dragging the sphere boundary. Similar work modes are applicable with the box-shaped selectors. A box can be resized by left-dragging any of its front-facing faces or the vertices or edges on them. In addition, congruent scaling by

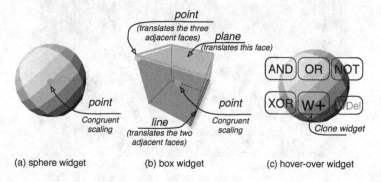

Fig. 3. A set of widgets appears on a box or sphere when the mouse cursor hovers over the box or sphere widget

right-dragging around the center of a front-facing face is also available for quicker re-sizing, while right-dragging the eccentric area of the face is mapped to panning. Users can also move a box along the depth dimension, as is necessary in exploring a 3D visu-alization, by right-dragging an edge of a front-facing face.

As in other multiple-view systems, the actions are synchronized in views. Rotating in one view can induce simultaneous rotation in other views. Selection and sculpturing are also synchronized. When the user operates in any of the views, all interactions with mouse inputs are mirrored in all other views. This synchronized-actions mechanism can be useful in comparing datasets, e.g., a patient's DMRI captured at different times. The multiple-view interface is also useful when combined with sculpturing. In the case where a normal brain is to be compared with one with ACC, for instance, the doctor used InShape to put the two datasets side by side, cull the peripheral fiber bundles using the sculpturing, and then fully engage in the fiber bundles around CC; the doctor was able to confirm an hypothesis using our visualization with the InShape interface.

3.3 Sculpture-Based Exploration

Neurologists would like to quickly sculpt away pathways not relevant to their tasks. They comment that their tasks are often related to a small region of a brain that might be too deep and therefore they would like to sculpt out TOIs from dense tubes so the inside can be seen. Fig. 4 illustrates a task scenario where a neurologist can examine the corticospinal tracts and the corpus callosum areas. He or she starts with a whole-brain DMRI fiber model and uses a sculpting sphere to gradually sculpt away the periph-eral fibers irrelevant to the current task to expose the two fiber boundles for further interaction.

The sculpture widget (a box or a sphere) excludes undesirable fibers that would oth-erwise occlude the view. This sculpture action is continuous when the left-mouse button is hold, resembling an eraser or a sculpting approach in that all fibers touched by the widget are removed. Sculpting can be particularly useful when the selection targets are known and are located in the inner regions. In addition, this action differs from the

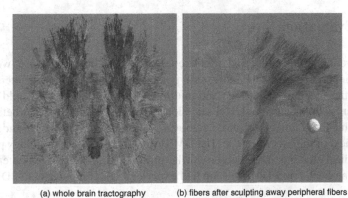

(a) whole brain tractography   (b) fibers after sculpting away peripheral fibers

Fig. 4. Sculpture-based interaction

"NOR" operator in the widget interface in Section 3.4 because of the different operation modality and results: this widget operates alone and can erase currently selected fibers. Users can simply remove those fibers that are blocking their views. Sculpting can be handy in removing large chunks of irrelevant fibers and helping the neurologists reach their tracts of interest.

3.4   User Interface

Our interface also supports associative logics, such as those in [20], CINCH [2], and BrownApp [1]. The sphere or box can work in either the "ADD" or "OR" or "NOR" mode. Table 1 shows InShape selector behaviors in each mode. Fig. 5 demonstrates the results from the multiple-widget interaction.

Table 1. Associative logic among multiple selectors

associative relation	selector behavior
AND	Fibers passing all selectors are selected
OR	Fibers passing one or more selectors are selected
NOR	Fibers passing all selectors are removed

(a) a AND b   (b) a OR b   (a) b NOT a

Fig. 5. Boolean logic operators.

4   User Evaluation

In order to verify the usability and effectiveness of our InShape interface, we invited two participant groups to test the tool. We hypothesized that the sculpting interface empowered by the shape-driven design and multiple synchronized views would enhance users' efficiency and accuracy of TOI selections. Our goal was to check whether users perceived value in the mixed selectors, removal-based design, hover-over widgets, and the intuitiveness of the user interface. The DMRI experts included a radiology professor and two medical students, and the non-DMRI group contained two computer science students. None of the non-DMRI group had experience in computer-based DMRI exploration.

4.1 Tasks

The participants were allowed to explore the data freely. They were also asked to select the white-matter bundles of their personal interest using InShape. A think-aloud protocol was enforced. With the non-DMRI group, it was our intent to collect data on the usability issues of the interface design. The tasks were replicated from an earlier study [3] for selecting a set of fiber bundles: corpus callosum, bilateral cingulate bundles, corticospinal tracts, superior longitudinal fasciculus, and cingulum and uncinate fasciculus. Since this group of participants did not know the DMRI domain, we highlighted these bundles in yellow. Both the single-view and multiple-view (dual-view) conditions were tested, with one view showing a normal dataset and the other view an ACC dataset.

4.2 Dataset

Two streamtube models of human brains were used in the study: a normal brain model in voxel resolution $0.9375mm \times 0.9375mm \times 4.52mm$ and a diseased model (agenesis) in resolution $1.71875mm \times 1.71875mm \times 3.00mm$. Whole brain DMRI tractography was performed using streamline tracing and culling techniques [1]. The normal case had 9,635 fibers and the agenesis case had 8,379 fibers.

4.3 Procedure

The experiment began with training in the concept of DMRI and the functionality of the InShape interface. Participants were asked to practice until they felt comfortable using InShape. The experiment for the three-expert group lasted about 2.5 hours and that for the other group about an hour.

We collected task completion times and measured selection accuracy, defined by bounding tightness (false positive, indicating how many non-TOI fibers are selected) and fiber coverage (false negative, indicating how many TOI fibers that should not be selected are selected). Computer-based log data were collected to record all participant actions such as mouse click, rotation, selection, zoom, and drop-down menu use.

5 Results and Discussion

The sample size in this pilot test was too small for formal quantitative measures, and only anecdotal results are reported here. The non-DMRI participants seem more efficient, perhaps because they were computer experts. The average task completion time was 2.2 minutes for the non-DMRI participants with relatively high selection accuracy, but we must interpret this result carefully in comparison to other tools in experiments with different sampling population, sample size, and test settings. In general, participants found that the interface was useful in examining the white-matter pathways and anatomical structures in the brain, and helpful in making accurate fiber selections.

Participants were inclined to use 1D input with the box widget. Our log data showed that the participants mostly tended to resize the box using 1D face movement ($> 30\%$), perhaps due to its low movement cost. Participants do not have to modify two or more planes simultaneously when a higher-dimensional manipulation is activated.

Participants also reported that the sphere shapes made selection easier and the scene less occluded. DMRI expert participants suggested several other extensions. One participant suggested situating a sphere widget in polar coordinates, so that the selection can be made by placing the center of the sphere within the TOI; coupling this method with the sculpturing could make the initial selection more efficient. Increasing the size of the sphere would enlarge the selection context to show more fibers of interest. In addition, participants reported that while neurosurgeons would examine the whole brain, they themselves would not. This implies that loading a good initial default (such as a subvolume of the brain) could further simplify TOI selection.

The removal mode was reported to be effective and the sculpting approach was judged a good fit to the way neurologists examine brain models in practical tasks. Additionally, the shape-driven design was considered useful because using boxes for TOIs of regular shape and spheres for those of curvilinear shape enhances selection accuracy. Participants also appreciated the in-the-flow interaction since it let them focus fully on their TOIs. While the multiple-view synchronization helps comparative study of multiple brain models, as reported, matching anatomical regions among brains (rather than merely calibrating on the geometry level, as currently) would be more helpful. As expected, the participants reported that they did not need to look away from the models with the hover-over widget.

6 Conclusion

We have presented InShape, an in-situ interface for interactive exploration of dense DMRI tubes. InShape extends many well-known techniques (box-based selection, brushing- and-linking, multiple views, and widget-based interfaces) to allow precise selection with spheres and sculpting away fibers within the model context. It also presents a prototype of shape-widgets, sculpting-based interaction, and multiple synchronized views for more precise selection. We have reported results from a pilot study. A formal study with more brain experts is planned to compare our interface with the 2D separated views in order to learn if our interface helps reduce the cognitive barriers between examining in 3D while selecting in 2D. We are also interested in comparing the shape style of selection and removal against more common methods like brushing and box selection in speed, comprehension, and user experience in general.

Acknowledgment. The authors wish to thank the participants for their time and effort and the anonymous reviewers for their helpful remarks. This work was supported in part by NSF IIS-1018769, IIS-1016623, IIS-1017921, EPS-0903234, DBI-1062057, CCF-1785542, and OCI-0923393 and NIH (RO1-EB004155-01A1). The authors would also like to thank Drs. Juebin Huang and Judy James at University of Mississippi Medical Center for discussion of the tasks and data analysis and Katrina Avery for her editorial support.

References

1. Zhang, S., Demiralp, C., Laidlaw, D.H.: Visualizing diffusion tensor MR images using streamtubes and streamsurfaces. IEEE Transactions on Visualization and Computer Graphics 9, 454–462 (2003)
2. Akers, D.: CINCH: a cooperatively designed marking interface for 3d pathway selection. In: Proceedings of the 19th Annual ACM Symposium on User Interface Software and Technology (UIST), pp. 33–42 (2006)
3. Chen, W., Ding, Z., Zhang, S., MacKay-Brandt, A., Correia, S., Qu, H., Crow, J., Tate, D., Yan, Z., Peng, Q.: A novel interface for interactive exploration of DTI fibers. IEEE Transactions on Visualization and Computer Graphics, 1433–1440 (2009)
4. Jianu, R., Demiralp, C., Laidlaw, D.H.: Exploring 3D DTI fiber-tracts with linked 2D representations. IEEE Transactions on Visualization and Computer Graphics 15, 1449–1456 (2009)
5. Diepenbrock, S., Prassni, J., Lindemann, F., Bothe, H., Ropinski, T.: 2010 IEEE Visualization Contest Winner: Interactive Planning for Brain Tumor Resections. IEEE Computer Graphics and Applications 31, 6–13 (2010)
6. Sherbondy, A., Akers, D., Mackenzie, R., Dougherty, R., Wandell, B.: Exploring connectivity of the brain's white matter with dynamic queries. IEEE Transactions on Visualization and Computer Graphics 11, 419–430 (2005)
7. Blaas, J., Botha, C.P., Peters, B., Vos, F.M., Post, F.H.: Fast and reproducible fiber bundle selection in DTI visualization. IEEE Visualization, 59–64 (2005)
8. Nowinski, W., Chua, B., Yang, G., Qian, G.: Three-dimensional interactive and stereotactic human brain atlas of white matter tracts. Neuroinformatics, 1–23 (2011)
9. Roberts, J.: On encouraging multiple views for visualization. In: Proceedings of Information Visualization, pp. 8–14 (1998)
10. Tufte, E., Goeler, N., Benson, R.: Envisioning information, vol. 21. Graphics Press (1990)
11. North, C., Shneiderman, B.: A taxonomy of multiple window coordinations. Technical report, Human-Computer Interaction Laboratory, University of Maryland (1997)
12. Shneiderman, B.: The eyes have it: A task by data type taxonomy for information visualizations. In: Proceedings of IEEE Symposium on Visual Languages, pp. 336–343 (1996)
13. Stolte, C., Hanrahan, P.: Polaris: A system for query, analysis and visualization of multi-dimensional relational databases. IEEE Transactions on Visualization and Computer Graphics 8, 52–65 (2002)
14. Keefe, D., Ewert, M., Ribarsky, W., Chang, R.: Interactive coordinated multiple-view visualization of biomechanical motion data. IEEE Transactions on Visualization and Computer Graphics, 1383–1390 (2009)
15. Bavoil, L., Callahan, S.P., Crossno, P.J., Freire, J., Vo, H.T.: VisTrails: Enabling interactive multiple-view visualizations. IEEE Visualization, 135–142 (2005)
16. Chen, J., Forsberg, A., Swartz, S., Laidlaw, D.: Interactive multiple scale small multiples. IEEE Visualization 2007 Poster Compendium (2007)
17. Wang, S., Kaufman, A.: Volume sculpting. In: Proceedings of the Symposium on Interactive 3D Graphics, pp. 151–156 (1995)
18. Baerentzen, J.: Octree-based volume sculpting. IEEE Visualization (late breaking hot topics) 98, 9–12 (1998)
19. Ferley, E., Cani, M., Gascuel, J.: Resolution adaptive volume sculpting. Graphical Models 63, 459–478 (2001)
20. Wakana, S., Jiang, H., Nagae-Poetscher, L.M., van Zijl, P.C.M., Mori, S.: Fiber tract-based atlas of human white matter anatomy. Radiology 230, 77–87 (2004)

Surface Construction with Fewer Patches

Weitao Li[1], Yuanfeng Zhou[1], Li Zhong[1], Xuemei Li[1] and Caiming Zhang[1,2]

[1] School of Computer Science and Technology, Shandong University, Jinan, China
[2] Shandong Province Key Lab of Digital Media Technology,
Shandong University of Finance and Economics, Jinan, China

Abstract. We present an algorithm to generate an interpolation or approximation model consisting of many patches from a triangle mesh, and each patch is a weighted combination of the three surfaces associated with the vertices of a triangle. Moreover, to make the whole surface include fewer patches, mesh simplification is introduced into the process of surface construction. The algorithm takes a triangle mesh and a given error as input, and iteratively deletes vertex whose distance to the surface model constructed from the simplified mesh is less than or equal to the given error until convergence. Since the method is based on surface approximation and vertex deletion, it allows us to control the error between the generated model and the original mesh precisely. Furthermore, many experimental results show that the generated models approximate the original models well.

1 Introduction

In computer graphics, objects are often represented by triangle meshes. As the size of geometric data set used in graphic systems grows very rapidly, automatic simplification techniques play an important role in data management. On the one hand, complex models which consume more rendering time and storage space maintain a convincing level of realism. On the other hand, a complex model can be approximated by a simply model under a given error. Therefore, in practical applications, complex models are usually replaced by simple models consisting of fewer patches.

In this paper, we discuss the problem of surface construction with fewer patches. Furthermore, a given error is regarded as the decimation criterion to decide whether a vertex should be deleted. And our approach can ensure that all distances from deleted vertices to the model constructed from the simplified mesh are less than or equal to the given threshold.

1.1 Related Works

There are two research areas related to our work closely: surface construction and mesh simplification.

In the field of surface construction: [1] presented an efficient algorithm to construct low-degree algebraic surfaces from any given collection of points having

G. Bebis et al. (Eds.): ISVC 2012, Part II, LNCS 7432, pp. 716–725, 2012.

derivative information. For fitting implicit algebraic surface of low degree, [2] described a new computational model based on the method mentioned in [1]. [3] employed implicit surfaces to approximate the given data sets. And [4] proposed a variational method for extracting general quadric surfaces from a 3D mesh.

Globally supported radial splines were employed to approximate a set of points or polygonal data in [5,6,7,8,9]. Generally, the resulting surfaces matched the input data roughly. However, some details of the original models were smoothed away. [10,11,12] used locally supported function for fitting an implicit surface to point clouds. But some spurious surface sheets away from samples might be generated. With a moving least-squares formulation with constraints over the polygons, [13] generated a surface that exactly interpolated the polygons, or approximated the input by smoothing away smaller features. But the computational cost was very high.

According to the research that about 90% parts of manufactured models can be represented by specific quadric surface - plane, cylinder and parabolic, [14,15,16,17,18,19,20] proposed many reconstruction methods with those surfaces.

In the field of mesh simplification: [21] proposed a decimation algorithm which used the distance from a vertex to the plane which was fitted from its 1-ring neighborhood vertices as a criterion to decide whether the vertex could be deleted or not. But the chief shortcoming of this algorithm was that it could not provide a global error between the original model and the simplified mesh. A mesh simplification approach ,which was regarded as one of the best simplification algorithms, was presented in [22]. In this algorithm, the quadric error metric was used to choose the edge which should be simplified and to compute the position of a vertex after contraction.

2   Surface Construction with Fewer Patches

In this section, we first give an overview of the surface construction with fewer patches (SCFP) algorithm, and then we elaborate on each step in detail.

2.1   SCFP Algorithm Overview

Let $\mathbf{M} = (\mathbf{V}, \mathbf{E}, \mathbf{F})$ be a triangle mesh equipped with vertex normal. To precisely control the global error between the deleted vertices and the surface model, we assume that each vertex can be deleted, and the distance from the deleted vertex to the surface model is less than or equal to the global error. The SCFP algorithm starts by deleting a vertex and triangulating its 1-ring neighborhood vertices. It then constructs a local surface from the region whose topological relations have changed, and determines whether the distance from the local surface to the current deleted vertex meets the criteria. It is obvious that the local surface construction method and vertex decimation approach are the most important parts of our algorithm, and we discuss them in the following parts.

2.2   Theoretical Foundation of Patch Construction

The local surface model generated from a local region $Nr = \{T_i\}, T_i \in \mathbf{F}$ is composed of several patches, and each patch is constructed from a triangle by a weighted combination of the three surfaces generated at its vertices.

For each triangle $T_i = \{v_i, v_j, v_k\} \in N_r$, we construct three implicit surfaces $f_i(\mathbf{x})$, $f_j(\mathbf{x})$, and $f_k(\mathbf{x})$ for its vertices, and transform the implicit surfaces into their parametric representations $P_i(s,t)$, $P_j(s,t)$, and $P_k(s,t)$ by a simple projection. Then the weighted patch which approximates or interpolates the vertices of the triangle is defined by

$$P_t(s,t) = \omega_i(s,t)P_i(s,t) + \omega_j(s,t)P_j(s,t) + \omega_k(s,t)P_k(s,t)$$
$$\omega_i(s,t) = \{10u - 15s^2 + 6s^3 + 30t(1-s-t)(Q_k(1-s-t)/D_{ik} + Q_jt/D_{ij})\}s^2$$
$$\omega_j(s,t) = \{10t - 15t^2 + 6t^3 + 30(1-s-t)s(Q_is/D_{ij} + Q_k(1-s-t)/D_{kj})\}t^2$$
$$\omega_k(s,t) = \{10(1-s-t) - 15(1-s-t)^2 + 6(1-s-t)^3 + 30st(Q_jt/D_{jk}$$
$$+Q_is/D_{ik})\}(1-s-t)^2$$

$$(1)$$

where

$$Q_i = \overrightarrow{v_iv_j} \cdot \overrightarrow{v_iv_k} \qquad D_{ij} = \| \overrightarrow{v_iv_j} \|_2^2$$
$$Q_j = \overrightarrow{v_jv_i} \cdot \overrightarrow{v_jv_k} \qquad D_{ik} = \| \overrightarrow{v_iv_k} \|_2^2$$
$$Q_k = \overrightarrow{v_kv_i} \cdot \overrightarrow{v_kv_j} \qquad D_{jk} = \| \overrightarrow{v_jv_k} \|_2^2$$

$\omega_i(s,t), \omega_j(s,t)$ and $\omega_k(s,t)$ are proposed in [23] and denote weight functions. Unlike [23], which parameterizes all points to a plane, we generate the local surface by putting all patches together without a global mesh parameterization. More details can be found in [23].

2.3   Implicit Quadric Surface Construction

Based on our theoretic analysis of the patch construction, we arrive at the following conclusion: for a triangle, the shapes of the implicit surfaces constructed at its vertices decide the shape of the local patch which is a weighted combination of those surfaces. Since implicit quadric surface is very natural for least-square approximation to data set without higher-order derivative information, we employ it to fit the adjacent region of a vertex.

The implicit surface fitting problem has been researched for about twenty years. However, those methods are unstable. To solve this problem, [24] proposed a 3L algorithm for fitting arbitrary degree implicit surfaces. [25] presented a Gradient One algorithm which added the gradient constraint, and [26] extended it to fit an implicit curve robustly. Specially, [27,28] proposed algebraic functions to measure the distance between two polygons. [29] used PIEM to compute the algebraic distance between an implicit surface and a polygon.

When an implicit surface is fitting to a region consisting of some triangles, we find that PIEM is robust. Thus, it is employed to construct an implicit surface which approximates a vertex's 1-ring adjacent triangles without sharp-features or corners. However, sharp-features and corners are very important for

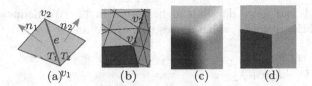

Fig. 1. Sharp-feature edge and corner

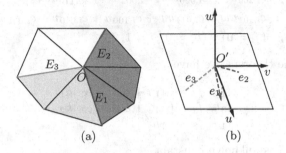

Fig. 2. Piecewise surface fitting

machine parts; using PIEM to deal with those regions, such characteristics will be smoothed away (Figure 1(c)). Therefore, we use the algorithm proposed in [30] to detect the characteristics, and employ a piecewise surface consisting of $n + 1$ implicit quadric surfaces to fit a vertex's 1-ring adjacent triangles having n sharp-feature edges. The resulting model preserves characteristics better (Figure 1(d)).

Take into vertex O and its 1-ring adjacent region shown in Figure 2(a) account, we assume that E_1 and E_2 are sharp-feature edges. To preserve the characteristics, we divide this region into three parts R_1(yellow), R_2(lime), and R_3(white) by an auxiliary edge E_3 and the sharp-feature edges, then construct three implicit quadric surfaces f_1, f_2, and f_3 from those parts. Meanwhile, f_1 meets f_2 with C^0 continuity along C_1. Similarly, f_2 meets f_3 with C^0 continuity along C_2. As E_3 is not a characteristic edge, f_1 meets f_3 with C^1 continuity along C_3–C_1, C_2, C_3 are the lines of intersection of those adjacent surfaces.

Let us introduce a local coordinates system (u, v, w) with the origin at O such that u is the unit projection vector of E_1 on O's tangent plane, and the positive direction of w is coincide with the direction of the normal $\mathbf{n}$ of O, moreover, $v = w \times u$. As illustrated in Figure 2(b), E_1, E_2, and E_3 are represented by $e_1(e_{1,u}, 0, e_{1,w})$, $e_2(\lambda_2, \lambda_1, e_{2,w})$, and $e_3(\lambda_4, \lambda_3, e_{3,w})$ in the local coordinates system.

Theorem 1. *Let g and h be distinct, irreducible polynomials. If the surface $g = 0$ and $h = 0$ intersect transversally in a single irreducible curve C, then any algebraic surface $f = 0$ that meets $g = 0$ with C_k re-scaling continuity along C must be the form $f = \alpha g + \beta h^{k+1}$. If $g = 0$ and $h = 0$ share no common components at infinity, then the degree of $\alpha g \leq$ degree of f and the degree of $\beta h^{k+1} \leq$ degree of f.*

The theorem 1 and more details can be found in [2]. We define the implicit surface f_i and the plane h_i which passes through w and e_i by

$$
\begin{aligned}
h_1 &= v \\
h_2 &= \lambda_1 u + \lambda_2 v \\
h_3 &= \lambda_3 u + \lambda_4 v \\
f_i &= c_{i,200} u^2 + c_{i,020} v^2 + c_{i,002} w^2 + c_{i,110} uv + \\
& \quad c_{i,101} uw + c_{i,011} vw + c_{i,100} u + c_{i,010} v + c_{i,001} w \\
C_i &= (f_i = 0) \cap (h_i = 0), i = 1, 2, 3
\end{aligned}
\tag{2}
$$

According to the theorem, we have

$$
\begin{aligned}
f_1 - f_2 &= (\alpha_1 u + \alpha_2 v + \alpha_3) h_1 \\
f_3 - f_2 &= (\alpha_4 u + \alpha_5 v + \alpha_6) h_2 \\
f_1 - f_3 &= \alpha_7 h_3^2
\end{aligned}
\tag{3}
$$

where $\alpha_i, i \in 1..7$ is unknown constant.

Solving the equation 3, then the quadratic surface can be written as

$$
\begin{aligned}
f_1 =\ & c_{1,200} u^2 + c_{1,020} v^2 + c_{1,002} w^2 + c_{1,110} uv \\
& + c_{1,101} uw + c_{1,011} vw + c_{1,100} u + c_{1,010} v + c_{1,001} w \\
f_2 =\ & c_{1,200} u^2 + (c_{1,020} - \alpha_5 \lambda_2 + \frac{\alpha_4 \lambda_1 \lambda_4^2}{\lambda_3^2}) v^2 \\
& + c_{1,002} w^2 + (c_{1,110} - \alpha_5 \lambda_1 - \alpha_4 \lambda_2 + \frac{2\alpha_4 \lambda_1 \lambda_4}{\lambda_3}) uv \\
& + c_{1,101} uw + c_{1,011} vw + c_{1,100} u + c_{1,010} v + c_{1,001} w \\
f_3 =\ & (c_{1,200} + \alpha_4 \lambda_1) u^2 + (c_{1,020} + \frac{\alpha_4 \lambda_1 \lambda_4^2}{\lambda_3^2}) v^2 \\
& + c_{1,002} w^2 + (c_{1,110} + \frac{2\alpha_4 \lambda_1 \lambda_4}{\lambda_3}) uv \\
& + c_{1,101} uw + c_{1,011} vw + c_{1,100} u + c_{1,010} v + c_{1,001} w
\end{aligned}
\tag{4}
$$

The unknown vector $[c_{1,200}, c_{1,020}, c_{1,002}, c_{1,110}, c_{1,101}, c_{1,011}, c_{1,100}, c_{1,010}, c_{1,001}, \alpha_4, \alpha_5]$ can be determined by PIEM.

3   Fewer Patches

We generate a patch from each triangle, and put them together to form a surface which approximates the triangle mesh. In this section, we describe how to reduce the number of patches used to form the surface under a given precision in detail.

As we mentioned before, our method ensures that all distances from deleted vertices to the surface constructed from the simplified mesh are less than or equal to the given threshold. To precisely control the global decimation criterion, we have to check whether all distances from deleted vertices to the surface still match criterion when we delete a vertex. Furthermore, our surface is composed of many patches. If we assign the deleted vertices to the residual triangles, we can check the criterion efficiently. Hence, a vertex list $\mathcal{L}_t$ is attached to each residual triangle, and it stores the deleted vertices located in the triangle. Then for each vertex v_i, we assume it can be deleted, and transform its 1-ring adjacent vertices $\{v_i^l(x, y, z)\}, l = 1..d$ to a local coordinates system whose origin and Z direction are v_i and its normal, respectively. After that, they are triangulated by Constrained Delaunay Triangulation, and a local mesh $M = [V, E, F]$ is generated. When we delete v_i and triangulate its 1-ring neighborhood, only the region adjacent to v_i is influenced. Therefore, if we guarantee that all distances between the surface model generated from $M = [V, E, F]$ and deleted vertices near $v_i \leq \varepsilon$, then the distances from all deleted vertices to the surface model $\leq \varepsilon$.

Algorithm 1 Fewer Patches($M = [V, E, F], \varepsilon$)

1: **for** each $t \in F$ **do**
2:    $\mathcal{L}_t \Leftarrow \Phi$
3: **end for**
4: **for** each $v_i \in V$ **do**
5:    Triangle $\{v_i^l(x, y, z)\}, l = 1..d$, Generate $M = [\mathbb{V}, \mathbb{E}, \mathbb{F}]$
6:    $\mathbb{L} \Leftarrow \bigcup_{t \in L_{v_i}} \mathcal{L}_t$
7:    **for** each $\mathcal{F} \in \mathbb{F}$ **do**
8:        $\mathcal{L}_{\mathcal{F}} \Leftarrow \Phi$
9:    **end for**
10:    **for** each $\mathfrak{L} \in \mathbb{L}$ **do**
11:        $\mathcal{L}_{LOCATION(\mathfrak{L})} \Leftarrow \mathfrak{L}$
12:    **end for**
13:    **for** each $\mathcal{F} \in \mathbb{F}$ **do**
14:        Construct $P_{\mathcal{F}}$
15:        **for** each $\ell \in \mathcal{L}_{\mathcal{F}}$ **do**
16:            **if** $distance(\ell, P_{\mathcal{F}}(\ell)) > \varepsilon$ **then**
17:                GOTO 4
18:            **end if**
19:        **end for**
20:    **end for**
21:    Delete v_i, Update $M = [V, E, F]$ and its $\mathcal{L}$
22: **end for**

In order to efficiently determine whether all distances meet the criterion, we use the deleted vertices stored in the triangle set L_{v_i} consisting of v_i's 1 and 2-ring neighborhood to form a new vertex list $\mathbb{L}$. Then for each vertex in $\mathbb{L}$, we find out which triangle of $M = [\mathbb{V}, \mathbb{E}, \mathbb{F}]$ contains it and put it into the vertex list $\mathcal{L}_t$ of the triangle. Therefore, if all distances from all deleted vertices belong to the local mesh meet criterion, v_i can be deleted. As a result, we propose a new

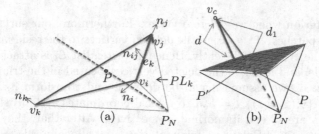

Fig. 3. Estimate the shortest distance(the weighted patch is drawn in wireframe)

method which can provide a global error metric, and distances from all deleted vertices to the whole surface constructed on the simplified mesh are less than or equal to the given threshold. Our method can be summarized by algorithm 1.

To use our algorithm, the distances from all deleted vertices to the weighted surface should be calculated first. Since the shortest distance d from a vertex v_c to the weighted surface is so difficult to calculate exactly, we propose a new approximate calculation method.

As shown in Figure 3, assume the projection point of v_c is in triangle $T = \{v_l\}, l = i, j, k$. Let $\{P_l(s,t)\}, l = i, j, k$ denote parametric surface constructed associated with each vertex of T. n_l , $e_l, l = i, j, k$ are v_l's normal and its opposite edges, respectively. As illustrated in Figure 3(a), plane PL_k is formed by e_k and n_{ij}, where $n_{ij} = 0.5(n_i + n_j)$. Similarly, plane PL_i, PL_j can be formed too. P_N is the intersection point of the three planes, and the intersection point of T and the line produced by connecting P_N and v_c is denoted by $P(s_{vc}, t_{vc})$, where (s_{vc}, t_{vc}) is its barycentric coordinates in triangle T. According to the approach described above, the corresponding point of P on the weighted surface is calculated by

$$
\begin{aligned}
P' =& \omega_i(s_{vc}, t_{vc})P_i(s_{vc}, t_{vc}) + \omega_j(s_{vc}, t_{vc})P_j(s_{vc}, t_{vc}) \\
&+ \omega_k(s_{vc}, t_{vc})P_k(s_{vc}, t_{vc})
\end{aligned}
\tag{5}
$$

As illustrated in Figure 3(b), it is manifest that $d \leq d_1 = \| v_c - P' \|_2$. If $d_1 \leq \varepsilon$, $d \leq \varepsilon$.

4   Experimental Results

Figure 4 shows different multi-resolution control meshes and their corresponding surfaces obtained by different thresholds. For the same model, the larger the threshold is, the more vertices are deleted, the simpler the simplified mesh is, and the fewer patches are contained in the weighted surface. The less the threshold is, the fewer vertices are deleted, and the more patches are used to form the weighted surface. The meshes and their corresponding surface models are depicted respectively in odd and even lines of Figure 4. The first and

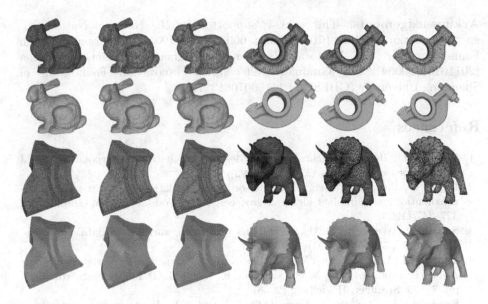

Fig. 4. Multi-resolution models we get using different precision

fourth columns demonstrate the original mesh and their corresponding weighted surface models. When $\varepsilon = 1.0e - 4$, about 58%, 60%, 26%, 37% triangles are deleted from bunny, fan-disk, rock-arm and triceratops, separately. The simplified meshes and the weighted surfaces constructed on them are shown in the second and fifth columns. Meanwhile, the third and sixth columns show the simplified meshes and the weighted surfaces obtained by deleting 78%, 77%, 75%, 78% triangles with $\varepsilon = 1.0e - 3$. Take the case of fan-disk into account, when $\varepsilon = 1.0e - 3$, it only needs 2966 patches to form the whole weighted surface which approximates all deleted vertices under the given threshold and interpolates all residual vertices, furthermore, the sharp-feature and corner characteristics are preserved. According to the analysis, it is manifest that the smooth surface model constructed from simplified mesh approximates the original model with a certain precision.

5   Conclusions and Future Works

We present a new algorithm to construct an approximation model from a mesh with fewer patches, develop a vertex deletion algorithm and propose a distance approximation formula to accelerate the mesh simplification process. The weighted surface model constructed from the mesh approximates the original one and all distances from deleted vertices to the weighted surfaces are less than or equal to the given threshold.

Acknowledgements. This work is supported by the National Nature Science Foundation of China (61020106001, 60933008,61103150). National Research Foundation for the Doctoral Program of Higher Education of China (20110131130004) and Graduate Independent Innovation Foundation of Shandong University (GIIFSDU) 11150070613227.

References

1. Bajaj, C.L., Ihm, I.: Algebraic surface design with hermite interpolation. ACM Transactions on Graphics 11, 61–69 (1992)
2. Bajaj, C., Ihm, I., Warren, J.: Higher-order interpolation and least-squares approximation using implicit algebraic surfaces. ACM Transactions on Graphics 12, 327–347 (1993)
3. Subaihi, I., Watson, G.: Algebraic ftting of quadric surfaces to data. Commun. Appl. Anal., 539–548 (2005)
4. Yan, D.-M., Liu, Y., Wang, W.: Quadric Surface Extraction by Variational Shape Approximation. In: Kim, M.-S., Shimada, K. (eds.) GMP 2006. LNCS, vol. 4077, pp. 73–86. Springer, Heidelberg (2006)
5. Savchenko, V., Pasko, E.A., Okunev, O.G., Kunii, T.L.: Function representation of solids reconstructed from scattered surface points and contours. Computer Graphics Forum 12, 181–188 (1995)
6. Turk, G., Brien, J.F.: Shape transformation using variational implicit functions. In: Proceedings of ACM SIGGRAPH, pp. 335–342 (1999)
7. Carr, J.C., Beatson, R.K., Cherrie, J.B., Mitchell, T.J., Fright, W.R., McCallum, B.C., Evans, T.R.: Reconstruction and representation of 3d objects with radial basis functions. In: Proceedings of ACM SIGGRAPH, pp. 67–76 (2001)
8. Turk, G., Brien, J.F.O.: Modelling with implicit surfaces that interpolate. ACM Transactions on Graphics 21, 855–873 (2002)
9. Yngve, G., Turk, G.: Robust creation of implicit surfaces from polygonal meshes. IEEE Transactions on Visualization and Computer Graphics 8, 346–359 (2002)
10. Muraki, S.: Volumetric shape description of range data using blobby model. In: Proceedings of ACM SIGGRAPH, pp. 67–76 (1991)
11. Morse, B., Yoo, T.S., Rheingans, P., Chen, D.T., Subramanian, K.: Interpolating implicit surfaces from scattered surface data using compactly supported radial basis functions. In: Proceedings of Shape Modelling International, pp. 89–98 (2001)
12. Ohtake, Y., Belyaev, A., Alexa, M., Turk, G., Seidel, H.P.: Multi-level partition of unity implicits. ACM Transactionson Graphics 22, 463–470 (2003)
13. Shen, C., Brien, J.F., Shewchuk, J.R.: Interpolating and approximating implicit surfaces from polygon soup. ACM Transactions on Graphics 23 (2004)
14. Beardsley, P.: Pose estimation of the human head by modelling with an ellipsoid. In: Proc. of IEEE Conf. on Automatic Face Gesture Recognition, Nara (1998)
15. Fitzgibbon, A., Pilu, M., Fisher, R.B.: Direct least square ftting of ellipses. IEEE Transaction on Pattern Analysis and Machine Intelligence 21, 476–480 (1999)
16. Li, Q., Griffiths, J.: Least squares ellipsoid specific fitting. In: Proc. of the IEEE CS GMP, pp. 335–340 (April 2004)
17. Allaire, S., Jacq, J.J., Burdin, V., Couture, C.: Type-constrained robust fitting of quadrics with application to the 3d morphological characterization of saddle-shaped articular surfaces. In: IEEE 11th International Conference on Computer Vision, ICCV 2007, pp. 1–8 (2007)

18. Dai, M., Newman, T.S., Cao, C.: Least-squares-based fitting of paraboloids. Pattern Recogniyion 40, 504–515 (2007)
19. Benk, P., Ks, G., Vrady, T., Andor, L., Marin, R.: Constrained fitting in reverse engineering. Computer Aided Geometric Design 19, 173–205 (2002)
20. Vanco, M., Hamann, B., Brunnett, G.: Surface reconstruction from unorganized point data with quadrics. Computer Graphics Forum 27, 1593–1606 (2008)
21. Schroeder, W.J., Zarge, J.A., Lorensen, W.E.: Decimation of triangle meshes. Computers Graphics 25, 175–184 (1991)
22. Garland, M., Heckbert, P.: Surface simplification using quadric error metrics. In: Proceedings of SIGGRAPH 1997, ACM Press ACM SIGGRAPH, Computer Graphics Proceedings. Annual Conference Series, pp. 209–216. ACM (1997)
23. Gao, S.-S., Zhang, C.-M., Zhong, L.: Interpolation by Piecewise Quadric Polynomial to Scattered Data Points. In: Bebis, G., Boyle, R., Parvin, B., Koracin, D., Remagnino, P., Nefian, A., Meenakshisundaram, G., Pascucci, V., Zara, J., Molineros, J., Theisel, H., Malzbender, T. (eds.) ISVC 2006. LNCS, vol. 4292, pp. 106–115. Springer, Heidelberg (2006)
24. Blane, M.M., Lei, Z., Civi, H., Cooper, D.B.: The 3l algorithm for fitting implicit polynomial curves and surfaces to data. IEEE Transactions on Pattern Analysis and Machine Intelligence 22, 298–313 (2000)
25. Tasdizen, T., Tarel, J., Cooper, D.: Improving the stability of algebraic curves for applications. IEEE Transactions on Image Processing 9, 405–416 (2000)
26. Helzer, A., Barzohar, M., Malah, D.: Stable fitting of 2d curves and 3d surfaces by implicit polynomials. IEEETransactions on Pattern Analysis and Machine Intelligence 26, 1283–1294 (2004)
27. Garland, M., Willmott, A., Heckbert, P.S.: Hierarchical face clustering on polygonal surfaces. In: Proc. ACM Symposium on Interactive 3D Graphics 2001, pp. 49–58. ACM Press, New York (2001)
28. Steiner, D.C., Alliez, P., Desbrun, M.: Variational shape approximation. ACM Transactions on Graphics 23, 905–914 (2004)
29. Takashi, K., Yutaka, O., Kiwamu, K.: Hierarchical error-driven approximation of implicit surfaces from polygonal meshes. In: Proceedings of the Symposium on Geometry Processing, vol. 256, pp. 21–30 (2006)
30. Cazals, F., Pouget, M.: Topology driven algorithms for ridge extraction on meshes. Technical Report RR-5526, INRIA (2005)
31. Baldonado, M., Chang, C.C., Gravano, L., Paepcke, A.: The stanford digital library metadata architecture. Int. J. Digit. Libr. 1, 108–121 (1997)

Interactive Control of Mesh Topology in Quadrilateral Mesh Generation Based on 2D Tensor Fields

Chongke Bi[1], Daisuke Sakurai[2], Shigeo Takahashi[2], and Kenji Ono[1]

[1] RIKEN Advanced Institute for Computational Science, Japan
[2] Graduate School of Frontier Sciences, The University of Tokyo, Japan

Abstract. Generating quadrilateral meshes is very important in many industrial applications such as finite element analysis and B-spline surface fitting. However, it is still a challenging task to design appropriate vertex connectivity in the quadrilateral meshes by respecting the shapes of the target object and its boundary. This paper presents an approach for interactively editing such mesh topology in quadrilateral meshes by introducing a 2D diffusion tensor field to the interior of the target object. The primary idea is to track the two principal directions of the tensor field first and then construct the dual graph of the quadrilateral mesh, so that we can control the mesh topology through the design of the underlying 2D diffusion tensor field. Our method provides interactive control of such mesh topology through editing the orientations of the tensor samples on the boundary of the target object. Furthermore, it also allows us to intentionally embed degeneracy inside the object to introduce extraordinary (i.e., non-degree-four) vertices according to user requirements.

1 Introduction

In industrial applications, such as finite element analysis (FEA), quadrilateral meshes have become more preferable to triangular meshes since they give better and more precise simulation results. However, existing methods do not provide an effective means of controlling the mesh topology that fully satisfy the requirements of industrial applications, in that they are more likely to produce unnecessary extraordinary vertices inside the target objects. Such extraordinary vertices tend to distort the shape of quadrilaterals contained in the mesh. On the other hand, several methods tried to generate quadrilaterals from the object boundary, while this inevitably incurs non-quadrilaterals such as triangles and pentagons around the center of the object.

In order to appropriately control the *mesh topology* in quadrilateral meshes, we have to retain the following two conditions: (1) The shape of the target object and its boundary should be taken into account when generating quadrilateral meshes. (2) The extraordinary vertices should be introduced only when non-quadrilaterals were generated. The first condition suggests that the generated quadrilateral mesh elements should be properly aligned with the boundary of the target object. The second condition requires us to minimize the number of extraordinary vertices generated in the quadrilateral mesh. However, existing methods can achieve either of them at most.

In this paper, we present an approach for interactively controlling such mesh topology of quadrilateral meshes in order to fulfill the requirements in industrial applications

G. Bebis et al. (Eds.): ISVC 2012, Part II, LNCS 7432, pp. 726–735, 2012.

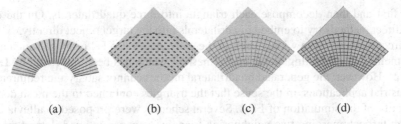

(a)   (b)   (c)   (d)

Fig. 1. Generating a quadrilateral mesh in the interior of a fan-shaped object. (a) Input fan-shaped object. (b) The generated 2D diffusion tensor field by referring to the shape of the object and its boundary, where each tensor sample is represented as an ellipse. (c) The streamlines along the two principal directions of the diffusion tensor field. (d) The quadrilateral mesh generated from its dual graph represented by the streamlines.

described above. This is accomplished by introducing a 2D diffusion tensor field in the target object, which is inspired by the dual relationship between a quadrilateral mesh and a 2D diffusion tensor field. Firstly, a 2D diffusion tensor field is generated in the interior of the target object by employing the Poisson equation. This way of introducing diffusion tensor field can fully respect the shape of the entire target object together with its boundary. We then transform the configuration of the diffusion tensor field to the mesh topology of the quadrilateral meshes. This is accomplished by tracking the streamlines along the two principal directions of the 2D diffusion tensor field to construct the dual graph of the quadrilateral meshes. Furthermore, our method also provides interactive control of the mesh topology through editing the orientations of the tensor samples on its boundary. We also equip our system with an interface for intentionally embedding degenerate tensor samples to control the positions of extraordinary vertices according to user requirements.

Figure 1 shows an example where a quadrilateral mesh of a fan-like object is generated using our approach. A 2D diffusion tensor field has been introduced by referring to the shape of the fan (Figure 1(b)). This is followed by tracking the streamlines along the two principal directions of the tensor field (Figure 1(c)). The final quadrilateral mesh is obtained by transforming the configuration of the streamlines to its dual, and thus maximally respects the shape of the fan and its boundary (Figure 1(d)).

The remainder of this paper is organized as follows: Section 2 surveys previous work related to ours. Section 3 explains the dual relationship between a 2D diffusion tensor field and a quadrilateral mesh. Section 4 describes how we introduced a 2D diffusion tensor field into the interior of the target object, which is followed by a scheme for constructing the dual graph of quadrilateral meshes through tracking the streamlines of the 2D diffusion tensor field in Section 5. Section 6 presents several experimental results to demonstrate the effectiveness of the proposed scheme. Section 7 concludes this paper and refers to possible future extensions.

2 Related Work

Existing schemes for quadrilateral mesh generation can be roughly classified into *indirect* and *direct* methods. Indirect methods generate triangular meshes in the target object first and then decompose each triangle into three quadrilaterals. On the other hand, direct methods try to embed quadrilaterals into the target object directly.

Indirect methods were widely used in the early stages. Lo [1] proposed such an indirect method to generating quadrilateral meshes, which has been improved by Lau et al. in [2]. However, the generated quadrilateral meshes cannot satisfy the requirements of industrial applications, in the sense that the triangles contained in the mesh degrade the quality of the simulation of FEA. Several schemes were proposed to alleviate this problem through merging two neighboring triangles to generate quadrilateral meshes. Owen et al. [3] have tried to minimize the number of triangles contained in the generated quad-dominant mesh. Repulsive forces were introduced to uniformly distribute the vertices of the triangular meshes over the target domains before merging pairs of neighboring triangles, including bubble packing methods [4] and square cells packing methods [5]. Nonetheless, all these schemes are still more or less likely to suffer from the same problem.

Recently, more attention has been paid to direct methods. The simplest one is the grid-based method [6], where the interior of the target object is filled with regular grids first and then the boundary region is partitioned into small quadrilaterals by using 2D operations. This method incurs the problems that the quadrilateral elements around the object boundary are distorted in shape when the boundary is not aligned with the grid horizontally or vertically. To alleviate this problem, Robert [7] employed an octree shape representation to improve the quality of the quadrilateral meshes. However, the grid-based methods cannot fully incorporate the shape of the object boundary into the generated mesh topology. In order to respect the boundary shape, several paving methods are proposed. Blacker et al. [8] developed a plastering method to extend quadrilateral mesh decomposition from the boundary to the center of the target object. However, it is unavoidable that irregular shapes are generated around the center in the last stage of the mesh generation. Transforming these irregular shapes to quadrilateral mesh elements requires us to change the entire mesh topology of the target object.

In all of the aforementioned approaches, no one focuses on the appropriate locations of extraordinary vertices for improving the quality of the overall mesh topology. Our approach employs a 2D diffusion tensor field, as an underlying configuration of the quadrilateral mesh, in order to conform the entire mesh topology to the shapes of the target object and its boundary. The details will be introduced in the next three sections.

3   Dual Relationship

Our proposed approach allows us to control the mesh topology of quadrilateral meshes consistently through editing the generated diffusion tensor fields in the target object. This approach is indeed supported by the dual relationship between 2D diffusion tensor fields and quadrilateral meshes, which will be described in this section.

The streamlines of diffusion tensor fields can be used to construct the dual graph of the quadrilateral meshes. Figure 2 shows such an example, where the streamlines in Figure 2(a) guide the dual graph of the quadrilateral mesh in Figure 2(b). As shown in Figure 2, an intersection between two streamlines corresponds to a quadrilateral face of the resulting mesh, since the four lines emanating from the intersection point correspond to the four side edges of the quadrilateral face.

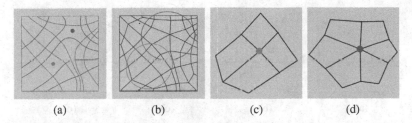

(a)   (b)   (c)   (d)

Fig. 2. Dual relationship between diffusion tensor fields and quadrilateral meshes. (a) The stream-lines in a 2D diffusion tensor field. (b) A quadrilateral mesh is generated from the dual graph consisting of the streamlines in diffusion tensor fields. (c) and (d) show the zoom-up views of the regions in the orange and blue circles of (b), respectively.

Note that *ordinary vertices* in quadrilateral meshes are matched with non-degenerate regions in the diffusion tensor fields. Figure 2(c) shows an ordinary vertex (in orange), whose degree is four. On the other hand, *extraordinary vertices* are generated in degenerate regions in the diffusion tensor fields. As shown in Figure 2(d), the degree of an extraordinary vertex (in blue) is not four. A degenerate point exists around the blue point marked in Figure 2(a).

4   Generating 2D Diffusion Tensor Fields

This section describes how we introduce a 2D diffusion tensor field into the interior of the target object. The diffusion tensor field can be used to respect the shape of the target object and its boundary, and to automatically locate degenerate points for generating extraordinary vertices in quadrilateral meshes.

4.1   Generating a Tensor Field along the Boundary

We begin with the generation of tensor samples on the boundary of the target object. For obtaining newly introduced tensor samples between the two endpoints of each boundary edge, we interpolate between the two tensor samples at the two endpoints along each boundary edge. The tensor samples at the corners of the object boundary are given as input, where the directions of the primary eigenvectors of the tensor samples are designed by referring to the boundary shape of the target object.

The interpolated tensor samples along the boundary should be consistent with the boundary edge direction of the input target object. For this purpose, we employ an eigenstructure-based interpolation method for diffusion tensor fields, because we can fully preserve the shape of the boundary by interpolating the eigenvectors and eigenvalues individually. Suppose that we calculate the interpolated tensor D^M at the ratio of $t : (1-t)$ in the range [0,1] between $D^S (t = 0)$ and $D^T (t = 1)$. The eigenvalues $\lambda_i^M (i = 1,2)$ and their corresponding normalized eigenvectors $e_i^M (i = 1,2)$ of D^M can be obtained as follows:

$$\lambda_i^M = t\lambda_i^S + (1-t)\lambda_i^T, \tag{1}$$

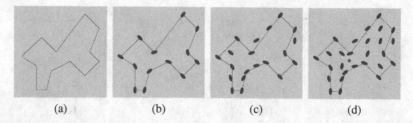

(a)   (b)   (c)   (d)

Fig. 3. Generating a 2D diffusion tensor field in a target object. (a) A target object. (b) The tensor samples on the vertices have been set as input. (c) The tensor samples along the boundary are interpolated. (d) The tensor samples in the interior of the target object are interpolated, where the blue point is a degenerate point.

$$(e_1^M, e_2^M) = R^t(e_1^S, e_2^S),$$
$$\text{where} \quad R = (e_1^T, e_2^T)(e_1^S, e_2^S)^{-1}. \tag{2}$$

Note that R is a 2×2 rotation matrix that transforms between the tensor samples D^S and D^T. Figure 3(c) shows the interpolated diffusion tensor field along the boundary by using this method. The interpolated tensor field retains the boundary shape since our first priority here is to reflect the boundary shape into the mesh topology to be generated inside the object, as already described in Section 1.

4.2 Generating Tensor Fields Inside the Objects

The diffusion tensor field in the interior of the target object should be interpolated for respecting the shape of the target object, as well as its boundary shape.

For this purpose, we employ the Poisson equation to generate the 2D diffusion tensor field. This is because the Poisson equation can continuously propagate the properties of the tensor field along the boundary into the interior. For applying the Poisson equation to such 2D diffusion tensor fields, it is preferable to represent the two eigenvectors of the diffusion tensor field as scalar values. In our approach, we use the polar angle of the tensor sample θ, where $-\pi < \theta < \pi$.

Now we can employ the Poisson equation to generate diffusion tensor field in the interior of the target object by using the polar angles. Suppose that we denote by $\theta_{i,j}^n$, the polar angle at a tensor sample $D_{i,j}$ in the n-th iteration for solving the Poisson equation. We can compute $\theta_{i,j}^{n+1}$ as follows:

$$\frac{\theta_{i,j}^{n+1} - \theta_{i,j}^n}{\Delta t} = \frac{d_x}{\Delta x^2}((\theta_{i+1,j}^n - \theta_{i,j}^n) - (\theta_{i,j}^n - \theta_{i-1,j}^n))$$
$$+ \frac{d_y}{\Delta y^2}((\theta_{i,j+1}^n - \theta_{i,j}^n) - (\theta_{i,j}^n - \theta_{i,j-1}^n)), \tag{3}$$

where Eq. (4) must hold for convergence.

$$d_x \frac{\Delta t}{\Delta x^2} + d_y \frac{\Delta t}{\Delta y^2} \leq \frac{1}{2}. \tag{4}$$

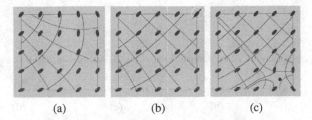

(a)   (b)   (c)

Fig. 4. Interactively controlling the mesh topology through editing the streamlines. (a) The streamline has been tracked in the diffusion tensor field. (b) User changed the direction of the primary eigenvector of the tensor samples (circled in yellow) by drawing a straight segment (in black). (c) The streamlines have been tracked in the diffusion tensor fields together with a degenerate point, which was inserted by the user through clicking the position.

Here, in our approach we use $\Delta t = 1/4$. Figure 3(d) shows the interpolated tensor field using the Poisson equation. The properties of the tensor samples along the boundary are successfully propagated into interpolated tensor samples inside the target object.

4.3   Locating Tensor Degeneracy

When aligning quadrilateral elements along the boundary of the target object, we may have to inevitably introduce non-quadrilateral elements such as triangles and pentagons in the interior of the target object. For generating a fully quadrilateral mesh over the object, our approach allows us to insert non-degree-four vertices to rearrange mesh connectivity. However, the number of such extraordinary vertices should be as small as possible while appropriately controlling their positions.

In our approach, the extraordinary vertices can be fully controlled through editing the degenerate points in the generated diffusion tensor fields, as described in Section 3. We locate the degenerate points in the diffusion tensor fields by employing a minimum spanning tree-based algorithm proposed in [9]. In Figure 3(d), a degenerate point represented by the blue point was introduced for generating an extraordinary vertex. How to introduce extraordinary vertices will be detailed in Section 5.

5   Generating a Quadrilateral Mesh through Constructing Its Dual Graph

In this section, we will explain how to transform the set of streamlines to its dual graph for constructing the quadrilateral mesh. An interface for interactively controlling the mesh topology through editing diffusion tensor fields will also be introduced.

5.1   Tracking Streamlines for Constructing the Dual Graph

In our approach, we track the primary and secondary eigenvectors over the generated 2D diffusion tensor fields by employing the algorithm proposed by Basser et al. [10]. In their method, a seedpoint is selected first to start the tracking process. The streamlines

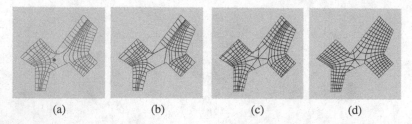

(a)   (b)   (c)   (d)

Fig. 5. Generating a quadrilateral mesh. Constructing the dual graph (b) of the quadrilateral mesh (c) from the streamlines (a). (d) Rearranging the geometry of the generated quadrilateral mesh.

are then pursued forward and backward from the selected seedpoint with a small step size, where the forward and backward directions can be calculated by interpolating the eigenvectors of the generated diffusion tensor field. The above subprocess will be continued until we reach the boundary of the target object.

In the above subprocess, it is very important to select appropriate seedpoints for obtaining an evenly-spaced streamlines from the diffusion tensor fields. For achieving this task, we employ the seedpoints selection method proposed by Jobard et al. [11]. Their method was designed for creating evenly-spaced streamlines by setting a threshold to spare enough space between a new seedpoint and the existing streamlines.

Figure 5(a) shows the streamlines tracked by using the aforementioned two methods. The figure clarifies that the tracked streamlines can fully respected by the shape of the target object and its boundary. Figure 5(b) is the dual graph of quadrilateral mesh in Figure 5(c).

Finally, in order to improve the geometric quality of the generated quadrilateral mesh, we employ the *Laplacian smoothing* method to rearrange the positions of the vertices in the quadrilateral mesh. Figure 5(d) shows the result after the geometric smoothing has been carried out. The geometry of the quadrilateral mesh has been well rearranged in the sense that the vertices are more uniformly distributed.

5.2   Interactive Control of the Mesh Topology

Our approach also provides an interactive control of the mesh topology through editing the tensor samples. Figure 4 shows such an example where the initial streamlines are extracted from the interpolated 2D diffusion tensor field. The configuration of streamlines can be further modified by users according to their requirements. This is accomplished by editing the primary directions of the diffusion tensor samples. In Figure 4(b), the tensor sample in the yellow circle has been rotated by drawing a straight segment (in black), which is used as the primary direction of the tensor sample.

Furthermore, our approach also allows users to intentionally embed degeneracy into the interior of the object, so as to introduce extraordinary vertices specifically on demand. This is useful in many applications for avoiding non-quadrilateral mesh elements to be generated. We perform this by changing the directions of the four tensor samples around the user-specified region. Figure 4(c) shows an example where we generate a degenerate point in a specific position indicated by the blue point. Here, the directions of the four tensor samples around the blue point are changed.

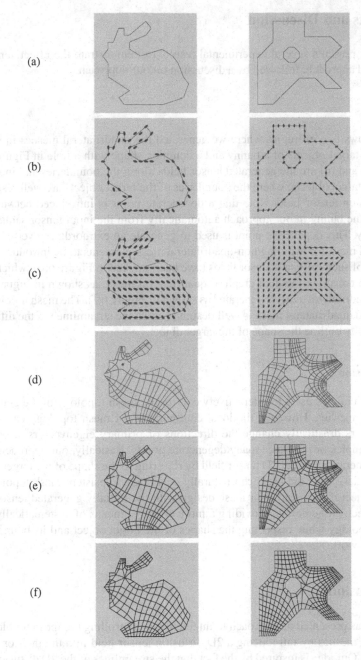

Fig. 6. Generating a quadrilateral mesh in the interior of the target objects. (a) Boundary shape of target objects. (b) The diffusion tensor field along the boundary of the target objects. (c) The generated diffusion tensor field in the interior of the target objects. (d) The streamlines along the two principal directions of the diffusion tensor fields. (e) The dual graphs of the quadrilateral meshes obtained from the streamlines in (d). (f) The quadrilateral meshes obtained from their dual graphs in (e).

6   Results and Discussion

This section presents several experimental results to demonstrate the effectiveness of the proposed approach, followed by a discussion on our approach.

6.1   Results

Figure 6 shows two examples where we generated the quadrilateral meshes in the interior of the target objects of a bunny and a complex shape with a hole in Figure 6(a). Figures 6(b) and (c) are the generated tensor fields along the boundaries and in the interior of the target objects, where the boundaries of the target objects are well respected in the diffusion tensor field. Note that a degenerate point is introduced between the two ears of the bunny in our approach automatically from the input tensor samples on the boundary. This degenerate point is used to generate an extraordinary vertex in the quadrilateral mesh for avoiding non-quadrilateral mesh elements to be introduced. The streamlines of such diffusion tensor fields have been tracked in Figure 6(d), which have been used to construct the dual graph of quadrilateral mesh, as shown in Figure 6(e). Finally, a quadrilateral mesh is generated as shown in Figure 6(f). The mesh topology of the generated quadrilateral mesh is well designed since the streamlines of the diffusion tensor can fully outline the shape of the target object.

6.2   Discussion

Our approach allows users to interactively control the mesh topology of the generated quadrilateral meshes. However, the local configuration of mesh topology can be degraded if users drastically change the directions of primary eigenvectors at original boundary samples or introduce many degenerate points. Basically, our approach automatically generate an initial 2D tensor field by referring to the shape of the target object and its boundary, thus the users can naturally generate a consistent mesh topology in the target object if they start their mesh design with the initially generated tensor field over the object. Nonetheless, providing a more effective means of systematically alter the mesh topology while respecting the shapes of the target object and its boundary is left as future work.

7   Conclusion

This paper has presented an approach to interactively controlling the mesh topology of quadrilateral meshes by introducing a 2D diffusion tensor field into the interior of the target object. Our idea is inspired by the fact that the streamlines of the 2D diffusion tensor field constitute the dual of the quadrilateral mesh in general. Our approach can fully respect the shape of the target object and its boundary by interactively controlling the mesh topology in quadrilateral meshes through editing the streamlines of the diffusion tensor field. Furthermore, our method also allows us to introduce extraordinary vertices while consistently rearranging the mesh topology by embedding tensor degeneracy into the underlying diffusion tensor field. Several results are presented to demonstrate the effectiveness of the proposed approach.

Our future extension includes the challenge to extend our framework to 3D cases, i.e., all-hexahedral mesh generation by introducing 3D diffusion tensor fields in the target volume, which will significantly improve the simulation quality of FEA.

Acknowledgements. This work has been partially supported by JSPS under Grants-in-Aid for Scientific Research (B) No. 22300037, and challenging Exploratory Research No. 23650042 and No. 23650052.

References

1. Lo, S.H.: Generating quadrilateral elements on plane and over curved surfaces. Computers and Structures 31, 421–426 (1989)
2. Lau, T.S., Lo, S.H., Lee, C.K.: Generation of quadrilateral mesh over analytical curved surfaces. Finite Elements in Analysis and Design 27, 251–272 (1997)
3. Owen, S.J., Staten, M.L., Canann, S.A., Saigal, S.: Q-Morph: an indirect approach to advancing front quad meshing. International Journal for Numerical Methods in Engineering 44, 1317–1340 (1999)
4. Shimada, K., Gossard, D.C.: Bubble mesh: Automated triangular meshing of non-manifold geometry by sphere packing. In: Proceedings of ACM Symposium on Solid Modeling and Applications, pp. 409–419 (1995)
5. Shimada, K., Liao, J., Itoh, T.: Quadrilateral meshing with directionality control through the packing of square cells. In: Proceedings of the 7th International Meshing Roundtable, pp. 61–75 (1998)
6. Ho-Le, K.: Finite element mesh generation methods: A review and classification. Computer Aided Design 20, 27–38 (1988)
7. Schneiders, R.: A grid-based algorithm for the generation of hexahedral element meshes. Engineering with Computers 12, 168–177 (1996)
8. Blacker, T.D., Stephenson, M.B.: Paving: A new approach to automated quadrilateral mesh generation. International Journal for Numerical Methods in Engineering 32, 811–884 (1991)
9. Bi, C., Takahashi, S., Fujishiro, I.: Interpolating 3D Diffusion Tensors in 2D Planar Domain by Locating Degenerate Lines. In: Bebis, G., Boyle, R., Parvin, B., Koracin, D., Chung, R., Hammoud, R., Hussain, M., Kar-Han, T., Crawfis, R., Thalmann, D., Kao, D., Avila, L. (eds.) ISVC 2010. LNCS, vol. 6453, pp. 328–337. Springer, Heidelberg (2010)
10. Basser, P.J., Pajevic, S., Pierpaoli, C., Duda, J., Aldroubi, A.: In vivo fiber tractography using DT-MRI data. Magnetic Resonance in Medicine 44, 625–632 (2000)
11. Jobard, B., Lefer, W.: Creating evenly-spaced streamlines of arbitrary density. In: Visualization in Scientific Computing, pp. 43–56 (1997)

A New Visibility Walk Algorithm for Point Location in Planar Triangulation

Roman Soukal, Martina Malková, and Ivana Kolingerová

University of West Bohemia, Plzeň, Czech Republic

Abstract. Finding which triangle in a planar triangle mesh contains a query point is one of the most frequent tasks in computational geometry. Usually, a large number of point locations has to be performed, and so there is a need for fast algorithms resistant to changes in triangulation and having minimal additional memory requirements. The so-called walking algorithms offer low complexity, easy implementation and negligible additional memory requirements, which makes them suitable for such applications. In this paper, we propose a walking algorithm which significantly improves the current barycentric approach and propose how to effectively combine this algorithm with a suitable hierarchical structure in order to improve its computational complexity. The hierarchical data structure used in our solution is easy to implement and requires low additional memory while providing a significant acceleration thanks to the logarithmic computational complexity of the search process.

1 Introduction

The point location problem is often solved in computational geometry tasks, such as triangulation construction and deformation, morphing and terrain editing. In this text, we focus on point location algorithms for triangle meshes, which are the most common geometry representation. These algorithms can also be used for terrain models represented by triangle meshes without any preprocessing, only by omitting the height information during the location. The input mesh is expected to be convex and without holes, other types of data should be triangulated first.

The point location problem is defined as follows. For a given planar triangle mesh and a query point, the task is to find which triangle from the mesh geometrically contains the query point. Algorithms solving this problem can be divided into two groups: algorithms with and without additional data structures. The former concentrate on having the lowest computational complexity possible, in this case $O(\log n)$ per query point (n is the number of vertices in the mesh). However, these algorithms have additional memory demands, they are more difficult to implement, and their modification to cover adding or removing vertices is often problematic. The latter group tries to avoid these disadvantages, but has a slightly higher, but still sublinear complexity. The most important representatives of this group are walking algorithms.

Walking algorithms use the triangle neighborhood relations to go (*walk*) via the triangles between the starting triangle and the one containing the query

G. Bebis et al. (Eds.): ISVC 2012, Part II, LNCS 7432, pp. 736–745, 2012.

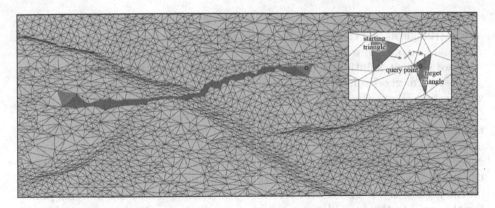

Fig. 1. Example of the point location with a walking algorithm on the part of a terrain model

point (see Figure 1). The starting triangle may be arbitrary, however, its clever selection may radically shorten the length of the walk.

The so-called barycentric walk [1] is an easily implementable algorithm which is independent of the triangles orientation. Although, in its original form, it is also one of the slowest walking algorithms, but has a high potential for speed-up, because its walk is the shortest among the existing walking algorithms. Our approach significantly improves the speed of this algorithm, making it more interesting for further use in complex algorithms and applications. We also propose its combination with a suitable additional data structure in order to improve its computational complexity. The hierarchical data structure used in our solution is easy to implement and requires low additional memory, but it provides a significant acceleration thanks to the logarithmic computational complexity of the search process.

The rest of the paper is organised as follows. Section 2 presents the existing walking algorithms and several approaches for a sophisticated selection of the starting triangle for the walk. Section 3 describes our new proposed algorithm, Section 4 shows experiments comparing our solution with the existing algorithms. Section 5 summarizes the paper.

2   State of the Art

Point location by walking algorithms usually works in two steps: (1) selection of the initial triangle for the walk, and (2) using the neighborhood relationships between the triangles to find the triangle containing the query point *(walking)*.

Clever selection of the initial triangle may radically improve the speed of the process. Some approaches provide solutions using additional information about the data, such as the range of the mesh vertices [2], take advantage of sorted vertices [3] or sort them properly prior to the location ([4],[5], [6], [7]).

Without any additional memory, we can speed up the process by selecting the initial triangle as the closest one from a randomly chosen subset of triangles [8]. An ideal size of such a set is $O(\sqrt[3]{n})$ [9] for a random input.

More efficient solutions lead to some additional memory consumption. [10] proposed a method simplifying the mesh and locating the point in the simplified version first. From the triangulation T with n vertices, only $m = k \cdot n$ vertices (where $k \in (0, 1)$) are randomly selected, triangulated and so a higher layer for the location is created. The number of triangles is much smaller in the new layer and it radically improves the speed of a walk in it. If m is still bigger than a chosen size, other layers are constructed in the same way. The point location then runs in several steps. First, the triangle containing the query point is found on the highest layer. The closest vertex of this triangle defines the starting point for the walk in the lower layer, until the triangle in the lowest layer is found. In each layer, the walk is short and therefore fast. [11] analyses this algorithm, specifies the computational complexity as $O(\log n)$ for any input and proposes the optimal value of $k = 0.025$ which is valid for random input and leads to the best rate between speed and memory usage.

[12] introduce a bucketing method, which uses a uniform grid to quickly find a proper initial triangle. Empty cells slow down the algorithm, therefore it is suitable mainly for uniformly distributed vertices. Some algorithms (e.g., [13], [14]) try to avoid the sensitivity of the original bucketing method to data uniformity by using adaptive structures instead of a uniform grid. However, on highly non-uniform data, the dynamic hierarchy algorithm mentioned above [10], [11] still provides better results with lower additional memory.

When we know the initial triangle, the walk may proceed. There exist several algorithms solving this step, and according to the style how they determine the way of the walk, they can be divided into three groups: visibility, straight and orthogonal walks.

Visibility walks use local "visibility" tests to determine the way of their walk. These tests look for such an edge that defines a line separating the query point and the third vertex of the triangle. The walk then moves across this edge to the neighborhood triangle.

The first visibility walk algorithm is called Lawson's oriented walk [15]. The algorithm starts in the initial triangle and uses this 2D orientation test to move to its neighbors until it reaches the query point:

$$orientation2D(\mathbf{t}, \mathbf{u}, \mathbf{v}) = \begin{vmatrix} u_x - t_x & v_x - t_x \\ u_y - t_y & v_y - t_y \end{vmatrix}, \tag{1}$$

where points $\mathbf{t}, \mathbf{u}$ define an oriented line and $\mathbf{v}$ is the tested point.

The algorithm tests the edges of the current triangle in a deterministic order, leading to the fact that the walk may loop for non-Delaunay triangulations (see Figure 3a). [16] proposed an algorithm avoiding the loops by choosing the edges of the current triangle in a random order. This modification is called *stochastic*. Furthermore, since it is not necessary to test the edge incident to the previous triangle, the process was speeded up by remembering this edge and skipping the test. The stochastic walk has been shown in [17] to need $O(\sqrt{n} \cdot \log n)$ expected time for uniform data.

[1] proposed a visibility walk algorithm which uses barycentric coordinates instead of the 2D orientation test. Barycentric coordinates b_0, b_1, b_2 describe the position of a point $\mathbf{q}$ with respect to a triangle $\tau_{t_0 t_1 t_2}$ (see Figure 2). The point $\mathbf{q}$ is an affine combination of $\mathbf{t_0}, \mathbf{t_1}, \mathbf{t_2}$:

$$\mathbf{q} = b_0 \cdot \mathbf{t_0} + b_1 \cdot \mathbf{t_1} + b_2 \cdot \mathbf{t_2} \tag{2}$$

where $b_i \in R$, $i = 0, 1, 2$, $b_0 + b_1 + b_2 = 1$.

b_i can by computed using the 2D orientation tests:

$$b_i(\mathbf{t_0}, \mathbf{t_1}, \mathbf{t_2}, \mathbf{q}) = \frac{orientation2D(\mathbf{t}_{[(i+1) \mod 3]}, \mathbf{t}_{[(i+2) \mod 3]}, \mathbf{q})}{orientation2D(\mathbf{t_0}, \mathbf{t_1}, \mathbf{t_2})} \tag{3}$$

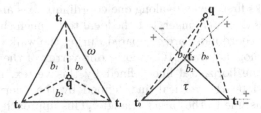

Fig. 2. Barycentric coordinates of $\mathbf{q}$ inside a triangle ω, $\mathbf{b}(\omega) = (0.25, 0.35, 0.4)$, and outside a triangle τ, $\mathbf{b}(\tau) = (-0.75, -0.25, 2)$

Each barycentric component b_i corresponds to one edge of the triangle, defining on which side of this edge the point $\mathbf{q}$ lies: either inner (positive value), or outer (negative value) side of the triangle, disregarding the orientation of the triangle. The higher the value of the component is, the bigger the triangle defined by the edge and the query point is. The walk via bigger triangles is usually shorter, therefore, the algorithm takes advantage of it by crossing such an edge of the triangle that corresponds to the component with the highest negative value.

The barycentric components of $\mathbf{q}$ are computed for each visited triangle. Since the third component can be computed from the other two components, we need three 2D orientation tests for each visited triangle. Although not stated by [1], the algorithm may loop in some rare cases - see Figure 3b. In the triangle τ, the area s_2 is greater than s_1, so the walk does not cross the edge leading to the triangle containing $\mathbf{q}$. A similar case happens in other thin triangles.

Straight walk algorithms walk along an oriented line $\overrightarrow{pq}$, connecting one point $\mathbf{p}$ (its choice depends on the particular solution) of the starting triangle with the query point $\mathbf{q}$ and then pass all triangles intersected by this line. This way, the walk is short.

The standard straight walk algorithm [16,19] chooses $\mathbf{p}$ as any of the starting triangle vertices and moves around it until it finds the triangle intersected by

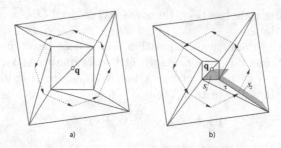

Fig. 3. a) Loop of Lawson's oriented walk [18] and b) Loop of the Barycentric walk

the line segment $\overrightarrow{\mathbf{pq}}$. Then it follows this line segment, using one orientation test to choose the proper edge, and another for checking if $\mathbf{q}$ is still on the outer side of this edge.

Orthogonal walks first navigate along one coordinate axis and then along the other, which makes the walk longer, but the local tests much cheaper, since only components of the coordinates are compared during the walk.

The original orthogonal walk [16] chooses one vertex of the starting triangle and then follows the horizontal line defined by this vertex, until it finds the triangle intersected by the vertical line defined by the query point. Then it continues along this line to the target triangle. This approach has a significant drawback: the border of a triangulation may be crossed during the walk, in which case a special modification is needed, resulting in a slower location process and additional implementation effort.

A speed-up of this walk was proposed in [2], where fewer tests are done during the location for a price that the walk may not find the correct target triangle, but only a triangle in its neighborhood, and a visibility walk algorithm is used for the final short location. The algorithm does not need any modification for dealing with the crossing of the border, but the final location by a visibility walk in this case may be much longer.

3   Proposed Algorithm

Our algorithm is based on the original barycentric walk [1]. However, unlike [1] it does not use the barycentric coordinates, but only the orientation tests of the edges, with respect to their relation to barycentric coordinates.

To decide which edge to cross to the next triangle, we compute the same orientation test as the Lawson's oriented walk [15], but we use not only its sign, but also the values. Let us call the proposed algorithm *direct walk* because the purpose of this change is to maintain the shortest walk of [1] while speeding up the algorithm by performing less orientation tests in each triangle.

If we denote the value of the orientation test for the point $\mathbf{q}$ and an edge opposite to vertex $\mathbf{v_i}$ as c_i (let us call it *orientation coordinate*), we can derive interesting properties from its relation to the barycentric coordinate b_i:

$$c_i(\mathbf{t_0}, \mathbf{t_1}, \mathbf{t_2}, \mathbf{q}) = b_i(\mathbf{t_0}, \mathbf{t_1}, \mathbf{t_2}, \mathbf{q}) \cdot orientation2D(\mathbf{t_0}, \mathbf{t_1}, \mathbf{t_2}) \tag{4}$$

where $c_i \in R$, $i = 0, 1, 2$ and $c_0 + c_1 + c_2 = orientation2D(\mathbf{t_0}, \mathbf{t_1}, \mathbf{t_2})$.

Our algorithm uses the properties of the orientation coordinates in the following way. In each triangle, it uses the orientation coordinate c_{prev} corresponding to the edge crossed to this triangle. The new coordinate c_i for this edge can be obtained as $c_i = -c_{prev}$. The coordinate $c_{[(i+1) \mod 3]}$ is computed using the orientation test (Eq. 1), and the last coordinate $c_{[(i+2) \mod 3]}$ is derived from the knowledge of the area A of the current triangle as $c_{[(i+2) \mod 3]} = 2A - c_i - c_{[(i+1) \mod 3]}$ (the barycentric property), where $2A = orientation2D(\mathbf{t_0}, \mathbf{t_1}, \mathbf{t_2})$.

As a result, for each visited triangle we have to compute two orientation tests: one for an edge and one to get the doubled area of the triangle (three for the first triangle, where we have to compute c_{prev} as well). In applications working with areas of the triangles of the input mesh, one may prefer to store them. The use of these values during the location then results in the need of only one orientation test per visited triangle (two for the first triangle).

Just as the barycentric walk, our algorithm is resistant to variable orientation of the triangles. If there is a possibility of such a case, we need to test the sign of the triangle area and in case of the negative sign, we reverse the signs of all the orientation coordinates. The orientation information can be easily obtained from the $2A$ value which produces a signed value. For more details, see the pseudo-code in Algorithm 1.

Input: the query point $\mathbf{q}$, the chosen starting triangle $\alpha \in T$
Output: the triangle ω which contains $\mathbf{q}$

integer i;
edge ϵ;
triangle $\tau = \alpha = \mathbf{t_0 t_1 t_2}$;
double min, c_0, c_1, c_2, c_a;

$c_0 = orientation2D(\mathbf{t_1}, \mathbf{t_2}, \mathbf{q})$;
$c_1 = orientation2D(\mathbf{t_2}, \mathbf{t_0}, \mathbf{q})$;
$c_a = $ stored double area of τ; // or $c_a = orientation2D(\mathbf{t_0}, \mathbf{t_1}, \mathbf{t_2})$
$c_2 = c_a - c_0 - c_1$;

// the following condition is necessary for non-uniform triangle orientation
// if $c_a < 0$ then begin $c_0 = -c_0$; $c_1 = -c_1$; $c_2 = -c_2$; end

$min = $ minimal c_j where $j \in \{0, 1, 2\}$;
$\epsilon = $ edge corresponding to minimal c_j;

while $min < 0$ **do**
    $\tau = $ neighbour of τ over ϵ;
    $i = $ index of ϵ in τ;
    $c_i = -min$;
    $c_{[(i+1) \mod 3]} = orientation2D(\mathbf{t}_{[(i+2) \mod 3]}, \mathbf{t}_{[(i+3) \mod 3]}, \mathbf{q})$;
    $c_a = $ stored double area of τ; // or $c_a = orientation2D(\mathbf{t_0}, \mathbf{t_1}, \mathbf{t_2})$
    $c_{[(i+2) \mod 3]} = c_a - c_i - c_{i+1}$;

    // the following condition is necessary for non-uniform triangle orientation
    // if $c_a < 0$ then begin $c_0 = -c_0$; $c_1 = -c_1$; $c_2 = -c_2$; end

    $min = $ minimal c_j where $j \in \{0, 1, 2\}$;
    $\epsilon = $ edge corresponding to minimal c_j;
end
return τ;

Algorithm 1. Direct walk

Often, there is a need for a faster algorithm for a price of some memory consumption, in which case, we offer to use our walking algorithm in a combination with a hierarchical structure proposed by [11] favoured for its low memory requirements, an easy modification (in terms of adding or deleting vertices from the input triangle mesh) and excellent results for non-uniform data. For the detailed description of the hierarchical structure, see Section 2. Here we will describe its combination with our method.

Our algorithm is suitable for such a structure thanks to its fast initialization step, which is performed in each layer. Also, when moving to the lower layer, we do not have to compute which vertex of the target triangle is closest to the query point - we already know its orientation coordinates and can therefore easily select the vertex with the highest orientation coordinate c_i and use it as the starting for the walk in the lower layer. This way, we save three distance computations in each descend. Also, since there is only a fragment of the original number of the triangles in the higher layers, we can store the area information there and thus speed up the algorithm even more for a low additional memory consumption.

4   Experimental Results

We selected the most popular and also the fastest of the mentioned algorithms to compare with our method: the remembering walk (RW) [15], the remembering stochastic walk (RSW) [16], the barycentric walk (BW) [1], the straight walk (SW) [16], and the orthogonal walk (OW) [2].

For the test purposes, we implemented the specified algorithms in C++ and tested on Intel Q6600 2,40GHz. SSE2 random generator was used for RSW algorithm since is declared as up to five times faster then the standard C random generator [21].

The tests were performed on the triangulations on three types of datasets: randomly distributed points in a unit square, LIDAR data and data from a cadastre. On each dataset, we constructed four types of triangulations: Delaunay (DT), Greedy, MWT and Min-max angle. The results were similar, so we present them on the most popular triangulation type, DT.

In each case, we performed 10^7 location processes and computed the average number of the tested quantities. The following qualities were examined for each algorithm: the average number of visited triangles ($\#\Delta$), the average number of tests ($\#tests$) and the average time per one location ($t[\mu s]$). Note that the number of tests done by each algorithm is presented only to measure how many tests per triangle the particular algorithm does on average, not to compare the performance of the algorithms - for that, we should use the time values, since the speed of their tests differ among the algorithms. The properties $\#tests$ and $\#\Delta$ consists of two values for OW: the former value concerns the walk, the latter concerns the final location performed by RW.

Table 1 shows the results of the tests for a random initial triangle and a random target point. Table 2 presents the performance of the algorithms when they are used in the hierarchical approach by [11]. The hierarchical approach was tested on randomly generated rectangular datasets, the datasets for testing the original algorithms were bounded by a rectangle and retriangularized to obtain a fair measurement of the orthogonal walk, which is slower for non-rectangular data, where it often crosses the border of the triangulation.

Two versions of our algorithm were tested, DW1 uses the precomputed area information, DW2 does not. For the hierarchical approach, DW2 stores the area information only for the higher layers, not for the original mesh.

Algorithms that performed remarkably worse with the selected hierarchical structure, such as the original barycentric walk and the straight walk, were not included in the final tests.

The results in Table 1 confirm that the shortest length of the barycentric walk was maintained and its speed was improved. Although the barycentric walk without the precomputed areas (DW2) is still slightly slower than the RW algorithm, it can be preferred when the triangle orientation varies throughout the triangulation. If the intended application uses the triangle areas for other purposes, DW1 can be used to obtain the best performance.

When combined with the hierarchical structure (see Table 2), our algorithm becomes the fastest, thanks to the short walk and avoiding the computation of

Table 1. Comparison of the walking algorithms with randomly chosen α (#Δ represents the number of visited triangles, #*tests* represents the number of performed tests, $t[\mu s]$ represents the time per one location)

	#Δ	#*tests*	$t[\mu s]$	#Δ	#*tests*	$t[\mu s]$	#Δ	#*tests*	$t[\mu s]$
	ϕ per located point			ϕ per located point			ϕ per located point		
	Cadastre data								
	4897 vertices (9774 Δ)			15824 vertices (31642 Δ)			70437 vertices (140868 Δ)		
RW	94.4	129.5	3.48	160.5	213.2	6.52	321.3	417.9	14.90
RSW	92.1	122.2	5.89	158.1	208.5	10.71	312.9	414.8	23.04
BW	87.4	262.2	5.92	149.6	448.5	10.72	283.1	849.2	21.87
SW	2.8+88.2	3.3+175.4	4.75	2.7+149.6	3.2+298.3	8.63	2.8+294.8	3.3+588.9	18.72
OW	94.2+4.3	192.6+7.0	2.01	176.5+2.7	357.3+5.2	4.32	319.0+3.5	642.0+6.4	9.50
DW2	87.4	175.8	4.34	149.6	300.2	8.00	283.1	567.2	17.43
DW1	87.4	88.4	2.87	149.6	150.6	5.48	283.1	284.1	12.63
	LIDAR								
	34932 vertices (69858 Δ)			313348 vertices (626690 Δ)			3722068 vertices (7444130 Δ)		
RW	205.5	275.4	8.98	613.5	823.3	38.18	2569.1	3444.2	181.51
RSW	202.6	265.6	14.26	608.5	792.6	54.09	2543.3	3305.4	248.08
BW	179.3	537.9	13.34	509.1	1527.3	47.24	2203.3	6609.8	222.76
SW	2.7+180.8	3.2+360.6	10.94	2.7+542.8	3.2+1084.6	42.98	2.8+2316.4	3.3+4631.8	202.65
OW	240.7+1.7	485.8+3.8	6.60	695.0+1.7	1394.3+3.8	31.88	2741.8+2.2	5487.8+4.5	148.45
DW2	179.3	359.6	10.20	509.1	1019.2	38.41	2203.3	4407.6	184.06
DW1	179.3	180.3	7.23	509.1	510.1	29.90	2203.3	2204.3	147.73
	Randomly distributed points in the unit square								
	10^4 vertices (19994 Δ)			10^5 vertices (199994 Δ)			10^6 vertices (1999994 Δ)		
RW	118.1	159.2	4.34	366.1	491.5	19.43	1144.7	1533.0	81.39
RSW	115.8	152.5	7.40	362.5	474.5	29.02	1130.5	1476.5	110.94
BW	103.2	309.6	7.08	326.8	980.1	27.60	1028.2	3084.5	105.12
SW	2.8+105.5	3.3+210.0	5.71	2.7+335.6	3.2+670.2	23.56	2.7+1065.4	3.2+2129.8	93.96
OW	137.8+1.8	279.9+3.9	2.84	433.3+1.7	870.8+3.8	15.71	1336.3+1.8	2676.9+3.9	72.42
DW2	103.2	207.4	5.22	326.8	654.6	21.71	1028.2	2057.4	86.89
DW1	103.2	104.2	3.46	326.8	327.8	16.16	1028.2	1029.2	69.37

Table 2. Comparison of the walking algorithms with hierarchical structure ($\#\Delta$ represents the number of visited triangles, $\#tests$ represents the number of performed tests, $t[\mu s]$ represents the time per one location)

	$\#\Delta$	$\#tests$	$t[\mu s]$	$\#\Delta$	$\#tests$	$t[\mu s]$	$\#\Delta$	$\#tests$	$t[\mu s]$
	ϕ per located point			ϕ per located point			ϕ per located point		
	10^2 vertices (2 layers)			10^3 vertices (2 layers)			10^4 vertices (3 layers)		
RW	7.8	10.8	0.40	13.4	18.7	0.60	24.0	32.0	1.06
RSW	7.6	11.2	0.60	13.2	18.4	0.96	22.6	29.0	1.63
OW	10.8 + 2.3	25.9 + 3.3	0.63	13.8 + 3.5	31.7 + 7.4	0.74	19.5 + 4.8	43.1 + 10.6	1.08
DW2	6.3	13.5	0.35	11.8	19.8	0.55	20.7	30.1	0.87
DW1	6.3	8.3	0.22	11.8	13.8	0.40	20.7	23.7	0.71
	10^5 vertices (3 layers)			10^6 vertices (4 layers)			10^7 vertices (4 layers)		
RW	30.9	41.2	1.36	24.5	30.8	1.21	35.5	47.9	1.84
RSW	27.0	36.7	1.95	23.3	32.1	1.81	36.3	48.8	2.88
OW	25.8 + 5.8	58.2 + 12.4	1.36	23.8 + 5.9	51.9 + 14.2	1.43	30.5 + 9.27	65.3 + 18.25	1.87
DW2	22.7	29.3	0.90	22.0	30.4	0.94	31.3	45.0	1.58
DW1	22.7	25.7	0.81	22.0	26.0	0.84	31.3	35.3	1.34

distances in each descend. Note that between the location time for 10^5 and 10^6 vertices, there is not much difference. The explanation is simple - 10^5 vertices is just below the limit for the creation of a new layer, while for 10^6 vertices the new layer is created.

Although we identified a situation where both the barycentric walk and our improvement can loop, during our thorough tests on the types of triangulations listed above, neither of the algorithms looped.

5 Conclusion

We presented a modification of the barycentric approach for the point location and proposed how to combine it with a popular hierarchical structure to gain more speed-up than in combination with other walking algorithms.

We compared the performance of our algorithm with the most popular and also the fastest of the existing walking algorithms. We compared both the original algorithms and their combination with the selected hierarchical data structure. Our approach proved to be faster than the original, while maintaining its advantages. When combined with the hierarchical structure, our algorithm becomes the fastest, thanks to its suitability for the structure.

Acknowledgement. This work has been supported by the Czech Science Foundation under the project P202/10/1435 and by University of West Bohemia under the project SGS-2010-02.

References

1. Sundareswara, R., Schrater, P.: Extensible point location algorithm. In: International Conference on Geometric Modeling and Graphics, pp. 84–89 (2003)

2. Soukal, R., Kolingerová, I.: Star-shaped polyhedron point location with orthogonal walk algorithm. Procedia Computer Science 1, 219–228 (2010)
3. Purchart, V., Kolingerová, I., Beneš, B.: Interactive sand-covered terrain surface model with haptic feedback. In: GIS Ostrava 2012 - Surface Models for Geosciences (2012)
4. Sloan, S.W.: A fast algorithm for constructing Delaunay triangulations in the plane. Advanced Engineering Software 9, 34–55 (1987)
5. Zhou, S., Jones, C.B.: HCPO: an efficient insertion order for incremental Delaunay triangulation. Information Processing Letters 93, 37–42 (2005)
6. Amenta, N., Choi, S., Rote, G.: Incremental constructions con brio. In: SCG 2003: Proceedings of the 19th Annual Symposium on Computational Geometry, pp. 211–219. ACM, New York (2003)
7. Buchin, K.: Incremental construction along space-filling curves. In: EuroCG 2005: Proceedings of the 21th European Workshop on Computational Geometry, pp. 17–20 (2005)
8. Mücke, E.P., Saias, I., Zhu, B.: Fast randomized point location without preprocessing in two and three-dimensional Delaunay triangulations. In: Proceedings of the 12th Annual Symposium on Computational Geometry, vol. 26, pp. 274–283 (1996)
9. Devroye, L., Mucke, E.P., Zhu, B.: A note on point location in Delaunay triangulations of random points (1998)
10. Mulmuley, K.: Randomized multidimensional search trees: Dynamic sampling. In: Proceedings of the 7th Annual Symposium on Computational Geometry, pp. 121–131 (1991)
11. Devillers, O.: The Delaunay hierarchy. International Journal of Foundations of Computer Science 13, 163–180 (2002)
12. Su, P., Drysdale, R.L.S.: A comparison of sequential Delaunay triangulation algorithms. In: Proceedings of the 11th Annual Symposium on Computational Geometry, pp. 61–70 (1995)
13. Zadravec, M., Žalik, B.: An almost distribution independent incremental Delaunay triangulation algorithm. The Visual Computer 21, 384–396 (2005)
14. Žalik, B., Kolingerová, I.: An incremental construction algorithm for Delaunay triangulation using the nearest-point paradigm. International Journal of Geographical Information Science 17, 119–138 (2003)
15. Lawson, C.L.: In: Mathematical Software III; Software for C1 Surface Interpolation, pp. 161–194. Academic Press, New York (1977)
16. Devillers, O., Pion, S., Teillaud, M.: Walking in a triangulation. In: Proceedings of the 17th Annual Symposium on Computational Geometry, pp. 106–114 (2001)
17. Zhu, B.: On Lawson's Oriented Walk in Random Delaunay Triangulations. In: Lingas, A., Nilsson, B.J. (eds.) FCT 2003. LNCS, vol. 2751, pp. 222–233. Springer, Heidelberg (2003)
18. Weller, F.: On the total correctness of Lawson's oriented walk. In: Proceedings of the 10th International Canadian Conference on Computational Geometry, pp. 10–12 (1998)
19. Mehlhorn, K., Näher, S.: Leda: A platform for combinatorial and geometric computing. Communications of the ACM 38, 96–102 (1995)
20. Soukal, R., Kolingerová, I.: Straight walk algorithm modification for point location in a triangulation. In: EuroCG 2009: Proceedings of the 25th European Workshop on Computational Geometry, Brussels, Belgium, pp. 219–222 (2009)
21. Owens, K., Parikh, R.: Fast random number generator on the intel pentium 4 processor. Intel Software Network (2009)

Real-Time Algorithms Optimization Based on a Gaze-Point Position

Anna Tomaszewska

West Pomeranian University of Technology in Szczecin,
Faculty of Computer Science,
Żołnierska 49, 71-210, Szczecin, Poland
atomaszewska@wi.zut.edu.pl

Abstract. In the paper, we present a real-time algorithm optimization based on a gaze point position. The data is provided by the eye tracker and allows the effective rendering of algorithms in real-time where their accuracy depends on the distance from the point-of-regard. We present the approach based on the algorithm simulating the subsurface scattering effect. The model complexity depends on the number of parameters that should be taken into account to simulate a real effect. In order to specify model parameters correctly dependent on the gaze point position, a series of perceptual experiments must be conducted. The quality of the results will be evaluated on the basis of the single stimulus perceptual metrics.

1 Introduction

The realism of virtual reality is playing an increasingly important role, as people expect the computer-generated world to successfully imitate the one known from their real life. Achieving a true physical accuracy of a complex scene in real-time is still beyond the capabilities of current standard desktop computers. Authentically looking visual effects can be efficiently worked out with the help of algorithms that simplify the natural phenomena. Despite the progress in this field that has been heavily accelerated with the introduction of programmable graphics cards, the execution of many algorithms of high complexity is still problematic.

In the paper, we focus on a real-time algorithm optimization based on the information concerning the gaze point position. In reaction to the progress in the field of eye tracking technology, which can be observed in the recent years [4], we have tried to prepare for the idea of a gaze-controlled effect for real time graphics which relies on the data provided by an eye tracker.

The presented approach is based on our algorithm for subsurface scattering [20]. In the translucent materials the light emitted from a point on the surface depends on the type of translucent material and the direction of light incident on the surface in the surrounding area. The Bidirectional Subsurface Scattering Reflectance Distribution Function (BSSRDF) takes all these parameters into consideration. Unfortunately, nowadays it is not possible to compute the radiance in real time on the basis of the complex BSSRDF equations. Hence, a

G. Bebis et al. (Eds.): ISVC 2012, Part II, LNCS 7432, pp. 746–755, 2012.

number of approximations have been proposed to speed up the computations of subsurface scattering phenomenon. The techniques focus on simplifying calculations that enable the algorithm execution in real time. In the presented algorithm we propose to use complex algorithms in dependence on the gaze point position. The more distant an object or phenomenon is from the point of interest, the worse quality of rendering can be used. The effect is fully generated only if the user concentrates directly on the effect or a scene close to it.

The paper is organized as follows. First, we provide a brief overview of eye tracking technologies in Section 2. It is followed by the description of our approach in Section 3. Section 4 shows achieved results. Finally we conclude the paper in Section 5.

2 Previous Work

Eye tracking is a technique of gathering real-time data concerning gaze direction of human eyes. In particular, position of the point, called *point-of-regard*, that a person is looking at is captured [1]. This information is acquired in numerous ways encompassing intrusive and remote techniques.

Intrusive eye trackers require some equipment to be put in physical contact with the user. In early works a coil embedded into contact lens was used [2]. The eye gaze was estimated from measuring the voltage induced in the coil by an external electro-magnetic field. In the electro-oculogram technique (EOG) [3] electrodes are placed around the eye. The eye movement is estimated by measuring small differences in the skin potential. In general, intrusive techniques are very accurate and often used in scientific experiments but rather not favourable in immersive VR applications.

More suitable for VR systems are remote techniques that use cameras to capture the image of the eye. Even if they require some intrusive head mounted devices [4, Sect. 6], they are still acceptable for many VR applications. The most common remote eye trackers apply the *corneal reflection* (CR) method. The eyes are exposed to direct invisible infra-red (IR) light, what results in appearance of Purkinje image with a reflection in the cornea (see Fig. 1, left). The reflection is accompanied by an image of the pupil. Captured by a video camera sensitive to the infra-red spectrum, the relative movement of both the pupil and corneal reflections are measured, what permits to estimate an observer gaze point. Commercial eye trackers can achieve the accuracy below 0.5 degree [5], [6]. The CR eye trackers require calibration to estimate the position of the head relatively to the screen plane. Then, it is possible to calculate the estimated screen-space gaze point coordinates with frequency higher than eye saccades [1] [7]. In our experimental approach we use the corneal reflection eye tracker manufactured by SMI Company (SMI RED250 [5]).

There are eye trackers that simultaneously process more than one corneal reflection [8],[9]. The most popular eye trackers can estimate gaze point with

---

[1] The saccades are defined as rapid movement of the gaze-point, characteristic for the human visual system (HVS).

Bright pupil   Corneal reflection

Fig. 1. The corneal reflection in the infra-red light, relative location of the pupil and the corneal reflection are used to estimate the observer's gaze-point

very high accuracy of about 1 min of arc. Their drawback is a need of using the chin rest and/or the bite bar for the head stabilization [4, Sec. 5.4].

Non-intrusive eye tracking can also base on tracking the limbus (boundary between the sclera and the iris) [10] or detection of the pupil position in an image. The latter method suffers from low contrast between the pupil iris and its boundary but the contrast can be enhanced with the use of IR light source lighting an eye [11]. Besides expensive commercial eye trackers, many solutions based on cheap web cameras often combined with IR photodiodes are available [12,13]. The development of inexpensive eye trackers can play a crucial role in the popularization of gaze-dependent technologies. Detailed reviews of eye tracking techniques are presented in [4,14].

3   Eye Tracker Based Approach for Optimization

The presented approach is based on our algorithm [20] which simulates the effect of subsurface scattering in a realistic way. The quality of the effect depends on Spherical Harmonics function that defines the object density for which the effect is simulated. The aim of the project is the calculation optimization by introducing the dependence between the results quality and the gaze point position.

It will allow the use the complex algorithm in real time without the necessity of its simplification when the effect appears in the Region of Interest (ROI), defined as surrounding the gaze point area. The ROI is assumed to have a rectangular shape to simplify the condition to an alternative of primitive relations. Simplifications are used only when the ROI is distant enough so that the lack of complexity will not be noticeable and hence they will not affect the quality of simulated realism perception. Therefore to reduce the negative impact on frame rate, we have decided to take further advantage of the fact, that the effect is intended to be controlled with an eye tracker. Bearing in mind the limitations of the human visual perception related to the peripheral vision inacuity by Duchowski in [4], we have tended to narrow the accurate subsurface scattering computation only to the object fully or partially located inside area, where it may actually be seen by the user.

In the example subsurface scattering algorithm, the reproduction of the object depth is the most time consuming process at the stage of visualization. The knowledge on the human visual system enables the selection of proper number of Spherical Harmonics band which will be used in projection to evaluate the object depth (see Fig. 2). They affect the complexity of calculations for different display regions. The quality level of the simulated effect is dependent on the angle towards gaze

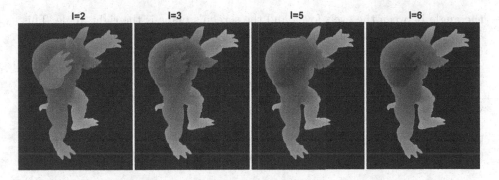

Fig. 2. Subsurface scattering effect generated for different SH bands number

direction. It results from the construction of the human eye, in particular the distance between rods and cones in the retina. Hence, optimization can be obtained by simplifying the level of precision according to the distance from the fixation point. To prevent the changes between the levels of details, linear interpolation was introduced. The idea is illustrated in Figure 3.

For the best efficiency, the size of the ROI has to be computed according to the user's distance to screen d and their angle of clear vision α. The first value can be provided in real time by the most modern eye trackers, while the latter we have assumed to be constant. It should cover the biological limitations of the human visual system and have a tolerance for the varying angle of eye-to-screen relation, which results in the non perpendicular alignment of the viewing axis towards the screen. In our formula, however, we use a simplification in which

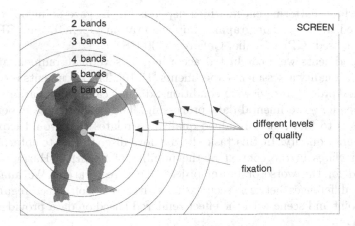

Fig. 3. Regions constitute subsequent levels of precision. Linear interpolation is made between them. A given level depends on the angle of measured towards the gaze direction. It represents the number of Spherical Harmonics bands. The radii of presented regions are computed in pixels.

the angle between the eye's fovea and the screen's plane is the right angle. This allows us to use basic trigonometry in order to calculate the approximate radius (r) of user's clear, foveal vision (Equation 1).

$$r = d * tan(\alpha) \tag{1}$$

where d is the distance between the eye and the screen, and α is the angle of acute vision. To convert the result value into screen space (r'), we need the knowledge of physical screen dimensions (w) and their size in pixels (w'). The final formula for calculating the radius in pixels is given by Equation 2.

$$r' = \frac{w' dtan(\alpha)}{w} \tag{2}$$

The correct angle value was selected and established in an experimental way to 15 degrees. To provide a convincing evidence that a new method or set of parameters used in the model is better than the state-of-the-art, image processing projects are often accompanied by user studies, in which a group of observers rank or rate the results of several algorithms. Such user studies are known as subjective image quality assessment experiments. The subjective video quality assessment methods originate from a wider group of psychometric scaling methods, which were developed to measure psychological attributes [19]. Image quality is an attribute that describes preference for a particular image processing algorithm. Psychometric methods are not new in computer graphics. Recent SIGGRAPH courses, such as [18], [21], [22], demonstrate increasing interest in them.

4   Results

The subsurface scattering effect, which depends on the gaze-point position, was implemented in C++. For graphics implementation OpenGL and GLSL languages were used. GPU: Nvidia, GeForce 8400 GS.

The experiments were conducted according to several recommendations for the design of quality assessment experiments [15,16]. The documents recommend experimental procedures, viewing conditions, display calibration parameters and the methods for experimental data processing. The goal of these experimental procedures is to find a scalar-valued quality correlates that would express the level of overall quality. In this task the authors discuss how to interpret such quality correlates in the context of the number of Spherical Harmonics (SH) bands used for the worse or better object depth restoration. We analyze the perceptive differences between scene with switched on/off effect regardless to the gaze point and scene with the effect rendered based on data provided by the eye tracker.

To achieve that, we rendered a scene showing an interior of a room. In the room a flying helicopter and a bunny sculpture were placed. The subsurface scattering effected the bunny sculpture (see Fig. 4 Bottom). The scene was rendered with 3 different combinations of the number of Spherical Harmonics bands and with

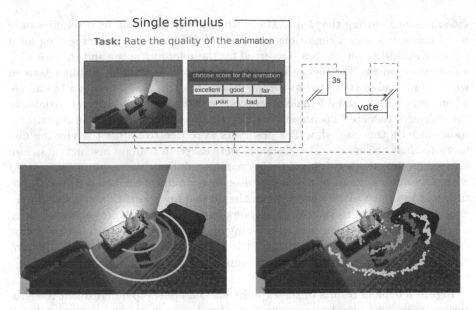

Fig. 4. Top: Overview of the single stimulus comparison method. The diagram shows the timeline of the approach and the corresponding screen. Bottom: The scene tested through the experiment. The colour dots on the images reconstruct, registered thought the animation, Left: the helicopter position and Right: the eye position.

3 different settings of the distance between bunny and flying helicopter influencing the rendering quality. The reference scene included the full effect regardless of the gaze point. A scene without the simulated effect was prepared as well. To conduct the experiment properly and to force the subjects to follow the same, required trajectory, they were asked to follow a moving object (helicopter). The location of the helicopter and the gaze point position were registered in real time (see Fig. 4 Bottom Right). It enabled to validate if the subject really followed the object. It is essential because of the fact that humans unintentionally move rapidly their eyes to other places than required (the saccades). Such situations were caught. If the subject focused in 90% on an object, the experiment was considered successful. In other case it was rejected as unreliable.

In the experiment Single Stimulus method was used which is dominant in video quality assessment [15,16]. The metric represents a categorical ranking, in which the observers judge the quality of displayed video for a short and fixed duration on a fixed 5-point scale containing five categories: excellent, good, fair, poor or bad (see Fig. 4 Top). Although 5-10 s presentation time is recommended for video, we found in the pilot study that 3 s presentation is sufficient to assess the video quality, yet it does not slow-down the experiment too much.

The videos were shown at random and include the reference scene, where the full effect is displayed independently on the ROI and scene rendered without subsurface scattering effect. There is no time limit at the voting stage but no video is shown during that time. The method is quite efficient as it requires only $n + 2$ trials to assess n conditions - number of different parameters setting for a scene. Two additional trials are reserved for the reference scene and the one with switched off effect. The algorithms were assessed by computer graphics laymen who had normal or corrected to normal vision. The age varied between 18 and 23. There were 9 male and 3 female observers. To estimate inter-observer variability the subjects repeated each session three times, but neither of the repetitions took place on the same day. The observers were free to adjust the viewing distance to their preference. The observers were asked to read an instruction before every experiment. According to [15] recommendation, the experiment started with a training session, in which the observers familiarized themselves with the task and interface. After that session, they asked questions or started the main experiment. To ensure that, the observers fully attended the experiment, three random trials were shown at the beginning of the main session without recording the results. The videos were displayed in a random order and with a different randomization for each session. Two consecutive trials showing the same scene were avoided if possible. All sessions were preceded by subject calibration.

Figure 5 depicts results of the experiment. The participants disliked the subsurface scattering simulation without eye tracker control. The scenes with eye

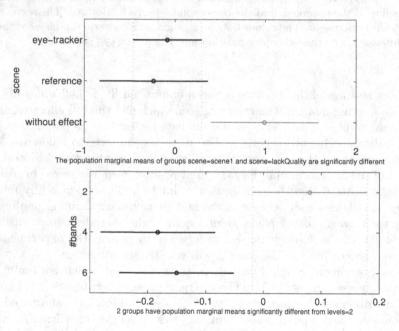

Fig. 5. ANOVA results. Top: Comparison of the quality between switched off effect and the effect simulated based on gaze-point and regardless to that point. Bottom: Comparison of the quality of the rendered scene in relation with the number of bands used for the object depth restoration. To stabilize results the quality is denoted by z-scores, computed as difference between score and average score per observer given to the scene and divided by standard deviation computed by observer.

tracker and the reference scene were assessed in a similar way, what narrowing the effect rendering to the ROI area has not bad impact on the final effect. Figure 5 Bottom showed an optimal dependence between the effect quality and the number of SH bands used for the object depth restoration. The eye-tracking control with number of SH bands equal to 2 significantly differ from the results received for the higher SH bands number e.g. 4 and 6. It received the highest z-value what corresponds to the lowest quality of rendering. The results for 4 and 6 bands were assessed similarly. Therefore the number of 4 SH bands appears sufficient for the simulation. Using more bands in the object depth calculation do not influence on the quality of simulated effect. To stabilize the results between the subjects, the tests were performed on the z-scores computed according to [17] based on the difference mean opinion score defined as the difference between reference and the test animation. The less difference the better quality.

During the experiment we found that subjects preferred the scenes with the subsurface scattering effect switched on. Moreover there is no significant difference between the scene rendered with highest quality regardless of the gaze point and the one optimized based on the data provided by the eye tracker. It proves that optimization of the real-time algorithm based on the gaze-point does not influence the perceived quality of simulated effect.

Additionally, it was found that abrupt changes are visible and annoying even if the eyes look far from the examined graphic model. The introduction of interpolation between the SH bands considerably improved the results. For the evaluation of the computational time, we tested the approach for different models for different SH coefficients parameters (see Figure 6). Results indicate the dependence between efficiency and number of SH coefficients used in the effect simulation. The results received for *bunny* model for Intel GMA X3100 are almost constant because of relatively small number of vertices to influence computational efficiency. A frame size was fixed to 1280 x 800 pixels.

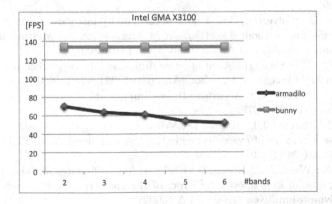

Fig. 6. FPS measured for two models: *armadilo* (129720 vertices) and *bunny* (9564 vertices) in dependence on different SH coefficients number. The results achieved for Intel GMA X3100.

5   Conclusions and Future Work

In this paper, we presented the approach for optimization in the usage of real-time complex algorithm on the basis of the subsurface scattering approach contained in the gaze point information. The data was received from the eye tracker. Instead of rendering of the effect with full quality, we made the approach dependent on the distance to the ROI. The full quality approach is rendered only when the object with a simulated effect is included in the ROI. Beyond the ROI, the algorithms of lower quality are generated. It decreases the complexity of calculations. Perceptual experiments were conducted in order to select an optimal set of algorithm parameters.

Acknowledgements. This work was supported by the Polish Ministry of Science and Higher Education through the grant no. N N516 193537.

References

1. Poole, A., Ball, L.J.: Eye Tracking in Human-Computer Interaction and Usability Research: Current Status and Future Prospects. Encyclopedia of Human-Computer Interaction. C. Ghaoui, Idea Group, Inc., Pennsylvania (2005)
2. Robinson, D.A.: A method of measuring eye movements using a scleral search coil in a magnetic field. IEEE Trans. Biomed. Eng. 10, 137–145 (1963)
3. Kaufman, A., Bandopadhay, A., Shaviv, B.: An eye tracking computer user interface. In: Proc. of the Research Frontier in Virtual Reality Workshop, pp. 78–84. IEEE (1993)
4. Duchowski, A.T.: Eye Tracking Methodology: Theory and Practice, 2nd edn. Springer, London (2007)
5. RED250 Technical Specification. SensoMotoric Instruments GmbH (2009)
6. Tobii T/X series Eye Trackers. Product Description. Tobii Technology AB, 2nd edn. (2009)
7. Loschky, L.C., Wolverton, G.S.: How late can you update gaze-contingent multiresolutional displays without detection? ACM Transactions on Multimedia Computing, Communications, and Applications 3(4), 1–10 (2007)
8. Cornsweet, T., Crane, H.: Accurate two-dimensional eye tracker using first and fourth Purkinje images. J. Opt. Soc. Am. 63(8), 921–928 (1973)
9. Crane, H., Steele, C.: Accurate three-dimensional eyetracker. J. Opt. Soc. Am. 17(5), 691–705 (1978)
10. Reulen, J., Marcus, J.T., Koops, D., Vries, F., Tiesinga, G., Boshuizen, K., Bos, J.: Precise recording of eye movement: the iris technique. Med. Biol. Eng. Comput. 26(1), 20–26 (1988)
11. Nguyen, K., Wagner, C., Koons, D., Flickner, M.: Differences in the infrared bright pupil response of human eyes. In: Proc. of the Eye Tracking Research and Applications Symposium, New Orleans, LA (2002)
12. Yoo, D.H., Chung, M.J., Ju, D.B., Choi, I.H.: Non-intrusive Eye Gaze Estimation using a Projective Invariant under Head Movement. In: Proc. of the Internat. Conf. on Automatic Face and Gesture Recognition, Washington, DC, pp. 94–99 (2002)

13. Hennessey, C., Noureddin, B., Lawrence, P.: A Single Camera Eye-Gaze Tracking System with Free Head Motion POG on Monitor Pupil Cornea. In: Proceedings of the 2006 Symposium on Eye Tracking Research & Applications (2006)
14. Morimoto, C.H., Mimica, M.: Eye gaze tracking techniques for interactive applications. Computer Vision and Image Understanding 98(1), 4–24 (2005)
15. ITU-R.Rec.BT.500-11, Metodology for the Subjective Assessment of the Quality for Television Pictures (2002)
16. ITU-T.Rec.P.910, Subjective audiovisual quality assessment methods for multimedia applications (2008)
17. Wang, Z., Bovik, A.C., Sheikh, H.R., Simoncelli, E.P.: Image quality assessment: from error visibility to structural similarity. IEEE Trans. on Image Processing 13(4), 600–612 (2004)
18. Ferwerda, J.A.: Psychophysics 101: how to run perception experiments in computer graphics. In: SIGGRAPH 2008: ACM SIGGRAPH 2008 Classes, pp. 1–60 (2008)
19. Torgerson, W.S.: Theory and methods of scaling. Wiley (1985)
20. Tomaszewska, A., Markowski, M.: Dynamic Scene HDRI Acquisition. In: Campiho, A., Kamel, M. (eds.) ICIAR 2010, Part II. LNCS, vol. 6112, pp. 345–354. Springer, Heidelberg (2010)
21. Tomaszewska, A.: Blind Noise Level Detection. In: Campilho, A., Kamel, M. (eds.) ICIAR 2012, Part I. LNCS, vol. 7324, pp. 107–114. Springer, Heidelberg (2012)
22. Tomaszewska, A.: Blind Noise Level Detection. In: Campilho, A., Kamel, M. (eds.) ICIAR 2012, Part I. LNCS, vol. 7324, pp. 107–114. Springer, Heidelberg (2012)

Depth Auto-calibration for Range Cameras Based on 3D Geometry Reconstruction

Benjamin Langmann, Klaus Hartmann, and Otmar Loffeld

ZESS - Center for Sensor Systems, University of Siegen
57068, Siegen, Paul-Bonatz-Str. 9-11, Germany
{langmann,hartmann,loffeld}@zess.uni-siegen.de

Abstract. An approach for auto-calibration and validation of depth measurements gained from range cameras is introduced. Firstly, the geometry of the scene is reconstructed and its surface normals are computed. These normal vectors are segmented in 3D with the Mean-Shift algorithm and large planes like walls or the ground plane are recovered. The 3D reconstruction of the scene geometry is then utilized in a novel approach to derive principal camera parameters for range or depth cameras. It operates based on a single range image alone and does not require special equipment such as markers or a checkerboard and no specific measurement procedures as are necessary for previous methods. The fact that wrong camera parameters deform the geometry of the objects in the scene is utilized to infer the constant depth error (the phase offset for continuous wave ToF cameras) as well as the focal length. The proposed method is applied to ToF cameras which are based on the Photonic Mixer Device to measure the depth of objects in the scene. Its capabilities as well as its current and systematic limitations are addressed and demonstrated.

1 Introduction

In recent years interest in 3D imaging gained considerable momentum and today 3D imagers are utilized in many different applications. Several techniques capable of providing depth measurements at video frame rate and with reasonable resolution are on the market. These devices are based on structured light (SL), on the Time of Flight (ToF) principle or utilize stereo vision. The former camera types are active sensors and usually illuminate the scene with infrared light. When high frame rates are desired, the light emitted by SL approaches differs typically in space and is constant over time.

On the other hand the light emitted by ToF approaches does not exhibit such a structure – but its characteristics obviously depend on the lighting devices. ToF cameras change the intensity of the emitted light over time (amplitude modulation) and the light is usually either continuously modulated or pulsed. In the first case the phase difference between the emitted and the received light is determined and it directly corresponds to the distance of the reflecting object. Again different technologies are on the market, one of those is the Photonic Mixer Device (PMD), see [1] for an overview.

G. Bebis et al. (Eds.): ISVC 2012, Part II, LNCS 7432, pp. 756–766, 2012.

ToF cameras operating with pulsed light measure directly the time until the light reaches the imaging chip. But to achieve an accuracy of a few centimeters extremely short time differences need to be measured, which is very challenging. To illustrate this: It takes the light about $10ns$ to reach an object in 3 meters distance to the camera and 3 centimeters correspond to $100ps$.

Depth imaging devices utilizing stereo vision are considered passive and consist of at least two cameras. Their respective images are compared and the disparities are calculated of all pixels to gain a depth map. A precise calibration of the cameras is required to achieve depth maps of sufficient quality. Nevertheless, this approach has several limitations, e.g. low textured objects or background light.

All of these cameras are used in a wide range of applications. Some of them are currently being researched, while others are already on the market. In some areas accurate depth measurements are not that important, i.e. in some video games or video chats range cameras are only used to perform highly reliable foreground segmentation. But when an application aims at the reconstruction of the observed scene, i.e. for maneuvering in the scene or to correlate the observations of several cameras, the depth measurements delivered by any range camera need to be as precise as possible. However, range cameras are typically subject to systematic measurement errors, which in turn make a calibration of the measurements necessary. Time-of-Flight cameras for example usually have a significant constant measurement error due to signal propagation delays. Additionally, errors caused by different light intensities and the actual distance to the reflecting object occur.

To account for the deterministic measurement errors range cameras are calibrated. This is often done by taking a number of images of a calibration object or pattern with known distances and then to use these to compensate for the measurement errors. The distances should cover the whole measuring range. Depth values not being part of the set of images taken are interpolated. Due to the fact that this procedure is relatively complex this calibration is usually only performed once. This poses the question whether the calibration is accurate for the complete lifetime of the camera and for all imaging parameters such as the modulation frequency of continuous wave ToF cameras. Obviously, the calibration can only be valid for one type of illumination system.

Therefore, self-calibration approaches for range cameras have been developed as an extension of self-calibration techniques for normal video cameras. The methods work by taking a number of images of a calibration object, e.g. a checkerboard, without known distance to the camera. Features of the calibration object are automatically detected and an optimization algorithm then determines the most likely camera parameters based on 2D and 3D point correspondences. These approaches have a significantly lower accuracy and it is very difficult to cover the whole measuring range. Range camera calibration methods will be discussed further in the next section.

Taking this one step further so-called auto-calibration methods are able to derive camera parameters without the need of any calibration objects and without known distances. These methods are extremely useful if just a video or a single image is available and not the calibration data of the camera or the camera in the used configuration itself. Additionally, if the calibration is needed quickly or if the camera parameters are frequently changed these auto-calibration methods can be applied. Certain image features such as parallel lines or planes are usually used in these methods. To our knowledge auto-calibration methods able to calibrate the distance error for range cameras have not been published up until now.

In this paper we firstly perform a segmentation of surface normals in 3D to obtain initial estimates of planes in the scene and then use a novel auto-calibration approach to determine the distance offset and the focal length of a range camera. Its capabilities, especially its accuracy, as well as limitations and possible applications of this approach are discussed.

This paper is structured as follows. In section 2 the previous work dealing with calibration and self-calibration of range cameras is reviewed. In section 3 the camera model is detailed. Afterwards, we show how the Mean-Shift algorithm can be utilized to detect planes in section 4 and the proposed depth auto-calibration approach is explained in section 5. The experiments to confirm the capabilities of the method are discussed in section 6 and this paper ends with a conclusion of this work in section 7.

2   Related Work

Camera calibration is a widely studied research topic. The book Multiple View Geometry in Computer Vision [2] gives an excellent overview and serves as a good starting point. Specifically, the auto-calibration of standard cameras, i.e. the estimation of camera parameters without special calibration objects or recording procedures, is covered. Therefore, papers aiming at the calibration of range cameras will be reviewed in the following.

In the past, several papers presented methods to calibrate range cameras. Typically error models are introduced and estimated in order to compensate for the measurement errors afterwards. In [3] B-Splines were used to model the intensity related distance error, and the standard intrinsic camera parameters are determined conventionally using image features and bundle adjustment. This work was revisited and refined in [4] and in [5] the distance calibration is fused with the standard intrinsic parameter estimation using a polynomial error model.

Another error source of ToF imaging was discussed in [6]. They try to compensate for multiple reflections of the light emitted by ToF cameras. An elaborate calibration device is used in [7] to compensate for different intensities due to different exposure times with a look-up table. This was also attempted in [8]. Moreover, multiple ToF cameras are studied and fused in [9] and in [10] the monocular combination of PMD and color imaging is addressed.

Table 1. Camera Parameters and Notations

Symbol	Quantity	Meaning
x, y	pixel	Coordinates of the pixel observing the object
d	$\in [0, 1]$	Raw depth measurements
w_P, h_P	meter	Width and height of a pixel
c_x, c_y	pixel	Position of the optical center on the imaging chip
$\delta(\cdot)$	meter	Function mapping a raw depth measurement to the distance of the object
ν	Hz	Modulation frequency
o	rad	Constant distance error (phase offset)
x_c, y_c	meter	2D coordinates of the pixel on the image plane
f	meter	Focal length of the camera
X, Y, Z	meter	3D camera coordinates

3   Range Camera Model

In the following the pinhole camera model for range cameras is described in detail. The camera model maps a pixel and a depth measurement to a 3D point. The parameters of this model and the notations are summarized in table 1. A pixel (x, y) is mapped to metric 2D coordinates (x_c, y_c) using a principal point (c_x, c_y) on the image plane with

$$x_c = (x - c_x)w_p \qquad y_c = (y - c_y)h_p \, . \tag{1}$$

The z-coordinate of the 3D point (X, Y, Z) with the $X - Y$ plane being identical to the image plane can be calculated based on these 2D coordinates and the distance $\delta(d)$ with

$$Z = \frac{\delta(d)}{\sqrt{1 + \frac{(x_c^2 + y_c^2)}{f^2}}} \tag{2}$$

and this leads to the x- and y-coordinates

$$X = \frac{x_c Z}{f} \qquad Y = \frac{y_c Z}{f} \tag{3}$$

with the focal length f. Range cameras often do not provide depth measurements directly, or only measurements, which need to be corrected. In this range camera model a function $\delta(d)$ maps a raw depth measurement to a depth in meters. For PMD based cameras, which utilize phase shift information, a simple distance function is given by

$$\delta(d) = \frac{c(2\pi d + o)}{4\pi\nu} \tag{4}$$

with c being the speed of light, ν the modulation frequency used and o a constant measurement error. Errors of higher order or special function have been used in the past, cf. [4].

4   Segmentation of Surface Normals

In order to recognize planes in the scene we perform a segmentation or feature space clustering of surface normals in 3D. The normal vectors are gained from range images and the Mean-Shit algorithm [11] is used for the segmentation. The depth values for each pixel are firstly transformed into 3D points with the range camera model and a set of default camera parameters. Afterwards, the surface normal $\underline{n} = (n_x, n_y, n_z)$ at each 3D point $\underline{p} = (p_x, p_y, p_z)$ can be estimated by averaging over the 8 normals of triangles spanned by $\underline{p}$ and combinations of its neighbor points. This will lead to interpolation errors at the borders of objects and therefore boundary points of detected planes will be neglected later on.

The Mean-Shift algorithm is a feature space approach and consists of two steps. In the first one (filtering) the mean-shift vectors are calculated iteratively and the feature points are moved accordingly until a convergence is reached. These vectors are determined by calculating the weighted average of neighboring feature points. Let $P_i = (\underline{p}_i, \underline{n}_i)$ and $P_j = (\underline{p}_j, \underline{n}_j)$ be points of the feature space. For the weight $g(P_i, P_j)$ between two points we use the product of two Gaussian kernels

$$
g(P_i, P_j) = \exp\left\{ -\frac{\left(\underline{p}_i - \underline{p}_j\right)^T \left(\underline{p}_i - \underline{p}_j\right)}{\sigma^2_{space}} \right\} \cdot \exp\left\{ -\frac{1 - \left(\frac{n_i \cdot n_j}{\|n_i\| \|n_j\|}\right)^2}{\sigma^2_{angle}} \right\} . \quad (5)
$$

The first term is based on the squared Euclidean distance between the two points and the second uses the squared sine to measure differences between normal vectors. σ_{space} and σ_{angle} are bandwidth parameters to control the influence of the different subspaces.

Given a point $P_i^{(t)}$ in the feature space at iteration t the next point can be computed with

$$
P_i^{(t+1)} = \frac{\sum_{P \in N(P_i^{(t)})} P \cdot g(P_i^{(t)}, P)}{\sum_{P \in N(P_i^{(t)})} g(P_i^{(t)}, P)} , \quad (6)
$$

where the set $N(\cdot)$ describes a spatial neighborhood. The Mean-Shift vector is the difference $P_i^{(t+1)} - P_i^{(t)}$ and for each point these iterative calculations are done independently. Once the Mean-Shift filtering is finished, the feature points are merged using a distance threshold.

5   Auto-calibration of Range Cameras

The camera imaging equations 2-4 make it obvious that X, Y, Z are nonlinear in all intrinsic camera parameters c_x, c_y, f, o with the frequency ν being an exception. Therefore, the usage of wrong parameters in the reconstruction of the scene geometry will result in deformed objects, e.g. even small differences will give planes a significant curvature. Another result of choosing wrong parameters are modified angles between objects in the scene, see figure 1a. In the experiments it was discovered that this property can in fact be reliably used to estimate the constant distance error.

Since many man-made objects are planar (walls, floors, tables), their reconstruction can be used to find the correct parameters by refining them until the object in question is actually planar. In theory all non-linear parameters can be estimated using a single plane in the scene.

5.1 Motivation

Uncorrected depth measurements using four different range cameras all based on PMD chips are depicted in figure 1b. These measurements were conducted using a precise linear translation unit.

The measurement errors can roughly be classified as follows: constant distance errors, errors due to over-exposure or insufficient lighting and an error of higher order sometimes called wobbling. Obviously, the constant distance error or phase offset needs to be corrected, whereas the exposure related errors cannot be compensated for and the higher order errors are minor. Moreover, it cannot be expected to compensate for these errors without elaborate measurements.

Related to the normal depth calibration of range cameras is the validation of range measurements. Small changes to the camera setup, e.g. different cables or lighting devices used in conjunction with PMD based cameras, lead to changes of the phase offset, but also different modulation frequencies as is demonstrated in figure 5. Aging or wear may also yield impaired depth measurements as does the relative positioning of lighting devices and camera. This is a concern for camera stereo setups as well.

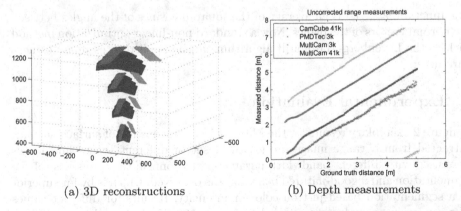

(a) 3D reconstructions   (b) Depth measurements

Fig. 1. 3D reconstructions of a simple scene consisting of three orthogonal walls while applying different offsets and uncorrected range measurements of four different ToF cameras

5.2   Proposed Range Auto-calibration

The proposed auto-calibration relies on the existence and successful detection
of at least one plane of significant size. The segmentation of surface normals is
used with an initial set of camera parameters to retrieve candidate planes in
the scene by selecting the largest segments, which are additionally larger than
a given threshold. Then a plane is fitted to each detected plane in the scene.
Here a simple least squares method could be applied, but we use a RANSAC
[12] based algorithm instead, which itself uses the well known eigenvalue plane
fitting method (also a type of least squares optimization). This algorithm is able
to handle outliers for example at border of objects and therefore yields good
results.

Now a measure $\rho_{curv}(\cdot)$ is utilized to judge the curvature of a plane and
thereby the accuracy of the camera parameters. $\rho_{curv}(\cdot)$ is simply given by the
square root of the average squared distance between all points of the plane to
the constructed plane. Let $\underline{n}_1, \underline{n}_2, \ldots, \underline{n}_m$ be the normal vectors of unit length
of the m constructed planes with points on the plane $\underline{p}_1, \underline{p}_2, \ldots, \underline{p}_m$ and let
$M_1, M_2, \ldots, M_m$ sets of the 3D points used in the construction. We can now
write

$$\rho_{curv}(M., \underline{n}., \underline{p}.) = \frac{1}{m} \sum_{i=1}^{m} \sqrt{\sum_{\underline{q} \in M_i} \left(\left(\underline{q} - \underline{p}_i \right) \cdot \underline{n}_i \right)^2}. \tag{7}$$

The angle between the reconstructed planes is used in a second measure to judge
the applied camera parameters. Here at least two pairwise orthogonal planes of
significant size have to be detected successfully in the scene. Then the measure
to rate the orthogonality of the detected planes in the scene is given by

$$\rho_{ortho}(\underline{n}_1, \underline{n}_2, \ldots, \underline{n}_m) = \frac{2}{m(m-1)} \sum_{i=1}^{m-1} \sum_{j=i+1}^{m} \left(\frac{\underline{n}_i \cdot \underline{n}_j}{\|\underline{n}_i\| \|\underline{n}_j\|} \right)^2. \tag{8}$$

The function $\rho_{ortho}(\cdot)$ is an average of the squared cosines of the angles between
the normal vectors of the planes. Now a standard non-linear optimization method
such as the Levenberg-Marquardt algorithm can be applied to estimate camera
parameters.

6   Experimental Evaluation

In figure 2 exemplary results for the Mean-Shift segmentation of surface normals
extracted from a range image are given. A set of different scenes and lenses
were used with different bandwidth parameters to confirm the robustness of the
segmentation and a good-natured behavior was observed, which is by far superior
to a segmentation based just on color or intensity. Results for different values
of σ_{angle} with normal vectors scaled to length 100 and a spatial bandwidth of
$\sigma_{space} = 5cm$ are shown.

In order to evaluate the curvature measure $\rho_{curv}(\cdot)$ a simple scene consisting
of just a single wall was acquired with the PMDTec CamCube 41k and a set of

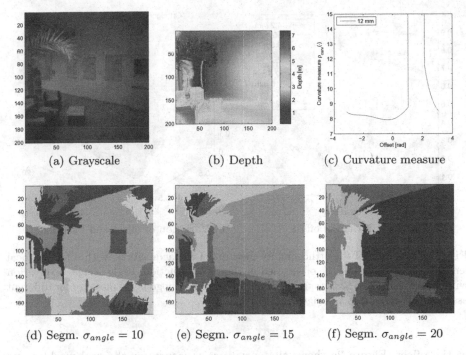

(a) Grayscale   (b) Depth   (c) Curvature measure

(d) Segm. $\sigma_{angle} = 10$ (e) Segm. $\sigma_{angle} = 15$ (f) Segm. $\sigma_{angle} = 20$

Fig. 2. Mean-Shift segmentation based on surface normals of a real world scene for different bandwidth parameters

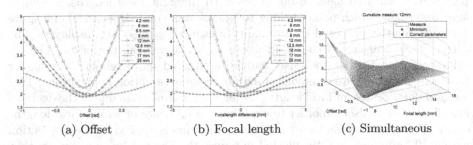

(a) Offset   (b) Focal length   (c) Simultaneous

Fig. 3. Curvature measure evaluated for a number of acquisitions with different lenses of a single wall with a size of 100×65 pixels

different lenses. An area of about 100×65 pixels was marked and the curvature measure was computed while assuming different offsets and focal lengths. Correct values were chosen for the fixed parameters. The accurate offset $o = -0.06$ was determined with a photogrammetric method. The results in figure 3 demonstrate that the offset as well as the focal length can be reliably estimated under these good conditions, but estimating both parameters at the same time is not possible. Additionally, large focal lengths cause problems due to the diminishing effect

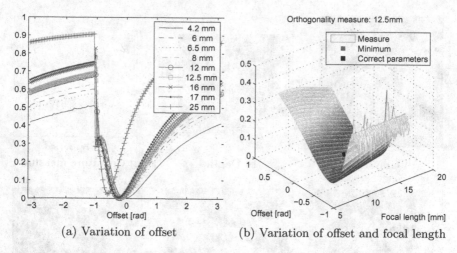

(a) Variation of offset   (b) Variation of offset and focal length

Fig. 4. Evaluation of the orthogonality measure. Left: acquisitions with different focal lengths and variation of the offset used in the reconstruction. Right: single acquisition with a focal length of 12.5mm and evaluation of the orthogonality measure for different assumed offsets and focal lengths.

of the offset. Figure 2c shows the curvature measure for the real world scene captured using a $12mm$ lens and applied to the largest plane for different offsets.

For the orthogonality measure a test scene consisting of three orthogonal planes was captured again using lenses with different focal lengths. A simple map describing which points belong to planes was created per hand. In figure 4 results for the proposed orthogonality measure $\rho_{ortho}(\cdot)$ are displayed. The correct focal length was chosen in figure 4a while varying the offset. The results demonstrate that the offset can be estimated quite reliably for smaller focal lengths which are commonly used for scene observation applications. This method deteriorates for focal lengths larger than $25mm$ again due to the decreasing effect of the offset. In order to examine the influence of wrong focal lengths on the 3D reconstruction the orthogonality measure is applied on the acquisition with a $12.5mm$ lens and it is plotted in figure 4b. The results demonstrate that the correct focal length cannot be estimated based on the proposed measure and by no means can both camera parameters be estimated. For an complete camera auto-calibration approach standard methods based on the recognition of lines etc. are therefore required. The previous results were confirmed using a scene containing two orthogonal walls which was captured at larger distances (3 to 6 meters depending on the focal length and with a real world scene (fig. 2).

In another experiment a scene displaying a table in front of a wall and some objects was acquired while applying a set of different modulation frequencies. The orthogonal walls were automatically recognized via segmentation. The focal length used was $17mm$ and the orthogonality measure was evaluated for the whole range of possible offsets. Previously, it was assumed that the modulation

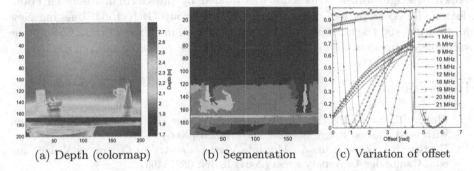

(a) Depth (colormap)   (b) Segmentation   (c) Variation of offset

Fig. 5. Acquisitions of a second test scene using different modulation frequencies. Segmentation of surface normals with Mean-Shift and evaluation of the orthogonality measure using different offsets in the reconstruction.

frequency would have little effect on the phase offset, but it was discovered that different frequencies yield severely different offsets. This was confirmed by visually inspecting the reconstructed planes and by applying photogrammetric methods.

7 Conclusion

A novel approach towards auto-calibration and validation of range cameras was presented in this paper. The method is applicable to all types of range or depth cameras, but in this paper it was tested with ToF cameras based on the Photonic Mixer Device. In general, a scene can only be reconstructed correctly from depth images if the right camera parameters are used in the process. The proposed approach utilizes that planes are reconstructed with a curvature and that angles between planes are modified if wrong parameters are applied. It was demonstrated that large planes like the ground plane or walls can be reliably detected with the Mean-Shift segmentation of surface normals. The geometry of these planes can then be used to estimate the constant distance measuring error as well as the focal length of the camera.

Two measures to judge the reconstruction of a scene are defined: the first one measures how planar the points of a detected plane are and requires a single plane of significant size in the scene. For the second measure angles between detected planes are computed while assuming that the detected planes are orthogonal, which is in many practical cases satisfied. It was not possible to estimate errors of higher order or to estimate camera parameters simultaneously. Nevertheless, other well known auto-calibration methods also relying on the successful detection of scene geometry could be applied in conjunction with the proposed method for a complete auto-calibration of any range camera. In addition to auto-calibration the introduced measures may be useful to constantly validate the depth measurements of a camera as well as when a zoom lens is used and the focal length is changed repeatedly.

Acknowledgements. This work was funded by the German Research Foundation (DFG) as part of the research training group GRK 1564 'Imaging New Modalities' and the authors would like to thank all members of the ZESS for the valuable discussions.

References

1. Möller, T., Kraft, H., Frey, J., Albrecht, M., Lange, R.: Robust 3d measurement with pmd sensors. Range Imaging Day, Zürich (2005)
2. Hartley, R.I., Zisserman, A.: Multiple View Geometry in Computer Vision, 2nd edn. Cambridge University Press (2004) ISBN: 0521540518
3. Lindner, M., Kolb, A.: Lateral and Depth Calibration of PMD-Distance Sensors. In: Bebis, G., Boyle, R., Parvin, B., Koracin, D., Remagnino, P., Nefian, A., Meenakshisundaram, G., Pascucci, V., Zara, J., Molineros, J., Theisel, H., Malzbender, T. (eds.) ISVC 2006. LNCS, vol. 4292, pp. 524–533. Springer, Heidelberg (2006)
4. Lindner, M., Schiller, I., Kolb, A., Koch, R.: Time-of-flight sensor calibration for accurate range sensing. Computer Vision and Image Understanding 114, 1318–1328 (2010)
5. Schiller, I., Beder, C.: Calibration of a pmd-camera using a planar calibration pattern together with a multi-camera setup. In: Proc. XXXVII Int. Soc. for Photogrammetry (2008)
6. Falie, D., Buzuloiu, V.: Further investigations on tof cameras distance errors and their corrections. In: 4th European Conference on Circuits and Systems for Communications, pp. 197–200 (2008)
7. Kahlmann, T., Remondino, F., Ingensand, H.: Calibration for increased accuracy of the range imaging camera swissrangertm. Image Engineering and Vision Metrology (IEVM) 36, 136–141 (2006)
8. Radmer, J., Fuste, P.M., Schmidt, H., Kruger, J.: Incident light related distance error study and calibration of the pmd-range imaging camera. In: IEEE Computer Society Conference on Computer Vision and Pattern Recognition Workshops, pp. 1–6 (2008)
9. Kim, Y., Chan, D., Theobalt, C., Thrun, S.: Design and calibration of a multi-view tof sensor fusion system. In: IEEE Conf. on Computer Vision & Pattern Recogn.; Workshop on ToF-Camera based Computer Vision, pp. 1–7. IEEE (2008)
10. Prasad, T., Hartmann, K., Weihs, W., Ghobadi, S.E., Sluiter, A.: First steps in enhancing 3d vision technique using 2d/3d sensors. In: Computer Vision Winter Workshop, Telc, Czech Republic, Citeseer, pp. 82–86 (2006)
11. Comaniciu, D., Meer, P.: Mean shift: A robust approach toward feature space analysis. IEEE Transactions on Pattern Analysis and Machine Intelligence 24, 603–619 (2002)
12. Fischler, M.A., Bolles, R.C.: Random sample consensus: a paradigm for model fitting with applications to image analysis and automated cartography. Commun. ACM 24(6), 381–395 (1981)

Author Index